Mixed Carbonate - Siliciclastic Sequences

Organized and Edited

by

Anthony J. Lomando

and

Paul M. Harris

SEPM Core Workshop No. 15
Dallas, April 7, 1991

Copyright 1991 by SEPM (Society for Sedimentary Geology)

SEPM gratefully acknowledges financial support from Reservoirs, Inc., Houston, Texas for the publication of this volume.

ISBN #0-918985-87-0

Additional copies of this publication may be ordered from SEPM. Send your order to:

SEPM
Post Office Box 4756
Tulsa, OKlahoma 74159-0756
U.S.A.

© Copyright 1991 by

SEPM (Society for Sedimentary Geology)
Printed in the United States of America

PREFACE

The study of carbonate-siliciclastic mixed sequences has seen an increase in the number of investigations that focus on mixed settings as part of the continuum between the carbonate and clastic end members. The level of interest in such mixed sequences has been reflected in recent years by sessions at national meetings that dealt specifically with the topic, and also by recent publications (Doyle and Roberts, 1988; Budd and Harris, 1990). Cyclic deposition in mixed basins by reciprocal sedimentation has become one of the foundation blocks for sequence stratigraphy (Van Siclen, 1958; Wilson, 1967; Meissner, 1972; Sarg, 1988). In addition, these mixed sequences have a variety of distinctive petroleum reservoir characteristics, important for both exploration and development programs. The emphasis now is on reevaluating ancient sequences in the light of a more dynamic understanding of spatial and temporal variations and controls on these sequences.

Mixtures of siliciclastic and carbonate rocks can occur due to lateral facies mixing (spatial variability) or by sea-level changes and/or variations in sediment supply, causing a vertical variation in the stratigraphic succession (temporal variability) (Budd and Harris, 1990). Mixing can occur through a wide range of scales, from millimeters to kilometers, respond to a wide range of processes, and can be influenced by all orders of cyclicity. Trends of mixing can vary along both depositional strike and dip, ranging from coastal environments to deep basinal settings. Major basinward transport of siliciclastics is controlled by rates of change of sea level and subsidence that together create accommodation space. Other significant factors include climate, supply volume and transport mode, shelf width and paleobathymetry (ramp, distally steepened ramp, rimmed platform), and current patterns.

We have subdivided the papers within this volume (see Fig. 1) under the following headings: **Shelf Wide, Coastal and Inner Shelf, Middle to Outer Shelf, Slope to Basin,** and **Paleokarst**.

In the **Shelf Wide Section** are two papers that are contrasting examples of the spectrum of mixing on a regional scale. Borer and Harris[1] present a Guadalupian example from the Permian Basin that is a regional treatment of temporal mixing where siliciclastics repeatedly alternate with carbonates across the entire width of the shelf in environments ranging from inner shelf to shelf margin. The paper by Shew[2] is a Jurassic example from Mississippi that is dominated by spatial mixing where fluvial input into a coastal environment is reworked by shelf processes into a carbonate-dominated regime and addresses mixing processes from nearshore to outer-ramp environments.

The **Coastal and Inner Shelf Section** is a group of five papers that describe environments ranging from fluvial and eolian to coastal marine. Dewet and others[3] present a Pennsylvanian example from the southern margin of the Illinois Basin where a terrigenous dominated system contains periodic carbonate marine incursions exemplifying both spatial and temporal mixing. Handford and Francka's[4] paper from the Mississippian of southwest Kansas is an example of shallow marine and carbonate eolian mixing and, in theory, would be a spatial mixing type. Lomando and Walker[5] present a Cenomanian example from offshore Angola consisting of a spatially mixed sequence of shoreface, beach, and lagoonal environments. Cuzella and others[6] report on a Pennsylvanian example from the Cherokee group in west-central Kansas. Here temporal mixing dominates where transgressive-regressive cycles are characterized by widespread carbonate deposition during major transgressive pulses. The final paper in this section, by Shaw

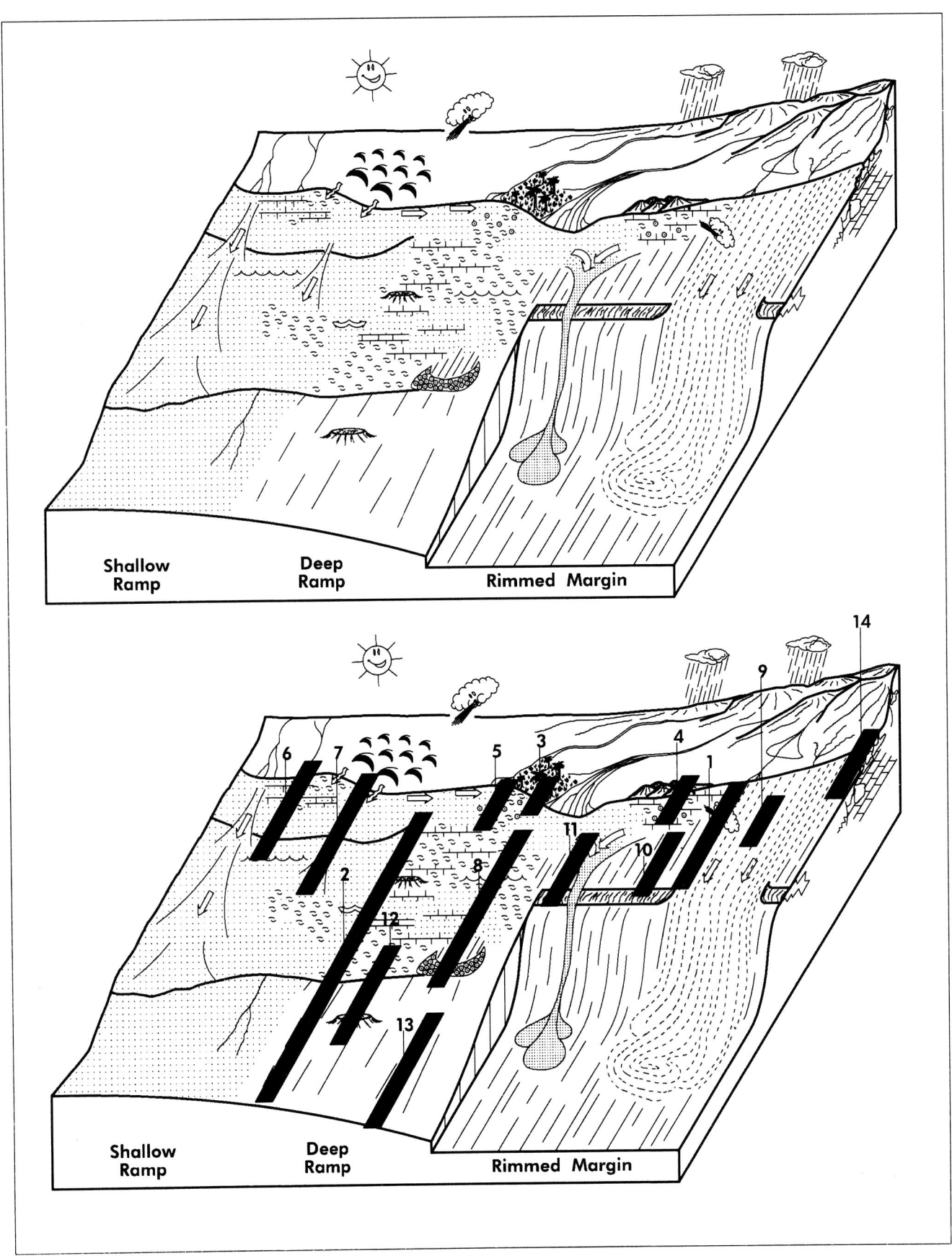

Fig. 1.--Block diagrams showing variety of mixed settings across both ramp and rimmed margin profiles (top) and settings described in papers of this volume (bottom). Numbers refer to preface text and also to table of contents.

and Schreiber,[7] is a middle Ordovician example from the Ancell Group of the Illinois Basin where shallow subtidal to supratidal environments contain a temporally mixed sequence controlled by both eustatic and tectonic processes.

Papers in the **Middle to Outer Shelf Section** are dominated by examples of subtidal deposition. This group of four papers begins with Harwood and Sullivan[8] who present a Mississippian example from the Irish Midlands. This sequence is an example of tectonically controlled spatial mixing. Tucker and Chalcraft[9] present a temporal mixing example driven by sea-level fluctuations during the Guadalupian in the Permian Basin. Bebout[10] presents another example from the Guadalupian of the Permian Basin, where carbonate-dominated packages are separated by siltstones and sandstones that are introduced during lowstands and subsequently reworked during initial phases of transgressions. The final paper in this section by Bush[11] presents a modern example along the Puerto Rican shelf of a storm-dominated inner-shelf system typified by spatial mixing processes.

The **Slope to Basin Section** contains two examples of the deepest-water environments covered in this volume. Smosna and Patchen[12] present an upper Ordovician example from the Trenton Limestone in the central Appalachian Basin. This basinal sequence is an example of temporal mixing during a tectonically driven transgressive phase of deposition. Foreman and others[13] present an upper Cambrian example from eastern Tennessee. Spatial mixing occurs along a distally steepened ramp-slope to basin system that is characterized by episodic input of shallow-water carbonate into a shale-dominated basinal setting.

The final paper is an example of unconventional mixing in the **Paleokarst Section**. Sieverding and Harris[14] present a Mississippian example from the southwest Wyoming thrust belt. In this temporal mixing example, the mixing process results from the downward percolation of siliciclastics into an underlying karst system during a major transgressive event.

Many mixed sequences have been described in the literature, but understanding the controls of these sequences from a process approach is now in an adolescent stage. We have endeavored to compile examples from a broad range of environments and ages in order to provide participants in the workshop, and subsequent readers of this volume, with a body of literature focused on mixing problems. We hope that this will aid and encourage both our academic and industrial colleagues to continue their efforts to understand the complexities of mixing processes and stimulate ideas and approaches to explore for and develop new oil and gas reserves.

The core workshop was made possible with the help of many people. The SEPM staff and continuing education committee supported the concept of the workshop and handled the logistical preparations. We thank Chevron Overseas Petroleum Inc. and Chevron Oil Field Research Company for supporting our efforts in organizing and presenting the workshop, and editing and preparing this volume.

A. J. LOMANDO   P. M. HARRIS

REFERENCES

BUDD, D. A., AND HARRIS, P. M., eds., 1990, Carbonate-siliciclastic mixtures: SEPM (Society for Sedimentary Geology) Reprint Series No. 14, 272 p.

DOYLE, L. J., AND ROBERTS, H. H., eds., 1988, Carbonate-clastic transitions: Developments in Sedimentology No. 42, Elsevier, Amsterdam-Oxford-New York-Tokyo, 304 p.

MEISSNER, F. F., 1972, Cyclic sedimentation in Middle Permian strata of the Permian Basin, *in* Elam, J. G. and Chuber, S., eds., Cyclic Sedimentation in the Permian Basin, Second Edition: West Texas Geological Society, Publication 72-60, p. 203-232.

SARG, J. F., 1988, Carbonate sequence stratigraphy, *in* Wilgus, C. K., Hastings, B. S., Kendall, C. G. St. K., Posamentier, H. W., Ross, C. A., and Van Wagoner, J. C., eds., Sea-Level Changes: An Integrated Approach: Society of Economic Paleontologists and Mineralogists, Special Publication No. 42, p. 155-182.

WILSON, J. L., 1967, Cyclic and reciprocal sedimentation in Virgilian strata of southern New Mexico: Geological Society of America Bulletin, v. 78, p. 805-818.

VAN SICLEN, D. C., 1958, Depositional topography - examples and theory: American Association of Petroleum Geologists Bulletin, v. 42, p. 1897-1913.

MIXED CARBONATE-SILICICLASTIC SEQUENCES

TABLE OF CONTENTS

Shelf Wide

1 DEPOSITIONAL FACIES AND MODEL FOR MIXED SILICICLASTICS
AND CARBONATES OF THE YATES FORMATION, PERMIAN BASIN
By J. M. Borer and P. M. Harris 1

2 UPWARD-SHOALING SEQUENCE OF MIXED SILICICLASTICS AND
CARBONATES FROM THE JURASSIC SMACKOVER FORMATION OF
CENTRAL MISSISSIPPI
By R. D. Shew 135

Coastal and Inner Shelf

3 DEPOSITION AND DIAGENESIS OF A MARINE-SWAMP MARGIN;
THE PROVIDENCE LIMESTONE AND ADJACENT COALS,
WESTERN KENTUCKY
By C. B. Dewet, S. O. Moshier, J. C. Hower,
and S. M. Rimmer 169

4 MISSISSIPPIAN CARBONATE-SILICICLASTIC EOLIANITES IN
SOUTHWESTERN KANSAS
By C. R. Handford and B. J. Francka 205

5 WAMBA FIELD, PEOPLE'S REPUBLIC OF ANGOLA, A CENOMANIAN
MIXED CARBONATE-SILICICLASTIC RESERVOIR
By A. J. Lomando and T. L. Walker 245

6 DEPOSITIONAL ENVIRONMENTS AND FACIES ANALYSIS OF THE
CHEROKEE GROUP IN WEST-CENTRAL KANSAS
By J. J. Cuzella, C. P. Gough, and S. C. Howard 273

7 LITHOSTRATIGRAPHY AND DEPOSITIONAL ENVIRONMENTS OF THE
ANCELL GROUP IN CENTRAL ILLINOIS: A MIDDLE ORDOVICIAN
CARBONATE-SILICICLASTIC TRANSITION
By T. H. Shaw and B. C. Schreiber 309

TABLE OF CONTENTS (contd.)

Middle to Outer Shelf

8 SEDIMENTARY HISTORY OF THE MOYVOUGHLY AREA, COUNTY WESTMEATH, IRELAND: EVIDENCE FOR SYNSEDIMENTARY FAULT MOVEMENTS IN A MIXED CARBONATE-SILICICLASTIC SYSTEM OF COURCEYAN AGE
By G. M. Harwood and M. Sullivan 353

9 CYCLICITY IN THE PERMIAN QUEEN FORMATION - U.S.M. QUEEN FIELD, PECOS COUNTY, TEXAS
By K. E. Tucker and R. G. Chalcraft 385

10 INTERRELATIONSHIP OF PLATFORM DOLOSTONE AND SILICIC SILTSTONE FACIES - GRAYBURG FORMATION (PERMIAN), CENTRAL BASIN PLATFORM AND OZONA ARCH, PERMIAN BASIN, WEST TEXAS
By D. G. Bebout 429

11 MIXED CARBONATE/SILICICLASTIC SEDIMENTATION: NORTHERN INSULAR SHELF OF PUERTO RICO
By D. M. Bush 447

Slope to Basin

12 ORDOVICIAN LIMESTONE AND SHALE IN THE CENTRAL APPALACHIAN BASIN: EARLY SEDIMENTARY RESPONSE TO PLATE COLLISION
By R. Smosna and D. G. Patchen 485

13 SLOPE AND BASINAL CARBONATE DEPOSITION IN THE NOLICHUCKY SHALE (UPPER CAMBRIAN), EAST TENNESSEE: EFFECT OF CARBONATE SUPPRESSION BY SILICICLASTIC DEPOSITION ON BASIN-MARGIN MORPHOLOGY
By J. L. Foreman, K. R. Walker, L. J. Weber, S. G. Driese, and R. B. Dreier 511

Paleokarst

14 MIXED CARBONATES AND SILICICLASTICS IN A MISSISSIPPIAN PALEOKARST SETTING, SOUTHWESTERN WYOMING THRUST BELT
By J. L. Sieverding and P. M. Harris 541

DEPOSITIONAL FACIES AND MODEL FOR MIXED SILICICLASTICS AND CARBONATES OF THE YATES FORMATION, PERMIAN BASIN

J. M. BORER AND P. M. HARRIS
Chevron Oil Field Research Company
P. O. Box 446, La Habra, CA 90633-0446

ABSTRACT

Subsurface and outcrop data show the distribution of siliciclastics, carbonates, and evaporites across the inner, middle, and outer portions of the shelf for the Yates Formation (Permian, Late Guadalupian) of the Permian Basin.

The evaporitic inner shelf consists of thick intervals of anhydrite and minor halite interbedded with anhydrite-cemented, Red Argillaceous Siltstone/ Sandstone. The sheet-like geometry of the beds and lack of evidence for channels, as shown by log correlations, suggest that sheetflood and eolian processes may have been the dominant modes of sediment transport.

The middle shelf was dominated by siliciclastic deposits and separated the evaporitic inner shelf from the carbonate-rich shelf margin. Siliciclastic facies in the middle shelf consist of alternating Light Brown Arkosic Sandstones (shoreline deposits), Dark Gray Argillaceous Siltstones (wet mud flats transitional to shallow lagoons), and Red Argillaceous Siltstones similar to that of the inner shelf. Carbonate lithologies consist predominantly of structureless to algal-laminated, peloidal dolomudstones that locally show signs of erosion (ripped-up clasts). Green-Gray Dolomitic Subarkosic Siltstones/Sandstones occur in the middle- to outer-shelf region and are associated with algal-laminated dolomudstones and minor, thin-bedded, pisolitic packstones (tidal flat to shallow lagoon deposits).

The shelf margin consists of Red Anhydritic Siltstone/ Sandstone that passes downdip to Gray Bioturbated, Kaolinitic Dolomitic Quartz Sandstone. These siliciclastics occur with dolomites of the shelf margin pisolite shoal complex, with the Gray Sandstones deposited in a more shallow marine environment than the subaerially deposited Red Siltstones.

Vertical stacking of the various facies is interpreted to be the result of cyclic sea-level variation and is thus a shelf-wide phenomenon. Silts and sands were transported across the shelf and to the shelf margin during lowstands of sea level, and outer-shelf clastics were reworked during subsequent sea-level rises. Cyclostratigraphic analyses suggest that the siliciclastic deposition and stacking of facies occurred during three orders of relatively low-amplitude, sea-level fluctuations. A depositional model summarizes the effect of these sea-level fluctuations on facies distribution for shelf areas with different slopes and/or subsidence (i.e., Central Basin Platform vs Northwest Shelf). An understanding of the sea-level fluctuations and the variable shelf profiles enhances facies correlations in a mixed siliciclastic and carbonate system like the Yates Formation.

INTRODUCTION

The Yates Formation is a siliciclastic, carbonate, and evaporitic sequence that was deposited on a broad, very shallow shelf rimming the Delaware Basin in Late Guadalupian time (Fig. 1A). The Yates is one of the shelf equivalents to the Capitan Reef Complex and the basinal siliciclastics of the Bell Canyon Formation (Fig. 1B). The underlying Seven Rivers and overlying Tansill formations differ from the Yates in that they are predominantly dolomites and evaporites with only minor siliciclastics.

Facies belts within the Yates Formation trend parallel to the shelf margin, resulting in a relatively high degree of facies continuity along strike but a high degree of variability across dip (King, 1948; Motts, 1962; Neese and Schwartz, 1977). North-south, dip-oriented cross section of the Yates Formation on the Northwest Shelf are similar to east-west, dip-oriented cross sections for the Central Basin Platform.

Although descriptions of carbonate facies in the Yates Formation are relatively thorough, little work has been done in describing the distribution and depositional mode of the siliciclastic rocks. These deposits are generally described as being of tidal flat, shallow marine, and/or lagoon origin with little or no attention given to the high degree of variability along depositional dip. Cores, logs, and outcrops show that variability in the clastic rocks is similar in scope to that of the carbonate rocks and is largely dependent on the depositional position on the shelf. Sandstones, siltstones, and minor shales of the Yates Formation exhibit both depositional (grain size, sorting, bed thickness, and sedimentary structures) and diagenetic (authigenic cements and clays) characteristics that are controlled primarily by the position of the deposits along depositional dip and, to a lesser extent, along depositional strike.

PREVIOUS WORK

World-renowned outcrops and giant subsurface accumulations of oil and natural gas have led numerous workers to study the sedimentary rocks of the Permian Basin. Important work detailing stratigraphic relations and facies distribution in the shelf carbonate complex includes that of King (1948), Newell and others (1953), Dunham (1972), Hileman and Mazzullo (1977), Pray and Esteban (1977), Scholle and Halley (1980), Ward and others (1986), and Harris and Grover (1989). These studies place the Yates Formation within a regional stratigraphic framework.

The Yates Formation was initially defined by Cartwright and Adams (in Gester and Hawley, 1929) for a 50-ft (15 m) sandstone bed of Late Permian age in the subsurface of the Yates Field, on the Central Basin Platform in Pecos County, Texas. Later, Mear and Yarbrough (1961) described an uncored, 104-ft (32 m) sequence of the Yates Formation in the Sumrall #5 Douglas

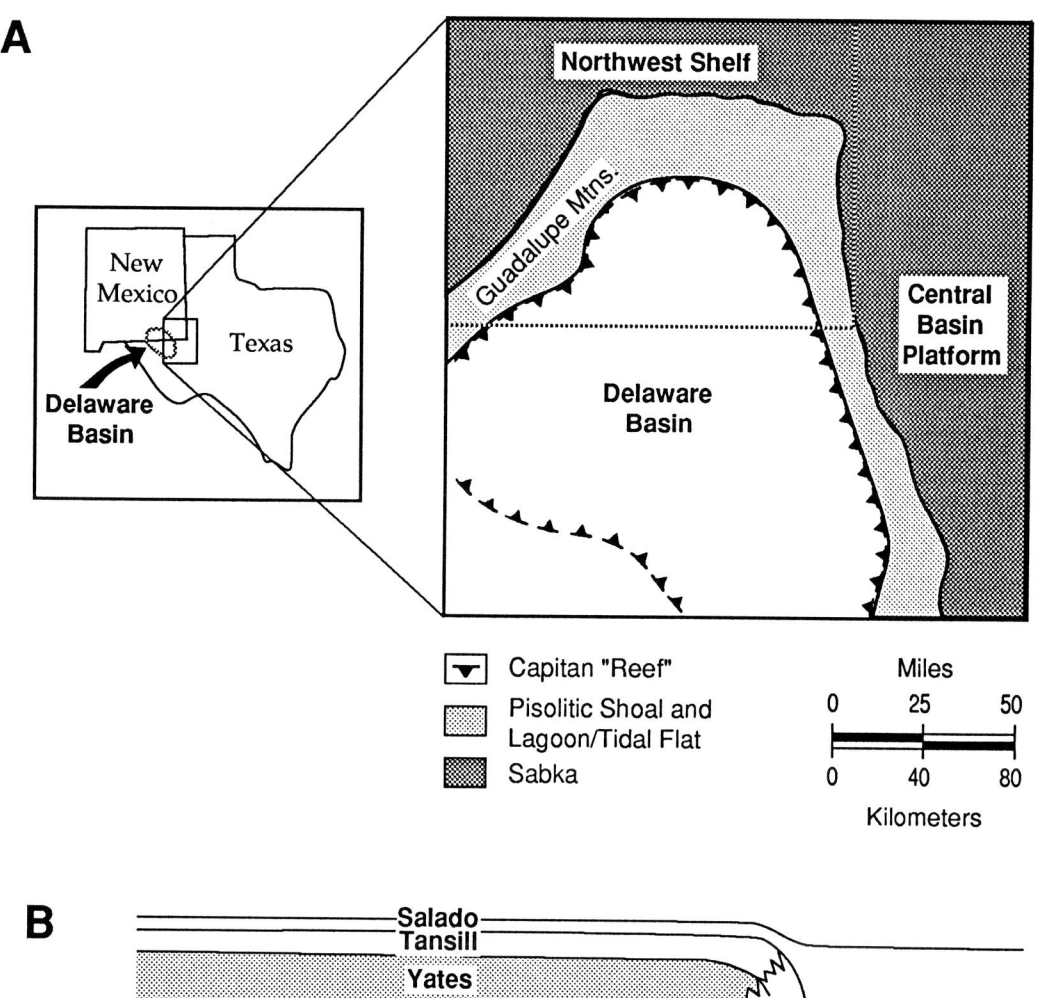
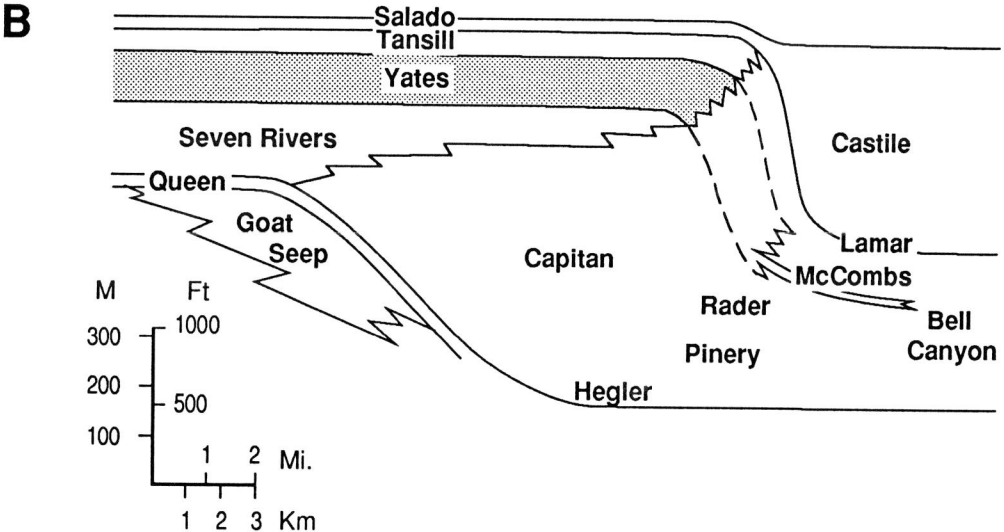

Figure 1.
(A) Location of the Delaware Basin in west Texas and New Mexico. The mixed clastic, carbonate, and evaporitic Yates Formation was deposited on a broad, shallow shelf that rimmed the Delaware Basin in Late Guadalupian (Late Permian) time. (B) Schematic cross section showing Upper Guadalupian stratigraphy of the Delaware Basin. The Yates Formation is a shelf equivalent to the Capitan Reef and basinal siliciclastics of the Bell Canyon Formation.

well in the Yates Field and designated this well as the subsurface "type section" for the Yates Formation. A subsurface type section for the Yates Formation on the Northwest Shelf was also designated by Tait and others (1962) when they defined the Artesia Group. More recently, Lanphere (1972) has proposed a surface type section for the Yates Formation in lower Dark Canyon, Guadalupe Mountains, New Mexico (Sec. 27, T23S, R25E).

Detailed work describing facies distribution for the Yates Formation in outcrop, and to a lesser extent in the subsurface, has been done by many workers. Most studies, however, have concentrated on carbonate lithologies (especially pisolite-rich beds and the occurrence of tepee structures) and have only briefly described the distribution and character of the associated siliciclastic rocks. Some workers (Mear and Yarbrough, 1961; Zinz, 1971; Franco, 1973; Smith, 1974; Candelaria, 1982; Thorkelson, 1983; and Mazzullo and others, 1985) have given more consideration to the siliciclastic rocks. In general, however, these descriptions and interpretations are incomplete or localized to only a relatively small area of study and thus provide no uniform model for siliciclastic deposition during Yates time.

Mear and Yarbrough (1961), for example, thought sandstones and siltstones in the Yates Field at the southern edge of the Central Basin Platform to be subaqueous, shallow lagoon sediments that were deposited by longshore currents and storm-induced waves. They considered the backreef lagoon to have a low-relief sand shoreline that was periodically inundated by north-directed storm waves that spread red siltstone over the eastern Central Basin Platform. Kendall (1969) briefly described Yates siltstones and sandstones from outcrops in the Guadalupe Mountains of New Mexico and formulated a contrasting interpretation. Due to the lack of traction current features in these rocks and the downwind relationship to Permian, red, eolian sands of Oklahoma and Kansas, he considered the Yates siliciclastics to be subaerial wind deposits. He envisioned the migration of dunes and sandflats across a broad supratidal shelf during low sea-level stands.

Silver and Todd (1969) developed a regional model, based on cyclic sea-level changes, for most of the Permian (late Wolfcampian, Leonardian, and Guadalupian) shelf deposits in the Permian Basin. In their model and that of Meissner (1969, 1972), sheet-like shelf sandstones and siltstones such as those in the Yates Formation were thought to have been deposited as coalescing eolian and fluvial sediments that prograded across supratidal flats to the shelf edge during the low sea-level stands. Smith (1974) gives a general description of Yates sandstones and siltstones in the Guadalupe Mountains and interprets the red siliciclastics and interbedded evaporites of the northwestern part of the Northwest Shelf to have been deposited in semicontinental conditions on a prograding coastal plain or sabkha. Smith (1974) after Silver and Todd (1969) considered the clastic rocks to be deposited across the shelf during low sea-level stands but showed that the clastic rocks of the outer shelf contain traction features that are evidence of water-laid deposition. He suggests that this was a product of marine reworking during subsequent sea-level highs.

Studies by the University of Wisconsin have suggested that the clastic shelf deposits of the Yates Formation have a shallow marine origin. Pray (1977) proposed an "All Wet," constant sea-level hypothesis for the shelf strata. In this model, the siliciclastic rocks were considered to have been uniformly deposited across the shelf by shallow seas during minor (1 to 2 m), gradual rises of sea level and/or episodic subsidence. Several M.S. theses by Pray's students have focused on outcrops in the Guadalupe Mountains and have further developed his "All Wet" model. Most pertinent is work done by Candelaria (1982, 1983, 1988, 1989) describing upper Yates carbonate and siliciclastic depositional patterns of the outermost shelf, but other studies include Hurley's (1978, 1979, 1989) descriptions of the lower Seven Rivers, outer-shelf carbonate and siliciclastic rocks in the North Fork of McKittrick Canyon; Neese's (1979, 1989) work on along-strike facies patterns of lower Tansill and upper Yates strata in Walnut Canyon; and Schwartz's (1981) studies of dip-oriented facies patterns of lower Tansill and upper Yates strata in Rattlesnake Canyon. Neese and Schwartz published preliminary results in Hileman and Mazzullo (1977).

Candelaria's stratigraphic sections encompass an area of 69 square miles (179 km^2), but all are within 3 miles (4.8 km) of the youngest Capitan shelf edge. Candelaria suggested the rare sedimentary structures present are evidence for subaqueous deposition with hydraulic energy diminishing shelfward. He considered the sandstone/carbonate couplets as shoaling-upward hemicycles. The clastic sediments occur as basal, deeper-water deposits and are overlain by shoal-water carbonate sediments.

Outcrop studies (e.g., Hurley, 1978, 1989; Borer and Harris, 1989) indicate that shelf sand deposition is in part synchronous with shelf carbonate deposition. These studies show that the siliciclastics extended seaward of the shelf margin pisolite shoal facies, the topographically highest part of the shelf (Dunham, 1972), either as migrating sheets or through channels that breached the shoals. The siliciclastics entered the Delaware Basin through gullies and channels in the reef, as evidenced by a thick, siltstone-filled channel within the Capitan reef facies of the PDB-04 core (Garber and others, 1989). The Capitan reef remained submerged throughout, because of its deep-reef nature. It continued to grow more or less unobstructedly, and allochthonous carbonate sediment was shed into the basin.

SILICICLASTIC FACIES AND ASSOCIATED CARBONATES

The Yates shelf can be divided into inner-, middle-, and outer-shelf (shelf margin) regions that each have a distinct association of siliciclastic and carbonate/evaporite facies. Platforms rimming the Delaware Basin have long been shown to contain an evaporitic inner shelf and an outer carbonate-rich shelf margin (King, 1948; Dunham, 1972; Hileman and Mazzullo, 1977). This study suggests that the clastic-rich belt separating the inner shelf and shelf margin is equally important and should be called the "middle shelf." This is an important distinction because of both the concentration of Yates oil production in the middle-shelf region and also the role the clastic-rich belt may play in separating the evaporite regime from the carbonate regime.

Cores from the Yates Formation in the middle and outer shelf regions of the Central Basin Platform and the Northwest Shelf are the primary basis for the lithologic interpretation presented in this paper (Fig. 2). A dip-oriented cross section (Fig. 3) depicts the relative position of four cores that were described in detail, and also the Dollarhide #7-15-C core. The Dollarhide #7-15-C core, described by Crawford and Dunham (1983), is of particular interest since it provides data from the updip, evaporitic, inner-shelf region some 12 to 14 miles (19 to 22 km) behind the Yates shelf edge. The two HSA cores are from the southern portion of Chevron's North Ward Estes Field. HSA #1187 is near the eastern edge of the field, approximately 8 to

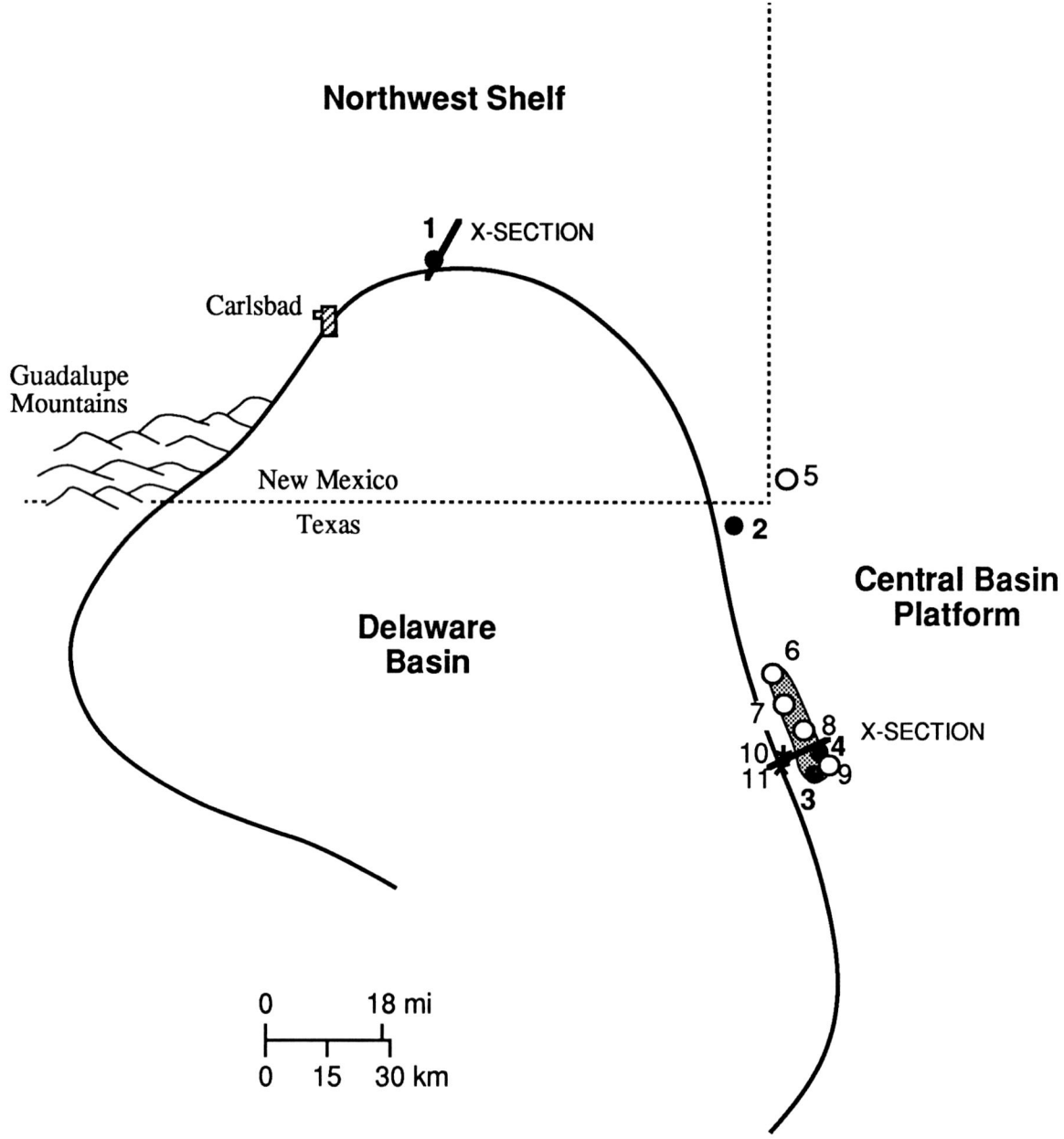

Figure 2
Map of Delaware Basin showing the location of cores and cross sections used in study. Solid dots represent Yates wells with cores and thin sections that were described in detail. Open circles represent Yates cores that were described in less detail, and asterisks show location of wells for which only cuttings were available. Heavy lines locate cross sections that were used to compare Yates stratigraphy between the Northwest Shelf and Central Basin Platform.

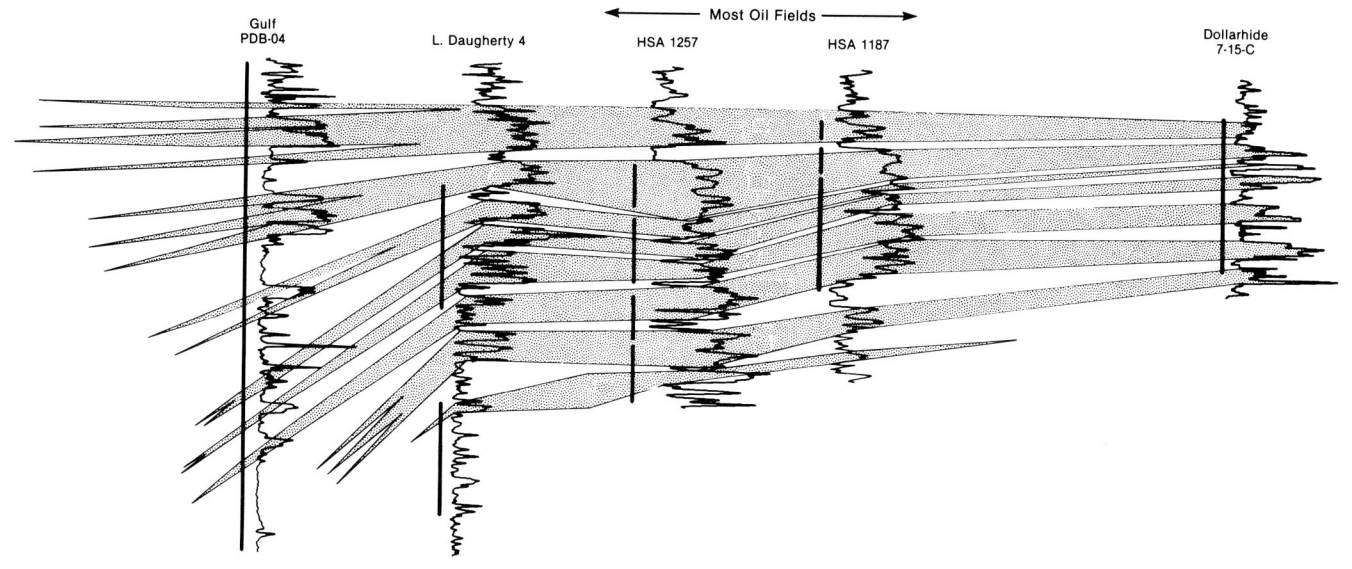

Figure 3.
Composite dip-oriented cross section showing the relative positions of Yates cores used in study and the correlations of laterally continuous siliciclastic and carbonate packages between wells. The wells are from various positions on the Central Basin Platform and the Northwest Shelf (see Fig. 2 for locations of the wells) and have been projected along strike into the cross section. Correlations are based on lithology, gamma-ray and density logs, and comparison to more detailed subsurface cross sections. Sandstones have high gamma-ray (API) values due to the presence of abundant feldspar; whereas, the carbonate intervals show low gamma-ray values. The heavy bars to the left of each gamma-ray log show where core was available.

Well	Distance from Capitan Margin m (km)
Gulf PDB-04	1-2 (1.6 - 3.2)
L. Daugherty 4	3-4 (4.8 - 6.4)
HSA 1257	6-7 (9.6 - 11)
HSA1187	8-9 (13 - 14.5)
Dolarhide 7-15-C	12-14 (19 - 22)

9 miles (13 to 14.5 km) behind the latest Yates shelf margin; whereas, HSA #1257 is in the central part of the field, 6 or 7 miles (9.6 to 11 km) behind the shelf edge. The L. Daugherty #4 core is from the Henderson Field and is 3 to 4 miles (4.8 to 6.4 km) behind the Yates shelf edge. The PDB-04 well, described in detail by Garber and others (1989), is on the Northwest Shelf about 1 to 2 miles (1.6 to 3.2 km) behind the Yates shelf margin.

Although variable, the siltstones and very fine sandstones throughout the Yates Formation are generally similar, in that they are lithic subarkoses to arkoses, with variable amounts of detrital clay, authigenic clays, and carbonate and anhydrite cements. These observations are in good agreement with those of Franco, 1973, and Candelaria, 1982. We subdivide the Yates into six siliciclastic lithofacies based on texture, mineralogy, log response, and the associated carbonates. The striking variation in these lithologies over the outer 15 miles (24 km) of the Yates shelf is reflected on Figure 3 by the basinward thickening of the Formation and the dramatic changes in gamma-ray response. The facies are defined as end members, and it should be kept in mind that there are transitions between facies. The six lithofacies and their depositional settings are:

1. Red Argillaceous Siltstone/Sandstone of the middle and inner shelf.

2. Light Brown Arkosic Sandstone of the middle shelf.

3. Dark Gray Argillaceous Siltstone of the middle shelf.

4. Green-Gray Dolomitic Subarkosic Siltstone/Sandstone of the middle to outer shelf.

5. Red Anhydritic Siltstone/Sandstone of the outer shelf.

6. Gray Bioturbated, Kaolinitic Dolomitic Quartz Sandstone of the outer shelf.

A cross section (Fig. 4) shows the general distribution of the siliciclastic lithofacies. The cross section is based on numerous, but widely spaced, point sources of rock data (Table 1) that are projected into relative positions along dip. A major limitation of the cross section is that the outer-shelf facies distribution is based on data from the Northwest Shelf; whereas, the distributions of middle- and inner-shelf facies are based largely on data from the Central Basin Platform. Outcrops (discussed later) in the Guadalupe Mountains provide some data on middle-shelf facies on the Northwest Shelf (Borer and Harris, 1989). However, most outcrops are for outer-shelf facies (Franco, 1973; Candelaria, 1982, 1989). No core data were available for the shelf-margin region of the Central Basin Platform, and therefore the distribution of both siliciclastic and carbonate facies in this region is poorly understood. Keeping these limitations in mind, the cross section is used to describe the general siliciclastic facies architecture of the Yates shelf.

The evaporitic inner shelf of Figure 4 was apparently an extensive low gradient area that began 6 to 8 miles (9.6 to 13 km) behind the shelf margin and extended for tens of miles to the north and east (across a filled Midland Basin) during Yates deposition (King, 1948; Ward and others, 1986). The inner-shelf region, as represented in this study by the Dollarhide #7-15-C well, consists of thick intervals of anhydrite and minor halite interbedded with anhydrite-cemented, Red

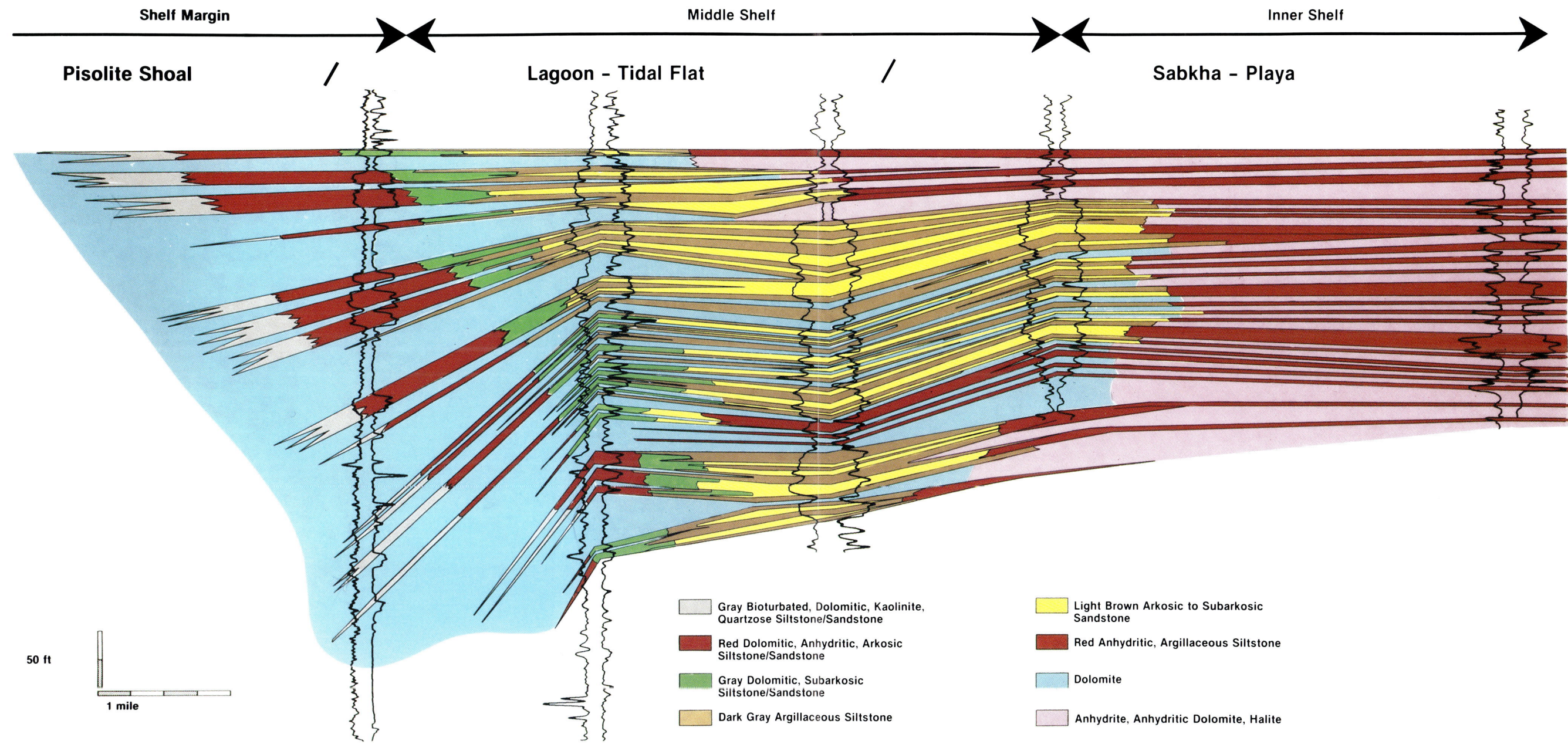

Figure 4
Schematic cross section showing the distribution of major siliciclastic lithofacies in the Yates Formation. The cross section is the same as in Figure 3; however, lithofacies have been identified for the cored intervals and interpreted for the uncored intervals using wireline log response. Interpolation between cored wells was aided by the use of logs from uncored wells. Note the gamma-ray log is plotted on the right and density log on the left for each well to produce the "Butterfly" appearance and accentuate the cyclic patterns apparent in the logs.

Shelf position	Environment	LOCATION	LITHOLOGY	ASSOC. CARB./EVAP.	TEXTURE	MINERALOGY	RESERVOIR QUALITY	CYCLICITY
EVAPORITIC INNER SHELF	SABKHA/ PLAYA	**DOLLARHIDE 7-15-C** NO. CENTRAL BASIN PLATFORM	- **RED ARGILLACEOUS SILTSTONE FACIES (WITH MINOR RED AND GRAY SANDSTONE).** - MEDIUM TO THICK-BEDDED (1- 10 FT. THICK). - SHARP BASAL CONTACTS AND GRADATIONAL UPPER CONTACTS WITH LAMINATED AND CHAOTIC ANHYDRITE. - STRUCTURELESS TO WISPY-LAMINATED, CHAOTIC ANHYDRITE-RICH ZONES, ABUNDANT DOLOMITE/ANHYDRITE RIP-UP CLASTS.	- **MASSIVE, LAMINATED, AND CHAOTIC ANHYDRITE WITH MINOR HALITE AND RESIDUAL DOLOMITE.** - THIN INTERLAMINATED DOLOMITE AND ANHYDRITE, AND MASSIVE OR NODULAR ANHYDRITE ARE PROBABLE ORIGINAL DEPOSITIONAL TEXTURES. - CHAOTIC AND GHOST TEXTURES SUGGEST THE ANHYDRITE IS LOCALLY A REPLACEMENT PHASE AFTER DOLOMITE.	- **SILTSTONE AND VERY FINE SANDSTONE** WITH MINOR, WELL-ROUNDED, FINE- TO MEDIUM-GRAINED QUARTZ SAND. - MODERATELY WELL-SORTED. - ABUNDANT **RED (HEMATITE-STAINED) DETRITAL CLAY/MICA** OCCURS AS WISPY LAMINAE, GRAIN RIMS, AND DIFFUSE CLAY-RICH ZONES. - TANGENTIAL TO STRAIGHT GRAIN CONTACTS; MINOR (ANHYDRITE) MATRIX-SUPPORTED TEXTURES.	- **SUBARKOSIC TO ARKOSIC** WITH COMMON LITHIC GRAINS AND ACCESSORY MINERALS. - **MICACEOUS.** - ABUNDANT **ANHYDRITE CEMENT.**	- **VERY POOR RESERVOIR QUALITY** DUE TO THE ABUNDANCE OF BOTH RED DETRITAL MATRIX AND ANHYDRITE CEMENT.	- **NOT ANALYZED FOR CYCLICITY IN ANY DETAIL.** - MM- TO CM-SCALE , VARVE-LIKE LAMINAE ARE LOCALLY WELL-DEVELOPED. - SILICICLASTIC - EVAPORITE PACKAGES MAY EXHIBIT CYCLICITY AND REQUIRE FURTHER STUDY.
SILICICLASTIC-RICH MIDDLE SHELF	TIDAL FLAT	**U.S. HWY 285 OUTCROP** GUADALUPE MNTS.	- **RED ARGILLACEOUS SILTSTONE FACIES (WITH ABUNDANT RED AND GRAY SANDSTONE).** - 20 TO 25 FT. THICK INTERVAL CONTAINING FOUR FINING-UP SEQUENCES OF SILTY SANDSTONE GRADING TO ARG. SILTSTONE. - SANDSTONES HAVE SHARP BASAL CONTACTS; GRADATIONAL CONTACT BETWEEN UPPER ARG. SLTST AND OVERLYING DOLOMUDSTONE. - WAVY-LAMINATED SILTSTONES; POSSIBLE ADHESION STRUCTURES IN SANDSTONES.	- **MASSIVE, ALGAL-LAMINATED, AND CLOTTED, DOLOMUDSTONE .** - DOLOMITE INTERVALS ARE NOT COMPLETE IN OUTCROP BUT WIRELINE LOGS SUGGEST THEY MAY BE 30 TO 40 FT. THICK. - FENESTRAL FABRIC IS COMMON. - ARGILLACEOUS INTERVALS COMMON.	- **SILTY VERY FINE-GRAINED SANDSTONE TO ARGILLACEOUS SANDY SILTSTONE.** - MODERATE- TO WELL-SORTED. - **ELONGATE PODS** (0.5 TO 1.5 CM. HIGH BY 5 TO 10 CM. LONG) OF **GRAY, WELL-SORTED VERY FINE-GRAINED SAND** SET IN A RED ARGILLACEOUS SILTSTONE/SANDSTONE MATRIX MAY BE **HORIZONTAL BURROWS OR SOME TYPE OF ADHESION STRUCTURES.**	- **SUBARKOSIC TO ARKOSIC** WITH ABUNDANT LITHIC GRAINS, ACCESSORY MINERALS, **MICA, AND RED (FE-STAINED) DETRITAL CLAY.** - **ABUNDANT FINE- TO MEDIUM-CRYSTALLINE DOLOMITE CEMENT.** - MINOR VERY FINE DOLOMITE MATRIX. - KAOLINITE AND SERICITE FROM PRESENT DAY WEATHERING OF OUTCROP.	- **POOR RESERVOIR QUALITY** DUE TO THE PRESENCE OF ABUNDANT (DETRITAL AND DOLOMICRITE) MATRIX AND DOLOMITE CEMENT (UP TO 40%). - <3% POROSITY.	- **WELL-DEVELOPED CYCLICITY.** - SMALL-SCALE (100 K.Y.) CLASTIC CYCLES ARE DEFINED BY MULTIPLE FINING-UP INTERVALS OF SANDSTONE AND ARG. SILTSTONE; BASAL SANDSTONES HAVE SHARP, EROSIVE CONTACTS WITH ARG. SILTSTONES OR DOLOMUDSTONE OF PRECEDING CYCLE. - LARGE-SCALE (400 K.Y.) CYCLE CONSISTING OF ALTERNATING THICK CLASTIC- AND DOLOMITE-RICH INTERVALS NOT COMPLETE DUE TO LIMITED OUTCROP EXPOSURE.
	LOW ENERGY SHORELINE	(east part of field) **HSA 1187** **HSA 1386** **HSA 1391** **NO.WARD ESTES FIELD** **CENTRAL BASIN PLATFORM** (west and central part of field) **HSA 1257** **G.W. O'BRIEN 1491** **G.W. O'BRIEN 1514** **HENDERSON FIELD** **NO. CENTRAL BASIN PLATFORM** **L. DAUGHERTY 4** (UPPER PART)	- **LT. BROWN ARKOSIC SANDSTONE FACIES , DK. GRAY ARGILLACEOUS SILTSTONE FACIES, AND RED ARGILLACEOUS SILTSTONE FACIES.** - THICK SILICICLASTIC INTERVALS (UP TO 50 FT. THICK) CONSISTING OF ALTERNATING LT. BROWN SS. (5-15 FT. THICK) AND DK. GRAY OR RED ARG. SILTST (3 TO 10 FT. THICK). - SANDSTONES ARE POORLY CONSOLIDATED AND STRUCTURELESS, WITH RARE WISPY-LAMINAE; ARG. SLTST ARE WISPY-, WAVY-, AND PLANE-LAMINATED WITH RARE RIPPLE CROSS-LAMINAE. - RED SILTSTONES ARE CONCENTRATED ALONG A SPECIFIC HORIZON IN THE LOWER YATES BUT ALSO OCCUR AT OTHER HORIZONS THROUGHOUT THE MIDDLE-SHELF REGION. - GRADATIONAL CONTACTS BETWEEN CLEAN SANDSTONES AND ARG. SILTSTONES; SHARP CONTACTS BETWEEN ARG. SLTST AND RIPPED-UP, ALGAL DOLOMUDSTONES.	- **MASSIVE, ALGAL-LAMINATED, AND CLOTTED, DOLOMUDSTONE AND INTRACLASTIC DOLOWACKESTONE/PACKSTONE.** - DOLOMITE INTERVALS ARE <1 TO 7 FT. THICK. - INTERVALS OF DOLOMUDSTONE RIP-UP CLASTS SET IN A DOLOMITIC ARGILLACEOUS MATRIX ARE COMMON AND OFTEN ASSOCIATED WITH RED ARG. SILTSTONES. - MINOR FENESTRAL FABRIC. - COMMON STYLOLITES. - SOLUTION-ENLARGED VUGS AND MICRO-FRACTURES TOWARD DOWN-DIP LIMIT.	**LIGHT BROWN SANDSTONE** - **SILTY, VERY FINE-GRAINED WITH MINOR WELL-ROUNDED FINE TO MED. QUARTZ SAND.** - **WELL-SORTED.** - SUBANGULAR TO SUBROUNDED. - TANGENTIAL GRAIN CONTACTS. - **POORLY CONSOLIDATED.** **RED AND GRAY SILTSTONES** - ARGILLACEOUS, VERY FINE SANDY SILTSTONE. - MODERATELY POOR TO MODERATELY GOOD SORTING. - ANGULAR TO SUBROUNDED. - TANGENTIAL TO STRAIGHT CONTACTS WITH ABUNDANT MATRIX SUPPORTED LENSES OR LAMINAE.	**LIGHT BROWN SANDSTONE** - **SUBARKOSIC TO ARKOSIC** WITH ABUNDANT LITHIC GRAINS AND ACCESSORY MINERALS. - VERY FINE- TO MEDIUM-CRYSTALLINE DOLOMITE CEMENT COMMON. - PATCHY ANHYDRITE CEMENT PARTICULARLY NEAR UPDIP TRANSITION TO EVAPORITIC INNER SHELF. - ABUNDANT GRAIN-COATING AUTHIGENIC CORRENSITE CLAY. - COMMON, BUT MINOR, AUTHIGENIC K-FELDSPAR CEMENT. **RED AND GRAY SILTSTONES** - **SUBARKOSIC TO ARKOSIC** WITH ABUNDANT LITHIC GRAINS, ACCESSORY MINERALS, MICA, AND DETRITAL CLAY. - ABUNDANT VERY FINE DOLOMITE MATRIX/ CEMENT. - ANHYDRITE CEMENT IN SOME RED SILT-STONE INTERVALS.	**LIGHT BROWN SANDSTONE** - **GOOD TO VERY GOOD RESERVOIR QUALITY . PRODUCTION IN NORTHWARD ESTES FIELD IS PREDOMINANTLY FROM THIS LITHOFACIES.** - 15 TO 30% POROSITY. - 10 TO 100 md. PERMEABILITY. - LOW PERMEABILITIES ARE RELATED TO PATCHY ANHYDRITE AND DOLOMITE CEMENTS AND AUTHIGENIC CORRENSITE. **DK. GRAY SILTSTONE** - **POOR TO MODERATE RESERVOIR QUALITY** DEPENDING ON THE AMOUNT DETRITAL CLAY AND DOLOMITE MATRIX. - 5 TO 15% POROSITY; HOWEVER, UP TO HALF MAY BE NONEFFECTIVE MICRO-POROSITY ASSOCIATED WITH DETRITAL AND AUTHIGENIC CLAYS. **RED SILTSTONE** - **POOR RESERVOIR QUALITY** DUE TO ABUNDANT MATRIX AND CEMENT.	- **MODERATELY WELL-DEVELOPED CYCLICITY.** - DEPOSITIONAL CYCLES ON THE CENTRAL BASIN PLATFORM ARE GENRALLY LESS DEVELOPED THAN ON THE NORTHWEST SHELF DUE TO A STEEPER TOPOGRAPHIC AND/OR SUBSIDENCE PROFILE . - ALTERNATING INTERVALS OF LT. BROWN , RESERVOIR SANDSTONE AND DK. GRAY, POOR RESERVOIR ARG. SILTSTONE FORM SMALL-SCALE (100 K.Y.) CYCLES SIMILAR TO THE WELL-DEVELOPED FINING-UP CYCLES FOUND IN THE U.S. HWY 285 OUTCROP. - DOLOMITE INTERVALS MAY REPRESENT THE HIGH SEA-LEVEL STAND PORTIONS OF LARGE-SCALE (400 K.Y.) CYCLES. - CHANGES IN THE LATTERAL POSITION AND QUALITY OF RESERVOIR SANDSTONES (I.E., SILICICLASTIC PORTIONS OF 400 K.Y. CYCLES) DEFINE A LOW-FREQUENCY (1.5-2 M.Y.) DEPOSITIONAL CYCLE.
	LAGOON	**HENDERSON FIELD** **NO. CENTRAL BASIN PLATFORM** **L. DAUGHERTY 4** (MIDDLE PART)	- **GREEN-GRAY, DOLOMITIC SILTSTONE/ SANDSTONE FACIES.** - MEDIUM- TO THICK-BEDDED (2-6 FT. THICK) - ASSOCIATED WITH THIN, TERRIGINOUS MUDSTONE BEDS. - SHARP BASAL CONTACTS, AND GRADATIONAL AND SHARP UPPER CONTACTS, WITH ALGAL DOLOMUDSTONES. - STRUCTURELESS, TO WAVY-LAMINATED.	- **ALGAL-LAMINATED, CLOTTED, AND MASSIVE PELOIDAL DOLOMUDSTONES, WACKE-STONES, AND PACKSTONES.** - RARE THIN (CENTIMETER-SCALE) PISOLITE BEDS - LOCALLY WELL-DEVELOPED FENESTRAL , INTERPARTICLE, AND MOLDIC POROSITY. - COMMON SOLUTION-ENLARGED VUGS AND MICROFRACTURES. - COMMON ARGILLACEOUS INTERVALS. - COMMON PYRITE AND STYLOLITES.	- **MUDSTONE AND ARGILLACEOUS SILT-STONE TO SILTY, VERY FINE-GRAINED SANDSTONE.** - COARSER TEXTURES EXHIBIT MODERATE TO MODERATELY GOOD SORTING; WHEREAS, FINER SAMPLES EXHIBIT POOR TO MODERATE SORTING WITH ABUNDANT DETRITAL CLAY MATRIX. - POINT OR STRAIGHT GRAIN CONTACTS PREDOMINATE; HOWEVER, **MATRIX SUP-PORTED TEXTURES ARE ALSO COMMON.**	- **SUBARKOSIC** WITH ABUNDANT LITHIC GRAINS (CHERT, VRF's, IRF's), ACCESSORY HEAVY MINERALS (ILMENITE, ZIRCON, HORNBLENDE, ETC.), AND MICA. - MINOR FELDSPAR DISSOLUTION AND KAOLIN-ITE GROWTH. - **ABUNDANT VERY FINE DOLOMITE MATRIX AND MEDIUM CRYSTALLINE PORE-FILLING CEMENT.** - MINOR FE-DOLOMITE.	- **SILICICLASTICS HAVE POOR RESERVOIR QUALITY** DUE TO ABUNDANT (DOLOMITE AND DETRITAL CLAY) MATRIX AND PORE-FILLING DOLOMITE CEMENT. - <6 % MEASURED POR.; <1 md. PERM. - **DOLOMITES ARE OIL-STAINED AND SHOW LOCAL PRODUCTION.** - MEASURED FENESTRAL, INTERPARTICLE AND MOLDIC POROSITIES FROM 3 - 8%; MEASURED PERM. FROM <1 TO 1000 md.	- **CYLICITY NOT READILY APPARENT DUE TO RAPID FACIES CHANGES IN LAGOONAL FACIES BELT.** - ALSO CYCLICITY ON THE CENTRAL BASIN PLATFORM IS LESS WELL-DEVELOPED THAN ON THE NORTHWEST SHELF DUE TO A STEEPER TOPOGRAPHIC AND/OR SUBSIDENCE PROFILE.
CARBONATE-RICH OUTER SHELF	PISOLITE SHOAL COMPLEX (BACK) (CREST)	**WALNUT CANYON HAIRPIN OUTCROP** GUADALUPE MNTS. **GULF PDB-04 UPPER YATES** NORTHWEST SHELF **DARK CANYON OUTCROP** GUADALUPE MNTS.	- **RED, ANHYDRITIC SILTSTONE/SANDSTONE FACIES** - THICK- TO VERY THICK-BEDDED (5 TO 15 FT. THICK). - COMMONLY SHOW FINING-UP SEQUENCE OF SILTY, VERY FINE SANDSTONE CAPPED BY DOLOMITIC, ARGILLACEOUS SILTSTONE. - SHARP BASAL AND GRADATIONAL UPPER CONTACTS WITH SHOAL CREST DOLOMITES. - WISPY- TO WAVY-LAMINATED. - **RED, ANHYDRITIC SILTSTONE/SANDSTONE FACIES** - THICK-BEDDED (5 TO 10 FT. THICK). - UNIFORM ,SILTY V.F. SANDSTONE BEDS WITH NO FINING-UP, SHALEY TOPS - SHARP BASAL AND GRADATIONAL UPPER CONTACTS WITH SHOAL CREST DOLOMITES. - STRUCTURELESS TO WISPY-LAMINATED.	- **PISOLITE SHOAL COMPLEX DOLOMITES WITH ABUNDANT ANHYDRITE CEMENT.** - MASSIVE AND ALGAL-LAMINATED DOLO-MUDSTONE; PELOIDAL DOLOWACKESTONE/ PACKSTONE; AND PISOLITIC DOLOPACK-STONE/GRAINSTONE. - SMALL (DECIMETER- TO METER-SCALE) SHOALING AND DEEPENING (AUTO?) CYCLES ARE COMMON. - SUPRATIDAL FEATURES INCLUDE VADOSE CRUSTS (MICRITIC AND PISOLITIC), RIP-UP HORIZONS, FENESTRAL FABRIC, AND TEPEE STRUCTURES (CHAOTIC BEDDING WITH LARGE ANHYDRITE- AND RED CLAY-FILLED VOIDS)	- **SILTSTONE AND VERY FINE SANDSTONE WITH MINOR, WELL-ROUNDED, FINE- TO MEDIUM-GRAINED QUARTZ SAND.** - TANGENTIAL TO STRAIGHT GRAIN CONTACTS; MINOR (DOLOMITE/ANHYDRITE) MATRIX-SUPPORTED TEXTURES. - MODERATELY WELL-SORTED. - ABUNDANT RED (HEMATITE-STAINED) DETRITAL CLAY/MICA OCCURS AS WISPY LAMINAE, GRAIN RIMS, AND THICK, DIFFUSE CLAY-RICH ZONES.	- **SUBARKOSIC TO ARKOSIC** WITH ABUNDANT LITHIC GRAINS (CHERT, VRF's, IRF's), ACCESSORY HEAVY MINERALS (ILMENITE, ZIRCON, HORNBLENDE, ETC.), AND MICA. - MINOR PORE-BRIDGING ILLITE (RECRYSTAL-LIZED DETRITAL CLAY?). - ABUNDANT, **EARLY (PRECOMPACTIONAL) DOLOMITE, ANHYDRITE, AND MAGNESITE CEMENTS.** - VERY RARE AUTHIGENIC K-FELDSPAR.	- **VERY POOR RESERVOIR QUALITY** DUE TO THE ABUNDANCE OF BOTH ORIGINAL MATRIX (DETRITAL AND CARBONATE) AND EARLY CARBONATE AND EVAPORITE CEMENTS. - CORE PLUG MEASUREMENTS WERE NOT MADE; HOWEVER, LOG POROSITIES RANGE FROM <0 TO 5%.	- **WELL-DEVELOPED DEPOSITIONAL CYCLES OCCUR AT SEVERAL SCALES AND EXHIBIT MILANKOVITCH BAND PERIODICITY.** - SMALLEST SCALE CYCLES (SHOALING AND DEEPENING CARBONATE SEQUENCES AND TEXTURAL VARIATIONS IN SILICICLASTIC INTERVALS) MAY BE AUTOCYCLIC OR RELATED TO HIGH-FREQUENCY (20-40 K.Y.) MILANKOVITCH SIGNALS. - SMALL-SCALE (100 K.Y.) MILANKOVITCH CYCLES ARE CARBONATE AND SILICICLASTIC COUPLETS THAT EXHIBIT REGULAR CHANGES IN THICKNESS (20 TO 50 FT) AND DOMINANT LITHOLOGY (BUNDLING) WHICH DEFINE LARGE-SCALE CYCLES. - LARGE-SCALE (400 K.Y.) MILANKOVITCH CYCLES ARE APPROXIMATELY 100 FT THICK AND CONSIST OF SUBEQUAL DOLOMITE- AND SILICICLASTIC-RICH INTERVALS . - CHANGES IN THE POSITION AND THICKNESS OF 400 K.Y. CYCLES DEFINE A LOW-FREQUENCY (1.5-2 M.Y.) DEPOSITIONAL CYCLE .
	(FRONT)	**GULF PDB-04 LOWER YATES** NORTHWEST SHELF **WALNUT CANYON MOUTH OUTCROP** GUADALUPE MNTS.	- **GRAY BIOTURBATED, KAOLINITIC DOLOMITIC QUARTZ SANDSTONE FACIES.** - MEDIUM TO THICK-BEDDED (3-10 FT THICK). - GRADATIONAL AND SHARP BASAL CONTACTS AND GRADATIONAL UPPER CONTACTS WITH SHOAL FRONT DOLOMITES. - PREDOMINANTLY BIOTURBATED (*CONICHNUS* AND *OPHIOMORPHA/THALASSINOIDES*). - LOCALLY STRUCTURELESS, WAVY-, AND PLANE-LAMINATED.	- **SHOAL-FRONT DOLOMITES AND LIMEY DOLOMITES THAT INCLUDE SUBTIDAL TO SUPRATIDAL FACIES .** - SUPRATIDAL FACIES ARE SIMILAR TO SHOAL-CREST LITHOLOGIES AND INCLUDE CHAOTIC PISOLITE INTERVALS, FENESTRAL DOLO-MUDSTONES, AND RIP-UP BEDS. - SUBTIDAL FACIES INCLUDE SKELETAL DOLO-PACKSTONES AND BOTH MASSIVE AND BURROWED DOLOMUDSTONES. - COMMON BRECCIATED DOLOMITES.	- **SILTY, VERY FINE SANDSTONES WITH INTERVALS OF FINE- TO MEDIUM-GRAINED LAMINAE.** - MODERATELY WELL- TO WELL-SORTED. - LITTLE OR NO DETRITAL CLAY. - ABUNDANT DOLOMICRITE MATRIX.	- **QUARTZOSE** (FELDSPAR AND LITHIC GRAINS DISOLVED). - ABUNDANT **EARLY DOLOMITE MATRIX AND CEMENT.** - **LATE DOLOMITE AND CALCITE CEMENT** - ABUNDANT **KAOLINITE** - AUTHIGENIC PYRITE, ANHYDRITE/GYPSUM, AND QUARTZ OVERGROWTHS ASSOCIATED WITH BURROWS. - OPAQUE GRAINS ALTERED TO LUECOXENE AND PYRITE.	- **POOR RESERVOIR QUALITY** DUE TO ABUNDANT DOLOMITE CEMENT. - 2 TO 10% POROSITY IS LARGELY NON-EFFECTIVE MICROPOROSITY ASSOCIATED WITH KAOLINITE, OR POORLY-CONNECTED SECONDARY PORES RESULTING FROM FELDSPAR DISSOLUTION. - PERMEABILITIES < 1md.	- **MODERATELY WELL-DEVELOPED CYCLICITY.** - CORE DATA SUGGESTS THAT SMALL-SCALE (100 K.Y.) CLASTIC-CARBONATE CYCLES THAT ARE 15 TO 25 FT. THICK IN THE SHOAL-CREST REGION ATTAIN THICKNESSES OF 30 TO 40 FT. IN THE SHOAL-FRONT REGION. - INCOMPLETE SECTIONS AND RAPID FACIES CHANGES MAKE THE SHOAL-FRONT CYCLES DIFFICULT TO ANALYZE.

Table 1 - SUMMARY OF YATES SILICICLASTIC CHARACTERISTICS FOR VARIOUS POSITIONS ON THE SHELF.

Argillaceous Siltstone/Sandstone. Inner-shelf evaporites exhibit textures suggestive of both subaqueous and subaerial (sabkha) deposition. Most of the evaporites, however, seem to have been deposited in shallow ponds (Sarg, 1977; Crawford and Dunham, 1983).

Red Argillaceous Siltstones/Sandstones occur as extensive sheets that may represent deflation surfaces on the broad, low-gradient coastal plain. The texture and composition of red detrital clays suggest that they were oxidized and infiltrated into the silt and sand deposits during stages of repeated wetting and drying (Walker, 1979; Singer, 1988). The sheet-like geometry of the beds and lack of evidence for channels, as shown by log correlations, suggest that sheetflood and eolian processes may have been the dominant modes of sediment transport.

The middle shelf of Figure 4 is defined here as a 2 to 4 mile-wide (3.2 to 6.4 km) region that is dominated by siliciclastic deposits and separates the evaporitic inner shelf from the carbonate-rich shelf margin. Most oil fields producing from the Yates Formation are aligned along the clastic-rich, middle-shelf region (see Ward and others, 1986). HSA #1187 represents the updip portion of the middle shelf and is very close to the transition to the evaporitic inner shelf; whereas, HSA #1257 is positioned near the center of the middle-shelf region. The upper Yates interval in L. Daugherty #4 represents the downdip limit of the middle-shelf region and is near the transition to the outer shelf. Siliciclastic facies in the middle shelf consist of alternating Light Brown Arkosic Sandstones, Dark Gray Argillaceous Siltstones, and Red Argillaceous Siltstones. Carbonate lithologies consist predominantly of structureless to algal-laminated peloidal dolomudstones that locally show signs of erosion (ripped-up clasts).

Good sorting and sparse detrital matrix suggest that the light brown sandstones were deposited in a relatively high-energy environment, and the continuous linear trend of the facies belt is suggestive of a shoreline setting. The lack of abundant upper flow regime sedimentary structures may suggest, however, that the shoreline was only of moderate energy. The winnowing process that led to deposition of the light brown sandstones could be related to the landward migration of a low-energy shoreline or tidal wedge during periodic sea-level rises. Alternatively the clean sandstones may simply represent a time when the sediment being supplied to the shelf was slightly coarser and better sorted.

Stratigraphic correlations and analyses of cyclic sequences for various positions on the shelf (to be discussed in a later section) suggest that intervals of Dark Gray Argillaceous Siltstones are correlative to thin carbonate wedges at the shelf margin (Fig. 4). The argillaceous siltstones are considered to have been deposited as wet mud flats that were transitional to shallow lagoons. During maximum flooding, thin carbonate mudstones were deposited across the middle shelf in a shallow lagoon.

Green-Gray Dolomitic Subarkosic Siltstones/Sandstones occur in the middle- to outer-shelf region and are associated with abundant algal-laminated dolomudstones and minor, thin-bedded, pisolitic packstones. The siltstone/sandstone intervals are argillaceous and contain numerous discrete thin beds of mudstone suggesting deposition in a relatively quiet, low-energy environment. The dark green-gray color of some siltstones and the presence of abundant pyrite nodules further suggest an organic-rich, reducing environment. The dolomitic siltstone/ sandstone facies is interpreted as having been deposited in a tidal flat to shallow lagoon just landward of a pisolite shoal complex.

The outer shelf or shelf margin of Figure 4 is represented by the lower Yates portion of the L. Daugherty #4 core and the PDB-04 core. The shelf margin is defined as the outer 3 to 4 miles (4.8 to 6.4 km) of shelf starting from the point where the formation begins to rapidly thicken and become dominated by carbonates, to the shelf break marked by the presence of Capitan reef facies (Dunham, 1972; Pray and Esteban, 1977; Garber and others, 1989). Outer-shelf siliciclastic facies are Green-Gray Dolomitic Subarkosic Siltstone/Sandstone that passes downdip to Red Anhydritic Siltstone/Sandstone, that in turn changes downdip to Gray Bioturbated, Kaolinitic Dolomitic Quartz Sandstone.

Red Anhydritic Siltstones/Sandstones are associated with dolomites of the shelf margin pisolitic shoal complex. Features in both the carbonates and the siltstones/sandstones suggest that the siliciclastics were deposited across subaerial shoal carbonates (marginal mound of Dunham, 1972) during sea-level lowstands. Supratidal carbonates with abundant desiccation features suggest that the pisolite shoal was periodically exposed above sea level. Furthermore, the presence of red silt and clay in tepee-related voids within carbonate intervals suggests that clastics were moving across the shoal while it was exposed. Also the sharp planar contacts that typically exist between the red siltstones/sandstones and the underlying dolomites suggest a period of erosion.

Features found within the siliciclastic rocks may also indicate deposition during a low sea-level stand. Red (FeO_2-stained) detrital clay plastered to sand and silt grains, as mentioned previously, is thought to be airborne dust that was oxidized and infiltrated into sediment during periods of repeated wetting and drying. The illitic clay, such as that present in the red anhydritic siltstone/sandstone facies, is believed to be characteristic of aridic soils formed during numerous cycles of deflation and deposition. The sheet-like bedforms of the shoal crest clastics are also suggestive of an erosional or deflation surface.

The sediment transport processes involved in moving sand and silt from the clastic-rich, mid-shelf region seaward across the carbonate shoal and to the shelf margin are not entirely clear. Evidence cited above suggests the transport took place during low stands of sea level and that eolian processes were likely active; however, unchanneled fluvial processes (sheetfloods) may also have been important. Interpretation of the transport mechanism is difficult since the outer-shelf clastics were probably reworked during subsequent sea-level rises.

Gray Bioturbated, Kaolinitic Dolomitic Sandstones were deposited along the seaward edge of the pisolite shoal complex in what is defined here as the shoal-front region. Associated carbonates consist of pisolite sequences that are more skeletal-rich than those found in the shoal-crest region. Thick beds of subtidal dolomudstones are also common, as are intervals of intraclastic dolowackestone/packstone. The gray sandstones are interpreted as having been deposited in a more marine environment than the subaerially deposited red siltstones. Abundant dolomicrite matrix and extensive bioturbation are suggestive of shallow marine conditions. Intervals with slightly coarser grain size (including well-sorted, fine- to medium-grained laminae) and only sparse detrital clay suggest slightly higher-energy currents than those which reworked the red siltstones. Pray (in Hurley, 1978) suggests that the basinward limit of the shelf was a site of high-energy winnowing currents. Concentrations of fine- to medium-grained clastic material and associated intraclastic dolomite may represent a high-energy shoreface; whereas, burrowed gray sandstone; local, thin-bedded detrital claystone; and medium- to thick-bedded, massive and burrowed dolomudstones may represent just offshore, shallow marine facies.

The siliciclastic facies and associated carbonates/evaporites are each described and illustrated in detail in the following sections in order to give a clear picture of the lithological variation that was present across the Yates shelf. After the descriptive sections, the facies distribution, especially the cyclical arrangement of siliciclastic and carbonate alternations in the Yates Formation, is explained with a depositional model that invokes low-amplitude sea-level fluctuations and perhaps associated climate variation.

Red Argillaceous Siltstone/Sandstone

Medium- to very thick-bedded, Red (and gray-red) Argillaceous Siltstone/Sandstone occurs in the inner-shelf (Dollarhide #7-15-C, Fig. 5) and middle-shelf (HSA #1187 and HSA #1257, Figs. 6 and 7) regions of the Central Basin Platform. In the Dollarhide #7-15-C core, nearly all siliciclastic intervals consist of this facies; whereas, in the HSA cores the red siltstones/sandstones occur 70 to 90 ft (21 to 27 m) above the base of the formation (Fig. 4). The red siltstones/sandstones are structureless to plane- and wavy-laminated. Rip-up clasts of massive and algal-laminated dolomudstone are common near the base and top of beds. Basal contacts and upper contacts with dolomudstone are sharp.

The Red Argillaceous Siltstones/Sandstones range from mudstones to argillaceous, silty sandstones (Fig. 8A through D). Samples are micaceous, lithic, subarkoses, and arkoses (Table 2), with abundant micaceous/illitic detrital clay and dolomicrite matrix/cement (Fig. 9A through C). The red color is due to very fine hematite(?) (not detectable in SEM) stain on the detrital clay and mica. The detrital clay occurs as laminae, grain coats, and as a pore-filling matrix intimately associated with dolomicrite matrix. Anhydrite cement is ubiquitous in the Dollarhide core and patchy to common in the HSA cores (Fig. 9D).

In Dollarhide #7-15-C, Red Argillaceous Siltstones/Sandstones are interbedded with massive and bedded anhydrite and dolomudstone of the inner shelf. Anhydrite is dominant and occurs with both primary textures and also replacement textures after dolomite (Fig. 5). A few thin beds of halite are also present. In the HSA wells the red siltstones are associated with algal, peloidal dolomudstones that commonly contain rip-up clasts and rarely have anhydrite-filled fenestrae (Fig. 6).

Light Brown Arkosic Sandstone

Light brown, poorly consolidated, lithic, arkosic sandstones (Figs. 10 and 11) are common in the HSA #1187 and HSA #1257 cores, and to a lesser extent in the L. Daugherty #4 core (Fig. 4). These sandstones are typically structureless to wispy- and wavy-laminated. Their poorly

Figure 5

Red argillaceous siltstones/sandstones from the Dollarhide #7-15-C well. The well is 15 to 20 miles (24 to 32 km) behind the Yates shelf margin in the evaporitic inner-shelf region. Siltstones have sharp basal contacts and gradational upper contacts with laminated and chaotic-bedded anhydrite. The siltstones are structureless to wispy-laminated. Disturbed and/or mottled beds are common. The siltstones are red due to the presence of hematite-stained, illitic(?), detrital clay and mica occurring as laminae, grain rims, and diffuse masses. Dolomite laminae and chaotic dolomite beds suggest some of the anhydrite is a replacement after (algal) dolomudstone. Halite beds are minor.

Photograph courtesy of G. A. Crawford, Unocal Science and Technology Company.

Figure 6

Red argillaceous siltstones/sandstones from the HSA #1257 well. The well is 6 or 7 miles (9.6 to 11 km) behind the Yates shelf margin in the clastic-rich middle-shelf region. Upper and lower contacts with dolomudstones are generally sharp. Dolomudstone rip-up clasts are abundant; algal-laminated dolomudstones are minor. Similar dolomites are replaced by anhydrite in the inner-shelf region.

Figure 7

Log response for the middle-shelf, Red Argillaceous Siltstones/ Sandstones; HSA #1257 well. Shaded regions are intervals of Red Argillaceous Siltstone. The facies is characterized by high bulk density/low porosity related to the presence of abundant detrital and carbonate matrix. Abundant detrital matrix results in high gamma-ray (API) values (thicker beds); however, the facies is commonly so dolomite-rich (thinner beds) that the gamma-ray log does not show much deflection.

Figure 5

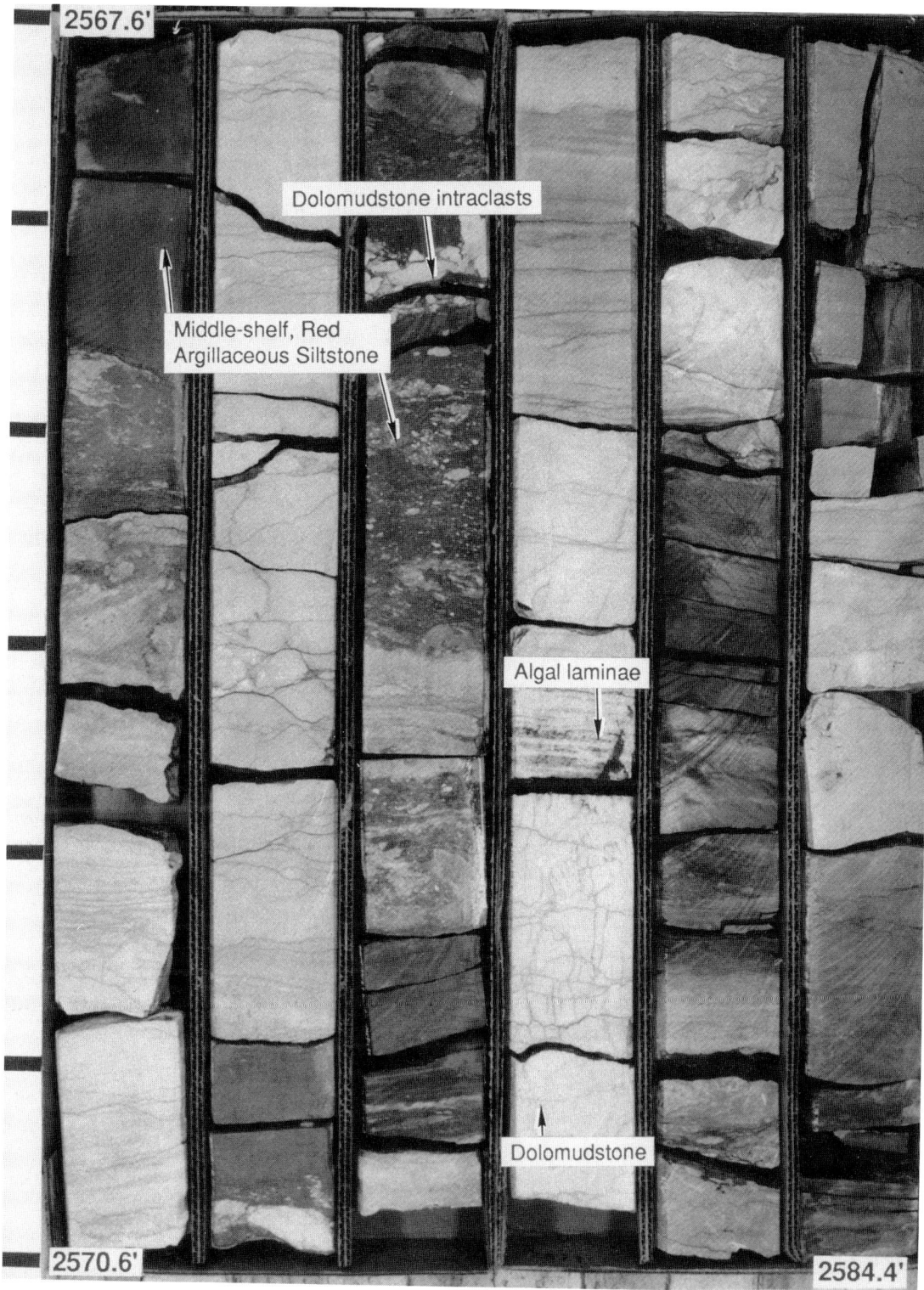

Figure 6

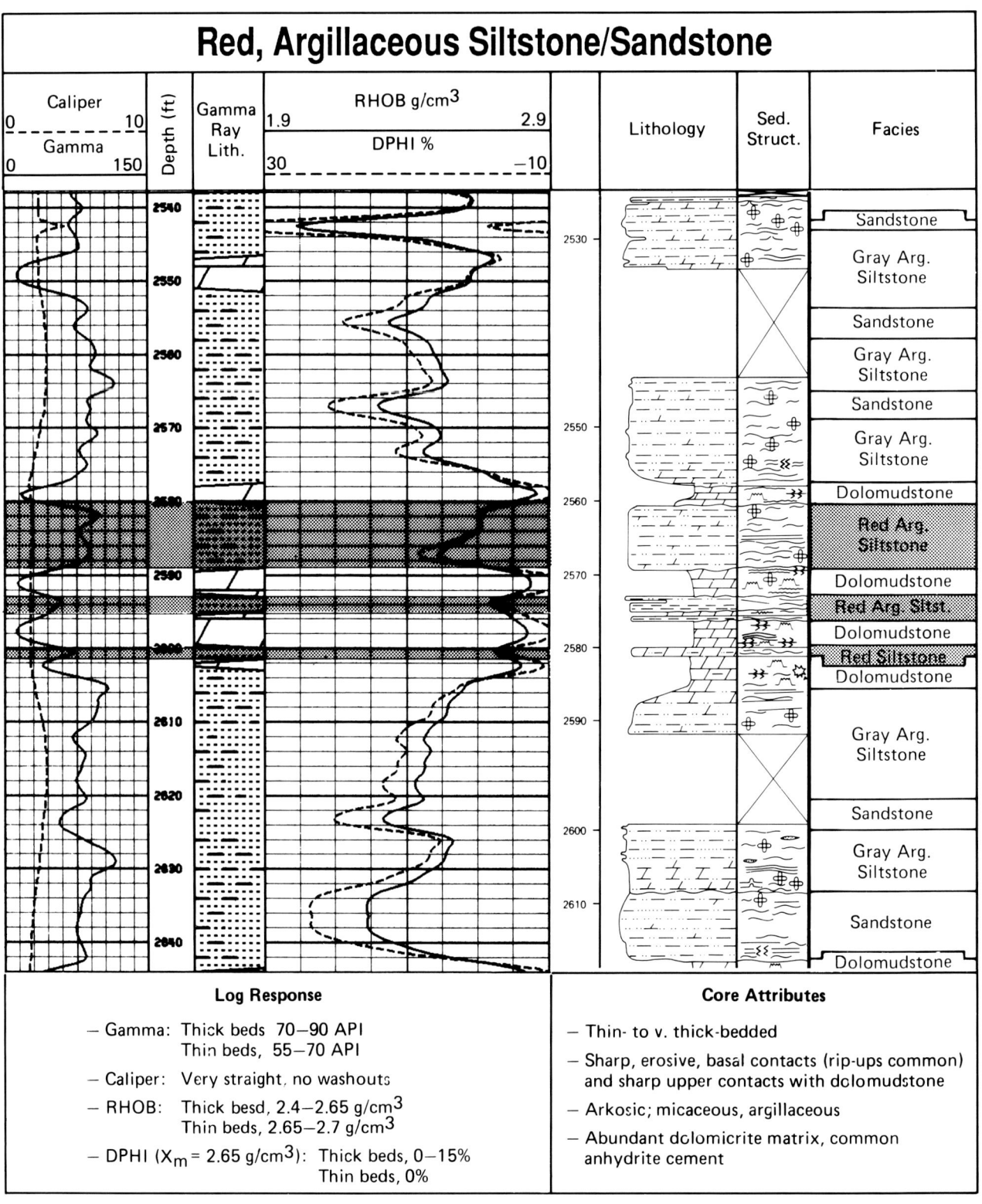

Figure 7

Table 2　　　　　RED ARGILLACEOUS SILTSTONE - PETROGRAPHIC DATA

<---SAMPLE INFO--->		<------FRAMEWORK GRAINS------>						<------INTERSTITIAL MATERIAL------>						<---TEXTURE--->	
WELL	DEPTH	QUAR	K-FE	PLAG	ROCK	HEAV	MICA	MATR	RED MATR	AUTH CLAY	DOLO	ANHY +GYP	PYR	SHR	POR	GRAIN SIZE φ	SORT φ
1187	2555.0	35.6	10.8	3.2	3.6	0.8	6.8	0	26	0	1.6	1.2	0.8	0	0	3.8	0.42
1187	2582.0	14.8	5.2	1.6	0	0	2.8	0	30.4	0	43.6	0	0.8	0	0	5.0	0.85
1187	2591.0	30.8	9.2	2.4	0.4	0.8	0.4	0	3.2	0	28.4	0	0	20	0	3.5	0.6
1257	2567.6	39.2	11.2	7.6	8.0	0	0	0	15.6	2.4	14.4	0	0	0	0	3.5	0.42

KEY

COLUMNS: QUAR=quartz; K-FE=Potassium feldspar; PLAG=plagioclase; ROCK=rock fragments; HEAV=heavy minerals; MICA=biot/musc; MATR=gray detrital matrix; RED MATR=red detrital matrix; AUTH CLAY=authigenic clay; DOLO=dolomite (* ferroan dolomite; Mag=magnesite; CAL=calcite; ANHY+GYP=anhydrite + gypsum; PYR=pyrite; SHR=solid hydrocarbon residue; POR=percent porosity; GRAIN SIZE (phi units); SORT=sorting (phi units)

Figure 8

A. Typical depositional texture of middle-shelf, Red Argillaceous Siltstones/Sandstones. Grain size ranges from silt to very fine sand. Red (hematite-stained) detrital matrix is abundant and forms up to 15% of the facies. This particular sample is moderately sorted. G. W. O'Brien #1491, 2443.5 ft, plane-polarized light. Note: <u>The first photomicrograph of each major lithofacies depicts the typical depositional texture and is shot at the same magnification for ease of comparison.</u>

B. The red detrital matrix occurs as irregular masses distributed unevenly throughout the sample. The red clay both fills pores and coats grains. Silt-sized mica is abundant. Also note the angular nature of the quartz silt grains. G. W. O'Brien #1491, 2897.4 ft, plane-polarized light.

C. Red detrital clay occurs in distinct wispy, micaceous laminae. HSA #1187, 2555.0 ft, plane-polarized light.

D. Clay- to silt-sized mica grains are common within the red matrix. The detrital clay and mica act as a cement, coating and binding framework grains and filling interparticle pores. HSA #1386, 2520.6 ft, plane-polarized light.

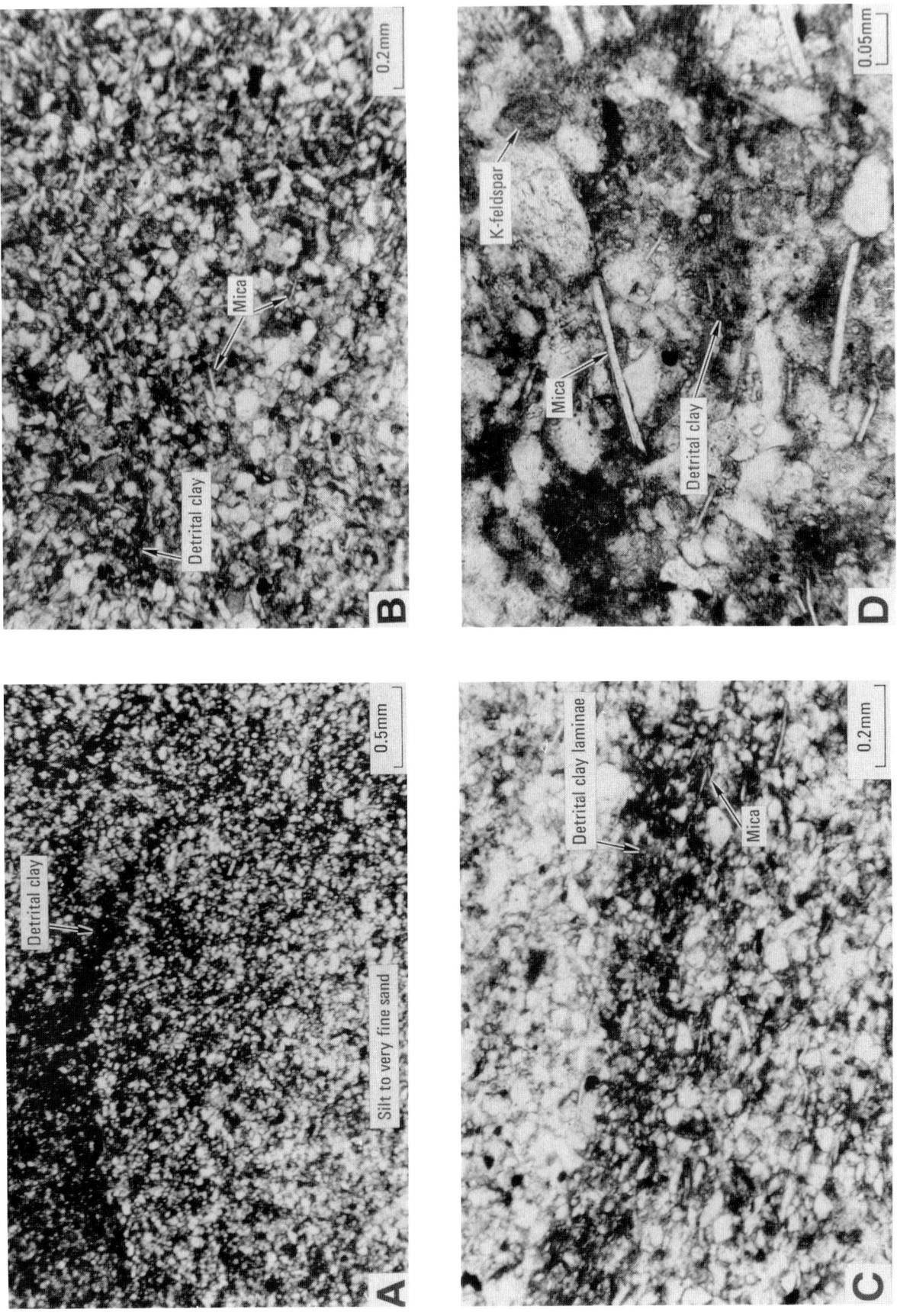

Figure 8

Figure 9

A. Very finely crystalline dolomite is common in the Red Argillaceous Siltstones/Sandstones. The dolomite occurs as small (<5 μm), equant crystals. Local areas of dolomicrite matrix-supported texture are common. HSA #1257, 2612.7 ft, cross-polarized light.

B. The red clay fills pores and coats grains in an uneven fashion suggestive of (airborne) dust oxidized and infiltrated between the silt grains during repeated wetting and drying of the sediment. Dolomite occurs both on grains and in the micropores of the detrital clay. Dolomite matrix-supported textures and the close association with the red detrital clay suggest the dolomite (or original micrite) may have formed about the same time the red clay was being infiltrated. HSA #1187, 2591.0 ft, cross-polarized light.

C. SEM photomicrograph of dolomitic Red Argillaceous Siltstones/Sandstones. Interparticle pore spaces are filled with very fine crystalline dolomite and red detrital clay and mica. Local dolomite matrix-supported textures suggest the detrital clay and mica are impurities that were trapped within early dolomicrite(?) matrix. Note detrital clay and mica plastered onto grain surfaces. HSA #1257, 2591.0 ft.

D. SEM photomicrograph of anhydritic Red Argillaceous Siltstones/Sandstones. The anhydrite cement is medium to coarsely crystalline and poikilotopic. Red (hematite-stained) detrital clay occurs only as very thin rims on the framework grains. The sparseness of the red clay suggests that the silt and sand grains were cemented early, before thick detrital clay rims could develop. This is further suggested by local, anhydrite-supported (precompactional) textures. G. W. O'Brien #1491, 2970.8 ft.

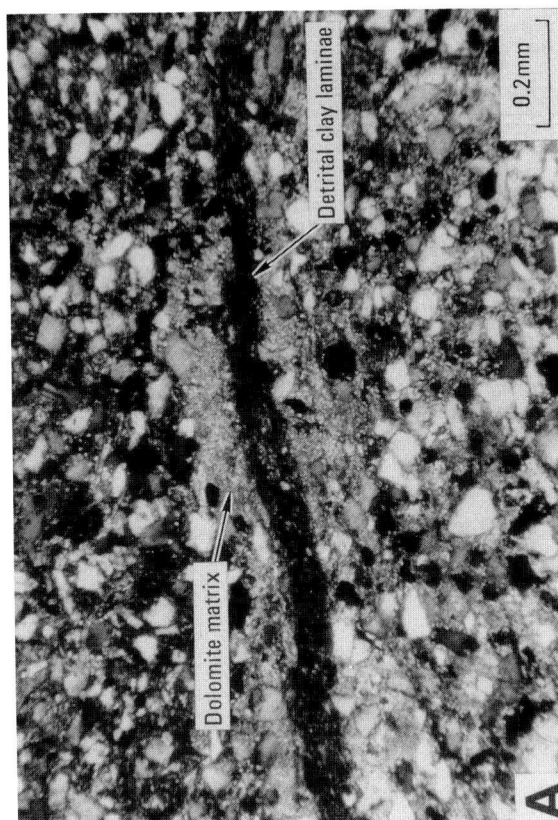
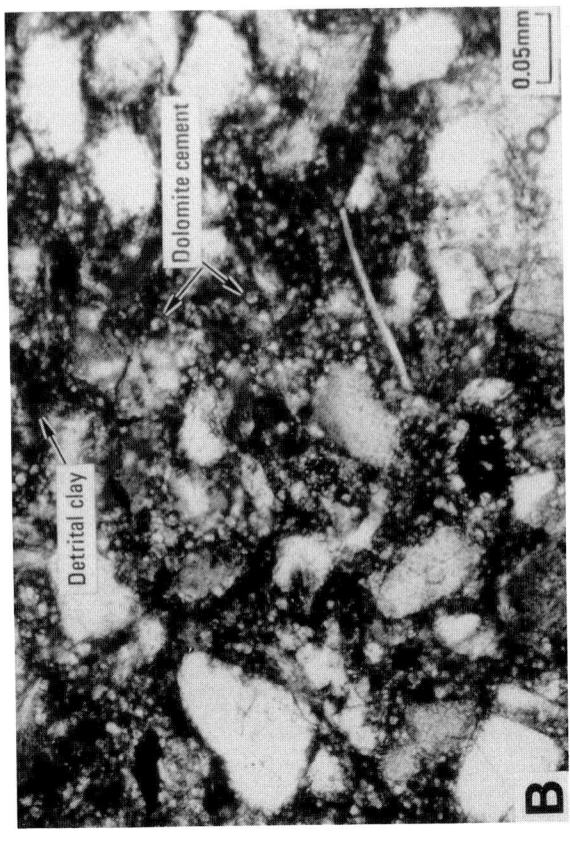
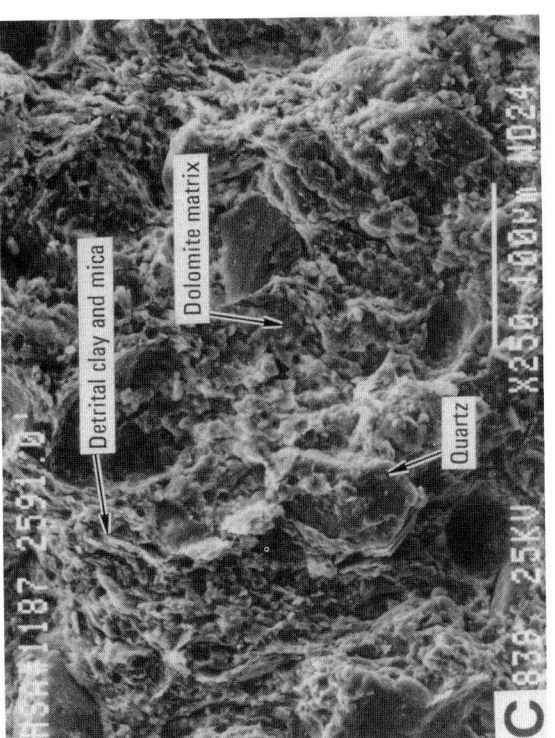
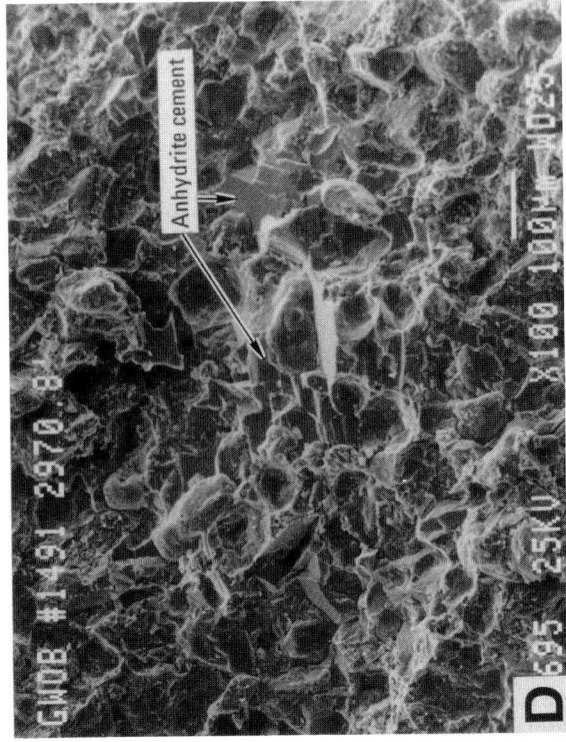

Figure 9

Figure 10

Light Brown Arkosic Sandstones from the clastic-rich, middle-shelf region of the Central Basin Platform. Core is from the HSA #1257 well. Structureless to wispy-laminated sandstones commonly grade into plane-, wavy-, and ripple-laminated Dark Gray Argillaceous Siltstones. Locally the sandstones and argillaceous siltstones are thinly interbedded. The sandstones are typically poorly consolidated, resulting in poor recovery, as seen in Figure 12.

Figure 11

Log response and core attributes of Light Brown Arkosic Sandstones; HSA #1257 well. The shaded intervals correspond to poorly consolidated arkosic sandstones. A minimal gamma-ray response occurs when there is little or no detrital clay in the arkosic sandstones. Low density/high porosity values result from the presence of open porosity due to the lack of both detrital matrix and authigenic cements. A 15% density log porosity value is the approximate log cutoff between Light Brown Arkosic Sandstones and Dark Gray Argillaceous Siltstones. The poorly consolidated nature of the sandstones is shown by the deflection in the caliper log and the poor core recovery in the lowest thick sandstone interval.

Figure 12

Poorly consolidated Light Brown Arkosic Sandstones from HSA #1257. The well is from the clastic-rich, middle-shelf region of the Central Basin Platform. The friable nature of the sandstones results in poor recovery using routine coring procedures. Whole-core pieces of sandstone exhibit little in the way of sedimentary structures. Crude, wispy-, to wavy-laminae and minor ripple-laminae are present. Upper and lower contacts with wavy-laminated, gray, argillaceous siltstones are gradational.

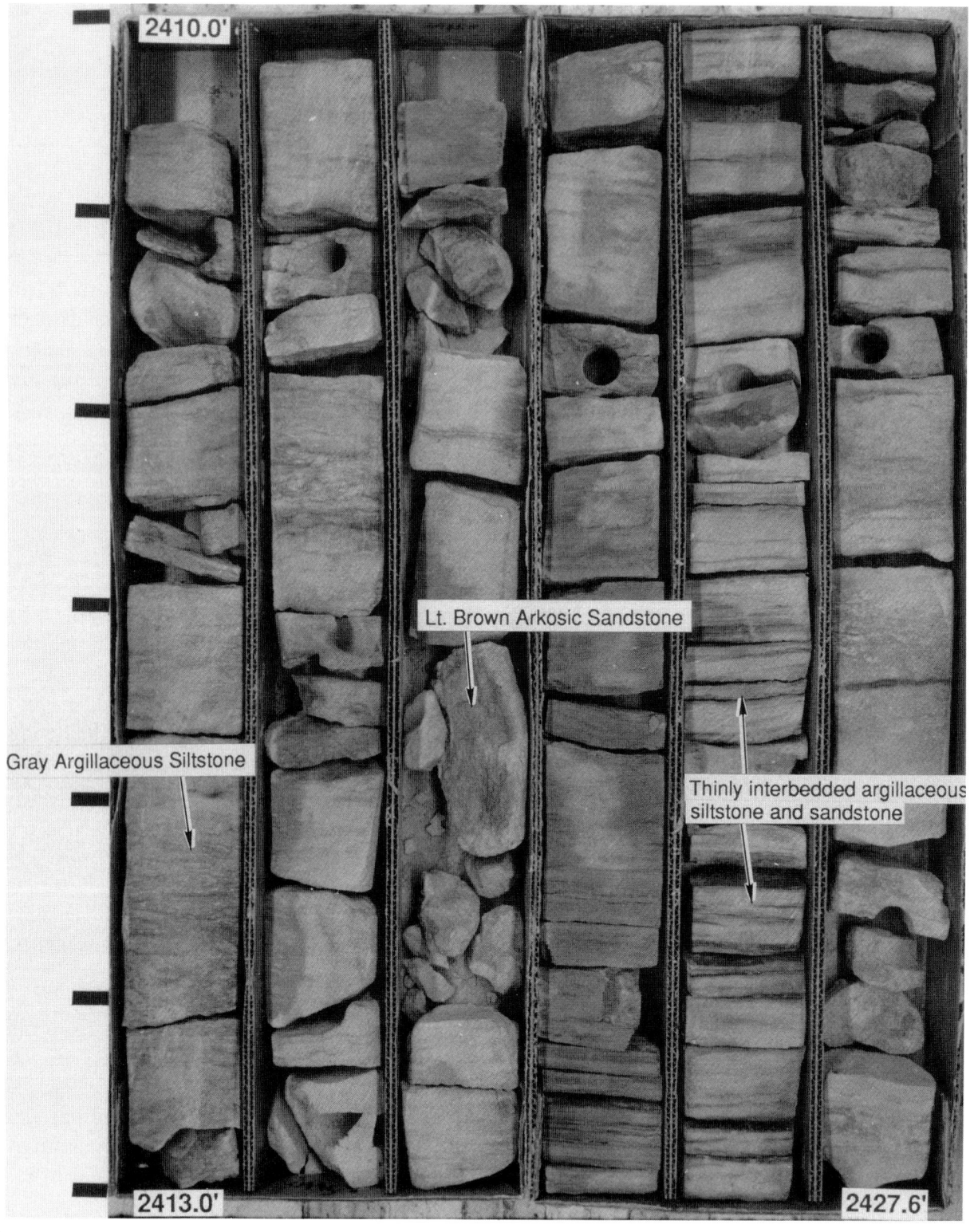

Figure 10

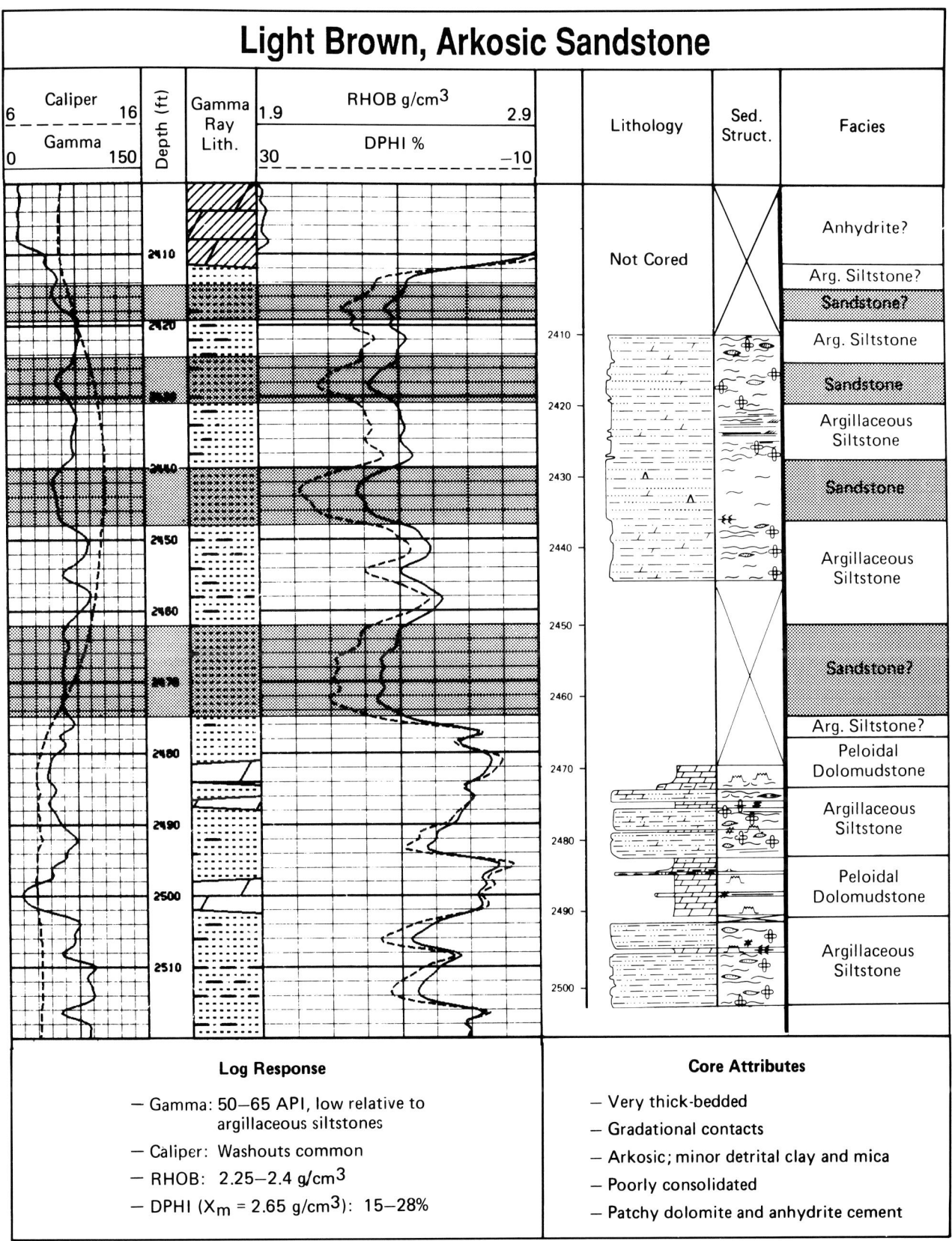

Figure 11

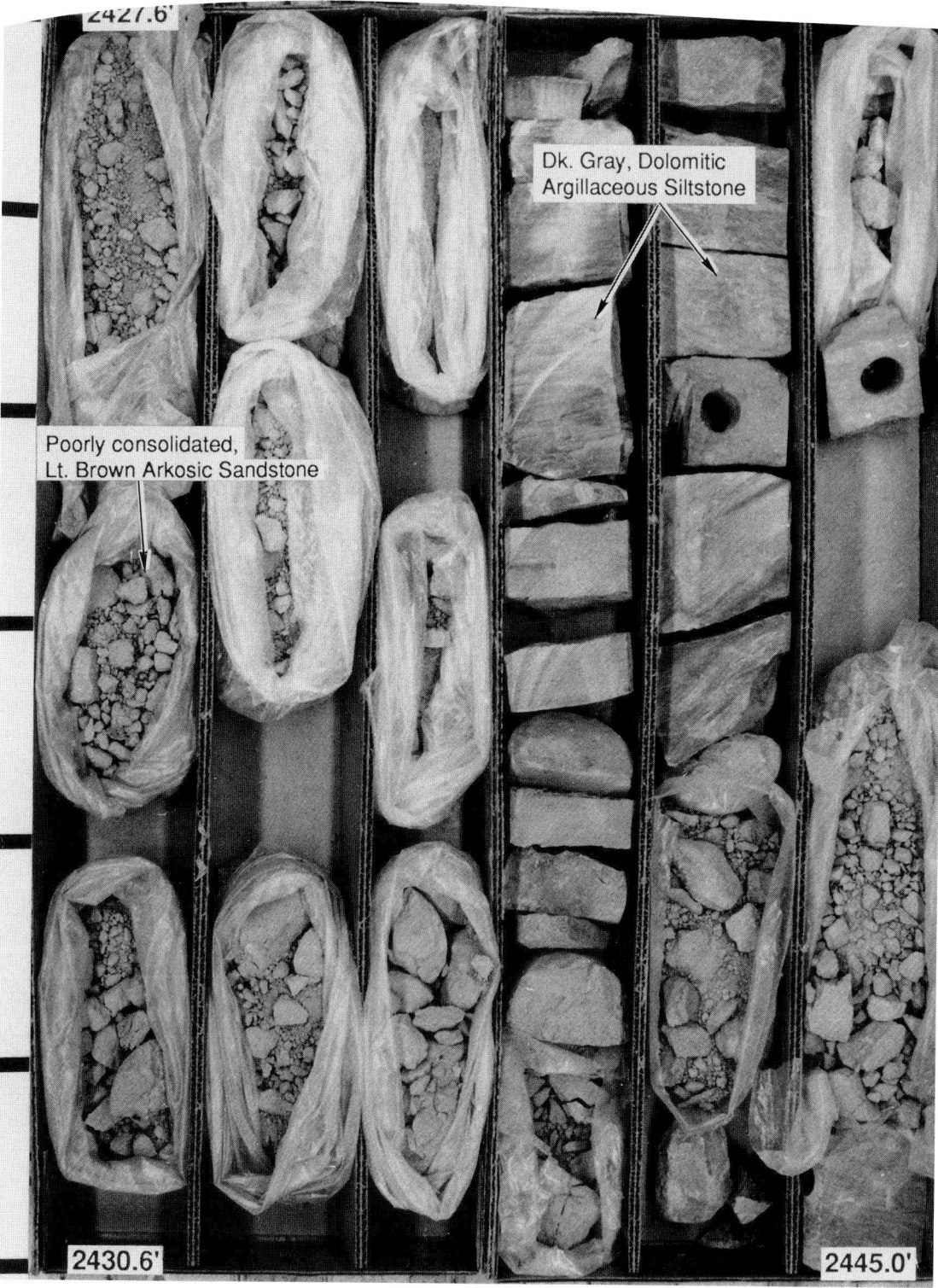

Figure 12

consolidated nature results in poor core recovery with standard coring techniques (Fig. 12). The absence of sedimentary structures in the sandstones may, in part, be due to depositional processes or to a simple lack of grain-size contrast. Some intervals of increased detrital clay content exhibit well-developed, wispy-, plane-, and ripple-laminae. In core, the clean sandstones typically have gradational lower and upper contacts with wispy- to wavy-laminated argillaceous siltstone/mudstone.

The sandstones are silty, very fine- to fine-grained, well- to very well-sorted subarkoses to arkoses (Fig. 13A and B; Table 3). Detrital grains are subangular to subrounded. In places grain dissolution is apparent, and secondary porosity is locally common (Fig. 13C). Solid hydrocarbon residue (dead oil) is locally very abundant (up to 20 percent) and may occlude primary intergranular porosity (Fig. 13D).

Diagenetic phases include patchy anhydrite and dolomite cements, abundant grain lining, authigenic corrensite clay, common but minor authigenic K-feldspar and pyrite, and rare quartz overgrowths (Fig. 14A through D). Anhydrite occurs as medium to coarse pore-filling crystal, locally poikilotopic, that are highly birefringent in cross-polarized light (Fig. 14A). Dolomite is present in most samples in small amounts as very fine to fine, subhedral to euhedral crystals that appear to be "floating" in pores or attached to grain surfaces (Fig. 14B). More rarely, dolomite occurs as a pervasive, pore-filling, medium crystalline cement.

Authigenic corrensite (mixed layer chlorite smectite) is very common and occurs as a crenulate, grain-coating clay (Figs. 14C and 15A through D). The size of the corrensite platelets increases toward the center of pores and locally gets sufficiently robust to clog pore throats (Fig. 15A). More typically, however, the corrensite forms a uniform thin (less than 10 µm) grain coat that does not occlude much primary porosity but greatly increases pore roughness (Fig. 15B).

Authigenic K-feldspar and quartz also occur as very fine euhedral crystals growing on detrital grains or grain-coating corrensite (Fig. 15C and D). Both of these cements are best viewed in SEM due to their very fine crystal size. The K-feldspar cement occurs as a minor phase in almost every sample; whereas, the quartz overgrowths are rare.

Carbonate rocks associated with Light Brown Arkosic Sandstones are gray, pale orange, and tan, structureless to wavy-laminated, peloidal dolomudstones. Although dolomites are associated with the light brown sandstones, the two lithologies are only rarely in direct contact. Instead, the clean light brown sandstones commonly grade into gray argillaceous siltstones that are in turn in direct contact with the dolomudstones. Clastic/carbonate ratios in the middle-shelf region range from 2.5 to 1 in most updip wells to 1.5 to 1 in downdip wells. The dolomites occur as variably thick beds that typically separate very thick clastic intervals. The dolomudstones are nondescript in core, exhibiting only rare textures suggestive of algal laminations and local thin zones of fenestral fabric. Intervals of dolomudstone intraclasts set in an argillaceous dolomitic matrix are also present. In thin section, the algal and peloidal nature of the dolomudstones is apparent.

Figure 13

A. Typical depositional texture of Light Brown Arkosic Sandstones. The sandstones are moderately well to well sorted, and grain size ranges from silt to lower fine sand. Detrital clay and mica are sparse to absent. HSA #1187, 2482.0 ft, plane-polarized light.

B. Well-sorted porous texture of the Light Brown Arkosic Sandstones. Grain contacts are tangential (point) to straight. Visually estimated porosity of this sample is 16%. Note abundant (8% to 10%) K-feldspar grains. Plagioclase grains are also common and can be distinguished from quartz in plane-polarized light by the presence of cleaved edges and turbid surfaces. In cross-polarized light plagioclase twins are apparent. HSA #1257, 2612.7 ft, plane-polarized light.

C. Isolated secondary porosity due to the partial dissolution of framework grains. SEM analyses suggest the dissolved grains are predominantly plagioclase; however, K-feldspar, lithic fragments, and accessory minerals also show some dissolution features. HSA #1257, 2612.7 ft, plane-polarized light.

D. Solid hydrocarbon residue (SHR) filling primary intergranular pores in the Light Brown Arkosic Sandstones. HSA #1257, 2527.5 ft, plane-polarized light.

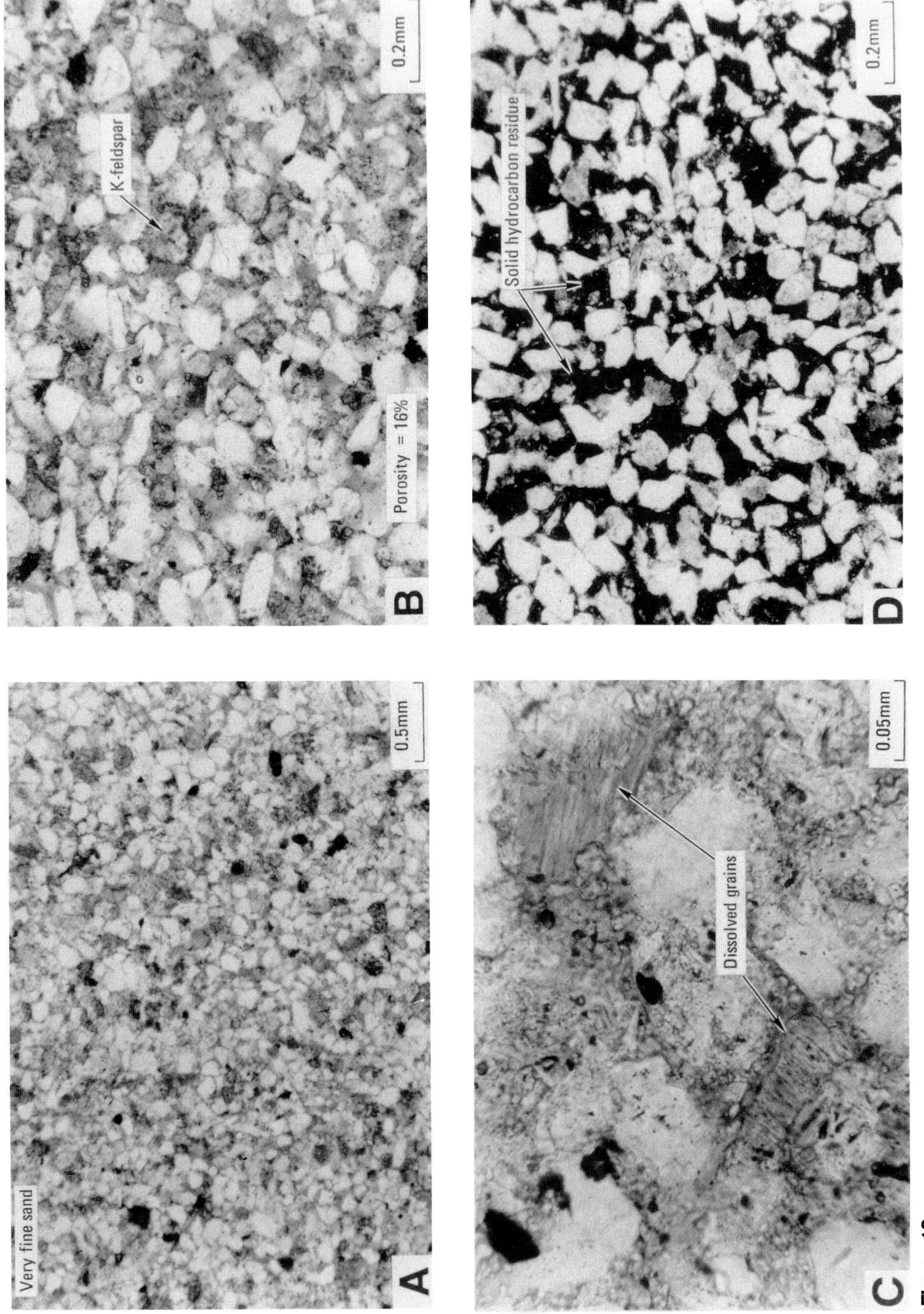

Figure 13

Table 3   LIGHT BROWN ARKOSIC SANDSTONE FACIES - PETROGRAPHIC DATA

<--SAMPLE INFP-->		<----------FRAMEWORK GRAINS---------->							<----------INTERSTITIAL MATERIAL---------->							<----TEXTURE---->	
WELL	DEPTH	QUAR	K-FE	PLAG	ROCK	HEAV	MICA	MATR	RED MATR	AUTH CLAY	DOLO	ANHY +GYP	PYR	SHR	POR	GRAIN SIZE φ	SORT φ
1187	2437.0	45.6	8	2.4	1.2	1.6	2.8	6.4	0	0	11.6	15.2	0.8	0	0.8	4.0	0.85
1187	2464.0	43.2	10.4	4	2	0	0	0.8	0	0	4	23.6	1.6	0	9.6	3.5	0.85
1187	2481.5	46.8	10.8	7.6	9.2	1.2	0	0.4	0	4.8	1.6	1.2	0	0	15.6	3.0	0.6
1187	2499.0	38.8	13.2	7.2	10	0.4	1.2	2.8	0	5.6	1.6	0	3.2	5.6	11.6	3.5	0.6
1257	2418.6	38.8	6.0	5.6	12.4	0.8	0.8	0	0	12.0	7.2	0.4	0.8	0	13.2	3.0	0.60
1257	2433.0	48.0	6.0	7.6	9.2	0.8	0.8	2.8	0	10.4	2.0	0.4	0.8	0	10.0	3.5	0.60
L.D4	2711.0	40	10	8	9	2	TR	2	0	6	15	0	TR	0	8	3.5	0.42
L.D4	2712.0	37	11	8	9	2	1	0	TR	8	9	TR	0	TR	11	3.25	0.35
1187	2518.0	48	12.4	4	3.2	0.4	2	7.2	0	5.2	0	2.4	0.4	6.8	7.6	3.5	0.35
L.D4	2748.0	36	10	9	11	2	TR	0	0	6	7	0	1	0	18	3.25	0.42
1187	2541.0	35.2	11.6	2.4	8	0.8	2.8	3.2	0	9.6	3.2	0.4	1.6	7.6	10	3.5	0.35
1257	2527.5	41.2	6.8	5.6	11.6	0.4	0	0	0	2.0	0.4	0.4	0	28.0	1.6	3.5	0.35
1257	2612.7	46.8	8.0	6.4	10.4	2.0	0	0.4	0	4.4	3.6	0	0	0	15.6	3.0	0.42
1257	2632.4	40.4	11.6	5.6	10.8	0.8	0	0	0	1.6	8.4	0	0	17.2	1.6	3.5	0.42
1257	2669.0	43.6	13.2	6.0	5.2	0.8	0	0.4	0	7.2	6.4	0	0	12.4	4.0	3.0	0.60

KEY

COLUMNS: QUAR=quartz; K-FE=Potassium feldspar; PLAG=plagioclase; ROCK=rock fragments; HEAV=heavy minerals; MICA=biot/musc; MATR=gray detrital matrix; RED MATR=red detrital matrix; AUTH CLAY=authigenic clay; DOLO=dolomite (* ferroan dolomite; Mag=magnesite; CAL=calcite; ANHY+GYP=anhydrite + gypsum; PYR=pyrite; SHR=solid hydrocarbon residue; POR=percent porosity; GRAIN SIZE (phi units); SORT=sorting (phi units)

Figure 14

A. Pore-filling, medium- to coarse-grained, poikilotopic anhydrite cement commonly found in the Light Brown Arkosic Sandstones. HSA #1391, 2485.0 ft, cross-polarized light.

B. Dolomite cement is common in Light Brown Arkosic Sandstones and occurs as scattered very fine and fine crystals and, to a lesser extent, as medium-crystalline, pore-filling cement. G. W. O'Brien #1514, 2876.5 ft, cross-polarized light.

C. Grain-coating and, to a much lesser extent, pore-filling authigenic corrensite is also common in the Light Brown Arkosic Sandstones. Noneffective microporosity is associated with the corrensite. For high-magnification SEM views of the grain-coating authigenic clay, see Figure 15A-D. HSA #1257, 2418.6 ft, plane-polarized light.

D. Authigenic K-feldspar is common throughout the arkosic sandstones. The K-feldspar cement (stained in thin section) occurs as very fine, euhedral crystals attached to clay-coated grains. Large stained patch in photomicrograph is a detrital K-feldspar grain. For a high-magnification SEM view of the authigenic K-feldspar, see Figure 15D. HSA #1257, 2418.6 ft, plane-polarized light.

Figure 15

A. SEM photomicrograph of grain-coating and pore-filling authigenic corrensite. The size of the crenulated clay platelets generally increases from the grain edge to the interior of the pores. Large, robust platelets locally fill pore space, resulting in noneffective secondary porosity. HSA #1187, 2481.5 ft, SEM.

B. Quartz grains coated by a uniform layer of crenulate, authigenic corrensite. Although the primary intergranular pore space has been only slightly reduced by the clay rim, the surface area and roughness of the pore walls have been greatly increased. HSA #1286, 2527.2 ft, SEM.

C. Euhedral quartz overgrowths are a minor diagenetic phase. The quartz cement occurs with dolomite and K-feldspar cements. The lack of clay on all the cements suggests that they are post-clay phases. HSA #1257, 2598.3 ft, SEM.

D. Euhedral authigenic K-feldspar growing on corrensite-coated grains. Lack of clay on the K-feldspar crystals shows that they are a post-clay phase. G. W. O'Brien #1491, 3106.0 ft, SEM.

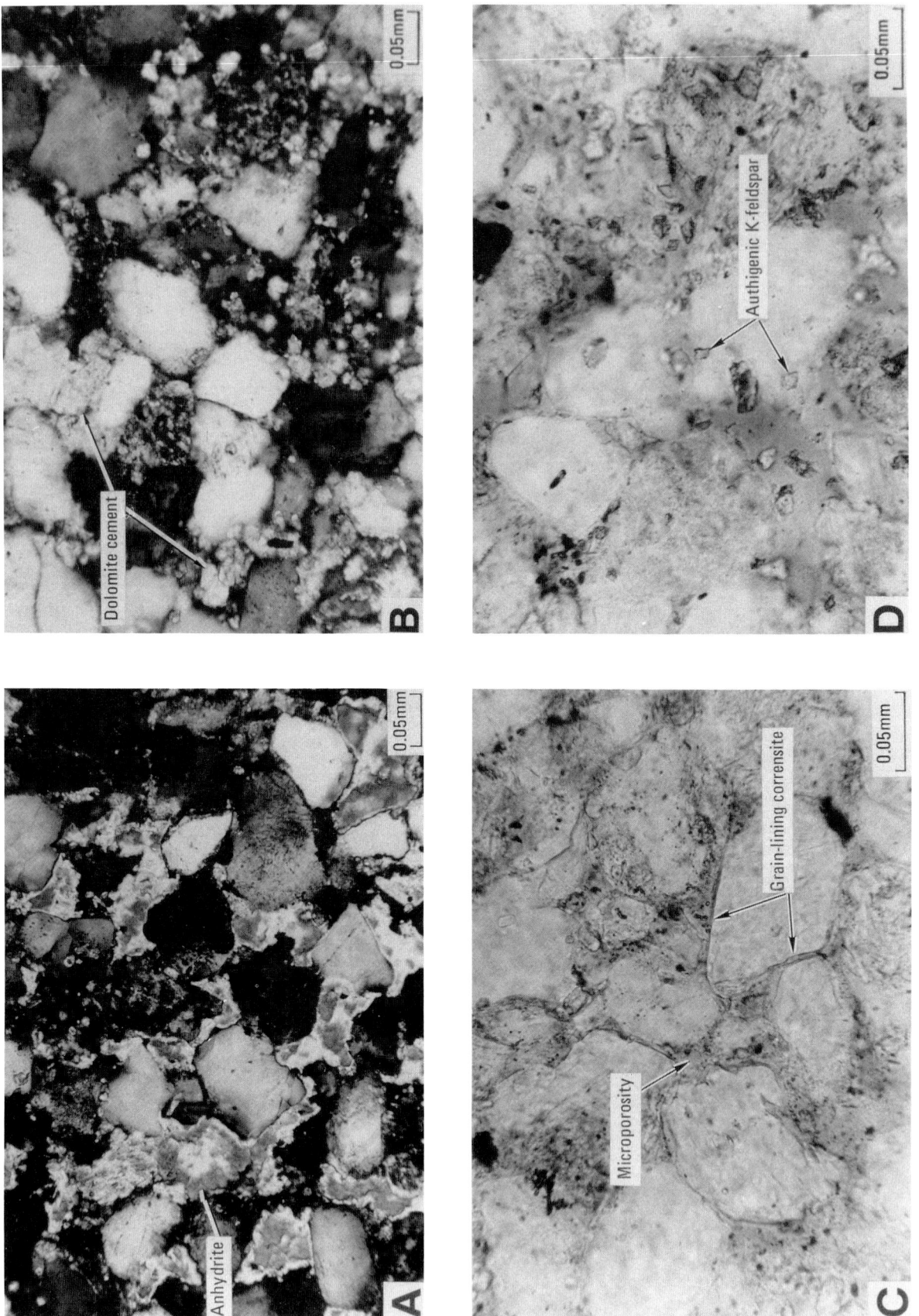

Figure 14

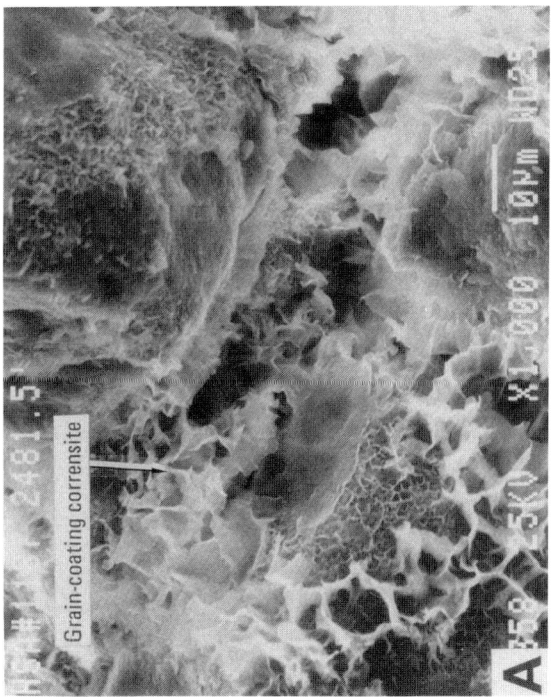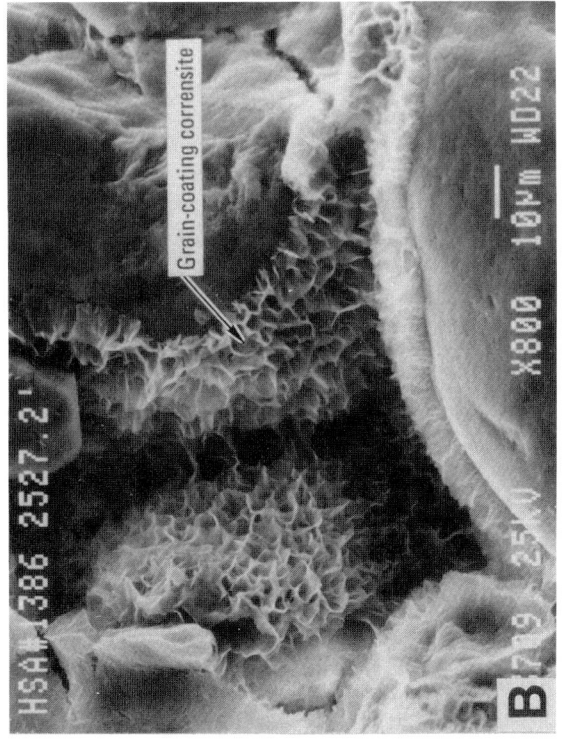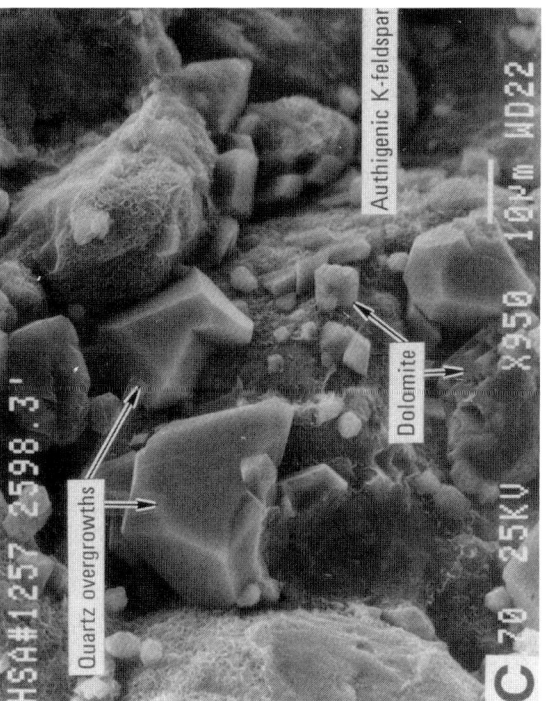

Figure 15

Dark Gray Argillaceous Siltstone

Medium to very thick beds of Dark Gray Argillaceous Siltstone are present in the HSA #1187, HSA #1257 (Figs. 16 and 17), and L. Daugherty #4 cores (Fig. 4). Argillaceous siltstones often occur at the base and top of clean, light brown sandstone intervals. The argillaceous siltstones are plane-, wispy-, and wavy-laminated, and they contain laminae, thin beds, and rip-up clasts of gray dolomudstone. Lower and upper contacts with dolomudstones are typically sharp; whereas, contacts with clean sandstones tend to be gradational.

Argillaceous Siltstones are texturally variable (Fig. 18A through D), being typically argillaceous sandy siltstones, but ranging from mudstone to argillaceous silty sandstone. The modal grain size falls between coarse silt and lower, very fine sand. Compacted terrigenous mudstone clasts also form local concentrated pods of clay matrix (Fig. 19A and B). Samples are argillaceous dolomitic, micaceous subarkoses to arkoses (Table 4).

Detrital clay ranges from 5 to 25 percent and occurs both as wispy to planar laminae and as a fine disseminated matrix. SEM analysis shows the detrital clay to have a composition similar to the authigenic corrensite in the Light Brown Arkosic Sandstones. Clay habits range from ragged detrital plates, through poorly crystallized, possibly authigenic forms, to clearly authigenic, crenulated, grain-lining forms. The variable clay habits suggest local recrystallization of the detrital clay. Very finely crystalline dolomite is typically associated with the detrital clay and mica (Fig. 19C). The very fine crystal size of the dolomite, intimate association with detrital matrix, and presence of dolomite laminae with matrix-supported textures suggest the dolomite is original matrix material. Carbonate rocks that are typically in direct contact with the gray argillaceous siltstones are algal, peloidal dolomudstones.

Green-Gray Dolomitic Subarkosic Siltstone/Sandstone

Medium- to thick-bedded Green-Gray Dolomitic Subarkosic Siltstones/Sandstones occur in the L. Daugherty #4 core in the lower part of the Yates Formation (Figs. 4 and 20 through 22). The siliciclastics are locally very argillaceous, and distinct, thin beds of green-gray terrigenous mudstone are common. The siltstones/sandstones are homogeneous to wavy- and wispy-laminated. Thin, wavy interlaminae of dolomudstone are common, and authigenic pyrite blebs occur along discrete laminae and thin beds. The base of the siliciclastic beds is typically very muddy and is in sharp contact with underlying dolomudstones. Upper contacts with dolomudstones are also often muddy but are slightly more gradational.

The dolomitic siltstones/sandstones range from mudstone and argillaceous siltstone, to silty, very fine-grained sandstone (Fig. 23A through D). The coarser textures have little or no detrital clay. Modal grain sizes are medium to upper very fine sand, and sorting is moderate to moderately good. Finer-textured samples contain up to 18 percent illitic/micaceous detrital matrix, have modal grain sizes in the silt to lower very fine sand range, and are poorly to moderately sorted.

Figure 16

Dark Gray Argillaceous Siltstone from the clastic-rich, middle-shelf region of the Central Basin Platform; HSA #1257 well. Sedimentary structures in the gray argillaceous siltstones are similar to those found in red argillaceous siltstones. Dark gray detrital clay commonly occurs as disseminated matrix and distinct planar, wavy, and wispy laminae. The detrital matrix is associated with dolomicrite matrix. Laminae, thin interbeds, and rip-up clasts of dolomudstone are also common. The gray siltstones are in contact with dolomudstones and arkosic sandstones. Contacts with arkosic sandstones are typically gradational, as shown in the photograph. Basal contacts with the dolomudstones are commonly sharp; whereas, the siltstones often become gradually more dolomitic before an abrupt upper contact with dolomudstones.

Figure 17

Log response and core attributes of Dark Gray Argillaceous Siltstones; HSA #1257 well. Using only wireline logs, it is difficult to distinguish between gray and red argillaceous siltstones. It is possible, however, to distinguish both the red and gray siltstones from arkosic sandstones and dolomites. The gray siltstones are characterized by large gamma-ray deflections due to the presence of abundant detrital clay. Density/porosity values are intermediate between dolomites and clean arkosic sandstones. A 15% porosity cutoff, using a density-derived porosity log (rock matrix = 2.65 g/cm^3), separates the argillaceous siltstones (<15%) from the clean arkosic sandstones (>15%). Although the density porosity log shows argillaceous siltstones with porosity values ranging from 7% to 15%, petrographic analyses show that the porosity is largely noneffective microporosity associated with the detrital clay matrix.

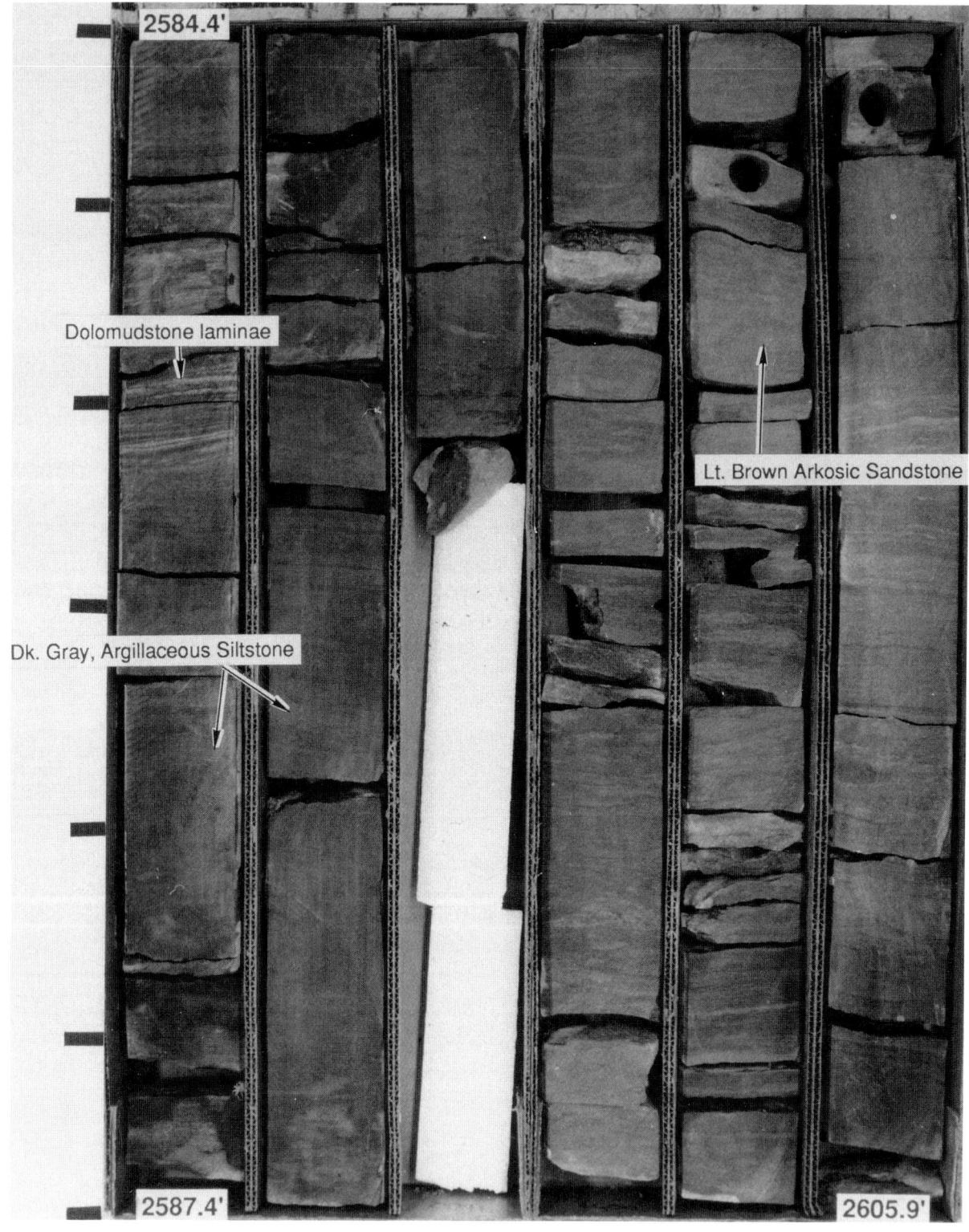

Figure 16

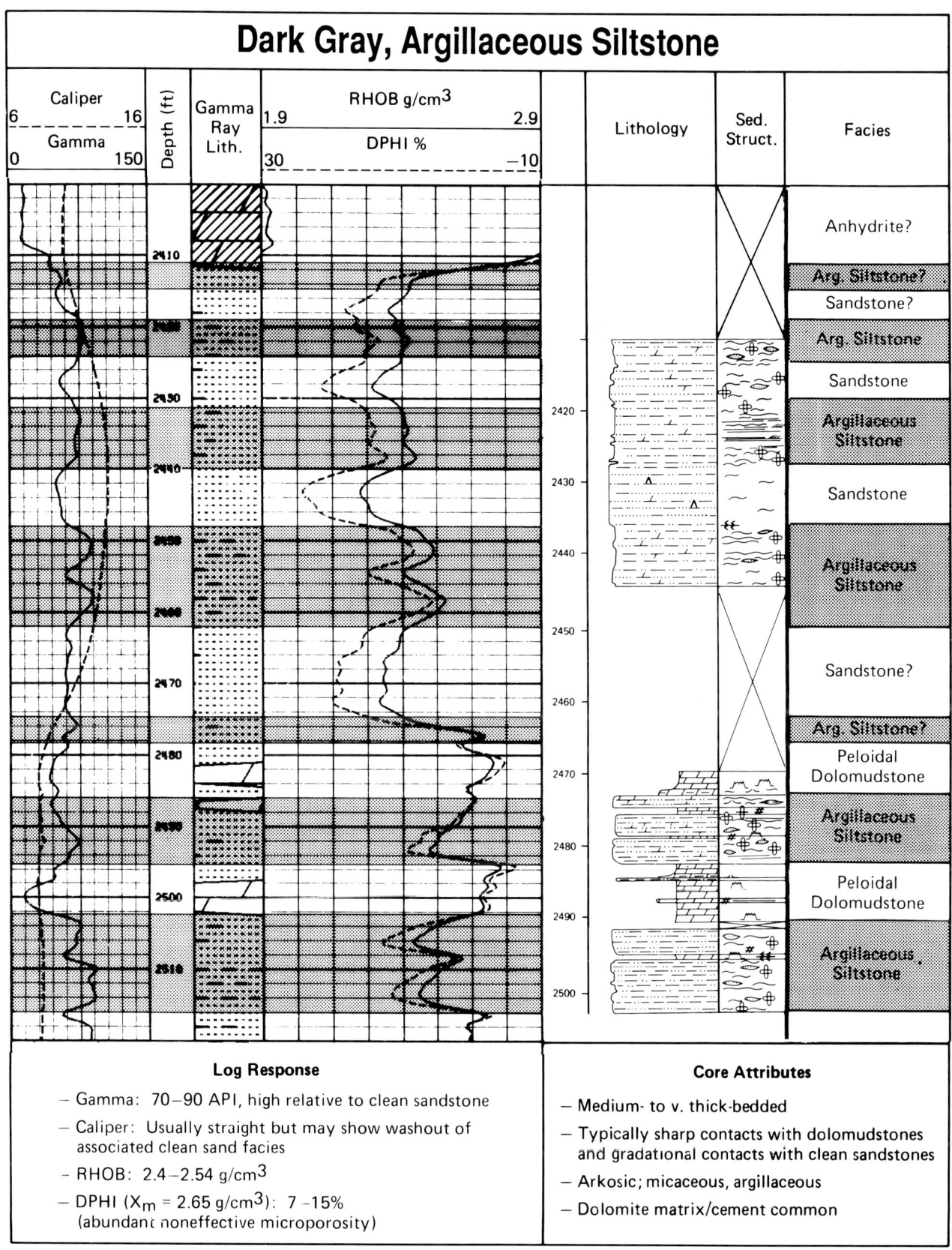

Figure 17

Figure 18

A. Typical petrographic texture of Dark Gray Argillaceous Siltstones. Grain size ranges from silt to very fine sand. Sorting ranges from poor to moderately good. Dark gray detrital clay is abundant and occurs as disseminated matrix and crude to well-defined laminae. HSA #1187, 2472.0 ft, plane-polarized light.

B. Abundant detrital clay gives the siltstones a dark gray color. Detrital clay occurs in diffuse patches rather than well-defined laminae. This lithofacies has little reservoir potential due to the abundant matrix and poor sorting, although some intervals that are transitional into the Light Brown Arkosic Sandstones exhibit moderate reservoir potential. HSA #1187, 2546.0 ft, plane-polarized light.

C. Micaceous, detrital clay laminae. Much of the clay-sized fraction is probably broken-down mica. The detrital clay and mica are plastered on some grain surfaces as uneven rims. Also note the very finely crystalline dolomite cement. G. W. O'Brien #1491, 2978.0 ft, plane-polarized light.

D. Same as Figure 18C in cross-polarized light.

Figure 19

A. Poorly sorted Dark Gray Argillaceous Siltstones. Detrital clay matrix fills intergranular pore spaces. Distinct patches of matrix are probably terrigenous mudstone intraclasts. HSA #1187, 2546.0 ft, plane-polarized light.

B. Same as A in cross-polarized light.

C. Typical SEM view of Dark Gray Argillaceous Siltstones. Framework grains are difficult to distinguish because they are covered by abundant detrital clay matrix and dolomite. The close association of the very finely crystalline dolomite with the detrital clay suggests that the dolomite is an early matrix phase. This is further supported by the presence of local dolomite-supported textures and thin laminae and lenses of dolomudstone within the argillaceous siltstones. HSA #1386, 2520.7 ft, SEM.

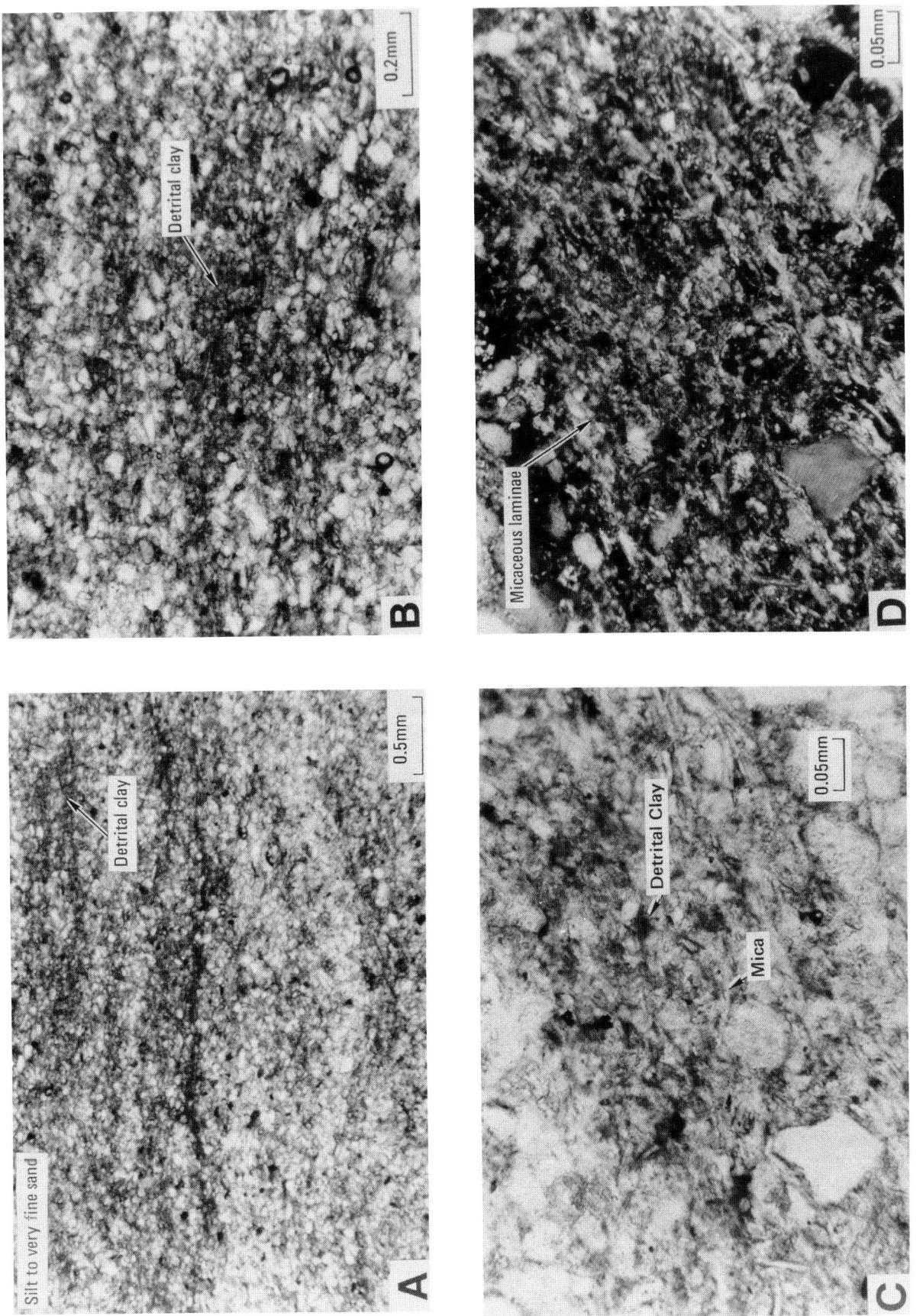

Figure 18

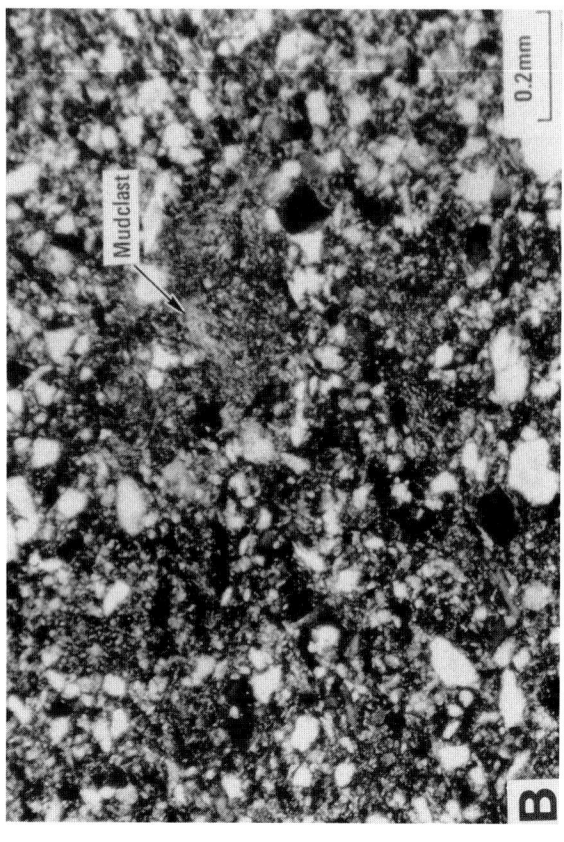
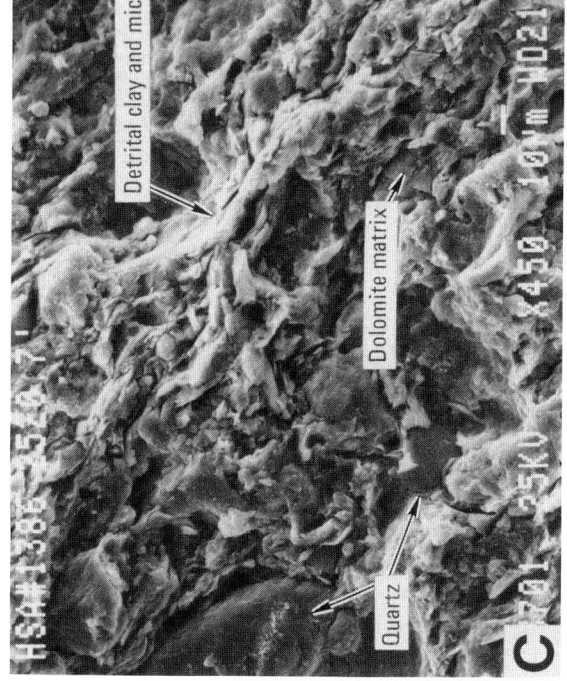
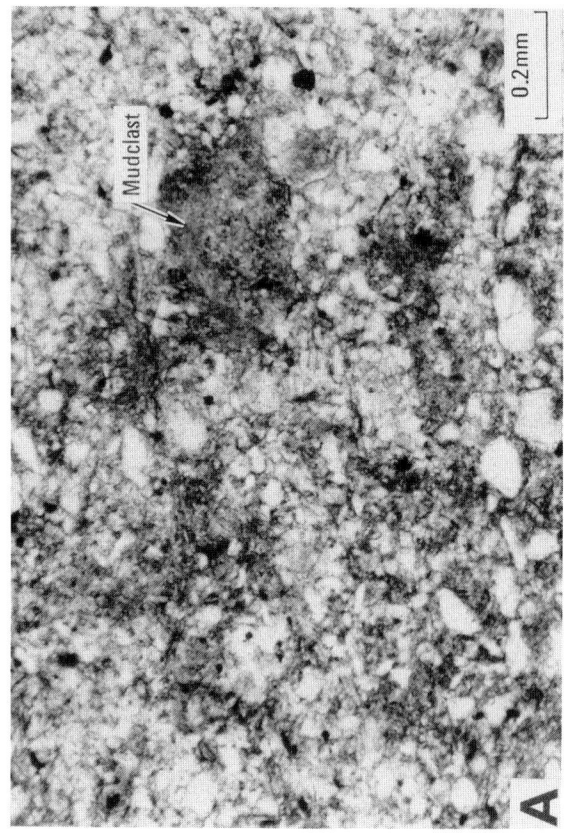

Figure 19

Table 4   DARK GRAY ARGILLACEOUS SILTSTONE FACIES - PETROGRAPHIC DATA

<---SAMPLE INFO--->		<----------FRAMEWORK GRAINS---------->						<----------INTERSTITIAL MATERIAL---------->						<----TEXTURE---->	
WELL	DEPTH	QUAR	K-FE	PLAG	ROCK	HEAV	MICA	MATR	RED MATR	AUTH CLAY	DOLO	ANHY +GYP	PYR	SHR	POR	GRAIN SIZE ϕ	SORT ϕ
1187	2472.0	36.4	12	5.6	2	0.4	6.8	24	0	0	7.6	0	0.4	0.4	0.4	4.5	0.6
1187	2491.0	32.8	11.6	2.8	3.6	0	4.4	37.2	0	0	3.6	0	1.6	0	0	4.5	1.5
1257	2438.0	35.2	8.8	6.8	6.4	7.2	5.6	8.4	0	2.8	21.2	0	0	0	1.2	4.0	0.60
L.D4	2708.0	35	9	9	9	1	4	18	0	6	8	0	0	1	1	3.75	0.8
L.D4	2710.5	37	10	8	8	1	3	10	0	6	14	0	0	0	3	3.75	0.8
L.D4	2715.0	38	8	9	8	2	2	6	0	6	21	0	TR	TR	4	3.75	0.8
L.D4	2717.2	22	5	4	3	1	3	32	0	3	26	0	1	0	0	4.5	1.5
L.D4	2747.3	32	7	7	5	2	3	22	0.8	2	12	0	2	0	TR	3.5	0.8
1257	2493.3	38.0	8.4	6.8	7.2	3.2	2.8	19.6	0	1.2	11.6	0	0	0	1.2	3.5	0.60
1257	2502.5	39.2	6.4	10.0	10.8	4.0	3.2	6.0	0	8.4	11.2	0	0	0	2.4	3.5	0.42
1187	2546.0	24.8	10.8	4.8	1.2	0.8	7.6	30.8	0	0	10.4	1.2	0	0	0.4	4.5	0.6
1187	2564.3	42	11.2	6	4	0	0	23.6	0	1.2	8.8	0.8	0.4	0	1.2	3.0	0.85
1257	2516.8	44.4	10.0	7.2	10.0	1.2	1.2	2.4	0	4.0	14.0	0	0	0.4	3.2	3.5	0.42
1187	2570.0	36.9	19.2	5.6	5.2	0.4	3.6	17.6	0	0	0	0	0	3.6	2.8	4.0	0.6
1187	2576.0	40.4	15.2	3.2	4.4	0.8	0.8	16.4	0	0.8	4.4	0.4	0.8	0	6	3.5	0.85
1257	2545.8	42.8	17.6	6.8	10.0	2.4	2	4.4	0	7.2	4.4	0	0.4	0	2.4	3.5	0.60
1257	2591.0	43.6	11.2	5.6	8.8	2.4	2	11.6	0	8.4	5.2	0	0	0	1.2	3.5	0.60
1187	2598.3	44.8	12.8	10.0	10.0	2.8	2	4.8	0	3.2	6.8	0	0.4	0.4	2.0	3.8	0.60
1257	2628.9	38.4	13.2	8.0	11.2	4.4	4	2.0	0	3.6	10.4	0	1.2	0	4.8	3.8	0.60

KEY

COLUMNS: QUAR=quartz; K-FE=Potassium feldspar; PLAG=plagioclase; ROCK=rock fragments; HEAV=heavy minerals; MICA=biot/musc; MATR=gray detrital matrix; RED MATR=red detrital matrix; AUTH CLAY=authigenic clay; DOLO=dolomite (* ferroan dolomite; Mag=magnesite; CAL=calcite; ANHY+GYP=anhydrite + gypsum; PYR=pyrite; SHR=solid hydrocarbon residue; POR=percent porosity; GRAIN SIZE (phi units); SORT=sorting (phi units)

Figure 20

Green-Gray Dolomitic Subarkosic Siltstones/Sandstones occurring in L. Daugherty #4. The well is located on the northern part of the Central Basin Platform near the transition between the middle-shelf, clastic-rich belt and the outer-shelf, carbonate-rich belt. Note the dolomitic and argillaceous nature of the siliciclastics. Basal contacts with gray algal dolomudstones are typically sharp and muddy; whereas, upper contacts may be sharp or gradational. A green-gray color and the presence of abundant authigenic pyrite, dolomite matrix, and detrital clay matrix suggest that the siliciclastics were deposited in a low-energy, reducing environment. The associated carbonate lithologies are suggestive of shallow lagoons that experienced periodic desiccation. Abundant algal structures and evaporites, and a paucity of fossils, suggest the lagoons were hypersaline. Desiccation features include fenestral fabric and sheetcracks. Stratigraphic relations suggest the lagoon(s) were positioned downdip of a predominantly tidal-flat environment and just updip of a pisolite shoal complex. Several thin beds (<3 in. or 5.5 cm thick) of peloidal, pisolitic, and intraclastic dolowackestone and packstone are probable intercalations of the pisolite shoal facies belt.

Figure 21

Continuation of cored interval from Figure 20. Green-Gray Dolomitic Subarkosic Siltstones/Sandstones are interbedded with algal dolomudstones and wackestones. See Figure 20 for a description of the core attributes and an interpretation of the depositional environment. The carbonate rocks associated with this lithofacies are of particular interest because they exhibit good porosity development and occasional production.

Figure 22

Log response and core attributes for Green-Gray Dolomitic Subarkosic Siltstones/Sandstones; L. Daugherty #4 well. Logs are characterized by minimal gamma-ray response and a high-density log response related to the large carbonate component of the siliciclastics. The low gamma-ray value may also be related to feldspar dissolution. Occasional, very high gamma-ray kicks correlate to thin, terrigenous mudstone beds. Density porosity logs suggest typical porosity values of 5% or less; but again the porosity is probably noneffective pore space within clay matrix or dissolved grains. Caliper logs exhibit no washouts, reflecting the cemented nature of both the siliciclastic and carbonate intervals.

Figure 20

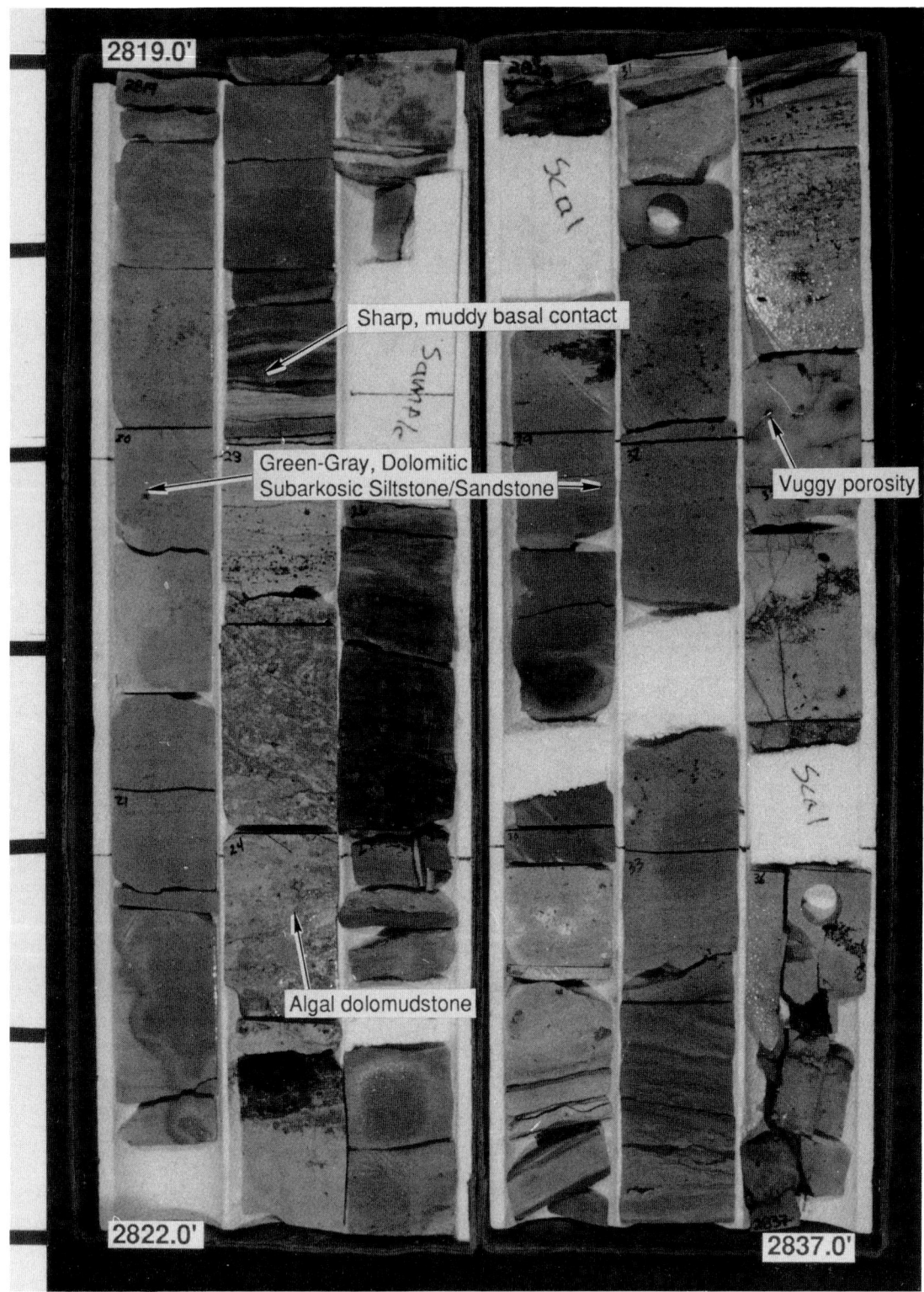

Figure 21

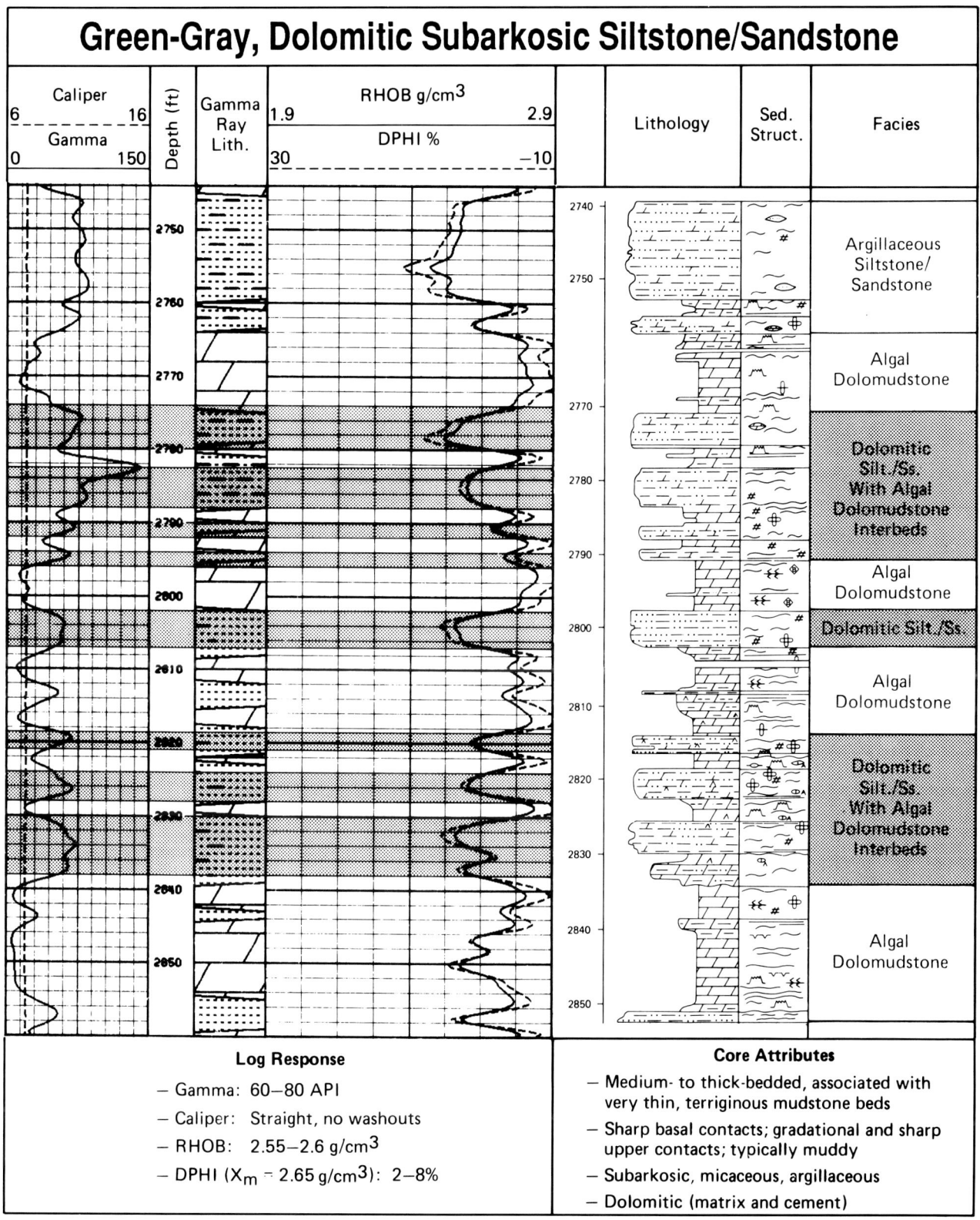

Figure 22

Figure 23

A. A typical texture for Green-Gray Dolomitic Subarkosic Siltstones/Sandstones shows the detrital grains consist of well-sorted silt. Matrix consists of both dolomicrite and detrital clay. Pyrite-filled structure is probably a burrow. L. Daugherty #4, 2787 ft, cross-polarized light.

B. Another typical texture is green-gray dolomitic siltstone to very fine sandstone. The line delineates one of many small oval burrows that are common. Well-sorted silt to very fine sand with abundant (about 50%) dolomicrite matrix occurs outside the burrow; whereas, the burrow is filled with poorly sorted silt to upper fine sand. Authigenic pyrite, gypsum, and quartz overgrowths are common in the burrow fill. L. Daugherty #4, 2827 ft, cross-polarized light.

C. Dolomicrite matrix and associated detrital clay in green-gray siltstones/sandstones. Detrital clay is usually common when there is abundant dolomicrite matrix. The clay occurs both as laminae and as submicroscopic platelets that are disseminated throughout the dolomite and give it a cloudy appearance. L. Daugherty #4, 2827 ft, cross-polarized light.

D. SEM photomicrograph showing the typical dolomite-rich nature of the green-gray siltstones/sandstones. The very fine crystal size, abundant detrital clay impurities, and loose packing of the detrital framework suggest that the dolomite is recrystallized, original micrite(?) matrix. Dolomite is so abundant (often over 50%) in many of these rocks that framework grains are almost completely obscured in SEM. L. Daugherty #4, 2800 ft, SEM.

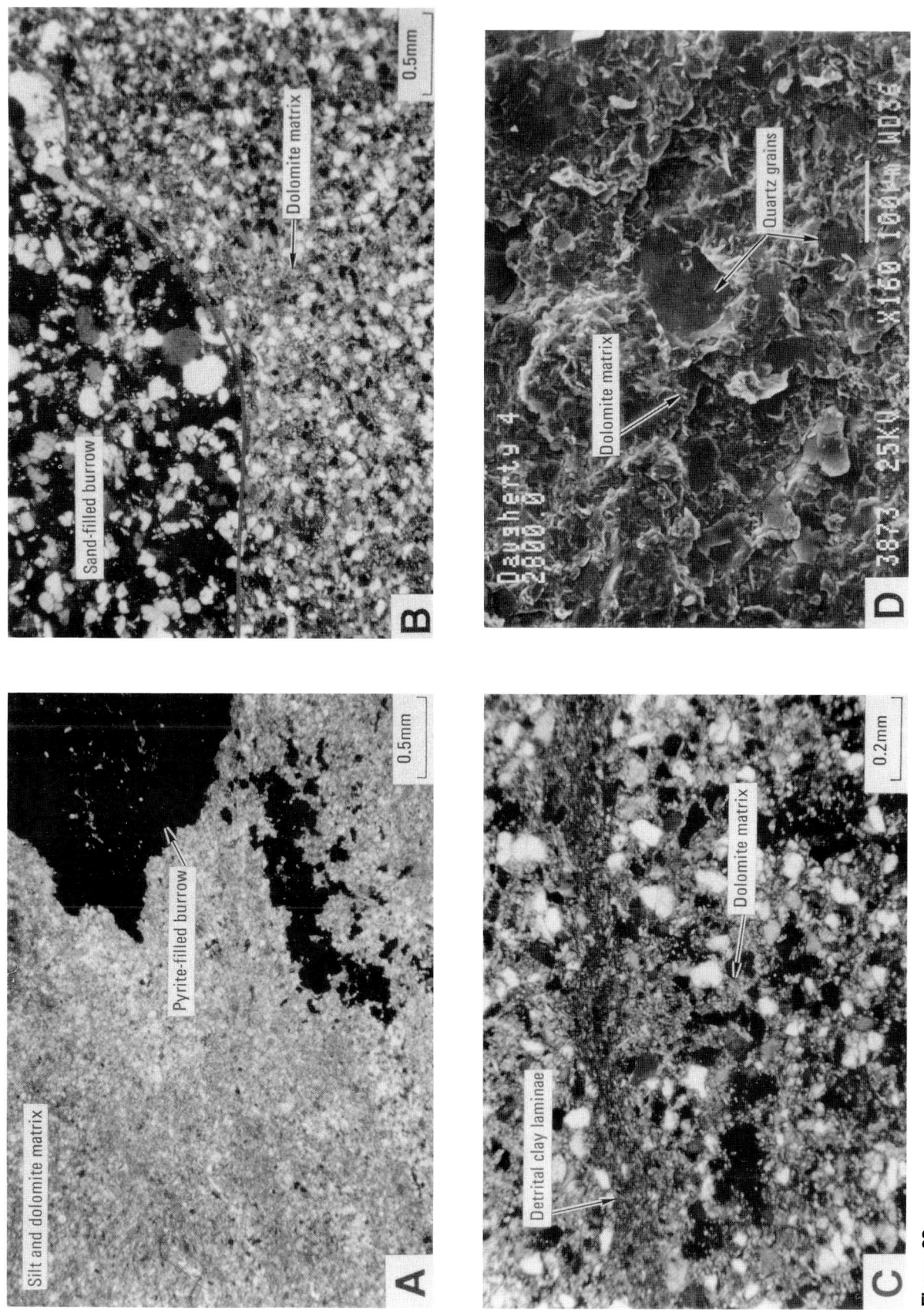

Figure 23

Dolomite ranges in abundance from 17 to 46 percent (Table 5), and dolomite occurs as both matrix and cement phases. The dolomite matrix is very finely crystalline and often associated with detrital clay and mica (Fig. 23C and D). Dolomite cement occurs as subhedral to euhedral, fine to medium, pore-filling crystals (Fig. 24A) that are only rarely Fe-rich.

Framework mineralogy is similar to the previously described facies, but is slightly less feldspar-rich due to grain dissolution and alteration. Samples are predominantly micaceous, lithic subarkoses with 5 to 15 percent feldspar as subequal amounts of K-feldspar and plagioclase. Pore-filling, microporous kaolinite occurs in some samples and is interpreted as the by-product of feldspar and lithic grain dissolution (Fig. 24A through C). Lithic grains are difficult to identify due to their very fine size, but chert and/or finely crystalline volcanic and metamorphic rock fragments appear to be most common.

Associated carbonates.--Medium to thick intervals of dolomite are interbedded with the Green-Gray Dolomitic Siltstones/Sandstones in the L. Daugherty #4 well (Figs. 20 and 21). The clastic-carbonate ratio is about 1:1 and is typical for the inner part of the outer shelf. Carbonate lithologies are much more abundant and diverse than those found updip in the clastic-rich, middle-shelf region, and some dolomite intervals exhibit well-developed porosity and are oil-stained; therefore, a more detailed description of the carbonate lithologies is warranted.

Carbonate facies are dolomudstone/wackestones that exhibit a wide variety of textures. Structureless, argillaceous, wispy-laminated, algal-laminated, and fenestral textures are common (Fig. 25A through D), but perhaps most common is a clotted algal texture. Intervals of dolopackstone and minor grainstone are also present. Allochems consist predominantly of algae, peloids, and very minor skeletal fragments. Thin (centimeter-scale) intervals of algal-coated pisolites and intraclasts are present (Fig. 25B).

Several types of porosity are locally well-developed in the dolomite intervals (Fig. 26A through D). Moldic and fenestral porosity are volumetrically the most important and attain values up to 10 percent or greater (Fig. 26B and D). Small (less than 0.1 to 0.5 mm), rounded molds are common but are of unknown origin; elongate molds of possible skeletal fragments are rare. Laminoid fenestral porosity is locally abundant and often solution-enlarged (Fig. 26C). Intra- and interparticle porosity can be substantial but is typically minor since most of the pores are filled with early cement (Fig. 26D). Large (5 to 15 mm) vugs and solution-enlarged microfractures are present in some intervals (Fig. 26A). All pore types are locally oil-stained.

Dolomite cement occurs with both very finely crystalline pore-lining and coarsely crystalline pore-filling textures (Fig. 27A and B). Cathodoluminescent (CL) microscopy of these same dolomite cements reveals three distinct cementation zones, each with internal microzonation (Fig. 27C and D). The first two CL zones only line pores; whereas, the third zone both lines and locally fills pores.

Table 5   GREEN-GRAY, DOLOMITIC SUBARKOSIC SILTSTONE/SANDSTONE FACIES - PETROGRAPHIC DATA

<---SAMPLE INFO--->		<---------FRAMEWORK GRAINS--------->						<-------INTERSTITIAL MATERIAL------->						<---TEXTURE--->	
WELL	DEPTH	QUAR	K-FE	PLAG	ROCK	HEAV	MICA	MATR	RED MATR	AUTH CLAY	DOLO	ANHY +GYP	PYR	SHR	POR	GRAIN SIZE φ	SORT φ
PDB4	1567.95	35	6	4	TR	TR	1	0	10	0	27	15	0	0	0	4.50	1.50
PDB4	1661.50	27	7	10	1	2	1	6	0	0	40	7	0	0	0	4.25	0.85
PDB4	1678.30	25	8	9	3	1	1	3	0	4	40	9	0	0	0	4.25	0.85
L.D4	2772.0	40	4	5	4	2	1	18	0	3	17	0	1	0	1	3.75	0.8
L.D4	2781.0	39	7	5	7	2	TR	0	3	8	29	0	0	0	0	3.5	0.6
L.D4	2787.0	23	3	2	1	1	4	18	0	TR	46	0	2	0	0	4.75	1.5
L.D4	2799.5	40	8	7	8	2	TR	0	3	9	22*	0	1	0	TR	3.25	0.6
L.D4	2816.5	38	8	7	8	3	0	TR	0	5	17*	11	2	0	TR	3.5	0.6
L.D4	2826.5	43	5	4	6	2	0	0	0	14	17	7	1	TR	TR	3.5	0.6
L.D4	2827.0	39	6	5	6	2	1	0	3	5	30	2	TR	0	1	3.25	0.8

COLUMNS: QUAR=quartz; K-FE=Potassium feldspar; PLAG=plagioclase; ROCK=rock fragments; HEAV=heavy minerals; MICA=biot/musc; MATR=gray detrital matrix; RED MATR=red detrital matrix; AUTH CLAY=authigenic clay; DOLO=dolomite (* ferroan dolomite; Mag=magnesite; CAL=calcite; ANHY+GYP=anhydrite + gypsum; PYR=pyrite; SHR=solid hydrocarbon residue; POR=percent porosity; GRAIN SIZE (phi units); SORT=sorting (phi units)

KEY

Figure 24

A. Typical subarkosic mineralogy of Green-Gray Dolomitic Siltstones/Sandstones. Some feldspars and lithic grains are partially to wholly dissolved, and the resulting void spaces are filled by kaolinite; whereas, other feldspars remain unaltered. Also note the patchy, subhedral to euhedral, fine- to medium-crystalline dolomite cement that is distinctly different from the dolomicrite matrix shown in Figures 23A through D. L. Daugherty #4, 2827 ft, cross-polarized light.

B. SEM photomicrograph showing partial feldspar dissolution and kaolinite growth in the Green-Gray Dolomitic Siltstones/Sandstones. Some medium crystalline dolomite cement is also present in the sample. Note the lack of open pores other than micropores associated with the kaolinite. The box represents a higher-magnification view shown in Figure 24C. L. Daugherty #4, 2799 ft, SEM.

C. Higher-magnification view showing micropores associated with the kaolinite platelets. Often the kaolinite blebs will be light brown when viewed with a plane-light microscope due to a slight oil stain within the micropores. L. Daugherty #4, 2799 ft, SEM.

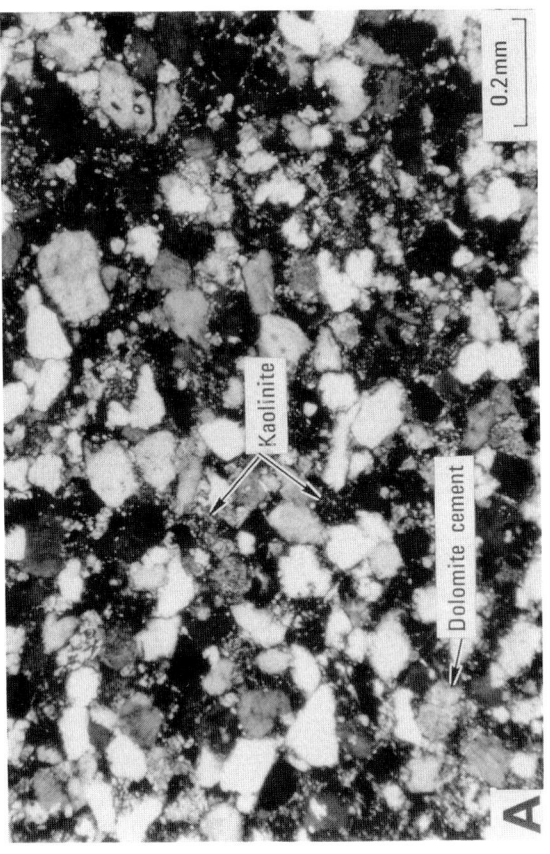
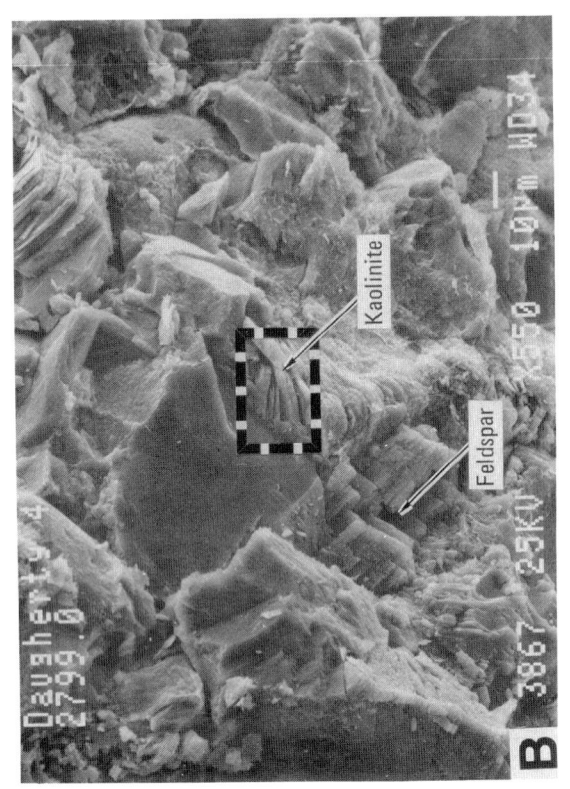

Figure 24

Figure 25

A. Dolomudstones and argillaceous dolomudstones are commonly associated with the Green-Gray Dolomitic Subarkosic Siltstones/Sandstones. The dolomudstone is structureless other than the swarms of detrital clay laminae. Massive dolomudstone beds are typically tan-colored and less than 1 or 2 ft (30 to 60 cm) thick. L. Daugherty #4, 2763 ft.

B. Peloidal, pisolitic dolopackstone overlying algal, pelletal/peloidal dolowackestone. The clotted, fine-grained dolowackestone is commonly associated with Green-Gray Dolomitic Subarkosic Siltstones/Sandstones. Thin (<4 in. = 10 cm) pisolitic and intraclastic beds with sharp upper and lower boundaries are rare but show the close proximity to the pisolite shoal complex. L. Daugherty #4, 2817 ft.

C. Peloidal dolomudstone with abundant laminoid, fenestral porosity that is possibly related to shrinkage during desiccation. Note the abundant laminae that may have had an algal origin, the dark oil stain, and small-scale vertical fractures. Locally these vertical fractures are solution-enlarged. L. Daugherty #4, 2847 ft.

D. Algal-laminated dolomudstone with incipient fenestral fabric. Vertical fractures are abundant and solution-enlarged in the lower part of the slab. Fractures and fenestral pores are oil-stained. L. Daugherty #4, 2848 ft.

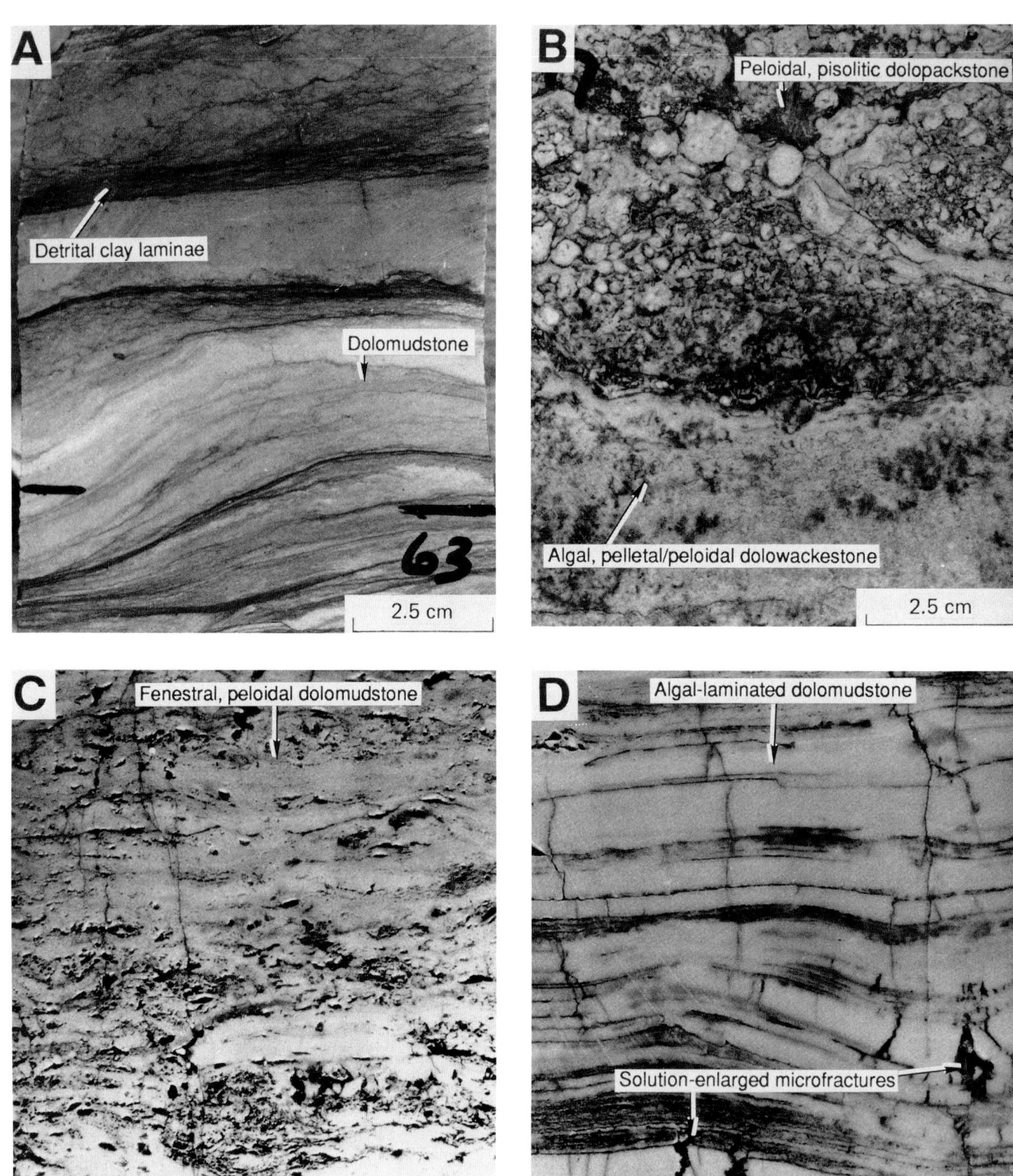

Figure 25

Figure 26

A. Large (3 to 15 mm) vugs lined by thin rims of dolomite cement. The vugs often exhibit oil staining. L. Daugherty #4, 2727 ft, cross-polarized light.

B. Well-developed moldic and interparticle porosity in a peloidal dolomudstone/wackestone. Small (0.25 to 0.5 mm), rounded molds are common; whereas, elongate (skeletal?) molds are rare. L. Daugherty #4, 2828 ft, plane-polarized light.

C. Fenestral porosity in algal-laminated dolomudstone. Pores are concentrated in distinct laminae. The small "birdseye" pores are probably formed during shrinkage and crinkling of an algal mat during early desiccation. The larger, more irregular-shaped pores may be related to solution enlargement of the "birdseye" structures. Also note the open microfracture that connects two solution-enlarged fenestral pores. L. Daugherty #4, 2849 ft, plane-polarized light.

D. Interparticle, moldic, and vuggy porosity in a pelletal dolowackestone/packstone. Most pores are rounded to slightly irregular and 0.25 to 0.5 mm wide. Locally these pores are solution-enlarged and connected, which results in a more vuggy appearance. Small, inter-particle pores are 0.1 mm or less in diameter and are typically filled with dolomite cement. Most of the larger pores are lined by dolomite cement. L. Daugherty #4, 2844 ft, plane-polarized light.

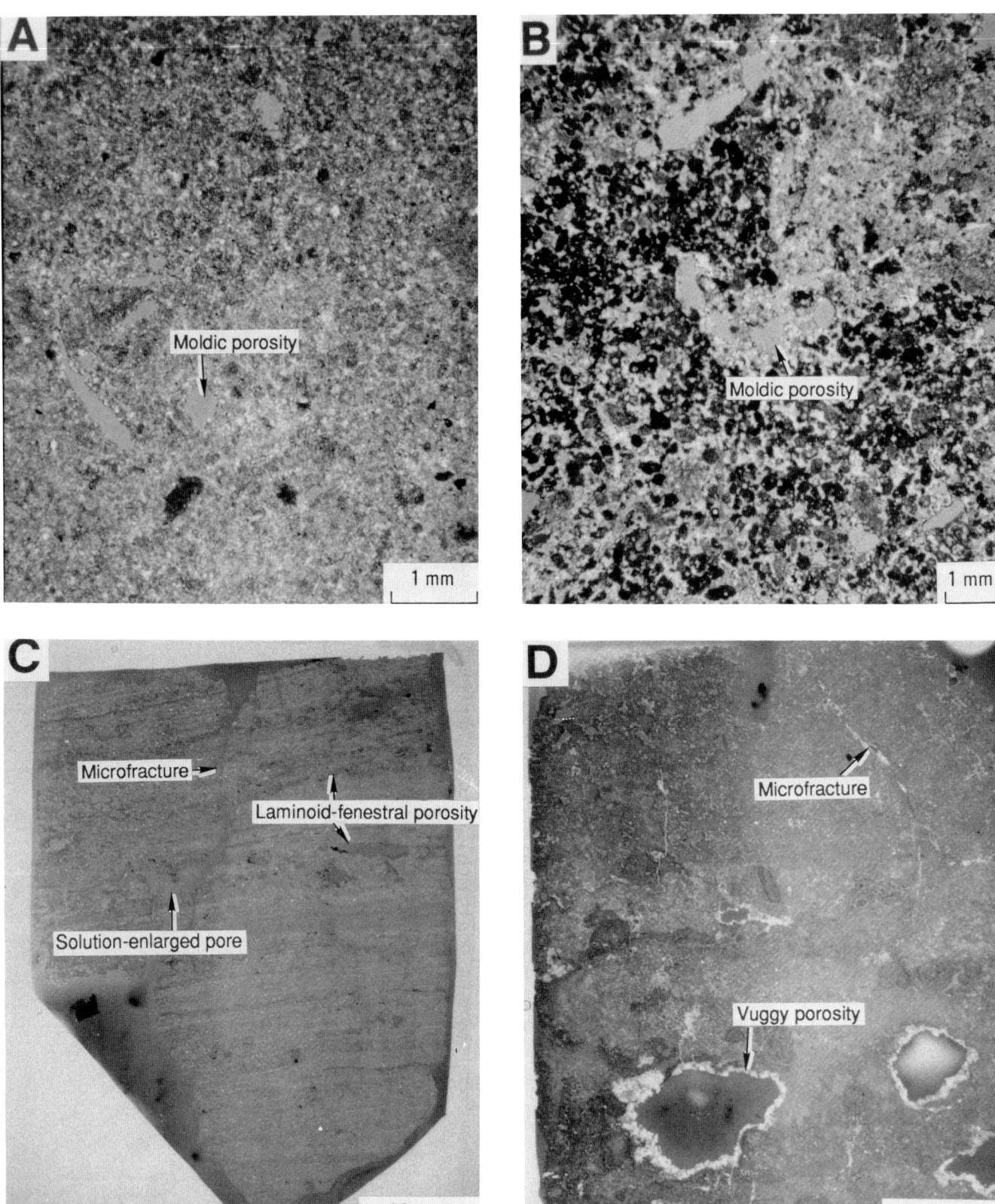

Figure 26

Figure 27

A. Two major dolomite cement types line and fill pores in the carbonate intervals of the Yates Formation in the L. Daugherty #4 well. Algal dolomudstone/wackestone shows the typical clotted texture of many dolomites associated with Green-Gray Dolomitic Subarkosic Siltstones/Sandstones. The cements consist of an early, finely crystalline, pore-lining phase and a later, coarsely crystalline, pore-filling phase. Figure 27B is a high-magnification view of the thin section. L. Daugherty #4, 2796.5 ft, cross-polarized light.

B. Finely crystalline dolomite cement lines both interparticle and intraparticle pores associated with clotted algal grains. A later phase of coarsely crystalline, euhedral dolomite partially fills some of the interparticle pores, growing from a substrate of the fine pore-lining dolomite. L. Daugherty #4, 2796.5 ft, cross-polarized light.

C. Cathodoluminescent characteristics of the grain-lining and pore-filling cement phases. Under cathodoluminescence, multiple phases of grain-lining cement are apparent. Major luminescent zones consist of (a) an initial, moderately bright, microzoned layer; (b) a middle, dark red layer; and (c) a microzoned bright layer. The late bright cement is locally coarsely crystalline and occurs as a pore-filling phase. L. Daugherty #4, 2727 ft.

D. Microzonation in the three major luminescent phases. Note the delicate light and dark zonation in the earliest cement phase. The second cement phase also shows some internal zonations under high magnification; a very thin, weakly luminescent, light pink layer was precipitated prior to the dark red layer that comprises the bulk of the zone. The later, bright, grain-lining and pore-filling cement also exhibits microzonation defined by alternating bright yellow-orange, orange, and dark orange bands. The microzonation within the bright cement can best be seen in Figure 28. L. Daugherty #4, 2796.5 ft.

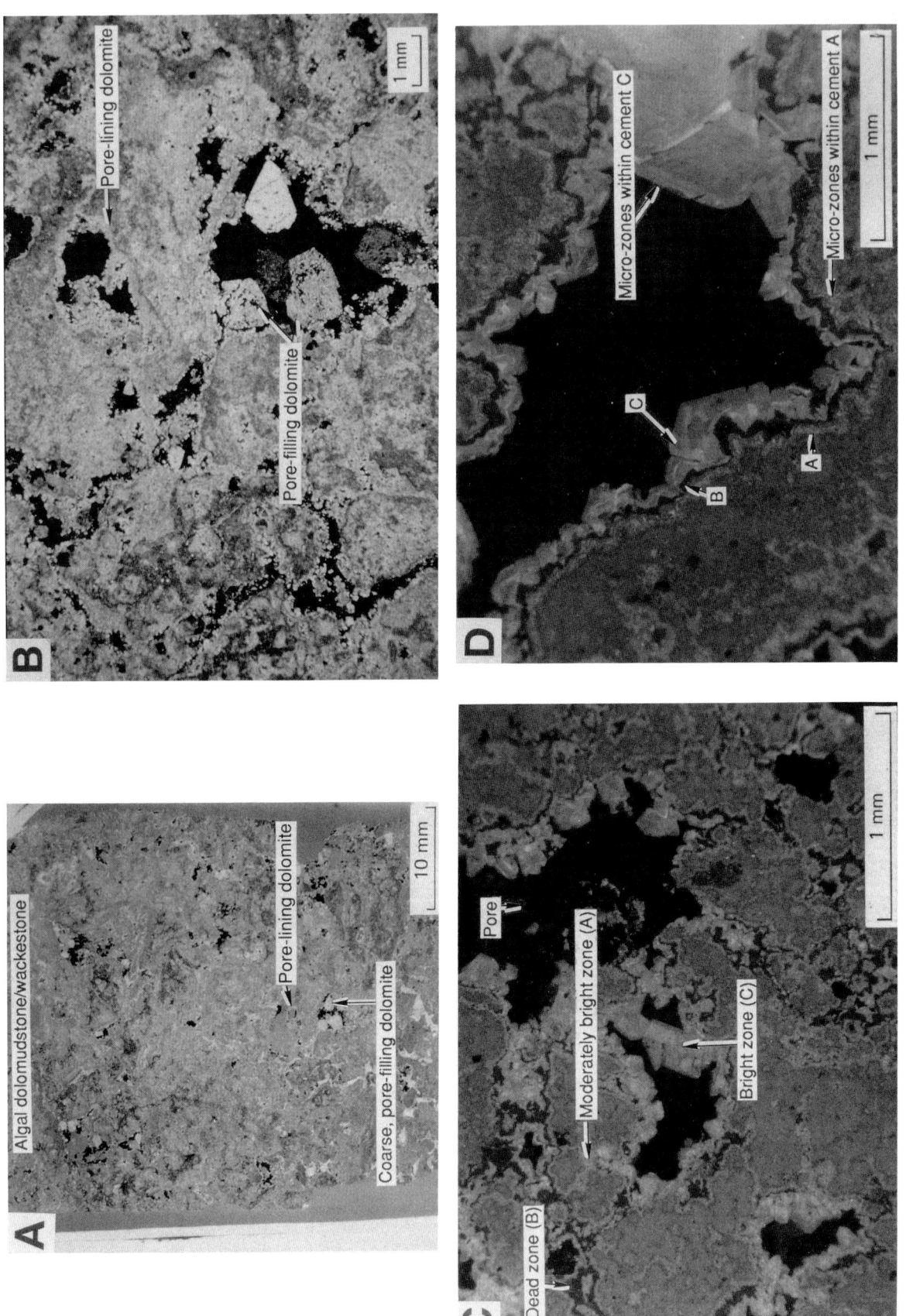

Figure 27

The first CL zone (A) exhibits moderately bright luminescence that, upon close inspection, is comprised of very fine (µm-scale) microzones of alternating bright and darker layers (Fig. 27D). Zone A occurs as a very finely crystalline pore-lining cement that grows directly on allochems and locally fills minor small intraparticle and interparticle pores.

The second CL zone (B) is a thin, dark red zone that lines pores and typically grows off of a substrate of zone A (Fig. 27D). Close inspection shows that zone B contains two microzones: (1) a thin, weakly luminescent pink layer directly overlying zone A; and (2) a thicker, dark red layer. In microfractures, zone A is absent and zone B is the first pore-lining phase occurring directly on matrix or allochems (Fig. 28).

The third cement zone (C) is a brightly luminescent cement that lines and locally fills pores (Figs. 27C and D and 28). When best developed, zone C occurs as coarse euhedral crystals. Microzones are readily apparent within the coarse crystals and consist of thick (0.1 to 0.4 mm) luminescent bands alternating with very thin (µm-scale), nonluminescent, dark bands (Figs. 27D and 28).

Although the relative timing of the various luminescent zones is clear from stratigraphic relations (A before B before C), the diagenetic environments of each cement phase are not entirely understood. A crossplot of carbon and oxygen isotope values obtained from various cements, matrix, and allochems in the Yates Formation exhibits a trend of presumed increasing temperature through time for the carbonate phases (Fig. 29). Dolomitized grains and matrix plot in a low-temperature field (or just slightly heavier in carbon). The coarse, brightly luminescent, euhedral pore-filling cement (C) plots in a higher temperature field and is clearly a burial cement. Results for cement zones A and B are ambiguous, however, because the isotopic values for these rim cements fall in the overlap field between high and low temperature dolomites. These values may unfortunately represent contaminated samples that contain cement zones A, B, and C, because all the cement phases can occur together within very thin rims that cannot be easily separated by microsampling.

Some information on the diagenetic environments for cements A and B can be inferred from the fact that cement zone A is absent in microfractures, where, instead, cement zone B is the initial lining phase (Fig. 28). This suggests that the fracturing clearly postdates cement zone A. Cement zone A is interpreted as an early phreatic cement due to its very fine crystal size, distribution directly on allochems, and even pore-lining nature. Cement A also fills small pores and was probably coincident with early lithification of the carbonates. The growth of cement B may have occurred any time after cement A and continued until sometime after fracturing.

Red Anhydritic Siltstone/Sandstone

Thick intervals of dark red, anhydritic siltstone/sandstone occur in the upper part of the PDB-04 core (Figs. 4, 30, and 31). These red siltstones/sandstones are typically wispy- to wavy-laminated, with only minor plane laminae. The most conspicuous structures are dark red, thin, discontinuous, wispy laminae of detrital clay that occur disseminated throughout the siltstones and also concentrated into local horizons which define crude bedding. Some intervals appear

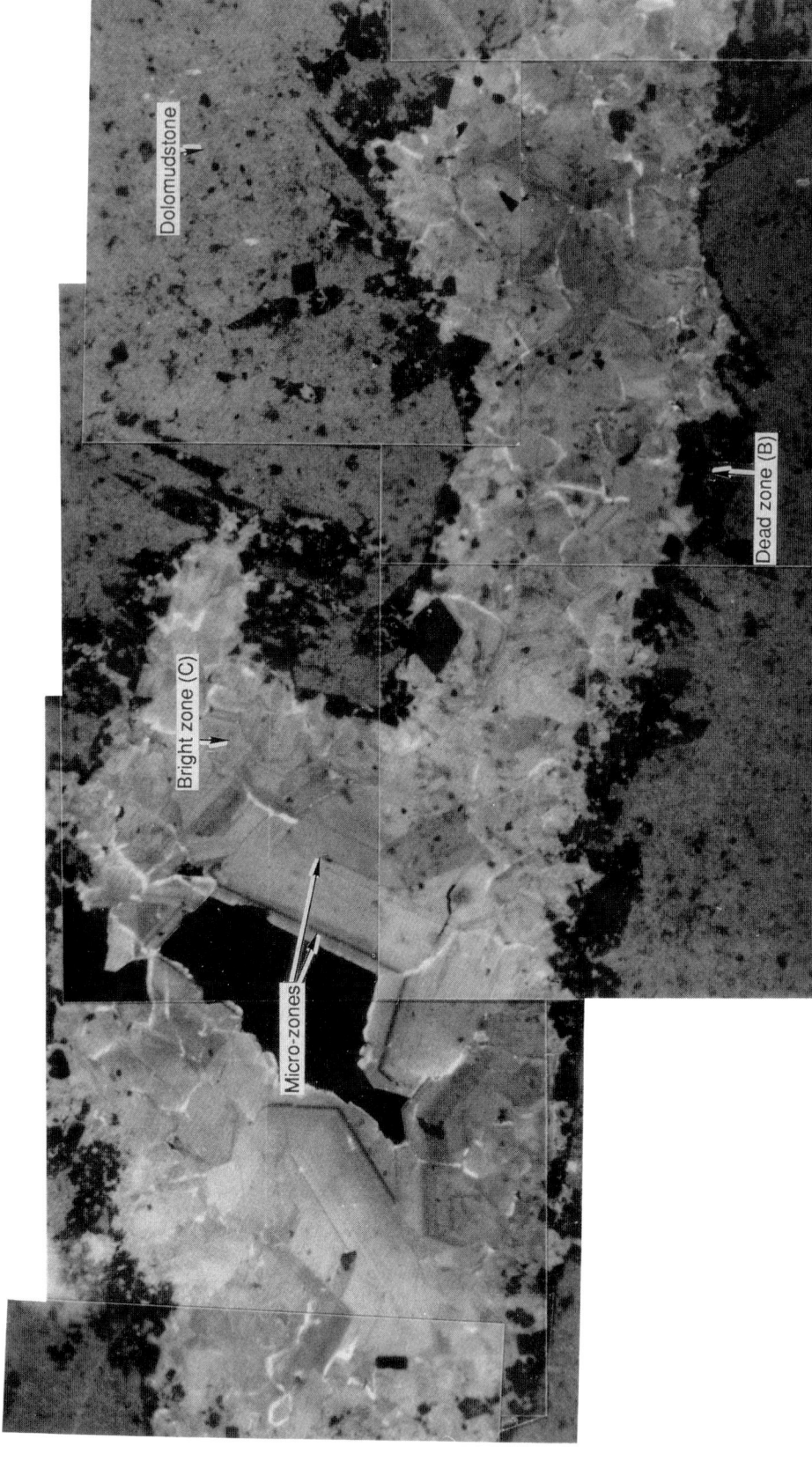

Figure 28
Cathodoluminescence photomicrograph showing two distinct zones of dolomite cement filling a microfracture in dolomudstone. The fracture is lined with a fine- to medium-crystalline, dark red cement that is the same as cement zone (B) in Figure 27C and D. The fracture is then nearly filled with a coarsely crystalline, microzoned, bright cement that is the same coarse, pore-filling cement zone (C) of Figure 27C and D. A lack of any early cement (A), which is well-developed in inter and intraparticle pores in Figure 27C and D, suggests the fracture postdates cement (A). L. Daugherty #4, 2791 ft.

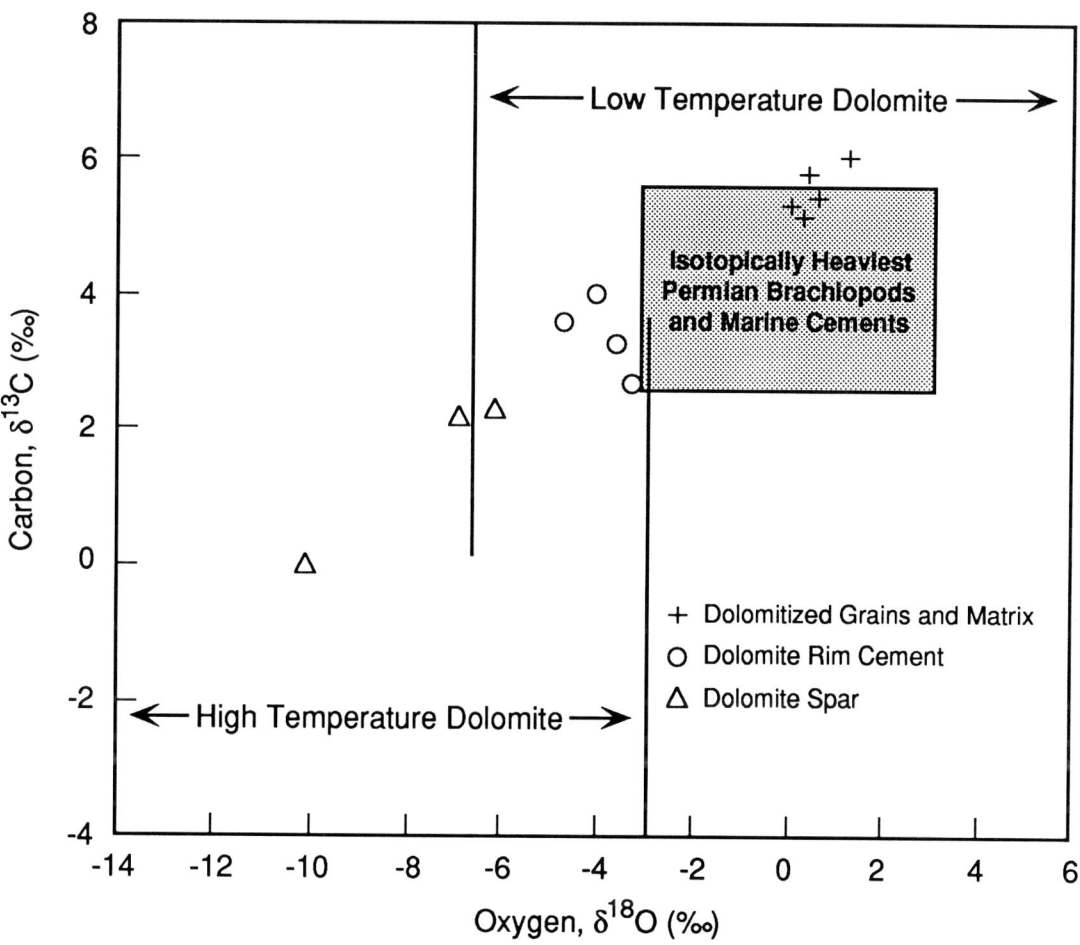

Figure 29
Crossplot of carbon and oxygen isotope values obtained from various dolomite cements, dolomite matrix, and grains from Yates carbonates in L. Daugherty #4. Dolomite matrix and grains have $\delta^{18}O$ values that plot within what we consider a low-temperature dolomite region and, in fact, within the data field for Permian seawater composition as estimated from brachiopods and marine cements. Dolomite rim cements (probably cathodoluminescent cement zones A, B, and C mixed together) have values that fall within the overlap between low- and high-temperature dolomite fields. Coarse, pore-filling dolomite spar cements (cathodoluminescent zone C) exhibit distinctly lighter (depleted) values suggestive of higher-temperature burial dolomite.

Figure 30

Red Anhydritic Siltstones/Sandstones from upper part of Yates interval in PDB-04 core. The well is positioned approximately 2 miles (3.2 km) behind the Yates shelf edge on the Northwest Shelf. The red siliciclastics typically have sharp basal contacts and gradational upper contacts with dolomites of the pisolite shoal complex. The dolomites shown here consist of fenestral dolomudstones and incipient pisolitic dolowackestones with abundant anhydrite-filled voids. The siliciclastics occur as thick intervals that exhibit little in the way of sedimentary structures other than crude, decimeter-scale interbeds defined by variations in cement type, abundant wispy detrital clay laminae, and rare ripple laminae. Intervals of small, rounded anhydrite blebs (fenestral fabric(?), burrows(?)) are also common. The dark color of the siliciclastics is due to a hematite stain on detrital clay that is present as grain rims, disseminated platelets, and laminae.

Figure 31

Log response and core attributes for Red Anhydritic Siltstones/ Sandstones; PDB-04 well. The outer-shelf red beds are characterized by high gamma-ray values due to the presence of abundant feldspar and detrital clay. High density-low porosity values and a straight caliper reflect the tight anhydrite-, magnesite-, and dolomite-cemented nature of the rocks.

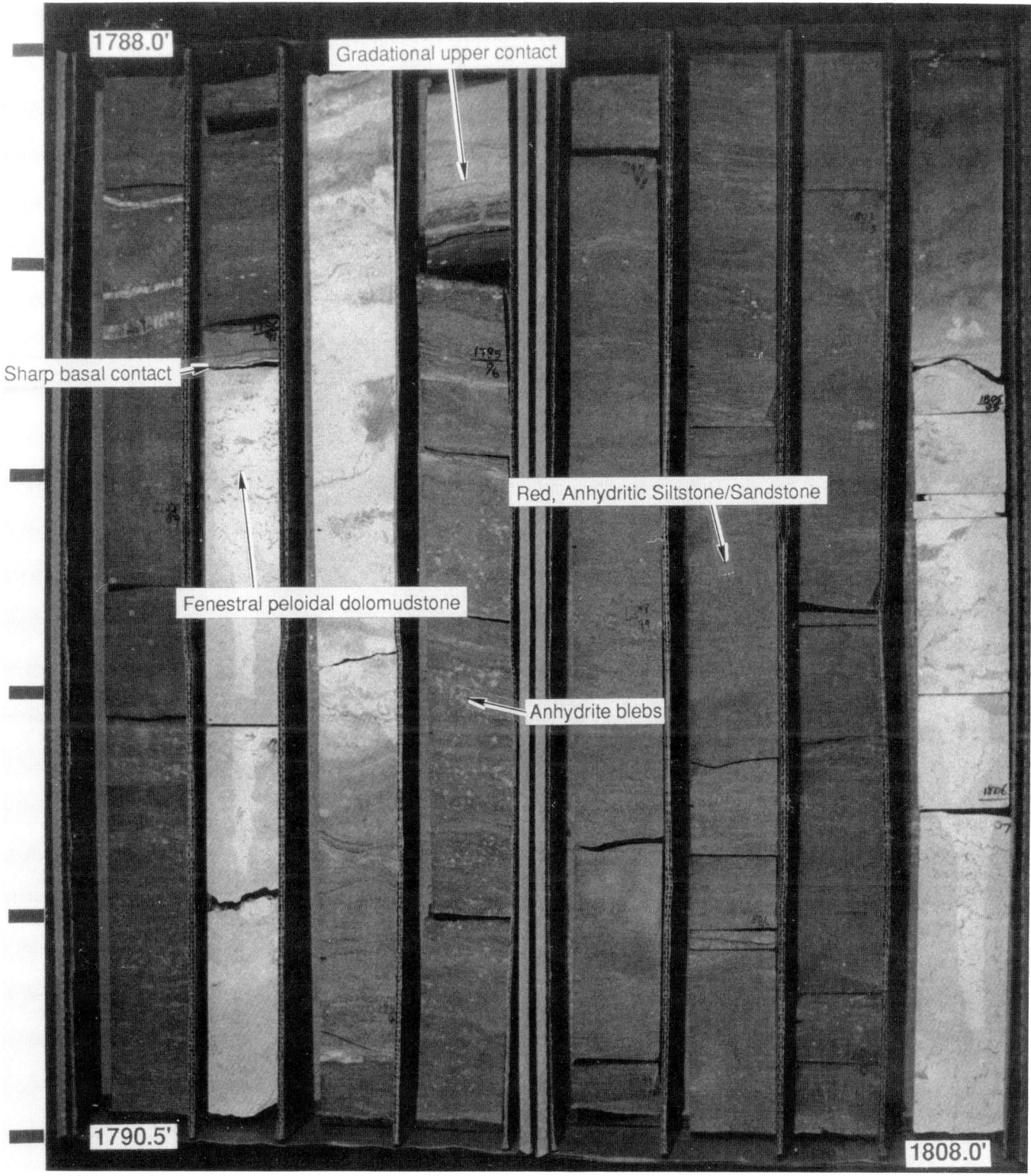

Figure 30

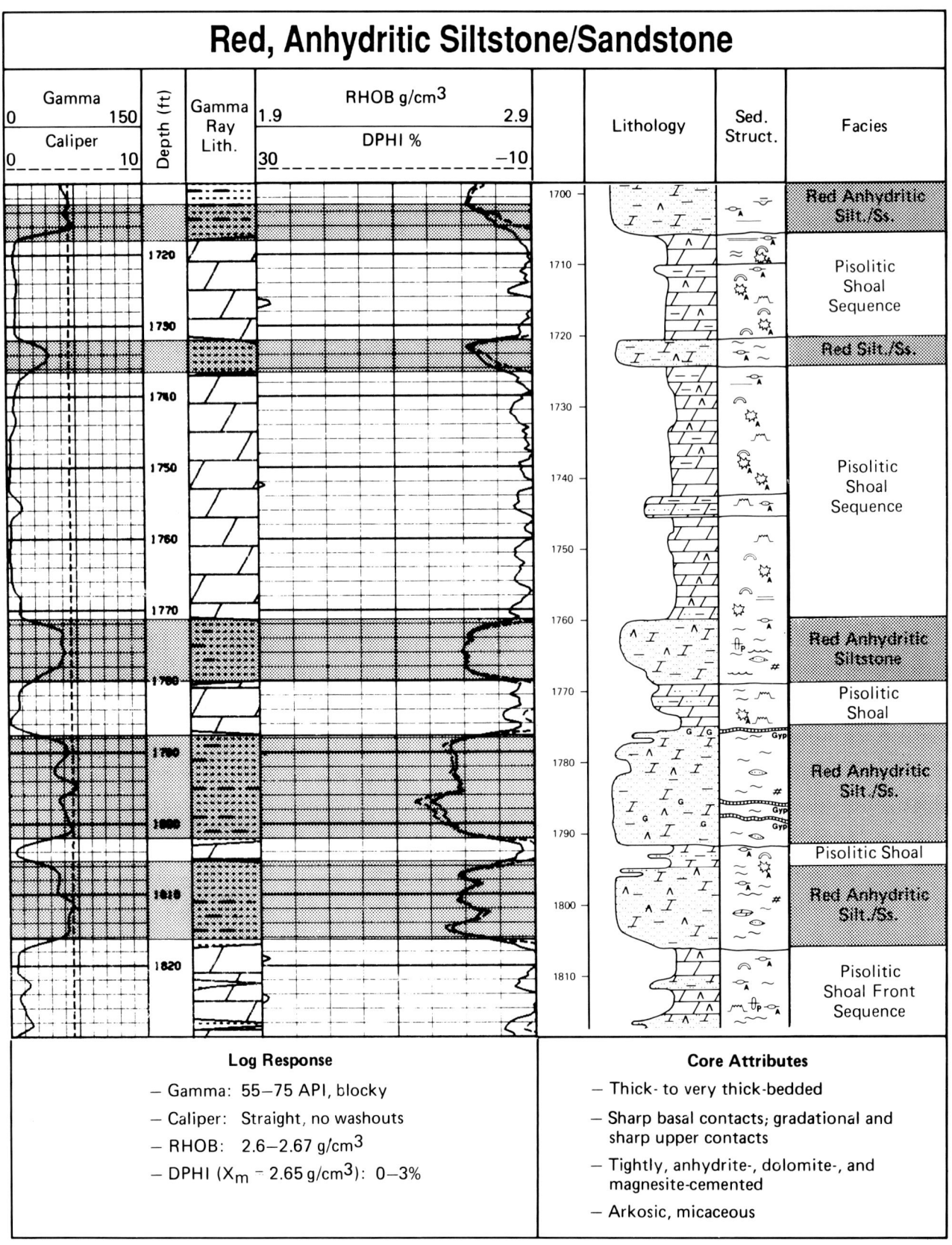

Figure 31

homogeneous and structureless in core, but show crude laminae in thin section. Laminae and thin beds of dolomudstone are common, as are anhydrite-cemented, fenestral(?) features. Crude bedding is also defined by variations in the type and amount of carbonate and evaporite cements. The siltstones/sandstones have sharp basal contacts and sharp to gradational upper contacts with dolomites.

Red Anhydritic Siltstones/Sandstones are very fine sandy, argillaceous siltstones and silty, very fine-grained sandstones (Fig. 32A through D). Grain size is variable with the modal grain size near the boundary between silt and very fine sand. Fine and medium sand is a minor constituent (less than 3 percent), occurring as very well-rounded, pitted or frosted quartz grains (LRFs = large, rounded, frosted grains)(Fig. 32A). Sorting is typically moderately good, but ranges from poor to good depending upon the amount of detrital silt and clay present in the sample. Grains other than the LRFs are angular to subrounded with a definite increase in roundness coinciding with increased grain size. Grain contacts are tangential (point) to straight with common matrix-supported (dolomicrite) lenses and pods (Fig. 32D). Small-scale sedimentary structures include normally graded laminae and lenses and pods of upper very fine sand. Heavy mineral trains composed primarily of detrital ilmenite are rare (Fig. 32C).

Red Siltstones/Sandstones are subarkoses and arkoses cemented by anhydrite and dolomite/magnesite (Table 6). Framework grains typically comprise 50 to 60 percent of the sample. Normalized to 100 percent, the detrital framework consists of 60 to 70 percent quartz, 20 to 30 percent feldspar, 1 to 5 percent lithics, 1 to 5 percent heavy minerals, and 1 to 3 percent mica (Fig. 33).

Red, illitic, detrital clay matrix comprises 5 to 20 percent of the red siltstone samples. The red (FeO_x-stained) clay occurs as thin wispy laminae, diffuse zones, and well-developed grain rims (Fig. 34A through D). Red clay in distinct laminae with abundant mica is detrital. Textural characteristics of the red grain-coating clay also suggest a detrital origin. Thin-section analyses show the grain-rimming clay is very similar to that in the detrital, red-clay laminae. Clay- and silt-sized mica platelets are plastered to grain surfaces and are suggestive of clastic deposition rather than authigenic growth (Fig. 34C and D). Also the oxidized nature and early emplacement of the clay rims (the clay underlies precompactional anhydrite cement) suggest a detrital origin.

The detrital dust may be eolian in nature since the grain-coating textures are like those described by others (Walker, 1979; Turner, 1980) and attributed to water-infiltrated airborne dust. The illite mineralogy is also common for "mature" eolian dusts, as Singer (1988) showed that illite forms pedogenetically in the surface horizons of arid and semi-aridic soils. Potassium required for the illitization process is introduced to the soils by the deposition of desert dusts that contain K-bearing minerals (e.g., mica, feldspar). The illitization process requires a long residence time and is enhanced by repeated wetting and drying cycles. Thus the illite-rich nature and pedogenetic features of Yates Formation red beds may record numerous cycles of deflation and deposition.

SEM analysis shows that authigenic clay is also a minor constituent of Red Anhydritic Siltstones/Sandstones. The most common is illite, which exhibits a delicate crenulate or hair-like growth habit that is clearly authigenic. EDX spectra, however, are nearly identical to those of ragged detrital illite (Fig. 35B). This suggests that some detrital illite has recrystallized

Figure 32

A. Common texture of the Red Anhydritic Siltstones/Sandstones. Framework grains are moderately well to well sorted and range from silt to very fine sand. Large (fine- to medium-grained), rounded, and frosted (LRFs) quartz sand grains are minor constituents. Note uniform, structureless nature of the siltstone/sandstone and patchy, detrital, and dolomicrite matrix-rich intervals. Gulf PDB-04, 1694.2 ft, plane-polarized light.

B. Wispy, red, detrital clay laminae abundant in the Red Anhydritic Siltstones/Sandstones. These sedimentary structures probably represent deposition in a low- to moderate-energy environment, with the wispy lamination being a type of clay-poor flaser bedding. Gulf PDB-04, 1694.8 ft, plane-polarized light.

C. Thick, detrital clay-rich laminae. Abundant hematite-stained detrital clay gives the shoal-crest siliciclastics their brick-red color. Also note the heavy mineral laminae at the contact between clay-rich and silt/sand-rich zones. This laminae is composed predominantly of unaltered detrital ilmenite grains. Gulf PDB-04, 1705.2 ft, plane light.

D. Dolomicrite matrix-supported laminae in red-gray siltstone/sandstone, at top of Yates interval in Gulf PDB-04, is transitional between Gray Dolomitic Subarkosic Siltstones/ Sandstones and Red Anhydritic Siltstones/Sandstones. The abundant matrix dolomite is suggestive of lagoonal siliciclastics; whereas, abundant anhydrite and a very weak, red stain are suggestive of shoal-crest siliciclastics. Gulf PDB-04, 1661.5 ft, plane-polarized light.

Figure 33

Mosaic of Back Scatter Electron (BSE) photomicrographs showing the arkosic mineralogy and tight anhydrite- and magnesite-cemented nature of typical Red Anhydritic Siltstones/Sandstones. This is the typical original framework mineralogy of Yates Formation siliciclastics as early cementation apparently protected framework grains from later diagenesis. Anhydrite cement appears as the washed-out white areas; whereas, magnesite appears as dark gray to black areas. All minerals labeled in the photo mosaic were identified using EDX analysis. Visible estimation suggests feldspars comprise 20% to 25% of the framework grains and occur as subequal amounts of plagioclase and K-feldspar. K-feldspar is easily identified due to its light gray color. Plagioclase and quartz have similar backscatter coefficients but can be differentiated by the altered nature and presence of cleavage in the plagioclase. Mica, lithic grains, and opaque minerals (often ilmenite) are also common constituents.

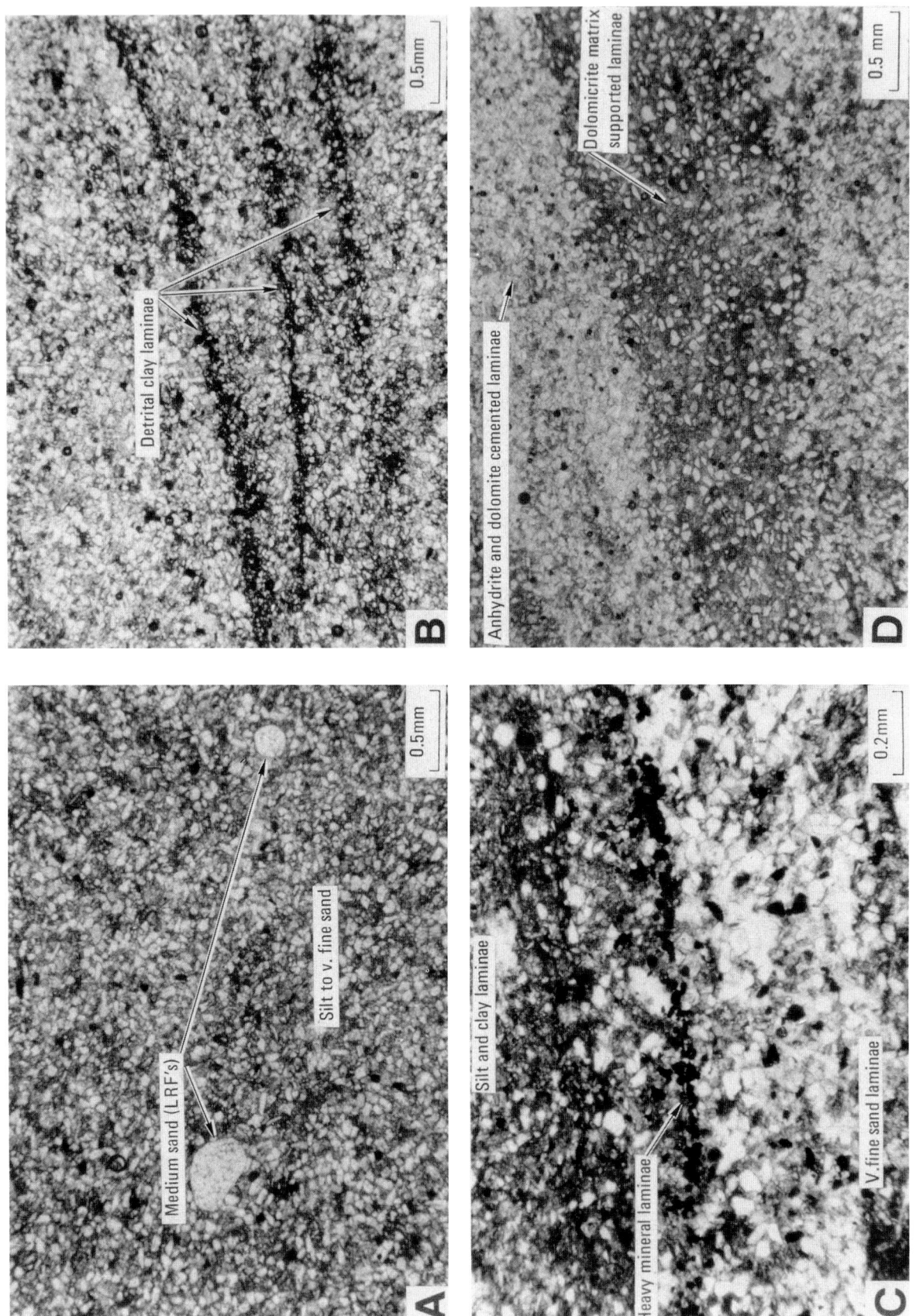

Figure 32

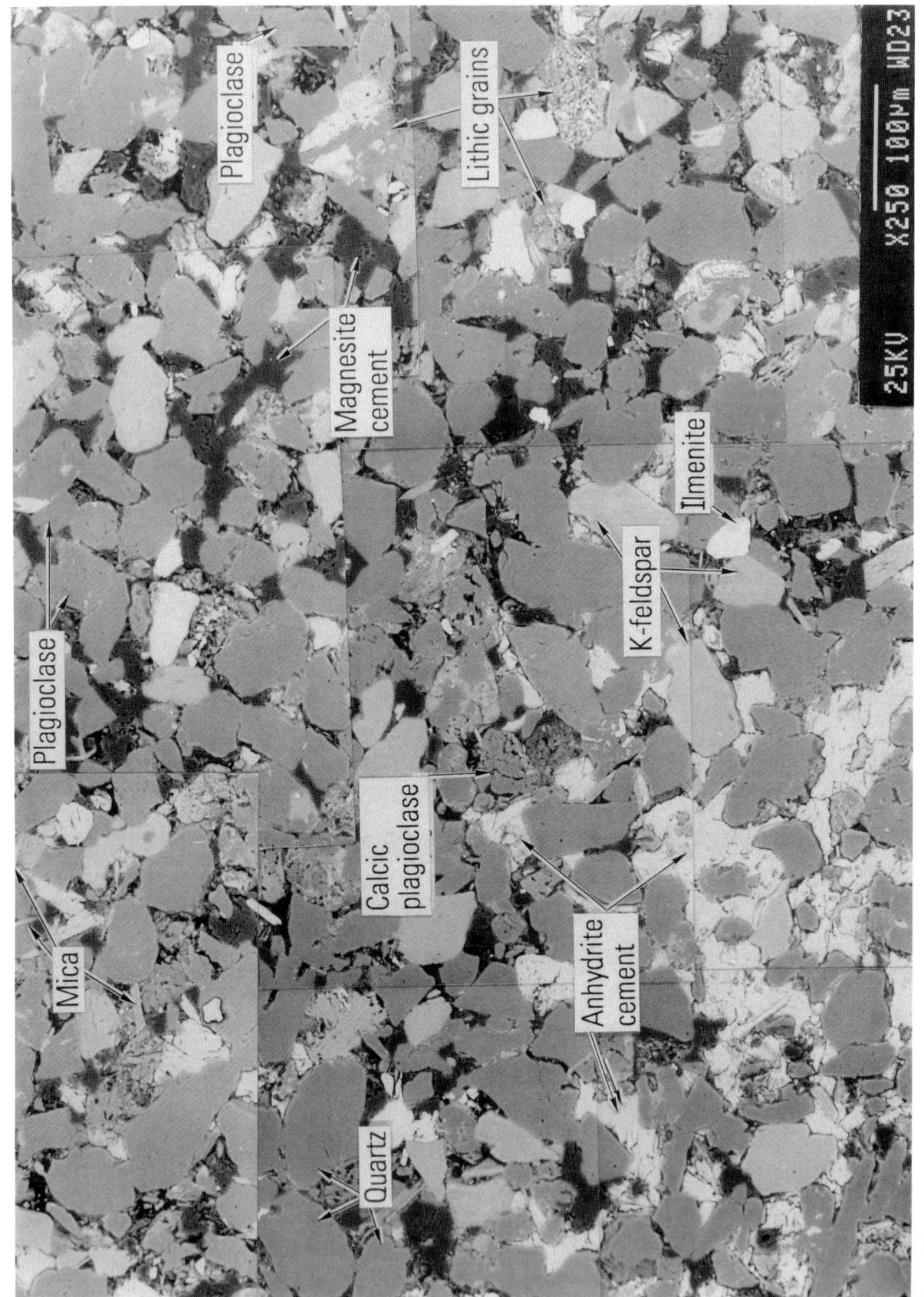

Figure 33

Table 6   RED, ANHYDRITIC SILTSTONE/SANDSTONE FACIES - PETROGRAPHIC DATA

<--SAMPLE INFO-->		<----------FRAMEWORK GRAINS---------->							<------INTERSTITIAL MATERIAL------>						<--TEXTURE-->	
WELL	DEPTH	QUAR	K-FE	PLAG	ROCK	HEAV	MICA	MATR	RED MATR	AUTH CLAY	DOLO /MAG	ANHY +GYP	PYR	SHR	POR	GRAIN SIZE φ	SORT φ
PDB4	1680.25	39	7	13	1	1	1	6	7	6	0/9	11	0	0	0	4.25	0.85
PDB4	1682.85	34	10	9	5	2	1	0	9	4	0/10	18	0	0	0	3.50	0.42
PDB4	1684.60	28	9	12	4	2	1	0	19	4	0/14	9	0	0	0	4.35	0.85
PDB4	1693.60	38	6	10	1	1	TR	0	17	0	20/0	7	0	0	0	4.00	1.50
PDB4	1694.15	34	6	10	3	2	TR	0	4	3	21/0	14	0	0	0	3.75	0.42
PDB4	1694.50	37	7	8	5	1	1	0	9	1	0/17	12	0	0	0	3.75	0.42
PDB4	1695.00	33	7	7	1	3	1	0	7	3	0/21	17	0	0	0	3.75	0.60
PDB4	1695.75	36	9	8	1	2	1	0	14	2	11/0	17	0	0	0	4.00	0.85
PDB4	1696.50	38	8	6	3	3	1	0	6	3	0/17	15	0	0	0	3.50	0.42
PDB4	1697.30	36	8	4	1	1	1	0	9	4	0/25	11	0	0	0	3.50	0.60
PDB4	1698.00	33	9	5	3	3	TR	0	3	2	0/5	36	0	0	0	3.25	0.42
PDB4	1698.75	30	7	9	TR	1	1	0	7	3	35/0	7	0	0	0	3.75	0.60
PDB4	1699.50	36	6	8	3	1	1	0	9	3	21/0	12	0	0	0	3.50	0.60
PDB4	1700.25	37	6	5	4	1	1	0	13	3	18/0	12	0	0	0	4.00	0.85
PDB4	1701.15	27	4	5	3	1	1	0	18	TR	0/30	11	0	0	0	4.00	0.85
PDB4	1701.95	34	4	7	3	1	1	0	10	TR	0/29	11	0	0	0	3.75	0.60
PDB4	1702.85	35	8	6	4	1	TR	0	9	2	0/14	26	0	0	TR	3.25	0.85
PDB4	1703.50	28	7	7	3	2	TR	0	6	2	0/15	30	0	0	0	3.25	0.42
PDB4	1704.35	24	4	4	1	1	TR	0	6	2	38/0	20	0	0	0	4.50	0.60
PDB4	1705.20	28	4	5	1	2	2	0	18	TR	25/0	16	0	0	0	4.50	0.85
PDB4	1764.30	32	6	6	4	1	1	0	9	TR	33/0	8	0	0	0	4.25	0.85
PDB4	1786.30	35	7	5	5	1	TR	0	4	TR	15/0	29	0	0	0	3.75	0.42
PDB4	1801.10	33	7	6	4	2	1	0	9	2	32/0	5	0	0	0	4.25	1.50
PDB4	1801.60	33	6	5	4	1	1	0	8	0	36/0	6	0	0	0	4.25	0.85

KEY

COLUMNS: QUAR=quartz; K-FE=Potassium feldspar; PLAG=plagioclase; ROCK=rock fragments; HEAV=heavy minerals; MICA=biot/musc; MATR=gray detrital matrix; RED MATR=red detrital matrix; AUTH CLAY=authigenic clay; DOLO=dolomite (* ferroan dolomite; Mag=magnesite; CAL=calcite; ANHY+GYP=anhydrite + gypsum; PYR=pyrite; SHR=solid hydrocarbon residue; POR=percent porosity; GRAIN SIZE (phi units); SORT=sorting (phi units)

Figure 34

A. Red detrital matrix can compose up to 20% of Red Anhydritic Siltstones/Sandstones. The color is interpreted as being due to an extremely fine hematite (or more poorly crystallized oxide or hydroxide) stain on detrital clay platelets and/or within the detrital or recrystallized clay lattice. Hematite has not been identified using any petrographic techniques. Note the abundant mica associated with the clay-rich horizon. Much of the detrital matrix is very fine silt- to clay-sized mica, as suggested by its illitic mineralogy. Gulf PDB-04, 1693.6 ft, plane-polarized light.

B. Back Scatter Electron (BSE) photomicrograph of argillaceous, micaceous laminae in Red Anhydritic Siltstones/Sandstones. In thin section the sample has red color and apparent matrix similar to the sample in Figure 34A; however, the matrix is much less apparent in the BSE image. This suggests visual estimations of red matrix in thin sections may be somewhat high. Most of the red clay occurs as thin coatings on framework grains and as inclusions in magnesite cement. Only small amounts of red clay are needed to give the rock a deep red color. Clay- to silt-sized mica platelets can also be seen plastered to the walls of framework grains. Gulf PDB-04, 1702.0 ft, BSE.

C. Very thin, red detrital clay coating framework grains. These red coatings are considered to be related to airborne dust infiltrated into the sediment and oxidized (forming iron-oxide pigments) during repeated wetting and drying events. Further reddening may also occur during intrastratal alteration of unstable framework silicates. Iron-bearing silicates (pyroxene, hornblende, biotite, and detrital ilmenite) in prolonged contact with oxygenated groundwater are susceptible to alteration by hydrolysis. Ions (including iron) are released into the groundwater, and authigenic mineral phases such as illite and hematite are precipitated (Turner, 1980). With continued reaction with oxygenated groundwater, the authigenic illite is likely to become red. Gulf PDB-04, 1694.5 ft, plane-polarized light.

D. SEM photomicrograph of red, illitic detrital clay. Note fine clay platelets are plastered on surface of quartz silt grain forming a thin grain coat. Other framework grains are coated by thick masses of detrital clay and dolomite. Gulf PDB-04, 1692.8 ft, SEM.

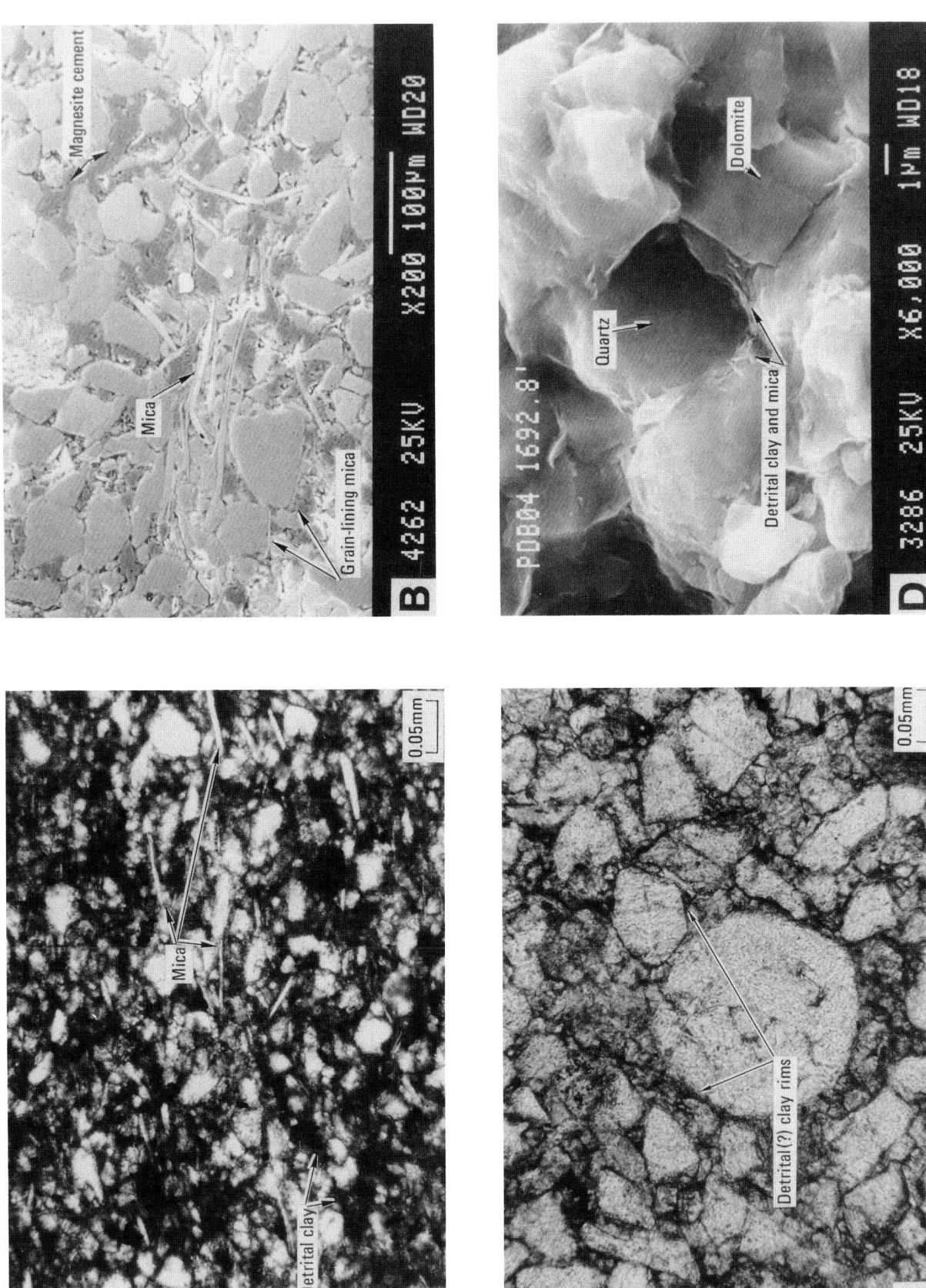

Figure 34

Figure 35

A. SEM photomicrograph and EDX spectrum of detrital clay. The clastic texture (irregular plates parallel to grain surfaces) and the occurrence in distinct laminae suggest the clay is detrital. The presence of K, Mg, and minor Fe, as seen from the EDX spectrum, is typical of illite. Illite is typically abundant in mature desert dusts that have undergone many cycles of deflation and deposition. Gulf PDB-04, 1680.4 ft, SEM.

B. SEM photomicrograph and EDX spectrum of very fine clay coating on detrital quartz grain. The delicate crenulate and hair-like morphologies are suggestive of an authigenic clay; however, the EDX spectrum is almost identical to that of clearly detrital illite in the same and other samples. This may suggest that some of the detrital illite has recrystallized or simply that the chemistry of the pore fluids from which the authigenic clay precipitated was controlled by the chemistry of the abundant detrital illite/mica present. Gulf PDB-04, 1702.9 ft, SEM.

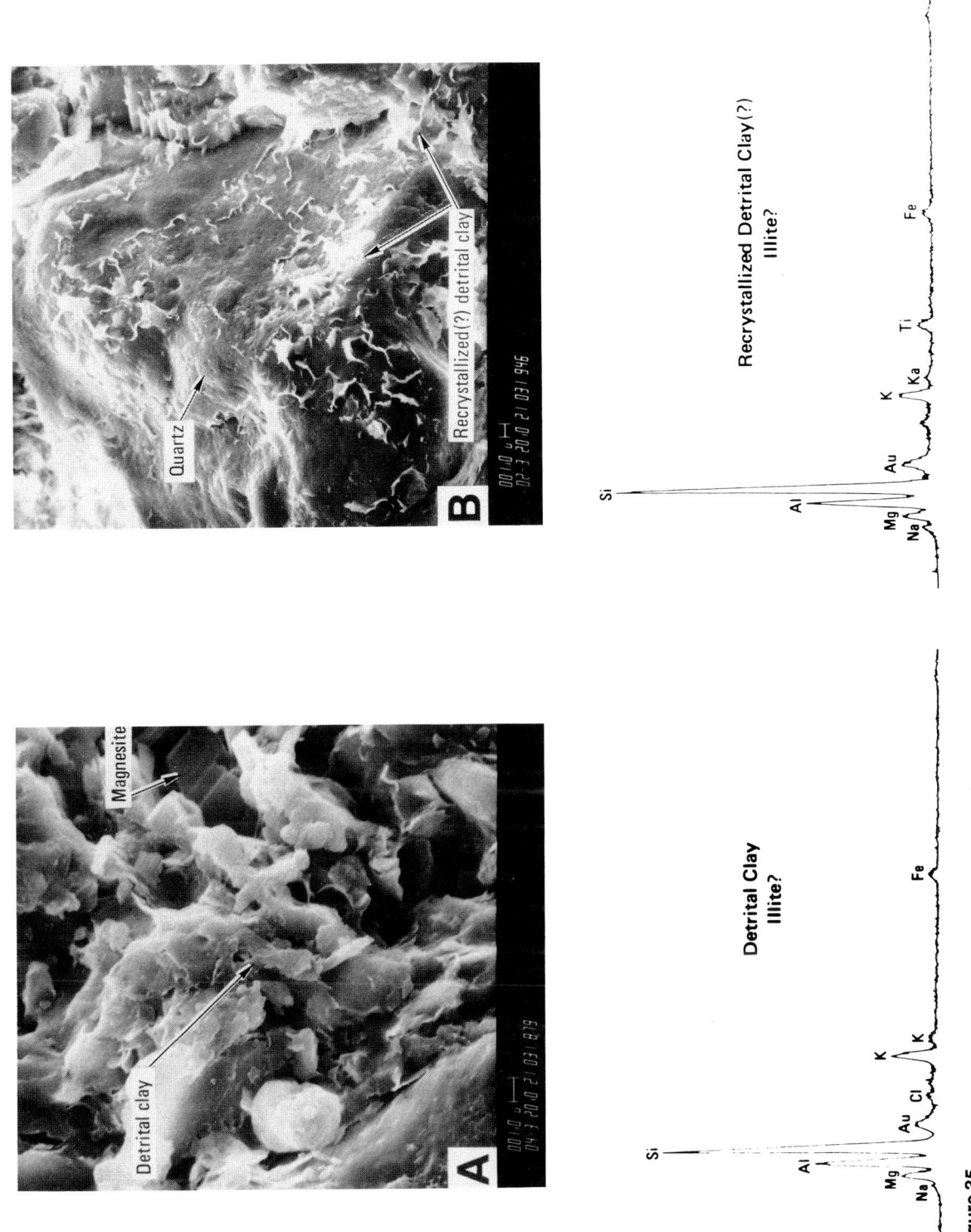

Figure 35

(compare Fig. 35A and B) or that the chemistry of the pore fluids from which the authigenic illite precipitated was dominated by the presence of the abundant illite/mica present as detrital matrix. Minor amounts of authigenic chlorite and mixed layer chlorite-smectite are also present as authigenic phases and occur predominantly as grain coatings and as small blebs after detrital grains.

Evaporite and carbonate cements are abundant in the Red Anhydritic Siltstones/Sandstones and typically fill primary intergranular porosity. Dolomite/magnesite and anhydrite cements typically comprise 25 to 40 percent of a sample. X-ray, microprobe, and SEM analyses show that magnesite and dolomite are generally mutually exclusive at the thin-section scale. Decimeter- to meter-scale intervals of dolomite cement alternate with intervals cemented by subequal amounts of anhydrite and magnesite. The alternating cement zones form crude bedding that is best seen in X-radiographs (Fig. 36).

Dolomite occurs both as a cement and as recrystallized(?) matrix (Fig. 37A through C). The matrix dolomite is very finely crystalline and often cloudy due to the presence of detrital clay and mica inclusions. Local, matrix-supported fabrics and dolomudstone laminae attest to the matrix origin of some dolomite. The matrix dolomite could be an original growth phase but is more likely a very early replacement phase after a metastable micrite matrix. In either case, stable carbon and oxygen isotopic compositions reported by Garber and others (1989) indicate that the dolomite formed in a near-surface environment, probably from waters with elevated salinities. Dolomite also occurs as pore-filling, very fine to medium crystalline cement (Fig. 37C). Dolomitized matrix is distinguished with difficulty from very finely crystalline cement in thin section. Both probably occur commonly together.

Magnesite typically occurs as very finely to finely crystalline pore-lining and pore-filling masses that are intimately associated with anhydrite cement (Fig. 38A). More rarely the magnesite occurs as subhedral to euhedral, medium crystals. In thin section and in SEM, the magnesite appears very similar to dolomite (Garber and others, 1990); however, EDX analyses (Fig. 38B) indicate the lack of calcium and abundance of magnesium that are characteristic of magnesite. Petrographic analyses show that when magnesite and anhydrite occur together in a pore, the magnesite lines the pore; whereas, anhydrite fills the pore center (Fig. 38C). These relations suggest that the magnesite is a relatively early primary cement phase that preceded at least some anhydrite cement. Alternatively the magnesite may have partially replaced pore-filling anhydrite, with the replacement occurring from the pore walls inward.

Anhydrite cement is also very abundant in Red Anhydritic Siltstones/Sandstones. Anhydrite-rich horizons occur both on laminae and thin-bed scales. Dolomite may occur with minor anhydrite; when the anhydrite is abundant, the associated carbonate cement is typically magnesite. Anhydrite occurs most commonly as medium to coarsely crystalline, subhedral to euhedral, blocky pore fills (Figs. 38A and C, and 39A and B) and, to a lesser extent, as very coarse, poikilotopic crystals (Fig. 40A). Euhedral, bladed, radiating crystals (displacive growth?) are rare. Some poikilotopic anhydrite, possibly after gypsum, is an early, precompactional phase supporting loosely packed grains with no grain or matrix replacement.

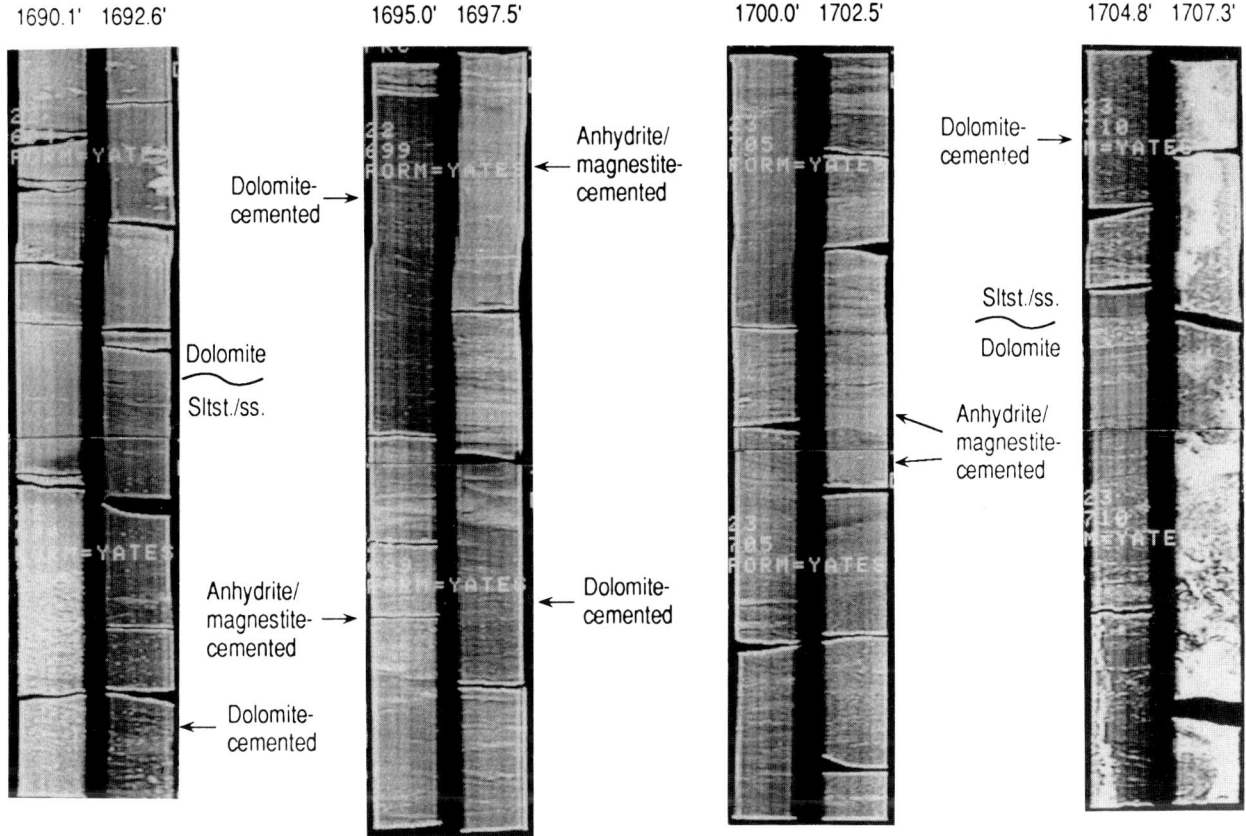

Figure 36
CT scan X-radiographs exhibiting decimeter-scale cement stratification in Red Anhydritic Siltstones/Sandstones from the PDB-04 core. Crude bedding is defined by light gray (dense), anhydrite- and magnesite-cemented intervals alternating with dark gray (less dense), dolomite-cemented intervals. The distinct cement horizons result in magnesite and dolomite typically being mutually exclusive on the scale of a thin section. Note anhydrite/magnesite intervals exhibit disrupted textures similar to fenestral fabric. Small, rounded bright spots are blebs of anhydrite. Dolomite-cemented zones typically have well-developed internal (wavy) stratification.

Figure 37

A. Dolomicrite matrix in red-gray siltstones/sandstones. The fine crystal size and matrix-supported texture suggest the dolomite is recrystalized, original (micrite?) matrix rather than cement. Dolomicrite matrix-supported laminae or lenses are common in Red Anhydritic Siltstones/Sandstones and are usually associated with fine silt and clay. Gulf PDB-04, 1661.5 ft, plane-polarized light.

B. SEM photomicrograph and EDX spectrum of dolomite in Red Anhydritic Siltstones/Sandstones. The presence of mica and detrital clay matrix suggests the dolomite is (recrystallized?) matrix. EDX analysis shows the dolomite is stoichiometric and contains no iron. Gulf PDB-04, 1767.5 ft, SEM.

C. Dolomite cement in Red Anhydritic Siltstones/Sandstones. Fine- to medium-crystalline, subhedral dolomite fills pores and completely occludes porosity. Red clay rims on the framework grains clearly predate the dolomite cement. Gulf PDB-04, 1694.5 ft, cross-polarized light.

Figure 38

A. Pore-filling magnesite and anhydrite cements in Red Anhydritic Siltstones/Sandstones. Due to their similar appearance, magnesite cannot be readily distinguished from dolomite in thin section. On the thin-section scale, however, the two mineral phases are mutually exclusive in the siliciclastic beds, and magnesite occurs only when there is abundant anhydrite present. Gulf PDB-04, 1703.5 ft, cross-polarized light.

B. SEM photomicrograph and EDX spectrum of magnesite cement in Red Anhydritic Siltstones/Sandstones. The magnesite occurs as very fine to medium, subhedral to euhedral, rhombohedral crystals. EDX analyses exhibit a strong magnesium (Mg) peak and a weak to moderate iron (Fe) peak. Red clay occurs as coatings on framework grains and within intercrystalline pore space. Gulf PDB-04, 1702.9 ft, SEM.

C. Back Scatter Electron (BSE) photomicrograph of polished thin section exhibiting textural relations between magnesite and anhydrite cements. Magnesite occurs as dark gray pore-lining cement; whereas, anhydrite occurs as washed-out white area in the pore center. This textural relationship is observed whenever the two mineral phases occur together in a pore and suggests that the pore-lining magnesite predated the pore-filling anhydrite. Gulf PDB-04, 1701.4 ft, BSE.

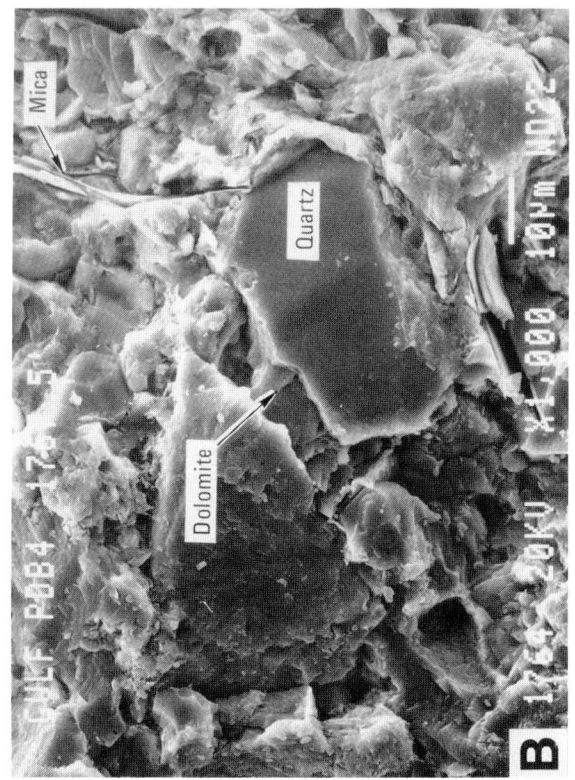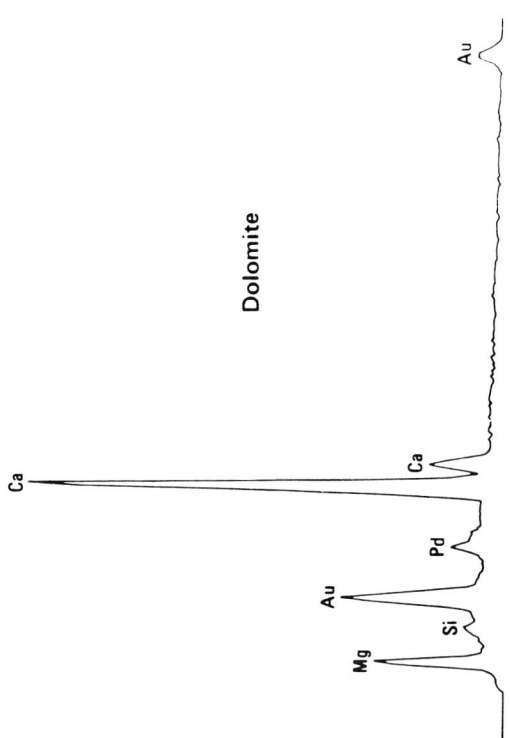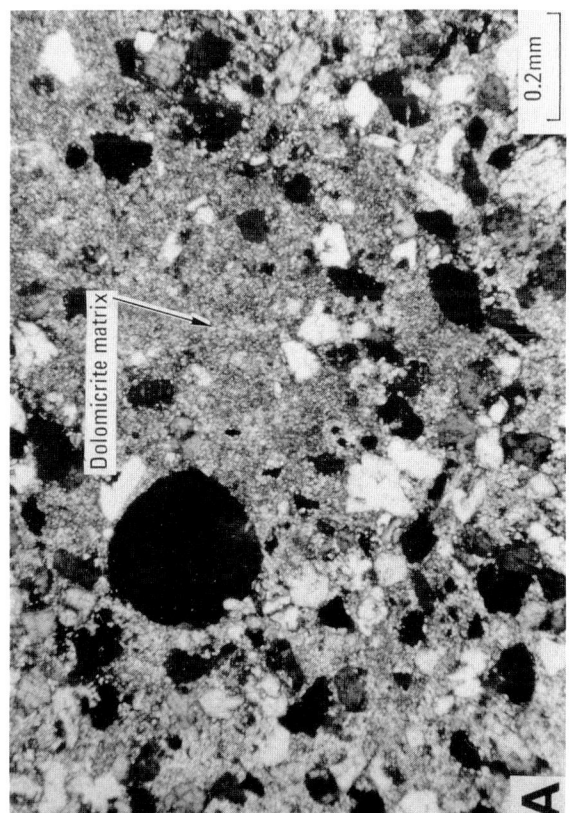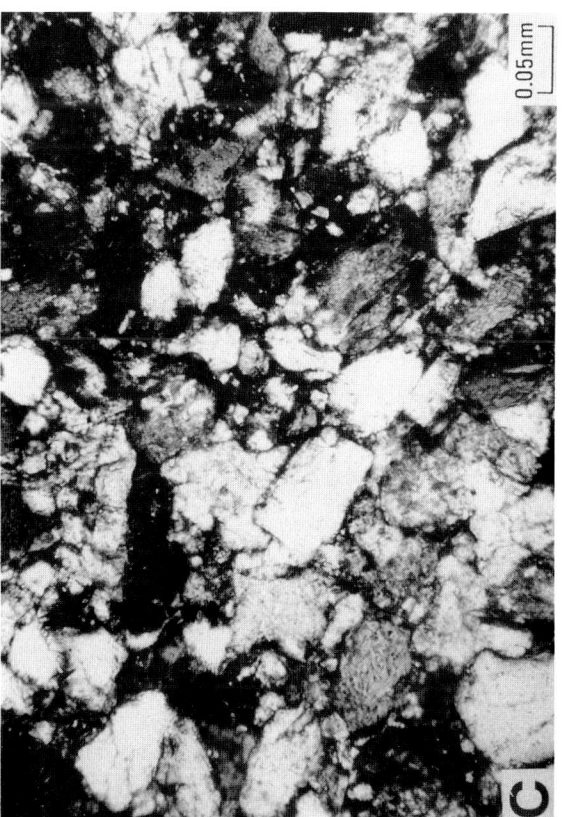

Figure 37

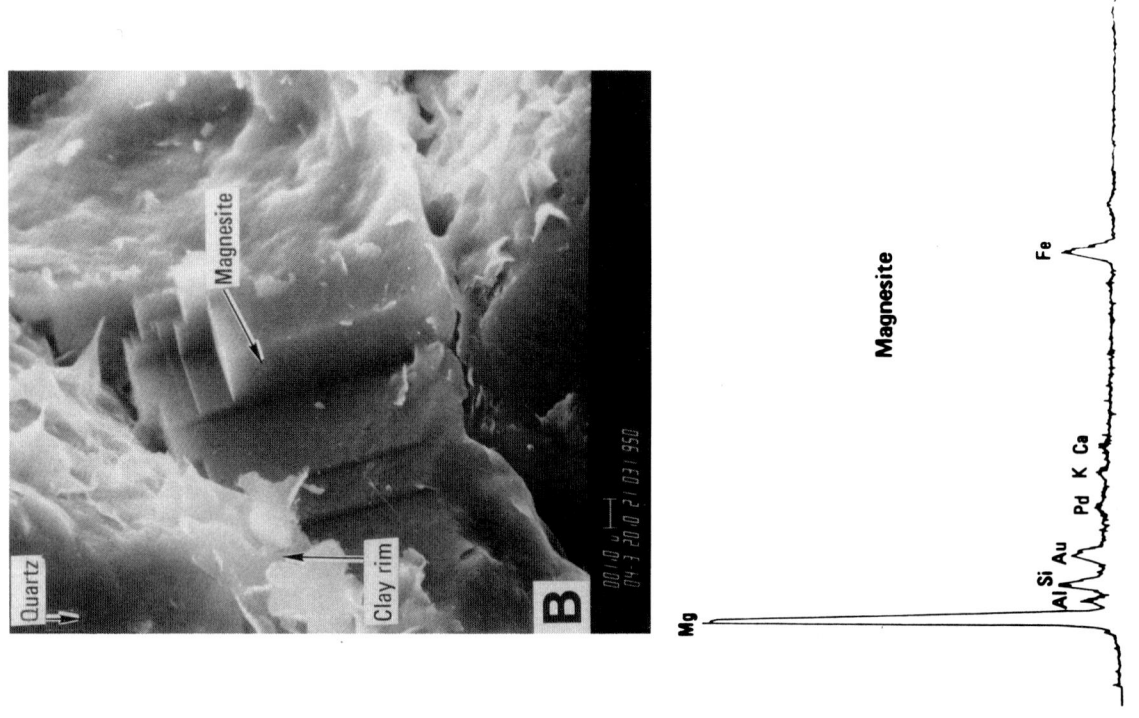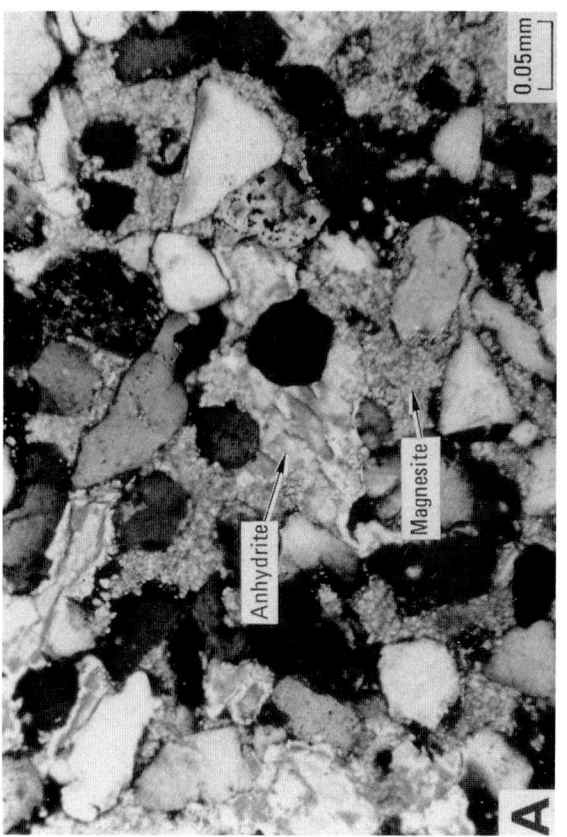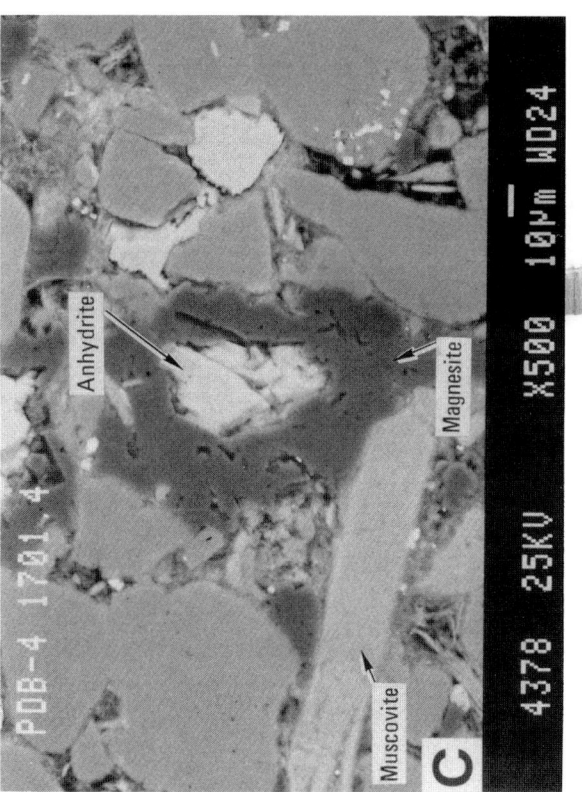

Figure 38

Figure 39

A. Textural relations between anhydrite and magnesite in Red Anhydritic Siltstones/Sandstones. Magnesite occurs as very finely crystalline masses that occur on the outer portion of pore-filling anhydrite crystals. Although the magnesite occurs at pore walls, it does not look like a typical primary pore-lining cement and instead may be replacing pore-filling anhydrite from the outside of the pore inward. Note the corroded outline of anhydrite crystals totally surrounded by magnesite in the lower left hand side of the photomicrograph. Gulf PDB-04, 1703.5 ft, cross-polarized light.

B. When magnesite and anhydrite occur together in a pore, the anhydrite typically appears etched and corroded compared to anhydrite in zones with no magnesite. In this sample the magnesite may have replaced pore-filling anhydrite along a front which moved from the right to left side of the pore. Gulf PDB-04, 1702.2 ft, BSE.

C. Fine- to medium-crystalline magnesite has clearly replaced a rounded detrital grain. Early red detrital clay rim outlines the ghost detrital grain. The well-rounded nature of the grain suggests it was a detrital carbonate grain or possibly an anhydrite nodule. Gulf PDB-04, 1702.9 ft, cross-polarized light.

D. BSE photomicrograph showing detrital grains replaced by magnesite. The fact that the magnesite blebs are the same size as detrital grains and also the presence of light gray, microporous rims (detrital clay coatings) on the magnesite blebs suggest a detrital grain precursor. Pore-filling magnesite present in this sample may have totally replaced original pore-filling anhydrite cement. Only several very small remnants of anhydrite are still present. Gulf PDB-04, 1701.6 ft, BSE.

Figure 40

A. SEM photomicrograph and EDX spectrum of anhydrite cement in Red Anhydritic Siltstones/Sandstones. Pore-filling anhydrite commonly occurs as medium to coarse poikilotopic crystals. The photo shows a coarse anhydrite crystal surrounding several quartz grains. Anhydrite is characterized by strong sulfur (S) and calcium (Ca) peaks on an EDX spectrum. Gulf PDB-04, 1694.2 ft, SEM.

B. SEM photomicrograph and EDX spectrum of authigenic K-feldspar. Very fine to fine, euhedral K-feldspar crystals occur in minor amounts and are associated with red (recrystallized?) illitic clay. Red illitic clay and associated euhedral, authigenic K-feldspars are common in red beds that have experienced intrastratal alteration of unstable framework grains due to prolonged contact with oxygenated waters (Turner, 1980). Gulf PDB-04, 1694.2 ft, SEM.

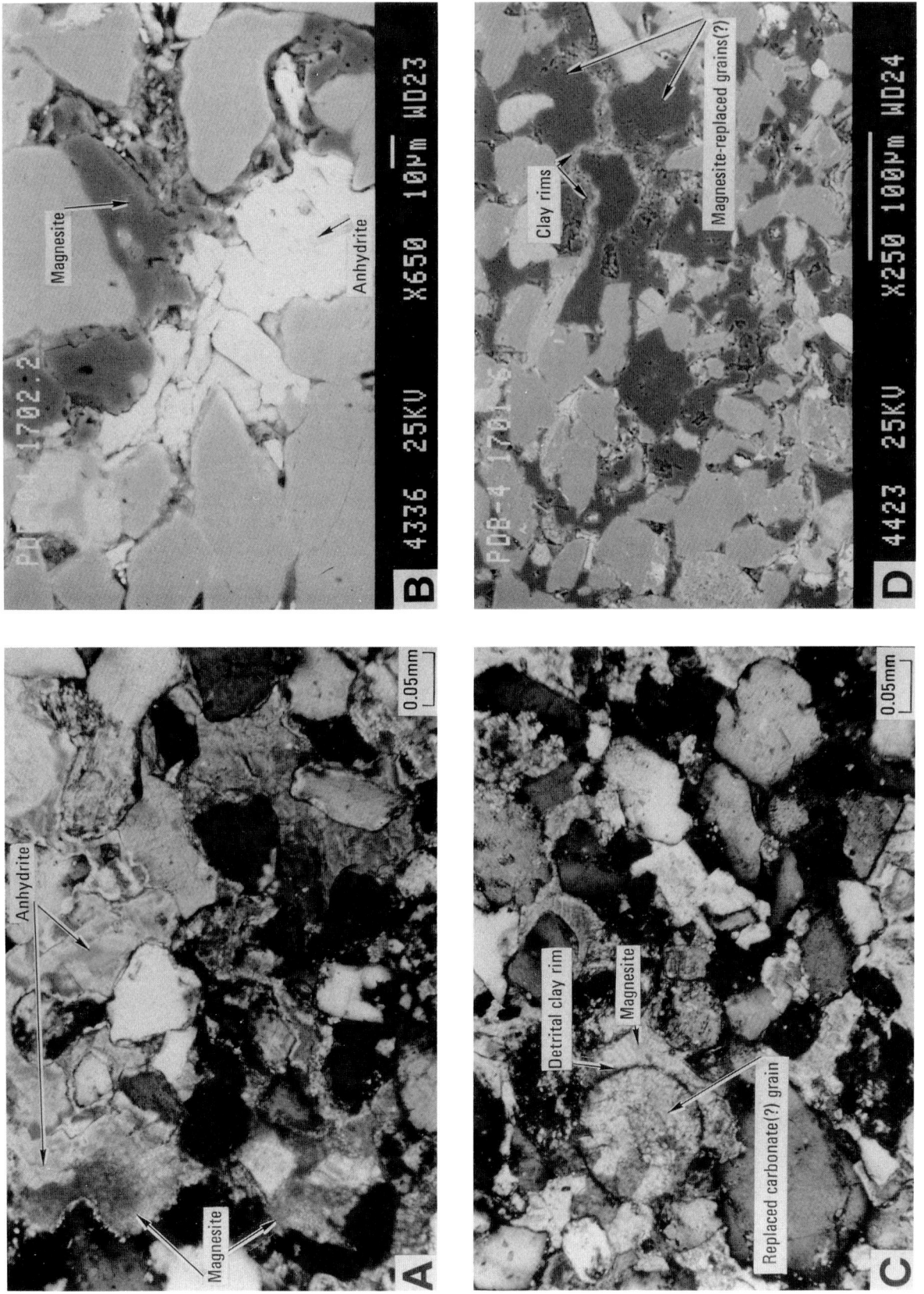

Figure 39

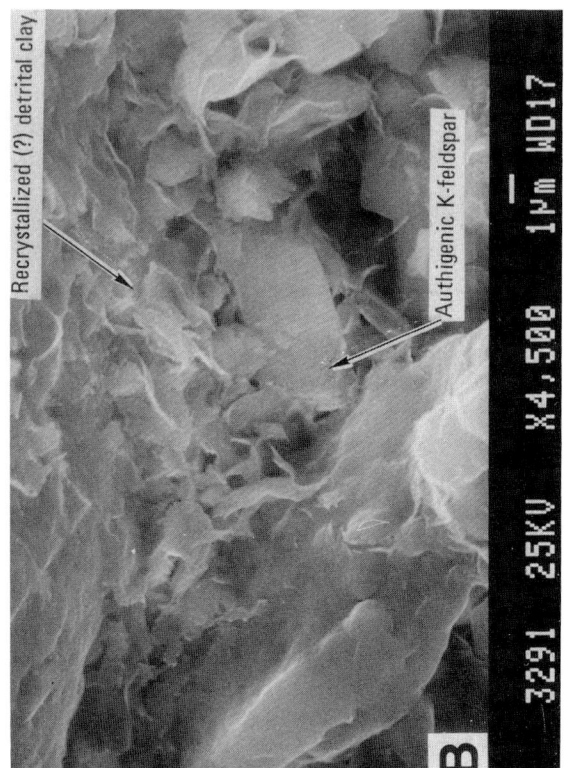
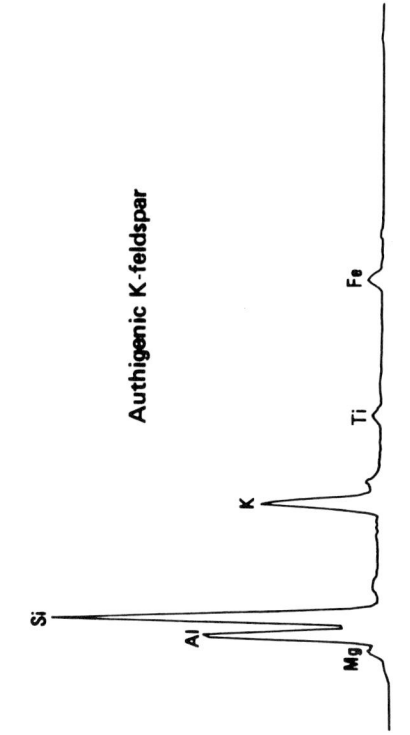
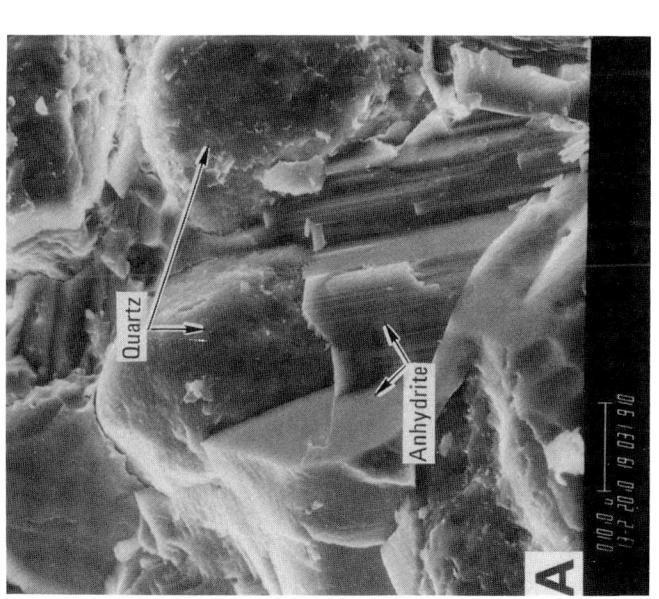
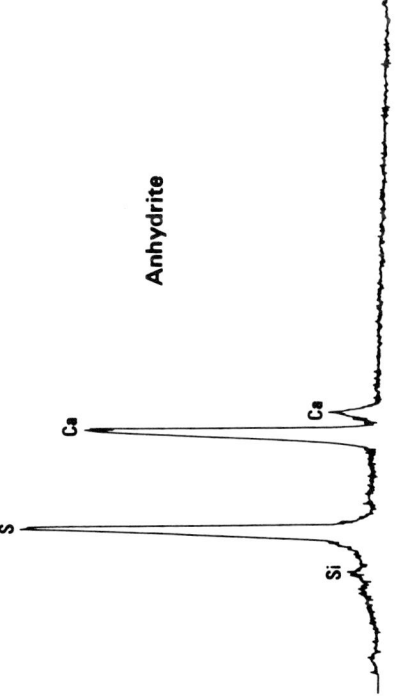

Figure 40

Minor amounts of very finely crystalline tabular authigenic K-feldspar occur in the red sand facies. The K-feldspar cement occurs as discrete crystals growing in association with recrystallized detrital and/or authigenic clay (Fig. 40B). Feldspar overgrowths were not found.

Red Anhydritic Siltstones/Sandstones are associated with very thick intervals of dolomite of the outer-shelf pisolite shoal complex. The clastic-carbonate ratio is about 1:2. The dolomite intervals consist of numerous well-developed shoaling cycles containing a basal-laminated dolomudstone that grades upward to peloidal and incipient pisolitic dolowackestone/packstone (with anhydrite-plugged fenestral porosity), that in turn grades to a capping unit of intraclastic, pisolitic dolopackstone/grainstone. Supratidal and desiccation features are common and include tepee structures with large anhydrite-filled voids, fenestral fabric, desiccation cracks, and brecciated dolomicritic crusts (Fig. 41).

Gray Bioturbated Kaolinitic Dolomitic Quartz Sandstone

Gray Bioturbated Kaolinitic Dolomitic Quartz Sandstones form medium to thick beds in the lower part of the Yates Formation in the PDB-04 core (Figs. 4 and 42 through 44). The gray sandstones generally are wavy-laminated with minor plane laminae; however, homogeneous, structureless intervals are common. Large (10 to 12 cm high and 2 cm wide) vertical burrows (*Conichnus*) and numerous (1 to 1.5 cm high and 1 to 3 cm wide), ovoid to elongate burrows (*Thalassinoides*) have authigenic pyrite associated with them. The gray sandstones have sharp to gradational basal contacts and gradational upper contacts with dolomites.

Grain size in the gray sandstone ranges from clay to medium sand with the average being lower very fine sand (Fig. 45A through D). The coarser clastic material occurs as well-rounded, pitted or frosted quartz grains that are dispersed throughout the samples and are also concentrated in distinct laminae or lenses (Fig. 45D). The sandstones are moderately to well sorted, and the finer sand fraction is subangular to subrounded. Grains range from being matrix-supported to having tangential (point) contacts.

In many samples detrital framework grains comprise only 35 to 50 percent of the rock (Table 7). The loose framework packing is related to both the presence of abundant dolomite matrix and the dissolution of original framework grains. In coarser-grained intervals grain packing is tighter and framework grains may comprise up to 70 percent of the rock. The detrital framework typically consists of 80 to 90 percent quartz, 1 to 10 percent feldspar, 1 to 10 percent rock fragments, 1 to 5 percent heavy minerals, 0 to 5 percent mica, and 0 to 5 percent carbonate intraclasts (Fig. 46A through D).

The gray dolomitic sandstones contain less feldspar than the other siliciclastic lithofacies. K-feldspar typically forms 1 to 3 percent of the rock; whereas, plagioclase is even less common. Both feldspar types exhibit common dissolution features (Fig. 47A through D). Secondary porosity related to the dissolution of feldspar grains is common. Most of the secondary porosity is noneffective, however, due to filling of the secondary pores by microporous kaolinite. Open secondary porosity attains values of 8 to 10 percent, but is largely noneffective due to the isolated (poorly connected) distribution of individual pores.

Figure 41

Pisolite shoal-crest dolomites associated with Red Anhydritic Siltstones/Sandstones in the PDB-04 core exhibit abundant desiccation features. These include fenestral fabric, pisolite crusts, sheetcracks, and small-scale erosion surfaces. The chaotic bedding with large voids filled by early marine (hypersaline) carbonate and anhydrite cements are tepee structures, possibly related to early desiccation and evaporite growth in the environment of deposition. The red color of some carbonate intervals also suggests exposure that allowed for the infiltration of airborne, hematite-stained clay.

Figure 42

Gray Bioturbated, Kaolinitic Dolomitic Quartz Sandstones in lower part of the Yates Formation in the PDB-04 well. The well is located at the shelf margin of the Northwest Shelf. The lower part of the cored Yates interval is positioned less than a mile from the time-equivalent shelf edge, and these gray sandstones exhibit both sharp and gradational upper and lower contacts with dolomites of the pisolite shoal front. The quartz sandstones are typically highly bioturbated, containing numerous *Ophiomorpha/Thalassinoides* and large *Conichnus* burrows. Other sandstone intervals are structureless or have wavy and planar laminae. Silt- to cobble-size dolomite intraclasts are found throughout the sandstones. Shoal-front carbonates exhibit some supratidal features similar to the shoal-crest dolomites (e.g., pisolite crusts, fenestral fabric, and possible tepee structures); however, subtidal facies such as massive and burrowed dolomudstones and skeletal dolopackstones/grainstones are also common.

Figure 43

Gray Bioturbated, Dolomitic Quartz Sandstones and associated brecciated carbonates in the lower part of the Yates Formation of the PDB-04 core. The gray sandstone is highly bioturbated and contains abundant dolomite clasts. Carbonate intervals are locally brecciated. The breccia zones may be tepee structures, rip-up horizons, or later dissolution features.

Figure 44

Log response and core attributes for Gray Bioturbated, Kaolinitic Dolomitic Quartz Sandstones; PDB-04 well. The shoal-front sandstones are characterized by minor gamma-ray deflections due to the low amounts of feldspar and detrital clay and the abundance of dolomite matrix. Abundant dolomite also results in high bulk densities, low density-derived porosities, and a straight caliper.

Figure 41

Figure 42

Figure 43

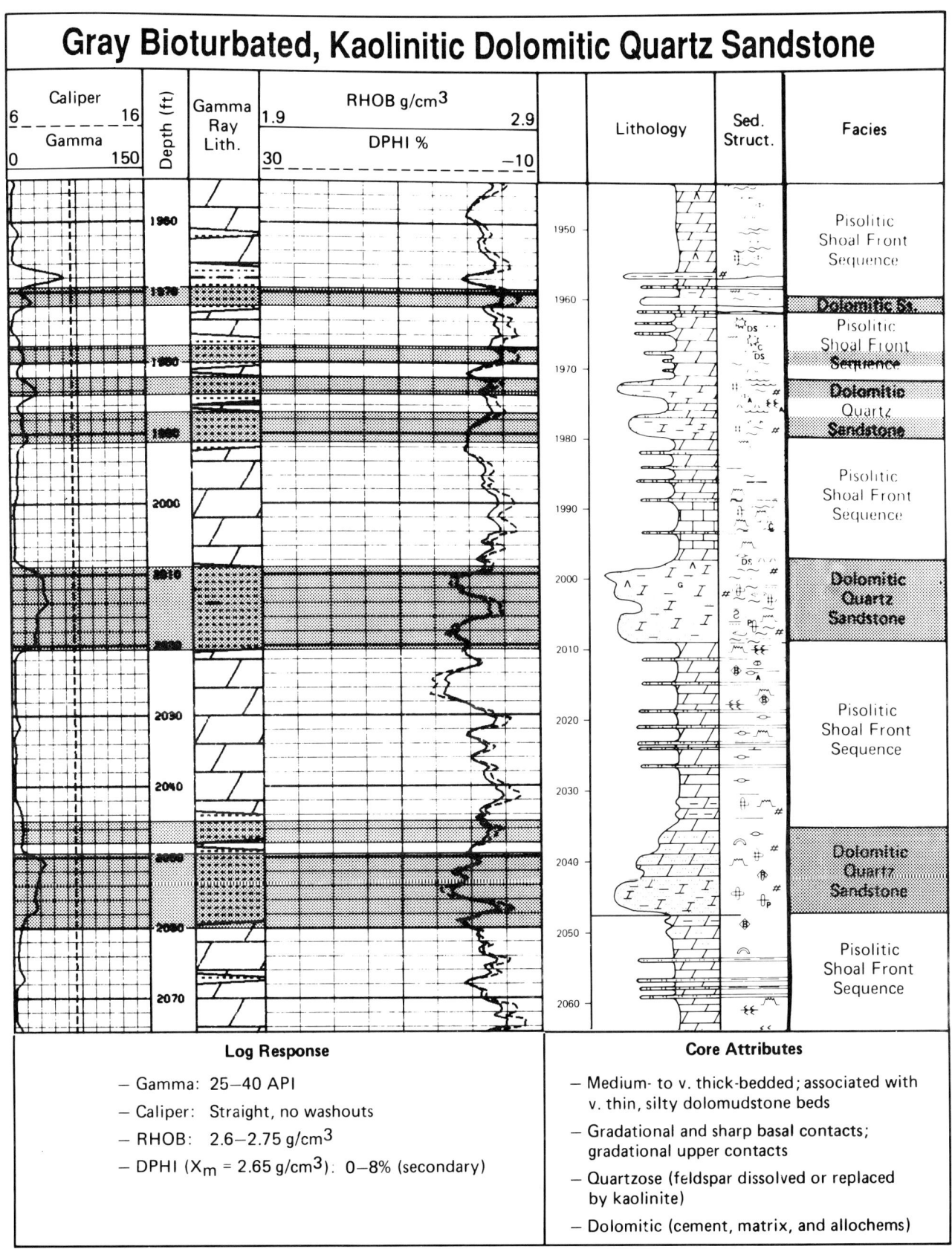

Figure 44

Figure 45

A. Typical texture of Gray Bioturbated, Kaolinitic Dolomitic Quartz Sandstones. Grain size ranges from silt to very fine sand and sorting is moderately good. Detrital clay matrix is less abundant than in Red Anhydritic Siltstones/Sandstones. Dolomite is very abundant and occurs as both recrystallized (micrite?) matrix and cement. Large (up to fine pebble gravel) carbonate rip-up clasts are common. Gulf PDB-04, 1979.0 ft, plane-polarized light.

B. Typical bioturbated, dolomite matrix-rich texture found in Gray Bioturbated, Kaolinitic Dolomitic Quartz Sandstones. Burrows typically contain coarser grains than in the surrounding matrix and are cemented by pyrite, gypsum/anhydrite, and quartz. The matrix shown here is actually best described as a silty, sandy dolomite since it contains greater than 50% dolomicrite matrix. Gulf PDB-04, 1973.0 ft, plane-polarized light.

C. Laminae consisting of intervals of very fine to fine sand alternating with intervals of silt to very fine sand. A very fine wispy laminae of pyrite and minor detrital clay marks the change in grain size. Gulf PDB-04, 2002.6 ft, plane-polarized light.

D. Laminae consisting of intervals of fine to medium sand alternating with intervals of silt to fine sand. Authigenic pyrite is concentrated at the change in grain size. Intervals with a fine- to medium-grained sand component and little or no detrital matrix suggest the shelf edge was periodically the site of winnowing currents. Gulf PDB-04, 1999.1 ft, plane-polarized light.

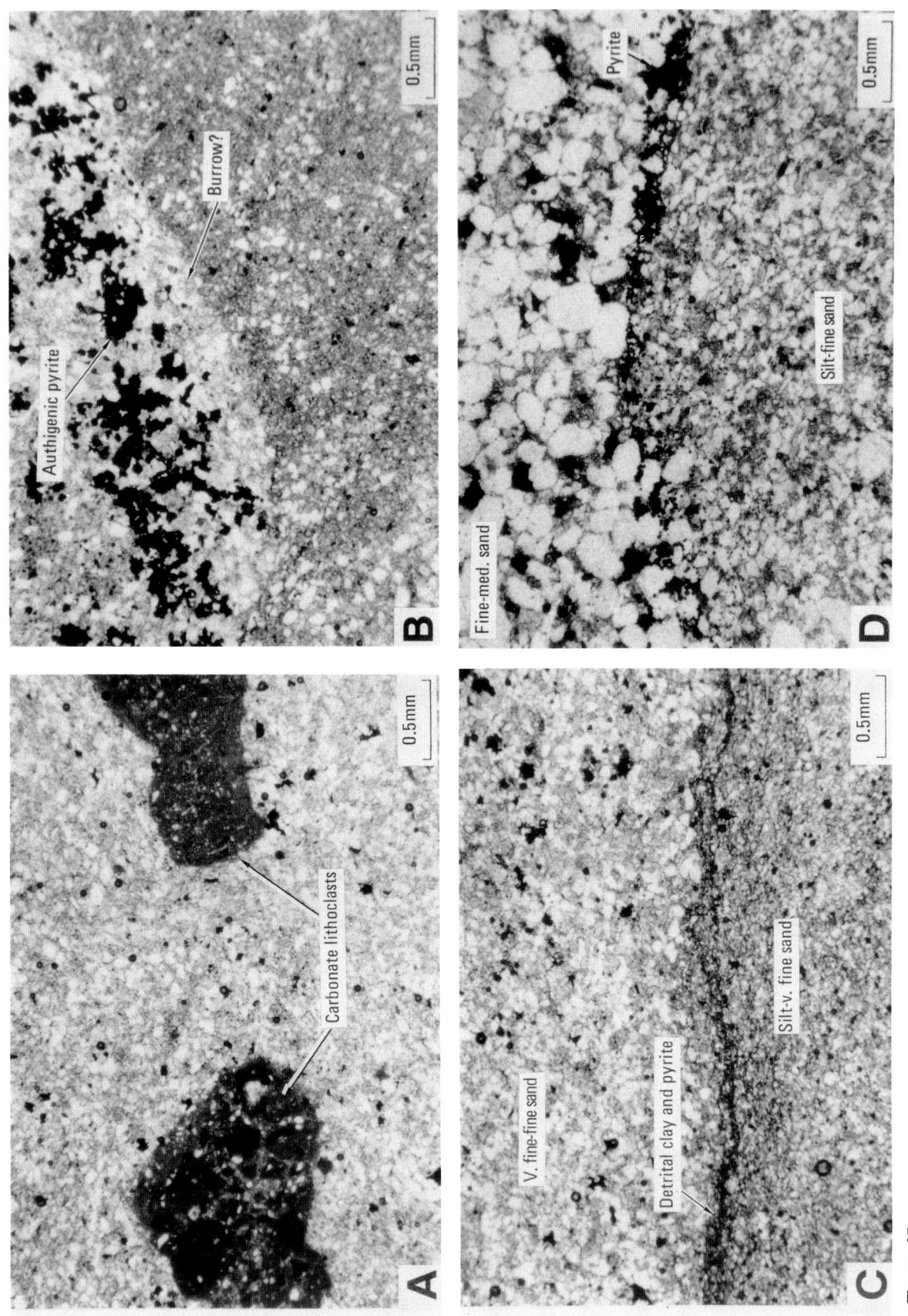

Figure 45

Table 7   GRAY, BIOTURBATED KAOLINITIC, DOLOMITIC QUARTZ SANDSTONE - PETROGRAPHIC DATA

<--SAMPLE INFO--> <----------FRAMEWORK GRAINS----------> <---------INTERSTITIAL MATERIAL---------> <---TEXTURE--->

WELL	DEPTH	QUAR	K-FE	PLAG	ROCK	HEAV	MICA	MATR	RED MATR	AUTH CLAY	DOLO /CAL	ANHY +GYP	PYR	SHR	POR	GRAIN SIZE φ	SORT φ
PDB4	1870.30	30	4	3	1	1	1	4	0	5	26/0	22	3	0	0	4.00	0.60
PDB4	1871.00	33	4	1	1	TR	0	3	0	6	33/0	16	3	TR	0	4.25	0.85
PDB4	1875.00	30	3	1	1	TR	1	3	0	5	44/0	8	2	0	TR	3.50	0.60
PDB4	1877.20	34	3	1	3	1	1	0	0	TR	48/0	7	3	0	0	4.25	0.85
PDB4	1898.70	31	4	TR	2	TR	0	0	0	6	42/0	12	2	0	TR	3.50	0.60
PDB4	1973.00	36	3	TR	1	1	TR	TR	0	5	47/3	0	2	TR	2	3.50	0.60
PDB4	1979.00	33	2	TR	1	1	TR	1	0	5	48/2	3	2	0	0	4.00	0.85
PDB4	1999.10	45	1	TR	5	2	0	0	0	8	17/8	0	8	0	6	2.00	0.85
PDB4	2000.30	45	3	1	TR	2	0	0	0	6	37/0	0	5	0	TR	3.25	0.42
PDB4	2001.00	35	2	0	2	2	TR	0	0	11	28/9	0	5	1	5	3.50	0.42
PDB4	2001.60	40	3	0	3	2	0	0	0	8	40/0	0	4	TR	TR	3.50	0.60
PDB4	2002.60	38	3	TR	2	1	3	7	0	8	32/0	0	5	0	TR	4.50	1.50
PDB4	2043.00	32	2	TR	2	1	TR	TR	0	9	45/0	0	3	0	6	3.50	0.60

KEY

COLUMNS: QUAR=quartz; K-FE=Potassium feldspar; PLAG=plagioclase; ROCK=rock fragments; HEAV=heavy minerals; MICA=biot/musc;
MATR=gray detrital matrix; RED MATR=red detrital matrix; AUTH CLAY=authigenic clay; DOLO=dolomite (* ferroan
dolomite; Mag=magnesite; CAL=calcite; ANHY+GYP=anhydrite + gypsum; PYR=pyrite; SHR=solid hydrocarbon residue;
POR=percent porosity; GRAIN SIZE (phi units); SORT=sorting (phi units)

Figure 46

A. Structureless texture found often in the Gray Bioturbated, Kaolinitic Dolomitic Quartz Sandstones. Silty, very fine- to fine-grained quartz sandstone with abundant dolomite cement, authigenic pyrite, and dolomite intraclasts. Gulf PDB-04, 2003.7 ft, plane-polarized light.

B. Same as Figure 46A in cross-polarized light. Abundant fine to medium crystalline, subhedral to euhedral dolomite is apparent under crossed polars. Much of the dolomite cement is filling voids that were probably formed by the dissolution of feldspars and lithic grains. This conclusion is based on the loose packing of the remaining framework quartz grains, the lack of feldspars, and the uniform silt to fine sand size of many of the dolomite crystals. Gulf PDB-04, 2003.7 ft.

C. Three-inch (7.5 cm) interval of laminated fine- to medium-grained sandstone that caps the top of a 10-ft (3 m) bed of Gray Bioturbated, Kaolinitic Dolomitic Quartz Sandstone. The sand grains are well to very well rounded and consist predominantly of quartz. Within individual laminae sorting is moderately good to good. Also note the presence of authigenic pyrite and secondary pores related to the dissolution of framework grains. Gulf PDB-04, 1999.1 ft, plane-polarized light.

D. Same as Figure 46C in cross-polarized light. Altered framework grains and patchy calcite cement are apparent under crossed polars. Five to ten percent calcite cement is common in the coarser-grained intervals. Gulf PDB-04, 1999.1 ft.

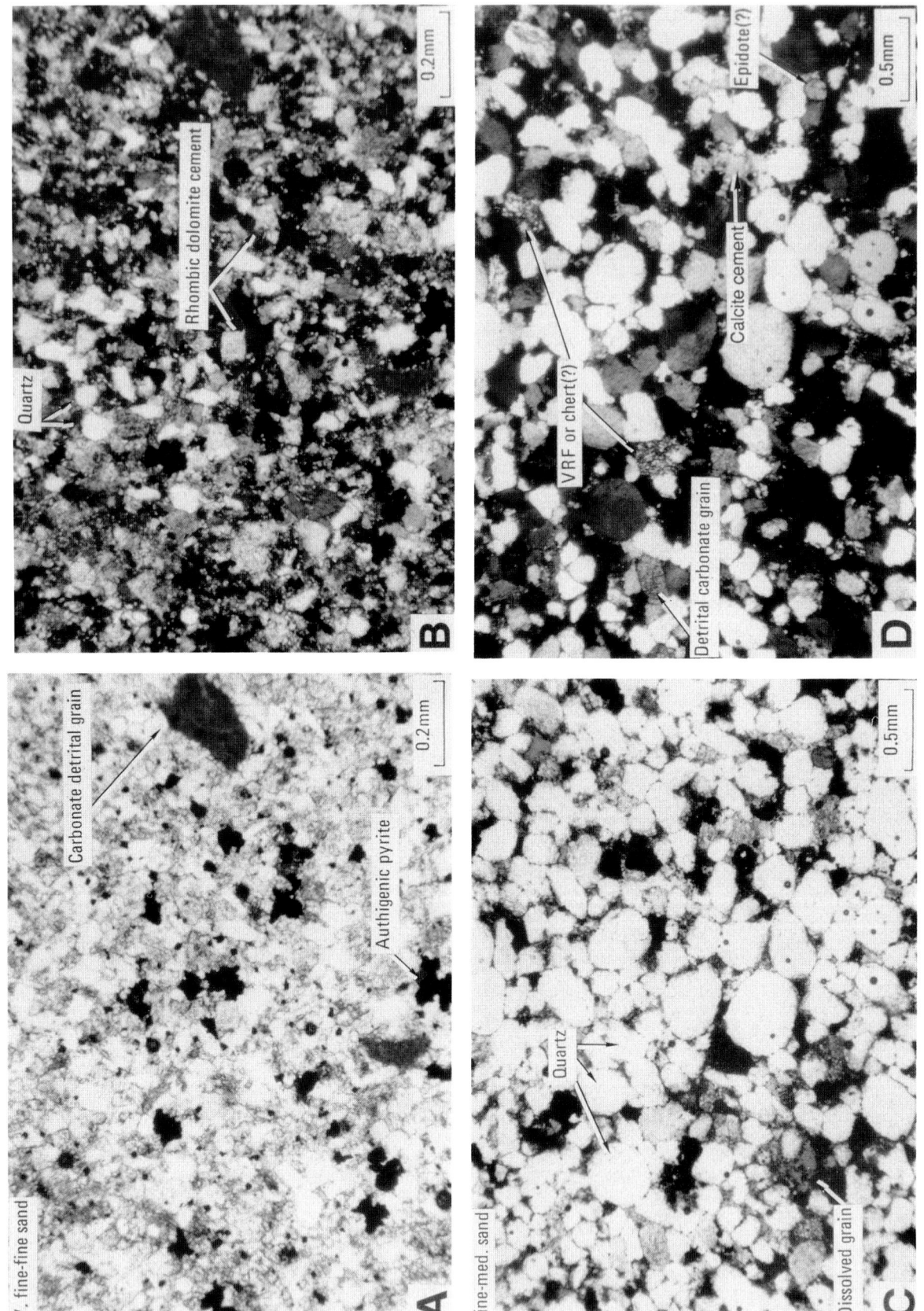

Figure 46

Figure 47

A. Partially dissolved (feldspar?) framework grains and authigenic, microporous kaolinite filling the secondary voids. Kaolinite after feldspar is not a replacement mechanism but rather involves dissolution of the feldspar and reprecipitation of kaolinite. Although nearly all the feldspar is dissolved, the resulting secondary porosity is largely noneffective due to poor connectivity and the extensive filling by microporous kaolinite. Gulf PDB-04, 1999.1 ft, plane-polarized light.

B. Same as Figure 47A in cross-polarized light. The authigenic clay is interpreted as kaolinite due to its microporous texture and low birefringence. SEM analyses (Fig. 49B and C) have further documented the presence of kaolinite. Gulf PDB-04, 1999.1 ft.

C. SEM photomicrograph showing partially dissolved K-feldspar in a sandstone from the middle part of the Yates Formation in Gulf PDB-04. The sandstone is transitional between Red Anhydritic Siltstones/Sandstones in the upper portion of the core and the Gray Bioturbated, Kaolinitic Dolomitic Quartz Sandstones in the lower portion of the core. A pale orange color and the presence of abundant anhydrite are typical for red siltstones/sandstones; whereas, the presence of abundant dolomite matrix, burrows, and authigenic pyrite are suggestive of gray sandstones. Feldspar is present but in lesser amounts (5% to 10%) than in Red Anhydritic Siltstones/Sandstones. Furthermore, most of the feldspar exhibits dissolution features. Gulf PDB-04, 1870.0 ft, SEM.

D. Open secondary porosity related to feldspar and/or lithic grain dissolution. Note shape and size of pores are similar to that of framework grains. Porosity may range from 5% to 8% but is typically poorly connected and therefore largely noneffective. More commonly the secondary pores are filled with microporous kaolinite. Gulf PDB-04, 1999.1 ft, plane-polarized light.

91

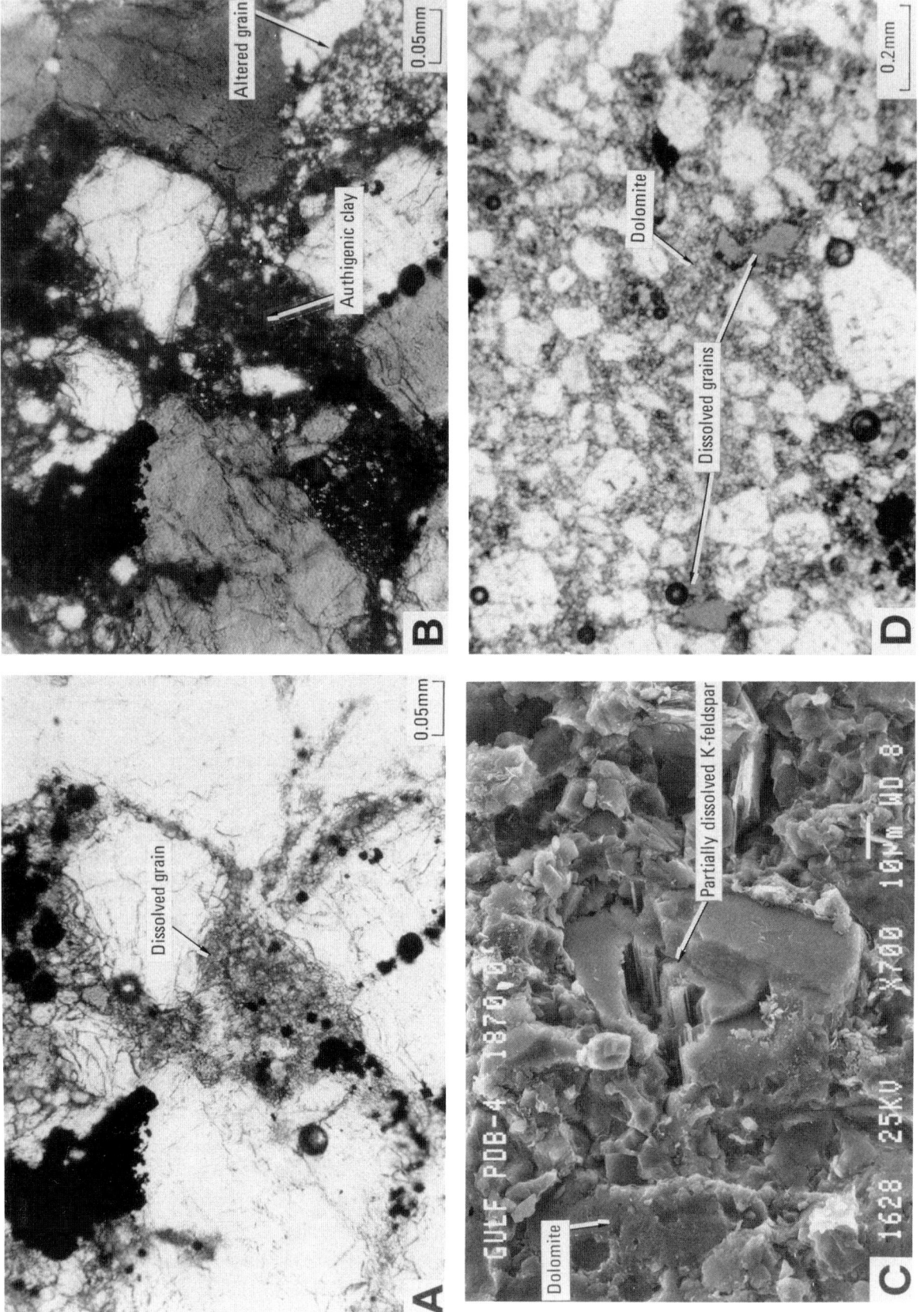

Figure 47

Chert and/or volcanic rock fragments(?) and local concentrations of carbonate mudstone and wackestone intraclasts form the chief lithic components. Detrital opaque minerals are common but typically highly altered in the gray sandstones (Fig. 48A and B). Detrital ilmenite is generally altered to leucoxene, pale yellow in reflected light. Other opaque grains are altered to pyrite. Accessory minerals include tourmaline, zircon, epidote, and rutile.

Detrital clay and mica are minor and usually comprise less than 5 percent of the rock. Dark gray, wispy laminae of detrital clay and silt-sized mica are common, but minor, constituents. The detrital clay is illite as in the red anhydritic siltstones; however, in the gray sands there is no red, FeO_2 stain on the clay platelets.

Dolomite comprises 25 to 55 percent of the gray sandstones and occurs both as recrystallized matrix (dolomicrite) and as a cement. Matrix dolomite is recognized by its cloudy, micro- to very finely crystalline nature and by the fact that it occurs in lenses and laminae with floating (matrix-supported) detrital silt/sand grains (Figs. 45A and B, and 48C and D). Detrital clay and mica are commonly associated with the dolomicrite-rich intervals. The abundance of dolomicrite is highly variable and ranges from 5 to 40 percent.

Dolomite is also present as fine to medium crystalline, subhedral to euhedral cement that may comprise 15 to 30 percent of a sample (Figs. 46A and B, and 49A and B). Several rhombic dolomite crystals have been found to exhibit a threefold growth zonation, with the middle zone being slightly more calcitic.

Moderate amounts (up to 9 percent) of pore-filling calcite cement occur in several of the gray sandstone samples. The calcite may be the only cement filling a pore (Fig. 46D), but is more commonly found intergrown with corroded dolomite rhombs. The calcite is typically anhedral to subhedral and fine to medium crystalline. The origin of the calcite is not entirely clear; however, petrographic textural relations suggest the calcite is probably replacing dolomite (dedolomitization).

Other cements include authigenic pyrite, quartz overgrowths, and gypsum (with minor anhydrite). These three authigenic phases are often associated together in burrows (Fig. 48C and D). Authigenic pyrite and quartz overgrowths are also associated with coarser-grained laminae (Fig. 49D). Pyrite is present in all of the gray sandstone samples.

Authigenic kaolinite (dickite) is a major constituent of the gray dolomitic sandstones and typically comprises 5 to 15 percent of a sample. The kaolinite generally fills secondary pores formed by the dissolution of detrital feldspar grains (Fig. 47A and B) and may occur as grain-sized blebs comprised of well-developed tabular "books" or individual platelets growing on grain surfaces and interspersed with pore-filling dolomite (Fig. 49B and C). A general decrease in feldspar and increase in kaolinite with depth (Fig. 50) suggests the kaolinite is forming at the expense of the feldspar (see discussion below).

The Gray, Bioturbated Kaolinitic, Dolomitic Quartz Sandstones are associated with carbonates similar to those found with the Red Anhydritic Siltstones/Sandstones; however, individual shoaling cycles are generally more mud-dominated and have smaller pisolites. Fenestrae and

Figure 48

A. Authigenic pyrite and leucoxene are common in the Gray Bioturbated, Kaolinitic Dolomitic Quartz Sandstones. The authigenic opaque minerals fill pores and occur as alteration products after iron-bearing detrital grains. Opaque minerals cannot readily be distinguished in plane-polarized light. Gulf PDB-04, 1999.1 ft, plane-polarized light.

B. Pyrite occurs as pore-filling and grain-replacing phases; whereas, leucoxene is predominantly an alteration product of detrital ilmenite grains. Gulf PDB-04, 1999.1 ft, reflected light.

C. Diagenetic phases that occur in burrows due to the presence of organic material and/or grain-size sorting (permeability) variations. The burrows typically contain loosely packed and slightly coarser framework grains than the surrounding sediment. The burrows are tightly cemented by gypsum/anhydrite, pyrite, and minor quartz overgrowths. Gulf PDB-04, 1976.8 ft, plane-polarized light.

D. Same as Figure 48C in cross-polarized light. Note low birefringent gypsum, high birefringent anhydrite, and euhedral dipyramidal quartz overgrowths. Burrows typically occur in dolomicrite matrix-rich host sediment. Gulf PDB-04, 1976.8 ft.

Figure 49

A. SEM photomicrograph of euhedral rhombic dolomite cement that is common in Gray Bioturbated, Kaolinitic Dolomitic Quartz Sandstones. Dolomite rhombs typically range from 0.02 to 0.075 mm (20 to 75 μm) in length. Gulf PDB-04, 2043.0 ft, SEM.

B. SEM view of authigenic kaolinite (dickite) and dolomite cement that is ubiquitous to Gray Bioturbated, Kaolinitic Dolomitic Quartz Sandstones. Kaolinite and dolomite are so abundant that they typically obscure framework grains in SEM. Kaolinite occurs as grain-shaped blebs with well-developed books and also as fine disseminated plates. Dolomite cement can be subhedral masses or euhedral rhombs. Gulf PDB-04, 2043.0 ft, SEM.

C. Cross-section view of broken quartz grain shows how framework grains are coated and surrounded by disseminated kaolinite plates and subhedral dolomite cement. Gulf PDB-04, 2000.9 ft, SEM.

D. SEM photomicrograph of authigenic pyrite and quartz overgrowths that are commonly found within burrows, but may also be found along laminae boundaries and within coarser-grained lenses or laminae. This sample is a medium-grained laminae from a zone of interlaminated, fine- to medium-grained sandstone and silty, very fine sandstone. Gulf PDB-04, 1999.1 ft, SEM.

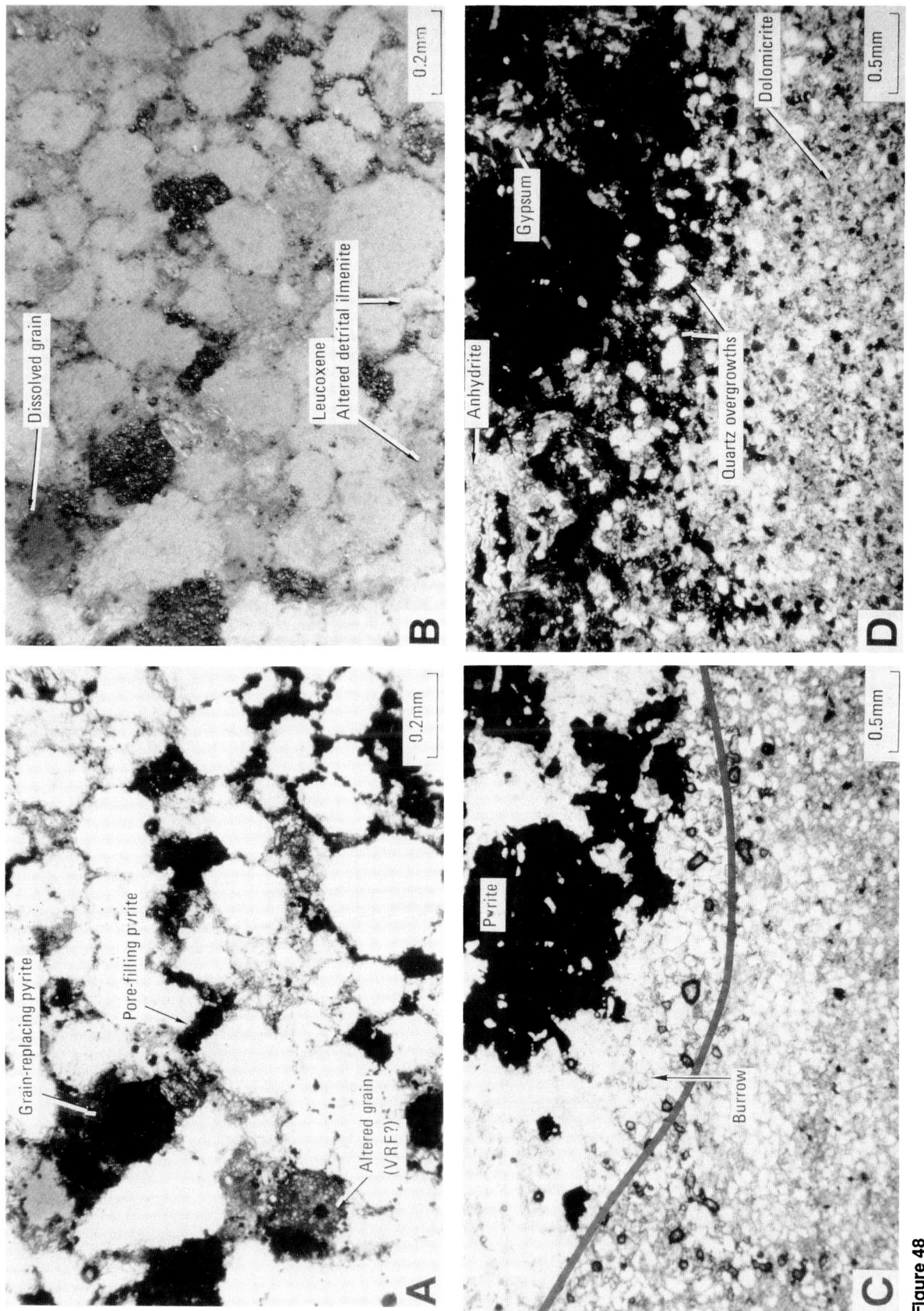

Figure 48

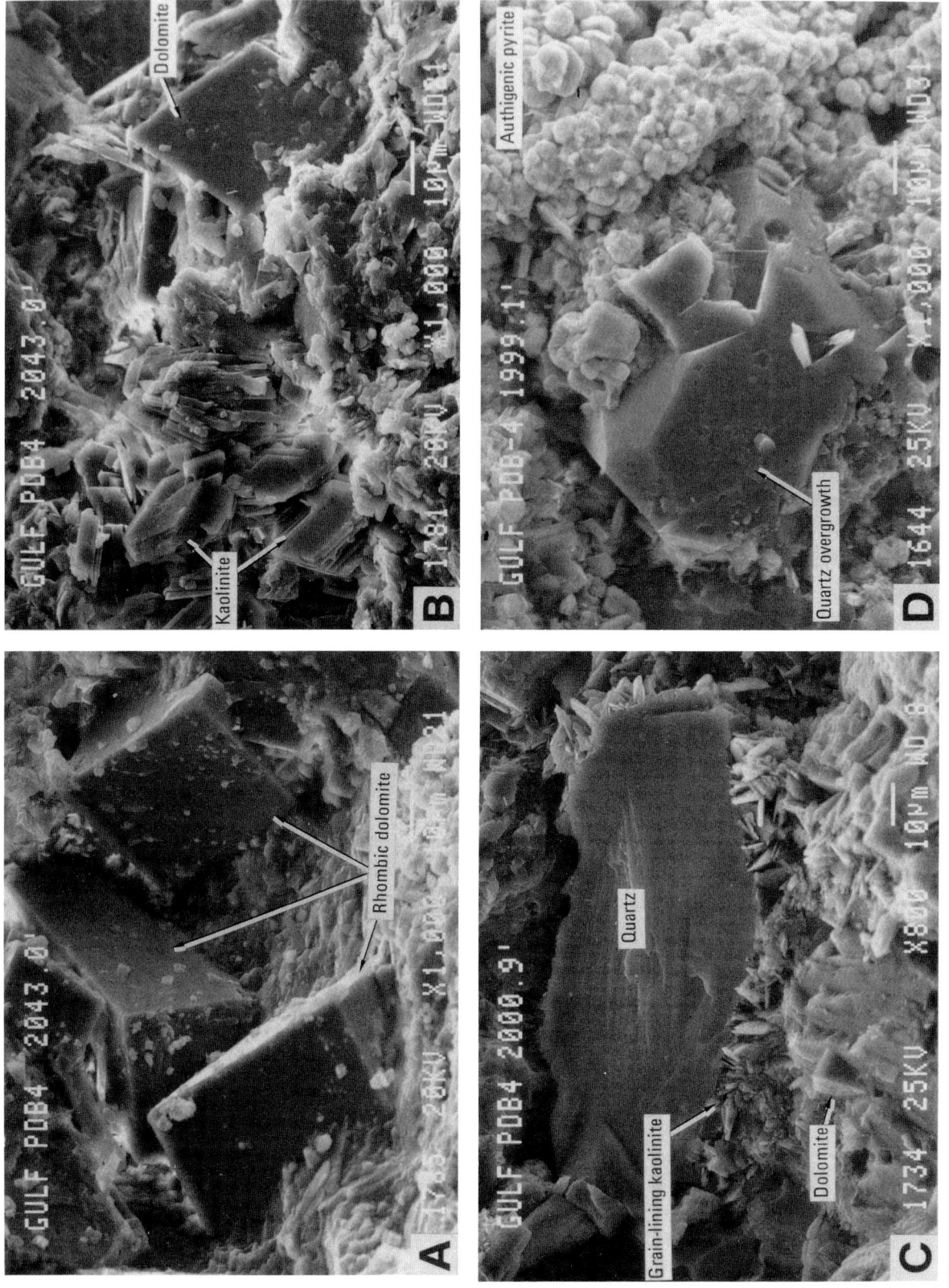

Figure 49

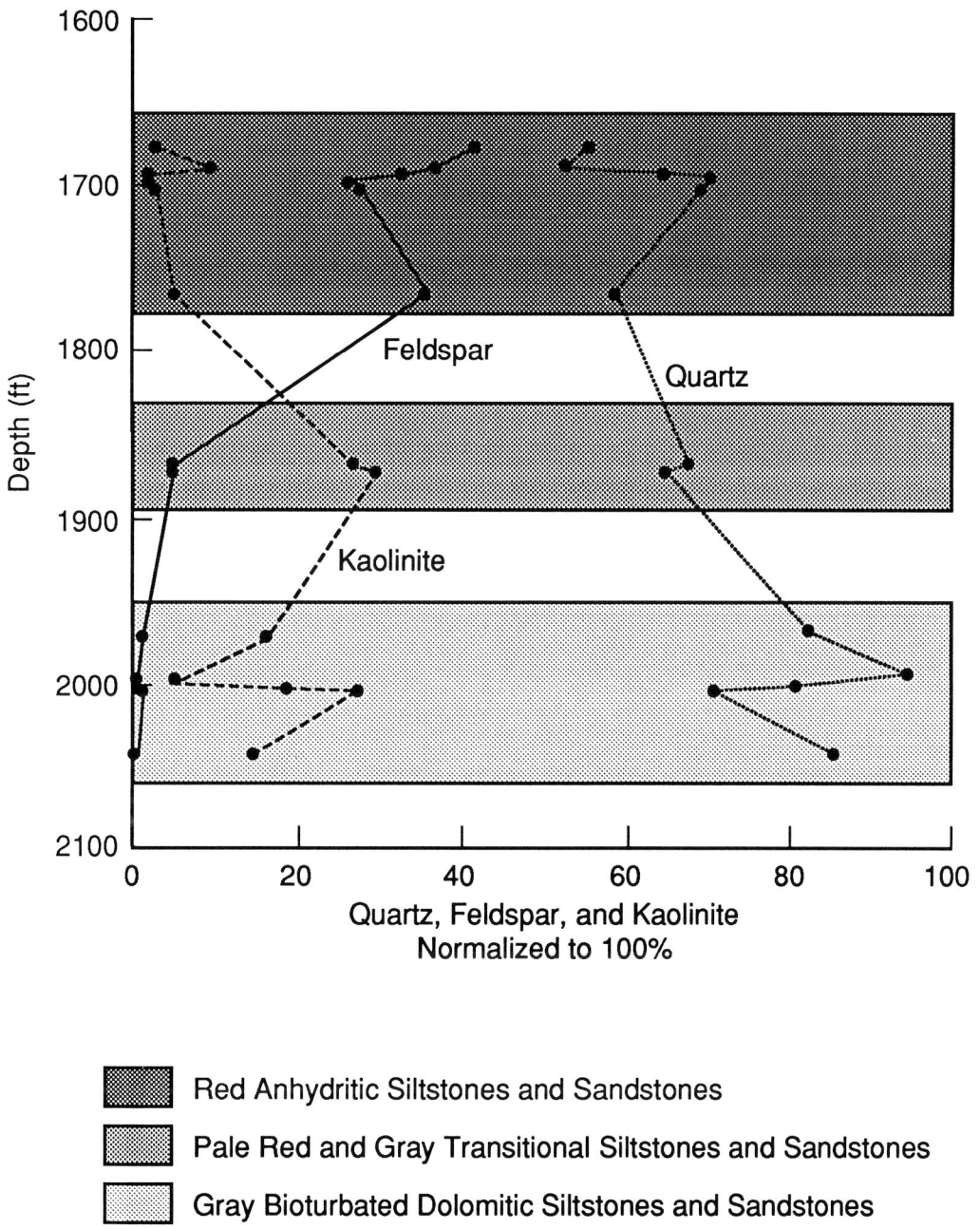

Figure 50
Total quartz, feldspar, and kaolinite normalized to 100% and plotted against depth for siliciclastic lithofacies in the PDB-4 core. Mineral percentages were acquired using X-ray diffraction techniques. The strong correlation between mineralogy and depth is controlled by siliciclastic lithofacies that vary stratigraphically due to the prograding nature of the Yates shelf. Shaded boxes correspond to distinct sandstone lithofacies found in the core. Facies belts get progressively closer to the Capitan aquifer with depth.

large voids are filled with laminated, silty, argillaceous, dolomicritic cement rather than anhydrite. Thick intervals of massive and mottled dolomudstone with no desiccation features are present, as are thin beds of fusulinid packstone. Carbonates interbedded with the gray sandstones are locally brecciated (Fig. 43). The brecciated intervals may be related to desiccation and/or resedimentation (rip-ups and/or tepee structures) in the depositional environment or to burial dissolution.

Feldspar Dissolution

Hull (1957) and Dunham (1972) suggested that a narrow belt of shelf margin sandstones lacking feldspar (basin sandstones and shelfward sandstones both contain abundant feldspar) was produced by early diagenesis within the environment of deposition. Hull (1957) attributed the lack of feldspar to differential abrasion, noting a decrease in feldspar content with increasing grain size. Dunham (1972) suggested chemical weathering and kaolinitization of feldspars occurred along a subaerially exposed paleotopographic high. Neither of these hypotheses are viable explanations for the lack of feldspar and abundant kaolinite in the gray sandstones studied here.

Facies relationships show that the dissolution of feldspars and resulting growth of kaolinite did not take place along a subaerially exposed paleotopographic high as proposed by Dunham (1972). In the PDB-4 core, feldspar dissolution and kaolinite growth occur in shoal-front rocks that formed seaward of the shoal crest (paleotopographic high) and are interpreted as intertidal to marginally subtidal deposits. Shoal-crest rocks, which are considered to be more subaerial (supratidal), contain abundant feldspar. If the feldspar dissolution and kaolinite growth were due to chemical weathering in the environment of deposition, it did not take place along the paleotopographic high but rather seaward and downdip of the shoal crest.

Several lines of evidence also suggest that feldspar dissolution was not a result of differential abrasion, as suggested by Hull (1957). On a sample-to-sample basis, the gray sandstones exhibit no relationship between grain-size and feldspar content. That is, although the shelf-edge sandstones may exhibit a slight overall coarser grain size than other siliciclastic facies, samples with variable textures within the facies exhibit unpredictable degrees of feldspar dissolution. Further evidence against the differential abrasion model is that siltstones and sandstones deposited in environments that would not have been subjected to winnowing currents (such as the toe of slope or channels/grooves within the deep reef) are also largely devoid of feldspars. Lastly, the occurrence of kaolinite blebs and open secondary pores may suggest that differential abrasion was not the cause of feldspar dissolution. If currents strong enough to mechanically weather grains were present, then clay alteration products would be winnowed and secondary pores would not be preserved.

A more plausible explanation is that feldspar dissolution and kaolinite growth are facies-specific, late-stage burial events related to meteoric flushing of the shelf margin during uplift of the Guadalupe Mountains. The Capitan aquifer has been present since uplift of the Guadalupe Mountains in the late Pliocene-Pleistocene (Hill, 1987). The proximity of the gray sandstones to porous carbonates of the Capitan Reef has likely resulted in extensive meteoric flushing and kaolinitization of the feldspars. This would also explain the paucity of feldspars in siltstones/sandstones deposited in channels and cavities within the reef and at the toe of slope.

On the Northwest Shelf, Red Anhydritic Siltstones/Sandstones deposited along the shoal crest did not experience meteoric flushing due to their greater distance from the porous reef tract and, more importantly, because they were tightly cemented by early anhydrite and dolomite/magnesite. The lack of interaction with meteoric waters preserved the original subarkosic to arkosic mineralogy in the red siltstones/sandstones.

On the Central Basin Platform, Green-Gray Dolomitic Subarkosic Siltstones/Sandstones found in the L. Daugherty #4 well exhibit minor feldspar dissolution and kaolinite growth (Fig. 51). These sandstones and siltstones were deposited in an algal lagoon-tidal flat setting, immediately updip (landward) of a pisolite shoal complex. The presence of associated porous carbonates (reef, backreef, pisolite shoal, and lagoonal/algal flat carbonates) may have allowed some meteoric flushing and kaolinitization in the siliciclastics. However, low permeabilities related to the original depositional texture (abundant clay and dolomite matrix) and the greater distance to the Capitan aquifer may have limited the kaolinitization. Some feldspars are dissolved and have been reprecipitated as kaolinite; whereas, other feldspars show little or no alteration. Authigenic K-feldspar is also common in the siliciclastics and may mark the updip limit of meteoric flushing.

CYCLICITY AND DEPOSITIONAL MODEL

The depositional model proposed here is based on (1) an examination of the cyclical arrangement of siliciclastic sands and carbonates/evaporites in logs, cores, and outcrops that capture the range of facies variation across the Yates shelf; and (2) an interpretation of the causal mechanism for the cyclicity being low-amplitude sea-level fluctuations that were operating at different periodicities and interacting to cause the mixing of the lithologies at different scales. The changes seen in Yates stratigraphy at any position on the shelf are a record of the significant variation through Yates time in depositional conditions, and the model here is a reasonable attempt at explaining these variations.

Cyclicity from Logs and Cores

Gamma-ray logs through a shelf margin sequence of the Yates Formation exhibit characteristic cyclic inflections related to alternating carbonate and siliciclastic intervals (Figs. 3, 4, and 52). Large-scale cycles are defined by 30 to 50 ft (9 to 15 m) thick intervals of low API unit dolomite coupled with 30 to 60 ft (9 to 18 m) thick intervals of high API unit arkosic siltstones/sandstones with interbeds of dolomite. Couplets composed of 3 to 16 ft (1 to 4.9 m) thick intervals of high API unit siltstone/sandstone and 3 to 14 ft (1 to 4.3 m) thick intervals of low API unit dolomite form small-scale cycles that give the siliciclastic intervals a characteristic bundled appearance.

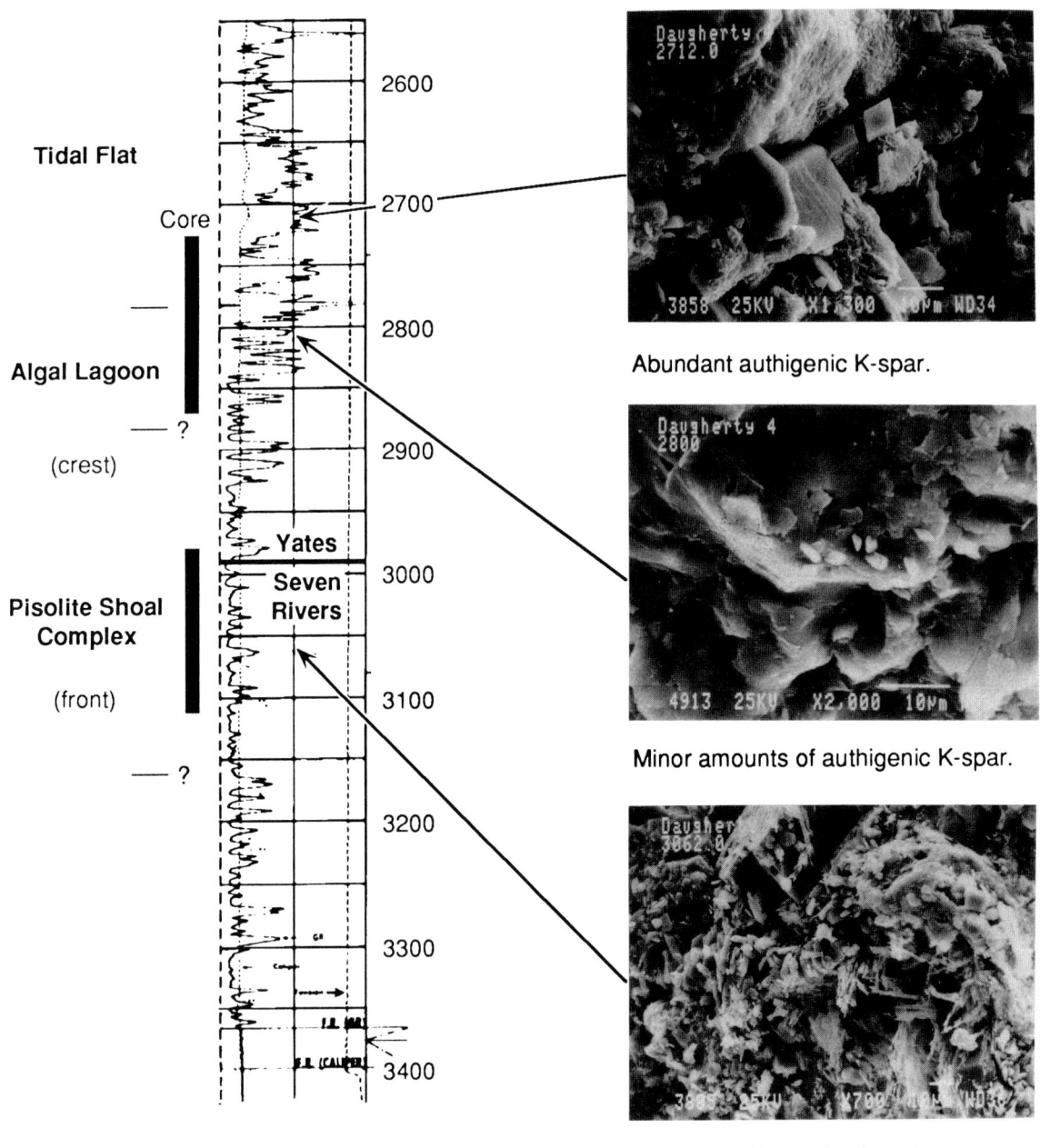

Figure 51
Authigenic and detrital feldspars decrease with depth in the L. Daugherty #4 well. The decrease in feldspar is actually a facies - dependent phenomenon that occurs systematically with depth due to the prograding nature of the formation. Facies belts get progressively closer to the Capitan aquifer with depth.

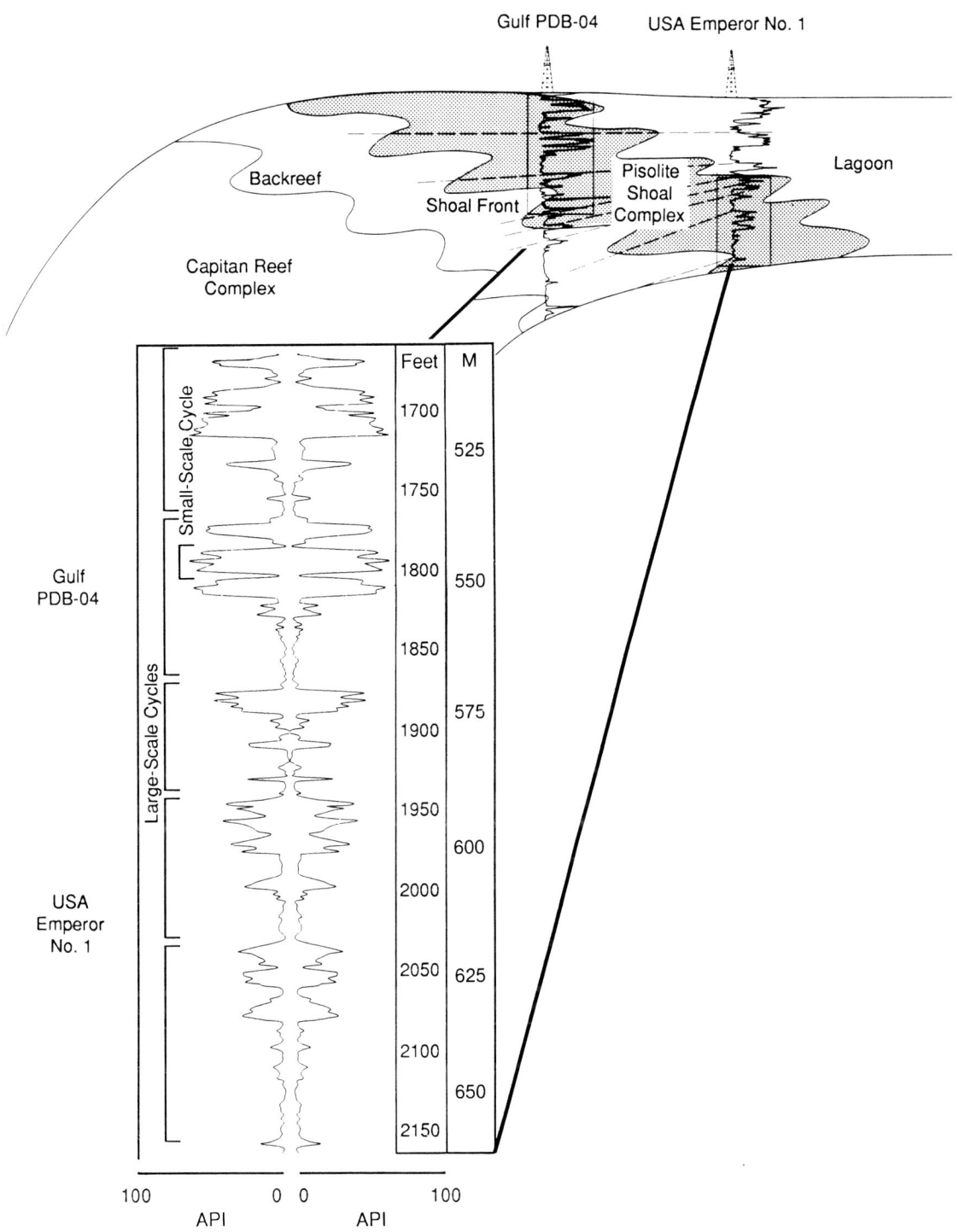

Figure 52
Rock sequences deposited at the shelf margin of the Northwest Shelf exhibit a characteristic log signature which is interpreted to be a very sensitive indicator of changes in sea level and/or climate. At the shelf margin, siliciclastics interfinger with carbonates that formed in a pisolitic shoal environment. In the shoal- crest region, clastic intervals consist of multiple (2, 3, or 4) sandstone/siltstone beds with thin carbonate interbeds. The distinct bundled fashion of the carbonates and siliciclastics results in a characteristic multi-spiked gamma-ray log response.

These various cycles represent alternations through time of the lithofacies and depositional settings that were previously described. An understanding of the cyclicity is critical before a depositional model can be presented in more detail.

Due to the prograding nature of the Yates Formation, a composite log constructed from parts of two wells is required to represent the entire formation through an interval that was deposited under uniform conditions (e.g., subsidence/accumulation rates, carbonate depositional rate, clastic depositional rate, and erosional rate)(Fig. 52). Five large-scale siliciclastic-carbonate cycles are present on Figure 52. A sixth cycle pinches out updip of the U.S.A. Emperor #1 well and may be considered as either the basal part of the Yates Formation or the upper part of the Seven Rivers Formation. On subsurface cross sections of the Northwest Shelf, Meissner (1969, 1972) clearly placed the lower siliciclastic-carbonate couplet in the Yates Formation; whereas, here the base of the formation is considered to be just below a well-developed triplet sandstone package (about 2080 ft on the composite log of Fig. 52). Because the formation top is picked at the top of a siliciclastic interval and the formation base at the bottom of a siliciclastic interval, the Yates Formation includes about 4.75 large-scale cycles, with the carbonate portion of the lowest cycle being placed in the Seven Rivers Formation.

A calculation of the average duration of the large-scale cycles is not straightforward, because the chronostratigraphy of the Upper Guadalupian section in the Permian Basin is poorly defined. Paleontologic data and eustasy cycles of Ross and Ross (1987) suggest the Yates and Tansill formations were deposited during a third-order sea-level fluctuation of 1 Ma duration. Sarg (1989) suggests the Yates Formation was deposited over a span of about 2 Ma during 2 third-order sea-level fluctuations of about 1 Ma each. These data suggest the duration of the Yates Formation may range from about 0.5 to 2 Ma and result in an average large-scale cycle duration ranging from approximately 105 ka (4.75 cycles/0.5 Ma) to 420 ka (4.75 cycles/2 Ma).

The 105- to 420-ka range for the durations of the five large-scale, clastic-carbonate packages suggests that if they are related to orbital forcing (Milankovitch cycles), then they are most likely to be 100- or 400-ka eccentricity cycles. In order to test the possibility of orbitally forced cycles and to constrain the cycle durations, the composite log of Figure 52 was correlated to Berger's (1978) plot of the caloric equator for the Quaternary for both 100- and 400-ka cycle scenarios (Fig. 53).

The caloric equator is an important measurement of astronomical insolation and is a perfect integrator of the orbital parameters (Berger, 1978). The maximum positive values on the caloric equator curve were considered by Milankovitch (1941) to correspond to the occurrence of the Ice Ages in the Northern Hemisphere. The Milankovitch model suggests that maximum positive values correspond with summers that were cool enough in high northern latitudes to allow for a positive annual snow budget and global cooling due to the surface albedo effect.

The caloric equator curve of Berger (1978) describes orbital forcing parameters for the Quaternary only. Little is known about probable changes in the forcing mechanism or the system response in the more geologic past. In the 100-ka scenario of Figure 53, the caloric equator curve is plotted at a scale such that 100 ka corresponds to a 100 ft-thick (30 m), large-scale, siliciclastic-carbonate cycle. The Yates Formation would thus have a duration of approximately

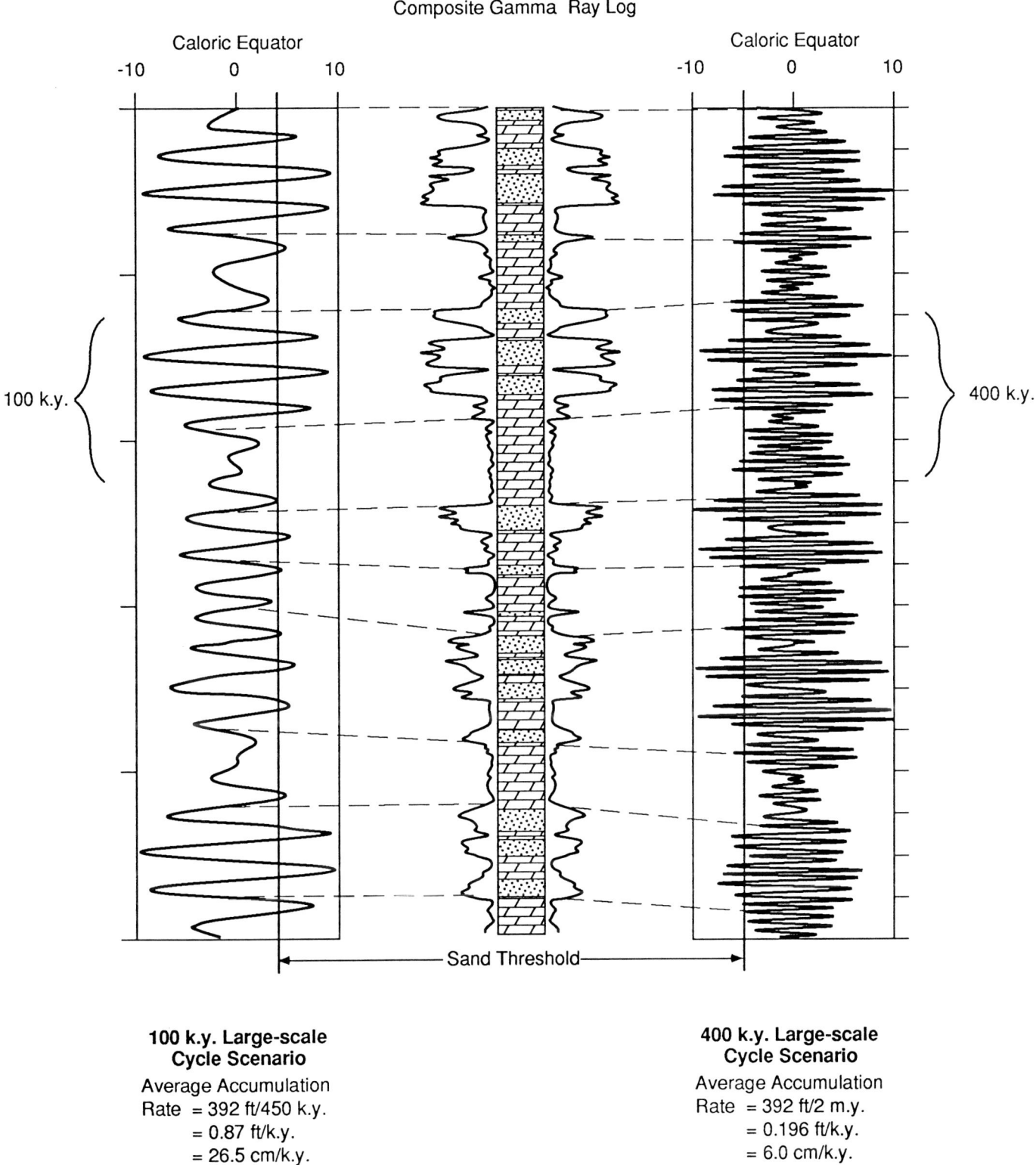

Figure 53
The composite gamma-ray log from Figure 52 is tuned to Berger's (1978) caloric equator curve in order to qualitatively test two scenarios of orbital forced cyclicity. The Berger curve describes the long-term variations of the caloric equator for the Quaternary and should only be considered schematic for the Permian. The latitude of the caloric equator is given in degrees: positive for northern latitudes; negative for southern latitudes.

0.48 Ma, which corresponds to an average accumulation rate of 0.87 ft/ka (26.5 cm/ka). In the 400-ka scenario the caloric equator curve is plotted such that a large-scale cycle on the composite log corresponds to 400 ka on the Berger curve. This results in a duration of about 2 Ma for Yates Formation with an average accumulation rate of 0.196 ft/ka (6 cm/ka).

A good correlation between lithology and amplitude variations on the caloric equator curve is present for both scenarios. Siliciclastic intervals correspond to maximum values (positive or negative); whereas, thick dolomite intervals correspond to times of low-amplitude oscillations. This suggests that some type of threshold (certain magnitude of sea-level/base-level drop, perhaps coupled with a particular climate regime) had to be reached before siliciclastic deposition took place. Bundling of the high-amplitude kicks on the Berger curve into packages of 2, 3, or 4 is similar to the bundling of siliciclastic intervals in the Yates. Although both scenarios provide reasonable fits, several lines of reasoning suggest that the 400-ka, large-scale cycles and 2-Ma Yates duration are the more likely. These include a lower, more reasonable average accumulation/subsidence rate (6.0 vs 26.5 cm/ka); common 4:1 cycle bundling; and the presence (seen best in core) of additional, very small depositional cycles (within the small-scale 100-ka cycles) that may be related to higher-frequency (precession and obliquity) orbital cycles.

Fischer plots are a technique used to evaluate changes in small-scale (high-frequency) cycle thickness over time in order to define longer-duration (lower-frequency) cycles or sea-level fluctuations (Fischer, 1964; Read and Goldhammer, 1988). The stratigraphic thicknesses of the small-scale cycles are corrected for regional linear subsidence and plotted against time. The plots define longer-duration cycles and infer relative sea-level curves because any long-term changes in sea level or subsidence will result in changes in the thickness of the small-scale cycles. During a long-term fall in sea level or decrease in subsidence, accommodation space is less and small-scale depositional cycles are thinner than expected from linear subsidence; whereas, during a long-term rise in sea level or increase in subsidence, accommodation space is greater and depositional cycles are thicker than expected from linear subsidence.

Cycles of uniform duration must be used when plotting the Fischer diagram. Typically the duration of the interval of interest is divided by the total number of cycles in order to get an average duration for the cycles. In the case of the Yates interval, however, both of the two possible durations (100 and 20 ka) are considered for the small-scale cycles (Fig. 54). The two horizontal axes on Figure 54 represent both possible scenarios for the duration of the small-scale cycles. In each case an average linear subsidence is calculated by dividing the thickness of the section by the total time represented. Although the shape of the plot does not change for the two scenarios, the linear subsidence values change due to the different cycle durations.

A major limitation in using the Fischer plot method is the ambiguity in picking some small-scale cycles. Cycles may be missing and/or condensed, particularly at points corresponding to large third-order sea-level lowstands. Missing and/or poorly identified condensed cycles may change the bundled pattern on a Fischer plot. The only difference between the plots of Figure 54A through C is the way small-scale cycles were picked within the third (middle) large-scale cycle. Although each is slightly different, the five large-scale cycles that comprise the Yates Formation are clearly depicted on all three plots. The plots suggest that the large cycles are asymmetric (quick rise and slow fall) and are characterized by 4:1 bundling.

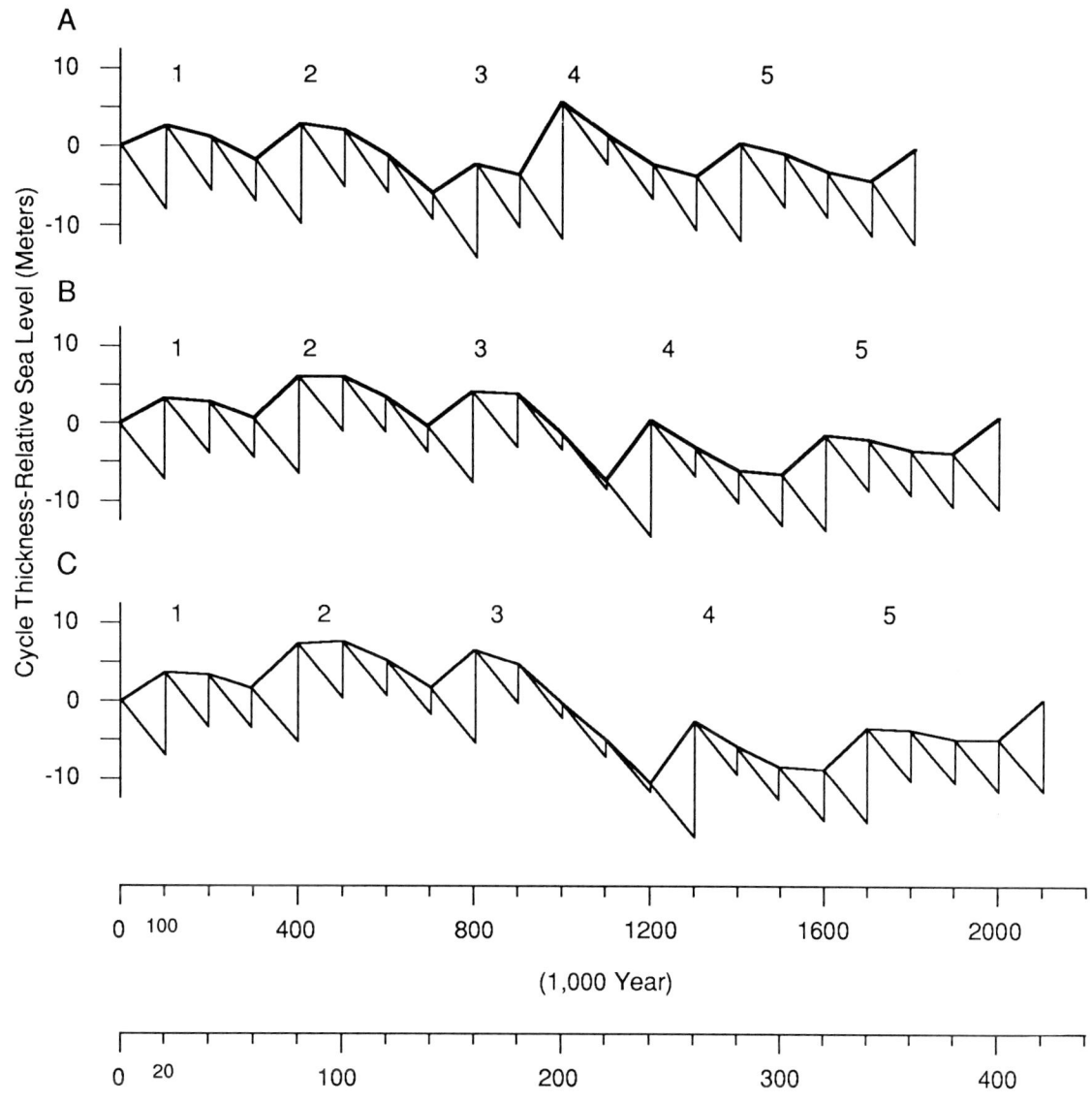

Figure 54
Several Fischer plot scenarios (A, B, and C) are possible for the Yates Formation, primarily due to the ambiguity in picking small-scale (high-frequency) cycles within a possibly condensed large-scale cycle (cycle no. 3). The two horizontal axis represent the possible Milankovitch scenarios for the duration of the small-scale cycles, i. e., 100,000 or 20,000 years duration. The calculated linear subsidence values change for the 20- and 100 ka scenarios, but the shape of the plot remains the same. Although there is ambiguity in picking some small-scale cycles, all plots exibit a dominate 4:1 (small- to large-scale) cycle bundling. This suggests the large - and small-scale cycles may be related to 400- and 100ka Milankovitch (eccentricity) rhythms. The Fischer plots further suggest that the five 400 ka sea-level fluctuations that took place during Yates deposition were asymmetric (fast rise and slow fall) and had magnitudes ranging from about 26 to 39 ft (8 to 12 m).

Although there is some ambiguity as to the number of small cycles comprising the middle large-scale cycle of Figure 54, several lines of evidence suggest the thick siliciclastic interval may actually represent several condensed small-scale cycles corresponding to a large third-order fall in sea level (Fig. 54B or C). Several thin dolomite interbeds occur within the siliciclastic interval. The thin dolomites are not apparent on the gamma-ray log but are present in core (Fig. 55) and may mark condensed cycle tops. No other individual (high-frequency) siliciclastic beds have such distinct carbonate interbeds. Furthermore, if the interval is treated as part of only one small-scale cycle (Fig. 54A), then the middle large-scale cycle does not exhibit the 4:1 bundling that is so well developed in the other large-scale cycles. For this reason the Fischer plots of Figure 54B or C are preferred since the middle cycles on these plots exhibit both bundling and asymmetry similar to the other cycles. These plots also place the condensed cycles on the falling limb of a third-order sea-level fluctuation (see below) where thinner cycles are expected due to the decrease in accommodation space.

The presence of very small cycles at the shelf margin of the Yates is best explained using a 2 Ma duration for the formation. In this case, small depositional cycles may be related to high-frequency orbital cycles, and analyses of these smallest cycles suggest they occur with frequencies similar to the 20-ka (precessional) and/or 40-ka (obliquity) Milankovitch cycles. If, instead, Yates deposition only lasted 0.5 Ma, then the smallest cycles must be locally produced (autocyclic) or related to subtle variations in shelf subsidence.

Further comparison of the composite gamma-ray log to the Berger caloric equator curve (Fig. 53) shows that the internal character in siliciclastic beds is best represented by 400-ka large-scale cycles. Gamma-ray inflections within an individual siliciclastic interval correspond to alternating detrital clay-rich and clay-poor zones and occur at about the same frequency as the 20-ka (precessional) kicks on the caloric equator curve. Decimeter-scale (early) cement stratification could also represent high-frequency orbital cycles, in that alternating zones of magnesite/anhydrite and dolomite cement (Fig. 36) could simply be following depositional textures or be the result of cyclic changes in the hydrologic conditions during deposition or early burial.

Numerous small-scale cycles are also present within the carbonate intervals of the Yates but cannot be recognized using the gamma-ray log. These cycles, seen in core and outcrop, are typically less than 6 ft (1.8 m) thick and are formed of shoaling and deepening pisolite shoal sequences that require short-term variations of relative sea level in order to form. A typical shoaling cycle consists of massive or laminated dolomudstone that grades to fenestral peloidal dolomudstone/wackestone, that is in turn capped by pisolitic and/or intraclastic dolopackstone/grainstone. Large anhydrite-filled voids are often associated with the pisolitic/intraclastic intervals and, based on a comparison with outcropping examples, are considered to be tepee structures and probably related to subaerial exposure.

In order to further evaluate these smallest cycles in core, the shoal-crest interval from the PDB-04 well was examined in more detail. The shelf-margin sequence in the core was subdivided into water depth-dependent lithologies (Fig. 55) in order to evaluate potential cyclicity. A lithologic (lith) log with uniform sample spacing was constructed by fitting a spline curve through the midpoint values of the lithology interval data, so the lithology data could be compared to the Berger caloric equator curve. The resulting log exhibits much more character in the thick

Figure 55

Graphic log (A) shows lithologic variation within a pisolite shoal-crest portion of the PDB-04 core. Lithologies are plotted in (B) so that those interpreted as the most subaerial lithologies are plotted on the left, and the most subaqueous lithologies are plotted on the right. Brief descriptions are provided in the key. Lith log (C) was built by fitting a spline curve through the midpoint values of the lithology interval data. The interpolated curve, with uniform sample spacing, is required for spectral analysis and also for easy comparison to Berger's caloric equator plot (D). Note the detailed character of the lith log compared to the composite gamma-ray log (Fig. 52). The lith log provides data within the thick dolomite sections that exhibit little log character on the gamma-ray log. With the 400-ka large-scale cycle scenario, kicks (cycles) on the lith log exhibit high frequencies similar to those expected from the precession (20 ka) and obliquity (40 ka) orbital signals. In the 100-ka large-scale cycle scenario (not shown), the high-frequency depositional cycles would have to be autocyclic or related to subtle variations in subsidence.

The lith log is constructed for just the pisolite shoal-crest facies interval. This entire interval was probably deposited during relatively uniform conditions (sedimentation rates, subsidence rate, and erosion rate) and should give a uniform response to any cyclic signal. The lagoonal and shoal-front facies, respectively, above and below the processed interval, were deposited under conditions different from those of the shoal crest and, therefore, are expected to exhibit different responses to the cyclic signal.

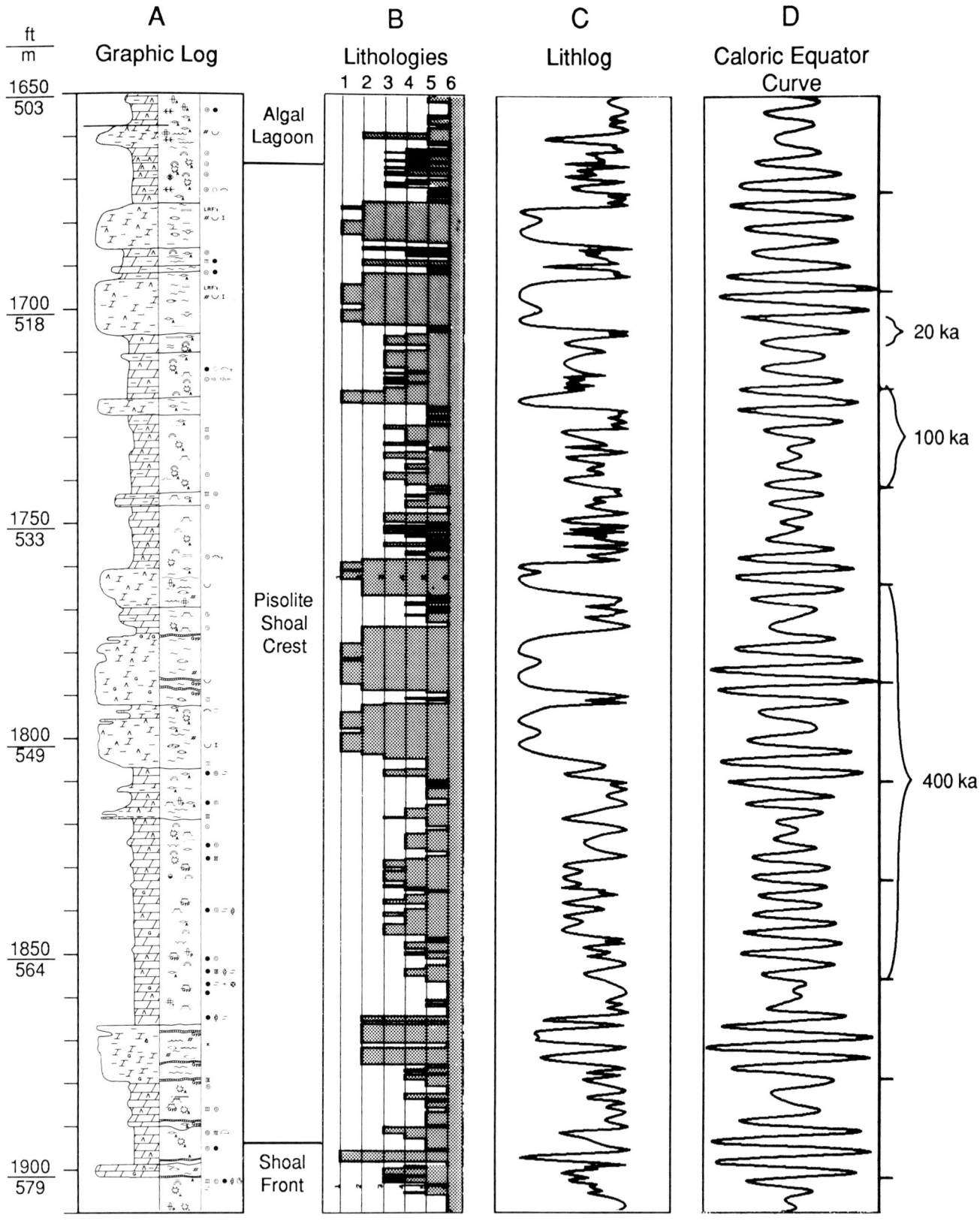

1 Red anhydritic siltstone/sandstone
2 Red, argillaceous or dolomitic siltstone/sandstone
3 Chaotic dolomite and anhydrite with desiccation features (teepee)
4 Pisolitic and intraclastic dolopackstone/grainstone
5 Fenestral dolomudstone
6 Massive or laminated dolomudstone

Figure 55

carbonate portions of 400-ka cycles than the gamma-ray log. Comparison of the lithologic log with the caloric equator curve shows that the small carbonate cycles and their associated variation of water depth occur with a frequency similar to the precessional (20 ka) and obliquity (40 ka) orbital signals.

Cyclicity and Facies Relations in
Guadalupe Mountains Outcrops

Four classic outcrops of the Yates Formation in the Guadalupe Mountains were also evaluated in terms of stratigraphic sequence and depositional cycles in order to (1) gain a better understanding of the small-scale cyclicity, and (2) illustrate lateral facies changes of both siliciclastics and carbonates on the Yates shelf (Borer and Harris, 1989). The outcrops are placed in their relative depositional and stratigraphic position along a dip-oriented cross section (Fig. 56) that spans the outer 8 to 10 miles (13 to 16 km) of the Yates Formation on the Northwest Shelf. A more detailed schematic cross section (Fig. 57) illustrates how the cyclic packages change across a portion of the shelf. Small-scale (20 to 100 ka) wedges of carbonate from the shelf margin pinch out into thick (100 to 400 ka) siliciclastic intervals in an updip direction (note that the suggested durations for the different scale cycles have been modified from Borer and Harris, 1989). This results in 10 to 25 ft-thick (3 to 7.6 m) siliciclastic-carbonate cycles at the shelf margin being equivalent to 5 to 10 ft-thick (1.5 to 3 m) fining-up siliciclastic intervals in the middle-shelf region. In other words, a sea-level rise that provided accommodation space for the deposition of 5 to 15 ft (1.5 to 4.5 m) of pisolitic dolomite at the shelf crest provided accommodation space for the deposition of only a few feet of dolomitic, argillaceous siltstone in the middle-shelf region. This has important implications for the distribution of siliciclastic lithofacies in the subsurface of the Northwest Shelf and Central Basin Platform and suggests that the deposition of argillaceous siltstones may be a cyclic shelf-wide phenomenon.

The four outcrops are each described in detail, beginning with the most updip depositionally. The location of each outcrop is described both in terms of its present location and its depositional position on the time-equivalent Yates shelf (distance from reef edge). Each outcrop is then further evaluated in terms of stratigraphic packaging and depositional cycles.

U.S. Highway 285 roadcut.--On U.S. Highway 285, about 2 miles (3 km) south of the junction with Rocky Arroyo Road (137), sandstones and siltstones of the Yates Formation crop out in low-relief roadcuts (T21S, R26E, Sec. 11, SE1/4, NW1/4). Structure contours drawn on the top of the Yates (Motts, 1962) suggest that these rocks are from the middle of the Yates Formation, about 100 to 150 ft (30 to 45 m) below the formation top. The outcrops are situated 9 to 10 miles (14 to 16 km) behind the youngest Capitan reef and 7 to 9 miles (11 to 14 km) behind their time-equivalent reef edge (Fig. 56). These siliciclastics represent the updip limit of the middle-shelf, clastic-rich facies belt.

The well-bedded clastic interval at the roadcut is 25 to 30 ft (7.5 to 9 m) thick and bounded by a sharp basal contact and slightly gradational upper contact with algal-laminated, fenestral dolomudstones. Four thinning- and fining-upward sequences are present within the siliciclastic

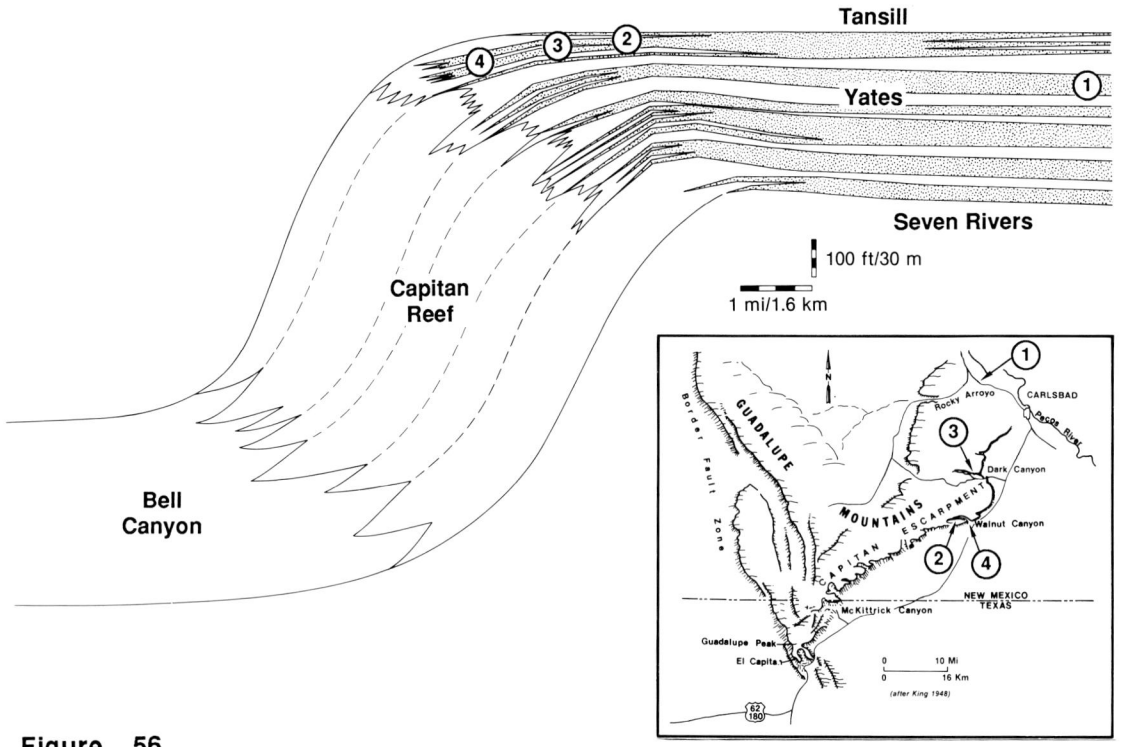

Figure 56
Schematic cross section of Yates Formation in the subsurface of the Northwest Shelf. Four outcrops in the Guadalupe Mountains have been placed in their relative depositional position on the shelf: (1) U.S Hwy. 285 roadcut, (2) Walnut Canyon above Hairpin Turn, (3) Dark Canyon, and (4) mouth of Walnut Canyon. The cross section is based on subsurface wireline log picks. Stippled region is siliciclastic, unstippled is carbonate. Six distinct clastic intervals are depicted on the cross section.

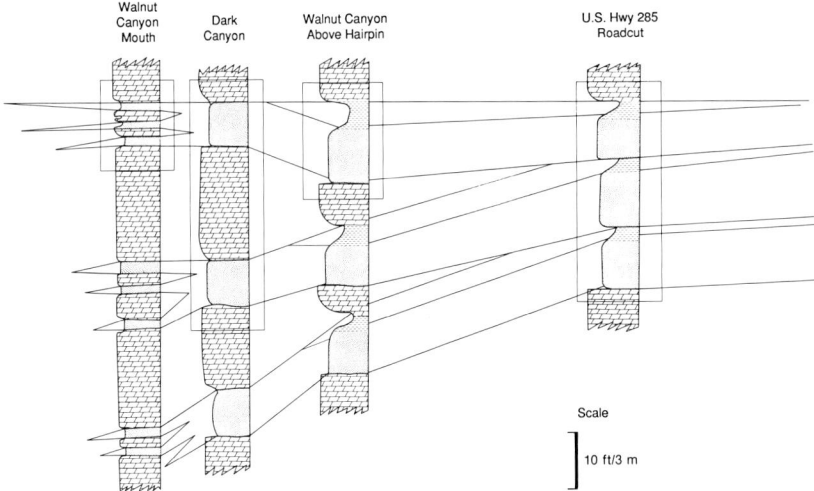

Figure 57
A schematic cross section illustrating depositional styles at each outcrop section. Changes in cycle packaging across the shelf are due to the updip pinch-out of small-scale (100 ka) carbonate wedges into the siliciclastic portion of large-scale (400 ka) depositional cycles. Thick (10 to 25 ft; 3 to 7.5 m) siliciclastic-carbonate cycles deposited at the shelf margin are equivalent to 5 to 15 ft- thick (1.5 to 4.5 m), fining-up siliciclastic intervals in the middle-shelf region. Sea-level rises that resulted in the deposition of 5 to 10 ft (1.5 to 3 m) of pisolitic dolomite at the shelf crest resulted in the deposition of only a few feet of argillaceous siltstone in the middle- shelf region.

interval (Fig. 58A). Each fining-up sequence consists of thick- to medium-bedded, silty, very fine-grained, subarkosic to arkosic sandstones (Fig. 58B and C) that grade upward to thin- to medium-bedded, argillaceous (illitic/micaceous) siltstones and mudstones. Boundaries between mudstones and overlying sandstones are sharp. Dolomicrite matrix increases toward the top of each sequence.

Two orders of cyclicity are evident in the outcrop. The clastic and upper carbonate interval form a couplet that defines a large-scale cycle that is only partially exposed. Small-scale, 5 to 10 ft (1.5 to 3 m) cycles are defined by the fining-up sequences in the clastic part of the larger cycle. Silty sandstones in the lower part of the small cycles are considered to have been deposited in a peritidal sandflat setting; whereas, the argillaceous siltstones and mudstones at the top of the small cycles must represent a relative sea-level rise and deposition in a shallow lagoon environment.

Walnut Canyon above Hairpin Turn.--Pale orange Yates sandstones and siltstones crop out in Walnut Canyon along New Mexico Highway 7 to Carlsbad Caverns. An excellent exposure of the uppermost Yates sandstone occurs on the western leg of a sharp hairpin turn, about 1.5 miles (2.4 km) from the cavern parking lot (T24S, R25E, Sec. 30, NW1/4, SE1/4). The outcrop is positioned 1.25 to 1.5 miles (2 to 2.5 km) updip from the youngest Yates reef edge (Fig. 56) and 6 to 8 miles (9.7 to 13 km) seaward of the previously described outcrop. The siliciclastics are thought to have been deposited across a very shallow to subaerially exposed pisolite shoal.

The 8 to 10 ft (2.4 to 3 m) thick clastic interval is one fining-up sequence consisting of a 5 ft (1.5 m) thick basal sandstone that grades up to wavy-bedded argillaceous siltstone and mudstone (Fig. 58D). The sandstone bed is very uniform and continuous for the entire length of the outcrop; whereas, the argillaceous siltstone/mudstone cap thins and thickens. The sandstone is a structureless to planar-laminated, silty, very fine-grained dolomitic subarkose (Fig. 58E and F). Siltstones are micaceous and have abundant detrital and dolomitic matrix. Silty beds toward the top of the interval show wavy and flaser bedding with possible ripple marks. The basal contact with thin- to medium-bedded cyclic dolomites of the pisolite shoal is planar and sharp. Upward in the unit the clastics become increasingly dolomitic toward a sharp upper contact with silty, argillaceous dolomudstone.

The fining-up sequence present in the siliciclastic part of the outcrop is similar to the fining-up cycles found in the siliciclastics exposed along the U.S. Hwy. 285 roadcut. The Walnut Canyon sequences, however, are not successive but rather interbedded with dolomites of the pisolite shoal complex. This suggests that small-scale rises of relative sea level that controlled the deposition of several feet of siltstone and mudstone in the midshelf region resulted in 5 to 15 ft (1.5 to 4.5 m) of dolomite deposition at the inner part of the shelf margin.

Dark Canyon.--Sheet-like, pale orange, dolomitic, silty sandstones near the top of the Yates Formation are exposed in the walls of lower Dark Canyon. These outcrops occur along the southern wall of the canyon about 1 mile (1.6 km) southwest of the canyon mouth (T23S,

Figure 58

- A. U.S. HWY. 285 ROADCUT. Three fining-up sequences in a red, midshelf, siliciclastics interval of the Yates Formation. Each fining-up cycle consists of thick- to medium-bedded, silty, very fine-grained sandstone that grades up to thin- to medium-bedded, argillaceous siltstone/mudstone. Dolomicrite matrix increases toward the top of the cycle. Cycle contacts are sharp and represent a time of nondeposition during low sea-level stands. Silty sandstones were trapped on the shelf in a peritidal-flat setting as sea level began to rise. The muddy cycle tops were deposited in a shallow lagoon during sea-level highstands. Algal dolomudstones were deposited during long-term sea-level highs.

- B. U.S. HWY. 285 ROADCUT. Silty sandstone from basal portion of lowest fining-up sequence. Sedimentary structures consist of nondescript wispy and wavy laminae and small-scale, irregular lenses of clean sand.

- C. U.S. HWY. 285 ROADCUT. Thin-section photomicrograph of silty, very fine-grained subarkose. Bottom portion of thin section contains abundant silt and red (iron-stained) detrital clay; upper portion is from clean sand lens similar to those in Figure 58B. Plane light, 79X magnification.

- D. WALNUT CANYON ABOVE HAIRPIN. Pale orange, fining-up clastic interval packaged between well-bedded, cyclic dolomites of the pisolite shoal complex. Plane-laminated, silty, very fine-grained sandstone grades up to wavy-laminated, dolomitic, argillaceous siltstone/mudstone. Basal contact of sandstone with Yates dolomite is sharp and planar; upper contact with Tansill dolomite is undulatory and slightly gradational.

- E. WALNUT CANYON ABOVE HAIRPIN. Sharp, planar contact between plane-laminated, silty, very fine-grained subarkose and underlying pisolitic dolopackstone/grainstone of the pisolite shoal complex.

- F. WALNUT CANYON ABOVE HAIRPIN. Thin-section photomicrograph showing alternating, planar siltstone and sandstone laminae in basal silty sandstone. Dolomicrite matrix is common, particularly in the siltstone laminae. Plane light.

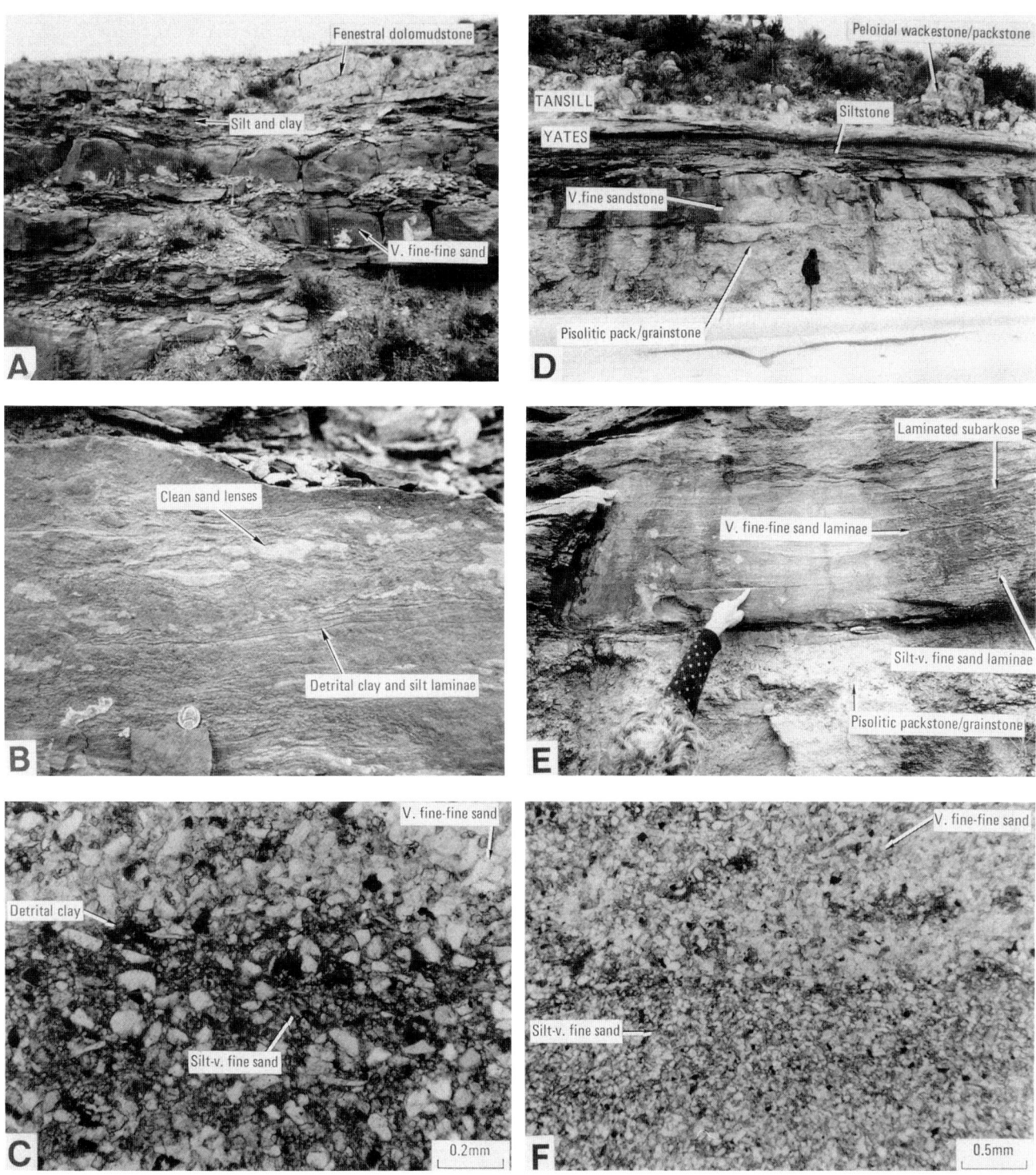

Figure 58

R25E, Sec. 23, SE1/4, SE1/4) and are positioned 0.75 to 1 mile (1.2 to 1.6 km) behind the youngest Yates reef edge (Fig. 56). This places the outcrops depositionally in a position about 0.5 miles (0.8 km) downdip from the previously described Walnut Canyon outcrop. Basal Tansill and upper Yates carbonates in this part of Dark Canyon consist of thick, cyclic, pisolite sequences with fenestral porosity and classic tepee structures characteristic of the pisolite shoal of the shelf margin. The upper Yates sediments were probably deposited slightly seaward of the shoal crest.

Two 5 to 10 ft (1.5 to 3 km) thick, very fine-grained subarkosic sandstones are separated by a 20 ft (6 m) thick section of thin- to medium-bedded dolomite of the pisolite shoal complex. Sandstones have sharp, planar, upper and lower contacts (Fig. 59A). Coarse sand and gravel-sized carbonate rip-up clasts are abundant at the basal contact of the upper sandstone. The sandstones (Fig. 59B and C) are structureless to wavy-bedded with planar laminae common near the base. Wispy detrital clay laminae are minor throughout; gypsum nodules and gypsum-filled fenestrae are abundant.

Small-scale, 20 to 30 ft (6 to 9 m) cycles consist of siliciclastic-carbonate couplets that are equivalent to the 5 to 10 ft (1.5 to 3 m) thick, fining-up siliciclastic sequences of the midshelf region. No fining-up sequences occur in the siliciclastic interval. Small-scale cycles can also be seen in the carbonate portions of large-scale cycles. These small-scale intracarbonate cycles consist of alternating well- and chaotic-bedded (tepee structures) dolomites of the pisolite shoal (Fig. 59A).

Mouth of Walnut Canyon.--Outcrops near the mouth of Walnut Canyon, about 1 mile (1.6 km) northwest of White City off New Mexico Highway 7 (Sec. 27, T24S, R25E, SE1/4, SW1/4), are intermittently exposed by flash floods. Low along the north canyon wall, several thin- to medium-bedded, discontinuous, gray siltstones/sandstones form small recessive ledges in resistant dolomite cliffs (Fig. 59D through F). The outcrops are positioned within a few tenths of a mile from the youngest Yates reef edge (Fig. 56). Here the Yates clastics are believed to have been transported to a position where they pinch out in the skeletal, intraclastic dolopackstone/grainstone facies immediately behind the reef near the shelf edge.

The gray clastics are followed with difficulty upstream approximately 100 ft (30 m) to recognizable outcrops of uppermost Yates, pale orange sandstones. The gray, dolomitic siltstone and very fine sandstones are planar- to wavy-bedded with local lenses and laminae of fine to medium sand. Basal contacts with imbricate clasts of skeletal, intraclastic dolopackstone/grainstones are sharp and erosive, suggesting deposition in a high-energy environment. Upper contacts with skeletal dolowackestones/packstones are also sharp. The sandstones are primarily composed of quartz and kaolinite derived from the alteration of feldspars and lithic fragments. They contain sparse detrital clay but abundant dolomite as lithoclasts, matrix, and cement. No distinct cyclic sequences are apparent in the limited outcrop area.

Figure 59

A. DARK CANYON. Sheet-like, pale orange, silty sandstone at top of Yates Formation. Sandstone is packaged between thick intervals of thin- to medium-bedded dolomite of the pisolite shoal complex. Basal contact between the sandstone and dolomite is sharp and planar; upper contact is slightly gradational. Also note tepee structures along distinct horizon in the upper dolomite interval.

B. DARK CANYON. Slightly gradational upper contact between wispy- to wavy-laminated, silty, very fine-grained sandstone and overlying dolomudstone of the pisolite shoal complex.

C. DARK CANYON. Thin-section photomicrograph of silty, very fine-grained subarkose. Red (iron-stained) detrital clay is sparse. Feldspars and lithic grains are often partially dissolved and have microporous kaolinite filling void spaces.

D. WALNUT CANYON MOUTH. Thin-bedded, discontinuous, gray, dolomitic, silty, very fine-grained sandstones at the top of the Yates Formation. The dolomitic siltstones/sandstones crop out just a few tenths of a mile (300 to 500 m) updip of the massive Capitan reef. Sandstones are interbedded with dolomudstones and skeletal, intraclastic dolopackstones and grainstones of the immediate back-reef facies belt. Basal contacts with intraclastic dolomites are sharp; upper contacts with dolomudstones and dolowackestones can be gradational or sharp.

E. WALNUT CANYON MOUTH. Sharp contact between dolomitic siltstone/sandstone and underlying skeletal, intraclastic dolopackstone/grainstone. Note large dolomite rip-up clasts at basal contact. A slightly more gradational contact with overlying dolomudstone/wackestone occurs near the top of the photograph.

F. WALNUT CANYON MOUTH. Plane- and ripple-laminated dolomitic siltstone/sandstone within immediate back-reef facies belt. Regions showing no internal stratification may be bioturbated.

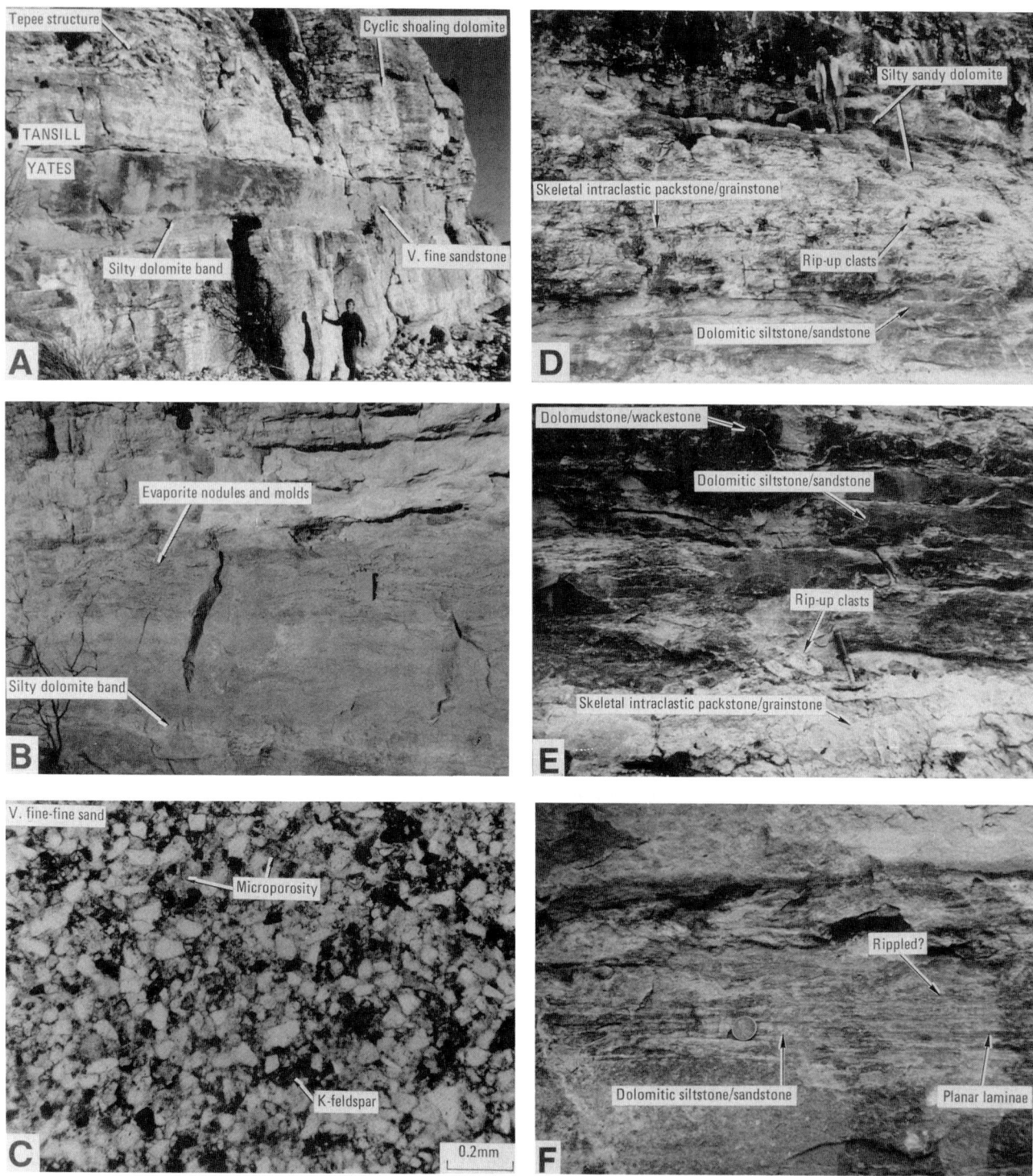

Figure 59

Third-Order Sea-Level Cycle

We believe an approximately 1.5- to 2-Ma duration, sea-level fluctuation coincided with deposition of the Yates Formation and is related to the character of shorter-duration cycles that are seen on logs, cores, and outcrops. The lower portion of the Yates Formation was deposited during the transgressive part of this "third-order" cycle; whereas, the upper portion of the formation was deposited during the regressive phase.

A Fischer plot of 100-ka cycles in the Yates Formation defines not only 400-ka bundles of cycles, but also the third-order sea-level fluctuation (Fig. 60). The Fischer plot suggests the third-order fluctuation had a low amplitude of 6 to 8 m and a wavelength of 1.5 to 2 Ma; however, the limited amount of stratigraphic section limits an interpretation of the duration of the third-order cycle. It is not entirely clear what produces a third-order sea-level fluctuation, in that it may be driven by local or global tectonics or may be a long-duration (low-frequency) expansion of the orbital signals. Read (1989), Osleger and Read (1989), and Bond and others (1989) have shown that 1- to 2-Ma eustatic sea-level cycles are common throughout the Cambrian and Ordovician, and Read (personal communication) suggests that many of these cycles are long-duration Milankovitch cycles. Recent work by R. K. Matthews (personal communication, 1990) suggests that low-frequency modulation of the precessional signal may be transformed into long period (millions of years) sea-level fluctuations by nonlinear geologic processes.

The third-order cycle coinciding with the Yates Formation controlled the character of the 400- and 100-ka depositional cycles. The Fischer plot (Fig. 60) and lithofacies cross section (Fig. 4) show that the thickness, dominant lithology, and internal packaging of these shorter-duration cycles are related to their position on the third-order curve. Although further study is required, it is apparent that 400-ka cycles deposited during the low sea-level portion of the third-order cycle are siliciclastic-rich; whereas, those deposited during the highstand portion are carbonate-rich. Condensed, siliciclastic-rich cycles may occur during a third-order sea-level fall.

400,000-Year Eccentricity Cycles

Siliciclastic rocks dominated the Yates shelf during the lowstand portions of asymmetric, 400-ka eccentricity cycles. Sea-level lowstands correspond to (glacial?) times of high-amplitude fluctuations of the caloric equator and account for one-half to three-fourths of the 400-ka cycles. Carbonates dominated the shelf during the remaining highstand portions of 400-ka cycles and occurred during (nonglacial?) times when variations in the caloric equator were at a minimum. At the margin of the Northwest Shelf, carbonate rocks accumulated at a higher rate than siliciclastic rocks, since a greater amount of time is represented by sea-level lowstands than by sea-level highstands (up to 3:1), yet the thicknesses of the highstand carbonate and lowstand siliciclastic intervals are the same. The character (thickness, internal packaging, depocenter, lateral extent, and dominant lithology) of 400-ka depositional cycles is controlled by both the position on a third-order sea-level cycle and the character of internal 100-ka cycles.

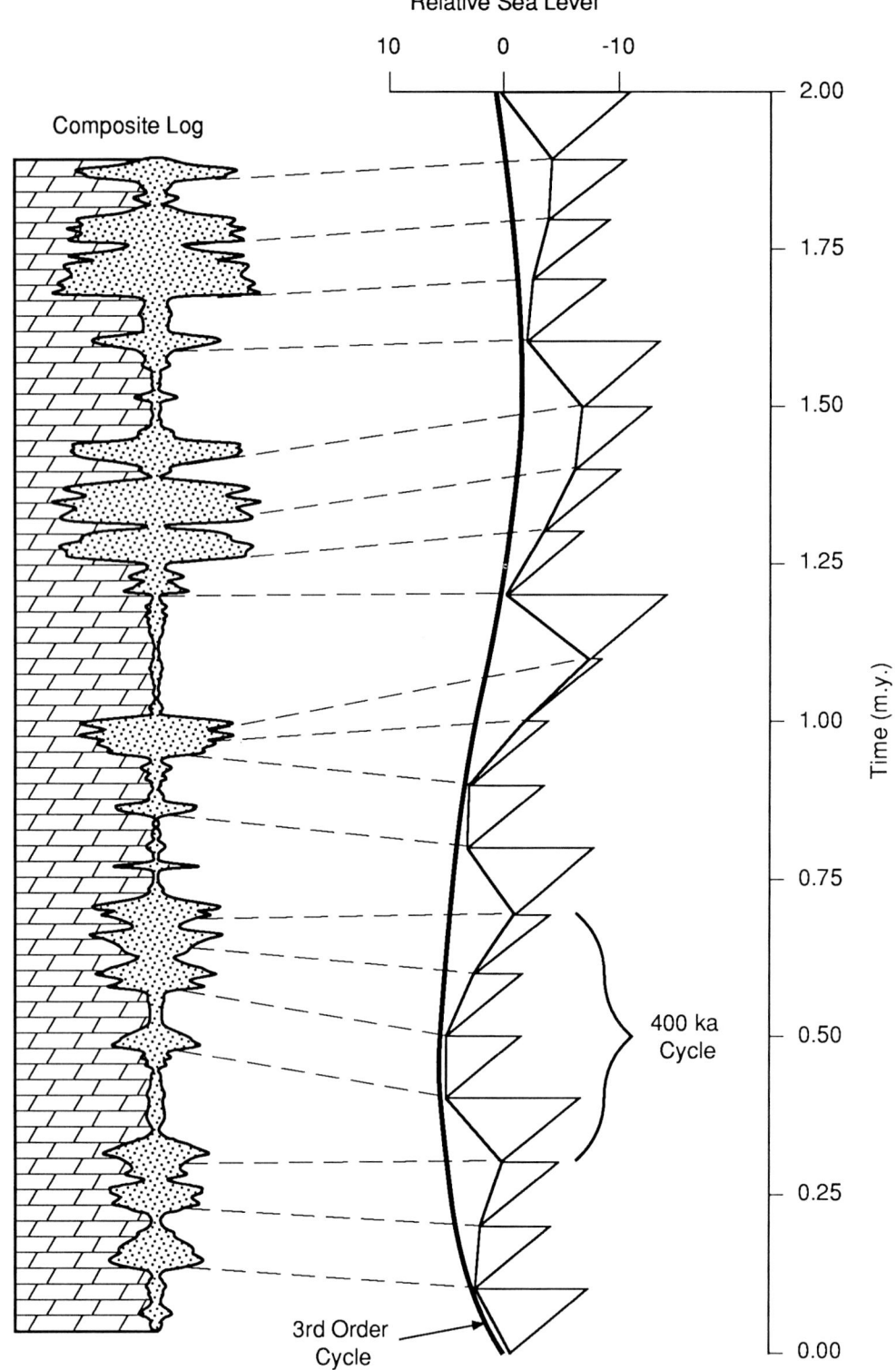

Figure 60
Fischer plot for the Yates Formation from Figure 55B illustrates how the changing thickness of 100 ka cycles defines 400 ka cycles that in turn delineate a 1.5- to 2 ma third-order sea-level fluctuation. The lower portion of the Yates Formation was deposited during the transgressive part of the third-order cycle; whereas, the upper portion was deposited during the regressive phase.

100,000-Year Eccentricity Cycles - Lowstands

Sea-level fluctuations related to 100-ka eccentricity orbital cycles control the internal packaging of the larger (400-ka and third-order) cycles within the Yates Formation. When 100-ka lowstands occurred in phase with 400-ka lowstands, sand and silt were transported across a very shallow to subaerially exposed shelf and pisolite shoal portion of the shelf margin, to be deposited in the basin. Outcrop studies summarized in Harris and Grover (1989) have shown that the shelf margin Capitan reef was actually a deeper-water reef and not a barrier reef. The reef was thus not exposed by these low-amplitude sea-level fluctuations, and it did not impede the transport of sand into the basin.

Sharp planar contacts between sheet-like siliciclastic beds and underlying dolomites of the pisolite shoal complex (with locally truncated tepee structures) suggest the low sea-level stands were a time of sediment bypass and erosion on the outer shelf. Siliciclastics were probably transported across the inner and middle shelf by a combination of sheetflood and eolian deflation processes, as there is little evidence for fluvial channels in the shelf strata. Near the shelf edge some channeling of sand through the pisolite shoal complex and reef is likely; however, much of the sand and silt probably crossed the peneplained shoal complex as sheets.

Wind transport of sediment across the shelf and into the Delaware Basin may have been very important during low sea-level stands. Fischer and Sarnthein (1988) show that older, basinal-equivalent siliciclastic units (Brushy Canyon and Cherry Canyon formations) have a distinct eolo-marine texture as compared to known Pleistocene eolo-marine sediments off the Saharan west coast. Furthermore, red detrital clay and mica found coating grains in shoal-crest siltstones/sandstones have textures very similar to those described by Walker (1979) in modern desert sands and attributed to the infiltration of airborne dust during repeated wetting and drying events. A deflation flat scenario during times of a lowered water table (low sea level and/or arid climate) is an attractive model to explain the peneplained shelf surface and the transport of abundant sand/silt across the shelf without evidence for fluvial channels.

When 100-ka lowstands corresponded with 400-ka highstands, shelf carbonates were periodically exposed and subjected to desiccation, but no siliciclastics were transported across the shelf. Marine pisolites deposited at the shelf margin during highstands (see below) may have been overprinted with (marine) vadose diagenetic features at this time. Intervals of anhydrite-cemented, chaotic-bedded dolomite (tepee structures) within the thick carbonate portion of 400-ka cycles are also considered to mark these 100-ka lowstands.

100,000-Year Eccentricity Cycles - Highstands

When 400-ka sea-level lowstands coincided with 100-ka sea-level highstands, siliciclastic sediments were not bypassed to the basin but instead were trapped on the Yates shelf. The change in base level related to the 100-ka sea-level highs increased accommodation space on the shelf and allowed for the trapping of silt and sand. The transgression reworked siliciclastics deposited along the shelf margin. This reworking or "whiskbroom" effect may have played a part in spreading the sand bodies into even sheets across the shelf or alternatively the siliciclastics may have been originally transported across the shelf as sheet-like bodies.

During continuation of the 100-ka sea-level rise, the shelf edge eventually became the site of carbonate deposition. The reef, forereef, and immediate backreef facies tracts expanded, and a thick wedge of pisolitic dolomite developed as a marginal mound (Dunham, 1972). The pisolitic shoal complex acted as a barrier and a quiet-water lagoon(s) formed in the middle-shelf region. In the lagoon(s) argillaceous siltstones were deposited as the tops of fining-up, siliciclastic sequences (such as those seen in the Walnut Canyon and U.S. Hwy. 285 outcrops). Discontinuous carbonates may have been deposited in some parts of the middle shelf as lagoon or tidal-flat deposits.

When (100 ka) sea level once again lowered, lagoons disappeared and silt and sand were transported across the shelf and out to the basin. Eventually the migrating sand and silt eroded the shelf to a peneplain and truncated both the pisolitic carbonates and the argillaceous siltstones. The reef margin again remained intact due to its deep-rim location. Erosion was probably limited considering the low shelf gradient and the low magnitude of the 100-ka sea-level variations (see below). Repetition of the 100-ka sea-level fluctuations resulted in the classic interfingering on the Northwest Shelf of carbonate wedges with siliciclastics.

When 400-ka sea-level highstands acted in phase with 100-ka highstands, the Yates shelf experienced maximum flooding. During these times massive subtidal dolomudstones were deposited far into the shelf interior. Locally the massive dolomites were desiccated and partially eroded during subsequent lowstands to form horizons of rip-up clasts.

Regional Variation

Cyclic depositional sequences within the Yates Formation are less well developed and therefore more difficult to analyze on the Central Basin Platform as opposed to the Northwest Shelf. The variation is due to a steeper slope and/or increased subsidence on the Central Basin Platform. A greater potential for missing cycles, related to either lack of deposition or subsequent erosion, exists for the steeper gradient shelf. The steeper gradient also results in cycle packages with thinner and less continuous carbonate portions of both 100- and 400-ka cycles (Fig. 61).

At the shelf margin of the Central Basin Platform, high-frequency (100 ka) carbonate wedges are poorly developed and project updip only about 0.25 miles (0.4 km) before pinching out (Fig. 61). In contrast, carbonate wedges on the Northwest Shelf project updip 1.5 miles (2.4 km) or more in response to the same sea-level rises. The carbonate portion of 400-ka cycles are thick at the shelf margin but thin rapidly to less than 10 ft (3 m) across the Central Basin Platform. On the Northwest Shelf the same carbonate portions of 400-ka cycles are 10 to 40 ft (3 to 12 m) thick and exhibit little thinning across the shelf.

Figure 61

The packaging of depositional cycles and the change in formation thickness across the outer shelf are considerably different for the Central Basin Platform and the Northwest Shelf. The differences are related to cyclic sea-level fluctuations acting on outer shelves with different topographic or subsidence profiles. On the Central Basin Platform, a steep shelf slope and/or large subsidence differential have resulted in rapid landward thinning of depositional cycles over the outer 2.7 miles (4.3 km) of shelf. Fifty feet (15 m) thick, 400-ka shelf-margin carbonates thin to less than 10 ft (3 m) along the outer middle shelf. Thin, 100-ka carbonates are poorly developed even near the shelf and pinch out on average about 0.25 miles (0.4 km) updip into siliciclastic intervals. In contrast, the lower shelf gradient and/or more uniform subsidence profile of the Northwest Shelf have resulted in much less updip thinning of depositional cycles. Thick (up to 50 ft; 15 m), 400-ka carbonates were deposited well into the interior of the shelf. Furthermore, 100-ka carbonate wedges are well developed and interfinger updip 1.5 miles (2.5 km) or more before pinching out into siliciclastic intervals.

Figure 62

A. Differences in formation thickness across the outer portions of the Northwest Shelf and the Central Basin Platform can be quantified by calculating the average slope/cycle for the two shelves. This calculation involves the change in formation thickness (t) across some distance (HD) among the total number of cycles. Using thicknesses and cycles picked from subsurface cross sections, the average slope/ 100-ka cycle is calculated for each shelf. The resulting slopes may represent topography and/or subsidence gradients. Results show that the outer Northwest Shelf has a much lower gradient (0.5 m/km or 2.6 ft/mile) than the outer Central Basin Platform (1.7 m/km or 9 ft/mile).

B. Graphical representations of the topographic/subsidence profiles of the Northwest Shelf and Central Basin Platform may be combined with stratigraphic data in order to estimate the average magnitude of both 100- and 400-ka sea-level fluctuations. Profiles are drawn as curves with the maximum slopes occurring along the outer 1 mile (1.6 km) of shelf so as to accurately represent the thickening seen in cross sections. The profiles show that a 6-ft (2 m) sea-level rise would result in a 1.5-mile (2.4 km) transgression and deposition of a well-developed, 100-ka carbonate wedge on the Northwest Shelf; whereas, only a minor transgression (less than 0.25 miles; 0.4 km) and poorly developed 100-ka carbonate wedge would be deposited on the Central Basin Platform. Similarly an approximately 37-ft (11 m) sea-level rise would provide accommodation space for only a 14-ft (4.3 m), 400-ka carbonate package on the Central Basin Platform; whereas, the same sea-level rise would result in extensive flooding of the Northwest Shelf and a much thicker, 32-ft (9.7 m), 400-ka carbonate interval if carbonate sedimentation kept up with sea-level rise.

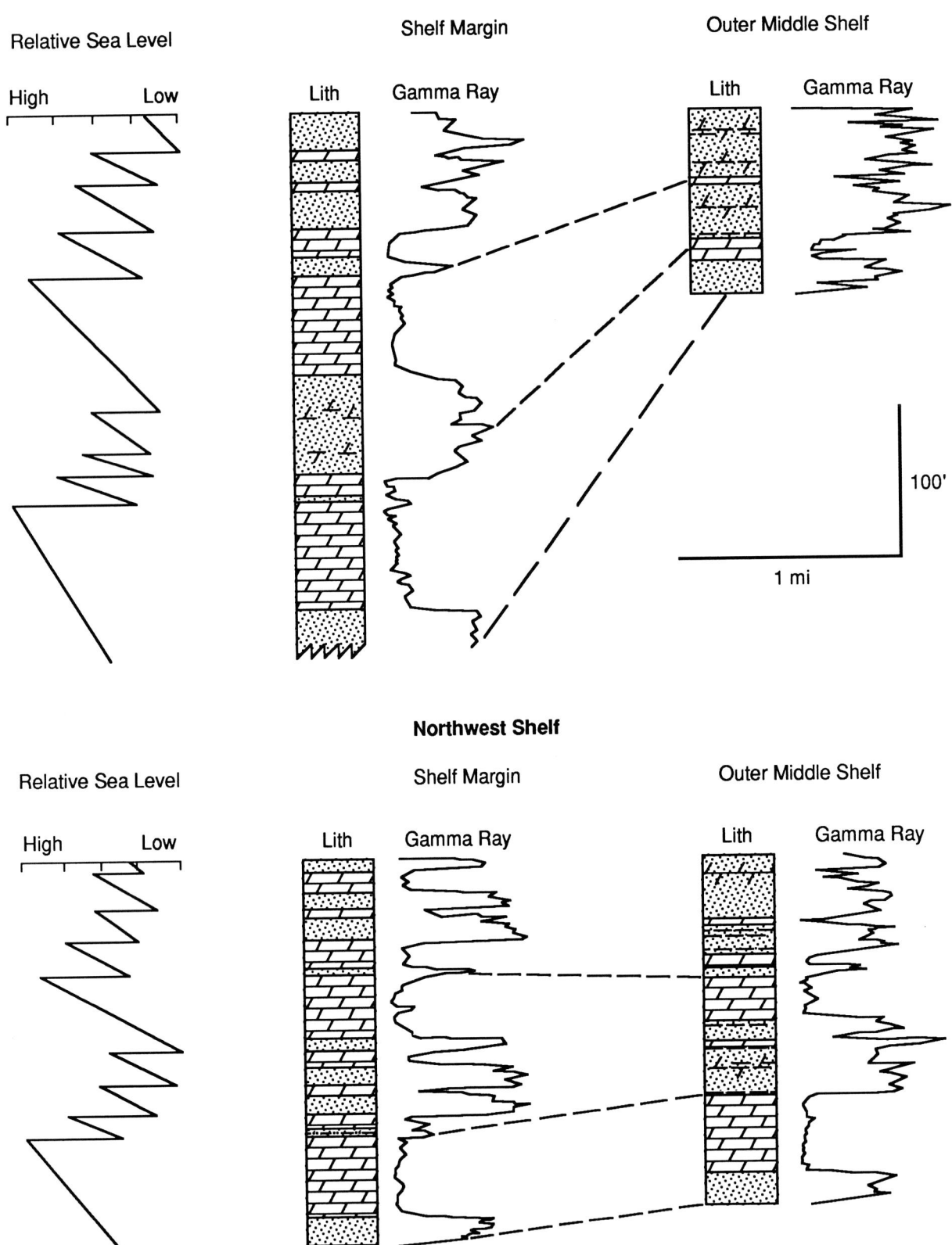

Figure 61

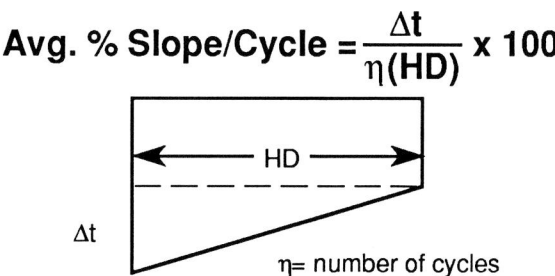

A.

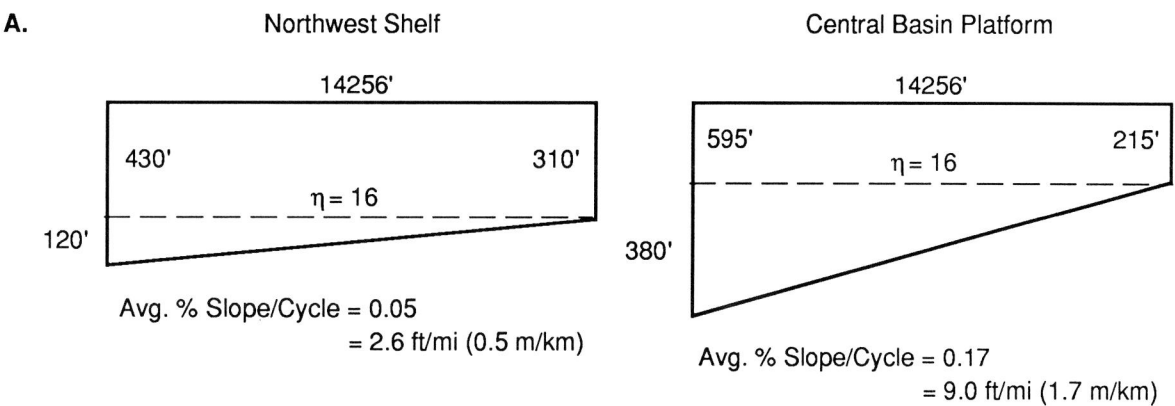

B.

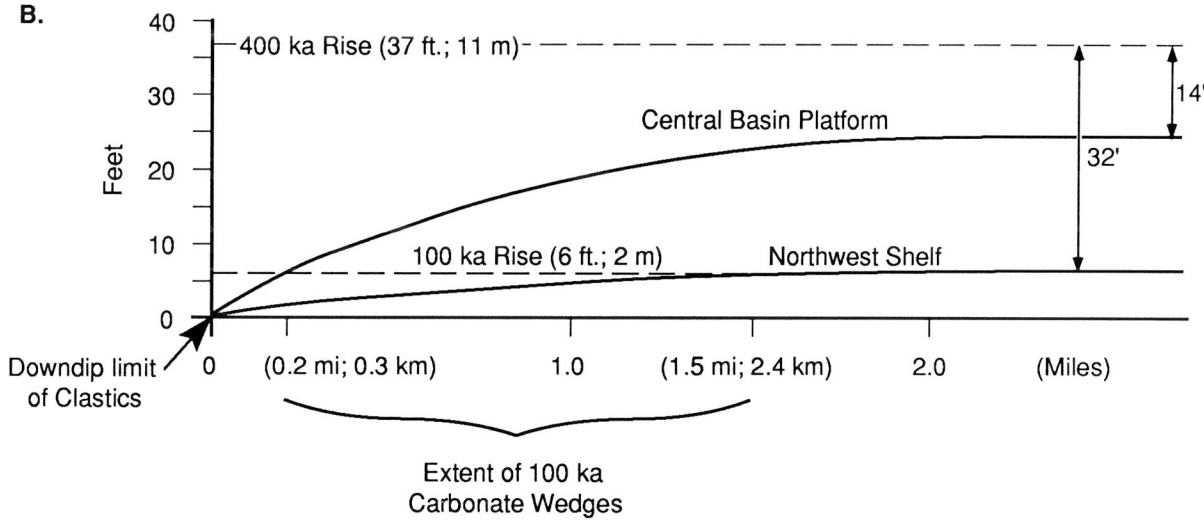

Figure 62

In the middle-shelf region of the Central Basin Platform, the siliciclastic portion of 400-ka cycles consist of 10 to 50 ft-thick (3 to 15 m) intervals of alternating Light Brown Arkosic Sandstone and Dark Gray Argillaceous Siltstone. The alternation between clean sandstones and argillaceous siltstones is interpreted as being the product of 100-ka sea-level fluctuations. The highstand carbonate portions of both 100- and 400-ka cycles are locally missing due to nondeposition (or erosion) and stacked siliciclastic intervals.

An average slope/cycle calculation (Fig. 62A) divides the change in stratigraphic thickness over some distance (a slope) among the total number of (small-scale) cycles present in the stratigraphic interval (Read, 1989). The calculated slope quantifies change in accommodation space across the shelf related to either a topographic or subsidence gradient and approximates the slope that would have to be transgressed by a sea-level rise. Results suggest that the outer 2.7 miles (4.3 km) of the Northwest Shelf had a slope of approximately 2.6 ft/mile (0.5 m/km); whereas, the Central Basin Platform had a slope of approximately 9.0 ft/mile (1.7 m/km) at a comparable location. Outcrop studies and the subsurface cross section of Figure 4 suggest that much of the formation thickening takes place near the shelf edge, so the average values given above are too low for the extreme shelf edge and too high for the updip portion of the outer shelf.

Work presented in this paper strongly suggests that sea level must have been fluctuating, albeit with only low-amplitude variations, to account for the stratigraphy of the Yates Formation. Climate effects, e.g., variations in wind regime or arid-humid alternations by themselves would not have produced the accommodation space on the shelf and the shoaling depositional cycles that are present in Yates cores. Crude estimates of the magnitude of sea-level fluctuations are made by combining the average slope/cycle data with information on the spatial distribution of 100- and 400-ka carbonates. The calculated average slope/cycle for each shelf is graphically represented in Figure 62B. A curve with increasing steepness toward the shelf edge is used to represent the outer-shelf gradient, since (1) the style of thickening seen in cross sections suggests that this is a more accurate representation of the shelf profile than a straight line, and (2) similar thickening has been documented for the Tansill Formation (Tyrrell, 1969) and Seven Rivers Formation (Hurley, 1989) on outcrop.

The 400-ka sea-level events are controlled by the modulation of 100-ka cycles. The magnitude of the 400-ka events can be crudely estimated by the magnitude of the maximum 100-ka events. Using an average thickness of 20 ft (6 m) for the thick 100-ka clastic-carbonate couplets on the Central Basin Platform and correcting for regional subsidence (driving and loading), a sea-level rise of 37 ft (11 m) is required to provide the accommodation space for the thickest 100-ka cycles on the Central Basin Platform. The same sea-level rise on the Northwest Shelf provides more than twice the accommodation space and allows for the deposition of laterally continuous, 20 to 40 ft (6 to 12 m) thick, "400-ka" carbonate intervals.

The magnitude of 100-ka sea-level fluctuations is estimated to be approximately 6 ft (2 m), based on the average 1.5-mile updip extent of carbonate wedges. Some wedges extend further updip on the Northwest Shelf but probably represent only slightly higher (a few feet) sea-level rises since the shelf gradient is so low. Calculated slope data suggest that the same sea-level rises would onlap less than 1/4 mile (0.4 km) on the Central Basin Platform. This is in agreement with log data from the Central Basin Platform that suggest the carbonate wedges are poorly developed and typically extend updip only 1/4 mile (0.4 km) from where siliciclastic beds pinch out near the shelf edge.

These estimates of sea-level fluctuations are crude and are limited by two factors that should be noted. First is the assumption that the downdip limit of sand deposition marks sea-level lowstands and can be used to tie the two shelves (slopes). This assumption should generally hold true for a depositional model in which siliciclastics are bypassed during lowstands and trapped during the subsequent highstand. The second limitation is not knowing how much of the accommodation space is provided by subsidence. This issue is complicated considering the interaction of driving subsidence, loading subsidence, and lag times. A simplified correction for subsidence is based on an accommodation rate of 4 cm/ka for the flat (updip) portion of the outer shelf on the Central Basin Platform. This was calculated by dividing thickness (215 ft or 65.5 m) by time (1600 ka) for this portion of the shelf (Fig. 62A). It is assumed that half of this accommodation (2 cm/ka) was provided by subsidence and therefore 6 ft (2 m) of subsidence-related accommodation needs to be subtracted from the average thick 100-ka cycle in order to estimate the 400-ka sea-level rise. Although these estimates of sea-level fluctuations are crude, they clearly show that the sea-level signal during the Upper Guadalupian was one of high-frequency, low-magnitude events.

Climate-Controlled Siliciclastic Sediment Supply

Climate-controlled variations in siliciclastic sediment supply may have also played a major role in the packaging of Yates stratigraphy. The rhythmic, abrupt alteration between siliciclastic and carbonate sediments at the 400-ka scale and formation scale (clastic-rich Yates, Queen, and Grayburg formations vs carbonate-rich Tansill, Seven Rivers, and San Andres formations) suggests that the siliciclastic sediment supply was repeatedly turned on and off. The cyclic nature of the siliciclastic sedimentation and the type of clastic sediment are both suggestive of a climatic rather than tectonic control. Although the arkosic nature of the sediments suggests a recently uplifted source terrain, the fine-grained texture suggests that the sediments were deposited far from any tectonic highland.

Furthermore, the depositional environment, paleogeography, and paleolatitude of the Yates inner shelf suggest a depositional system in which sediment transport was very sensitive to climatic change. The Yates inner shelf is interpreted as a carbonate/evaporite sabkha (with small lagoons) that was periodically overrun by a distal prograding sheet (coastal plain) of siliciclastics. The siliciclastic sediments exhibit many features (red beds, deflation surfaces, bimodal sorting, and sheet-sandstones) that suggest an arid desert environment and a component of eolian transport.

The ultimate source of Yates siliciclastics is unknown: approximately 100 miles north and northwest of the Delaware Basin, the Yates is truncated by an unconformity at the base of the Triassic, and, to the northeast, the Yates thins dramatically and cannot be reliably traced across the Amarillo uplift of Oklahoma (Tait and others, 1962). Paleogeographic reconstructions and paleocurrent analysis of late Permian eolian deposits in the Four Corners region suggest a general north-south to northwest-southeast dominant wind direction (Loope, 1981). With these wind directions in mind, the Yates shelf was positioned downwind from several late Permian uplifts and sand sea deposits that may have been a siliciclastic source. Granitic highlands that may have provided arkosic sediment during the late Permian include the Pedernal, Roosevelt, Amarillo, Sacramento, Uncompahgre, Calfax, Sierra Grande, and Bravo Highs (Hills, 1972; Newsom, 1989).

Milankovitch climate theory suggests that wind direction and wind strength would have changed through time as solar insolation varied with the earth's orbital parameters. During the Quaternary low latitude regions experienced an increase in aridity and wind strength that coincided with glacial maxima (Loope, 1981). The larger temperature gradient between the equator and the poles increased wind strength; whereas, decreased evaporation over cooled oceans would have reduced precipitation (Bowler, 1976; Fairbridge, 1976; Sarnthein, 1978; Lancaster, 1981).

Loope (1981, 1985) used a model of increased aridity and wind speed during glacial (low sea level) times to explain regional stacked deflation surfaces in the Wolfcampian to early Leonardian Cedar Mesa Sandstone of southeastern Utah. He suggests that extensive (traced for tens of kilometers) flat bedding surfaces with abundant burrows and rhizoliths represent periods when the eolian sand seas were deflated to the level of ground water table in response to increased wind strength and decreased sediment supply that coincided with glacial maxima and sea-level lowstands. During interglacial transgressions the deflation surfaces were locally transgressed by carbonate-rich seas or partially covered by shallow ephemeral lakes. Loope's model can be used to explain both a pulsed source for Yates siliciclastics and to describe attributes of siliciclastic deposition and transport on the Yates Shelf.

CONCLUSIONS

Work presented here describes the deposits and examines controls over deposition of a mixed siliciclastic-carbonate depositional system like the Yates. The detailed core studies show variations in depositional facies within any one of the siliciclastic beds between the inner, middle, and outer portions of the shelf, and also better evaluate the seaward shift of the facies tracts within the Yates. Considerable facies variation is indeed present, and this variation, in part, can go unnoticed during routine log correlations. Siliciclastics in the middle shelf consist of alternating Light Brown Arkosic Sandstones, Dark Gray Argillaceous Siltstones, and Red Argillaceous Siltstone/Sandstones. Carbonate lithologies consist predominantly of algal and peloidal dolomudstones. Vertical stacking of the various facies is interpreted to be the result of cyclic sea level and is thus a shelf-wide phenomenon.

Cyclostratigraphic analyses (i.e., the evaluation of the changes in the packaging of depositional cycles through time and at different positions on the shelf) suggest that the stacking of depositional sequences occurred during three orders of eustatic sea-level cycles. Two of the sea-level fluctuations were orbitally forced events with 100- and 400-ka durations (Milankovitch eccentricity cycles). A third-order sea-level cycle with a duration of 1.5 to 2 Ma may be related to either a low-frequency Milankovitch signal or tectonics. Fischer plots and stratigraphic analysis (using average slope/ cycle data calculated from regional cross sections) suggest that all the sea-level fluctuations had relatively low amplitudes ranging from about 2 m for the 100-ka cycles, 8 to 12 m for the 400-ka cycles, and 6 to 8 m for the 1.5- to 2-Ma cycle.

Regional facies variation occurred as the complex sea-level signal operated on portions of the shelf with different topographic and/or subsidence profiles. The steep profile of the Central Basin Platform enhanced the reworking of the sandstones and resulted in a strong vertical component to the stacking. In contrast, the lower gradient of the Northwest Shelf may have resulted in generally more argillaceous sandstones, a more updip location to the sand trend, and a greater horizontal component to stacking. The third-order sea-level fluctuation controlled the lateral position of successive sandstone depocenters and the general onlap/offlap (transgression/regression) configuration within the Yates. The thickness and vertical continuity of individual sandstones (i.e., siliciclastic portions of 400-ka depositional cycles) can be related to the position of the 400-ka sequences on the third-order (1.5 to 2 Ma) curve. Sequences deposited during the lowstand portions of the third-order cycle contain thick, stacked siliciclastic intervals; whereas, those deposited during the highstand portion are thinner and contain interbedded carbonates.

Results from this study suggest cyclostratigraphic analysis can be combined with the routine study of facies in order to better understand mixed siliciclastic and carbonate systems such as those of the Yates Formation. An interpretation of the frequency and magnitude of sea-level fluctuations and knowledge of the variability of shelf profiles enhances lithologic correlations and increases the predictability of facies correlations.

REFERENCES

BERGER, A., 1978, Long-term variations of caloric insolation resulting from the Earth's orbital elements: Quaternary Research, v. 9, p. 139-167.

BOND, G. C., KUMINZ, M. A., STECKLER, M. S., AND GROTZINGER, J. P., 1989, Role of thermal subsidence, flexure, and eustasy in the evolution of early Paleozoic passive-margin carbonate platforms, in Crevello, P. D., Wilson, J. L., Sarg, J. F., and Read, J. F., eds., Controls on Carbonate Platform and Basin Development: Society of Economic Paleontologists and Mineralogists Special Publication 44, p. 39-61.

BORER, J. M., AND HARRIS, P. M., 1989, Depositional facies and cycles in Yates Formation outcrops, Guadalupe Mountains, New Mexico, in Harris, P. M. and Grover, G. A., eds., Subsurface and Outcrop Examination of the Capitan Shelf Margin, Northern Delaware Basin: Society of Economic Paleontologists and Mineralogists Core Workshop No. 13, p. 305-317.

BOWLER, J. M., 1976, Aridity in Australia: age, origins and expression of aeolian landforms and sediments: Earth Science Reviews, v. 12, p. 279-310.

CANDELARIA, M. P., 1982, Sedimentology and depositional environment of upper Yates Formation siliciclastics (Permian, Guadalupian), Guadalupe Mountains, southeast New Mexico: Unpublished Masters thesis, University of Wisconsin, Madison, 267 p.

_____, 1983, Permian upper Yates Formation carbonate/siliciclastic depositional patterns, Northwest Shelf, Guadalupe Mountains, New Mexico: (Abstract) American Association of Petroleum Geologists Bulletin, v. 67(3), p. 435.

_____, 1988, Paleohydraulic interpretation of upper Yates Formation sheet sandstones, Guadalupe Mountains, New Mexico: (Abstract) American Association of Petroleum Geologists Bulletin, v. 72(2), p. 167.

_____, 1989, Shallow marine sheet sandstones, upper Yates Formation, Northwest Shelf, Delaware Basin, New Mexico, in Harris, P. M. and Grover, G. A., eds., Subsurface and Outcrop Examination of the Capitan Shelf Margin, Northern Delaware Basin: Society of Economic Paleontologists and Mineralogists Core Workshop No. 13, p. 319-324.

CRAWFORD, G. A., AND DUNHAM, J. B., 1983, Sedimentation in the Permian Yates Formation, Dollarhide Field, Central Basin Platform, Andrews County, West Texas: Permian Basin Cores - A Workshop: Permian Basin Section Society of Economic Paleontologists and Mineralogists, Core Workshop No. 2, p. 225-271.

DUNHAM, R. J., 1972, Capitan reef, New Mexico and Texas: facts and questions to aid interpretation and group discussion: Society of Economic Paleontologists and Mineralogists, Permian Basin Section, Publication 72-14, 270 p.

FAIRBRIDGE, R. W., 1976, Effects of Holocene climatic change on some tropical geomorphic processes: Quaternary Research, v. 6, p. 529-556.

FISCHER, A. G., 1964, The Lofer cyclothems of the Alpine Triassic: Kansas Geological Survey Bulletin, v. 169, p. 107-149.

_____, AND SARNTHEIN, M., 1988, Airborne silts and dune-derived sands in the Permian of the Delaware Basin: Journal of Sedimentary Petrology, v. 58, p. 637-643.

FRANCO, L. A., 1973, Deposition and diagenesis of the Yates Formation, Guadalupe Mountains and Central Basin Platform: Unpublished Masters thesis, Texas Tech University, Lubbock, Texas, 68 p.

GARBER, R. A., GROVER, G. A., AND HARRIS, P. M., 1989, Geology of the Capitan shelf margin-subsurface data from the northern Delaware Basin, in Harris, P. M. and Grover, G. A., eds., Subsurface and Outcrop Examination of the Capitan Shelf Margin, Northern Delaware Basin: Society of Economic Paleontologists and Mineralogists Core Workshop No. 13, p. 3-269.

GARBER, R. A., HARRIS, P. M., AND BORER, J. M., 1990, Occurrence and significance of magnesite in Upper Permian (Guadalupian) Tansill and Yates formations, Delaware Basin, New Mexico: American Association of Petroleum Geologists Bulletin, v. 74, p. 119-134.

GESTER, G. C., AND HAWLEY, H. J., 1929, Yates Field, Pecos County, Texas: structure of typical American oil fields, a symposium: American Association of Petroleum Geologists Bulletin, v. 2, p. 480-499.

HARRIS, P. M., AND GROVER, G. A., eds., 1989, Subsurface and outcrop examination of the Capitan Shelf Margin, Northern Delaware Basin: Society of Economic Paleontologists and Mineralogists Core Workshop No. 13, 481 p.

HILEMAN, M. E., AND MAZZULLO, S. J., eds., 1977, Upper Guadalupian Facies, Permian Reef Complex, Guadalupe Mountains, New Mexico and West Texas: 1977 Field Conference Guidebook, v. 1, Permian Basin Section Society of Economic Paleontologists and Mineralogists, Publication 77-16, 150 p.

HILL, C. A., 1987, Geology of Carlsbad Cavern and other caves in the Guadalupe Mountains, New Mexico and Texas: New Mexico Bureau of Mines and Mineral Resources Bulletin 117, 150 p.

HILLS, J. M., 1972, Late Paleozoic sedimentation in West Texas Permian Basin: American Association of Petroleum Geologists Bulletin, v. 56, p. 2302-2322.

HULL, J. P. D., JR., 1957, Petrogenesis of Permian Delaware Mountain sandstone, Texas and New Mexico: American Association of Petroleum Geologists Bulletin, v. 41, p. 278-307.

HURLEY, N. F., 1978, Facies mosaic of the lower Seven Rivers Formation (Permian, Guadalupian), Guadalupe Mountains, New Mexico: Unpublished Masters thesis, University of Wisconsin, Madison, 198 p.

_____, 1979, Seaward primary dip of fall-in beds, lower Seven Rivers Formation, Permian, Guadalupe Mountains, New Mexico (abstract): American Association of Petroleum Geologists Bulletin, v. 63, p. 471.

_____, 1989, Facies mosaic of the lower Seven Rivers Formation, McKittrick Canyon, New Mexico, in Harris, P. M. and Grover, G. A., eds., Subsurface and Outcrop Examination of the Capitan Shelf Margin, Northern Delaware Basin: Society of Economic Paleontologists and Mineralogists Core Workshop No. 13, p. 325-346.

KENDALL, C. G. ST. C., 1969, An environmental reinterpretation of the Permian evaporite-carbonate shelf sediments of the Guadalupe Mountains: Geological Society of America Bulletin, v. 80, p. 2503-2526.

KING, P. B., 1948, Geology of the southern Guadalupe Mountains, Texas: United States Geological Survey Professional Paper 215, 183 p.

LANCASTER, N., 1981, Paleoenvironmental implications of fixed dune systems in southern Africa: Palaeogeography, Palaeoclimatology, Palaeoecology, v. 33, p. 327-346.

LANPHERE, S., 1972, Proposed surface reference section for Yates Formation, Eddy County, New Mexico: American Association of Petroleum Geologists Bulletin, v. 56, p. 1534-1540.

LOOPE, D. B., 1981, Deposition, deflation and diagenesis of Upper Paleozoic eolian sediments, Canyonlands National Park, Utah: Ph.D dissertation, University of Wyoming, Laramie, Wyoming.

_____, 1985, Episodic deposition and preservation of eolian sands: a late Paleozoic example from southeastern Utah: Geology, v. 13, p. 73-76.

MAZZULLO, S. J., MAZZULLO, J., AND HARRIS, P. M., 1985, Eolian origin of quartzose sheet sands in Permian shelf facies, Guadalupe Mountains: Permian carbonate/clastic sedimentology, Guadalupe Mountains: Analogs for shelf and basin reservoirs, Symposium preceding 1985 Spring Field Trip, Permian Basin Section Society of Economic Paleontologists and Mineralogists, Midland, Texas, program with abstract, p. 3.

MEAR, C. E., AND YARBROUGH, D. V., 1961, Yates Formation in southern Permian Basin of West Texas: American Association of Petroleum Geologists Bulletin, v. 45, p. 1545-1556.

MEISSNER, F. G., 1969, Cyclic sedimentation in Middle Permian strata of the Permian Basin, West Texas and New Mexico (abstract), *in* Elam, J. G. and Chuber, S., eds., Cyclic sedimentation in the Permian Basin: West Texas Geological Society Symposium, p. 135.

_____, 1972, Cyclic sedimentation in Middle Permian strata of the Permian Basin, *in* Elam, J. G. and Chuber, S., eds., Cyclic Sedimentation in the Permian Basin: West Texas Geological Society, Midland, Texas, p. 203-232.

MILANKOVITCH, M., 1941, Kanon der Erdbestrahlung und seine Anwendung auf das Eiszeitenproblem: Akademie Royale Serbe, v. 133, 633 p.

MOTTS, W. S., 1962, Geology of the West Carlsbad Quadrangle, New Mexico: Geologic Quadrangle Maps of the United States, Published by the United States Geological Survey, Map GQ-167.

NEESE, D. G., 1979, Carbonate facies variation on Guadalupian shelf crest (upper Yates and lower Tansill formations), Guadalupe Mountains, New Mexico (abstract): American Association of Petroleum Geologists Bulletin, v. 63, p. 501.

NEESE, D. G., 1989, Peritidal facies of the Guadalupian shelf crest, Walnut Canyon, New Mexico, *in* Harris, P. M. and Grover, G. A., eds., Subsurface and Outcrop Examination of the Capitan Shelf Margin, Northern Delaware Basin: Society of Economic Paleontologists and Mineralogists Core Workshop No. 13, p. 295-303.

_____, AND SCHWARTZ, A. H., 1977, Facies mosaic of the upper Yates and lower Tansill Formations, Walnut and Rattlesnake canyons, Guadalupe Mountains, New Mexico, *in* Hileman, M. E. and Mazzullo, S. J., eds., Upper Guadalupian Facies, Permian Reef Complex, Guadalupe Mountains, New Mexico and West Texas: Field Conference Guidebook, v. 1, Permian Basin Section Society of Economic Paleontologists and Mineralogists, Publication 77-16, p. 437-450.

NEWELL, N. D., RIGBY, J. K., FISCHER, A. G., WHITEMAN, A. J., HICKOX, J. E., AND BRADLEY, J. S., 1953, The Permian reef complex of the Guadalupe Mountains region, Texas and New Mexico: Freeman and Company, San Francisco, 236 p.

NEWSOM, D. F., 1989, The lithology, environment of deposition, and reservoir properties of sandstones in the Upper Queen Formation (Guadalupian, Permian) at Concho Bluff Queen Field, Crane County, Texas: M.S. thesis, Texas A&M University, 187 p.

OSLEGER, D., AND READ, J. F., 1989, Comparative cycle stratigraphy of Cordilleran and Appalachian Late Cambrian passive margins: field and computer-modeling study: Abstract from American Association of Petroleum Geologists-Society of Economic Paleontologists and Mineralogists-Society of Exploration Geophysicists-Society of Professional Well Log Analysts Pacific Section 64th Annual Meeting Program, Palm Springs, p. 39-40.

PRAY, L. C., 1977, The all wet constant sea-level hypothesis of upper Guadalupian shelf and shelf edge strata, Guadalupe Mountains, New Mexico and Texas, *in* Hileman, M. E. and Mazzullo, S. J., eds., Upper Guadalupian Facies, Permian Reef Complex, Guadalupe Mountains, New Mexico and West Texas: Field Conference Guidebook, v. 1, Permian Basin Section Society of Economic Paleontologists and Mineralogists, Publication 77-16, p. 437-450.

_____, AND ESTEBAN, M., 1977, Upper Guadalupian facies, Permian reef complex, Guadalupe Mountains, New Mexico and west Texas (1977 Field Conference Guidebook), v. 2: Road logs and locality guides: Midland, Texas, Permian Basin Section, Society of Economic Paleontologists and Mineralogists Publication 77-16, 194 p.

READ, J. F., 1989, Controls on evolution of Cambrian-Ordovician passive margin, United States Appalachians, *in* Crevello, P. D., Wilson, J. L., Sarg, J. F., and Read, J. F., eds., Controls on carbonate platform and basin development, Society of Economic Paleontologists and Mineralogists Special Publication 44, p. 147-165.

READ, J. F., AND GOLDHAMMER, R. K., 1988, Use of Fischer plots to define third-order sea-level curves in Ordovician peritidal cyclic carbonates, Appalachians: Geology, v. 16, p. 895-899.

ROSS, C. A., AND ROSS, J. R. P., 1987, Late Paleozoic sea levels and depositional sequences, *in* Cushman Foundation for Foraminiferal Research, Special Publication 24, p. 137-149.

SARG, J. F., 1977, Sedimentology of the carbonate-evaporite facies transition of the Seven Rivers Formation (Guadalupian, Permian) in southeast New Mexico, *in* Hileman, M. E. and Mazzullo, S. J., eds., Upper Guadalupian Facies, Permian Reef Complex, Guadalupe Mountains, New Mexico and West Texas: Field Conference Guidebook, v. 1, Permian Basin Section Society of Economic Paleontologists and Mineralogists, Publication 77-16, p. 451-478.

_____, 1989, Middle-Late Permian depositional sequences, Permian Basin, West Texas-New Mexico, *in* Bally, A. W., ed., Atlas of Seismic Stratigraphy, v. 3: American Association of Petroleum Geologists Studies in Geology no. 27, p. 140-154.

SARNTHEIN, M., 1978, Sand deserts during glacial maximum and climatic optimum: Nature, v. 272, p. 43-46.

SCHOLLE, P. A., AND HALLEY, R. B., 1980, Upper Paleozoic depositional and diagenetic facies in a mature petroleum province (a field guide to the Guadalupe and Sacramento mountains of west Texas and New Mexico): United States Geological Survey Open-File Report 80-383, 191 p.

SCHWARTZ, A., 1981, Facies mosaic of the upper Yates and lower Tansill formations (Upper Permian), Rattlesnake Canyon, Guadalupe Mountains, New Mexico: Unpublished Masters thesis, University of Wisconsin, Madison, 155 p.

SILVER, B. A., AND TODD, R. G., 1969, Permian cyclic strata, northern Midland and Delaware basins, West Texas and southeastern New Mexico: American Association of Petroleum Geologists Bulletin, v. 53, p. 2223-2251.

SINGER, A., 1988, Illite in aridic soils, desert dusts, and desert loess: Sedimentary Geology, v. 59, p. 251-259.

SMITH, D. B., 1974, Sedimentology of upper Artesia (Guadalupian) cyclic shelf deposits of northern Guadalupe Mountains, New Mexico: American Association of Petroleum Geologists Bulletin, v. 58, p. 1699-1730.

TAIT, D. B., AHLEN, J. L., GORDON, A., SCOTT, G. L., MOTTS, W. S., AND SPITLER, M. E., 1962, Artesia Group of New Mexico and West Texas: American Association of Petroleum Geologists Bulletin, v. 46, p. 504-517.

THORKELSON, J. M., 1983, Depositional environments of select shelfal sands associated with the Permian Basin of West Texas and southeastern New Mexico: Unpublished Masters thesis, University of Southwestern Louisiana, 62 p.

TURNER, P., 1980, Continental red beds, *in* Developments in Sedimentology, v. 29: Elsevier, New York, 562 p.

TYRRELL, W. W., JR., 1969, Criteria useful in interpreting environments of unlike but time-equivalent carbonate units (Tansill-Capitan-Lamar), Capitan Reef Complex, West Texas and New Mexico, *in* Friedman, G. M., ed., Depositional Environments in Carbonate Rocks: Society of Economic Paleontologists and Mineralogists Special Publication 14, p. 80-97.

WALKER, T., 1979, Red color in dune sand, *in* McKee, D. E., ed., A study of global sand seas, United States Geological Survey Professional Paper 1052, p. 61-83.

WARD, R. F., KENDALL, C. G. ST. C., AND HARRIS, P. M., 1986, Upper Permian (Guadalupian) facies and their association with hydrocarbons - Permian Basin, West Texas and New Mexico: American Association of Petroleum Geologists Bulletin, v. 70, p. 239-262.

ZINZ, L. Z., 1971, Environmental framework and diagenesis of the Yates Formation, Apache Mountains, Culberson County, Texas: Unpublished Masters thesis, Texas Tech University, Lubbock, Texas, 132 p.

UPWARD-SHOALING SEQUENCE OF MIXED SILICICLASTICS AND CARBONATES FROM THE JURASSIC SMACKOVER FORMATION OF CENTRAL MISSISSIPPI

ROGER D. SHEW
Shell Development Company
P. O. Box 481, Houston, TX 77001

ABSTRACT

The Upper Jurassic Smackover Formation throughout most of the U.S. Gulf Coast is composed of a regressive sequence of carbonate ramp deposits. From central Mississippi to eastern Louisiana, however, carbonate deposition was influenced or interrupted by a large influx of siliciclastics derived from an ancestral Mississippi River source. Longshore currents and storm processes, combined with sea-level fluctuations, led to interfingering of the sandstones and carbonates in nearshore to basinal depositional environments.

In central Mississippi the sour gas trend (five fields at depths greater than 19,300 ft (5,880 m)) is characterized by mixed siliciclastic/carbonate deposits that occur in a general upward-shoaling sequence. The upward-shoaling sequence is evident both within fields as well as from downdip to more updip locations. In Thomasville, the most updip field, the upward-shoaling sequence is composed of (1) outer ramp bioturbated storm deposits interbedded with mudstones and wackestones, (2) lower shoreface and shelf ridge sand bodies that are gradational with pelletal packstones and minor ooid grainstones, and (3) shoreface and shoal sandstones with gradational contacts into high-energy ooid grainstone deposits with local skeletal buildups. Tidal-flat carbonate mudstones and Buckner Formation sabkha evaporites are the uppermost units of the upward-shoaling sequence and are the top seals to the hydrocarbon accumulation. Harrisville, the most downdip field, contains a basal sandstone that is interpreted to be a basinal turbidite deposit interbedded with microlaminated carbonates. Only the lower half of the upward-shoaling sequence is present in Harrisville, with the uppermost sandstones interbedded with middle to outer ramp skeletal and pelletal packstones and wackestones.

Thomasville and Harrisville are sour gas (H_2S concentrations of 35 to 41 percent), productive from low porosity and permeability sandstones that are interbedded primarily with nonreservoir carbonates. Even though a complex diagenetic history has strongly overprinted the original depositional fabric and reduced the reservoir quality, high geopressures, large drainage areas, and enough altered primary and secondary porosity remain for both high production rates and economic prospects in these hostile subsurface environments. Important diagenetic events include early compaction, cementation by both calcite and more importantly dolomite, quartz and feldspar overgrowths, hydrocarbon migration and calcite dissolution, bitumen formation, and thermochemical sulfate reduction to form the high H_2S concentrations. The original depositional setting and proximity of the carbonate interbeds to the sandstones have played major roles in defining the trap, reservoir quality, hydrocarbon accumulation, and final hydrocarbon mix in these reservoirs.

INTRODUCTION

The Upper Jurassic (Oxfordian) Smackover sediments in central Mississippi are anomalous to typical Gulf Coast Smackover ramp deposits. First, abundant siliciclastics occur in gradational and sharp contacts with the carbonate ramp deposits; second, the sandstones are the hydrocarbon-bearing intervals instead of the high-energy ooid grainstone and/or other carbonate facies. Cores from the sour gas area, Thomasville and Harrisville fields in particular, provide the opportunity to study these mixed siliciclastic/carbonate depositional environments in detail. The sediments are comprised of a general upward-shoaling sequence of basinal to nearshore sandstones and carbonates.

The fields are sour gas-productive (greater than 35 percent H_2S) from deep (greater than 19,000 ft (5,880 m)), geopressured (greater than 0.88 psi/ft (19.9 kPa/m)), high-temperature (greater than 350°F (177°C)) sandstone reservoirs. Field and well productivities, reservoir quality, and hydrocarbon presence and type are strongly influenced by the depositional setting, proximity to carbonate interbeds, and complex diagenetic history. Core descriptions, petrographic data, and petrophysical properties are presented to illustrate the facies and rock property variability that occur both within fields and from downdip to updip locations.

STUDY AREA

The informally designated sour gas trend (Fig. 1A), composed of five fields, is located in central Mississippi in Rankin and Simpson counties. Sour gas is produced from deep, low-quality Smackover sandstones interbedded with nonreservoir carbonates. Thomasville and Harrisville fields, respectively, are the most updip and downdip fields in the trend. Continuous core from the Crain No. 1-R, plus shorter selected cores from multiple wells in Thomasville, are used to illustrate outer ramp to nearshore and sabkha deposition. Limited core from Harrisville, in particular the Hilton No. 1, contains more basinal sandstone and carbonate deposits. Combining these two fields provides a relatively complete sequence of Smackover mixed siliciclastic/carbonate deposition in central Mississippi.

REGIONAL GEOLOGIC SETTING

Tectonic Setting

Harrisville and Thomasville fields are located within the Mississippi Interior Salt Basin (Fig. 1B). The basin was formed during the rifting of the Gulf of Mexico (Pindell and Dewey, 1982) and was separated from the true Gulf of Mexico by the South Mississippi and Wiggins uplifts. Repeated influx of oceanic waters into the shallow, arid basin and subsequent evaporation led to the accumulation of thick Louann Salt (Callovian) deposits (Fig. 2A). Early loading of this mobile substrate by the Upper Jurassic (Oxfordian) Norphlet eolian sandstones and Smackover sediments led to the formation of faulted and nonfaulted structural highs. These salt-generated highs are part of the salt-roller belt that parallels the Jurassic paleo-shoreline. The other dominant structural element in central Mississippi is the Jackson Dome complex (Fig. 1A). This

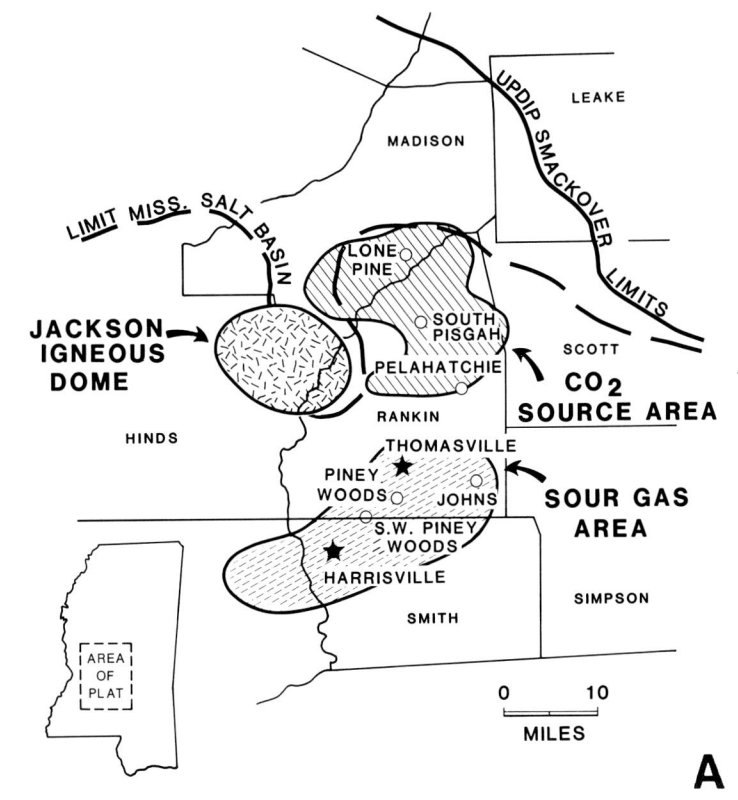

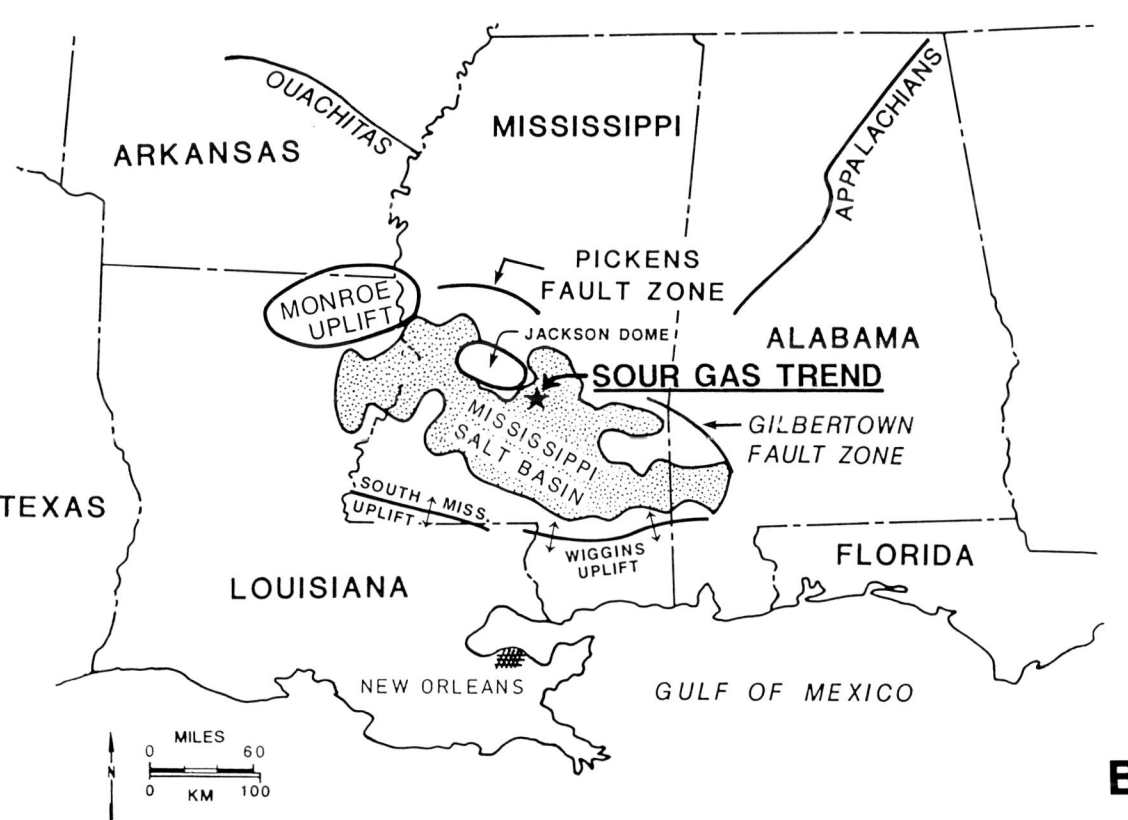

FIGURE 1 - Location map (A) of the sour gas area showing the major producing trends, fields, and tectonic elements. The sour gas trend is comprised of five fields. Figure 1B illustrates the regional tectonic setting of the Mississippi Interior Salt Basin and the sour gas area.

FGIRUE 2 - Generalized vertical sequence of the central Mississippi Jurassic section. Figure 2B provides details of the Smackover stratigraphic column in Thomasville Field. Also shown is the type log; the stippled pattern, cross-over of the FDC with the CNL, indicates the presence of porous sandstones. The FDC/CNL provides the most usable data for petrophysical evaluations (Modified from Shew and Garner, 1990).

late Cretaceous to early Paleocene igneous intrusion led to the thermal metamorphism of Smackover carbonates and generated large amounts of carbon dioxide that were trapped in previously formed structures (Studlick and others, 1990).

Stratigraphic and Depositional Setting

The generalized sour gas trend vertical sequence is shown in Figure 2A. With minor variation, predominantly the presence of siliciclastics in the Smackover, this sequence describes the Middle to Upper Jurassic stratigraphic sequence present throughout the U.S. Gulf Coast. The Norphlet overlies the Louann Salt and consists of eolian deposits that are locally prospective for hydrocarbons and carbon dioxide. The Smackover unconformably overlies the Norphlet and is characterized by carbonate ramp deposition (Ahr, 1973; Budd and Loucks, 1981; and Moore, 1984). A ramp profile is typified by belts of paralic sedimentation that are composed of higher-energy (ooid grainstone) deposits nearshore grading to middle ramp packstones and wackestones, and finally to below wave base or basinal mudstones. After an initial rapid sea-level rise and deposition of a thin basal Smackover unit, the remainder of the Smackover is composed of a regressive carbonate sequence formed during a slow relative sea-level rise. Therefore the vertical sequence mirrors all or part of the lateral depositional units and forms a general upward-shoaling sequence of basinal to nearshore deposits (Fig. 3). The Buckner Formation overlies and is in part laterally equivalent to the Smackover; it is the uppermost facies (sabkha) of the upward-shoaling sequence.

Variations of this upward-shoaling sequence occur near syndepositional salt highs. However, from central Mississippi to eastern Louisiana, the influx of abundant siliciclastics (Fig. 4) leads to a variable vertical sequence. The sandstones also exhibit a general upward-shoaling sequence. Although siliciclastics have been reported elsewhere in the Smackover (Chimene, 1976; Ahr and Palko, 1981; and Judice and Mazzullo, 1982), they predominate only in the above areas.

Determining the long-lived source of the abundant siliciclastic deposits is somewhat problematic because of sparse drilling at these great depths and the poor quality of seismic. Structural complications associated with salt tectonism are also important. Deltas (Dinkins and others, 1968) and ancestral river drainage systems (Mann and Thomas, 1968) have been suggested as the sediment sources. With modification, the theory of a nearby river system and other minor ones to supply sediments to the basin is believed to be the most attractive explanation. Burke and Dewey (1973) have shown that many large river systems form in the failed arms of triple junctions. And, as described by Ervin and McGinnis (1975), the ancestral Mississippi River is believed to have formed in a failed arm during the rifting of the Gulf of Mexico. This ancestral system could have delivered a large volume of siliciclastics to the interior basin, which could then be reworked by longshore currents and storm processes. It is notable that the siliciclastics have little or no detrital clay except for minor amounts in the more basinal areas. Extensive reworking and sediment bypass to the deeper basin are the likely processes of clay removal. Extreme mechanical weathering in the source area or drainage through a clay-poor source terrain are less likely possibilities.

FIGURE 3 - The composite stratigraphic sequence in the sour gas trend describes an upward-shoaling sequence both within fields and from downdip to updip locations. Interpreting the carbonates are important to an understanding of siliciclastic deposition. The four main types of carbonate deposits are shown.

A - Highly compacted ooid grainstone. The ooid grainstone interval is the uppermost unit (except for intertidal mudstones and evaporites) of the upward-shoaling sequence and is the prospective horizon throughout most of the Gulf Coast. The ooids are either highly compacted (sutured contacts -S) or cemented.

B - Mudstone to wackestone typical of the outer ramp. These often occur in sharp contacts with the interbedded sandstones.

B - Skeletal and pelletal (P) packstones and wackestones of the inner to outer ramp. These are often gradational with the sandstones. Minor silt is present in this sample.

D - Microlaminated carbonate (MLZ) represents low energy carbonate deposition. It is believed to be the source rock for the Smackover and Norphlet formations. An interesting relationship occurs in Harrisville where the lower sandstones are interbedded with the MLZ.

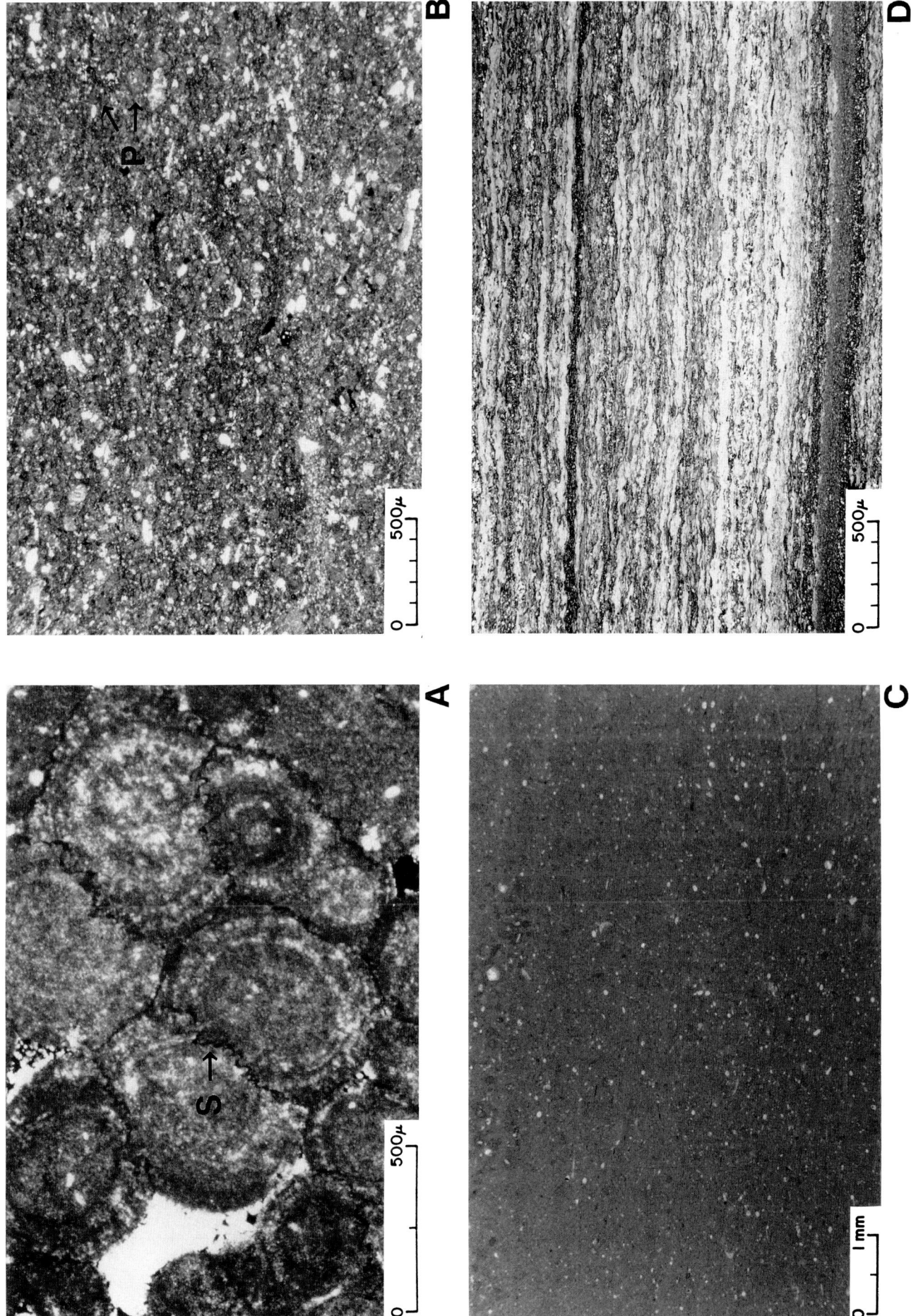

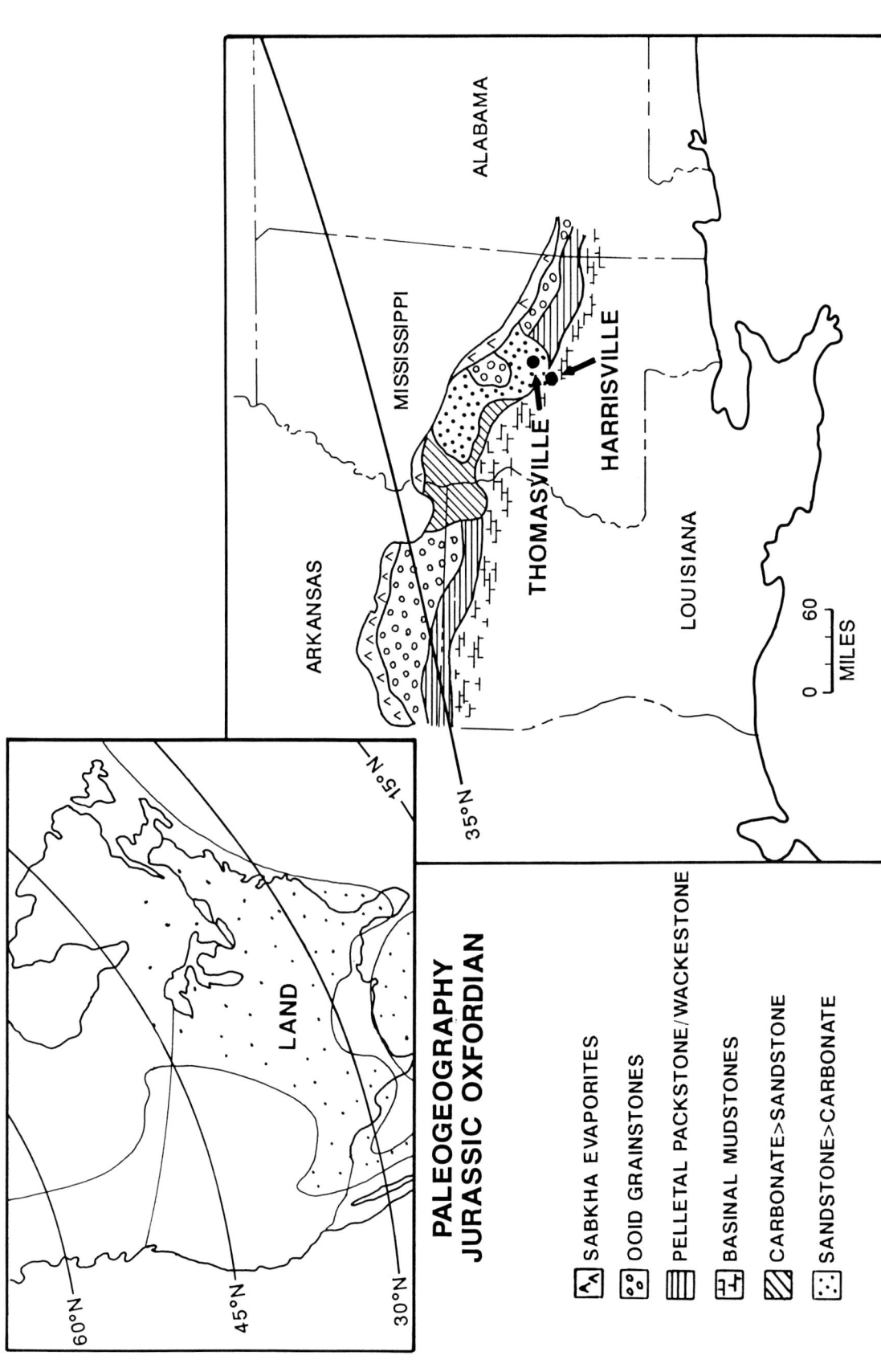

FIGURE 4 — Smackover facies distribution map. Typical paralic carbonate deposition is interrupted from central Mississippi to eastern Louisiana by the influx of abundant siliciclastics. Thomasville and Harrisville occur on the eastern edge of the siliciclastic wedge (modified from Bishop, 1968).

Sea-level fluctuations, though impossible to document because of the limited data mentioned previously, may be very important in explaining the frequency of mixed siliciclastic and carbonate deposition. Sea-level fluctuations, coupled with variable sediment input rates, may have led to cyclic interbedding, with the sandstone intervals representing the higher-order relative lowstands. Similar stacked deposits in mixed depositional settings have been described by Mack and James (1986), and Horne and others (1989).

SOUR GAS AREA

Structure and Field Development

The sour gas area is comprised of five fields: Thomasville (most updip), Johns, Piney Woods, S.W. Piney Woods, and Harrisville (most downdip). The top Smackover structure map, plus schematic cross section for each field except Johns, which is just east of Thomasville, are shown in Figure 5. Seismic reflection data were the basis for drilling these structures; simple and fault closure were recognized at the Smackover and Norphlet horizons. Although the Norphlet was found to be nonprospective, multiple sandstone reservoirs within the Smackover were found to be hydrocarbon-bearing. Recent improvements in the acquisition and processing of the seismic data have provided some opportunity for stratigraphic mapping of these porous sandstone packages interbedded with "hard" nonreservoir carbonates.

Exploration for and development of these fields have been long and problematic because of the hostile subsurface environment that includes great depths, high temperatures, hard geopressures, and corrosive gas mixtures (see Table 1). Thomasville Field (greater than 19,300 ft (5,880 m) in depth) was discovered in 1969 and was not fully developed until over a decade later. A replacement well, Crain No. 1-R, was drilled in 1985. Harrisville Field (greater than 22,500 ft (6,900 m) in depth) was discovered in 1980.

High flow rates and large reserves are possible even in these low porosity and permeability reservoirs. Hard geopressures (greater than 0.88 psi/ft (19.9 kPa/m) in Thomasville; greater than 0.98 psi/ft (20.5 kPa/m) in Harrisville) and thick, laterally extensive sandstones have yielded high flow rates and large reserves per well. The average per-well rate is 5 MMCFGPD, and some wells have maintained rates of 10 to 20 MMCFGPD for over 10 years. Development has been on 1280-acre (518 ha) spacing. The reservoir drive mechanism is depletion gas drive.

Depositional Model - Thomasville Field

Thomasville contains multiple sandstone layers (Fig. 6) interbedded with tight carbonates. Although the lower Smackover sequence was not cored in Thomasville, clear evidence of an upward-shoaling sequence is present, particularly in continuous core obtained in the Shell Crain No. 1-R (Fig. 2).

From the base to the top of the productive Smackover interval (core was not obtained in the lower Smackover in Thomasville), the sandstones and carbonates consist of:

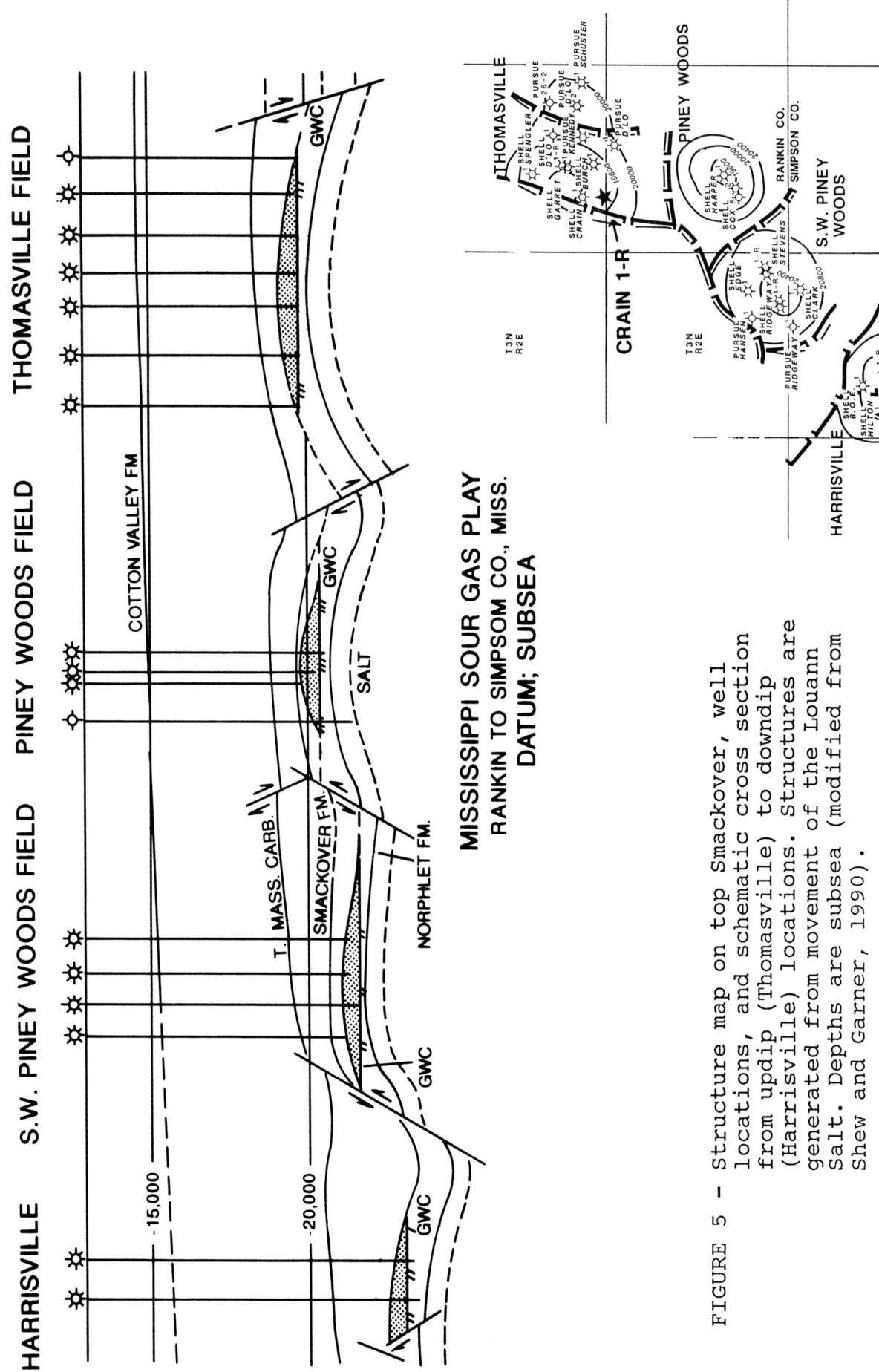

FIGURE 5 – Structure map on top Smackover, well locations, and schematic cross section from updip (Thomasville) to downdip (Harrisville) locations. Structures are generated from movement of the Louann Salt. Depths are subsea (modified from Shew and Garner, 1990).

TABLE 1 - RESERVOIR DATA

RESERVOIR PARAMETERS	THOMASVILLE	PINEY WOODS	SW PINEY WOODS	HARRISVILLE
DEPTH (FT.)	>19,3000	>20,000	>20,800	>22,900
BHT (°C)	182	190	196	207
PRESSURE (PSI/FT)	.88	.90	1.00	.98–1.01
GAS COMPOSITION %				
CH_4	56.1	51.0	70.1	56.8
H_2S	34.7	45.0	27.4	41.4
CO_2	9.2	4.0	2.5	1.8

TABLE 2 - SANDSTONE COMPOSITION AND DIAGENESIS

AVERAGE COMPOSITION (Bulk Volume Excluding Porosity) is:

FRAMEWORK GRAINS	%	CEMENTS	%
Quartz	61.5	Dolomite	15.4
K-feldspar	6.1	Calcite	2.7
Plagioclase	<1.5	Anhydrite	<1.0
Rock Fragments	<2.0	Qtz. Overgrowths	<2.0
Carb. Allochems	<2.0	K-spar Overgrowths	<2.0
Chert	<2.0	Bitumen	8.1
		Pyrite	<1.0

DIAGENETIC EVENTS	SHALLOW　　RELATIVE DEPTH　　DEEP
COMPACTION	
CALCITE CEMENT	
ANHYDRITE CEMENT	
DOLOMITE CEMENT AND REPLACEMENT	
K-SPAR DISSOLUTION AND OVERGROWTHS	
QUARTZ OVERGROWTHS	
CALCITE DISSOLUTION	
SPAR, CHERT, ALLOCHEMS DISSOLUTION EVENTS	
HYDROCARBON MIGRATION	
CALCITIZATION OF DOLOMITE	
PYRITE	
FRACTURE-FILL AND SADDLE DOLOMITE	
GENERATION OF H_2S, CO_2, AND SULFUR	
MINOR AUTHIGENIC CLAY	
ESTIMATED TEMPERATURE C	80　　100　　120　　140

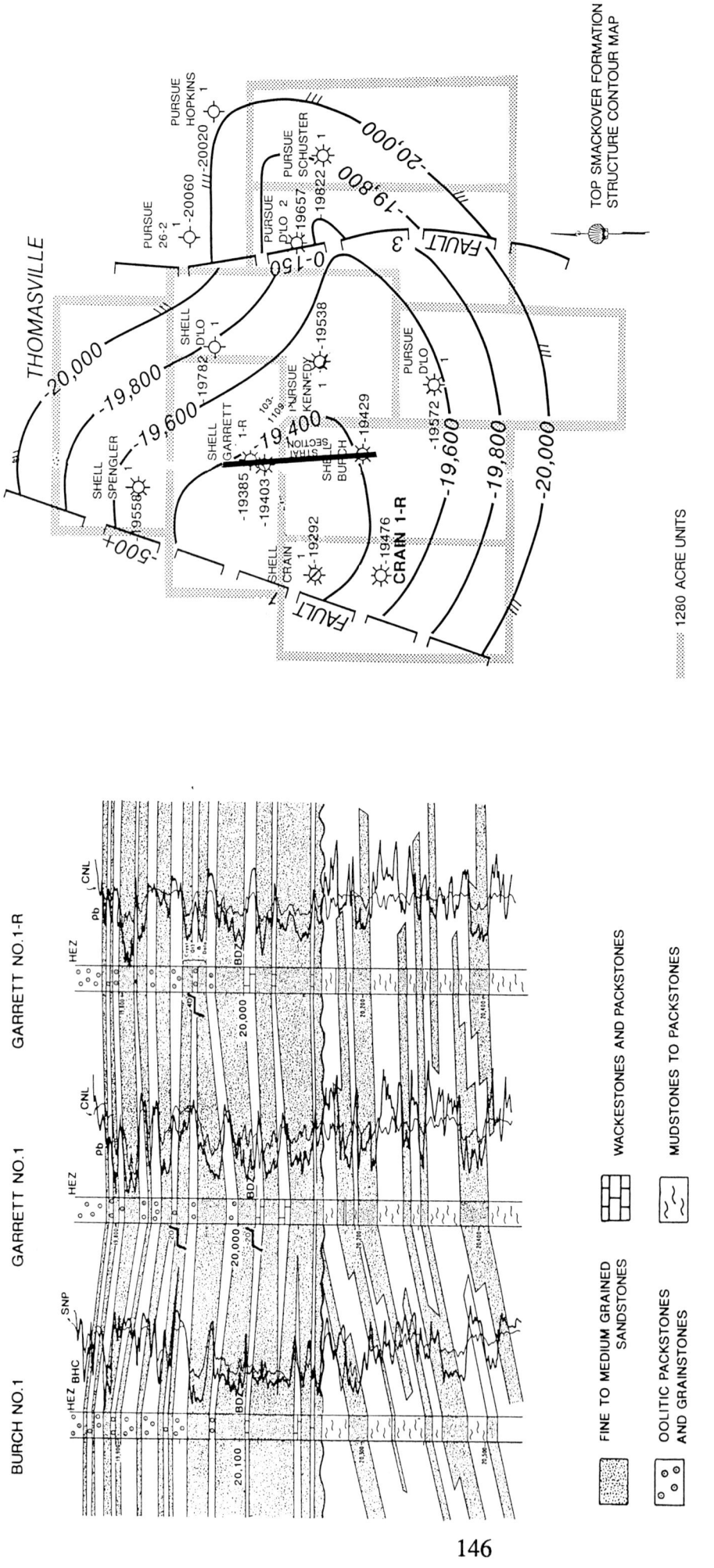

FIGURE 6 - Stratigraphic cross-section in central part of Thomasville Field illustrating the changes in thickness and continuity in various intervals of the Smackover sequence. Map shown is the top structure map with location of cross-section shown. Development has been on 1280-acre spacing (modified from Shew and Garner, 1990).

1. Outer ramp storm deposits that have punctuated facies contacts with the interbedded carbonate mudstones and wackestones. The sandstones are sharp-based and thin, with bioturbated tops (Fig. 7). Sandstones are moderately to poorly sorted, upper very fine to upper fine grain size, and have poor horizontal and poorer vertical permeabilities.

2. Middle to inner ramp sandstones composed of amalgamated, thick shoal/ridge sand bodies (similar to those described by Tillman and Martinsen, 1984) and bars (Fig. 8) that are interbedded and gradational with packstones and minor ooid grainstones. This middle zone contains the best reservoir quality and most of the reserves in Thomasville. In addition to using the interpretation of the carbonates to support middle to inner ramp deposition, the moderate- to high-energy sandstone deposits support this interpretation. The sandstones are characterized by a general absence of bioturbation; coarser grain size (upper fine to medium sand); undulatory bedding and numerous amalgamations; low-angle cross-bedding and hummocky cross-stratification; and parallel bedding.

   These sandstones are laterally extensive. However, another important reason for high reservoir quality is that sandstone quality consistently deteriorates proximal to carbonate interbeds (see Reservoir Quality section) because of increased cementation. The net/gross ratio is also much higher in this interval. Thicker sandstones of the middle interval have higher porosity and permeability than the lower or upper Smackover zones.

3. The uppermost prospective zone in Thomasville contains thin sandstones interbedded gradationally with ooid grainstones (Figs. 9 and 10). These are high-energy shoreface and shoal deposits that are gradational with the middle interval. Ooid grainstones are cross-bedded with thin interbeds of pelletal, skeletal, and oncolite grainstones. Skeletal buildups are small and patchy. The gradational contacts have alternating laminae of sands and ooids with low-angle cross-stratification and planar cross-stratification (Fig. 9). Reservoir quality is poorer in these thinner sandstones because of cementation (Fig. 10), and minor faulting has isolated some of the sand lenses.

4. The uppermost part of the upward-shoaling sequence is composed of intertidal carbonate mudstones with minor anhydrite laths. Capping the entire sequence are nodular, massive anhydrites of the Buckner Formation (Fig. 11).

Depositional Model - Lower Sandstone, Harrisville Field

The most striking change from north to south in the sour gas trend is the decrease in siliciclastics interbedded with the carbonates, further reflecting the more proximal source and longshore and storm origin of sandstones on the inner ramp. Limited core from Harrisville is important, however, as it contains the basal Smackover unit that is present throughout much of the Gulf Basin and which is most important as the hydrocarbon source. A basal sandstone occurs interbedded with the Microlaminated Zone (MLZ). Overlying and grading up from the basal Smackover unit are pelletal and skeletal wackestones and packstones. Elsewhere in the

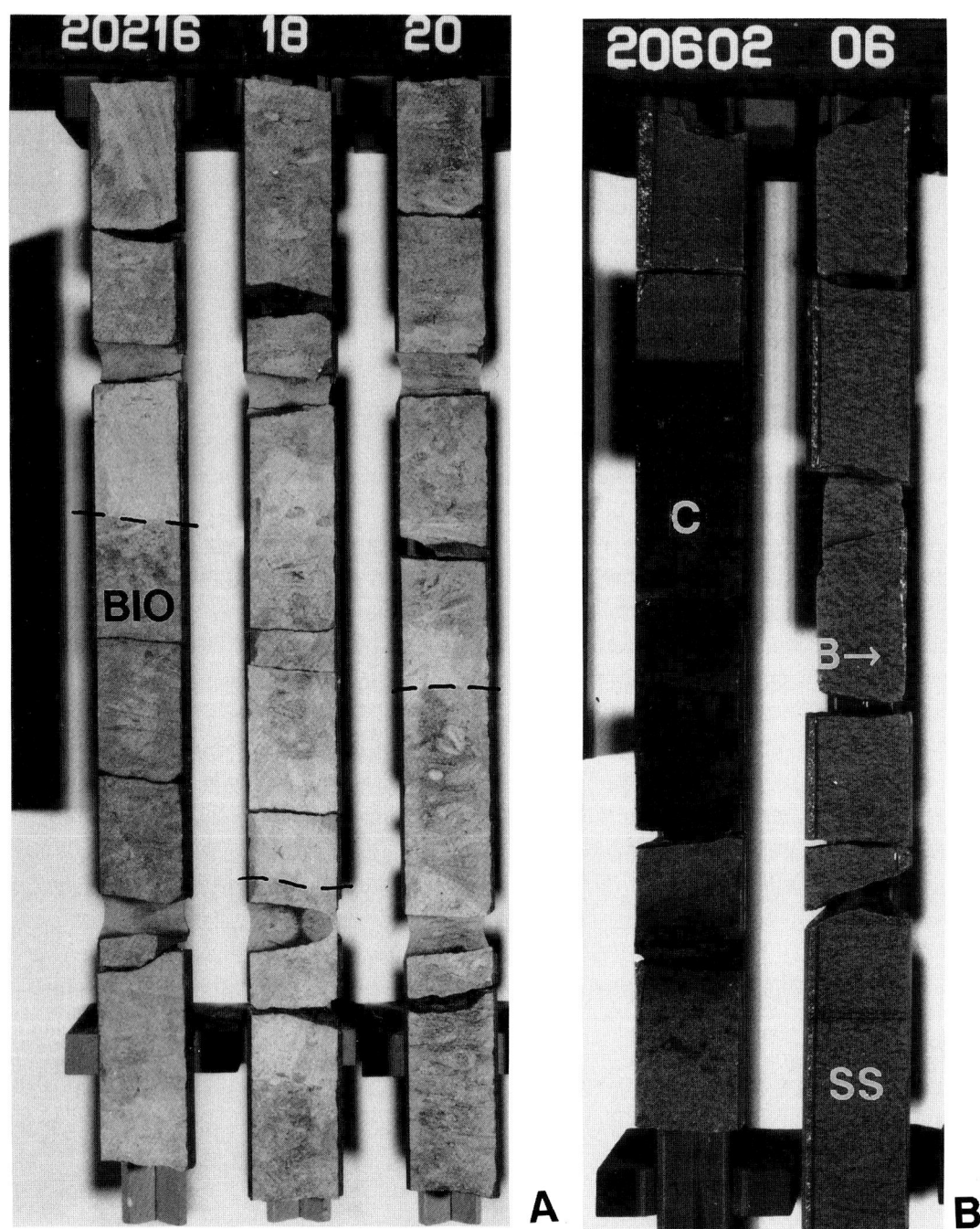

FIGURE 7 - Cores from the lower cored Smackover interval in Thomasville Field. Core B is located further off structure (Shell D'Lo No. 1) below the water level. Patchy anhydrite cement is present as is patches of bitumen (B) within the sandstone (SS). Tight carbonate (C) mudstones with sharp contacts are present. These sands are questionably large sand ridges and/or feeder channels to a more basinal setting. Core A (Shell Crain No. 1-R) has outer ramp, composite storm deposits with sharp bases (dashed line) and extensive bioturbation (BIO).

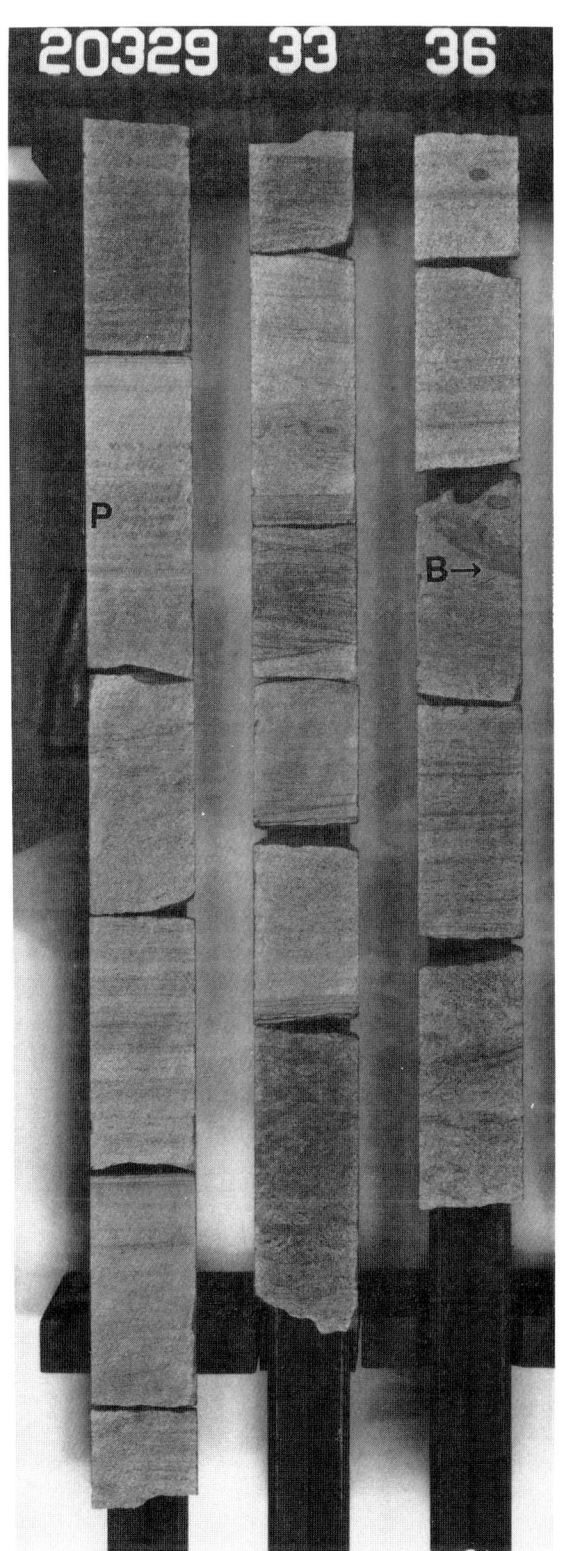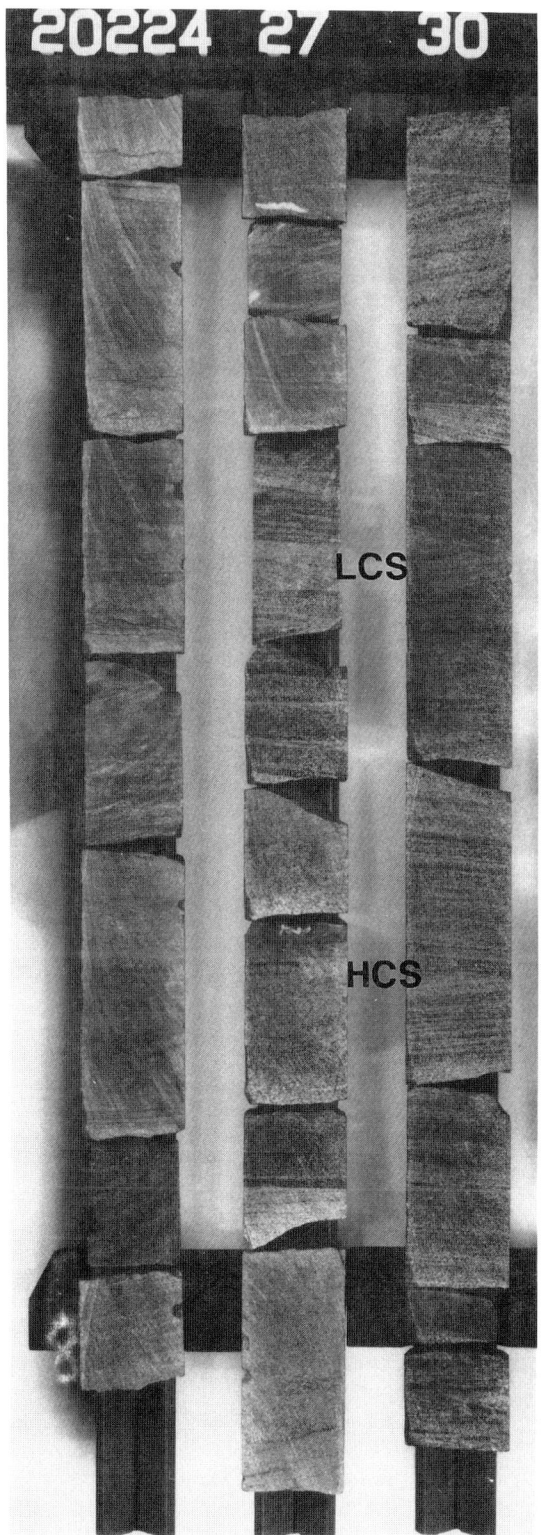

FIGURE 8 - Cores of the middle to inner ramp with planar (P), low-angle and trough (LCS), and hummocky cross-stratification (HCS) indicating a moderate to sometimes high energy depositional environment. Sandstones are composite, amalgamated beds with generally the best reservoir quality. Only minor bioturbation (B) occurs.

FIGURE 9 - The uppermost zone in Thomasville field is characterized by thin sandstones with gradational and interfingering contacts with moderate and high energy carbonate deposits. Alternating laminae of carbonate and siliciclastics are present in the transitional zones. An important point is the presence of better reservoir quality in the thicker sandstones (abundant bitumen, which darkens the sandstone, indicates more porous sands) such as at site A. Sandstones are often more heavily cemented adjacent to the carbonate interbeds and have lower porosities and permeabilities.

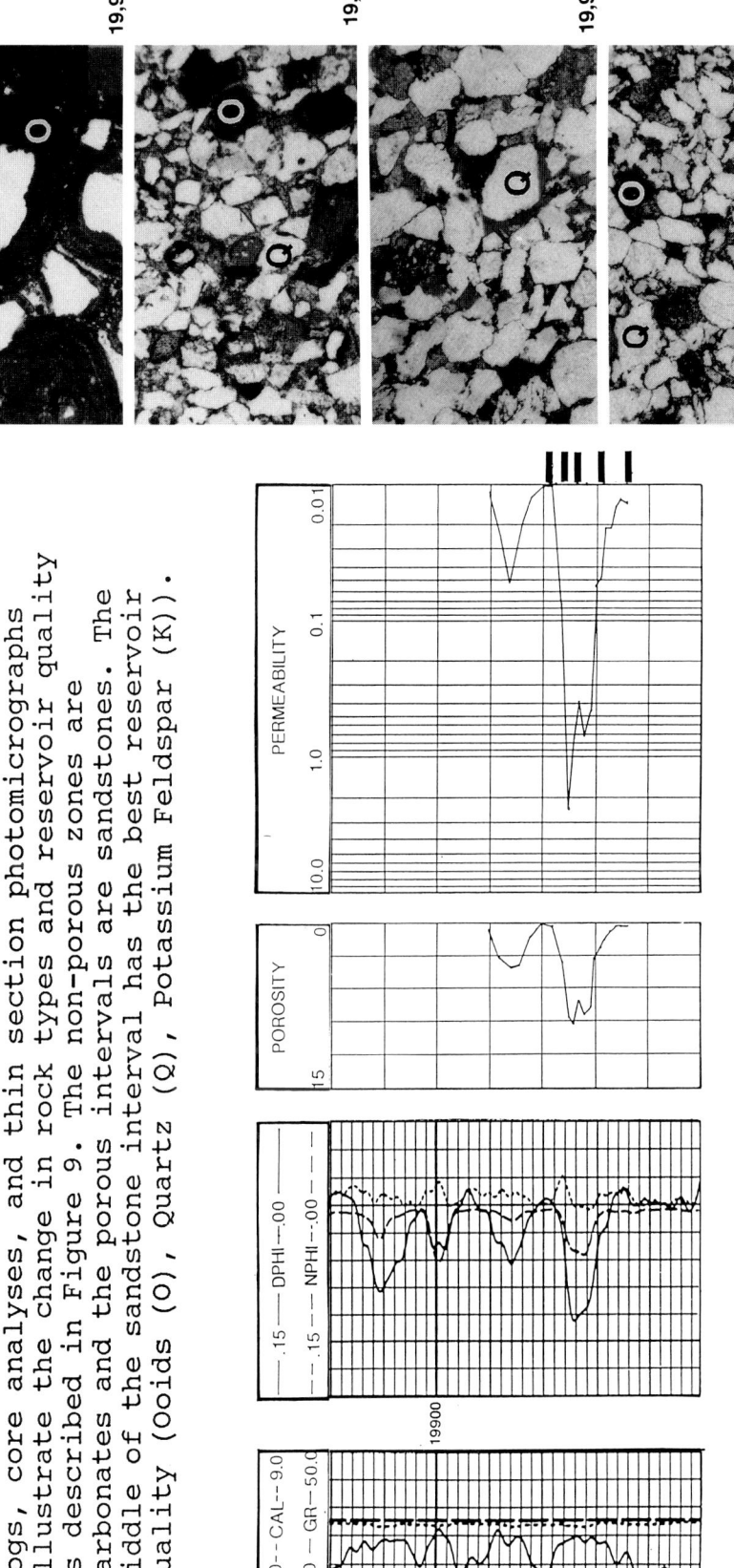

FIGURE 10

Logs, core analyses, and thin section photomicrographs illustrate the change in rock types and reservoir quality as described in Figure 9. The non-porous zones are carbonates and the porous intervals are sandstones. The middle of the sandstone interval has the best reservoir quality (Ooids (O), Quartz (Q), Potassium Feldspar (K)).

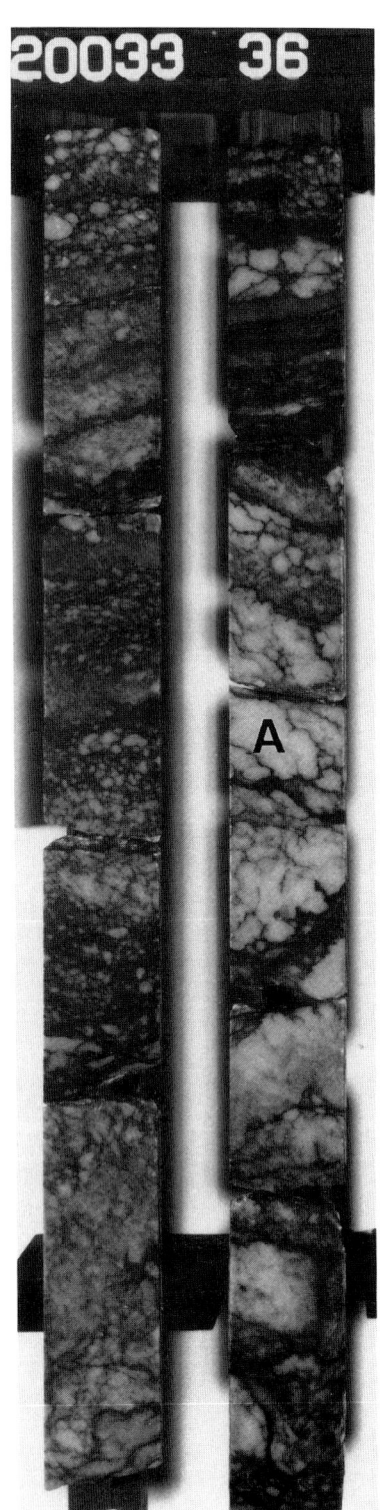
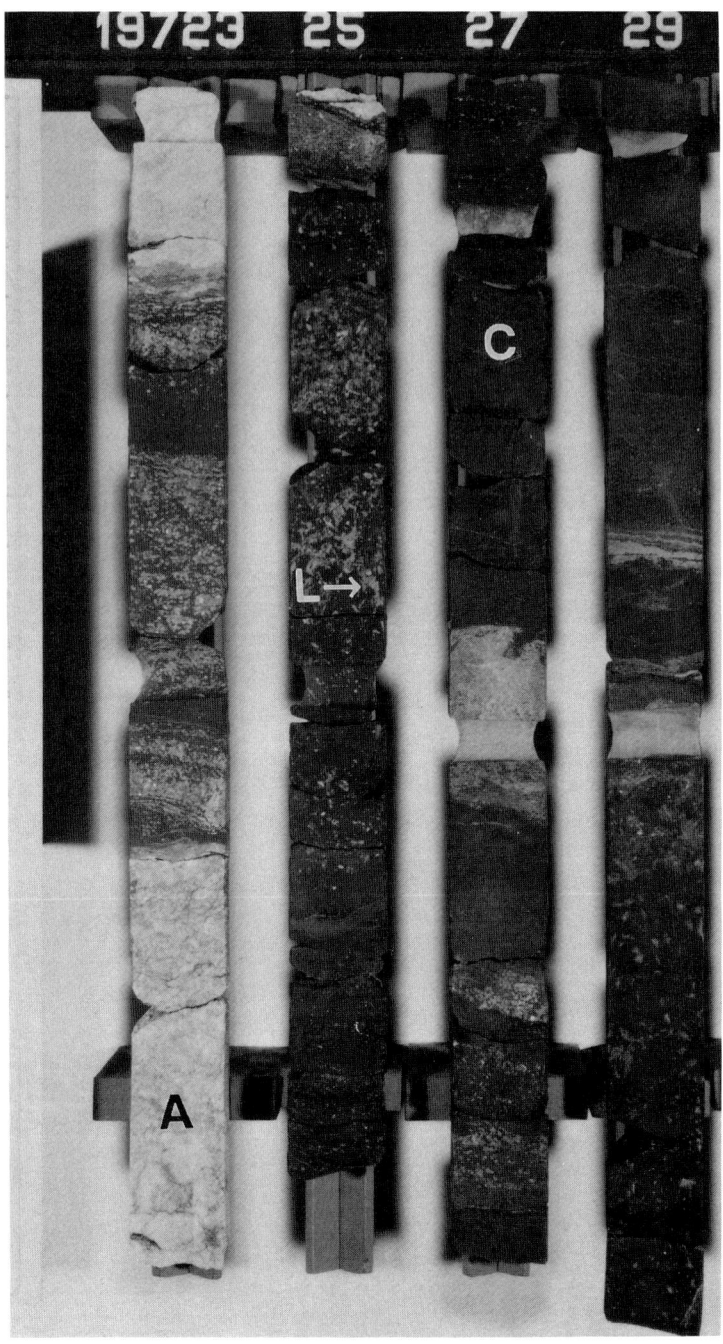

FIGURE 11 - The uppermost part of the upward-shoaling sequence is capped by intertidal carbonates (C) with occasional anhydrite laths (L), and massive to nodular anhydrite (A) of the sabkha (Shell Crain No. 1-R).

Smackover these sediments have been referred to as the Brown Dense Zone (BDZ), rarely hydrocarbon-bearing. The middle and upper sandstone intervals in Harrisville occur within the BDZ.

The MLZ is composed of parallel to crinkly laminae of dolomitic lime mudstones, with abundant bitumen parallel to bedding (see Fig. 3D). Dispersed quartz silt and silt laminae occur within the carbonate laminae. The sandstones are interpreted to be "basinal" (the Interior Salt Basin was probably never very deep) turbidite deposits. Evidence includes sharp-based sandstones, massive to poorly graded sand units, gradational contacts to thin shales and low-energy basinal carbonates, and paleogeographic position. The sandstones, based on seismic and well logs, are laterally extensive deposits. Log, core, and thin-section (representative lithologies) data from the lower sandstone in the Hilton No. 1 are shown in Figure 12.

Summary of Depositional Model

The depositional model for the sour gas complex must account for both the in-field and downdip to updip stratigraphic sequence between fields, and the abundance and distribution of sandstones and their relationship to the interbedded carbonates. The interpretations are based on sedimentary structures, macrofossils, trace fossils, vertical sequences, and on depositional environments interpreted from the carbonate interbeds. Figure 13 is a schematic representation of the ramp model for the Smackover interbedded siliciclastic/carbonate depositional sequence. The Thomasville area illustrates outer ramp to nearshore and sabkha deposition, and the lower sandstone and MLZ carbonates in Harrisville are more basinal deposits.

Reservoir Quality

Carbonate interbeds are generally nonreservoir deposits in the sour gas area. Only minor amounts of the reserves (less than 2 percent) occur in extensively dolomitized ooid grainstones. For the most part, the ooid grainstones are either highly compacted with sutured contacts (Fig. 3A), or porosity is occluded by early cementation. Little leaching and/or dolomitization of any of the carbonates have occurred to create the carbonate reservoirs typically found elsewhere in the Smackover trend.

The sandstones are fine- to medium-grained, poorly to moderately sorted, and subarkosic (see Table 2 for the sandstone properties). Although a complex diagenetic sequence (Table 2) has severely overprinted the original depositional fabric, both are important to the final reservoir properties. Depositional controls include proximity to carbonate interbeds, thickness of sandstone beds, reworking of the sediments, and environment of deposition, such as the bioturbated storm deposits having very low permeability. Important diagenetic events include compaction, numerous cementation and dissolution events, and hydrocarbon migration and alteration (bitumen formation) through cracking and/or thermal sulfate reduction (see section on sour gas origin). Porosity and permeability in the reservoir sandstones are approximately 7.0 percent and 0.3 md respectively.

FIGURE 12 - The lower zone in the Shell Hilton No. 1, Harrisville Field, contains thin sandstones interbedded with the microlaminated carbonates. The upper sandstones in Harrisville are interbedded with pelletal and locally skeletal wackestones and packstones. The lower interval sandstones are interpreted to be basinal turbidite deposits. Sandstone quality is less adjacent to the carbonate interbeds because of increased cementation, similar to the relationship in Thomasville. Three thin section photomicrographs illustrate the dominant sediments and reservoir properties present in the lower sandstone. Thin section A is an early-cemented (D = dolomite) sandstone (non-reservoir). Sample B is a subarkosic sandstone that is highly compacted with minor secondary (Sec) and altered primary porosity (quartz (Q) and potassium feldspar (K) and bitumen (B) are labeled). Sample C is the microlaminated zone with minor quartz silt (S) present.

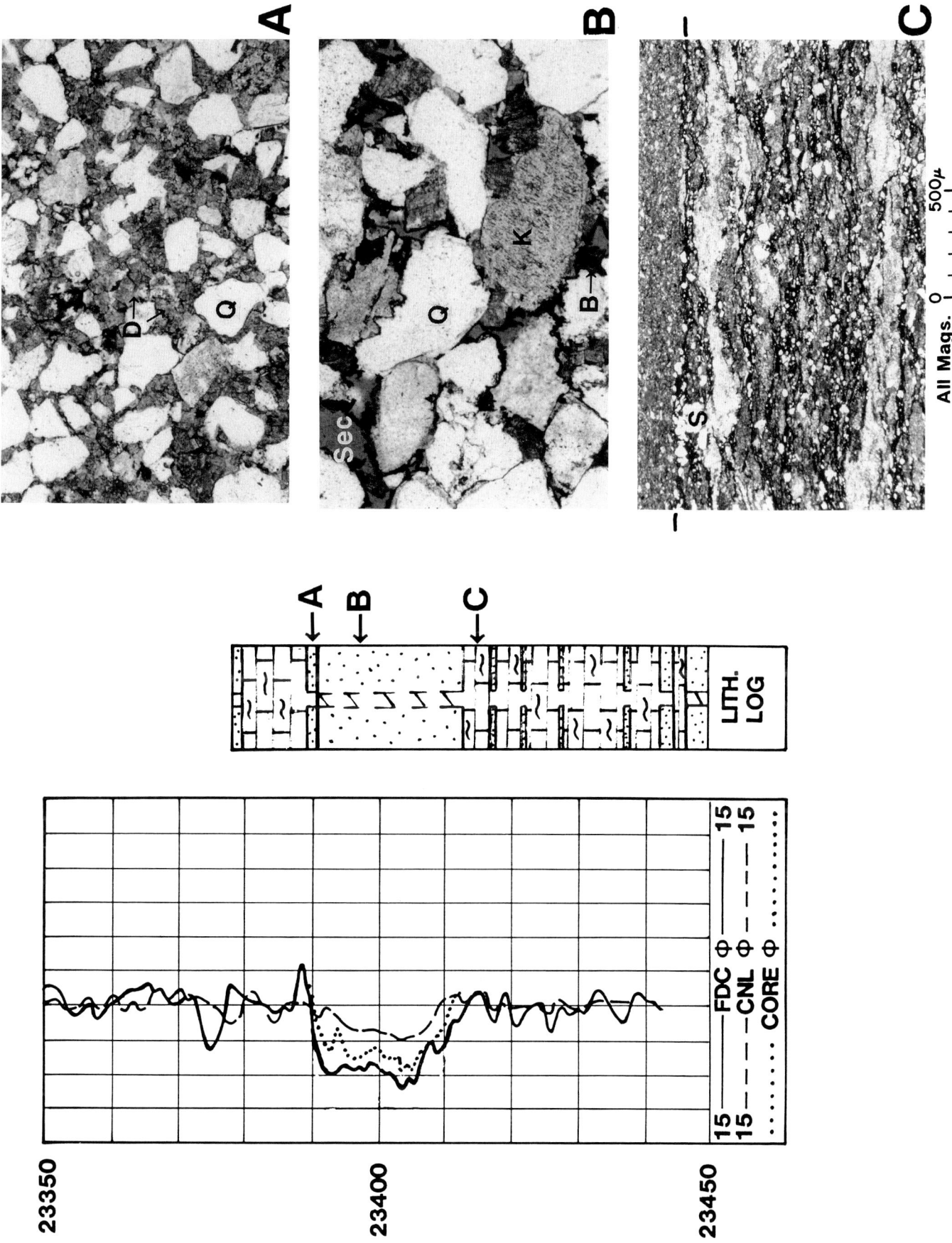

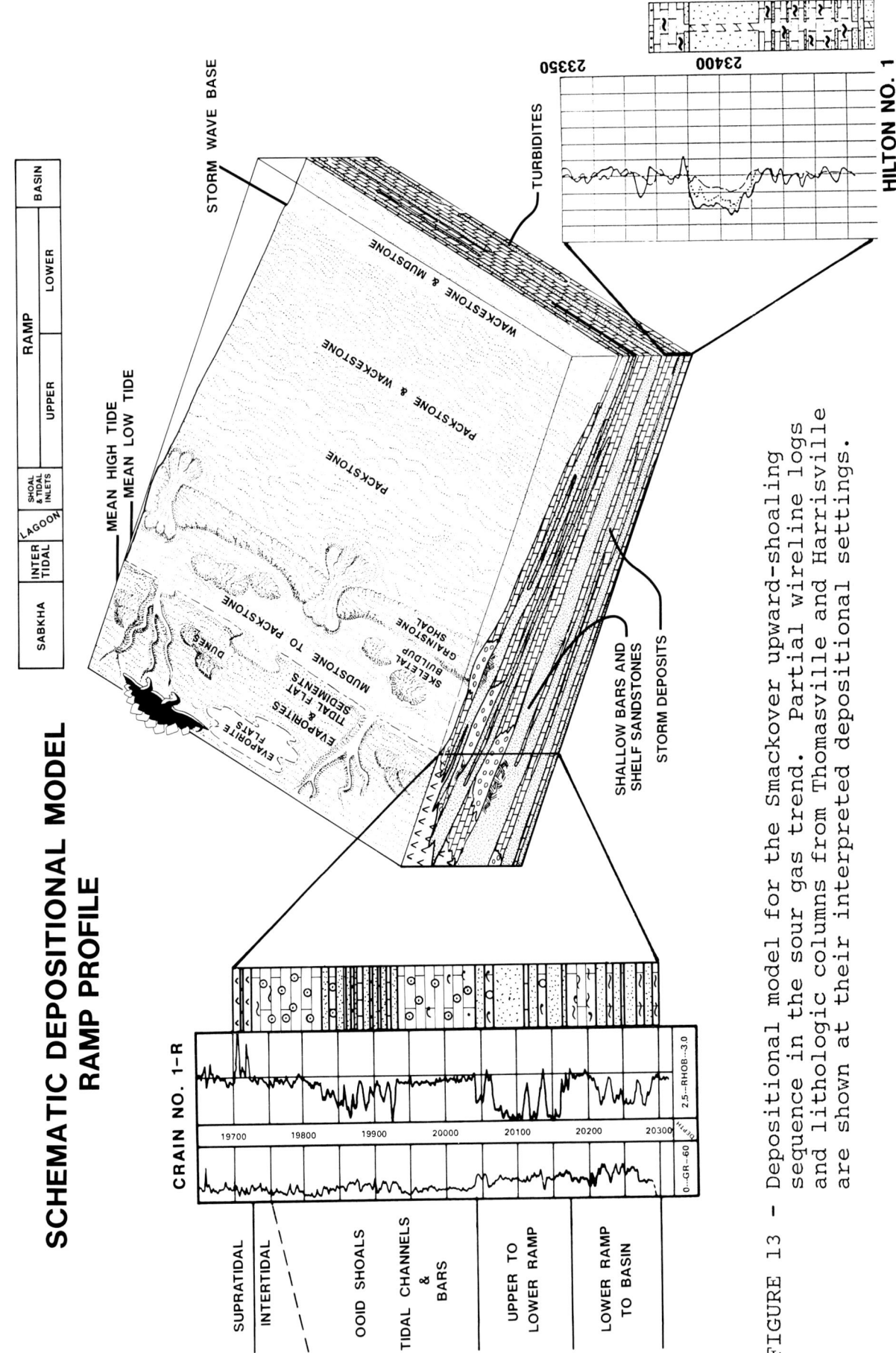

FIGURE 13 — Depositional model for the Smackover upward-shoaling sequence in the sour gas trend. Partial wireline logs and lithologic columns from Thomasville and Harrisville are shown at their interpreted depositional settings.

The carbonate and bitumen amount and distribution are most important in governing reservoir quality. An optimum amount (6 to 22 percent) of carbonate has been found to correspond with the higher porosities and permeabilities. Rhombic dolomite, in particular, seems to have helped inhibit the overcompaction of the sandstones. Exceptions include zones with entirely secondary vugular porosity, either from local dolomitization or feldspar dissolution, which may have high porosity but low permeability. Zones with both secondary and altered primary porosity are most important. Figure 14 illustrates variable reservoir quality associated with variable types and amounts of carbonate. One of the important controls, as seen in Figures 9 and 10 previously, is the often-observed cementation of the sandstone layers proximal to carbonate interbeds. Another example is shown in Figure 15, which contains core, logs, thin sections, and petrophysical properties showing corresponding increases in carbonate allochems and cement with decreasing porosity.

Bitumen (5 to 15 percent) both occludes porosity and reduces permeability. Leaching experiments have revealed a fine network of bitumen in both the pores and pore throats. Bitumen presence does usually indicate reservoir-quality sandstones and the presence of a former oil pool. Several of the other major diagenetic controls on reservoir quality include cementation and dissolution events, and locally fibrous illite (Fig. 16).

Origin of sour gas.--Fields in the sour gas trend contain gases with high concentrations of hydrogen sulfide (25 to 45 percent). It is unusual for siliciclastic reservoirs to contain such high H_2S concentrations because of (1) the usual presence of iron that may combine with and remove the H_2S, and (2) the usual absence of a proximal source of sulfate. However, sandstones in the sour gas trend "act" like carbonates in that they contain little clay and other Fe-bearing minerals and, more importantly, have a proximal source of anhydrite in both the overlying Buckner evaporites and as secondary pore-fill.

The origin of high H_2S concentrations has been explained by both thermal and bacterial sulfate reduction (Orr, 1977). Sulfur isotope, petrographic, and temperature relationships support a thermal sulfate reduction (TSR) origin in the sour gas area; TSR is the reduction of sulfate by hydrocarbons at high temperatures and the generation of H_2S and other products. The reaction and products are dependent on the reservoir's thermal history and on the availability of sulfate (predominantly pore-fill anhydrite) to react with the hydrocarbons.

Sulfur isotope data are the primary evidence for TSR. The isotopic values of the reactants (anhydrite) and products (H_2S, sulfur) should be similar in TSR because of little or no fractionation at high temperature. The isotopic values in the sour gas area all range from +16.1 to +17.3 per mil CDT (Shew, 1987). Petrographic evidence showing calcite replacing anhydrite and the local presence of sulfur (Fig. 17) also supports the TSR interpretation. Sassen and Moore (1988) have made similar interpretations for the high H_2S concentrations in Smackover carbonate reservoirs elsewhere in the Gulf Coast. TSR is one of the last and most important diagenetic events in the sour gas trend from an economic perspective. Special drilling and completion procedures are needed for these wells, and the H_2S is separated from the gas and processed into sulfur.

FIGURE 14 - Four sandstone samples illustrate the range and controls on reservoir quality, particularly of carbonate cement.

A) Early cementation as evidenced by floating grains in pervasive dolomite (D) and calcite (C) cement leads to poor quality.

B) Highly compacted sandstone sometimes occurs where little or no dolomite was present early. Quartz overgrowths (QO) and embayed (E) grains indicate pressure solution and compaction.

C) Minor porosity occurs where carbonate allochems (A) are present and locally dissolved. Too much carbonate cement (D) generally occurs in these areas. Cement and bitumen (B) have plugged most of the pores in this sample.

D) The best reservoir quality occurs where an optimum amount of carbonate cement, particularly rhombic dolomite (D), formed early to "prop" the framework grains, to prevent over-compaction. However, the best reservoir quality occurs where both altered primary (P) and secondary (S) porosity occur. Most of the secondary porosity is related to calcite dissolution.

FIGURE 15 - Reservoir quality of the sandstone is strongly influenced by proximity to the carbonate interbeds. Core, core analyses, and thin sections illustrate this relationship with poorer reservoir quality adjacent to the carbonates. Increased carbonate allochems (O) and cement (C) occurs in the upper sample. The darker core below sample A is heavily stained with bitumen (Bit) where porosity increases. In the thin section photos Q = quartz, K = potassium feldspar, O = ooid, D = dolomite, C = calcite, B = bitumen, and P = porosity.

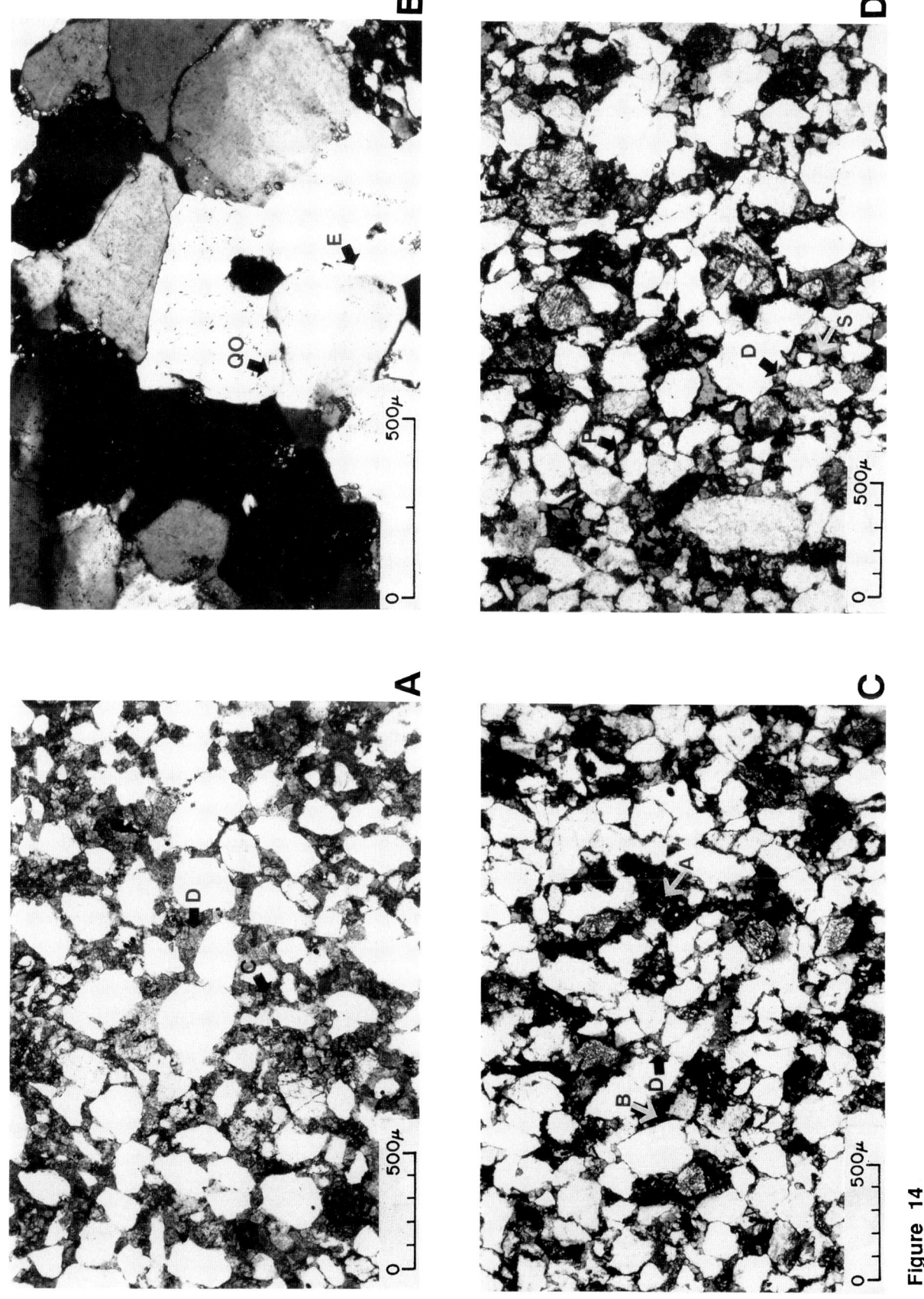

Figure 14

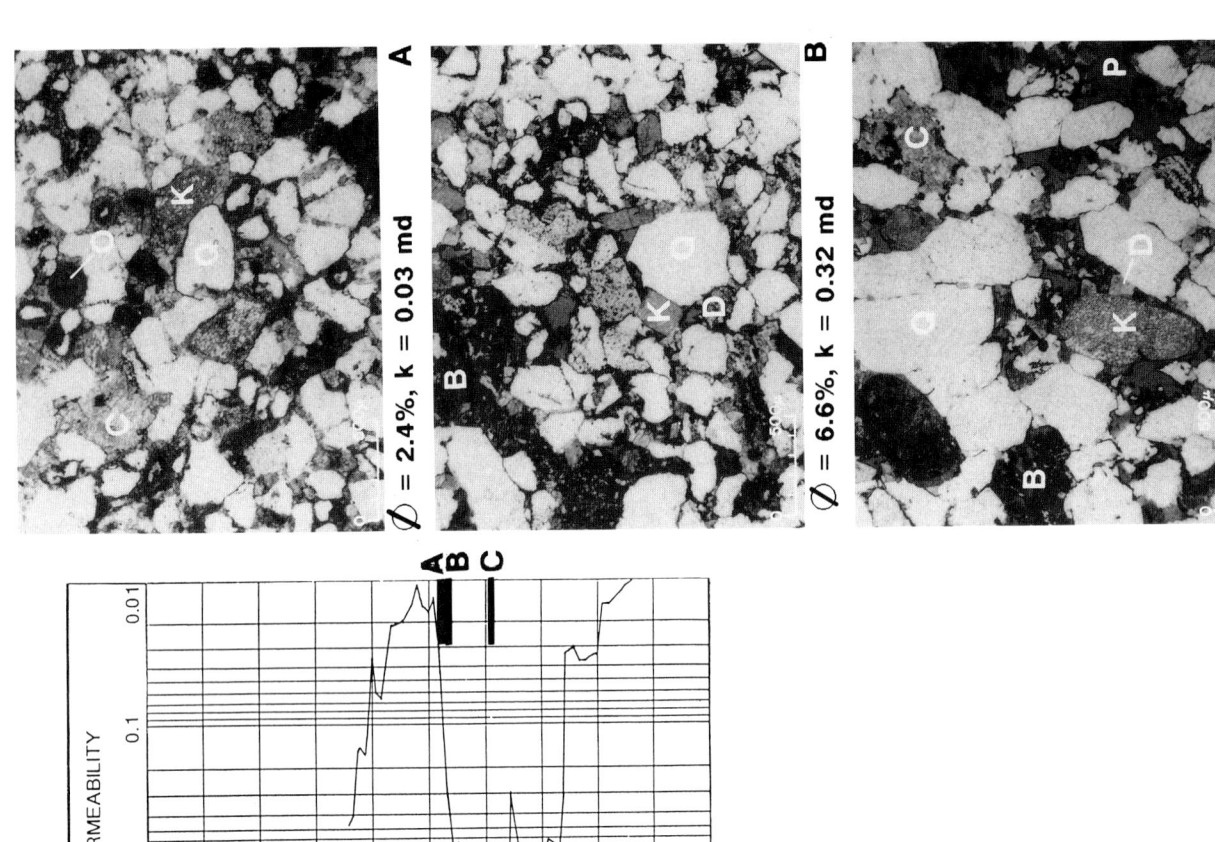

Figure 15

CONCLUSIONS

The sour gas area of central Mississippi is interesting because of the anomalous mixed siliciclastic/carbonate deposits, complex diagenetic sequence, and the hostile but highly productive subsurface environment. Important points include:

1. Smackover deposition is best described by overall regressive carbonate ramp sedimentation, with high-energy grainstones nearshore grading to lower-energy deposits offshore. The lowermost unit, MLZ, is the source rock.

2. Carbonate deposition was interrupted by a thick wedge of siliciclastic deposits from central Mississippi to eastern Louisiana. The clastic wedge is likely associated with the ancestral Mississippi River.

3. Siliciclastics were reworked by longshore and storm processes, which help explain both the absence of fines (clays) as well as the interbedded carbonate and sandstone deposits. Repeated and interbedded carbonate and siliciclastic deposition could be related to sea-level fluctuations, sediment-input fluctuations, and/or to syndepositional salt movement. Sandstones were also transported to the basin and occur as laterally extensive turbidite deposits.

4. Nearshore, inner to outer ramp, and "basinal" sandstone deposits are interbedded with tight nonreservoir carbonates that describe an upward-shoaling depositional sequence capped by the Buckner Formation. Interpretation of the carbonate depositional environments is important to the interpretation of the sandstones, as they occur in facies mixing and punctuated relationships. The upward-shoaling sequence is evident both within fields and from downdip to updip field locations.

5. Productivities, reservoir quality, and hydrocarbon presence and type are strongly influenced by the depositional setting, proximity to carbonate interbeds, and complex diagenetic history. Although reservoir quality is generally low, the high geopressures, laterally extensive sandstones, and enough secondary and altered primary porosity provide for large drainage areas and high flow rates. One of the most important diagenetic events is TSR, which has led to the formation of large amounts of H_2S. The sandstones "act" much like carbonate reservoirs in having few Fe-bearing minerals to remove the H_2S and abundant anhydrite to react with the hydrocarbons.

ACKNOWLEDGMENTS

The author thanks Shell Offshore Inc. and Shell Development Company for permission to publish this paper. The author also acknowledges the many workers in both Exploration and Production who have contributed to a better understanding of the central Mississippi Jurassic Smackover trend. Special thanks are extended to M. M. Garner for his recent work in establishing a structural and field framework for the interpretation of the depositional model and reservoir properties.

FIGURE 16 - Porosity and permeability in the sandstones are altered and severely reduced by the complex diagenetic overprint. Sample A is a thin section photomicrograph and B-D are SEM's.

A) Abundant bitumen (B) and poikilitopic calcite (C) have reduced the porosity but minor altered primary (P) and secondary (S) porosity are present. The secondary porosity in this sample is from dissolution of k-spar and calcite.

B) Quartz overgrowths (QO), dolomite (D), and bitumen (B) plug the pores and pore throats.

C) Abundant bitumen (B) encasing minor dolomite (D) in this sample is very important in reducing porosity in the sandstone. Its presence also illustrates the presence of the porous sandstones indicating these areas have remained the most porous sands through burial history. The bitumen may have formed either during thermal cracking and/or during the thermal sulfate reduction process. Leaching experiments to remove the carbonates and siliceous material indicate the presence of a network of bitumen in the pores and pore throats that are important in reducing porosity and permeability.

D) Fibrous illite, though not pervasive, is locally important in reducing permeability.

162

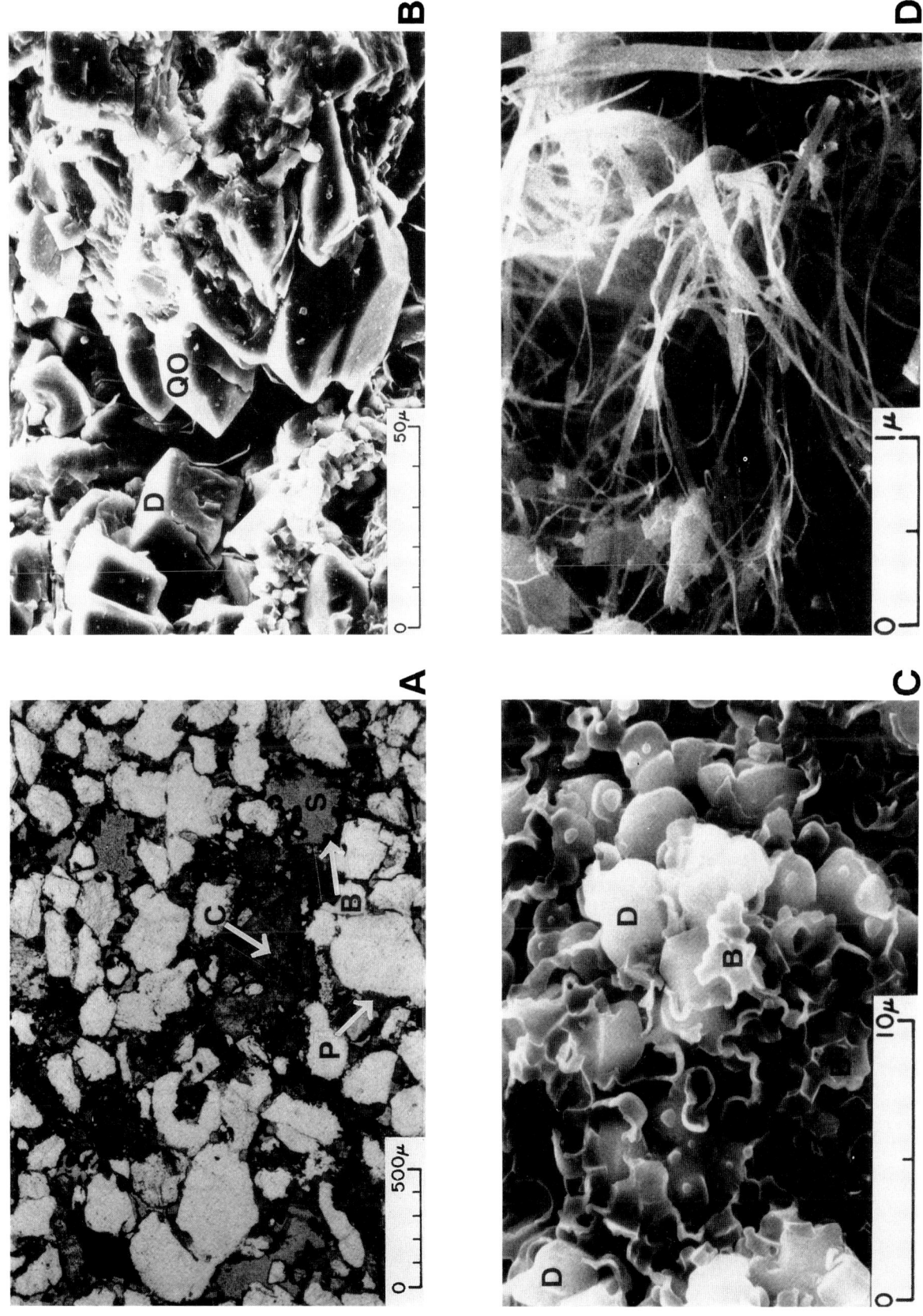

FIGURE 17 - The origin of the high H_2S contents in the sour gas area is related to thermochemical sulfate reduction (TSR). Evidence includes temperature relationships (increased H_2S with temperature increase), petrographic evidence of calcite (C) replacement of anhydrite (A) (anhydrite reacts with hydrocarbons), presence of sulfur (S), and most importantly sulfur isotope evidence where the H_2S, sulfur and anhydrite have similar isotopic values. These should be similar in TSR (little or no fractionation at high temperatures).

$$CaSO_4 + CH_4 \longrightarrow CaCO_3 + H_2S + H_2O$$

AN EXCESS OF SULFATE MAY PRODUCE ELEMENTAL SULFUR

$$CaSO_4 + 3H_2S + CO_2 \longrightarrow CaCO_3 + 4S + 3H_2O$$

$$\left[\begin{array}{l} \text{EQUATIONS FROM SIEBERT (1985)} \\ \text{THIS AND A PAPER BY ORR (1977) ARE THE BEST} \\ \text{SUMMARIES OF THE ORIGIN OF HIGH } H_2S \text{ COMPOSITIONS} \\ \text{AND EVIDENCE FOR THERMAL SULFATE REDUCTION} \end{array} \right]$$

THOMASVILLE SULFUR ISOTOPE VALUES

ANHYDRITE*	$\delta^{34}S$ = +16.3 TO 17.3‰ CDT
H_2S	$\delta^{34}S$ = +16.1 TO 17.3‰ CDT
SULFUR	$\delta^{34}S$ = 16.1‰ CDT

*INCLUDES BUCKNER MASSIVE ANHYDRITE AND PORE-FILLING ANHYDRITE

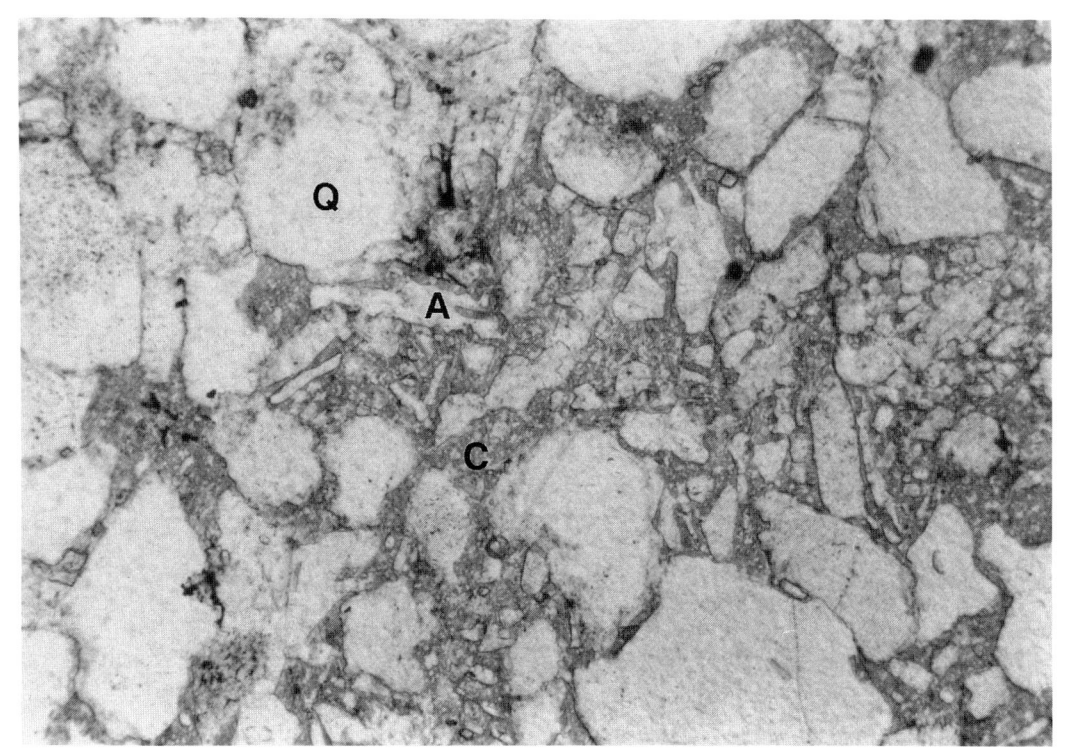
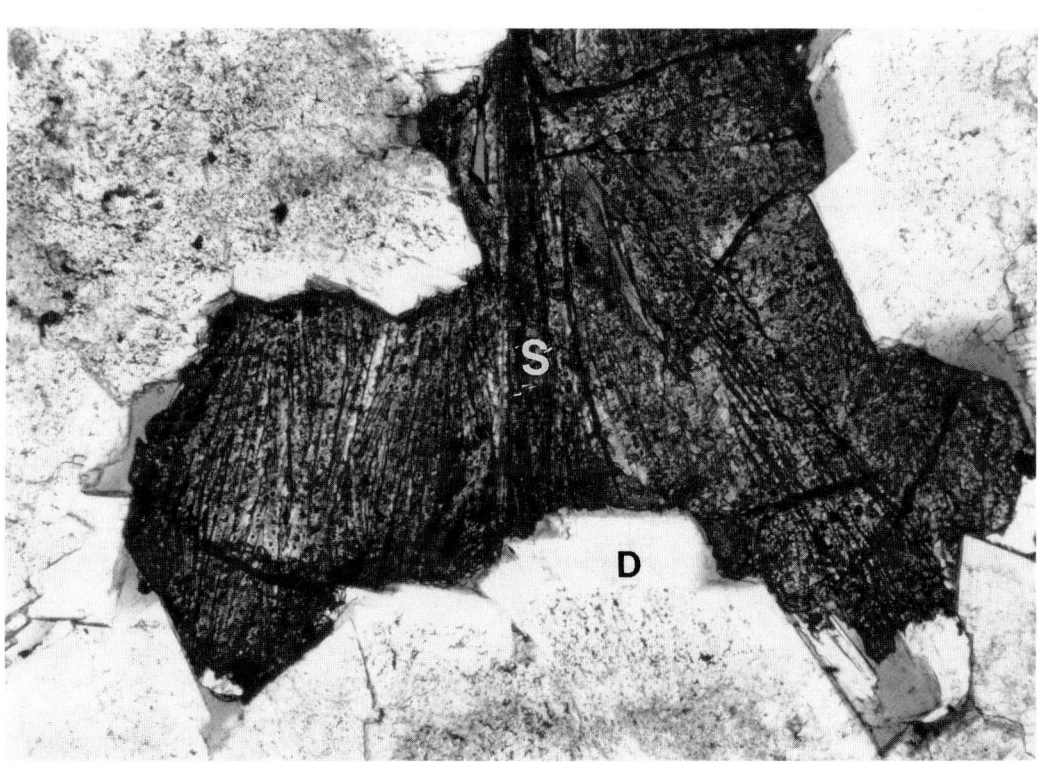
165

REFERENCES

AHR, W. M., 1973, The carbonate ramp - an alternative to the shelf model: Gulf Coast Association of Geological Societies Transactions, v. 23, p. 221-225.

_____, AND PALKO, G. J., 1981, Depositional and diagenetic cycles in Smackover limestone-sandstone sequences, Lincoln Parish, Louisiana: Gulf Coast Association of Geological Societies Transactions, v. 31, p. 7-17.

BISHOP, W. F., 1968, Petrology of Upper Smackover limestone in North Haynesville Field, Claiborne Parish, Louisiana: American Association of Petroleum Geologists Bulletin, v. 52, p. 92-128.

BUDD, D. A., AND LOUCKS, R. G., 1981, Smackover and Lower Buckner formations, South Texas: depositional systems on a Jurassic carbonate ramp: Bureau of Economic Geology, University of Texas, Austin, Report of Investigations 112, 38 p.

BURKE, K., AND DEWEY, J. F., 1973, Plume-generated triple junction: key indicators in applying plate tectonics to old rocks: Journal of Geology, v. 81, p. 406-433.

CHIMENE, C. A., 1976, Upper Jurassic reservoirs, Walker Creek Field area, Lafayette and Columbia counties, Arkansas, in North American Oil and Gas Fields: American Association of Petroleum Geologists Memoir 24, p. 177-204.

DINKINS, T. H., OXLEY, M. L., MINIHAN, E., AND RIDGEWAY, J. M., 1968, Jurassic stratigraphy of Mississippi: Mississippi Geology, Economic and Topographical Survey Bulletin 109, 77 p.

ERVIN, C. P., AND McGINNIS, L. D., 1975, Reelfoot rift: reactivated precursor to the Mississippi Embayment: Geological Society of America Bulletin, v. 86, p. 1287-1295.

HORNE, J. C., REEL, C. L., AND CUMMINS, G. D., 1989, Effects of sequence stratigraphy on distribution of Cambro-Ordovician siliciclastic hydrocarbon reservoirs in Michigan Basin (abstract): American Association of Petroleum Geologists Bulletin, v. 75, p. 1034.

JUDICE, P. C., AND MAZZULLO, S. J., 1982, The gray sandstones (Jurassic) in Terryville Field, Louisiana: basinal deposition and exploration model: Gulf Coast Association of Geological Societies Transactions, v. 32, p. 23-43.

MACK, G. H., AND JAMES, W. C., 1986, Cyclic sedimentation in the mixed siliciclastic-carbonate Abo-Hueco transitional zone (Lower Permian), southwestern New Mexico: Journal of Sedimentary Petrology, v. 56, p. 635-647.

MANN, M. C., AND THOMAS, W. A., 1968, The ancient Mississippi River: Gulf Coast Association of Geological Societies Transaction, v. 18, p. 187-197.

MOORE, C. H., 1984, The Upper Smackover of the Gulf Rim: depositional systems, diagenesis, porosity evolution and hydrocarbon production, *in* Ventress, W. P. S., Bebout, D. G., Perkins, B. F., and Moore, C. H., eds., The Jurassic of the Gulf Rim: Third Annual Research Conference, Gulf Coast Section, Society of Economic Paleontologists and Mineralogists, p. 283-307.

ORR, W. L., 1977, Geologic and geochemical controls on the distribution of hydrogen sulfide in natural gas, *in* Advances in Organic Geochemistry: Proceedings of 7th International Meeting on Organic Geochemistry, Madrid, Spain, p. 571-597.

PINDELL, J., AND DEWEY, J. F., 1982, Permo-Triassic reconstruction of western Pangea and the evolution of the Gulf of Mexico/Caribbean region: Tectonics, v. 1, no. 2, p. 179-211.

SASSEN, R., AND MOORE, C. H., 1988, Framework of hydrocarbon generation and destruction in eastern Smackover Trend: American Association of Petroleum Geologists Bulletin, v. 72, p. 649-663.

SHEW, R. D., 1987, Isotopic and petrographic evidence for a thermochemical sulfate reduction origin of high H_2S concentrations in central Mississippi: Society of Economic Paleontologists and Mineralogists, Annual Midyear Meeting, Abstracts with Programs, Austin, Texas, p. 77.

_____, AND GARNER, M. M., 1990, Reservoir characteristics of nearshore and shelf sandstones in the Jurassic Smackover Formation, Thomasville Field, Mississippi, *in* Barwis, J. H., McPherson, J. H., and Studlick, J. R. J., eds., Sandstone Petroleum Reservoirs, Casebooks in Earth Sciences: Springer-Verlag, New York, p. 437-464.

STUDLICK, J. R. J., SHEW, R. D., BASYE, G. L., AND RAY, J. R., 1990, A giant carbon dioxide accumulation in the Norphlet Formation, Pisgah Anticline, Mississippi, *in* Barwis, J. H., McPherson, J. H., and Studlick, J. R. J., eds., Sandstone Petroleum Reservoirs, Casebooks in Earth Sciences: Springer-Verlag, New York, p. 181-203.

TILLMAN, R. W., AND MARTINSEN, R. S., 1984, The Shannon shelf-ridge sandstone complex, Salt Creek Anticline area, Powder River Basin, Wyoming, *in* Tillman, R. W. and Siemers, C. T., eds., Siliciclastic Shelf Sediments: Society of Economic Paleontologists and Mineralogists Special Publication 34, p. 85-142.

DEPOSITION AND DIAGENESIS OF A MARINE-SWAMP MARGIN; THE PROVIDENCE LIMESTONE AND ADJACENT COALS, WESTERN KENTUCKY

CAROL B. DEWET
Department of Geosciences,
Franklin and Marshall College, Lancaster, PA 17604
STEPHEN O. MOSHIER, JAMES C. HOWER,
AND SUSAN M. RIMMER
Department of Geological Sciences,
University of Kentucky, Lexington, KY 40506

ABSTRACT

Marine limestones, terrigenous siltstones, shales, and coals exhibit complex stratigraphic relationships in the western Kentucky coal fields at the southern margin of the Illinois Basin. The Providence Limestone member of the Sturgis Formation (Pennsylvanian) is a mixed siliciclastic-carbonate unit bounded by the No. 11 (Herrin) and No. 13 (Baker) coal seams. The coals contain fractured and brecciated horizons, cemented by sparry calcite and dolomite. Shallow-marine limestones and clastic rocks are interbedded with coals; lateral thickness and lithologic changes are abrupt. Up to 6.6 ft (2 m) thick beds of limestone are brecciated, with abundant intraclasts and disrupted fabrics. Slickensided claystones containing plant debris (underclays) directly underlie the coal seam deposits. The coals are overlain by thin shales with limestone nodules, followed by normal marine limestones. The limestones are generally disrupted, with pedogenic features overprinting the original marine fabric. Hard claystones overlie the disrupted limestones.

The coals and Providence lithologies have complex diagenetic histories. Early lithification of brecciated coal fragments prevented complete organic decay, and the brecciated coals have well-preserved macerals. Petrographic examination of the Providence shows an original marine faunal assemblage in wackestones, packstones, and grainstones. Multiple episodes of carbonate cementation, silicification, and pedogenic alteration have altered the primary textures, creating a diverse suite of microfabrics.

INTRODUCTION

The 197 ft (60 m) thick interval between the No. 9 coal and the No. 13 coal in Hopkins County typifies the complex stratigraphic relationships between marine carbonates, detrital clastics, and coals in the western Kentucky coal field (Fig. 1). The interval includes the Providence Limestone (Fig. 2). Many horizons within the 'limestone' beds consist of silty, argillaceous wackestones and packstones, indicating that clastic and carbonate sedimentation was coeval. The Providence carbonate-clastic-coal sequence represents an important type of marginal marine setting - one of economic significance involving a fossil-fuel resource. Recently obtained exploration cores provide a unique opportunity to examine changes in carbonate-clastic-dominated

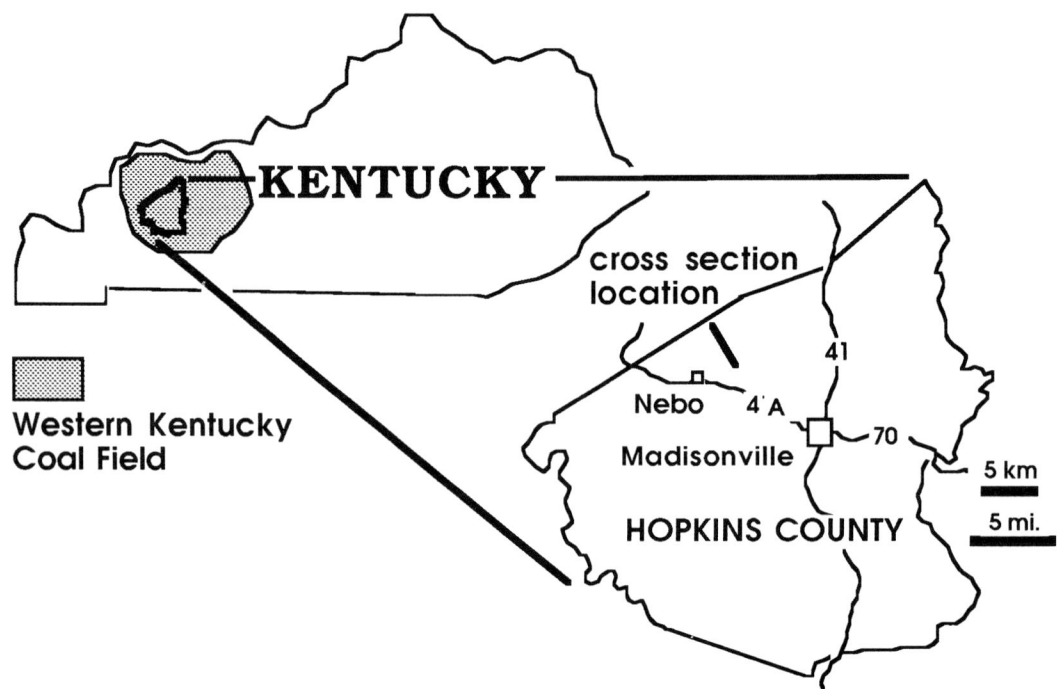

Figure 1. Location of cross-section connecting four cores through the Providence Limestone and No. 11 coal in northwest Hopkins County, KY.

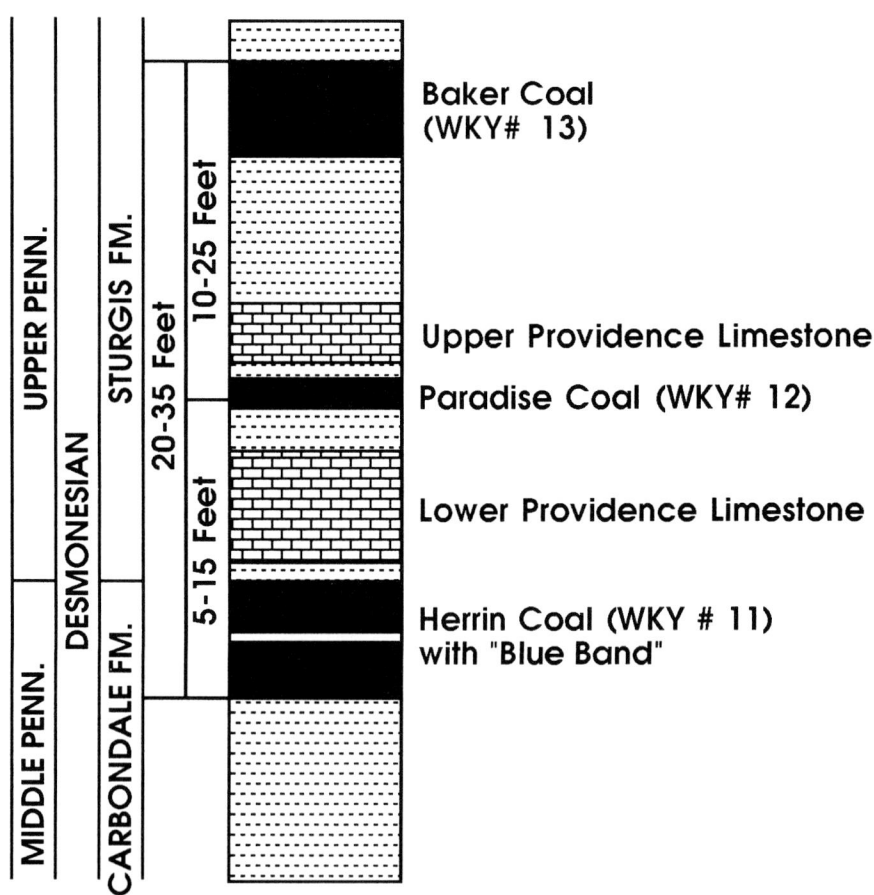

Figure 2. Stratigraphy of the Desmonesian section in western Kentucky that includes the Providence Limestone. In the study area, the No. 12 coal is absent and the Providence Limestone forms one unit.

depositional processes and subsequent diagenetic products. Laterally the sequence between the coals changes in thickness and lithology, reaching a maximum thickness of 33 ft (10 m), thinning to 6.6 ft (2 m). The lateral variability reflects an irregular depositional coastline coupled with synsedimentary tectonics.

STRATIGRAPHY AND DEPOSITIONAL PATTERNS IN THE COAL FIELD

Using interval isopach and lithofacies isopleth maps based upon correlations between hundreds of exploration core logs, Neuder (1984) demonstrated contemporaneous structural control on sediment distribution and coal thicknesses in eastern Hopkins County. A synthesis of the stratigraphic relationships, north to south across eastern Hopkins County, is shown in Figure 3. Coarsening-upward detrital deposits form the lower two-thirds of the sequence (from the No. 9 to the No. 11). The upper third is composed of coal, limestone, shale, and sandstone in equal proportions. An east-west trending depression in the northern area that parallels mapped fault zones is the focus of thick detrital sedimentation, reduction or loss of the Nos. 9 through 12 seams, and splitting of the No. 13 coal. Accumulation of thick detrital deposits and thin or split coals and their laterally equivalent limestones or black shales would have been accommodated in a structurally controlled topographic low. The Nos. 9, 11, and 12 coals are thicker on the structurally high area to the south. Lithofacies distributions across this area suggest the maintenance of a topographic high, and coal thicknesses appear to have been controlled by differential compaction of underlying sediments.

THE PROVIDENCE LIMESTONE - A MARINE CARBONATE IN A COAL MEASURE

The Providence Limestone is part of a carbonate depositional system of regional significance. In the coal measures of the Illinois Basin (to the northwest), the time-equivalent Brereton Limestone is a thin deposit (greater than 3.3 ft (1 m) thick) of muddy carbonates with prominent algal-bryozoan mud banks, up to 6.6 ft (2 m) in thickness (O'Connell, 1983; Khawlie and Carozzi, 1976). The unit represents the deposition of marine carbonates in shallow, open water over abandoned and transgressed deltas (O'Connell, 1983).

East of Hopkins County, in Muhlenberg and Ohio counties, the Providence consists of upper and lower limestone units situated between the three coals (Fig. 2). The lower unit consists of bioturbated skeletal wackestones with abundant bryozoans, forams, brachiopods, mollusks, blue-green algae (*Girvanella*), and phylloid algae (Perkinson, 1978; Perkinson and others, 1984). These deposits formed mounds and filled lagoonal ponds during rapid transgression of an irregular topographic surface above the No. 11 coal. Algal-laminated mudstones and sparse skeletal wackestones form the upper unit, representing the development of a tidal-flat setting over the No. 12 coal (Perkinson, 1978).

The Providence Limestone has not previously been examined in detail in Hopkins County, where it exhibits a complex, interfingering stratigraphic relationship with the Nos. 11 through 13 coals (Fig. 3). In eastern Hopkins County the lower Providence unit lies between the No. 11 and

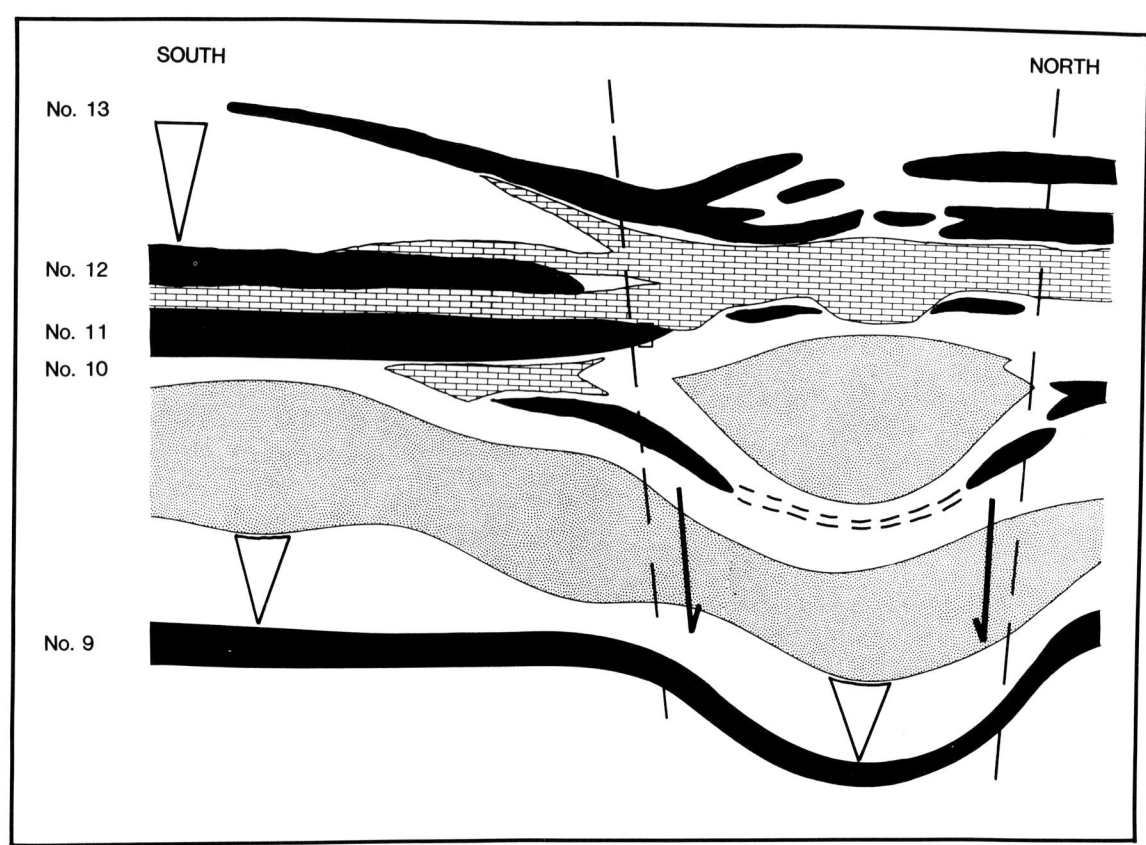

Figure 3. Stratigraphic relationships of coals (black), carbonates (brick) and detrital clastics (sandstones-stippled; shales, siltstones-blank) in eastern Hopkins Co. Triangles represent coarsening-upward intervals. Structural control on deposits is shown in the northern area. Vertical exaggeration is about 1000:1. (From Hower, et al., 1987)

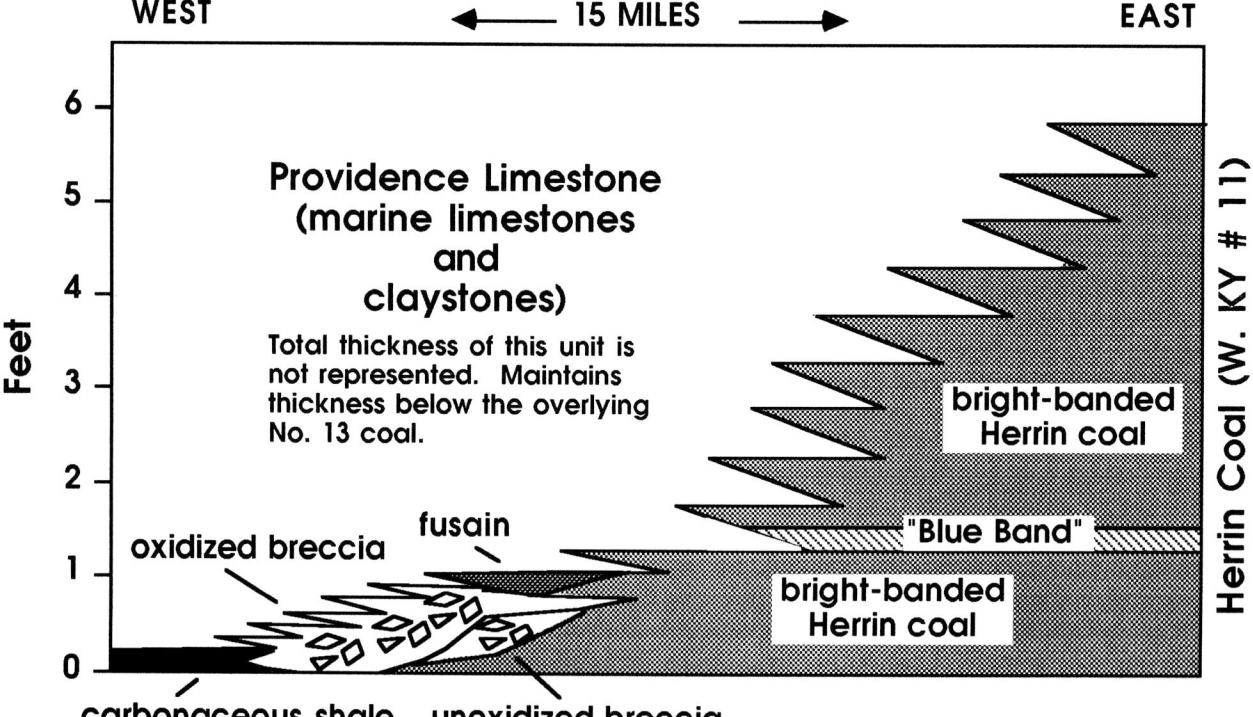

Figure 4. Observed characteristics of the No. 11 coal across Hopkins County including brecciation and pinchout.

No. 12, and also appears laterally adjacent to the No. 12, where it terminates to the north. In western Hopkins County the No. 12 coal is absent and the Providence consists of a single unit that would correlate to the lower unit to the east, or represent the upper and lower units merged.

New cores from the study area (and the subject of this presentation) show that the Providence is separated from the underlying No. 11 coal by a 1 to 2 ft (30 to 60 cm) thick shale. A typical underclay (1 to 3 ft (30 to 90 cm) thick) of green-gray to dark gray, fissile, slickensided claystone with abundant plant fragments lies between the Providence and the overlying No. 13 coal. The carbonate deposits include skeletal grainstones, wackestones, and mudstones that are highly biomottled. Fossil taxa include fragments of phylloid algae, crinoids, and brachiopods. Skeletal structures of the red "phylloid" algae, *Archaeolithophyllum*, are revealed in thin sections with cathodoluminescence (Moshier and Kirkland, 1990). Partial to complete dolomitization is common in the Providence, producing selective, mimicking, and nonmimicking fabric styles. In some locations the Providence reflects uninterrupted marine carbonate deposition, up to 11.5 ft (3.5 m) thick, with little alteration beyond typical recrystallization and dolomitization. At other locations the Providence includes thin mudstones and wackestones that are commonly brecciated and feature irregular bedding surfaces. These limestones are interbedded with and underlie beds that exhibit similar breccia fabrics but are composed of microcrystalline quartz and kaolinite.

THE "RAGGED EDGE" OF THE NO. 11 COAL

In eastern Hopkins County the No. 11 and No. 12 coals thin and pinchout over a short distance (less than 0.6 mile (1 km)) into expanded sections of the Providence Limestone (Fig. 3). Progressive loss of top benches in the No. 12 coal and corresponding increase in the thickness of the overlying Providence Limestone were documented by Austin (1979). The margin of the No. 11 coal involves thinning, progressive brecciation toward the "ragged edge," and diverse petrographic character toward the edge (Hower and others, 1987). A core from eastern Hopkins County, where the coal is 4 ft (1.2 m) thick, shows that the No. 11 peat was oxidized during the course of brecciation and is present as degraded vitrinite through fusinite. (Appendix A is a glossary of selected coal terms.) The fusinite is in the form of relatively large, "fusinized" wood fragments, not the fibrous fusain typically found in humic coals. Brecciated coals are cemented by calcite and dolomite, and contain thin, marine-limestone partings and numerous limestone lithoclasts. The skeletal content of the matrix and lithoclasts compares with that of the Providence Limestone.

New cores from western Hopkins County reveal previously unknown lateral and vertical transitions in the No. 11 coal (Fig. 4). The marginal coals vary from bright-banded, high vitrinite coal to carbonate-cemented breccias. In one core the coal is brecciated and reduced to 1 ft (0.3 m) thick, with indications that the peat was thoroughly oxidized during redeposition. The breccia contains only fusinite and semifusinite, with fungal sclerotinite (and marine fossils) in the carbonate matrix. Another core shows a vertical succession of increasing brecciation and oxidation of macerals similar to the lateral transition exhibited in the No. 11 coal across Hopkins County. The fusinized clasts preserve fragments of *Cordaites*, *Calamites*, and *Medullosa*; *Psaronius* and *Stigmaria* rootlets; and fungal sclerotia. Banded coals in these reduced beds include palynomorph assemblages dominated by arborescent lycopod, tree fern, sphenopsid, and *Cordaites* species (J. C. Hower, G. D. Wild, A. L. Raymond, and C. T. Helfrich, personal communication, 1990).

While oxidized to a semifusinite to fusinite level, the macerals in the marginal coals differ from the equivalent macerals in "normal" coals in terms of their size and degree of plant preservation structures. Inertinite macerals in the marginal coals have well-preserved textures that are normally associated with humic macerals from peats or lignites. Apparently calcite mineralization of the breccias prevented extreme fragmentation. These deposits should not be confused with the coal balls encountered in Illinois Basin coal measures.

DISTINCTIVE COLLATERAL FACIES CHANGES IN THE NO. 11 COAL AND PROVIDENCE LIMESTONE IN A MARGINAL SETTING

Lithofacies and Fabrics Observed in Core

Four cores from northwest Hopkins County show the variety of lithofacies associated with a coal-marine carbonate transition and the rock fabrics that appear to be distinctive of such a setting. The core locations are spaced along a nearly straight line oriented NW-SE, about 3 miles (4.8 km) in length (Fig. 1). The following stratigraphic changes are evident across this distance, from southeast to northwest (Fig. 5):

- The No. 11 coal thins from 5 ft to 7 in. (1.5 m to 18 cm) and exhibits the "ragged edge" characteristics discussed above (core photographs in Fig. 6).

- The No. 13 coal thickens from 2 to 6.9 ft (60 cm to 2.1 m)(core material not available for study; thicknesses measured from wireline logs).

- The interval including the Providence Limestone and detrital clastics thickens from 13 to 20.3 ft (4 to 6.2 m)(core photographs in Figs. 7-10).

At the southeast end of the section (core A), the interval between the coals includes underclays associated with the No. 13 coal and thick-bedded carbonates. At the northwest end (core B), the interval includes underclays and thin-bedded or brecciated carbonates and kaolinite-rich cherts. Descriptions of the lithofacies and their characteristic fabrics, as evident in the cores, follow.

Characteristics of the No. 11 coal.--At the location of core A, the No. 11 coal is 5 ft (1.55 m) thick and consists of common banded coal with thin layers of ash, bone, and pyrite. The "Blue Band," a clay parting found in the coal throughout the Illinois Basin, is 2.8 in. (7 cm) thick at this location.

The No. 11 coal in core B, at 4.2 ft (1.28 m) thick, is only about half the thickness of the mined coal. The coal is dominated by bright lithotypes commonly found in No. 11. Vitrains up to 0.8 in. (2 cm) thick are found. Two 0.3-in. (0.8 cm) pyrite lenses, along with cleat pyrite, are

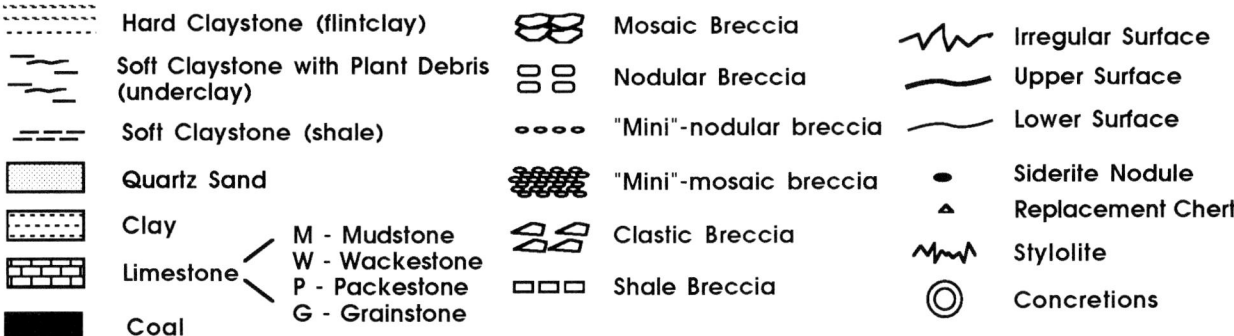

Figure 5B. Key to symbols in Fig. 5. See text for discussion

found in the central portion of the sequence and contribute to the 11.16 percent pyritic sulfur (dry basis; 12.44 percent total sulfur) of the 9.8-in. (24.9 cm) interval. The "Blue Band" is 3.1 in. (7.8 cm) thick at this location. The thickness of the coal below the "Blue Band" is similar to the thickness under the same parting in core A. Thus, it would appear that the reduction in thickness between the cores is from the top of the coal (Fig. 5A).

In core C the No. 11 coal is 13.9 in. (35.3 cm) thick and dominated by bright lithotypes (Fig. 6). The coal appears to have been truncated below the "Blue Band," which is absent in this location. The basal 3.6 in. (9.2 cm) of the coal contains a 0.2-in. (0.4 cm) pyrite band, the main contributor to the 6.32 percent pyritic sulfur (dry basis; 11.03 percent total sulfur) in that zone. The coal does exhibit a dulling-upward trend, abruptly terminated by a 0.4 in. (1.1 cm) vitrain found 10.6 in. (26.8 cm) above the base. The vitrain is overlain by a 0.1-in. (0.4 cm) dull clarain and a 0.2-in. (0.5 cm) bright clarain, followed by a 0.4-in. (1.1 cm) inertinite breccia, similar to the breccia found in well D. The breccia is overlain by a 0.2-in. (0.5 cm) bright clarain, 0.9-in. (2.4 cm) bone, a 0.5-in. (1.3 cm) bright clarain, and a 0.5-in. (1.2 cm) fusain (unmineralized).

The No. 11 coal in core D is only 7.1 in. (18.1 cm) thick, but is particularly interesting as it represents a vertical transition from bright-banded coal to breccia (Fig. 6). The lower 2.1 in. (5.3 cm) of the coal consists of bright clarain, vitrain, and a pyritized fusain. The next 1.3 in. (3.3 cm) is a breccia of coal fragments of a high volatile, B/C bituminous vitrinite and liptinite. The overlying 2.4 in. (6 cm) of breccia is dominated by semifusinite and fusinite clasts in a clay and carbonate matrix. The inertinite fragments are relatively large, up to 0.6 x 0.4 in. (1.5 x 1 cm), and preserve the wood structure. The top 1.4 in. (3.5 cm) of the core is a relatively unmineralized fusain.

Soft claystones and sandstones.--Underlying the No. 13 coal seam across the section is a thick layer of claystone that is soft and friable, green-gray to dark gray in color, slickensided, containing abundant plant fragments. In cores B, C, and D, the deposit is about 6.4 ft (2 m) thick, as is the No. 13 coal (Fig. 5A). In contrast, the claystone is only 29.9 in. (76 cm) thick in core A and the No. 13 coal is correspondingly thin. A thin horizon of tabular shale clasts

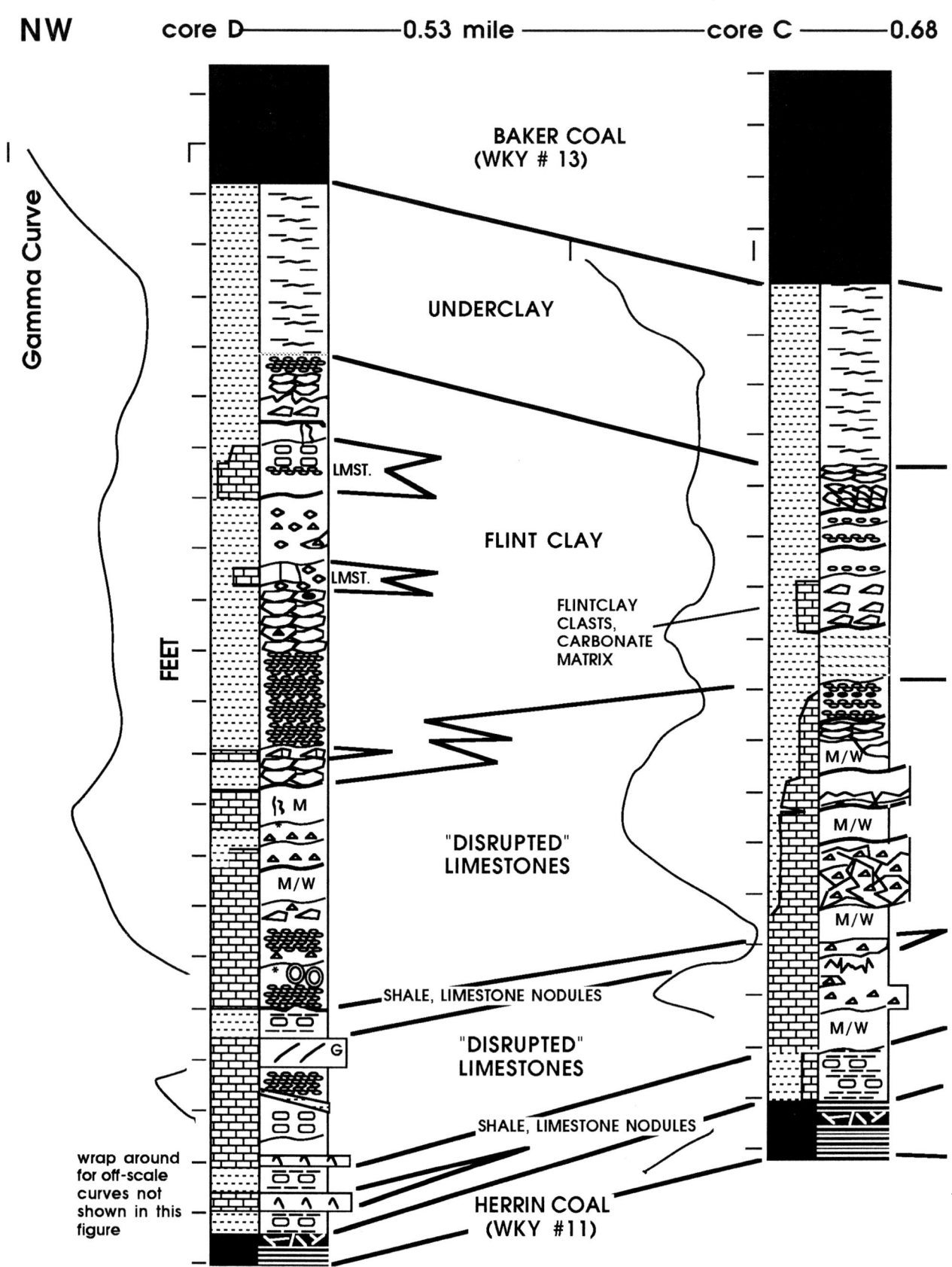

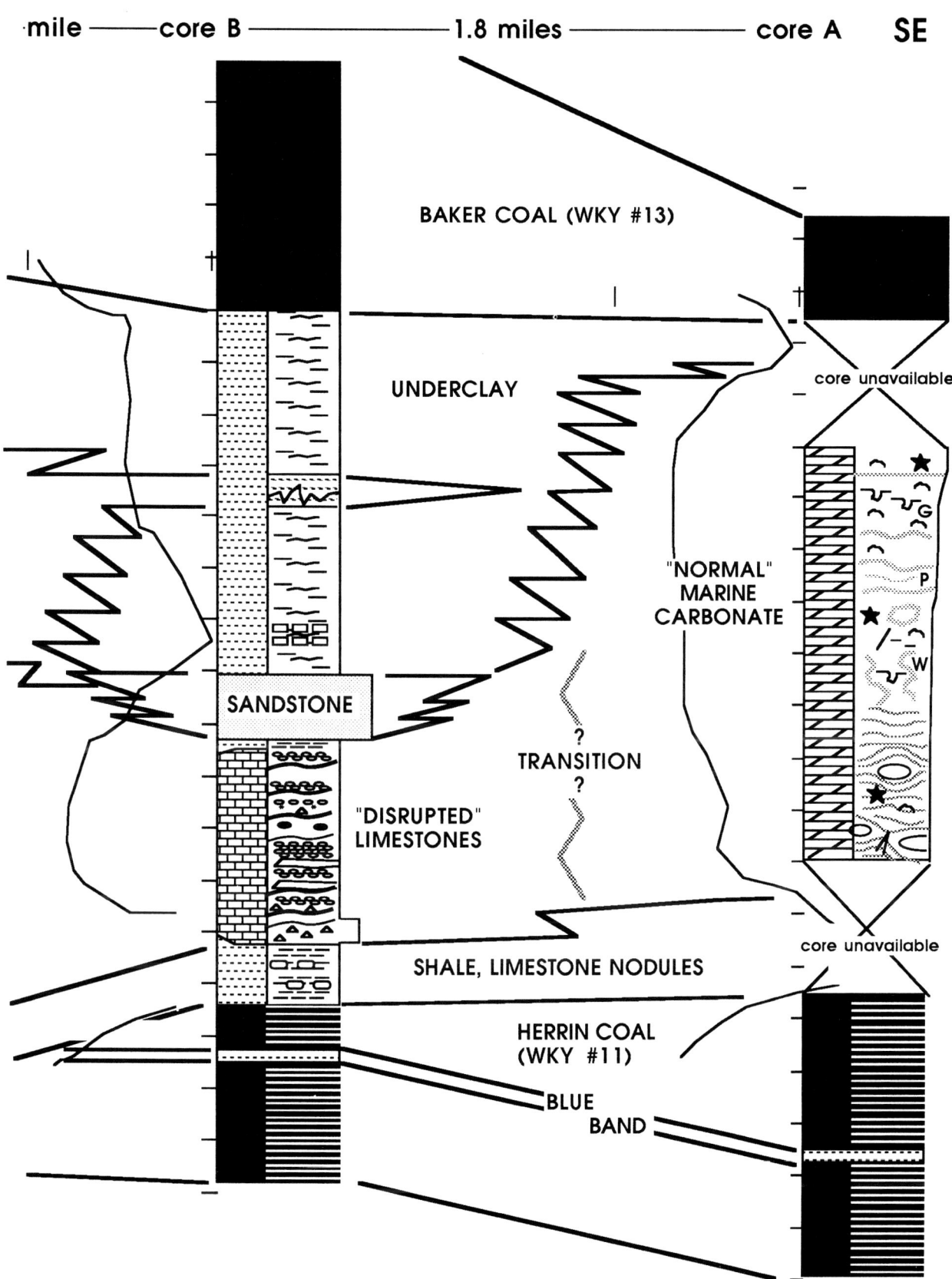

Figure 5A. Cross-section with correlations of facies between cores. Key to symbols is in Fig. 5 B.

Figure 6. Photographs of the No. 11 coal intervals in core C and core D. Samples were impregnated in epoxy, slabbed and polished as blocks. See text for discussion.

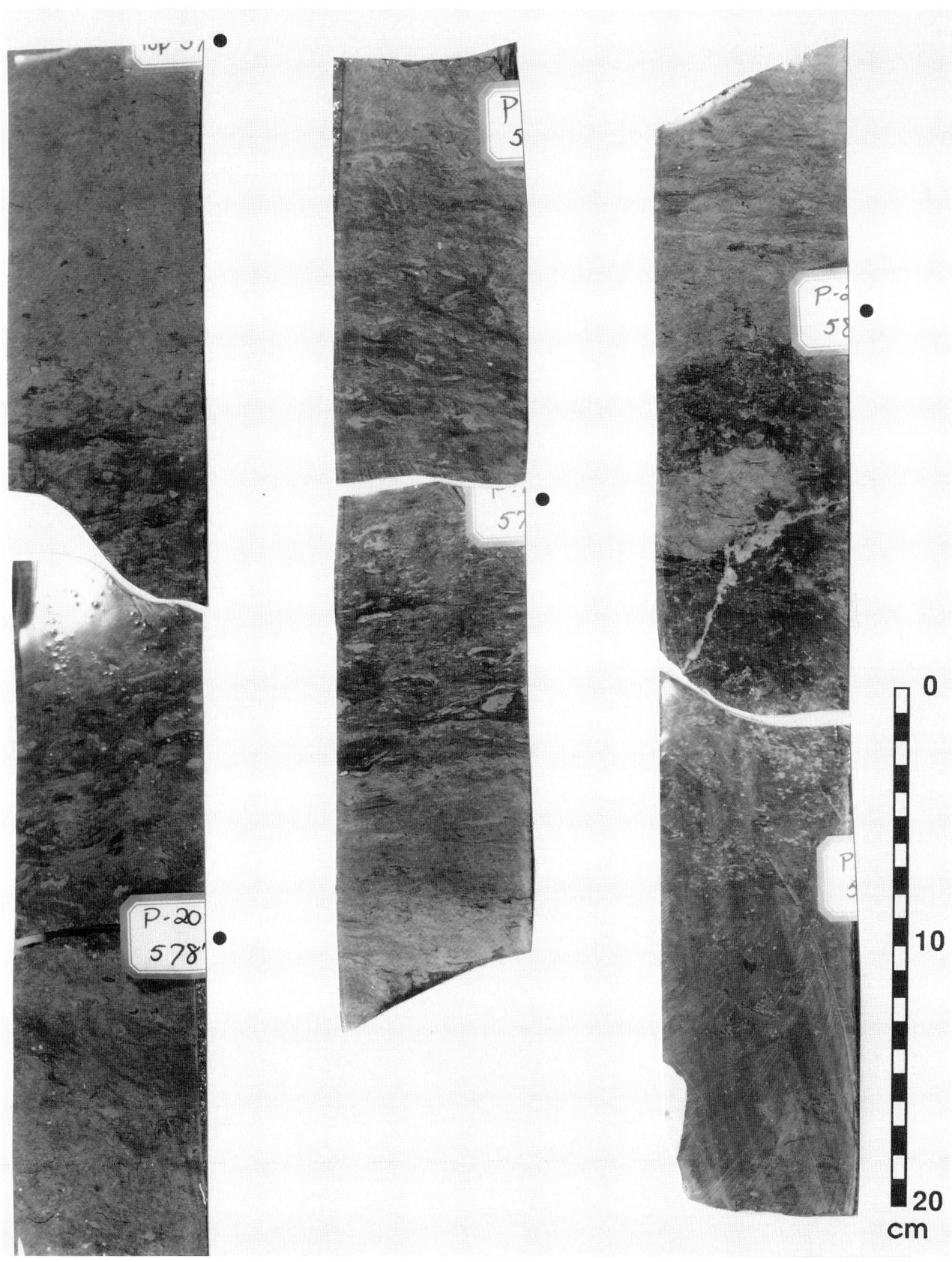

Figure 7. Photographs of the Providence Limestone in core A. This interval represents normal marine deposition. Wetted slabs. Dots mark footage divisions that match those in Fig. 5A.

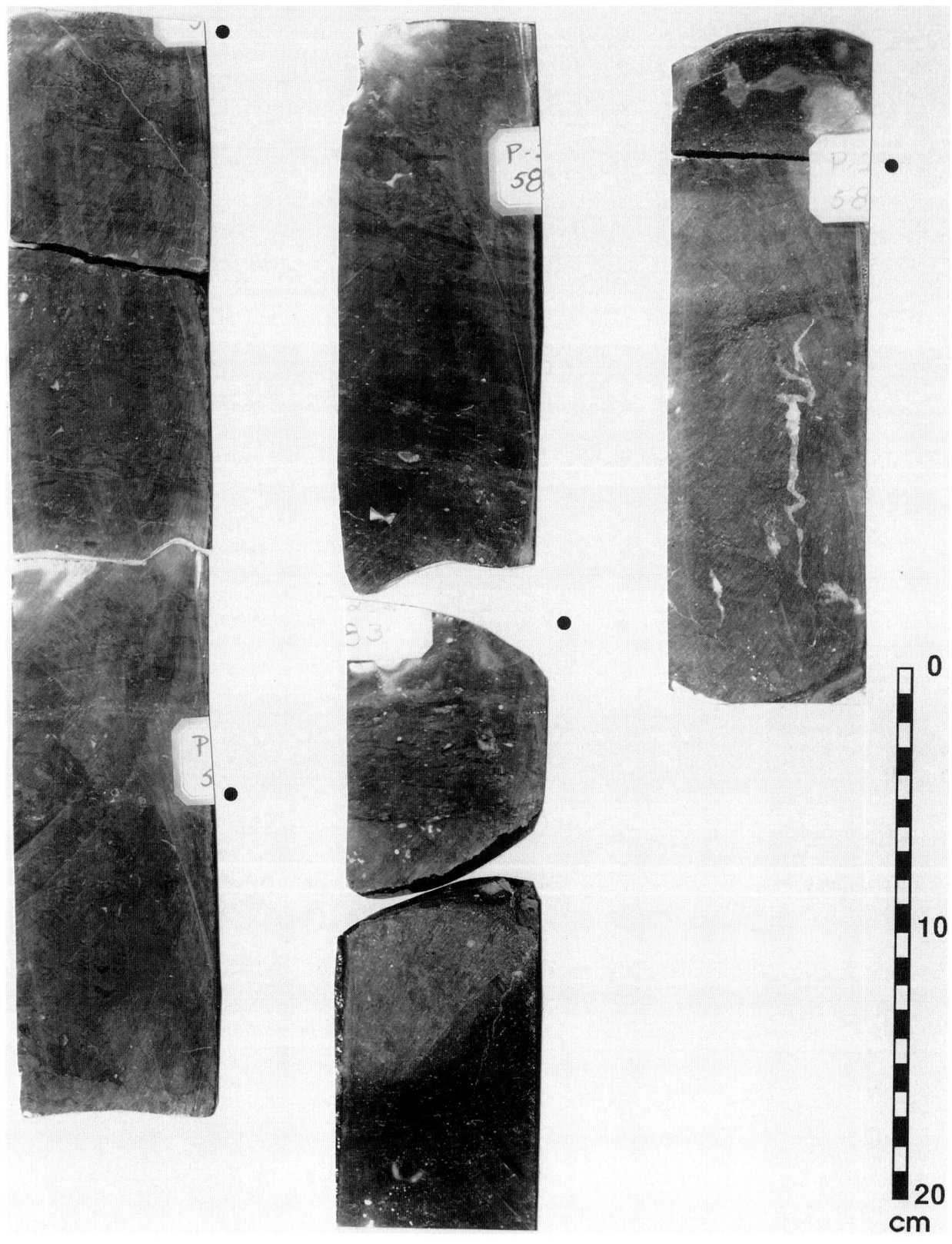

Figure 7 (continued).

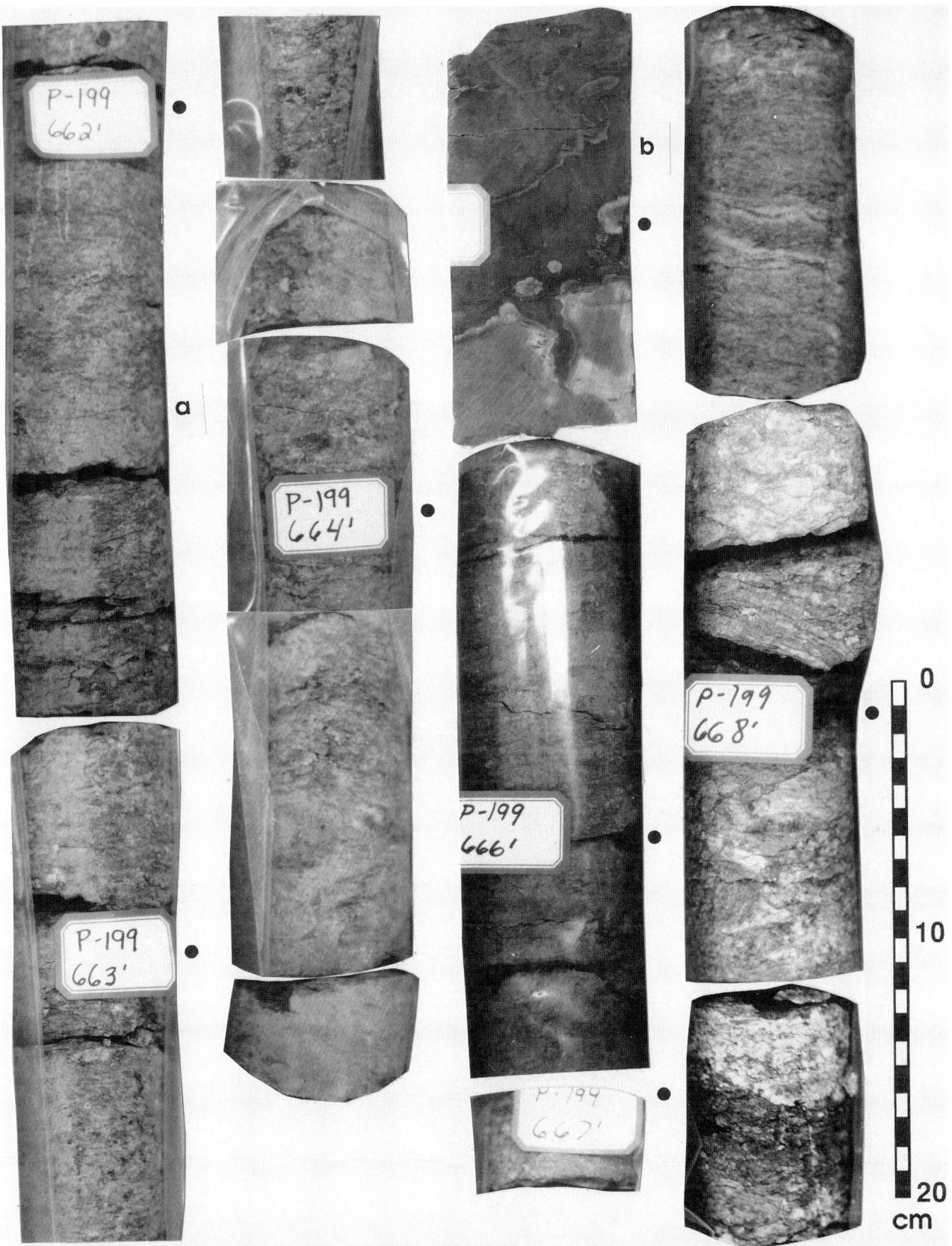

Figure 8. Photographs of the Providence Limestone in core B. This interval includes (from the top): the underclay beneath the No. 13 coal (a); thin interbedded flint clay within the underclay (b); fine sandstone (c); disrupted limestone with concretions (d). Wetted slabs. Dots mark footage divisions that match those in Fig. 5A.

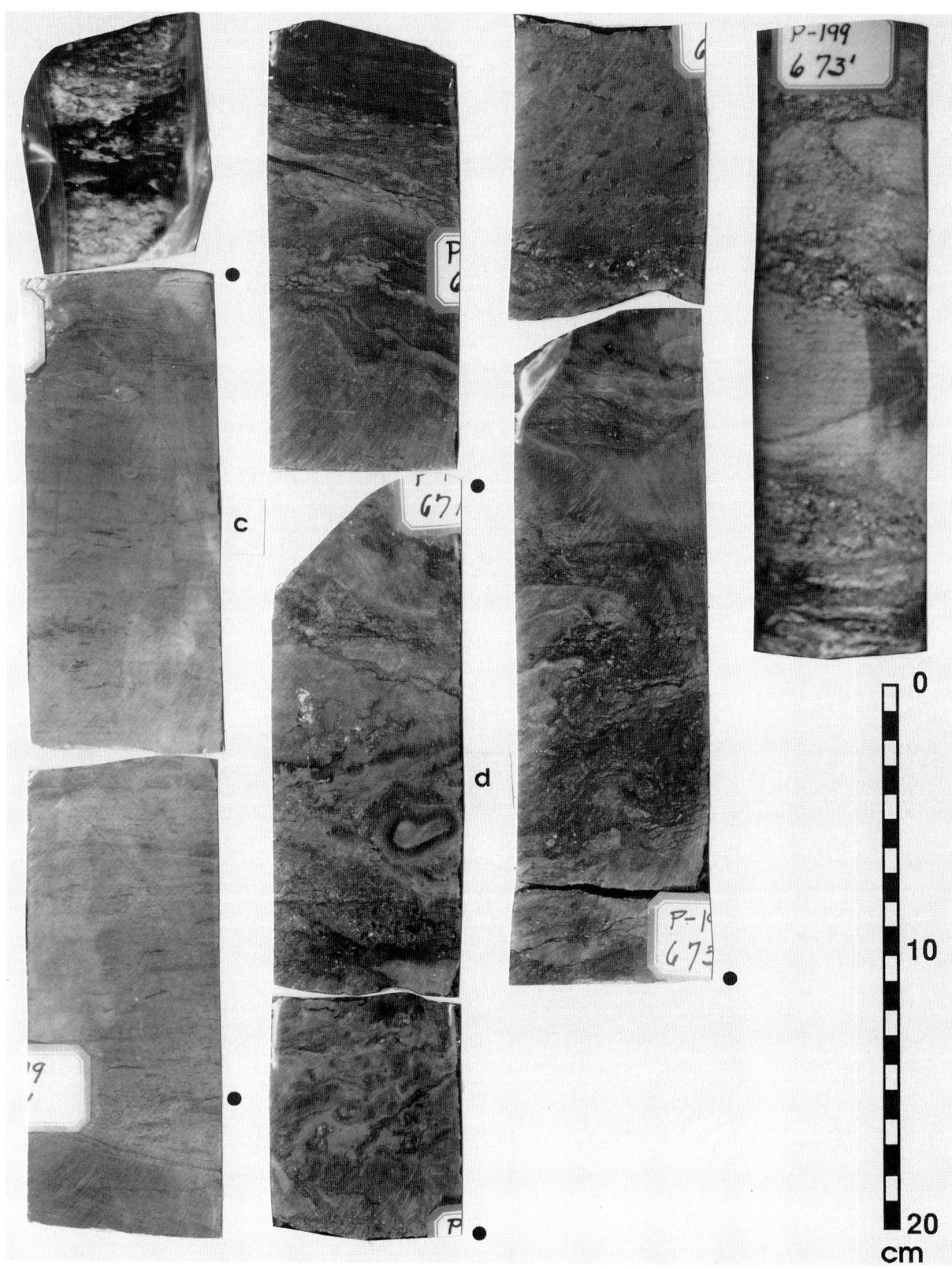

Figure 8 (continued).

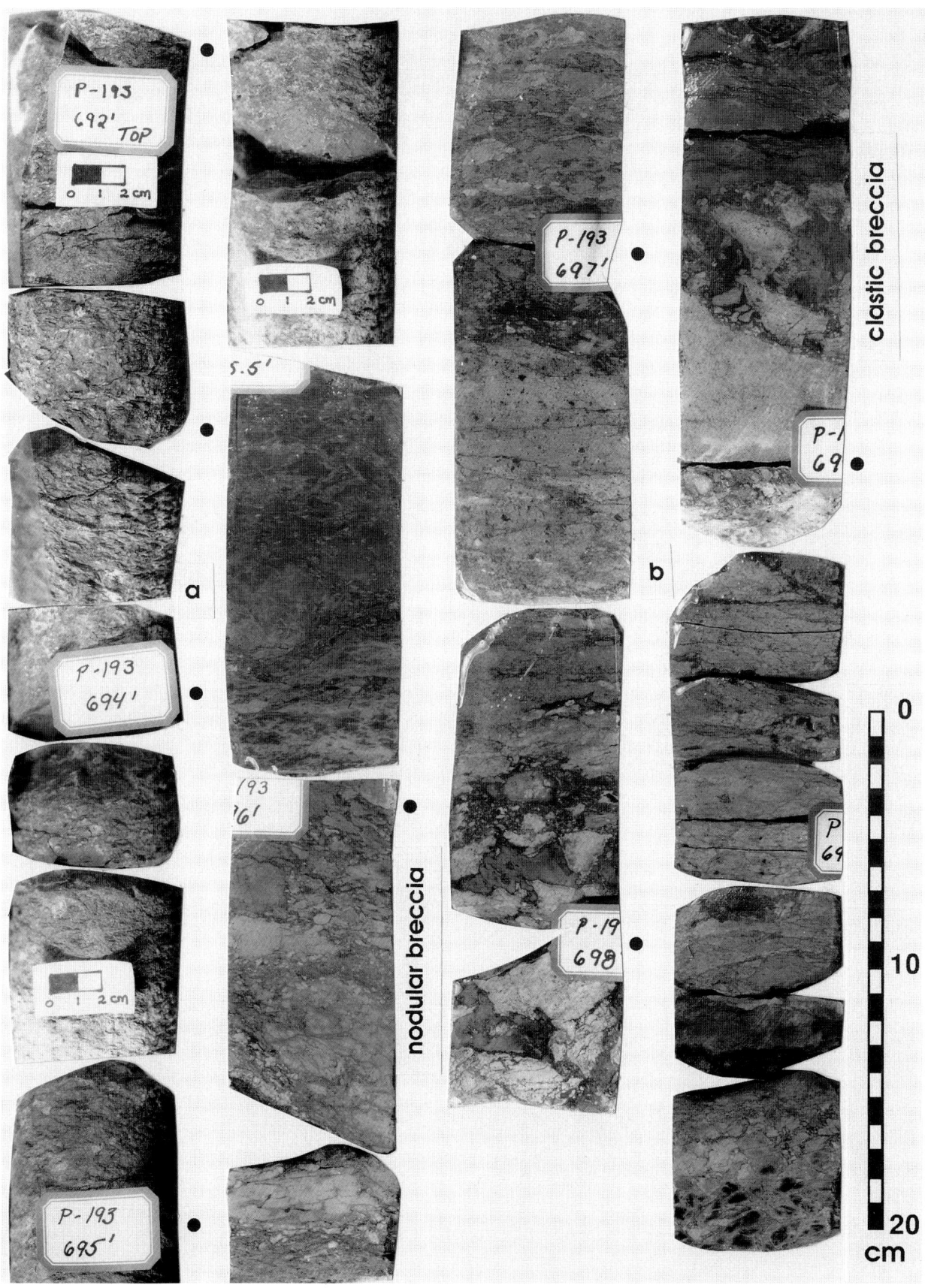

Figure 9. Photographs of the Providence Limestone in core C. This interval includes (from the top): the underclay beneath the No. 13 coal (a); flint clays (b); disrupted limestones (c). Wetted slabs. Dots mark footage divisions that match those in Fig. 5A.

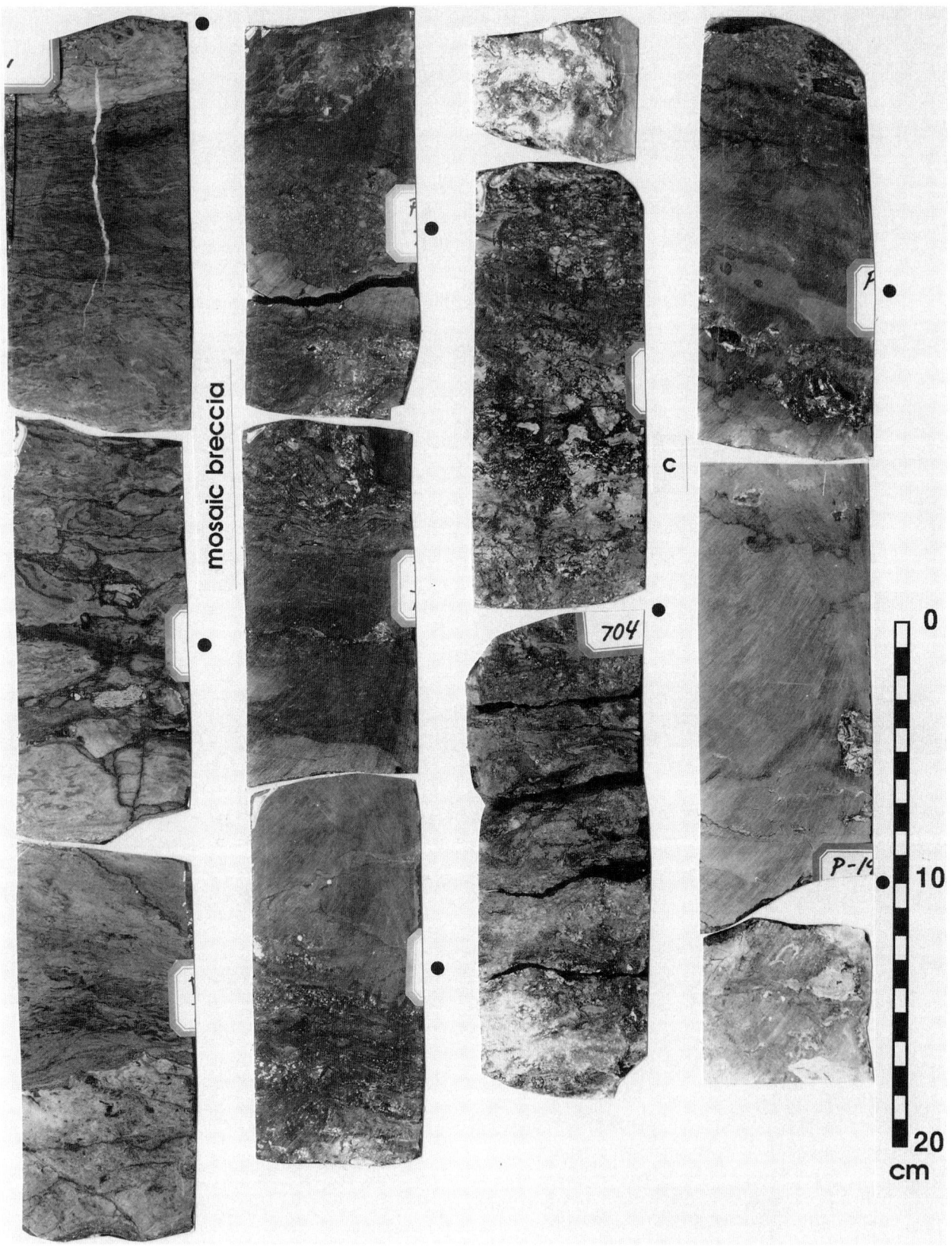

Figure 9 (continued).

Figure 10. Photographs of the Providence Limestone in core D. This interval includes (from the top): the underclay beneath the No. 13 coal (a); flint clays (b); normal limestones (c); disrupted limestones (d); shales with limestone clasts above the No. 11 coal. Wetted slabs. Dots mark footage divisions that match those in Fig. 5A.

Figure 10 (continued).

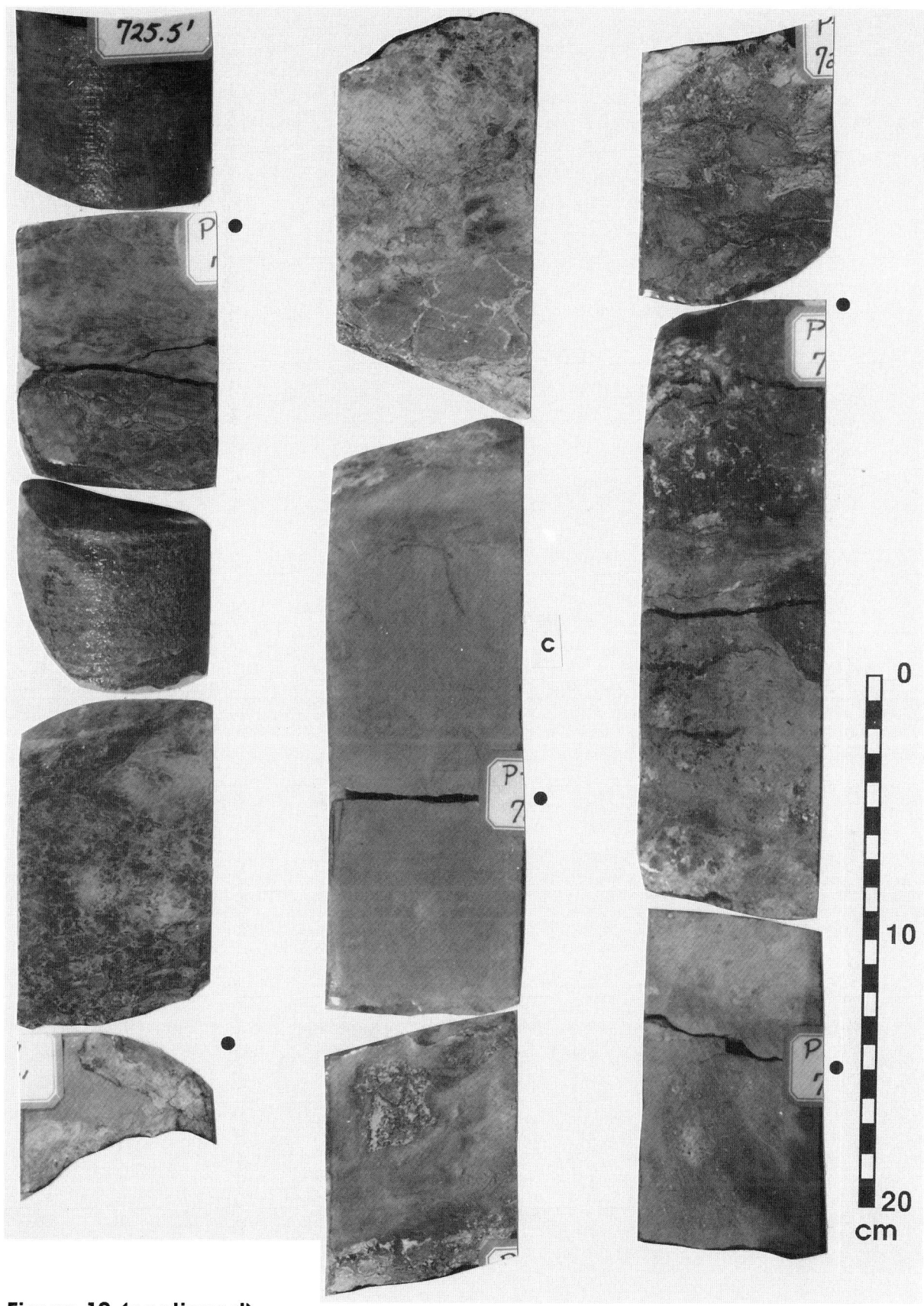

Figure 10 (continued).

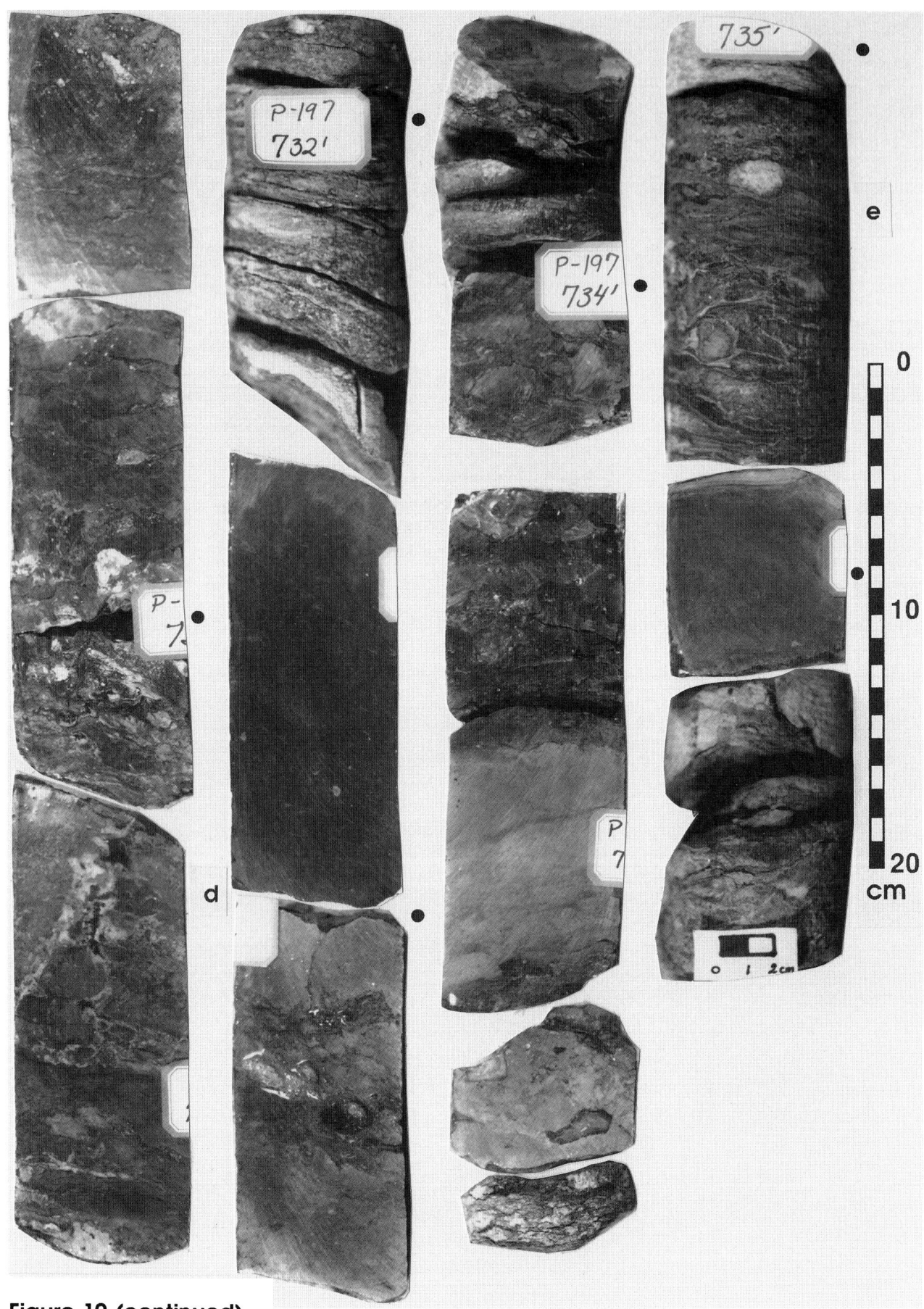

Figure 10 (continued).

occurs in core B near the base of the soft claystone (Fig. 8). The claystone is identified as a typical underclay, representing the old soil upon which the swamp was initiated and early plants were rooted. This underclay is composed of quartz, kaolinite, and/or illite (plus a minor amount of a mixed-layer clay), and trace amounts of calcite and pyrite.

A fine-grained sandstone, 14.6 in. (37 cm) thick, underlies the underclay in core B (Figs. 5A and 8). The sandstone contains weak, subhorizontal, discontinuous laminations. Short, *Skolithos*-type burrows are common.

A shale bed, 11.8 to 19.7 in. (30 to 50 cm) thick, overlies the No. 11 coal and is interbedded with the base of the Providence Limestone in the northwest (cores C and D). The shale contains thin carbonate nodules or clasts and exhibits anomalously high gamma radiation (Fig. 5A). The mineral composition of the shale includes kaolinite, illite (plus a mixed-layer component), and variable amounts of calcite, pyrite, and siderite.

Carbonates and kaolinite-rich cherts.--The carbonates and kaolinite-rich cherts between the No. 11 coal and underclay below the No. 13 coal exhibit remarkable variation in character across the study area (Fig. 5A). The Providence Limestone is 10 ft (3 m) thick at the southeast end of the section (core A), sandwiched between the underlying shale and overlying underclay (Figs. 5A and 7). Partially dolomitized wackestones and mudstones at the base of the unit are dark gray, slightly argillaceous, and highly bioturbated. Sponge spicule molds are abundant. They are filled with sparry calcite and/or chert but were originally composed of Opal A (Lane, 1981). Additional shell fragments represent echinoderms, brachiopods, bivalves, gastropods, bryozoans, and fusulinid foraminifera. Differential compaction is evident around early lithified concretionary portions of the carbonate. The center of the Providence consists of dark, bioturbated, skeletal wackestones with abundant crinoid fragments and whole brachiopods. The upper portion consists of lighter beds of fine-grained skeletal packstones that are weakly laminated and bio-mottled.

Toward the northwest, in cores B, C, and D, the carbonates become thin-bedded and are commonly brecciated (Fig. 5A). A thin bed of hard "claystone" (composed of quartz, calcite, kaolinite, and a minor amount of illite) is found in the middle of the underclay in core B (Fig. 8). The hard clay resembles typical flint clays (kaolinitic mudstones), mudstones that are widely associated with coal measures. The thickness of beds composed of kaolinite-rich chert progressively increases to the northwest in cores C and D, forming a 4 to 8.5 ft (1.2 to 2.6 m) thick deposit beneath the underclay (Figs. 5A, 9, and 10). The chert beds grade nearly imperceptibly into purer carbonate beds that represent the "disrupted" Provident Limestone. The chert and carbonate beds share many of the same components and fabrics, such as thin beds or various styles of brecciation and marine fossils. Thus, it appears that the kaolinite-rich cherts are an alteration product of the carbonates (see section below).

Descriptive (nongenetic) terms have been used to describe the breccia fabrics evident in the carbonate and kaolinite-rich chert (flint-clay) in the Providence Limestone interval in cores B, C, and D (Figs. 8-10). These are listed in Figure 5B and represented by symbols in the lithofacies cross section (Fig. 5A). "Clastic breccia" describes coarse (up to 2.4 in. (6 cm) long), angular clasts, supported in a mud to sandy matrix (see label in Fig. 9). The clasts are often derived from underlying beds. "Mosaic breccia" describes fragments that are closely fitted, as from in-situ disruption of a single bed (label in Fig. 9). "Nodular breccia" describes lumpy fragments that resemble mosaic breccia, except that there is more space between the fragments (see label in Fig. 9). Horizons of mosaic and nodular breccias include centimeter- or millimeter-scale fragments (the latter referred to as mini-mosaic or mini-nodular breccias). Brecciated horizons are interbedded with nonbrecciated horizons. The nonbrecciated horizons are commonly fractured and feature sharp upper contact surfaces.

The most likely cause of brecciation in the "disrupted" limestones was penetration by roots during periodic subaerial exposure of the carbonates or development of an extensive root system beneath the No. 13 coal swamp. Episodic development of paleosols is, however, more consistent with the observation of interbedded brecciated and nonbrecciated beds that have sharp upper surfaces. More detailed examination of the breccia fabrics will be necessary to more fully understand the nature of pedogenesis in these rocks.

Interpretation

Depositional patterns in the study area reflect rapid changes in topography and relative sea level, probably controlled by local structures, differential compaction, the influx of terrigenous sediments, and eustatic sea-level fluctuations. The succession from swamps to subtidal carbonates and back to swamps records a eustatic transgression-regression. Yet local structures must also have enhanced subsidence and influenced sedimentation and diagenetic patterns. In core A the thick No. 11 coal formed on a topographic high, but, throughout the transgression, the overlying thick marine limestone must have been deposited in a topographic (subtidal) low, relative to equivalent "disrupted" limestones in cores B, C, and D (pedogenically altered during periodic exposure).

Alternatively, it may be significant that the pedogenically altered limestones and "flint clays" occur where the No. 13 coal is thick. There are no pedogenic fabrics or flint clays in the limestone at core A where the No. 13 is thinner. If pedogenesis and flint clay development in the Providence were solely related to the thick overlying swamp, depositional conditions for the carbonates would have been uniform across the study area, without the development of structurally controlled topography. In either scenario regression involved the influx of terrigenous clastics (clays and sands) over marine carbonates and the reestablishment of a swamp (No. 13 coal).

DIAGENESIS

The Providence Limestone has undergone extensive diagenetic alteration in marine, freshwater, subaerial, and burial environments. Marine diagenesis included the precipitation of fibrous and peloidal cements in primary cavities. Subsequent subaerial exposure resulted in dissolution, fabric disruption, clay infiltration, and the formation of numerous pedological features. Silicification occurred early in the paragenetic sequence and again later during shallow burial. Dolomitization also occurred during burial diagenesis. The emphasis here will be on the features indicative of subaerial exposure, as they are particularly interesting in the cores and thin sections.

Subaerial Exposure Features

Cements.--The first cements are all nonferroan calcite, although peloidal, acicular, and fibrous forms probably had a high-Mg calcite or aragonite precursor mineralogy. Cement within the normal marine limestones (core A) is predominantly equant, nonferroan calcite (Fig. 11). Cement and allochems were partially replaced by weakly ferroan dolomite.

In cores B, C, and D, micritic peloids, rimmed with equant calcite spar crystals, are abundant in shelter pores and primary voids. The peloids are homogeneous in shape and size (subspherical, average 40 µm), occurring in clusters. In contrast, pelleted micrite is diffuse, forming mudstone and wackestone matrix sediment.

Fans of fibrous calcite form botryoidal crusts and pore-occluding cements in large cavities (Fig. 12). The acicular crystals are composed of radiating bundles of fibers. Growth zones are defined by alternating inclusion-rich and -poor internal zones. They may be overlain by marine micrite and shell debris. They are now composed of nonferroan, low-Mg calcite, but their original mineralogy has not yet been determined (Macintyre, 1983; Halley and Scholle, 1985; Kendall, 1985).

Equant calcite spar, both nonferroan and ferroan, forms the dominant pore-occluding cement within the Providence carbonate facies. It lines both primary and secondary pores and fills veins and fractures. Larger crystals exhibit compositional zonation, originating as an iron-free cement (nonferroan), followed by an iron-rich (ferroan) zone. Ferroan, nonferroan subzones may be present. Echinoid grains generally have syntaxial overgrowths of nonferroan sparry calcite.

Microfabrics of the disrupted limestones and silicified bed dissolution.--Partial to complete leaching of allochems and early cements is common. Originally aragonitic shells, such as gastropods, are preserved only as spar-filled molds. High-Mg grains, such as echinoderm plates, generally exhibit pitted and etched margins, although complete dissolution of the original skeletal material did occur. Such fragments are now preserved as spar-filled molds.

Early cements, particularly the fibrous calcites described above, show evidence of dissolution and fragmentation (Fig. 13). They are corroded, with serrated solution margins. Corroded and brecciated fragments of early sparry cements may also be floating in a clayey matrix (Fig. 14).

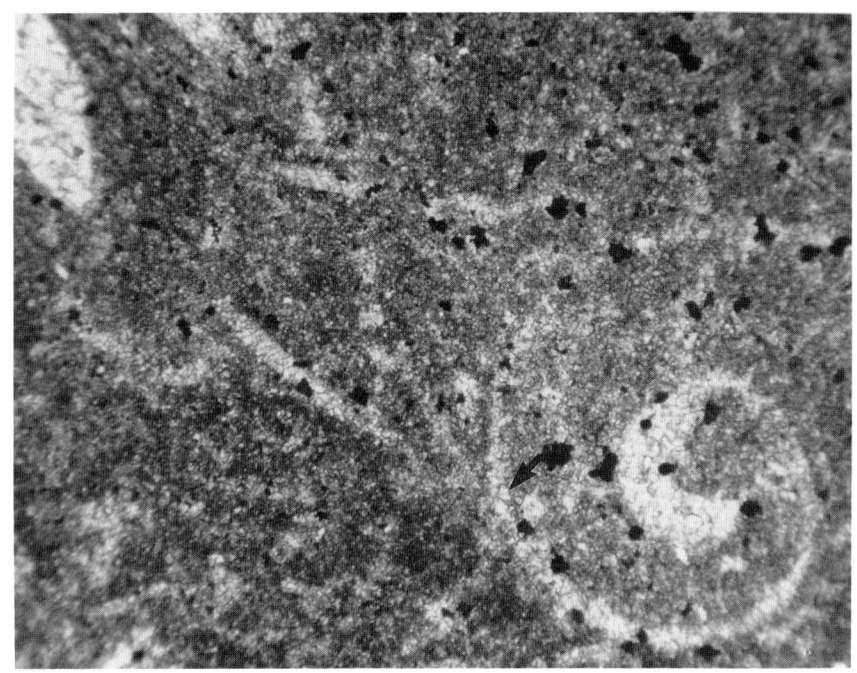

Figure 11. Typical marine wackestone lithology with sparry calcite filled shell molds in a micrite matrix. Minor quartz silt is present within the matrix (example at arrow). P-A-7-90. 4x.

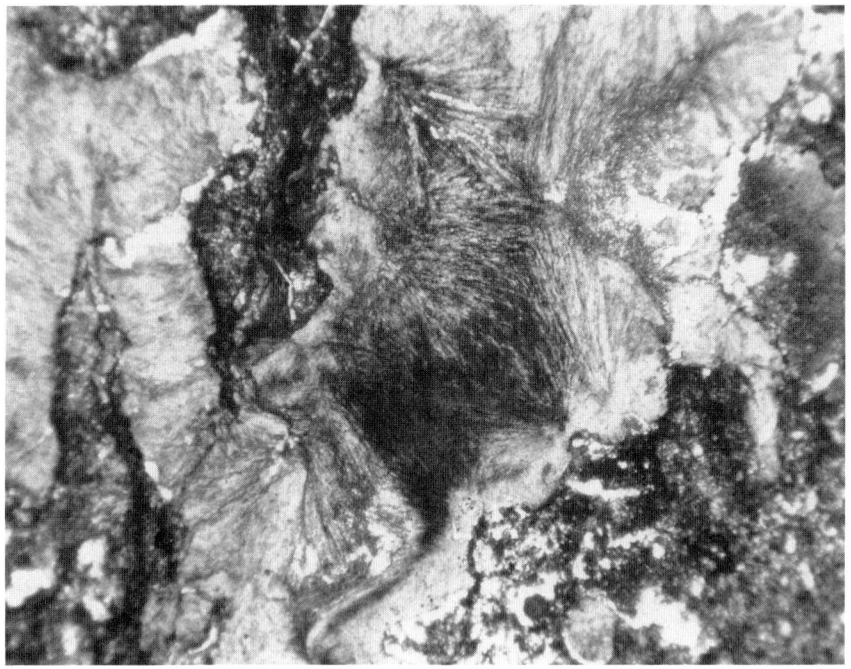

Figure 12. Coalescing fibrous cement fans. The concentration of inclusions varies from fan to fan. The cement is now composed of non-ferroan low-Mg calcite. P-D-13-90. 4x.

Figure 13. Fibrous cements have been fractured and reworked as clasts. They are preserved as relicts "floating" in an argillaceous, silty matrix. P-D-13-90. 4x.

Figure 14. Equant non-ferroan sparry calcite crystals are brecciated and corroded in an argillaceous matrix. This indicates that brecciation post-dates the precipitation of sparry cements which precipitated from meteoric porewaters. P-C-10-90. 4x.

Alteration.--Micritization, brecciation, and allochem degradation resulted in overall grain-size reduction and sediment homogenization (Fig. 15). Original marine packstones and wackestones may be altered to mudstones as primary features are obscured through fragmentation and micritization. Extensive alteration may produce a clotted fabric, which may be further modified by alveolar textures (Fig. 16).

Intra- and circumgranular cracking, brecciation, and differential cementation have led to the development of mosaic and nodular textures defined by peds and glaebules (Fig. 17). Ferran cutans (iron-oxide coatings) are locally abundant. Rhizoliths and crystallaria indicate the former presence of roots (Fig. 18A and B).

Replacement.--Allochems and clasts often are partially or completely replaced by an opaque, black, oxide clay coating (Fig. 19). Phosphate pebbles are locally abundant.

Silica has replaced much of the lime deposits as fabric-destructive chert and fabric-mimicking chalcedony. Silicification appears to have occurred at least twice; chert clasts, coated with ferran cutans, are common within carbonate facies, and secondary replacement chert has superseded all pre-existing fabrics and cements (Fig. 20). Chert also fills many late-stage fractures that cross-cut earlier cements.

Dolomitization.--Replacive dolomite is abundant only in core A where the upper two-thirds of the limestone have been replaced or partially replaced by fabric-obscuring sucrosic dolomite (Fig. 21). The dolomite is weakly ferroan. Larger, ferroan, baroque dolomite crystals occur in pores and fractures in all of the cores (Fig. 22).

Paragenesis and Interpretation

Paragenesis.--The first cements to precipitate within all of the limestones were peloidal and fibrous forms in primary interparticle pores. Micrite lithification and microspar development followed. Angular intraclasts of rigid micrite and microspar indicate an early origin. The first phase of silicification must also have occurred early during the diagenetic sequence, as chert intraclasts are common.

- Nonferroan, low-Mg calcite precipitated in primary and secondary voids, showing that it postdates initial aragonite dissolution and high-Mg calcite stabilization.

- Major sediment alteration occurred repeatedly, probably concurrent with phases of equant calcite spar precipitation.

- The second period of silicification, and the dolomitization, occurred during late diagenesis, after calcite cementation was essentially complete (Fig. 23).

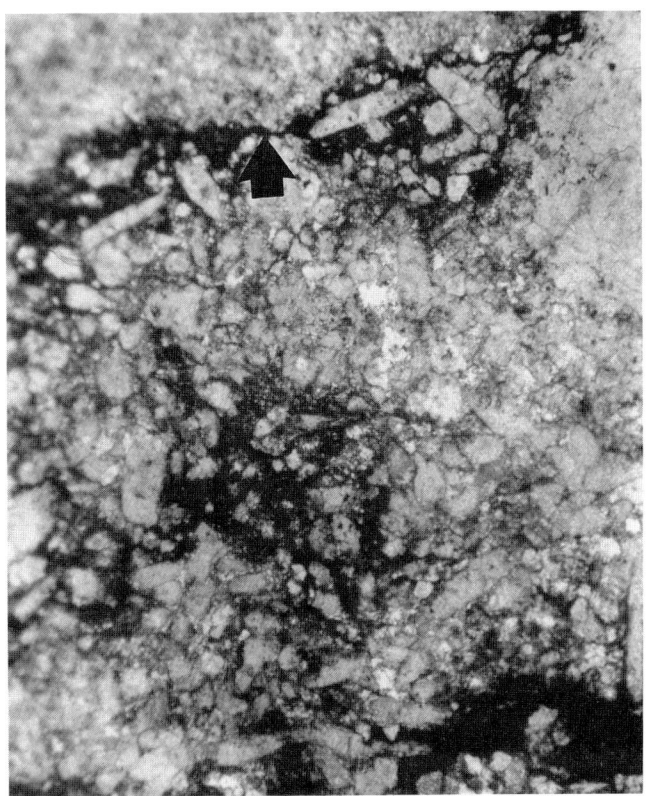

Figure 15. Pervasive brecciation and fragmentation has affected both calcite cement and matrix. The intertices are filled with dense, argillaceous micrite and quartz silt. A sharp boundary (arrow) marks the truncation of the lower surface. The overlying micrite is not brecciated. P-C-9-90. 4x.

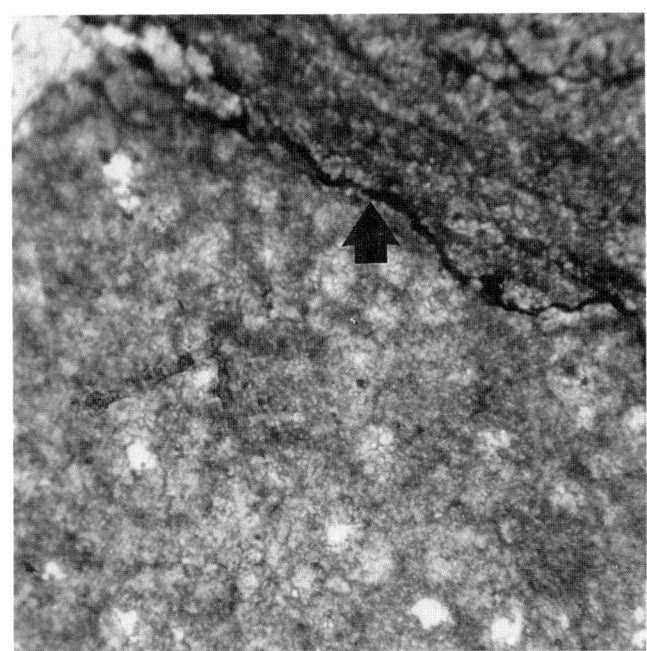

Figure 16. Alveolar texture in micrite with sparry calcite-filled crystallaria. The horizon is abruptly truncated by a stylolitic clay-lined irregular surface (arrow). P-B-8-90. 4x.

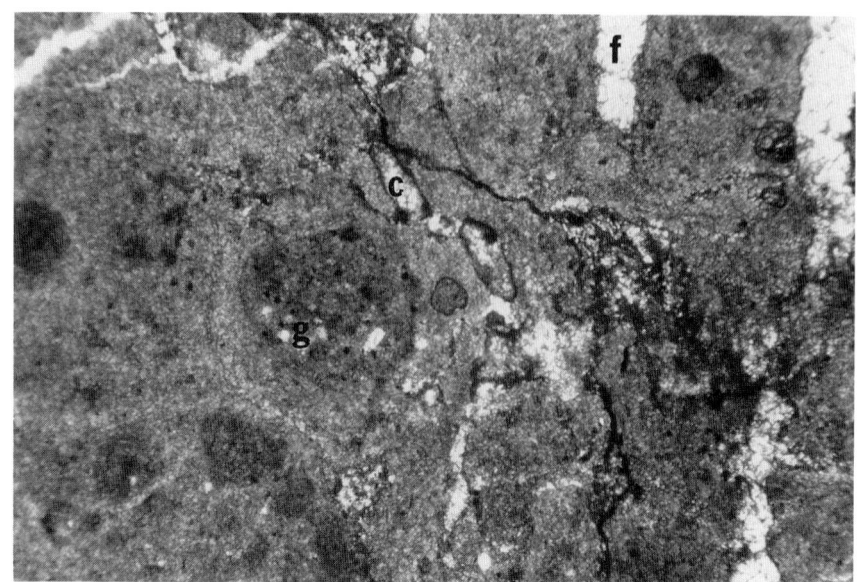

Figure 17. Glaebules (g), cutans (c) and sparry calcite-filled fractures (f) attest to pedogenic alteration of the original marine sediments. P-D-7-90. 4x.

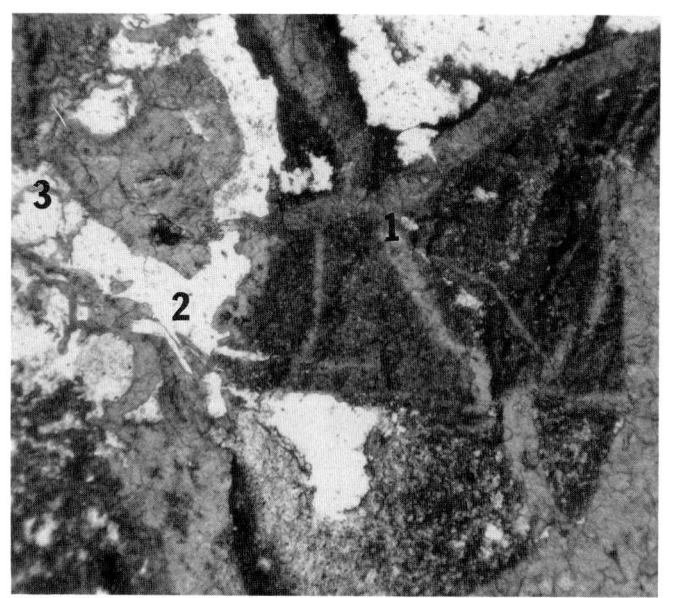 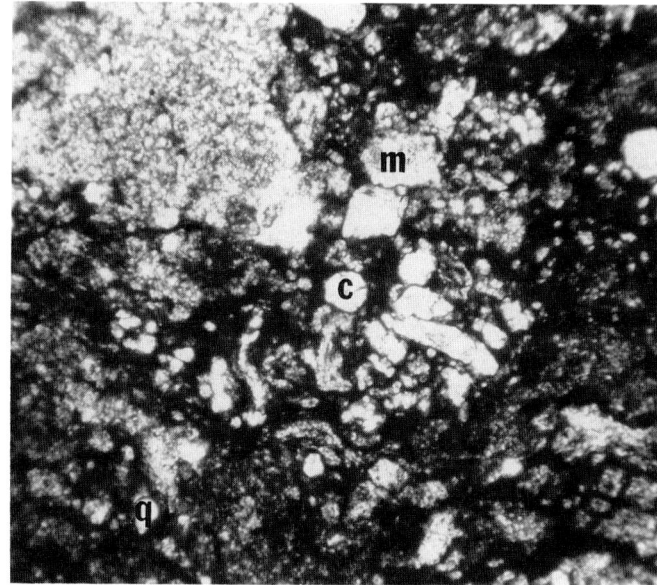

Figure 18.A. Numerous rhizoliths cross-cutting pre-existing fabrics and other root molds. Non-ferroan calcite filled root molds (1) are truncated by chert filled root molds (2) which are cut by small calcite filled rhizoliths (3). P-D-14-90. 4x.

Figure 18.B. Crystallaria (c), fragments of the original carbonate matrix (m) and quartz silt grains (q) are lithified in dense black mud and argillaceous micrite. P-B-10-90. 4x.

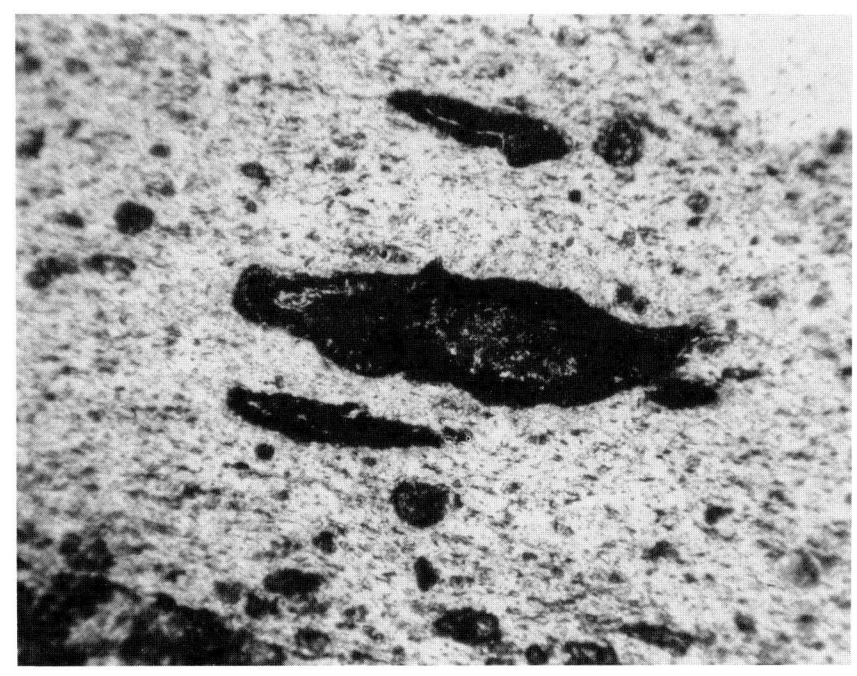

Figure 19. Intraclasts of lime mud and shell fragments are coated in an opaque film, creating blackened grains which are incorporated in a silicified matrix. P-D-4-90. 4x.

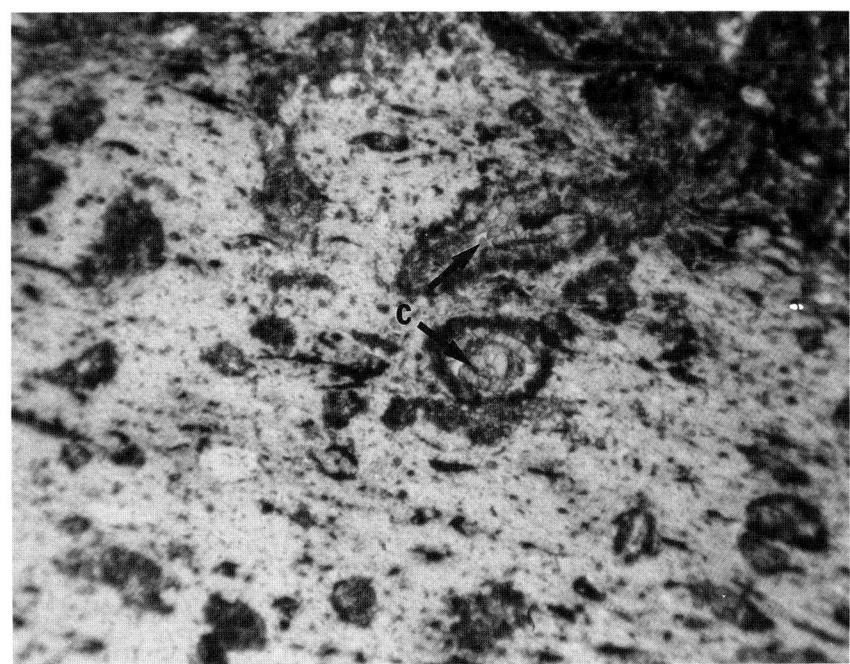

Figure 20. Relicts of non-ferroan sparry calcite cement (c) remain in some pores, although the remainder of the original carbonate has been replaced by chert. P-D-4-90. 4x.

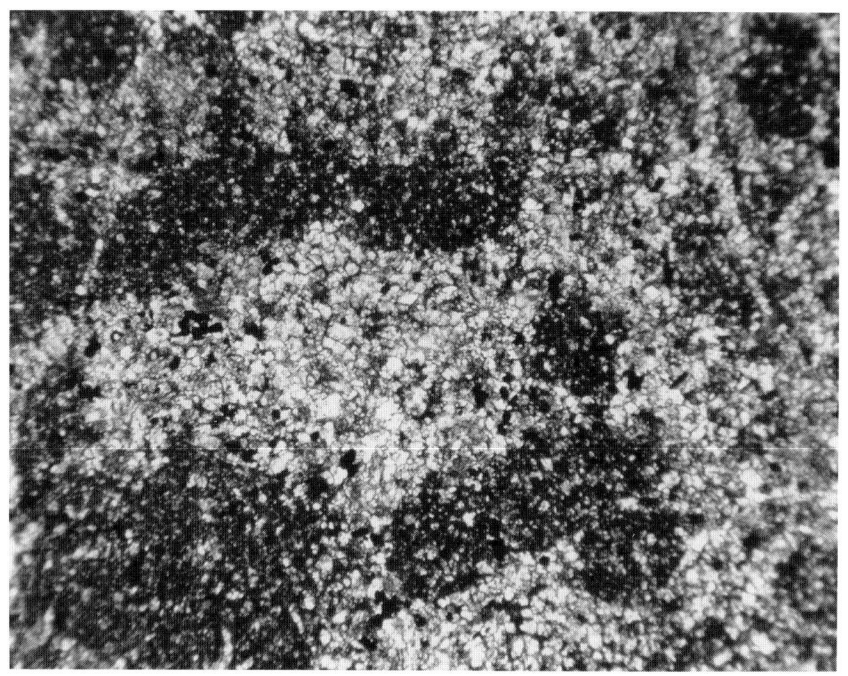

Figure 21. Partial dolomitization of the original mixed micrite and quartz silt (light particles in dark material) matrix. The sucrosic dolomite is weakly ferroan (light areas) and the matrix carbonate is non-ferroan (dark areas). P-A-7-90. 4x.

Figure 22. Weakly ferroan zoned baroque dolomite rhombs in clayey micrite matrix. P-C-10-90. 10x.

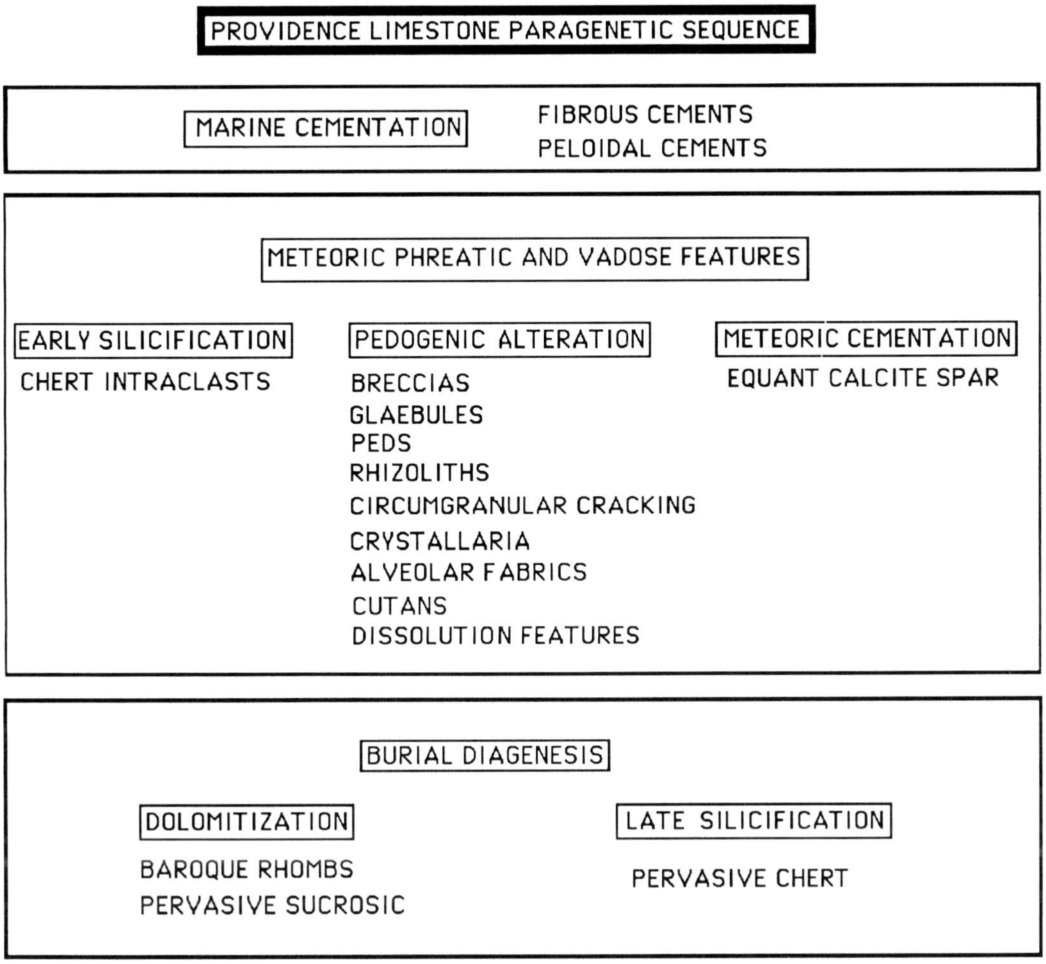

PROVIDENCE LIMESTONE PARAGENETIC SEQUENCE

MARINE CEMENTATION
FIBROUS CEMENTS
PELOIDAL CEMENTS

METEORIC PHREATIC AND VADOSE FEATURES

EARLY SILICIFICATION
CHERT INTRACLASTS

PEDOGENIC ALTERATION
BRECCIAS
GLAEBULES
PEDS
RHIZOLITHS
CIRCUMGRANULAR CRACKING
CRYSTALLARIA
ALVEOLAR FABRICS
CUTANS
DISSOLUTION FEATURES

METEORIC CEMENTATION
EQUANT CALCITE SPAR

BURIAL DIAGENESIS

DOLOMITIZATION
BAROQUE RHOMBS
PERVASIVE SUCROSIC

LATE SILICIFICATION
PERVASIVE CHERT

Figure 23. Schematic representation of the paragenetic sequence within the Providence Limestone member of the Sturgis Formation.

Interpretation.--The peloidal and fibrous cements precipitated in the marine phreatic environment in shelter pores and cavities. Their abundance, association with primary pores, and marine sediment overlays support this interpretation (Halley and Scholle, 1985; Kendall, 1985). Phylloid, algae-dominated crusts were sites for early lithification, although the fibrous cements also occur within sediments lacking obvious reef-building organisms. This indicates that there was sufficient seawater circulation to maintain supersaturation with respect to aragonite and/or calcite throughout various pore systems.

Dissolution and etching of early cements and allochems occurred in waters undersaturated with respect to carbonate. Dissolution is a common feature in the vadose zone, where undersaturated meteoric water moves readily through sediments. Brecciation, micritization, and localized cracking are processes commonly associated with pedogenesis and subaerial exposure (Arakel, 1982; Esteban and Klappa, 1983; Blodgett, 1988; Ettensohn and others, 1988; Goldstein, 1988; Retallack, 1988). The association of dissolution and alteration features strongly suggests that the marine sediments were repeatedly exposed to the subaerial environment.

This interpretation is supported by petrographic features including peds and glaebules leading to nodular textures, clotted and alveolar fabrics, rhizoliths, crystallaria, and cutans, features that are ascribed to pedogenic processes in the nearsurface environment (Arakel and McConchie, 1982; Wright, 1982; Warren, 1983; Wright and others, 1988).

The two phases of silicification probably represent a minimum of two periods of acidic porewater circulation through the lime sediments. The first early diagenetic silicification may have been the result of porewater dewatering from decaying organic matter in the present No. 11 coal seam. Silica from opaline sponge spicules was readily available and would be mobilized in acidic porewaters (Dapples, 1959; Degens, 1965; Engelhardt, 1977; Carson, 1987). As overburden pressure increased, acidic water expelled from the swamp sediments would have circulated through the overlying muds, silicifying them wherever silica saturation was attained (Williams and Crerar, 1985).

The second phase of silicification occurred during burial when increased overburden caused further porewater expulsion, driving both silicifying and dolomitizing fluids through the rocks (Gregg and Sibley, 1984; Hardie, 1987; Hesse, 1987; Thiry and others, 1988). Silica may have been sourced from remobilized spicular silica, organically complexed silica (Namy, 1979), and pressure-solution of quartz grains (Knauth, 1979; Hesse, 1987).

Magnesium for the dolomite may have been sourced from clay and/or feldspar degradation and high-Mg calcite stabilization. Slow-moving, warm burial fluids are capable of dolomitizing large quantities of carbonate (Hardie, 1987). This type of dolomitization tends to be concentrated along lithologic boundaries (Hardie, 1987). There is a major lithologic change between wells A and B, suggesting that such a change may have influenced the dolomitization pathway.

CONCLUSIONS

Interbedded limestones, mixed clastic-carbonate units, shales, and coals are the result of a complex interplay between depositional and local tectonic factors. The change from fully marine fauna to terrigenous plant remains within less than 9.8 ft (3 m) of section indicates a rapid and complete change in depositional setting.

Early diagenesis included marine and meteoric cementation. Subaerial exposure resulted in pedogenic alteration of the limestones in all sections except core A.

Subsequent diagenetic products include ferroan calcite cement, dolomite, and silica precipitation.

ACKNOWLEDGMENTS

This paper represents preliminary findings and interpretations from our continuing study of the Providence Limestone/Herrin Coal interval in western Kentucky. We wish to thank the following people who have helped us thus far by providing rocks to study or food for thought, especially Mike McClure and Island Creek Coal, John Ferm, Gerry Weisenfluh, and Buddy Justice.

REFERENCES

ARAKEL, A. V., 1982, Genesis of calcrete in Quaternary soil profiles, Hutt and Leeman lagoons, Western Australia: Journal of Sedimentary Petrology, p. 109-125.

_____, AND MCCONCHIE, D., 1982, Classification and genesis of calcrete and gypsite lithofacies in paleodrainage systems of inland Australia and their relationship to carnotite mineralization: Journal of Sedimentary Petrology, v. 52, p. 1149-1170.

AUSTIN, S. A., 1979, Depositional environment of the Kentucky No. 12 coal bed (Middle Pennsylvanian) of western Kentucky with special reference to the origin of coals: unpublished doctoral dissertation, Pennsylvania State University, University Park, 390 p.

BLODGETT, R. H., 1988, Calcareous paleosols in the Triassic Dolores Formation, southwestern Colorado, *in* Reinhardt, J. and Sigleo, W. R., eds., Paleosols and Weathering Through Geologic Time: Principles and Applications: Geological Society of America Special Paper No. 216, p. 103-122.

CARSON, G. A., 1987, Silicification fabrics from the Cenomanian and basal Turonian of Devon, England: isotopic results, *in* Marshall, J. D., ed., Diagenesis in Sedimentary Sequences: Geological Society Special Publication 36, p. 87-102.

DAPPLES, E. C., 1959, The behavior of silica in diagenesis, *in* Silica in Sediments: Society of Economic Paleontologists and Mineralogists Special Publication 7, p. 36-54.

DEGENS, E. T., 1965, Geochemistry of sediments: Prentice-Hall, New York, p. 73-80.

ENGELHARDT, W., 1977, The origin of sediments and sedimentary rocks, Part III: Halstead Press, p. 324-327.

ESTEBAN, M., AND KLAPPA, C. F., 1983, Subaerial exposure environment, *in* Scholle, P. A., Bebout, D. G., and Moore, C. H., eds., Carbonate Depositional Environments: American Association of Petroleum Geologists Memoir 33, p. 2-95.

ETTENSOHN, F. R., DEVER, G. R., JR., AND GROW, J. S., 1988, A paleosol interpretation for profiles exhibiting subaerial exposure "crusts" from the Mississippian of the Appalachian Basin, *in* Reinhardt, J. and Sigleo, W. R., eds., Paleosols and Weathering Through Geologic Time: Principles and Applications: Geological Society of America Special Paper 216, p. 49-80.

GOLDSTEIN, R. H., 1988, Paleosols of Late Pennsylvanian cyclic strata, New Mexico: Sedimentology, v. 35, p. 777-803.

GREGG, J., AND SIBLEY, D., 1984, Epigenetic dolomitization and the origin of xenotopic dolomite texture: Journal of Sedimentary Petrology, v. 54, p. 908-931.

HALLEY, R. B., AND SCHOLLE, P. A., 1985, Radiaxial fibrous calcite as early-burial, open-system cement: isotopic evidence from Permian of China (abstract): American Association of Petroleum Geologists Bulletin, v. 69, p. 261.

HARDIE, L. A., 1987, Dolomitization: a critical view of some current views: Journal of Sedimentary Petrology, v. 57, p. 166-183.

HESSE, R., 1987, Selective and reversible carbonate-silica replacements in Lower Cretaceous carbonate-bearing turbidites of the eastern Alps: Sedimentology, v. 34, p. 1055-1077.

HOWER, J. C., TRINKLE, E. J., GRAESE, A. M., AND NEUDER, G. L., 1987, Ragged edge of the Herrin (No. 11) coal, western Kentucky: Journal of Coal Geology, v. 7, p. 1-20.

KENDALL, A., 1985, Radiaxial fibrous calcite: a reappraisal, *in* Schneidermann, N. and Harris, P. M., eds., Carbonate Cements: Society of Economic Paleontologists and Mineralogists Special Publication 36, p. 59-77.

KHAWLIE, M. R., AND CAROZZI, A. V., 1976, Microfacies and geochemistry of the Brereton Limestone (Middle Pennsylvania) of southwestern Illinois, U.S.A.: Archives des Sciences, Geneva, v. 29, p. 67-110.

KNAUTH, L. P., 1979, A model for the origin of chert in limestone: Geology, v. 7, p. 274-277.

LANE, N. G., 1981, A nearshore sponge spicule mat from the Pennsylvanian of west-central Indiana: Journal of Sedimentary Petrology, v. 51, p. 197-202.

MACINTYRE, I. G., 1983, Growth, depositional facies and diagenesis of a modern bioherm, Galeta Point, Panama, *in* Harris, P. M., ed., Carbonate Buildups - A Core Workshop: Society of Economic Paleontologists and Mineralogists Core Workshop No. 4, p. 578-593.

MOSHIER, S. O., AND KIRKLAND, B. L, 1990, Generic identity and diagenesis of Pennsylvanian phylloid algae revealed by cathodoluminescence (abstract): American Association of Petroleum Geologists, v. 74, p. 726.

NAMY, J., 1979, Early diagenetic chert in the Marble Falls Group (Pennsylvania) of Central Texas, *in* Silica in Sediments: Nodular and Bedded Chert: Society of Economic Paleontologists Reprint Series No. 8, p. 106-112.

O'CONNELL, D. B., 1983, Paleoecology and environments of deposition of the Brereton Limestone Member (Pennsylvanian, Desmoinesian) in southwestern Illinois: unpublished M.Sc. thesis, Southern Illinois University, Carbondale, 120 p.

PERKINSON, M. C., 1978, Interpretations of the depositional environments for the Providence Limestone Member of the Sturgis Formation (Upper Pennsylvanian) in Muhlenburg and Ohio counties, Kentucky: unpublished M.Sc. thesis, Eastern Kentucky University, Richmond, 88 p.

PERKINSON, M. C., KUHNHENN, G. L., AND SMITH, A. D., 1984, Microfacies classification and its statistical verification for the Providence Limestone Member of the Sturgis Formation (Upper Pennsylvanian) in Muhlenburg and Ohio counties, Kentucky: The Compass, v. 61, p. 124-135.

RETALLACK, G. J., 1988, Field recognition of paleosols, *in* Reinhardt, J. and Sigleo, W. R., eds., Paleosols and Weathering Through Geologic Time: Principles and Applications: Geological Society of America Special Paper 216, p. 1-20.

THIRY, M., AYRAULT, M. B., AND GRISONI, J-C., 1988, Ground-water silicification and leaching in sands: example of the Fontainebleau Sand (Oligocene in the Paris Basin): Geological Society of America Bulletin, v. 100, p. 1283-1290.

WARREN, J. K., 1983, Pedogenic calcrete as it occurs in Quaternary calcareous dunes in coastal South Australia: Journal of Sedimentary Petrology, v. 53, p. 787-796.

WILLIAMS, L. A., AND CRERAR, D. A., 1985, Silica diagenesis II, general controls: Journal of Sedimentary Petrology, v. 55, p. 312-321.

WRIGHT, V. P., 1982, The recognition and interpretation of paleokarst: two samples from the Lower Carboniferous of South Wales: Journal of Sedimentary Petrology, v. 52, p. 83-94.

_____, PLATT, N. H., AND WIMBLEDON, W. A., 1988, Biogenic laminar calcretes: evidence of calcified root-mat horizons in paleosols: Sedimentology, v. 35, p. 603-620.

APPENDIX A - GLOSSARY OF SELECTED COAL TERMS

bone - (a) Coal that has a high ash content; hard and compact. (b) Argillaceous partings in coal.

clarain - A coal lithotype characterized macroscopically by semibright, silky luster, and sheet-like, irregular fracture. It is distinguished from vitrain by containing fine intercalations of a duller lithotype, durain. Its characteristic microlithtype is clarite.

cleat - In a coal seam, a joint or system of joints along which the coal fractures. There are usually two cleat systems developed perpendicular to each other.

flint clay - A smooth, flint-like, microcrystalline clay rock composed dominantly of kaolin, which breaks with a pronounced conchoidal fracture and resists slaking in water. It becomes plastic upon prolonged grinding in water.

fusain - A coal lithotype characterized macroscopically by its silky luster, fibrous structure, friability, and black color. It occurs in strands or patches and is soft and dirty when not mineralized. Its characteristic microlithtype is fusite.

fusinite - A maceral of coal within the inertinite group with intact or broken cellular structure, a reflectance well above that of associated vitrinite, and a particle size generally greater than about 50 microns, except when isolated from other macerals.

inertinite - A coal maceral group including micrinite, macrinite, sclerotinite, fusinite, semifusinite, and inertodetrinite. They are characterized by a relatively high carbon content and a reflectance higher than that of vitrinite.

liptinite (exinite) - A coal maceral group including sporinite, cutinite, alginite, resinite, and lipodetrinite, derived from spores, cuticular matter, resins, and waxes; is relatively rich in hydrogen.

semifusinite - A maceral of coal within the inertinite group having a reflectance intermediate between that of fusinite and that of associated vitrinite. It shows plant-cell structure, with cavities generally oval or elongated in cross section, but in some specimens less well defined than in fusinite; and it has a particle size generally greater than about 50 microns, except when isolated from other macerals.

underclay - A layer of fine-grained detrital material, usually clay, lying immediately beneath a coal bed or forming the floor of a coal seam. It represents the old soil in which the plants (from which the coal was formed) were rooted, and it commonly contains fossil roots (especially of the genus *Stigmaria*).

vitrain - A coal lithotype characterized macroscopically by brilliant, vitreous luster; black color; and cubic cleavage with conchoidal fracture. Its characteristic microlithtype is vitrite.

vitrinite - A coal maceral group that is characteristic of vitrain and is composed of humic material. It includes povitrinite and euvitrinite and their varieties, distinguished by a middle level of reflectance, higher than exinite and lower than inertinite in the same coal.

MISSISSIPPIAN CARBONATE-SILICICLASTIC EOLIANITES IN SOUTHWESTERN KANSAS

C. ROBERTSON HANDFORD
ARCO Oil and Gas Company,
Plano Research Center, Plano, TX 75075
BENJAMIN J. FRANCKA
ARCO Oil and Gas Company, Midland, TX 79702

ABSTRACT

Previous investigations of Mississippian St. Louis and Ste. Genevieve limestones in southwestern Kansas have suggested deposition in shallow marine environments exclusively. However, recent work has determined that these formations also contain a heretofore unrecognized eolian facies. Cross-bedded eolianites, made up of well-sorted ooid, skeletal, and peloid carbonate grains mixed with siliciclastic sands, reach 100 ft (30.5 m) thick in an area encompassed by seven southwestern Kansas counties. Cross-bed sets average about 1 to 3 ft (0.3 to 0.9 m) thick, but some reach 9 ft (2.7 m) thick. Foreset laminae, which are inclined approximately 16° to 20°, were formed by eolian grainfall processes. Thin, continuous laminae with inversely graded (texturally and compositionally) grains are abundant and interpreted as subcritically climbing translatent strata formed by migrating wind ripples. St. Louis subtidal grainstones commonly contain fragmented, large skeletal grains and are both cross-bedded to burrowed and moderately sorted. In contrast, the eolianites are well sorted and lack any coarse-grained bioclasts. All carbonate grains in the eolianite deposits are extremely abraded and well rounded. Indirect evidence suggesting an eolian origin includes features interpreted to represent rhizocretions, paleocaliche crusts, and a brecciated eolianite thought to have formed by surface weathering of a partially cemented dune ridge.

INTRODUCTION

Quaternary carbonate eolian dunes and eolianite deposits are abundant along many high-energy carbonate coastlines, but few, if any, are reported to occur in the pre-Quaternary geologic rock record. McKee and Ward (1983) implied that ancient carbonate eolianites may be present but recognition of such a deposit is "difficult, particularly in the subsurface, because of its similarity to clastic limestone deposited in high-energy, shallow-marine environments."

Although prior investigators (Goebel, 1968; Handford, 1988) correctly interpreted Mississippian (Meramecian) carbonates in Kansas to represent high-energy, shallow-marine grain shoals and open, muddy to bioclastic-rich shelf environments, this study will show that the upper St. Louis Limestone in southwestern Kansas also includes as much as 100 ft (30.5 m) of mixed carbonate-siliciclastic rocks deposited in coastal eolian dunes and associated interdune environments. An eolian interpretation is based upon the presence of (1) climbing translatent stratification of the type believed to be diagnostic of eolian sedimentation, (2) subaerially formed pedogenic features (laminated paleocaliche crusts, rhizocretions), and (3) textural and sorting features characteristic of eolian sediments.

REGIONAL SETTING

Southwestern Kansas lies well within the heart of the North American craton, and it has a relatively thin sedimentary cover. Of the approximately 9500 ft (2648 m) of Paleozoic-Mesozoic cover, as much as 1700 ft (518 m) of it consists of Mississippian rocks (Goebel, 1968). These rocks form an arcuate subcrop pattern beneath the pre-Pennsylvanian unconformity around the Hugoton embayment in western Kansas and dip gently southward (Fig. 1A). The western margin of the Hugoton embayment is marked by the Las Animas arch-Sierra Grande uplift and to the northeast lies the Central Kansas uplift, across which Mississippian strata are largely missing.

Upper Mississippian (Meramecian) rocks, which include the Warsaw, Salem, St. Louis, and Ste. Genevieve limestones, are up to 850 ft (259 m) thick in the Hugoton embayment (Fig. 2A). Except for the Ste. Genevieve and upper St. Louis limestones, these bioclastic, oolitic, and cherty limestones typically lack siliciclastic sediments (Goebel, 1968) and were, for the most part, deposited in shallow seas. The presence of evaporites, however, in the lower St. Louis may indicate that salina- and/or sabkha-type environments were also present. Maher and Collins (1949) suggested that Mississippian seas reached their northernmost extent in Meramecian time. The Central Kansas uplift and Las Animas arch exposed older Paleozoic rocks during late Meramecian time and shed siliciclastic sediments into the Hugoton embayment, thus marking the initial retreat of Mississippian seas from the craton (Goebel, 1968). Subsequently, retreating coastal environments deposited a succession of paralic carbonates, shales, and sandstones of the Chesterian interval, which disconformably overlie the largely carbonate Meramecian strata (Goebel, 1968).

LITHOFACIES

More than 160 ft (48.8 m) of core, representing the St. Louis Limestone, from three wells in Big Bow Field, Stanton and Grant counties, Kansas (Fig. 1), were described graphically (Figs. 3, 5, and 7; see Fig. 2B for key) and photographed (Figs. 4, 6, and 8). Five lithofacies are present and record deposition in environments ranging from open muddy-shelf, ooid shoal, tidal-flat, eolian dune, and interdune-wadi environments. These lithofacies are identical to those Handford (1988) described from the St. Louis and Ste. Genevieve limestones of Damme Field, Finney County, Kansas, which lies approximately 70 miles (113 km) northeast of Big Bow Field (Fig. 1).

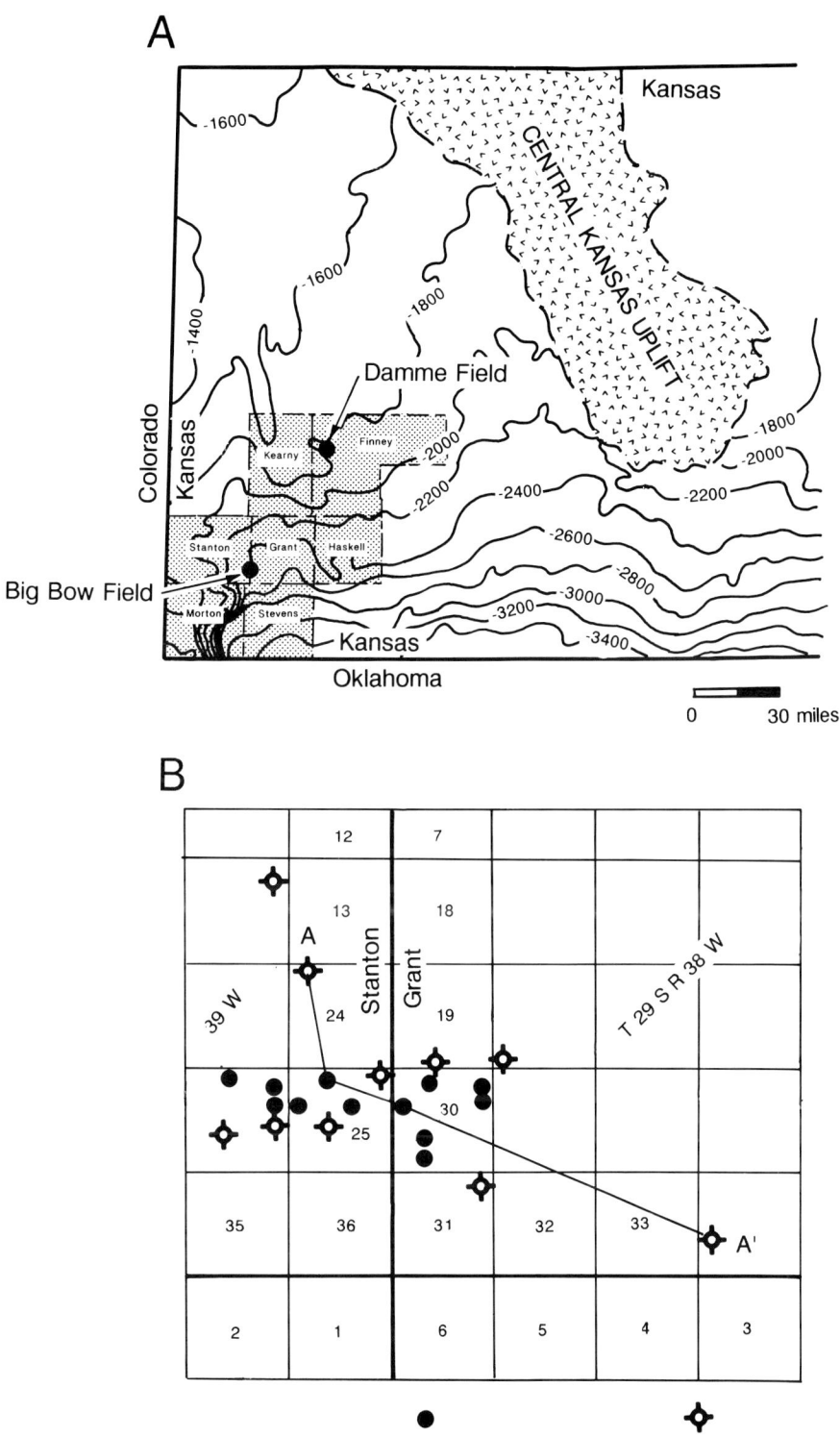

FIG. 1.--(A) Map of western Kansas showing subsea depth (in ft) to top of Mississippian System. Mississippian eolianites are present in the 7 shaded counties and in both Damme Field and Big Bow Field. Modified from Goebel (1968). (B) Map showing well control in Big Bow Field. Cores and logs used in this study are displayed in Figs. 3-8 and in the cross section A-A', shown in Fig. 9.

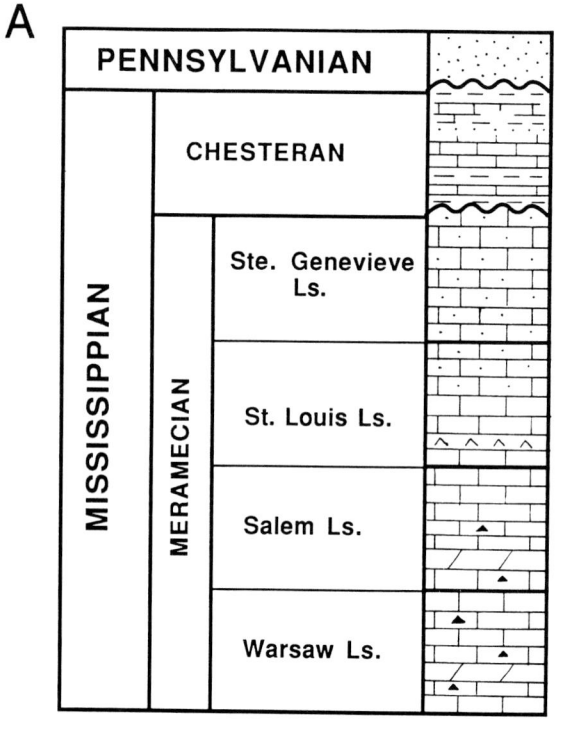

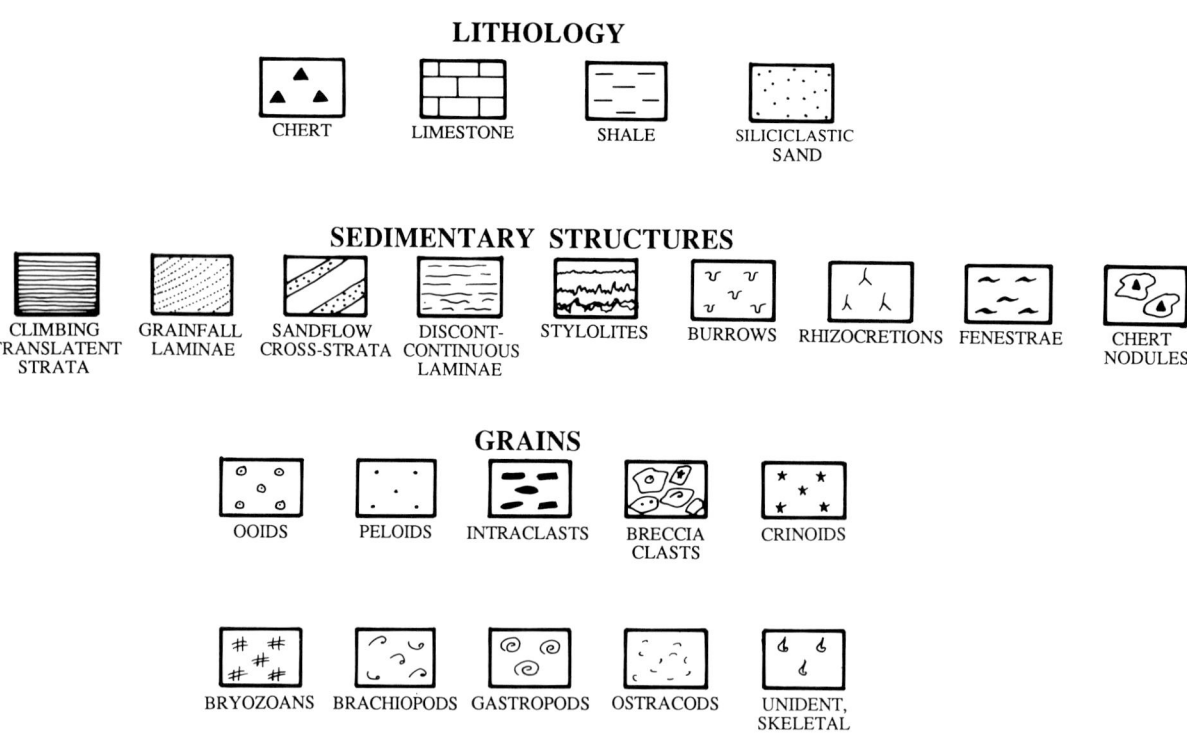

FIG. 2.--(A) Generalized stratigraphic column for upper Mississippian System in southwestern Kansas. (B) Key to lithology, sedimentary structures, and grains for Figs. 3, 5, and 7.

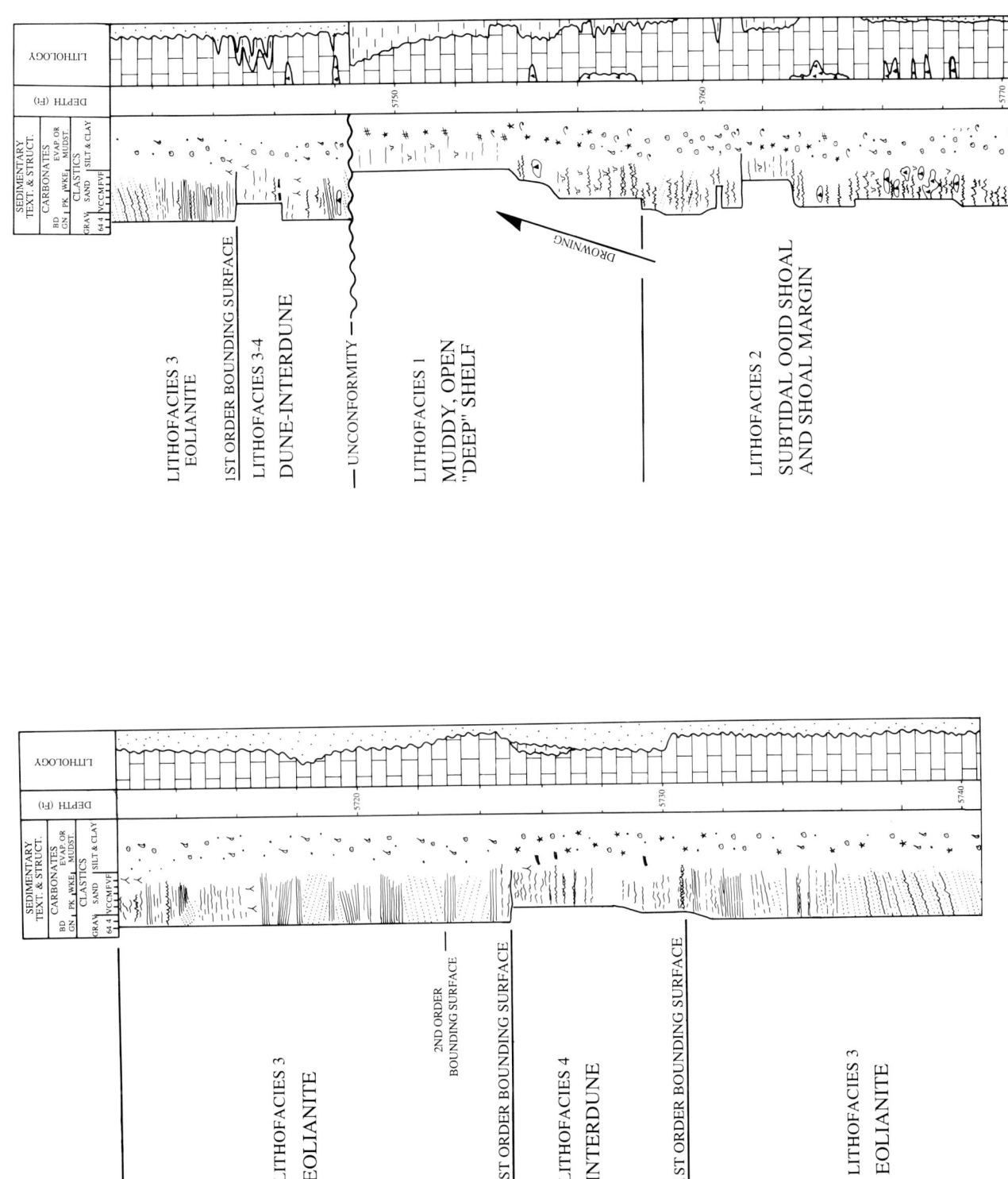

FIG. 3.—Graphic core description of St. Louis Limestone from ARCO M. M. Wilkerson #1, section 24, T. 29 S., R. 39 W., Stanton County, Kansas.

5712

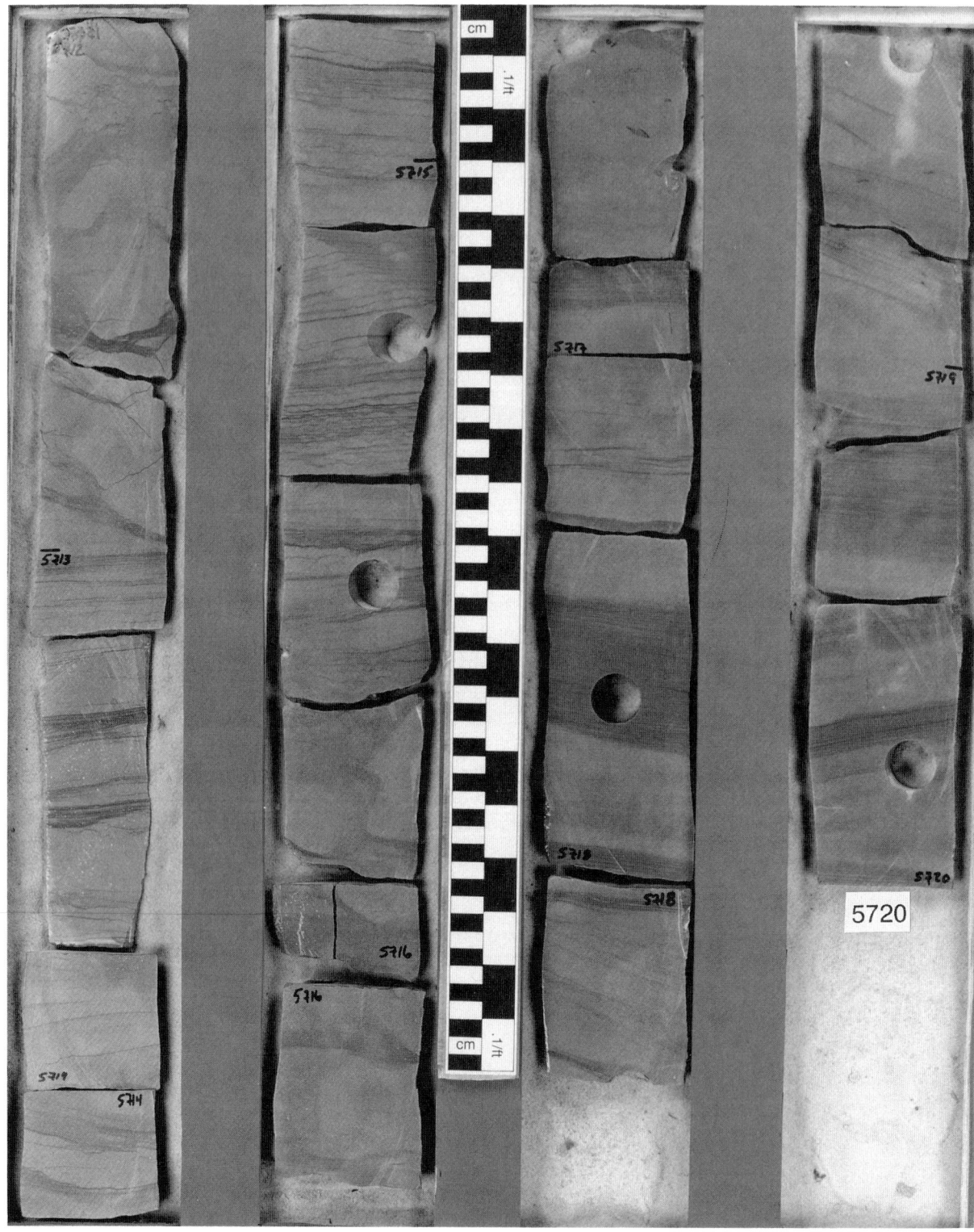

FIG. 4.--Core photographs of the ARCO M. M. Wilkerson #1, Sec. 24, T 29 S, R 39 W, Stanton County, Kansas.

5720

FIG. 4.--Continued

5728

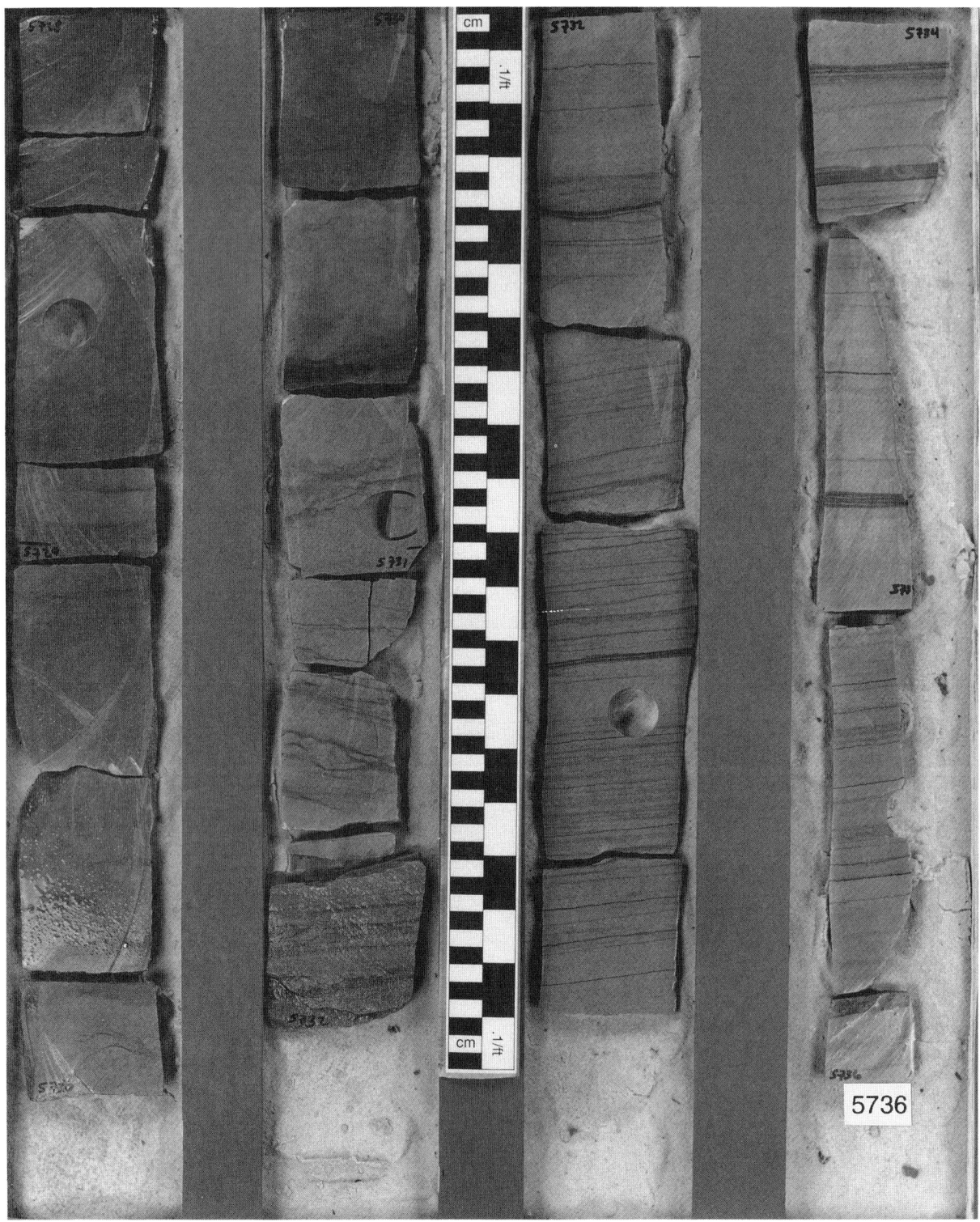

FIG. 4.--Continued

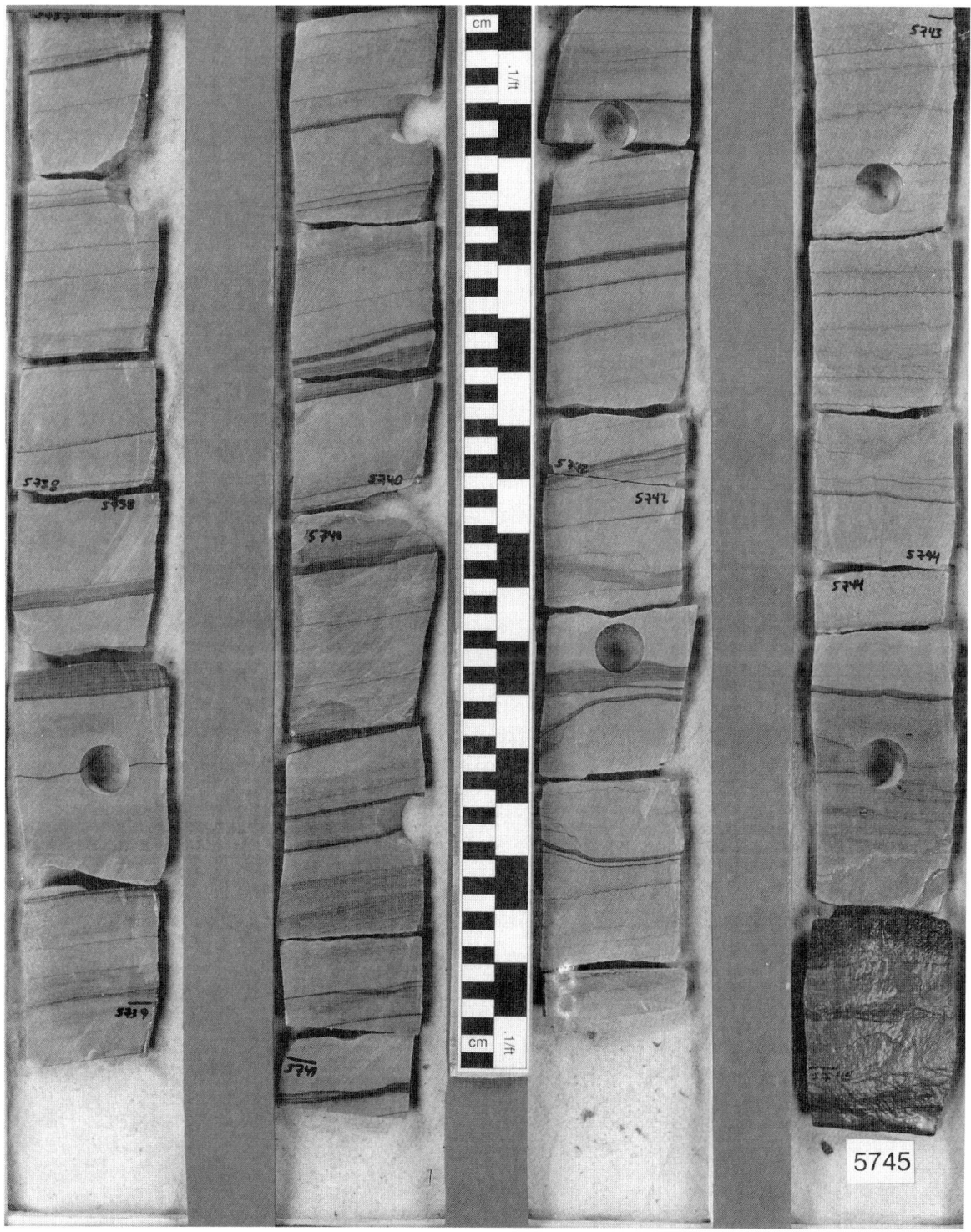

FIG. 4.--Continued.

FIG. 4.--Continued.

5755

FIG. 4.--Continued.

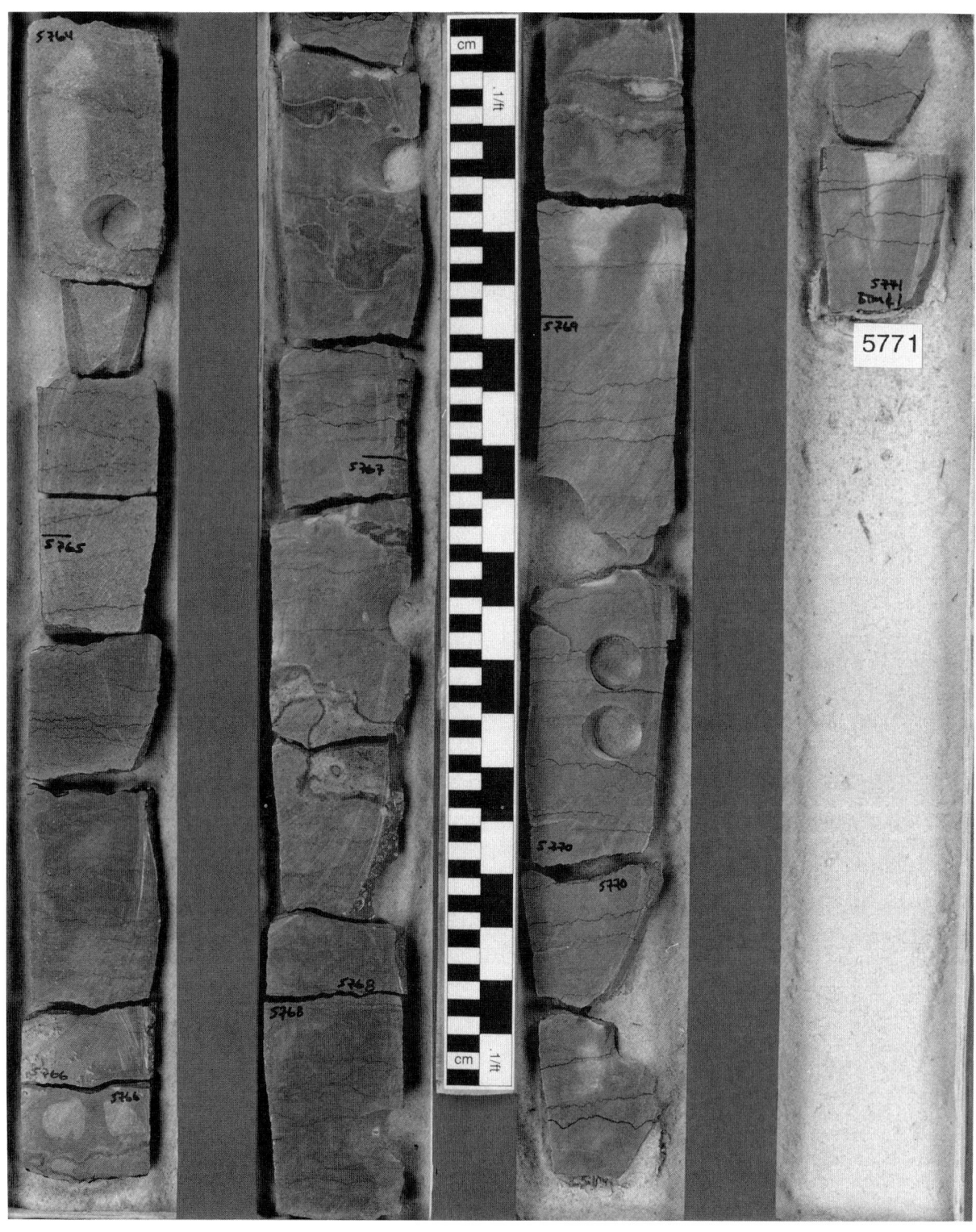

FIG. 4.--Continued.

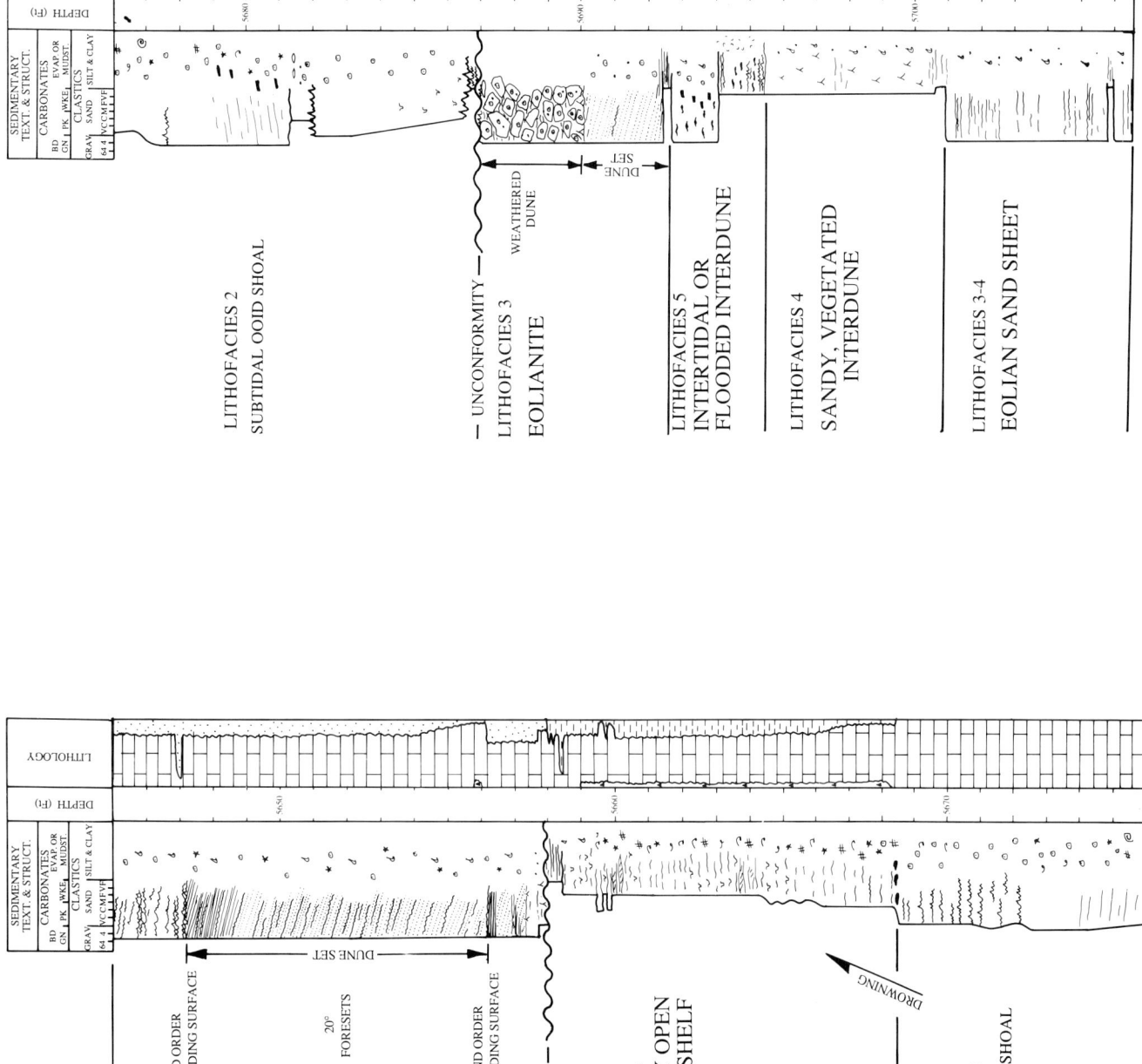

FIG. 5.—Graphic core description of St. Louis Limestone from ARCO P. E. Plummer #1, section 25, T. 29 S., R. 39 W., Stanton County, Kansas.

FIG. 6.--Core photographs of the ARCO P. E. Plummer #1, Sec. 25, T 29 S, R 39 W, Stanton County, Kansas.

FIG. 6.--Continued.

FIG. 6.--Continued.

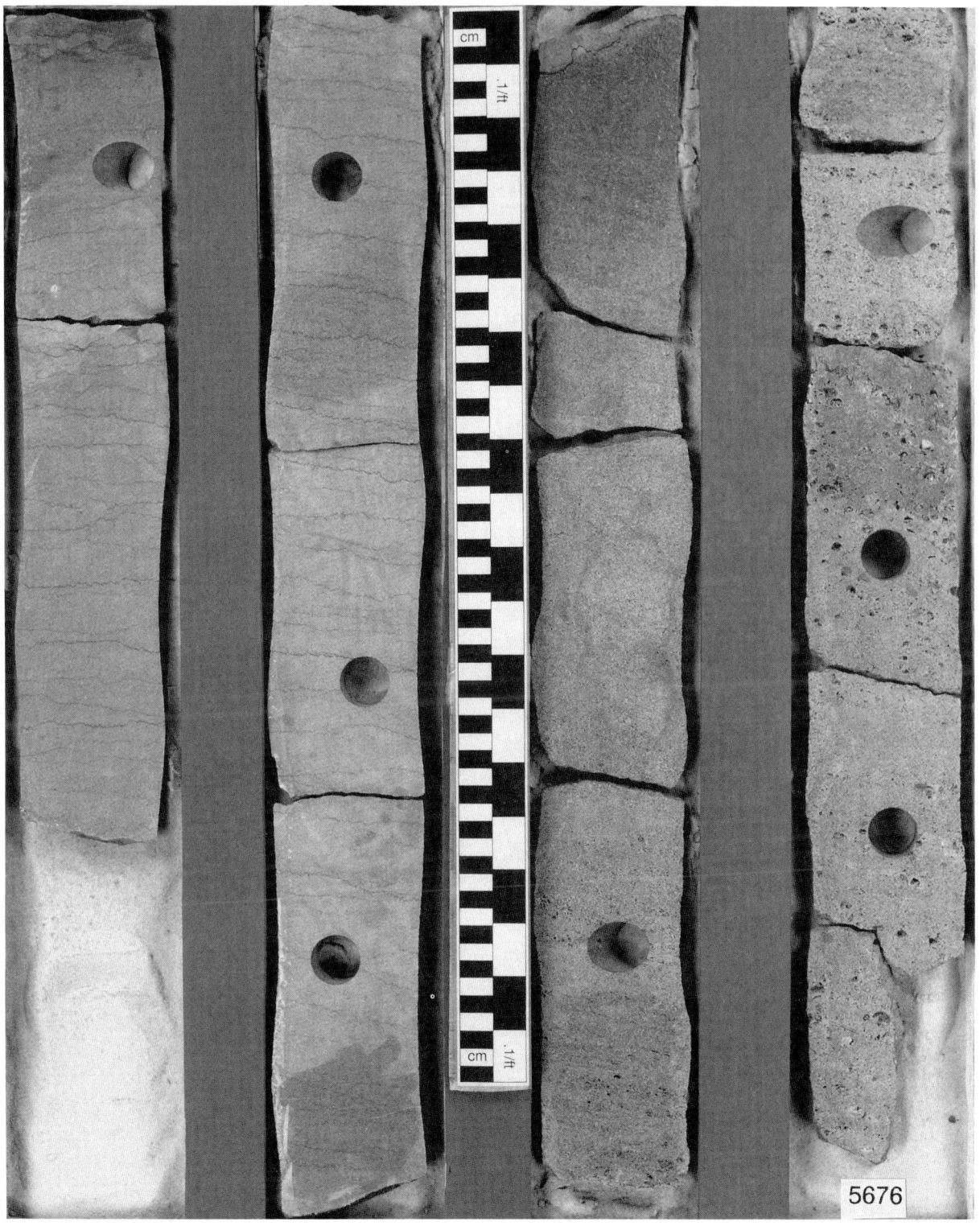

FIG. 6.--Continued.

FIG. 6.—Continued.

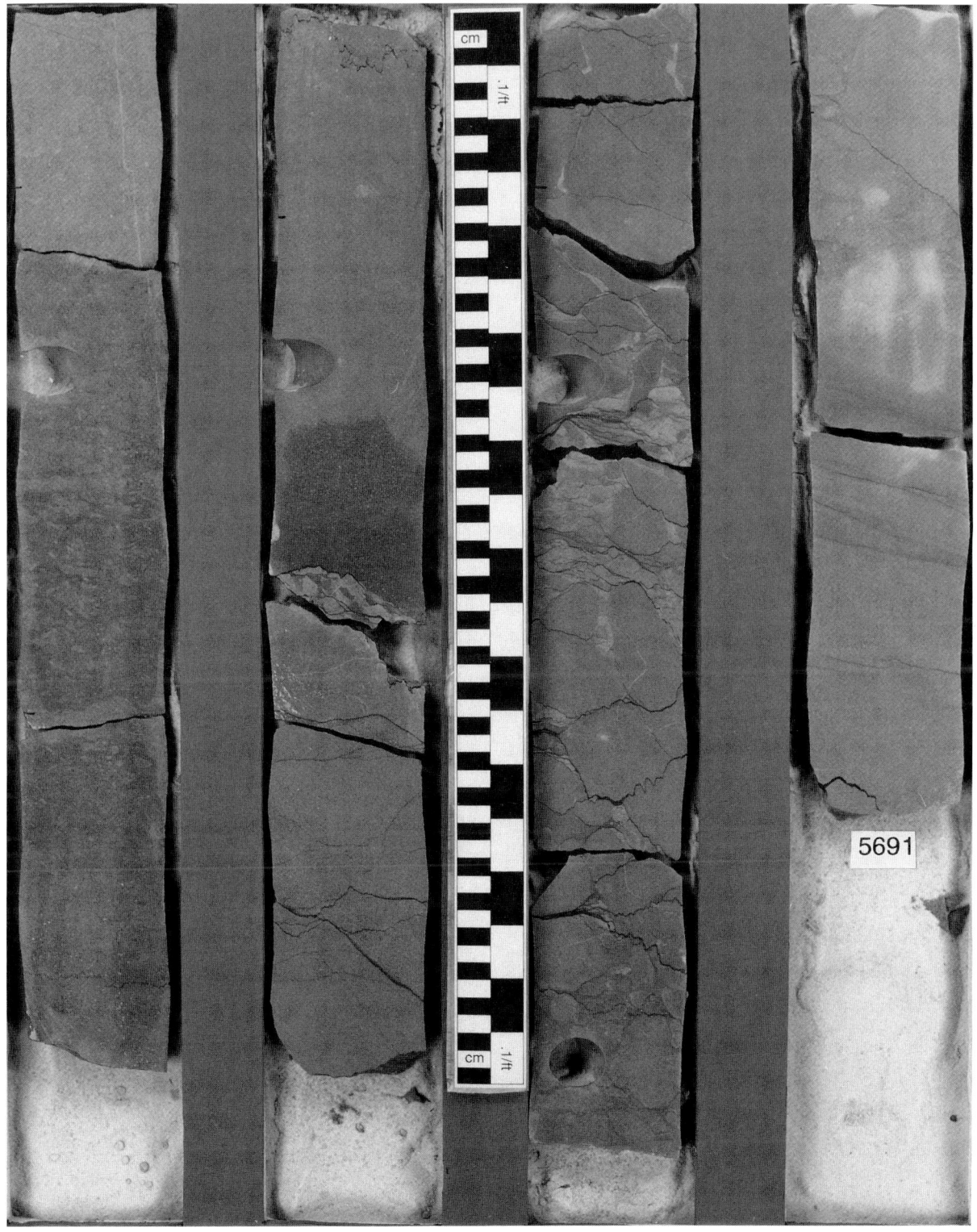

FIG. 6.--Continued.

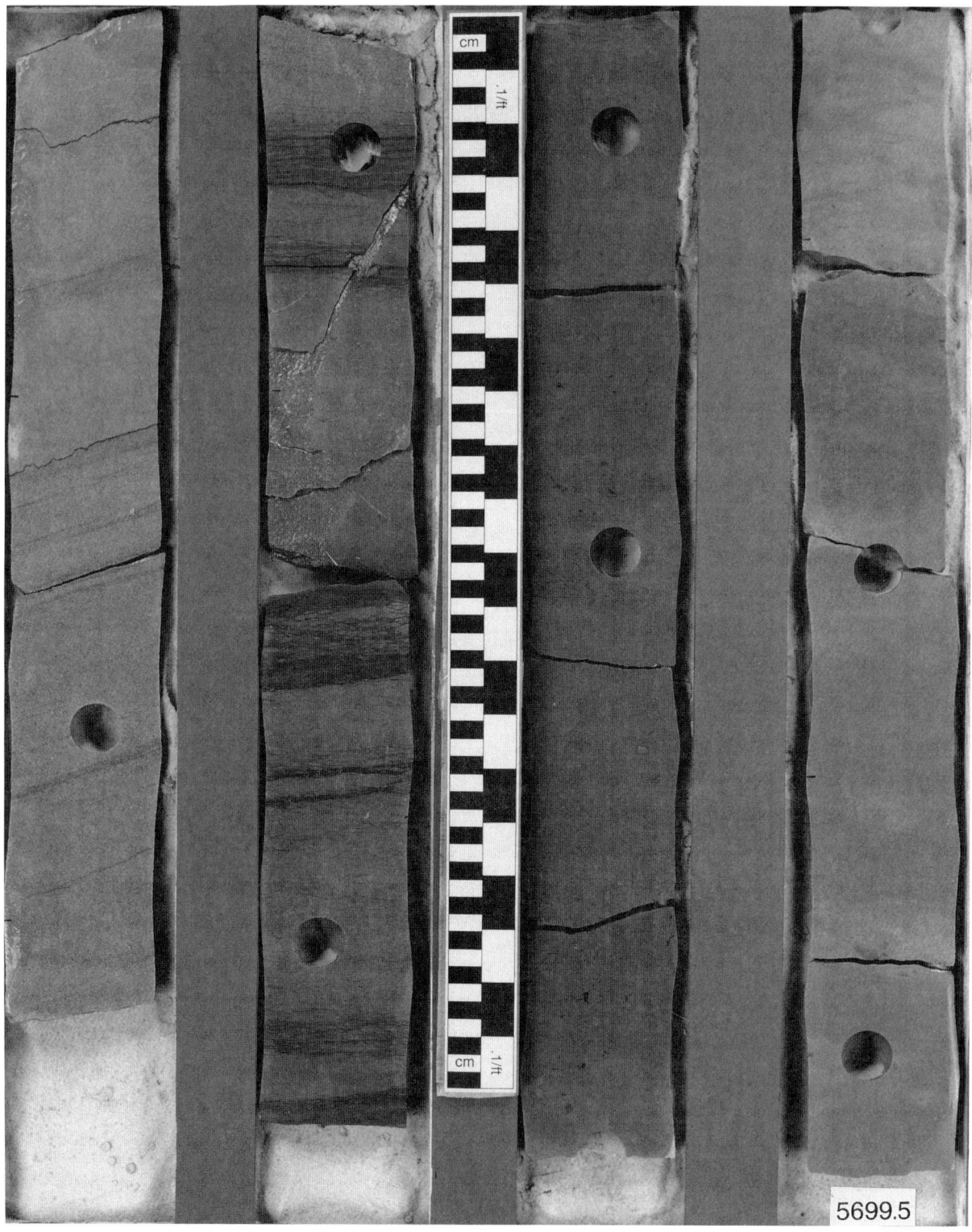

FIG. 6.--Continued.

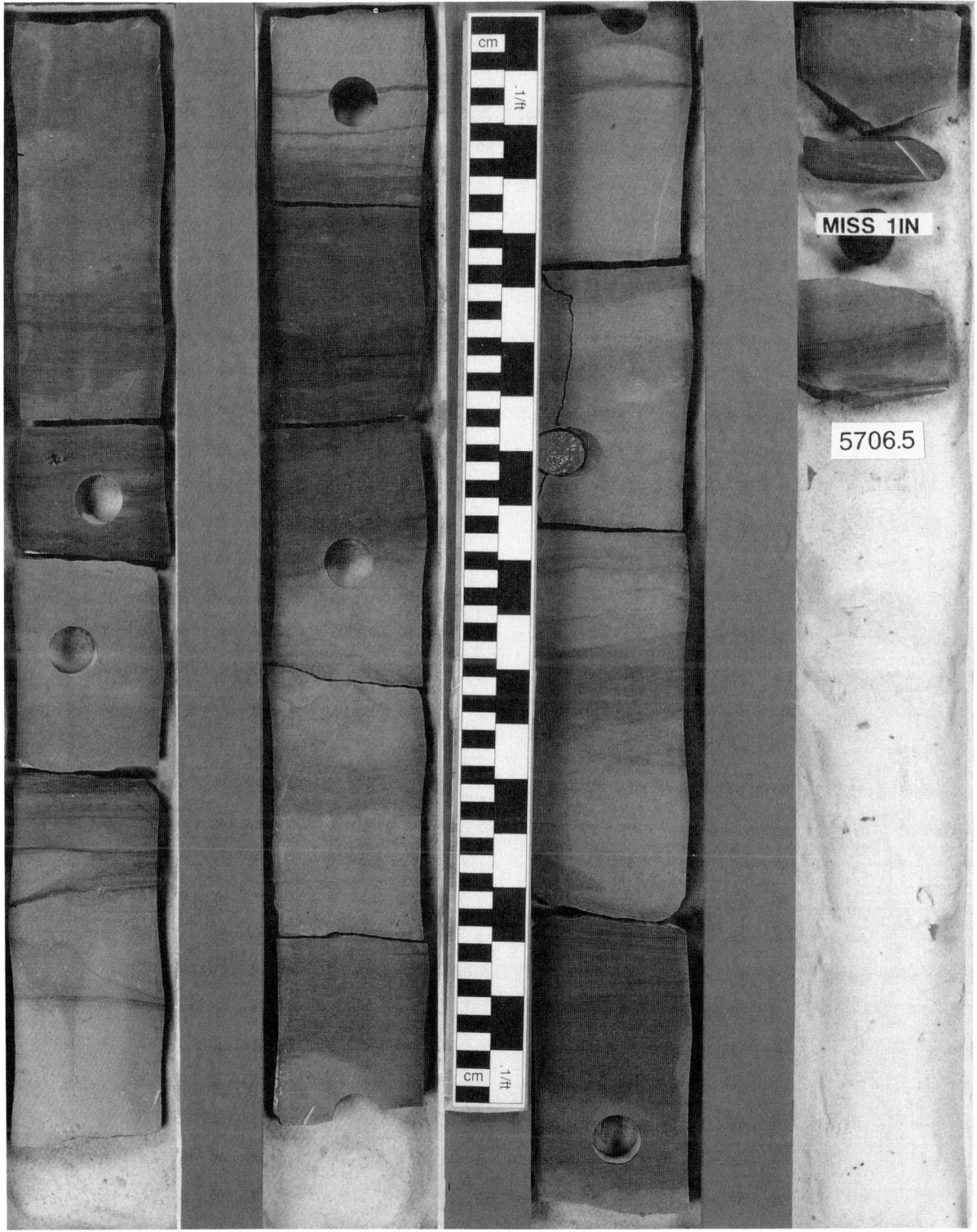

FIG. 6.--Continued.

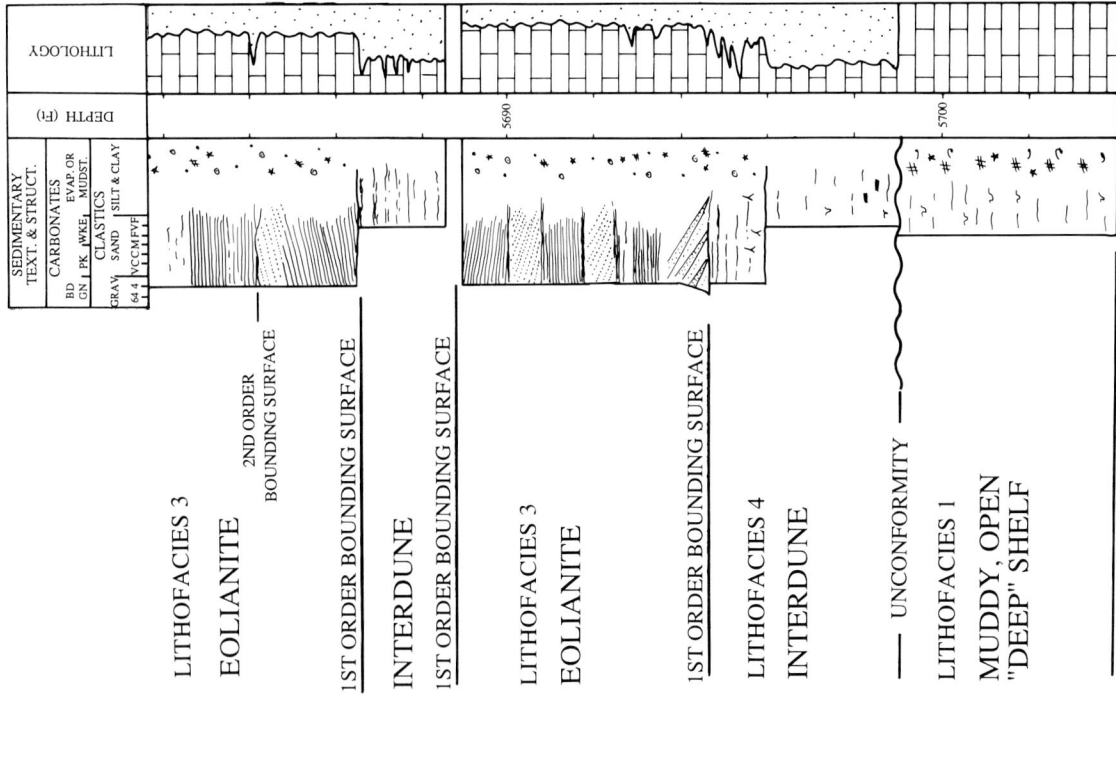
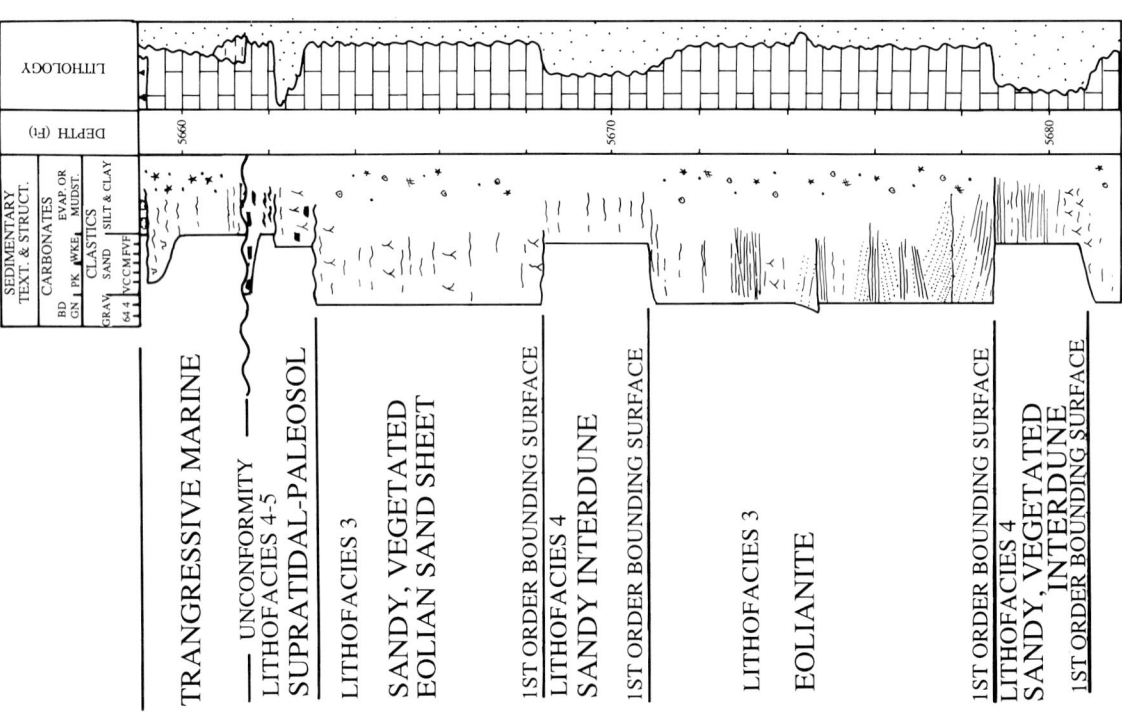

FIG. 7.—Graphic core description of St. Louis Limestone from ARCO B. Schmidt #3, section 34, T. 29 S., R. 38 W., Grant County, Kansas.

FIG. 8.--Core photographs of the ARCO B. Schmidt #3, Sec. 34, T 29 S, R 38 W, Grant County, Kansas.

FIG. 8.—Continued.

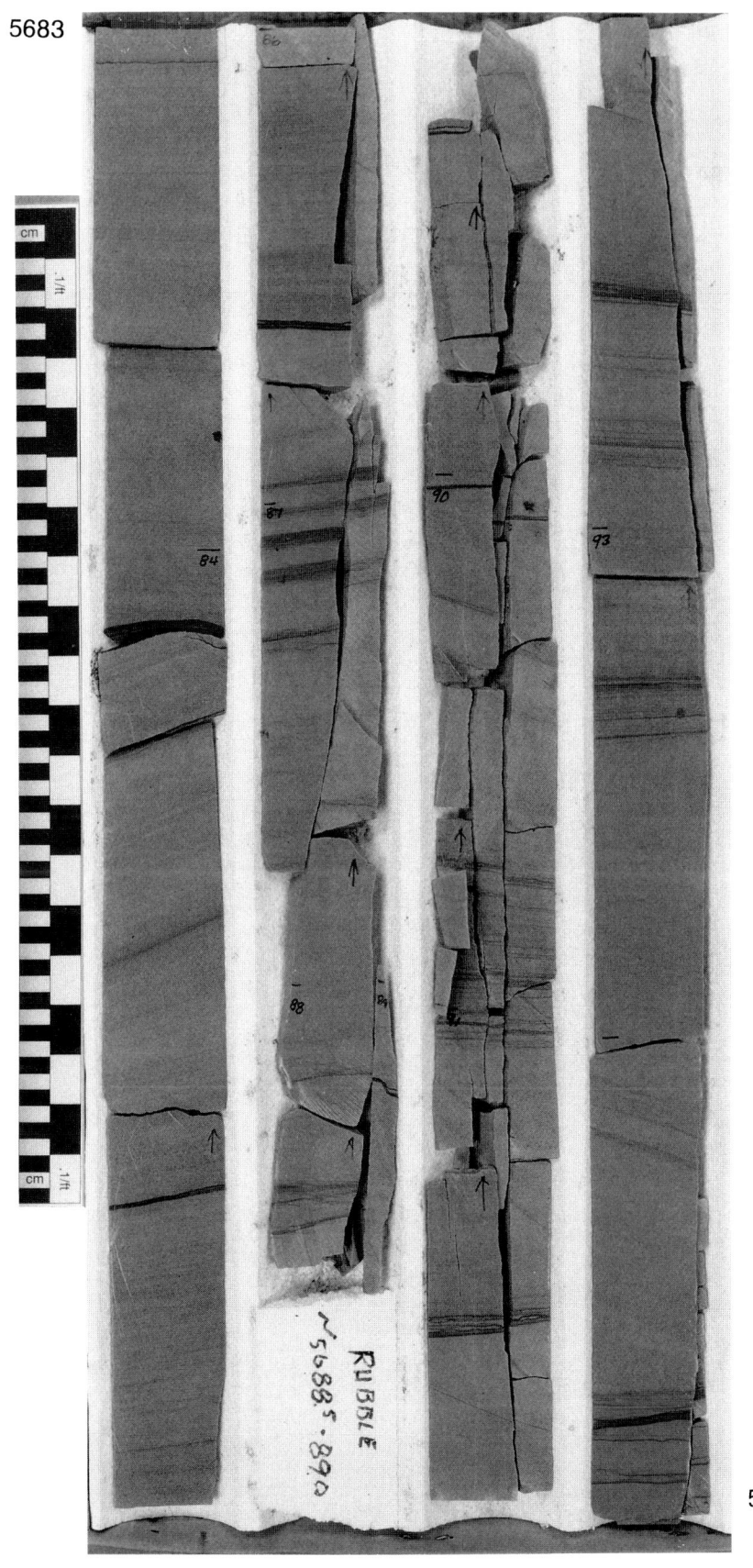

FIG. 8.--Continued.

FIG. 8.--Continued.

*Lithofacies 1 - Silty, Shaly, Skeletal
Wackestone-Packstone and Shale - Muddy-Open Shelf*

This lithofacies ranges from 7 to 10 ft (2 to 3 m) thick in Big Bow Field and consists of silty and shaly, fossiliferous limestones interbedded with thin shales (Figs. 3-9). It contains abundant and only slightly abraded fragments to nearly whole echinoderm, brachiopod, and bryozoan skeletal grains, ranging from 0.02 to 14 mm in diameter. Their presence in a micrite-supported shaly rock suggests open but quiet marine conditions. Shaliness of the lithofacies also points toward a quiet, muddy shelf. Lithofacies 1 has a relatively sharp basal contact in the Plummer core (5668 ft (1727.6 m) in Figs. 5 and 6) with the underlying ooid grainstones of Lithofacies 2, and a thin conglomerate of reworked ooid grainstone also marks the basal contact. This contact may represent a subaerial unconformity or a submarine hardground. The gradual increase in shale upward through the unit indicates either a deepening- or shallowing-upward history. Lithofacies 1 is unconformably overlain by Lithofacies 3.

Lithofacies 1 is similar to that at Damme Field (Handford, 1988), and it overlies porous ooid grainstones (Lithofacies 2) comprising the producing reservoir. At Damme Field, however, the lithofacies is also present below the ooid grainstones lithofacies. The sandwiching of ooid grainstone (Lithofacies 2) between an apparently widespread and consistently thin Lithofacies 1 is interpreted to represent first a shallowing-upward (wackestone overlain by ooid grainstone) followed by a deepening-upward (grainstone overlain by wackestone) history. Thus, the shaly, fossiliferous limestones and thin shales represent a muddy open-shelf with water depths of approximately 25 to 100 ft (7.6 to 30.5 m).

*Lithofacies 2 - Porous, Ooid-Skeletal Grainstone -
Marine Ooid/Skeletal Grain Shoal*

Mississippian reservoirs produce from ooid-skeletal grainstones in southwestern Kansas. This lithofacies is about 15 to 20 ft (4.6 to 6.1 m) thick in most producing fields. One can easily recognize the lithofacies in cores because of its oolitic and coarse-grained fossil content (Fig. 10), its large primary interparticle pores, and dark yellowish brown oil stain. Porous ooid-skeletal grainstones are also easily recognized by the electric-log signatures, because the resistivity, sonic, and density curves faithfully record this facies' interconnected porosity (Fig. 9). Porosity ranges from 10 to 15 percent, and permeabilities commonly reach 100 md.

Ooids are fine- to medium-grained and well sorted. They have radial-concentric cortices surrounding echinoderm, bryozoan, and peloid nuclei. Intermixed skeletal components include crinoids, bryozoans, brachiopods, and foraminifers. Crinoids are almost as abundant as the ooids in some parts of the lithofacies (Fig. 10). Most are medium-grained, but some reach very coarse and granule size. As skeletal content increases, overall sorting decreases. Skeletal grains range from nearly whole and barely abraded to broken, well-worn particles. Largest skeletal grains are up to 9.6 mm in diameter. Quartz sand is rarely present.

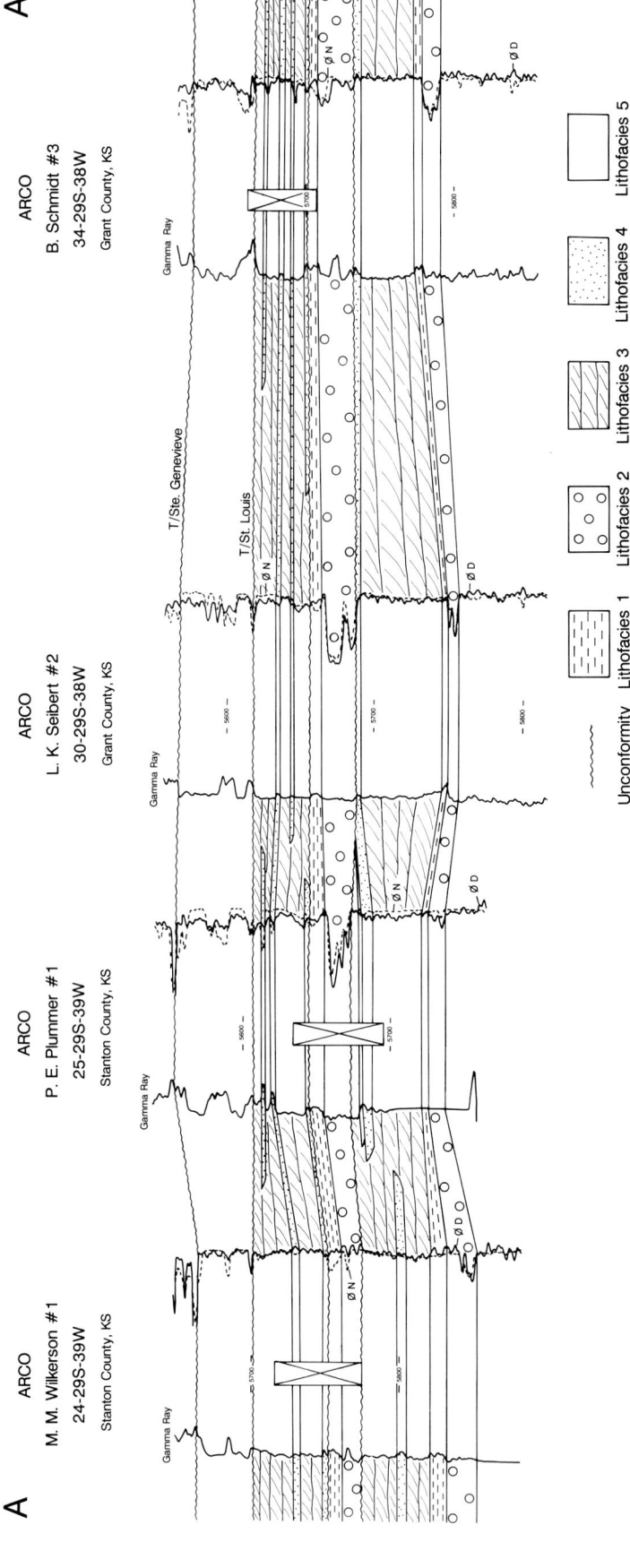

FIG. 9.--Stratigraphic cross section A-A' (see Fig. 1 for location) in Big Bow Field showing cored intervals and lithofacies distributions. Note that overlapping cores allows for nearly 100 ft of stratigraphic section for which logs can be calibrated to the cores.

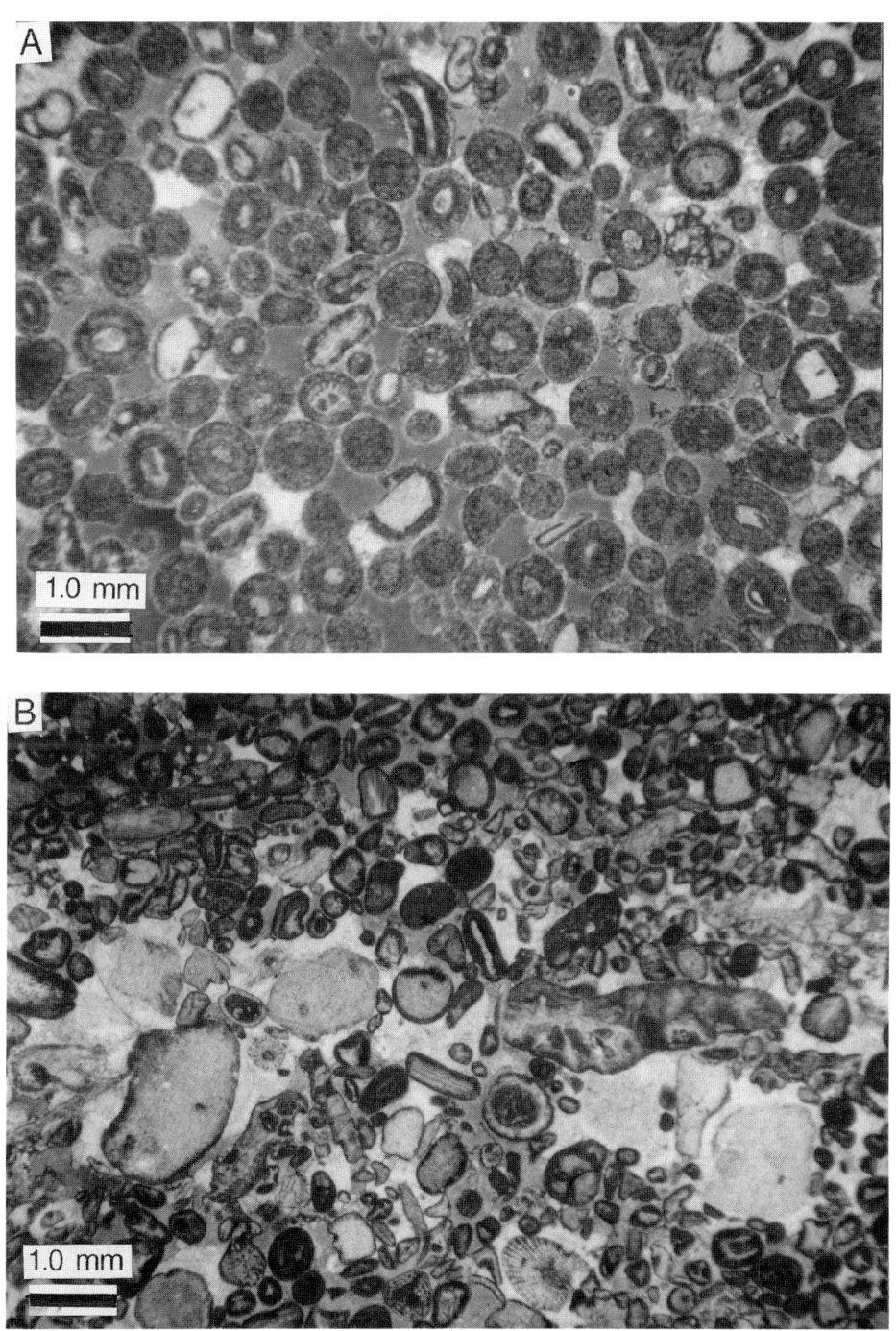

FIG. 10.--Thin-section photomicrographs of typical grainstones in lithofacies 2; (A) porous ooid grainstone and (B) porous, ooid, skeletal grainstone.

Cross-bedding is present but does not dominate the lithofacies. Cross sets are less than 1 ft (0.3 m) thick, and foreset laminae have dips of approximately 10°. The laminae are made up of alternating ooid-rich and skeletal-rich layers.

Characteristics of Lithofacies 2 indicate deposition in shallow-marine shoals, perhaps similar to Bahamian marine sand-belts (Ball, 1967; Handford, 1988). The presence of skeletal grains in these oolitic limestones indicates that open-shelf fauna either inhabited portions of the shoal or that the shoal was juxtaposed to an open shelf so that sediment mixing could occur, most likely during storms.

Lithofacies 3 - Siliciclastic, Peloid-Ooid, Skeletal Grainstones - Coastal Eolianite

This lithofacies forms stacked eolianite units averaging about 10.5 ft (3.2 m) thick in Big Bow Field, with each unit representing the depositional remains of migrating eolian dune bodies. Interdune-wadi deposits (approximately 4.5 ft (1.4 m) thick) of Lithofacies 4 (fine-grained sandstone) separate each of the eolianite units.

The coastal carbonate eolianites contain approximately 15 to 20 percent siliciclastic sand (quartz), with the balance made up of peloids, ooids, and skeletal grains (Fig. 11). Dominant skeletal grains include echinoderms, brachiopods, bryozoans, and foraminifers. All of the skeletal grains are very well rounded and spherical; even the normally platy or bladed brachiopods and bryozoans have been worn into spherical grains. Furthermore, some ooid coatings are partially worn.

In contrast to the marine-shoal grainstones, all of the ooids and skeletal particles in the eolianite lithofacies are fine- to medium-grained. Peloids, however, average 0.09 mm (very fine grain size). Thus, the carbonate fraction commonly has a bimodal size distribution. This is also true of the quartz fraction, which comprises two populations: subangular coarse silt and well-rounded coarse sand (Fig. 11). Both populations are either scattered throughout the lithofacies or they are present as discrete sedimentary laminae and insoluble concentrations along stylolites.

The eolianite lithofacies, as summarized above, contains both well-sorted and bimodal grain populations, generally in the fine to medium sand sizes; whereas, the marine shoal lithofacies is made up of medium- to very coarse-grained particles that are moderately to poorly sorted. These textural differences reflect the change from subaqueous to eolian physical processes of sedimentation present in marine to eolian environments. Marine carbonate environments, such as ooid-skeletal sand beaches, consist of various sand-size classes sorted by beach swash laminae. As a wave moves up the foreshore, its velocity quickly decreases to zero and then reaccelerates on its return down the foreshore. This rapid and constant fluctuation in current velocity allows for a variety of grain sizes (if available) to be transported and deposited. Thus, if calcareous skeletons that break down into coarse-grained sand- and gravel-size particles are present, they will mix with the ooids to form a coarse tail population. In contrast, carbonate dunes rarely, if ever, contain skeletal grains that are as coarse-grained as those comprising the source of dune sands in the adjacent marine environment. Day-to-day onshore winds do not exceed the shear

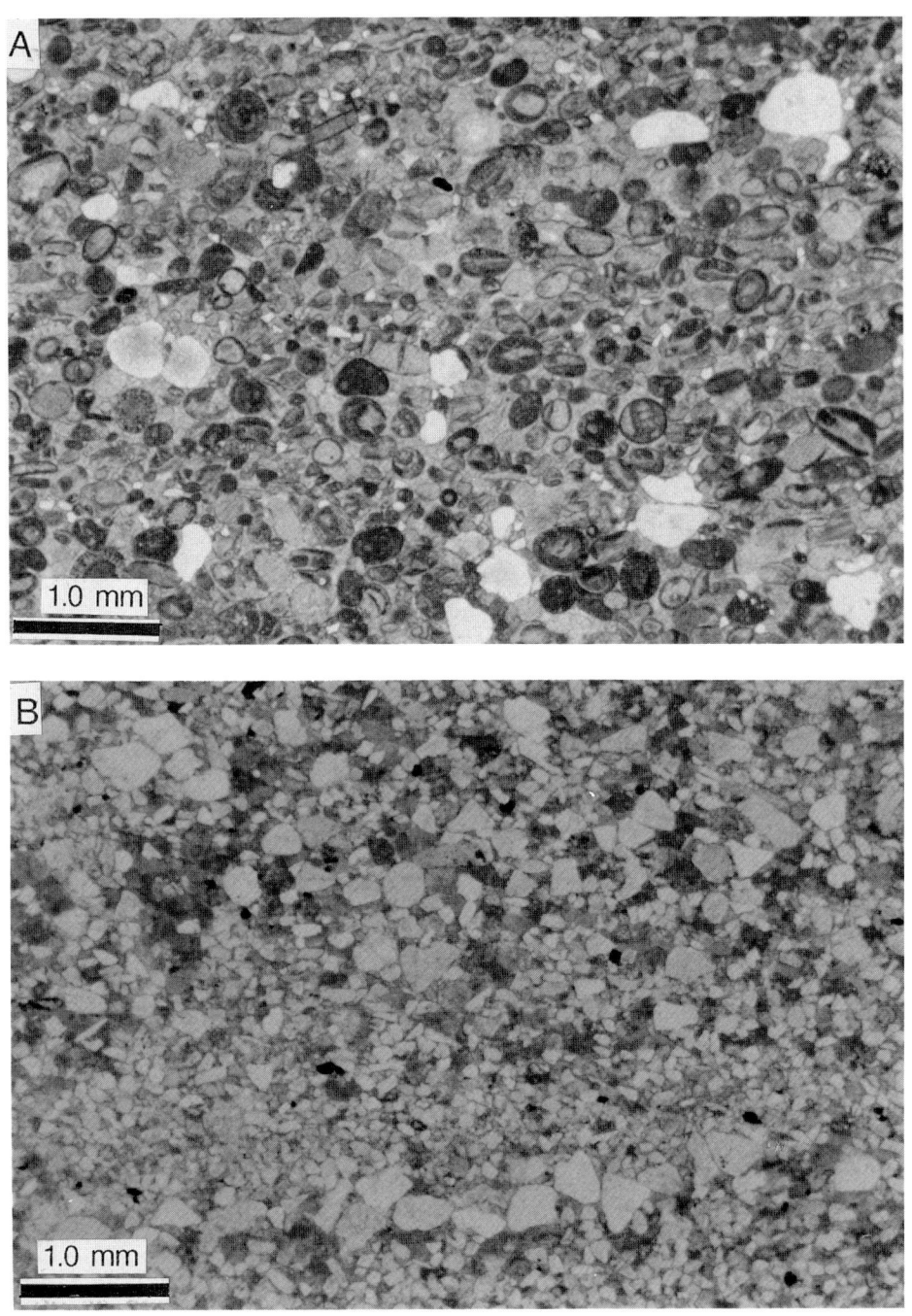

FIG. 11.--Thin-section photomicrographs of typical carbonate-siliciclastic eolianites (lithofacies 3), showing both (A) siliciclastic-poor and (B) siliciclastic-rich facies. Note the well-rounded and sorted carbonate grains in A and the presence of both silt- and sand-size quartz in A and B.

stresses needed to erode and transport coarse skeletal grains from the beaches and storm berms into the dunes. Thus, despite the abundance and availability of coarse skeletal grains in the marine shoals and perhaps the beach environments of the St. Louis seas, only the fine sand-size carbonate grains were transported into the dunes.

The eolianite lithofacies is almost pervasively cross-bedded. This is in contrast to the marine grainstone lithofacies, which is both cross-bedded and bioturbated. Eolianite cross-bed sets, which were determined by piecing together core slabs and observing foreset angles and bounding surfaces, range from less than 1 to 10 ft (0.3 to 3 m) thick and average approximately 1 to 3 ft (0.3 to 0.9 m). Foreset laminae are inclined about 16° to 20°, with dip angles locally increasing from bottom to top in individual sets.

Eolian stratification.--The basic genetic types of fine stratification formed by the deposition of windblown sand on dry surfaces in modern, siliciclastic eolian dunes are, following Hunter's (1977) terminology, plane-bed lamination, climbing translatent strata, grainfall lamination, and sandflow cross-strata (Fig. 12). Although plane-bed and grainfall lamination closely resemble each other, one can, with care, visually distinguish all of these types in most outcrops of eolianites.

Hunter (1977, 1981) recognized three types of climbing translatent strata in eolian sandstones: subcritically climbing, critically climbing, and supercritically climbing translatent strata, each depending on whether the angle of climb is less than, equal to, or greater than the angle of ripple climb (angle between the ripple stoss slope and the depositional surface). Of these types, only subcritically climbing translatent stratification is present in the Mississippian eolianites. According to Hunter (1981), this type of stratification can be recognized by its low dip angles, evenness and thinness of laminae, similarity in thickness to adjacent strata of the same kind, their sharp contacts, and the distinctness of laminae, which is due to inverse size grading (Fig. 12). Subcritically climbing translatent strata are present in all cores examined and make up as much as 50 percent of the bedding in an eolianite dune set (see 5683 ft (1732 m) in Figs. 7 and 8).

When windflow separation occurs in the leeside of dunes, previously saltating grains lose their forward momentum and fall onto the foreset slope to form grainfall lamination (Hunter, 1977, 1981). These slopes tend to be steeper than those that are rippled, but they cannot be so steep that sand avalanches if grainfall lamination is to be preserved (Hunter, 1977). Leeside slopes containing interpreted grainfall laminae in the Mississippian eolianites ranged from 16° to 20°. Foreset grainfall laminae are as thin as climbing translatent strata but are less distinctly laminated. They grade upslope and downslope into climbing wind translatent strata. Most of the inclined or foreset strata in the Mississippian eolianites formed as grainfall laminae. These laminae make up about 30 to 50 percent of the dune sets.

Sandflow cross-strata form when tongue- and cone-shaped masses of noncohesive sand grains avalanche down slipfaces whose initial inclination is greater than the angle of repose (approximately 30° to 32° measured for loose ooid sand at Isla Cancun). Hunter (1977) showed that sandflow cross-strata are steeply inclined, relatively thick (more than 0.4 in. (1 cm)), have sharp contacts, are lenticular, and have distinct pinch-outs or toes. Sandflow processes do not

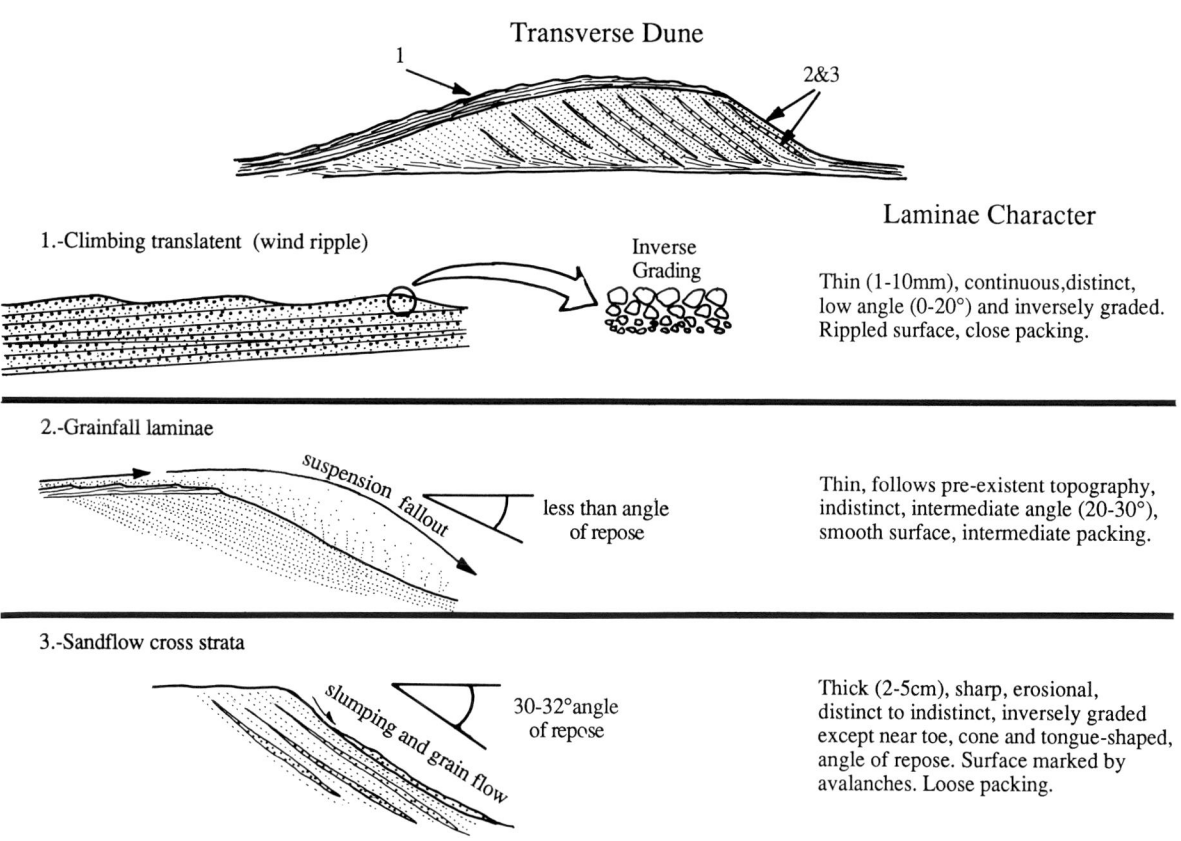

FIG. 12.--Idealized representation of eolian stratification types in small eolian dunes as described by Hunter (1977).

appear to have been important during deposition of the Mississippian eolianites because foreset dips do not exceed 22°, which is far less than the angle of repose. Although these rocks have been postdepositionally compacted by stylolitization, it is doubtful that this process decreased the dips by 8° to 10°. Inferred sandflow cross-strata were identified in only one set (5694 ft (1735.5 m) in Figs. 7 and 8).

Pedogenic features are not as common in the eolianite units as they are in the interdune deposits (Lithofacies 4), but their presence further indicates a subaerial origin for Lithofacies 3. Rhizocretions are present in some eolianites, but the most striking pedogenic feature is a laminated paleocaliche from Damme Field shown in Figure 13. An eolianite was cemented and pedogenically altered in the subaerial environment to form a caliche crust identical to those present in Pleistocene carbonate eolianites. The margin of the dune was eroded down to its caliche layer and subsequently onlapped by wadi or interdune sands.

Additional evidence for subaerial weathering is seen in an eolianite, which grades upward into a 3 ft (0.9 m) thick breccia of eolianite (5689 ft (1734 m) in Figs. 5 and 6). Clasts are poorly sorted, angular to rounded, and range from approximately 1 to 6 inches (2.5 to 15 cm) in diameter. Despite some rounding of clasts, this breccia shows no evidence of sediment transport, because some appear to be fitted, and they are barely separated from one another by a matrix of eolianite sand. A carbonate mudstone is the prevalent matrix at the top of the breccia. The clasts formed as a result of subaerial weathering (disaggregation) of a surface and near-surface crust of weakly cemented eolianite. Modern carbonate dunes are susceptible to meteoric cementation, which results in the formation of weakly cemented crusts at the surface. However, the crusts commonly break apart into blocky clasts because cementation is spotty and incomplete (Fig. 14). Wind erosion deflates loose sand and erodes cemented layers at the same time as roots penetrate and pry apart cemented layers. A similar process was probably responsible for the formation of the Mississippian eolianite breccia. The grainy sediment between clasts was deposited as loose dune sand, and the carbonate mud probably represents pond and/or initial marine sediment filtered in from above during a subsequent transgression. This transgression led to the deposition of marine-shoal ooid packstones and grainstones (Lithofacies 1) of the overlying porous reservoir.

Lithofacies 4 - Fine-Grained Siliciclastic Sandstone -
Interdune and Wadi

Several thin, calcareous sandstones, interpreted to represent interdune and wadi environments, are interbedded with the eolianite units (Figs. 3-9). They average 4.5 ft (1.4 m) thick. Although coarse-grained silt to fine-grained quartz sand make up the bulk of this lithofacies, peloids, ooids, and assorted skeletal grains are also present. Siliciclastic sand grains are subangular, generally well sorted, and range from 0.06 to 0.12 mm in diameter (coarse silt-very fine sand). However, some sandstones contain enough medium-grained (0.4 mm), rounded sand grains to render a bimodal texture.

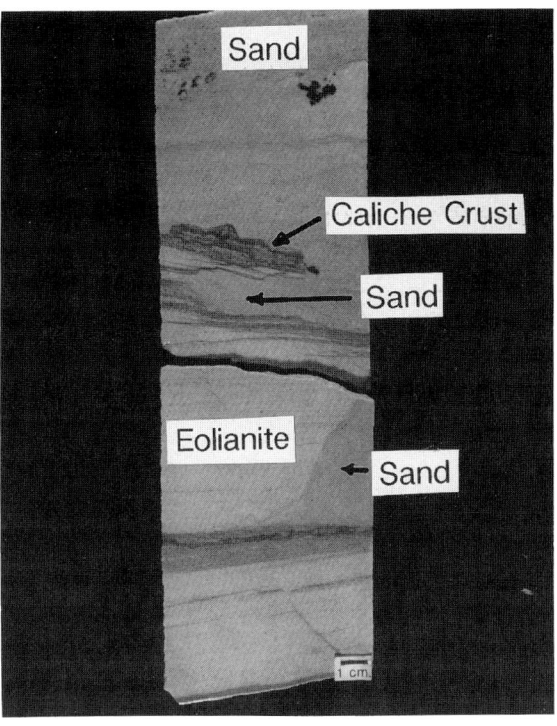

FIG. 13.--Mississippian core from Damme Field, Finney County, consists of a paleocaliche crust, coating an underlying eolianite, both of which have been eroded and onlapped by siliciclastic sand deposited in an interdune wadi.

FIG. 14.--Cemented clasts of modern carbonate eolianite at Isla Cancun, Mexico are scattered across the coastal dune ridge. Deflation of loose, carbonate sand from the dune ridge has exposed cemented, carbonate dune rock which is breaking apart by weathering into in-situ clasts.

Cross-bedding is present but is not nearly as common as structureless to bioturbated intervals. The most common structures are wispy or discontinuous laminae. There are also numerous tubular structures of laminated micrite surrounding a core of sparry calcite (5662.5 and 5680 ft (1725.9 and 1731 m) in Figs. 7 and 8). Diameters range from a couple of millimeters to approximately 0.4 in. (1 cm). They are similar to root structures (rhizocretions) described from Holocene-Pleistocene carbonate eolianites in Yucatan, the Bahamas, and from the Mediterranean (Klappa, 1980). The concentrically laminated part of the structure is the micritic carbonate that precipitated centripetally around living and/or decaying roots. Later, the organic matter comprising the roots decayed and sparry calcite cement precipitated in the resultant void.

Cross sections of Big Bow Field (Fig. 9) and Damme Field (Fig. 15 of Handford, 1989) indicate that the stratigraphically recurring sandstones of Lithofacies 4 are lenticular. Cores through this lithofacies show that most units have gradational upper and lower boundaries. Lithofacies gradations span a few inches to approximately 1 ft (0.3 m).

Handford (1989) previously interpreted the sandstone facies at Damme Field to be marine on the basis of the marine fossils it contained and its association with the siliciclastic, peloid-ooid, skeletal grainstones (Lithofacies 3) that Handford also interpreted to be marine. However, with the newly discovered evidence for an eolian environment of deposition for Lithofacies 3, the interbedded sandstones are now interpreted to represent interdune and wadi environments of deposition. All of the marine skeletal grains are worn and abraded, and none of the fauna, represented by these skeletal grains, lived in the sandstone depositional environment. They were mixed with the siliciclastics by eolian processes. As carbonate-rich dunes migrated across siliciclastic-dominated interdune areas, some carbonate grains were left behind as a deflation lag. These carbonate grains probably accumulated as a lag deposit where wadi channels migrated across interdune areas and eroded the margins of carbonate dunes. Some mixing may also have occurred as strong winds simply blew carbonate dune sands into the interdune areas.

Lithofacies 5 - Silty-Sandy, Fenestral, Ostracode
Wackestone and Intraclast, Skeletal-Ooid Grainstone -
Intertidal to Interdune Mudflat

This lithofacies is present in both the Plummer and Schmidt cores, but it is less than 3 ft (0.9 m) thick in both instances (5694 to 5696 ft (1735.5 to 1736 m) in Figs. 5 and 6, and 5662 ft (1725.8 m) in Figs. 7 and 8). The grainstones consist of intraclasts, skeletal grains, ooids, peloids, and quartz sand, and the skeletal grains are worn and abraded like the windblown material in Lithofacies 3. Abundant fenestrae and possibly keystone vugs are present and point toward an intertidal origin. The fenestral wackestones also contain windblown skeletal debris, but the most conspicuous grains are ostracodes. Their presence with the fenestrae is suggestive of a restricted, periodically exposed environment. Fine-grained terrigenous sand (0.12 mm) is fairly common in the wackestones and, like the fine skeletal grains, may be windblown sediment that was trapped and bound by algal mats in intertidal-supratidal areas between the dunes.

DEPOSITIONAL SUMMARY

Mississippian St. Louis strata record deposition in open, shallow-marine environments, but, as this study has shown, they also include over 100 ft (30.5 m) of heretofore unrecognized, mixed siliciclastic-carbonate eolianites and interbedded, calcareous wadi-interdune deposits. Unconformities punctuate the stratigraphic succession (Fig. 7, 5699 ft (1737 m)) so that "Waltherian" facies reconstructions must not juxtapose facies separated by these unconformities. Figure 15 is a depositional model of the St. Louis Limestone illustrating the hypothetical arrangement of environments if all were present during any one point in time. Porous, ooid-skeletal grainstones are transitional with shaly wackestones, which indicates that the open shelf passed laterally into ooid-skeletal shoal environments. Eolianites are inferred to have accumulated along the coast next to shallow grain shoals because the ooid and skeletal carbonate grains comprising the eolianites are the same type present in the shoal facies. Onshore winds blew loose carbonate grains from beaches and exposed shoals into subaerial backshore environments. The presence of interbedded sandstones and disseminated siliciclastic sand in the eolianites suggests that the eolian dunes and wadi-interdune environments lay adjacent to each other. The relative lack of siliciclastic sands in the offshore carbonates is attributed to limited fluvial input to the sea. Ephemeral streams emptied into the interdune environments, but the dunes probably blocked them from entering the sea. In addition, if paleowinds consistently blew onshore, little siliciclastic material could have been transported to the shoreline by eolian processes.

The presence of Mississippian eolianites in at least seven southwestern Kansas counties and 800 miles (1287 km) east in Indiana indicates that eolian carbonates are more common in the rock record than previously assumed. It further suggests that conditions favorable for their development in North America were a hallmark of Late Mississippian time. It is probably no coincidence that the widespread deposition of carbonate eolianites occurred during the second-order global cycle of relative sea-level fall near the end of the Mississippian Period (Vail and others, 1977).

ACKNOWLEDGMENTS

We thank ARCO Oil and Gas Company for giving permission to publish this study. Appreciation is also extended to Steve Johnson and Kelly Wilkins for core preparation and photography, and to Gary Garrett for graphics support.

REFERENCES

BALL, M. M., 1967, Carbonate sand bodies of Florida and the Bahamas: Journal of Sedimentary Petrology, v. 37, p. 556-591.

GOEBEL, E. D., 1968, Mississippian rocks of western Kansas: American Association of Petroleum Geologists Bulletin, v. 52, p. 1732-1778.

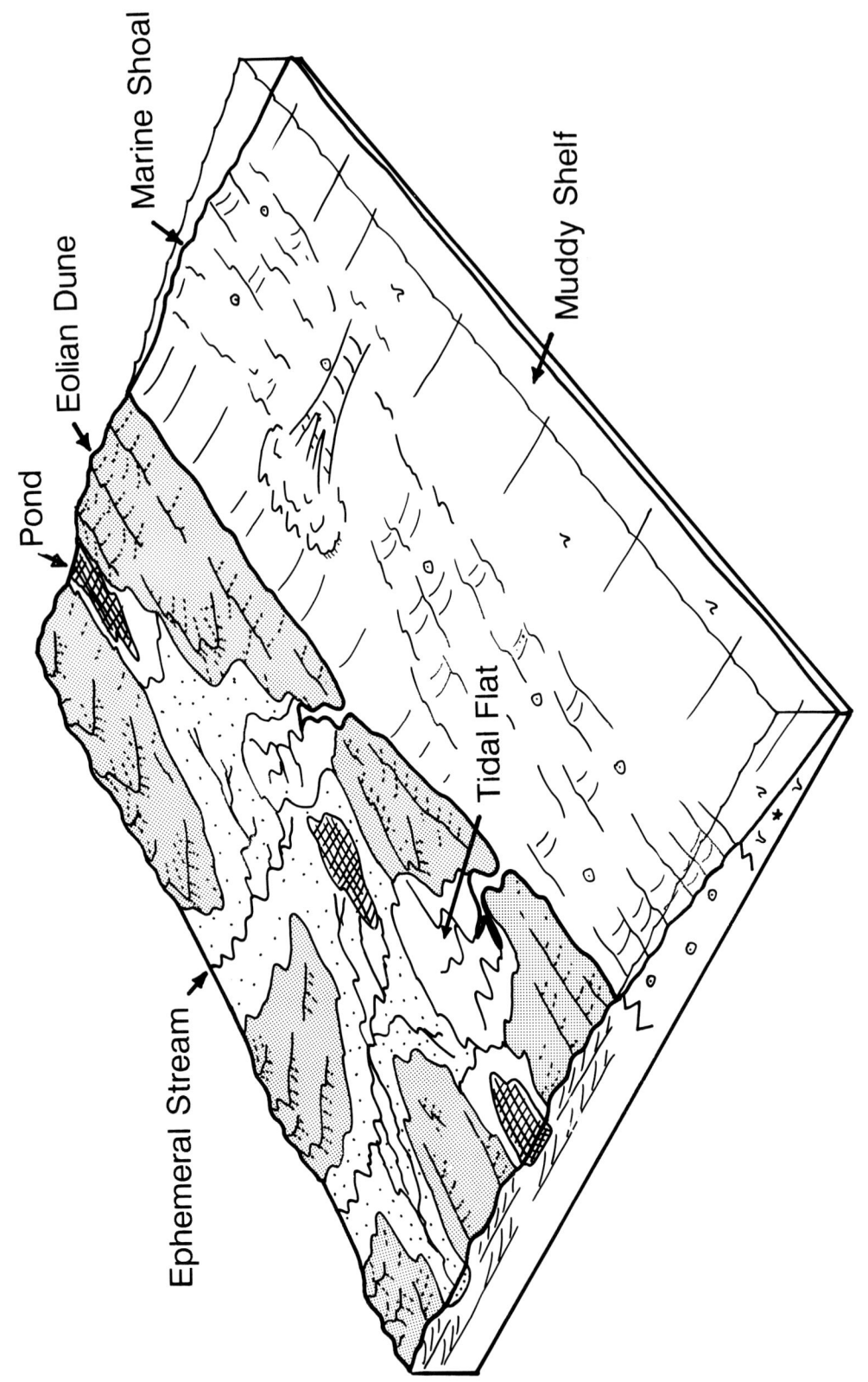

FIG. 15.—Schematic depositional model of coastal environments represented by the St. Louis Limestone at Big Bow Field.

HANDFORD, C. R., 1988, Review of carbonate sand-belt deposition of ooid grainstones and application to Mississippian reservoir, Damme Field, southwestern Kansas: American Association of Petroleum Geologists Bulletin, v. 72, p. 1184-1199.

HUNTER, R. E., 1977, Basic types of stratification in small eolian dunes: Sedimentology, v. 24, p. 362-387.

_____, 1981, Stratification styles in eolian sandstones: some Pennsylvanian to Jurassic examples from the western interior, U.S.A., *in* Etheridge, F. G., and Flores, R. M., eds., Recent and Ancient Nonmarine Depositional Environments: Models for Exploration: Society of Economic Paleontologists and Mineralogists Special Publication 31, p. 315-329.

_____, 1988, Eolianites in the Ste. Genevieve Limestone (Mississippian) of southern Indiana: Society of Economic Paleontologists and Mineralogists Annual Midyear Meeting Abstracts, v. 5, p. 26.

KLAPPA, C. F., 1980, Rhizoliths in terrestrial carbonates: classification, recognition genesis and significance: Sedimentology, v. 27, p. 613-629.

MAHER, J. C., AND COLLINS, J. B., 1949, Pre-Pennsylvanian geology of southwestern Kansas, southeastern Colorado, and Oklahoma Panhandle: U.S. Geological Survey Oil and Gas Investigations Preliminary Map 101.

MCKEE, E. D., AND WARD, W. C., 1983, Eolian environment, *in* Scholle, P. A., Bebout, D. G., and Moore, C. H., eds., Carbonate Depositional Environments: American Association of Petroleum Geologists Memoir 33, p. 132-170.

VAIL, P. R., MITCHUM, R. M., JR., AND THOMPSON, S., III, 1977, Seismic stratigraphy and global changes of sea level, Part 4: global cycles of relative changes of sea level, *in* Payton, C. E., ed., Seismic Stratigraphy - Applications to Hydrocarbon Exploration: American Association of Petroleum Geologists Memoir 26, p. 83-97.

WAMBA FIELD, PEOPLE'S REPUBLIC OF ANGOLA, A CENOMANIAN MIXED CARBONATE-SILICICLASTIC RESERVOIR

ANTHONY J. LOMANDO AND TRACY L. WALKER
Chevron Overseas Petroleum Inc.
San Ramon, CA 94583

ABSTRACT

Wamba Field, offshore Cabinda, People's Republic of Angola, produces significant quantities of oil from sandstones of the Upper Cretaceous Vermelha Formation. This sequence of mixed carbonate-siliciclastic sediments was deposited as a transgressive-regressive barrier island system. Approximately 320 ft (97.5 m) of conventional whole core was examined in order to define depositional facies and understand their relationships. Three principal facies were recognized: (1) inner shelf to shoreface sediments characterized by bioturbated, argillaceous, dolomitic siltstones, very fine-grained sandstones, and silty/sandy dolowackestones; (2) beach/barrier bar complex ranging from well-sorted, very fine- to fine-grained to poorly sorted, very fine- to very coarse-grained subarkoses and arkoses commonly containing planar and low-angle planar cross-laminations; and (3) lagoon and tidal-flat sediments composed predominantly of argillaceous dolomudstones and wackestones punctuated by dolomitic sandstones.

In this reservoir complex facies relationships and diagenesis cause extremely variable reservoir properties. Generally the sandstones of the beach/barrier bar complex exhibit the best overall reservoir quality resulting from sparse cementation, preserving intergranular pores and some secondary porosity development. The shoreface and lagoon facies exhibit the poorest reservoir quality and may act as local or regional permeability barriers. The mixing of siliciclastic and carbonate rocks occurs primarily by lateral facies mixing (spatial variability) along facies boundaries and transition zones, and by storm-punctuating processes. The larger-scale mixing variations in the stratigraphic succession (temporal variability) are caused by longer-term eustatic sea-level changes.

INTRODUCTION

The greater Takula area (Dale and others, 1990) is a trend of fields located 20 miles (32.2 km) offshore Cabinda, an enclave of the People's Republic of Angola (Fig. 1). The fields produce primarily from Cretaceous mixed carbonate-siliciclastic sequences (Fig. 2). Wamba Field was discovered in May 1982 by the Cabinda Gulf 44-5 well which was drilled as a step-out exploration well to appraise a structure north of the giant Takula Field. The field is in 210 ft (64 m) of water and is located 28 miles (45 km) WNW of the Cabinda Gulf Oil Company onshore base and terminal in Malongo. Production began in December 1986, with seven wells tied to one well jacket. The field has two reservoirs: the Vermelha Formation and the shallower and lower-permeability Mesa Formation. Current production is only from the Vermelha.

The Wamba structure is a northward-plunging structural nose separated from Takula Field by a growth fault with a throw ranging from 700 to 1100 ft (213 to 335 m; Figs. 3 and 4). The field has a maximum gross oil column of 375 ft (114 m) and is trapped by a seal formed from the

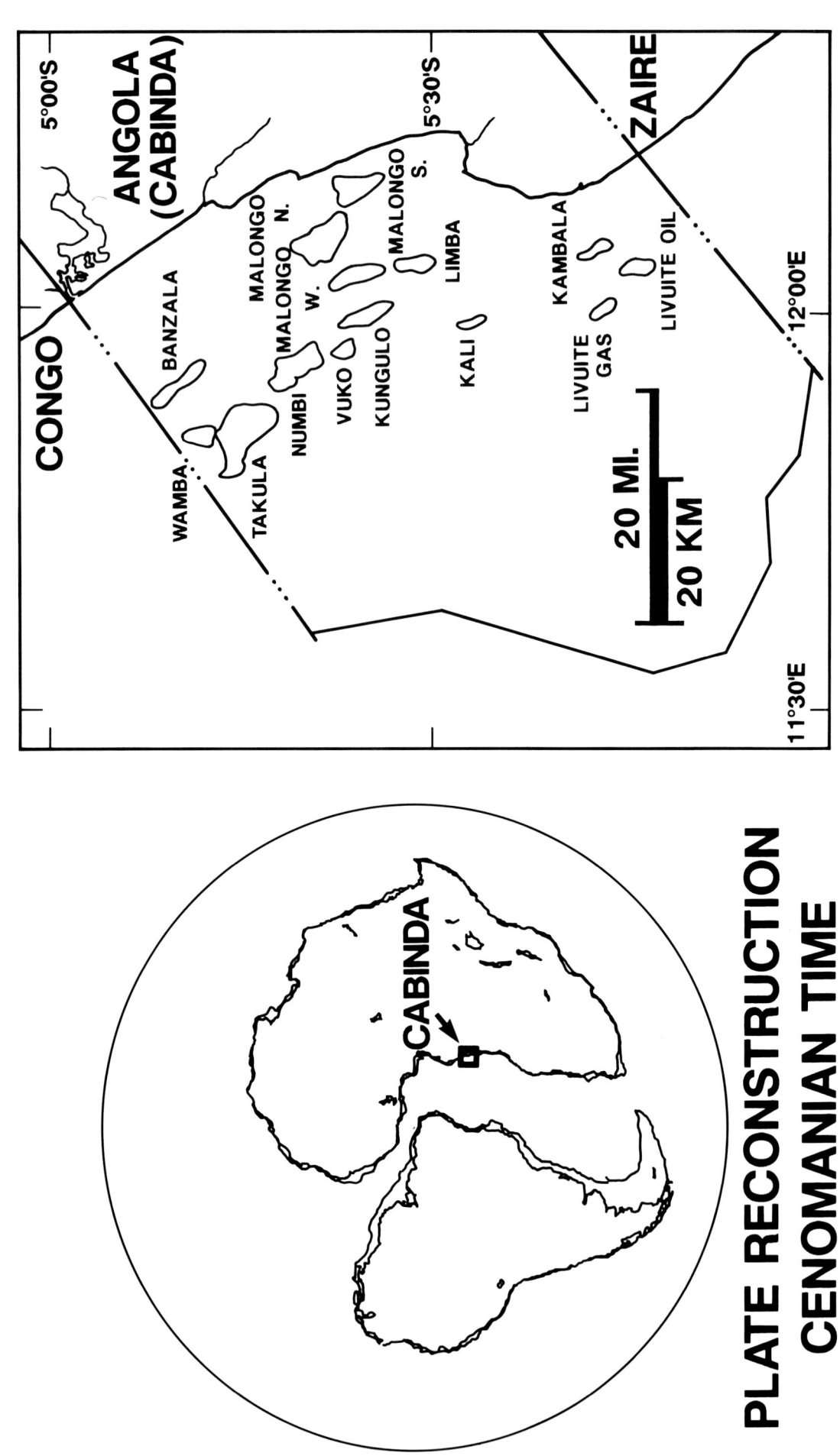

Fig. 1.—Location maps showing the location of Cabinda during the time of Vermelha deposition and the Cabinda concession and producing fields.

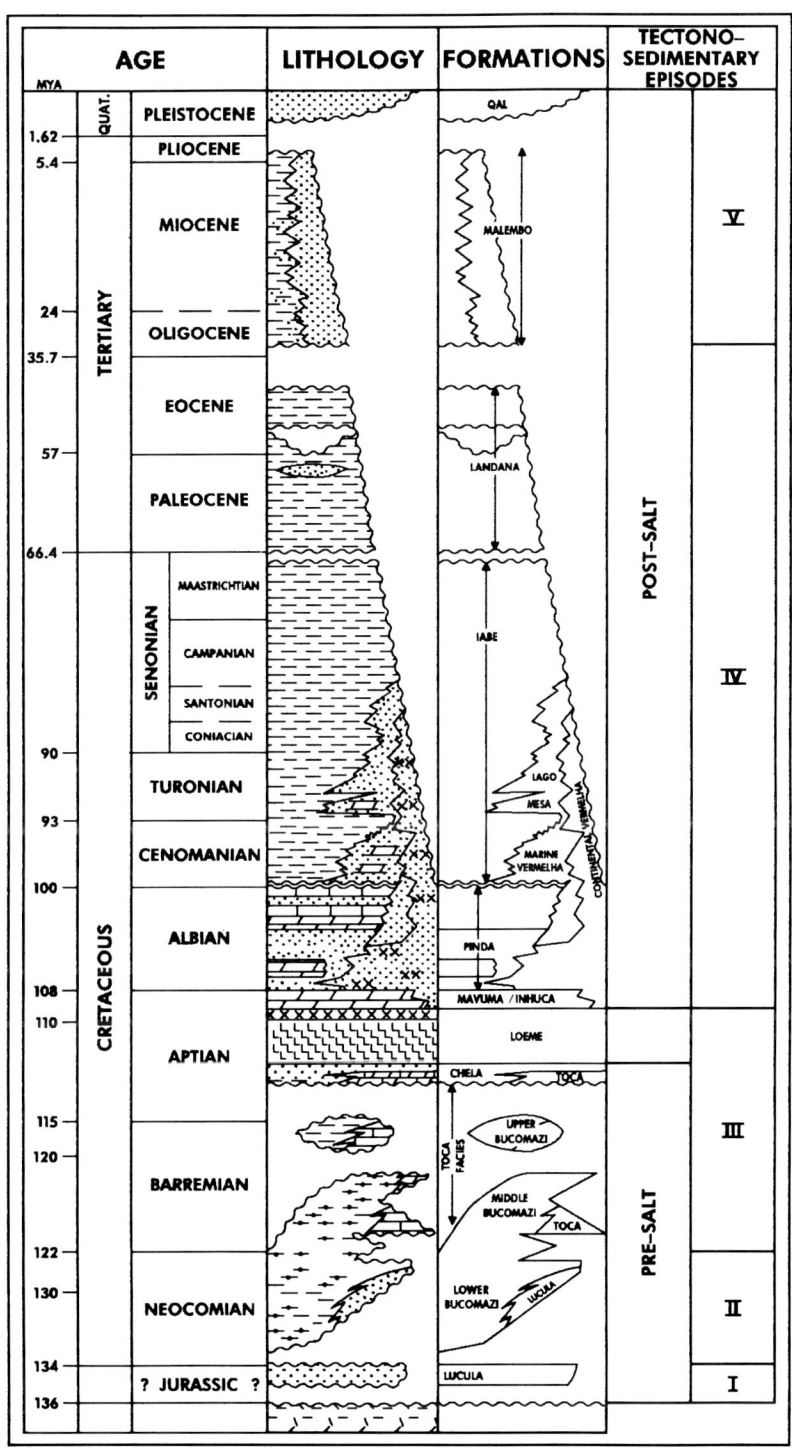

Fig. 2--Stratigraphic column for the Lower Congo Basin.

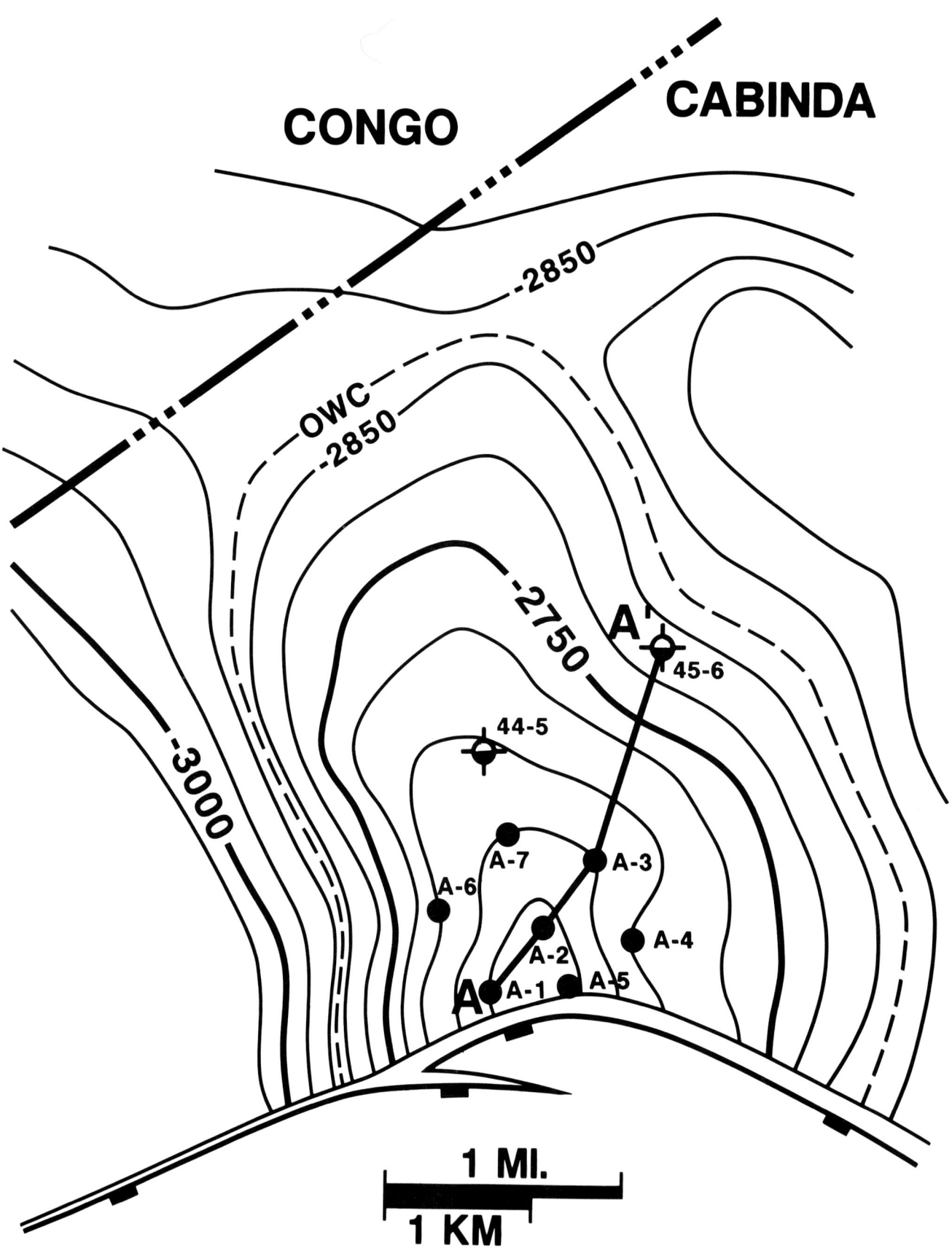

Fig. 3--Structure map drawn on the top Vermelha; contour interval equals 50 ft.

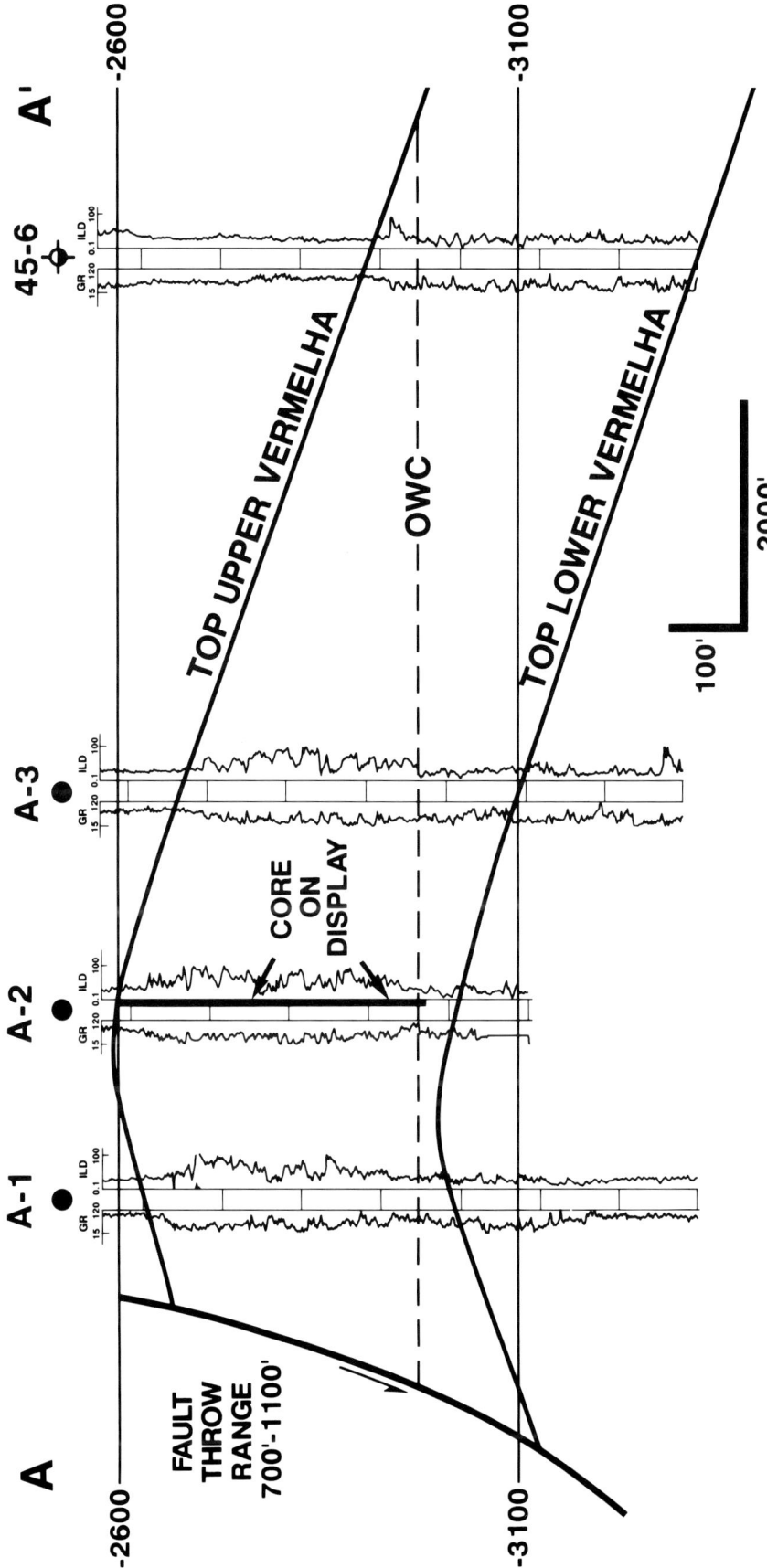

Fig. 4--Structural cross section across Wamba Field and stratigraphic position of the cores.

shales and siltstones of the overlying Lower Iabe Formation. Oil API gravity is 29.6°. The principal producing lithology is very friable (poorly cemented to uncemented) dolomitic sandstones that have average porosities of 29 percent and permeabilities as high as 5 darcys. Reservoir energy is provided by a very active water drive.

REGIONAL SETTING

The rifted margins of offshore West Africa evolved through a sequence of tectono-sedimentary episodes (Brice and others, 1982; Caflisch, 1979). In Cabinda the stratigraphic succession consists of five episodes that were described by Brice and others (1982) as: (1) Prerift Jurassic siliciclastics; (2) Synrift I-primarily Neocomian argillaceous sandstones and carbonates; (3) Synrift II-Barremian to Aptian evaporites, carbonates, and siliciclastics; (4) Postrift Albian to Eocene mixed carbonate/siliciclastics; and (5) Oligocene to Holocene sandstones and shales. This stratigraphic section in West Africa is often referred to by the simplified terms "pre-salt" containing episodes one, two, and three, and "post-salt" containing episodes four and five (Fig. 2). Post-salt Cretaceous rocks are characterized by two major transgressive-regressive sedimentation cycles that occur from the lower Congo Basin northward to the Douala Basin (Seiglie and Baker, 1984).

In Middle and Late Albian time, the rapidly opening South Atlantic epicontinental sea was still partially closed at its northern end. The latest Albian-early Cenomanian in the south witnessed the final separation of the Falkland Plateau from the S.E. African continental fragments (Reyment and Dingle, 1987). By Late Cenomanian-Turonian time, South America and Africa were completely separated, forming a long, continuous, proto-Atlantic seaway. During Albian-Cenomanian time the West African shelf was the site of very variable mixed carbonate-siliciclastic deposition from at least as far south as central Angola (Tillement, 1987) to as far north as Nigeria (Oti and Koch, 1990). This was the setting at the time of Vermelha deposition.

SEDIMENTARY FACIES

The Upper Vermelha consists of three principal facies types deposited in nearshore and coastal environments that range from inner shelf/shoreface to tidal flats (Fig. 5).

Shoreface

Shoreface facies in the Wamba A-2 well are typically characterized by bioturbated dark green to gray-green, argillaceous dolomitic siltstones and very fine sandstones (Fig. 6). Based on petrographic analysis the rocks of this facies are generally classified as dolomitic feldspathic arenites and dolomitic feldspathic wackestones. These rocks typically contain detrital quartz, feldspar, heavy minerals, and rock and skeletal fragments. Potassium feldspars are generally more abundant than plagioclase. Fine-grained matrix material consists of predominantly clay minerals and carbonate mud. Heavy minerals of detrital ilmenite and magnetite are often abundant locally. Rock fragments are rare and generally contribute less than 1 percent of the total composition. Carbonaceous material is locally present and occurs as fragments from 0.04 to 0.2 in. (1 to 5 mm) in size. Accessory minerals include garnet and detrital micas.

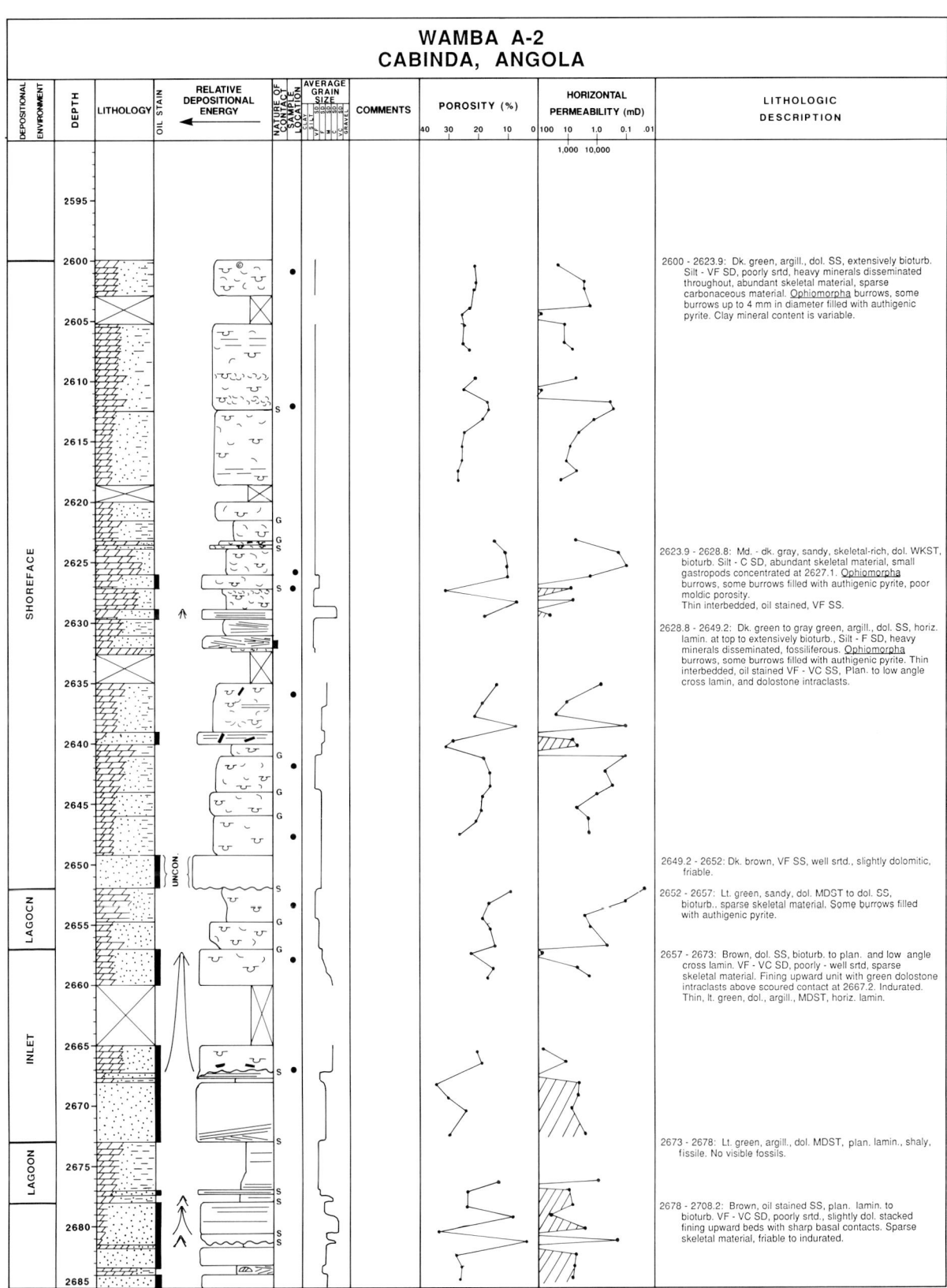

Fig. 5--Detailed core description of the Wamba A-2 cores.

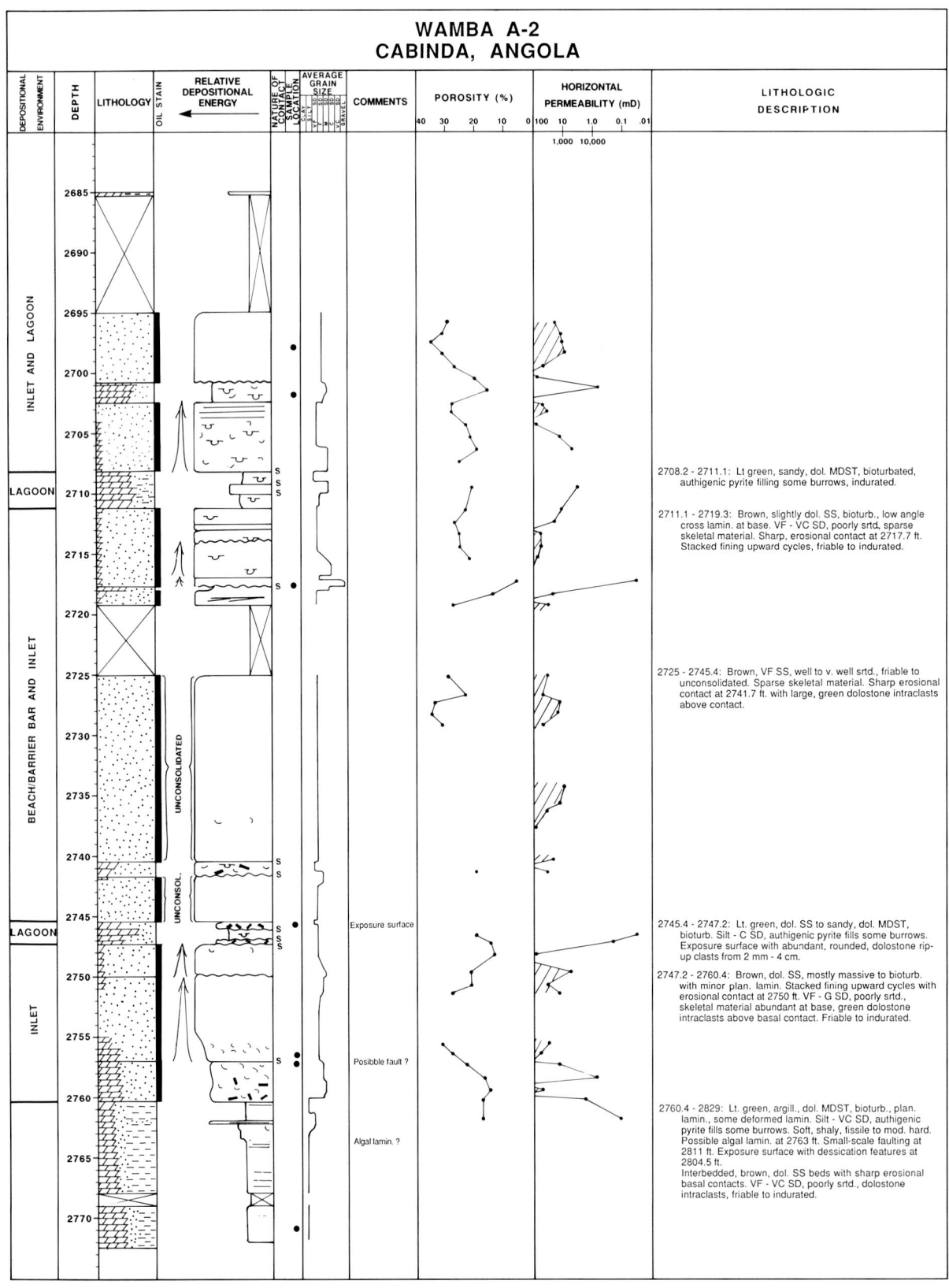

Fig. 5 cont.

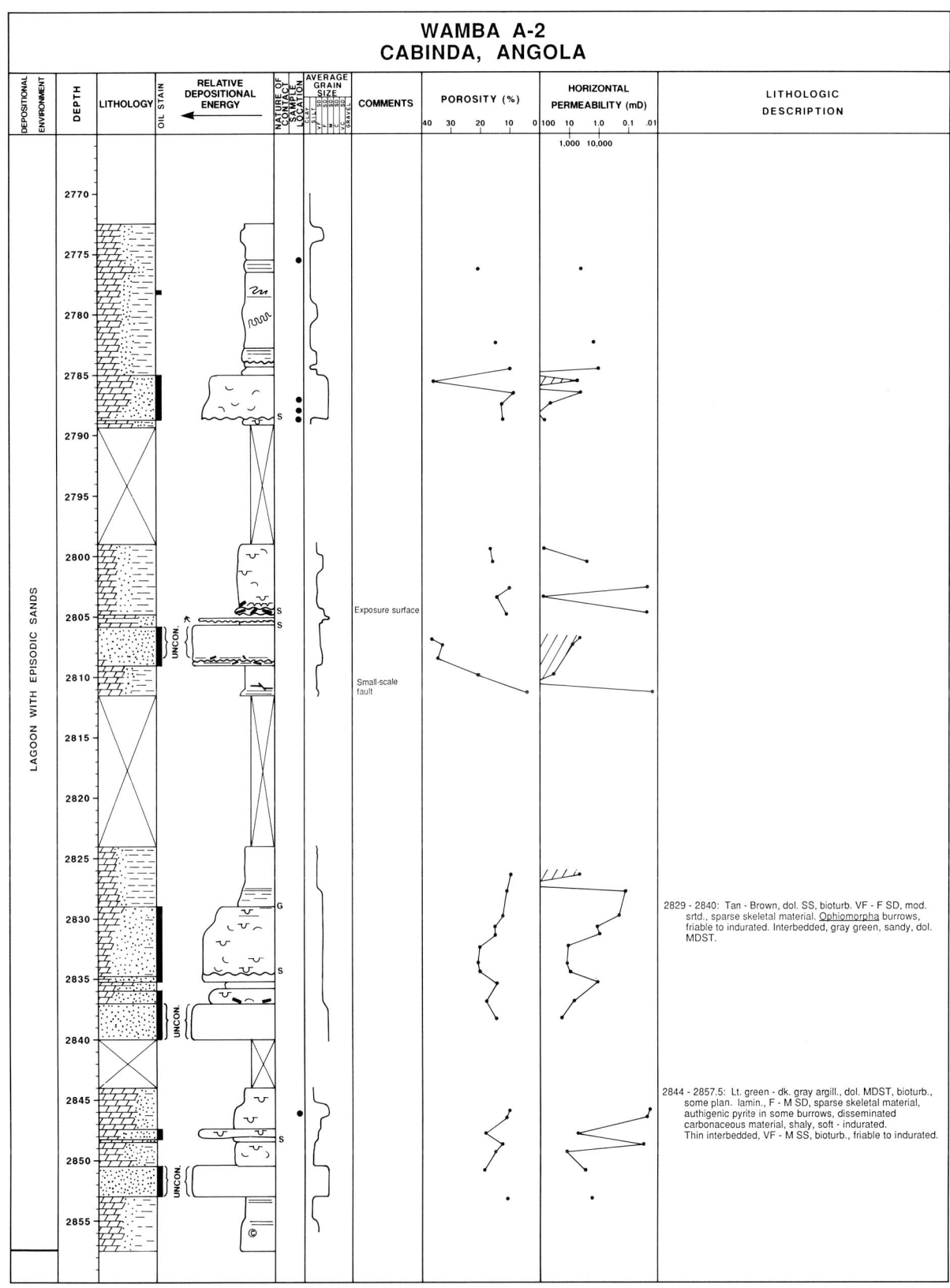

Fig. 5 cont.

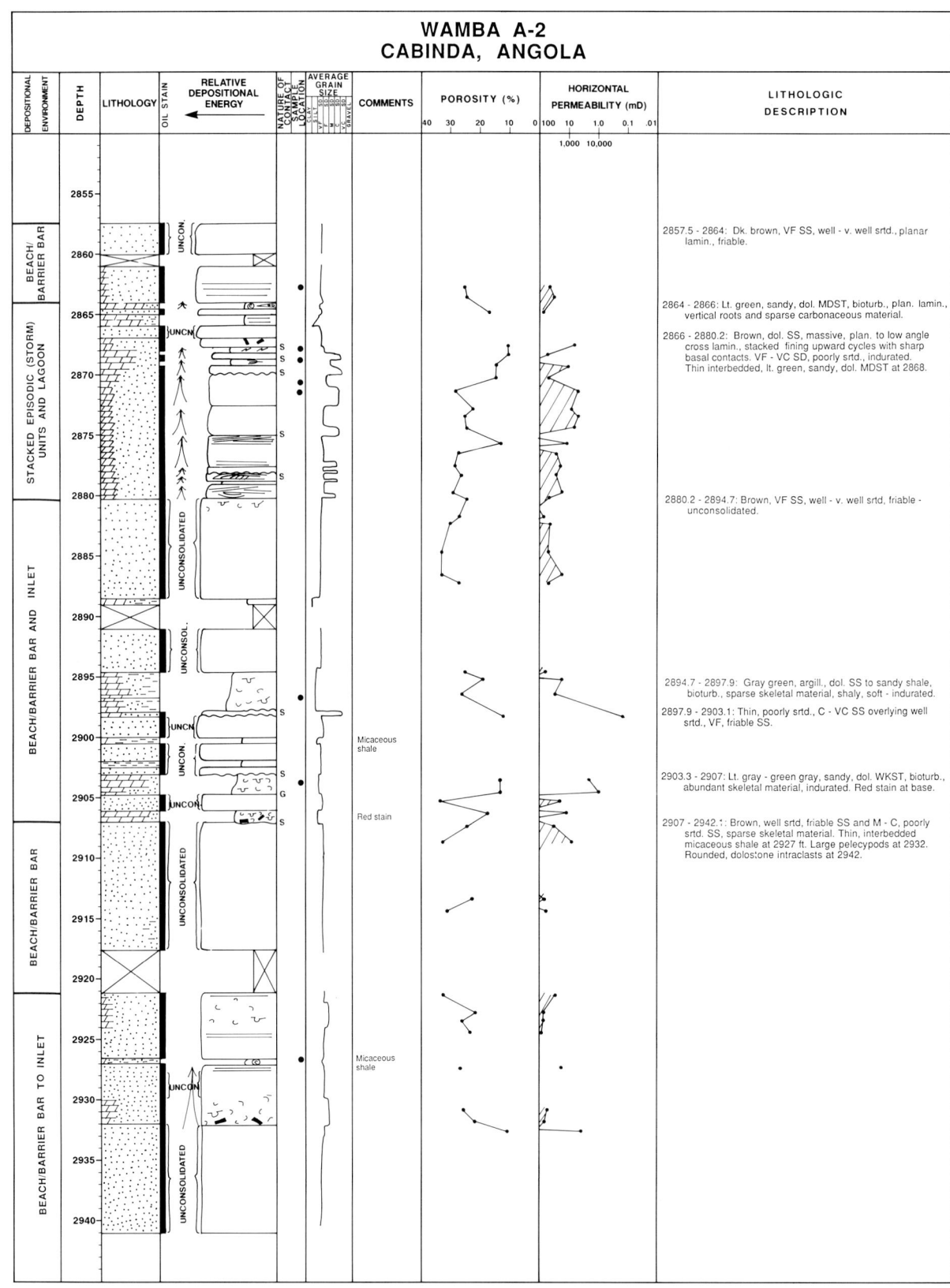

Fig. 5 cont.

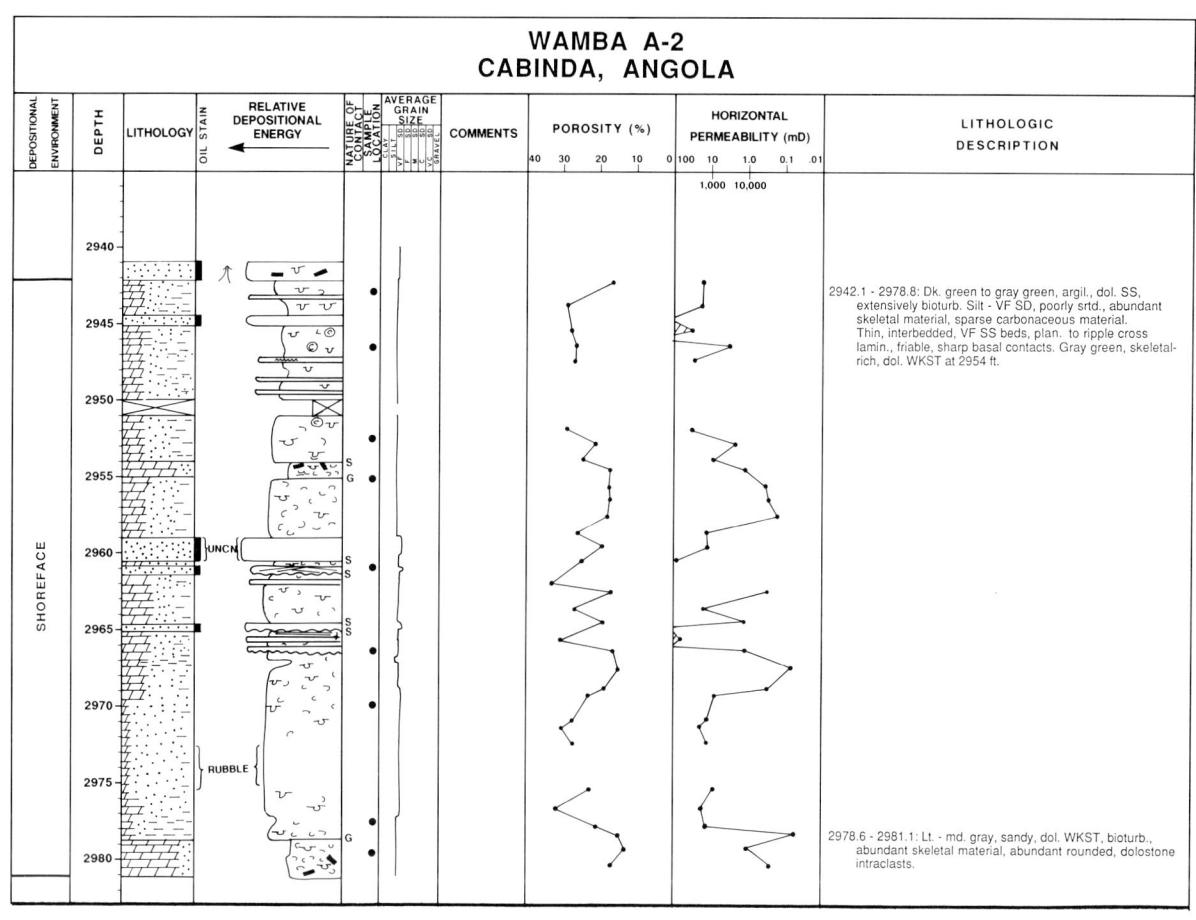

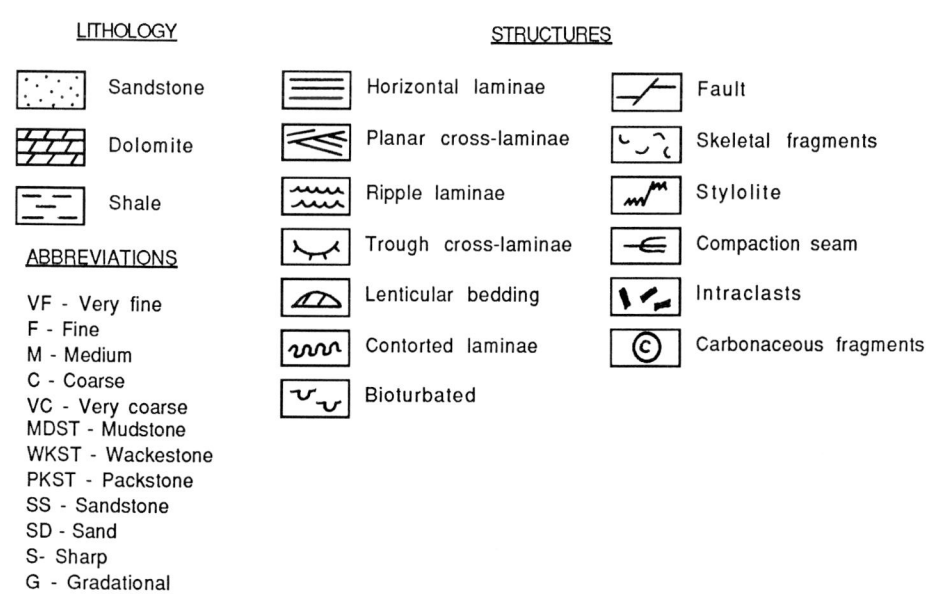

Fig. 5 cont.

The overall skeletal content of the shoreface facies varies considerably, but types are restricted to pelecypods and some gastropods that can be locally concentrated in thin, horizontal layers.

These deposits are often poorly sorted with siliciclastic grain sizes ranging from clay to very fine sand. The shoreface facies occurs in massive beds ranging in thickness from 3 to 10 ft (1 to 3 m) but have been extensively bioturbated so that primary bedding structures have been destroyed (Fig. 6A). Well-defined round to oval-shaped burrows (sometimes lined) range from 0.4 to 1 in. (1 to 3 cm) in diameter and are typically filled with very fine sand and pyrite. Thin laminations of heavy minerals are rarely preserved. Soft-sediment deformation occurs as contorted and folded laminae in several intervals.

In several places in the cored sequence, thin interbeds of true skeletal dolowackestone and dolopackstone occur (Fig. 6C). Colors vary from medium gray to greenish gray. Terrigenous sand and silt content can range from 1 to 20 percent, consisting of quartz and feldspar with accessory heavy minerals. In addition, dolostone intraclasts, ranging from 0.2 to 0.8 in. (0.5 to 2 cm) in size and generally subrounded to rounded, are often present. The dominant skeletal components are pelecypods and gastropods. Pelecypods make up the largest percentage of skeletal grains and occur most often as fragments. Both fragmented and whole gastropods are common and locally abundant, occurring as beds ranging from 1 to 3 ft (0.3 to 1 m) thick. The absence of primary bedding structures is due to extensive reworking by burrowing organisms to the extent that individual burrows are not identifiable. Diagenetic structures are common in these rocks and include stylolites, generally oriented subparallel to bedding, as well as compaction seams that may occur singularly or anastomosing.

The shoreface facies is generally associated with beach/barrier bar environments. Contacts with beach/barrier bar sandstones are generally sharp. Thin, interbedded sandstone units with sharp basal contacts are common (Fig. 6B). Contacts with the sandy skeletal dolowackestones are sharp to gradational.

Beach Barrier Bar

Sandstones of the beach/barrier bar facies do not vary greatly with respect to their siliciclastic constituents. Subarkosic to arkosic sands are the most common rock type of this facies, ranging in color from tan to dark brown as a result of oil staining. The detrital mineralogy of the beach/barrier bar sandstones in the Wamba A-2 well is dominated by quartz and feldspar, with potassium feldspars generally more abundant than plagioclase. Detrital ilmenite and magnetite are a significant component of the total composition, with lesser amounts of garnet and tourmaline present. Igneous rock fragments are less common in these better-sorted sandstones compared to the coarser sandstones in the tidal inlets and washover fans. Detrital micas are present in some samples and may be locally abundant, but skeletal fragments are rare. Textural characteristics of these sandstones are uniform, typically composed of very fine to fine, well-sorted terrigenous clastic grains. Sandstones with abundant dolomite cement are well indurated; whereas, sandstones with minor dolomite cement or completely lacking cement are friable to unconsolidated.

Fig. 6--Core slabs of the shoreface facies. A. Extensive bioturbation has destroyed primary sedimentary structures in this rock, which contains abundant skeletal material and intraclasts. B. Planar-laminated to unconsolidated, very fine-grained sandstone interbedded within bioturbated argillaceous and dolomitic, silt to very fine-grained sandstone. C. Sandy, skeletal dolowackestone in sharp contact above planar-laminated to burrowed, very fine-grained sandstone. Scale bar equals 2 cm.

Fig. 7--Core slabs of beach/barrier bar and tidal-inlet facies. A. Planar-laminated, very fine-grained sandstone. Dolomite is the major cement in the indurated zones. B. Unconsolidated, very fine-grained, well-sorted sandstone with interbedded dolomitic shale. C. Planar to low-angle cross-laminated, very fine-grained, dolomite-cemented sandstone. D. Poorly sorted sandstone with large dolostone rip-up clasts. E. Basal section of a massively bedded sandstone overlying planar-laminated dolomitic shale. Note the scoured contact at base of fining- upward sandstone. Scale bar equals 2 cm.

Fig. 8--Core slabs of the lagoon and tidal-flat facies. A. Dolomudstone with interbedded, poorly sorted, very fine-grained sandstone. B. Bioturbated mudstone. Authigenic pyrite concentrated in burrows. C. Dolomudstone with abundant dolostone rip-up clasts. D. Algal-laminated, argillaceous dolomudstone. Scale bar equals 2 cm.

Fig. 6

Fig. 7

Fig. 8

Sandstones of this facies occur in beds 1 to 15 ft (0.3 to 4.5 m) thick. Sedimentary structures are not visible in most intervals because of the friable nature of the sediments (Fig. 7A and B). Where visible, concentrations of heavy minerals are the most common form of lamination. Horizontal laminations are the most dominant bedding feature, and low-angle planar cross-bedding occurs locally (Fig. 7C). Bioturbation is rare; however, some individual lined burrows are preserved.

Lagoon and Tidal Flat

The restricted lagoon and tidal-flat facies makes up a significant portion of the Upper Vermelha in the Wamba A-2 well. This facies is characterized by silty to sandy, dark gray, argillaceous dolomudstones to wackestones and rare, silty to sandy dolomitic shales of pale green to gray-green color. Intervals dominated by low-energy lagoonal facies are often punctuated by 1 to 5 ft (0.3 to 1.5 m) thick episodic sandstone beds (Fig. 8A).

The dolomite and clay content of these deposits vary significantly in vertical section. Silty to sandy argillaceous dolomudstones typically contain a dolomite mud matrix ranging from 50 to 80 percent, with detrital clay and terrigenous clastic grains. Heavy minerals generally account for less than 2 percent of the total composition, but authigenic pyrite is common. The skeletal component in this facies is sparse and generally consists of small, thin-walled pelecypods and some gastropods, with ghosts of fine peloids locally present. The lagoonal facies occurs in beds from 1 to 25 ft (0.3 to 7.6 m) thick. Individual vertical and horizontal burrows are present and are often filled with authigenic pyrite. In rare cases bioturbation is extensive (Fig. 8B). Recognition of sedimentary structures is difficult due to the shaly and crumbly nature of these rocks, which results in numerous rubble zones in the cored sections. When recognizable, horizontal laminations of clay, silt, and very fine sand are the most common sedimentary structures in the mudstones (Fig. 8D).

Exposure surfaces are common in this facies and are recognized by mud cracks, the occurrence of rounded dolostone rip-up clasts above truncated surfaces, and vertical root zones (Fig. 8C).

The lagoonal facies is commonly punctuated by sandstone beds. These units range from 1 to 4 ft (0.3 to 1.2 m) thick, having sharp scoured bases with variable grain size. In many instances the sandstone beds are composed of well-sorted, very fine-grained sands with faint lamination preserved in the lower portions. In other cases these units are stacked, upward-fining sequences (tidal inlets) with scoured bases and basal lag deposits consisting of coarse skeletal fragments, green dolostone intraclasts, and medium- to coarse-grained sand that fine upward to very fine-grained sand (Fig. 7D and E). In several places in the cored sequence, exposure surfaces occur at the tops of either individual beds or stacked sequences. In most cases these units may represent episodic storm deposits developed when the beach/barrier bar sequences are breached during large storms generating washover fans. Some units appear to be clearly tidal channel in origin, with downcutting relationships at the base and channel abandonment features at the top culminating in exposure horizons.

DIAGENESIS

Diagenesis has played an important role in the development, preservation, and distribution of porosity and permeability in both the reservoir and nonreservoir rocks in the Upper Vermelha. Reservoir properties are extremely variable, largely due to the complex facies relationships and rapid facies changes within the section. Petrographic analysis of the facies described in the Wamba A-2 well has allowed recognition of a variety of diagenetic events including precipitation of interstitial carbonate cements, dolomitization, dissolution of pore-filling cement, carbonate grains and detrital framework grains, precipitation of overgrowth cements, and chemical compaction.

Sandstones of the beach/barrier bar and tidal-inlet facies exhibit the best overall reservoir potential, with measured porosities ranging from 10 to 35 percent and permeabilities ranging from 100 to 2000 millidarcys. Pore space in the well-sorted, very fine- to fine-grained sandstones of the beach/barrier bar facies is mainly intergranular (Fig. 9A). Minor amounts of intracrystalline porosity are visible within euhedral dolomite rhombs. Intragranular porosity is also recognized in minor amounts where feldspar and carbonate framework grains have been partially to completely dissolved (Fig. 9B through D). Dolomite, occurring as a mosaic of very fine to fine anhedral to subhedral crystals as well as euhedral crystals, is the dominant cement in these sandstones. Intergranular dolomite cement may be unevenly distributed in small, irregular patches or, in rare cases, may completely fill visible pore space. Other porosity-occluding mechanisms include precipitation of authigenic pyrite and well-developed overgrowths on detrital potassium feldspar grains.

Intergranular pores are the major pore type in the tidal-inlet sandstones. Minor amounts of intracrystalline porosity are recognized in zoned dolomite crystals displaying partial to complete dissolution of crystal cores and thin, partially dissolved outer zones. Intragranular pore space is visible where framework grains, mainly potassium feldspars, have been partially to completely dissolved. Moldic porosity is recognized in many samples and is sometimes partially to entirely filled with fine to medium, anhedral to euhedral dolomite crystals.

Dolomite is the major cement and is highly variable, ranging from 5 to greater than 35 percent. Some intervals within the tidal-inlet facies exhibit an apparent dolomite matrix-supported texture occurring as a mosaic of fine to medium anhedral to euhedral crystals. In grain-supported rocks dolomite cement may occur as unevenly distributed patches or may be more evenly distributed, thereby occluding much of the intergranular pore space.

The reservoir sandstones of the beach/barrier bar and tidal-inlet facies experienced similar diagenetic histories. Diagenesis began with early carbonate cementation, probably during shallow burial, which minimized the effects of compaction, resulting in a loosely packed, grain-supported framework. Potassium feldspar overgrowths probably occurred during early stages of diagenesis in the shallow subsurface prior to dolomite precipitation and replacement. Secondary intragranular and moldic porosity formation resulted from dissolution of framework grains and skeletal fragments, and minor intracrystalline porosity was created by selective dissolution of the cores as well as outer zones of euhedral dolomite rhombs.

Fig. 9--Thin-section photomicrographs of the beach/barrier bar facies in the Upper Vermelha. A. & B. Well-sorted, very fine-grained sandstone with minor dolomite cement and good intergranular porosity (P). Terrigenous clastic grains are mostly quartz and feldspar (F), with heavy minerals disseminated throughout. C. Poorly sorted sandstone with abundant dolomite cement/matrix. Note large pore created by leaching of a terrigenous clastic grain. D. Higher-magnification view of large pores created by leaching of terrigenous clastic grains. Dolomite is abundant and occurs as a mosaic of very fine anhedral crystals.

Fig. 10--Thin-section photomicrographs of the shoreface facies in the Upper Vermelha. A. The poor sorting in this sample is due to the admixing of carbonate mud, clay, and silt/very fine-grained sand by bioturbation. Note phosphatized skeletal fragment. B. Dolomite (D) is the primary matrix material in this sample, which contains minor, unevenly distributed intergranular pore space (P). C. Sandy dolowackestone with abundant phosphatized gastropods. D. Intracrystalline porosity created by selective dissolution of the cores of euhedral dolomite rhombs.

Fig. 11--Thin-section photomicrographs of lagoon facies in the Upper Vermelha. A. Dolomite is present as a mosaic of very fine, interlocking anhedral crystals in this laminated mudstone that contains abundant silt and clay. B. Very fine quartz and feldspar sand grains in a matrix of dolomite mud.

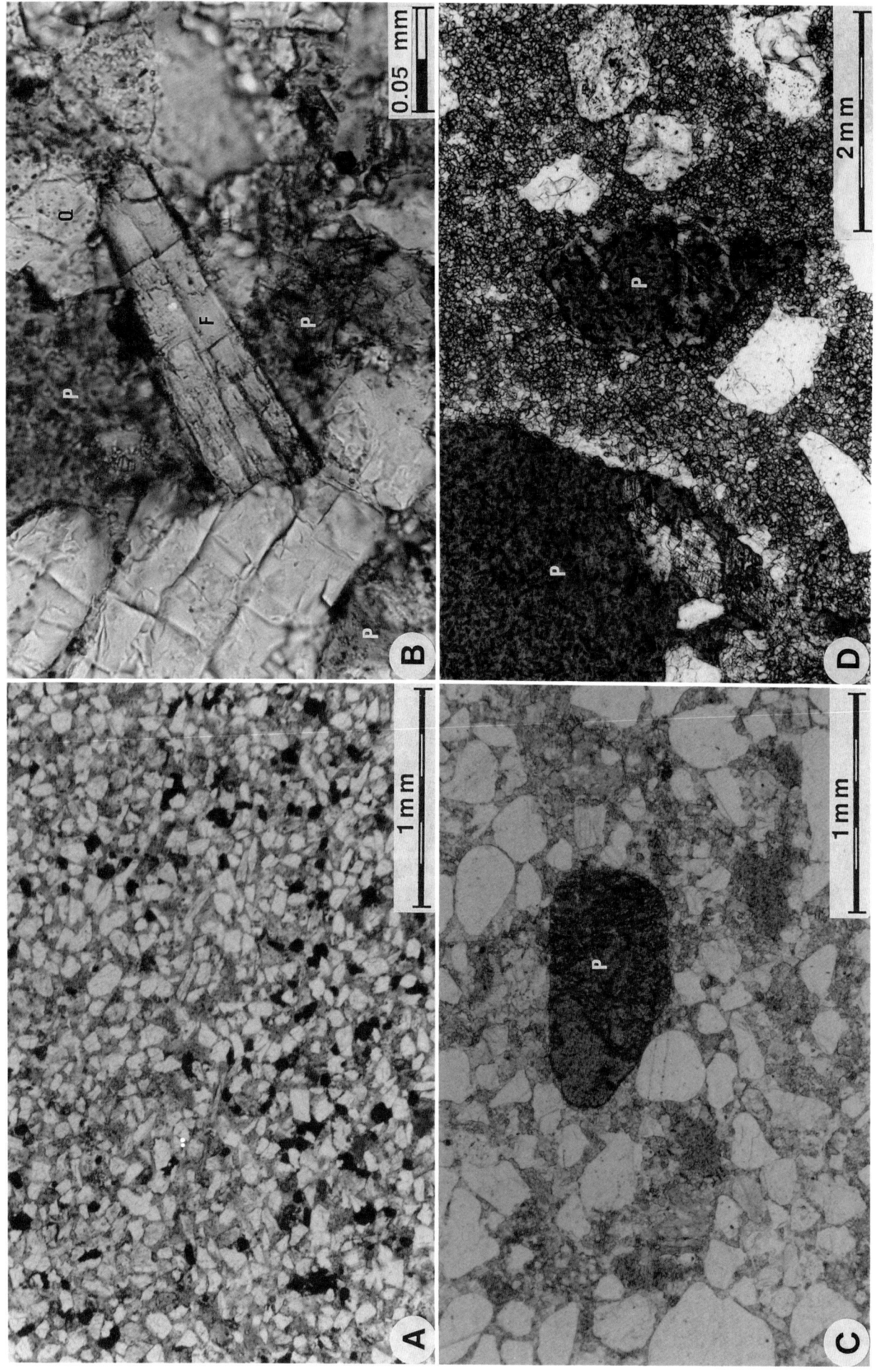

Fig. 9

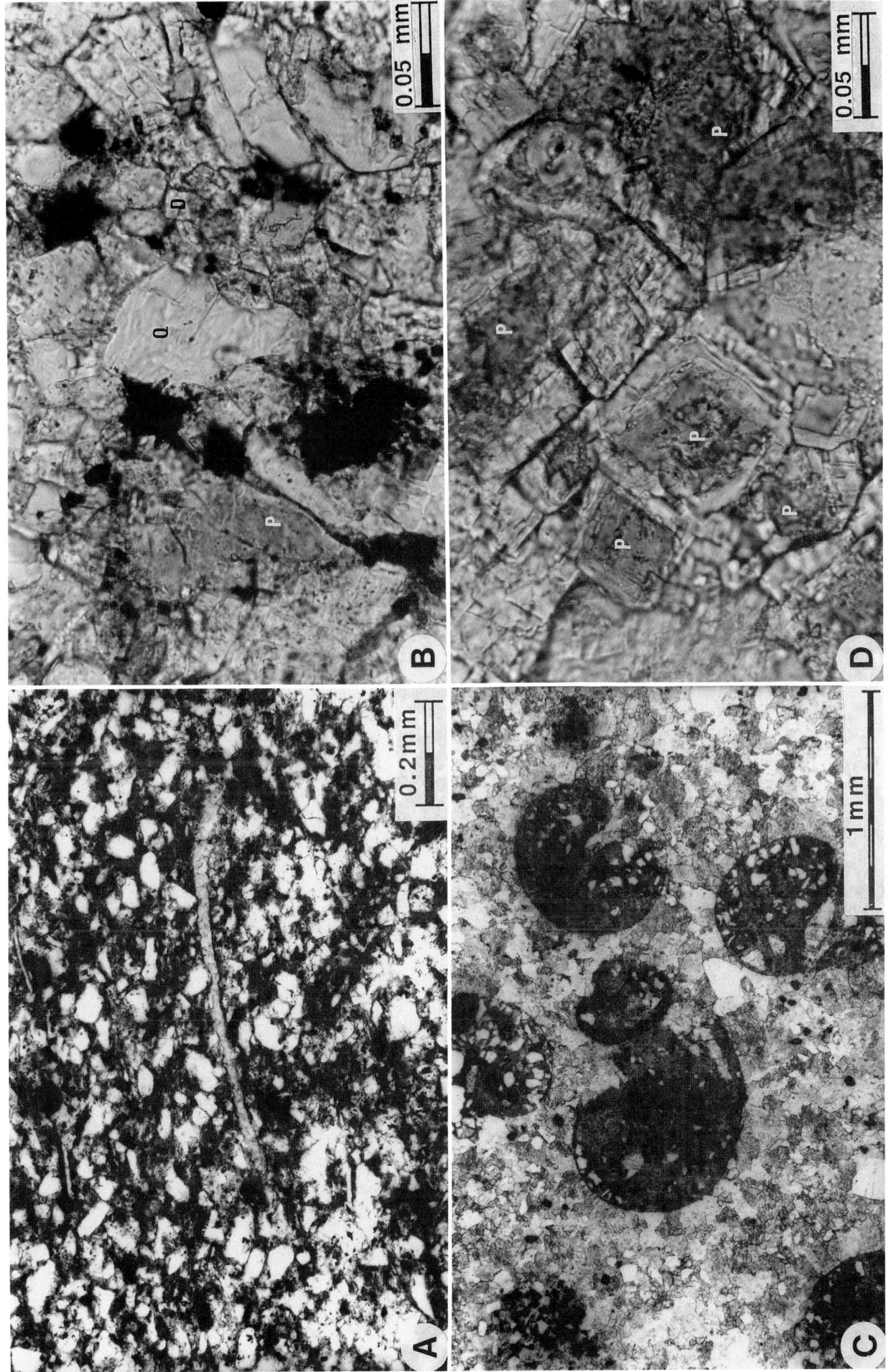

Fig. 10

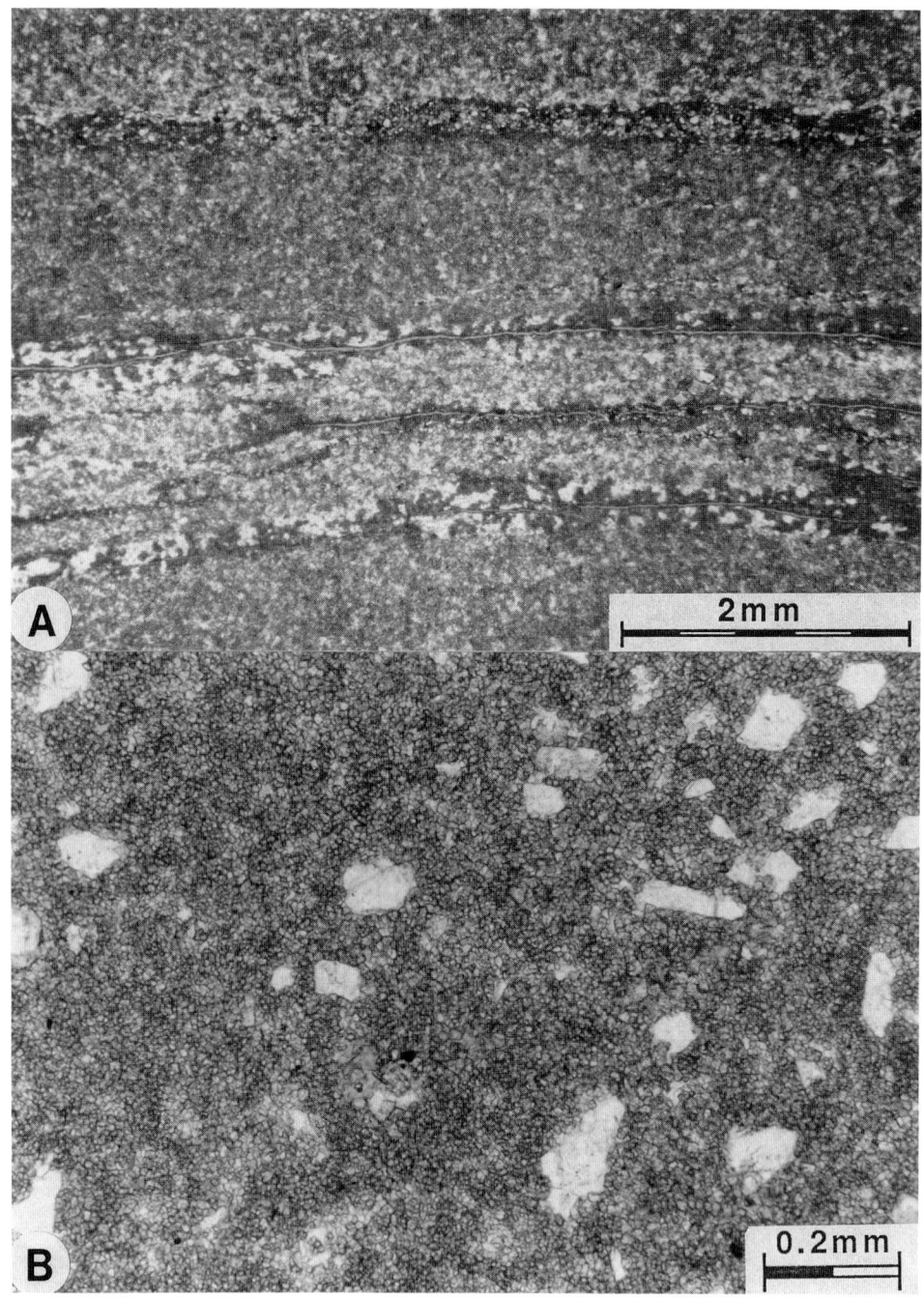

Fig. 11

The poorest reservoir quality occurs in the shoreface and restricted lagoon facies. Porosity and permeability values are greatly reduced in the shoreface and restricted lagoon facies in contrast to the beach/barrier bar and tidal-inlet facies. Because of their poor reservoir quality, these units may act as local or regional permeability barriers. In the shoreface facies both intergranular and moldic pore types are present. Much of the original pore space has been occluded by the admixing of carbonate and argillaceous muds through bioturbation (Fig. 10A). Moldic pores are the dominant pore type in the sandy dolowackestones of the shoreface facies but are unevenly distributed and poorly connected. Moldic, intercrystalline, and intragranular pore types are sparse in the restricted lagoon facies.

Dolomite is the major cement and matrix material in each of the lower-quality facies, occurring as a mosaic of very fine to medium anhedral crystals (Fig. 11). Fine to medium, subhedral to euhedral dolomite crystals are most often found partially or completely filling moldic pore space. Authigenic pyrite is common in each of these facies types but is more abundant in the restricted lagoon facies.

Chemical diagenetic events in the nonreservoir facies began with dolomitization of the original carbonate mud matrix. Dolomitization was probably accompanied, or closely followed, by selective dissolution of carbonate skeletal particles, creating a poorly developed network of moldic pores. Many of these moldic pores were partially to completely destroyed by precipitation of dolomite cement. A large percentage of the skeletal material was replaced by phosphate minerals before dolomitization, resulting in greater preservation of skeletal grains. Minor, late-stage dissolution of dolomite is indicated by corrosion of some dolomite crystals filling moldic pore space. Although sabkha and salina evaporites were not observed in Wamba Field, they are present in adjacent fields that produce from the Vermelha. In a regional sense these may have contributed to changes in pore fluid chemistry which influenced dolomitization, feldspar overgrowths, and dissolution of framework grains.

MIXING MECHANISMS

Facies suites encountered in the Upper Vermelha section can be viewed in a very simple fashion from a seaward to landward progression (Fig. 12). The most open marine facies type is interpreted as being deposited in inner-shelf to shoreface environments that grade into beach and barrier bar systems. These environments are dominated by very fine-grained quartz sand with clay, silt, and carbonate mud and skeletal fragments occurring in the bioturbated shoreface environment. Beach and barrier bar sediments are almost pure, very fine- to fine-grained sand. These exceptionally well-sorted sands are derived from a point source upcurrent, moved along the inner shelf by long-shore currents, and accreted along the foreshore/beach environment. The back-barrier environments of lagoons and tidal flats are a true, lithologically mixed environment. Background sediments appear to be carbonate-dominated, consisting of varying proportions of carbonate mud. Siliciclastic sand transport into the back-barrier system occurred by two mechanisms. Tidal channels cutting the barrier system provided avenues of sand transport into the lagoons. This sand was often reworked and admixed by bioturbation. Many of the sand beds found in the lagoonal sequences are episodic, with sharp, scoured bases and fining-upward character (Fig. 13). These were probably sourced as a result of storm surges through the tidal channels and as washover deposits. Variations in bed characteristics such as thickness, lamination, mud content, and homogenization due to bioturbation can be explained by proximity

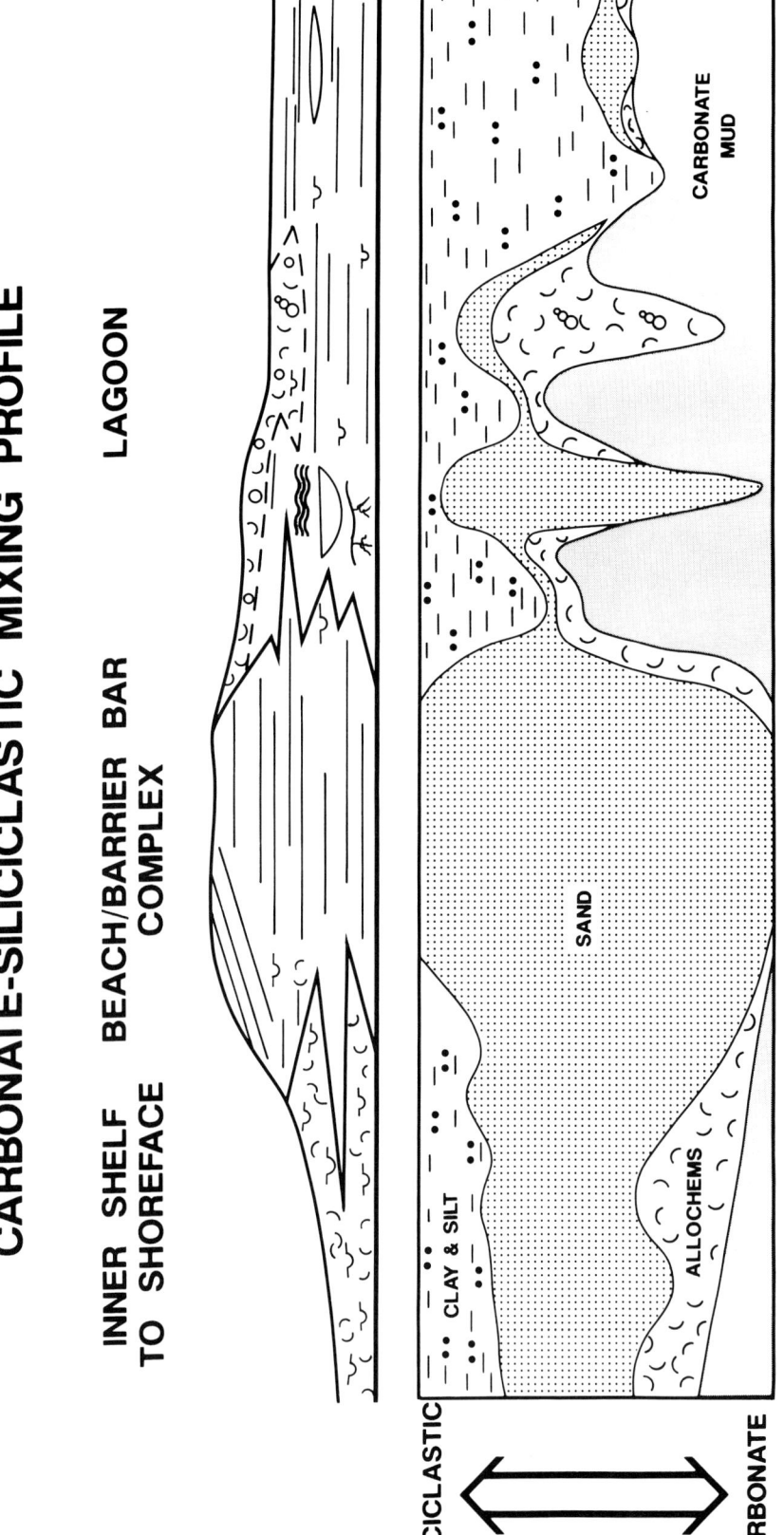

Fig. 12--Facies-mixing profile for Wamba Field.

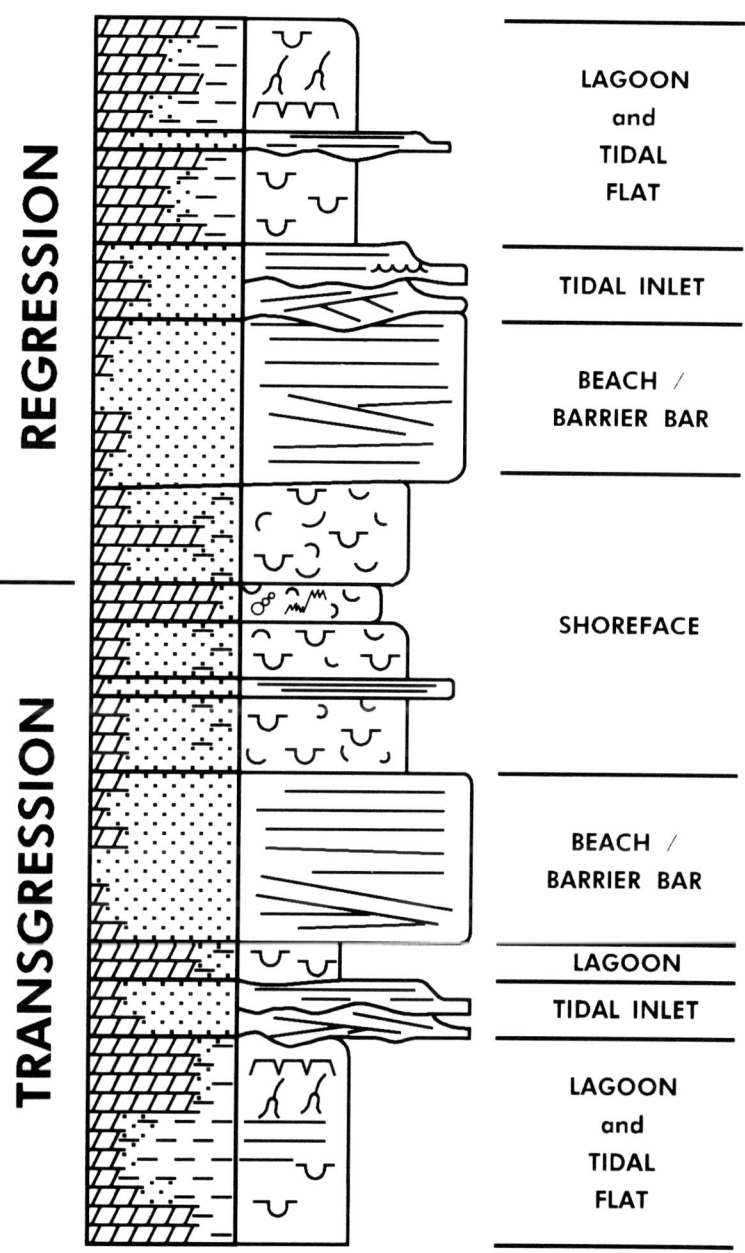

Fig. 13--Idealized transgressive-regressive sequence.

to the sand source. Thicker sands with better reservoir quality (low mud content) occur proximal to the barrier system, while thinner sands with more dolomite cement (higher mud content) occur in distal areas.

Mixtures of siliciclastic and carbonate rocks can occur due to lateral facies mixing (spatial variability) or by sea-level changes and/or variations in sediment supply causing vertical variations in the stratigraphic succession (temporal variability)(Budd and Harris, 1990). Using the mixing process approach of Mount (1984), three types of mixing occur in the Vermelha Formation. In-situ mixing dominates the inner-shelf/shoreface facies. Autochthonous carbonate mud and skeletal fragments (mostly mollusc) are mixed with very fine sand, silt, and clay primarily through bioturbation. Little mixing was evident in the beach/barrier bar facies. This may be controlled by the excellent winnowing and sorting associated with wave-dominated hydrodynamic processes in the foreshore/beach environment. In the lagoonal environment examples of both punctuated and facies mixing are observed. Punctuated mixing is observed in the episodic sandstone units, with scoured bases and fining-upward characteristics encased in sandy and argillaceous, low-energy dolowackestones and mudstones. Facies mixing is observed in portions of the section where carbonate- to siliciclastic-dominated transitions occur in the course of cyclic deposition.

REFERENCES

BRICE, S. E., COCHRAN, M. D., PARDO, G., AND EDWARDS, A. D., 1982, Tectonics and sedimentation of South Atlantic rift sequence: Cabinda, Angola, *in* Watkins, J. S., and Drake, C. L., eds., Studies in Continental Margin Geology: American Association of Petroleum Geologists Memoir No. 24, p. 5-20.

BUDD, D. A., AND HARRIS, P. M., eds., 1990, Carbonate-siliciclastic mixtures: Society of Economic Paleontologists and Mineralogists Reprint Series No. 14, 272 p.

CAFLISCH, L., 1979, Oil exploration in the pre-salt and post-salt sequence of West Africa - results in Cabinda: (Abstract) American Association of Petroleum Geologists Bulletin, v. 63, p. 427.

DALE, C. T., LOPES, J. R., AND ABILIO, S., 1990, Takula oil field and the greater Takula area, Cabinda, Angola: Offshore Technology Conference, Houston, Texas, p. 635-644.

MOUNT, J. F., 1984, Mixing of siliciclastic and carbonate sediments in shallow shelf environments: Geology, v. 12, p. 432-435.

OTI, M. N., AND KOCH, R., 1990, Mid-Cretaceous shelf carbonates: the Mfamosing Limestone, Lower Benue Trough (Nigeria): FACIES, No. 22, Erlangen, p. 87-101.

REYMENT, R. A., AND DINGLE, R. V., 1987, Paleogeography of Africa during the Cretaceous period: Paleogeography, Paleoclimatology, and Paleoecology, v. 59, p. 93-116.

SEIGLIE, G. A., AND BAKER, M. B., 1984, Relative sea-level changes during the middle and late Cretaceous from Zaire to Cameroon (Central West Africa), *in* Schlee, J. S., ed., Interregional Unconformities and Hydrocarbon Accumulation: American Association of Petroleum Geologists Memoir No. 36, p. 81-88.

TILLEMENT, B., 1987, Insight into Albian carbonate geology in Angola: Bulletin of Canadian Petroleum Geology, v. 35, No. 1, p. 65-74.

DEPOSITIONAL ENVIRONMENTS AND FACIES ANALYSIS OF THE CHEROKEE GROUP IN WEST-CENTRAL KANSAS

JEROME J. CUZELLA AND CHRISTOPHER P. GOUGH
National Cooperative Refinery Association,
1775 Sherman St., Ste. 3000, Denver, CO 80203
STEPHEN C. HOWARD
Consultant, 4 Nethergrange Court, Albert Road North,
Malvern, WORCS WR14 2TP, England

ABSTRACT

In west-central Kansas the Cherokee Group of early Desmoinesian, Pennsylvanian age is a mixed siliciclastic and carbonate sequence. It was deposited in environments that were transitional from continental to marginal marine as the sea transgressed over the Mississippian unconformity out of the Hugoton Embayment and onto the Central Kansas Uplift. Sandstones of the Cherokee Group are important oil reservoirs in west-central Kansas, but they are highly variable in properties and in aerial distribution; therefore, they are difficult to predict. Core studies and subsurface analysis using wireline logs reveal two persistent and widespread limestone beds that form useful stratigraphic markers within the Cherokee. They provide a framework for facies analysis and regional mapping that may be useful as a predictive tool for oil exploration. Six basic lithofacies are interpreted from lithologies and sedimentary structures observed in cores obtained from four wells in eastern Ness County: (1) basal Pennsylvanian conglomerate, (2) fluvial sands, (3) fine-grained tidal-flat deposits, (4) shallow marine limestones and shales, (5) shoreline sands and tidal-channel sands, and (6) braided-stream sandy conglomerates. Successions of these facies are equivalent to components commonly observed in Kansas cyclothems. Two transgressive-regressive cycles are identified in the Cherokee, and maximum transgression is correlated with the basal portion of the two widespread limestone beds.

Following burial of the Mississippian karstic surface, deposition of peritidal sediments occurred on a uniformly shallow platform. Sedimentation was punctuated by episodes of subaerial exposure and weathering. Clastics derived from the eroding Central Kansas Uplift were probably supplied episodically to the coastal plain by braided streams that were reworked by coastal processes.

INTRODUCTION

The Cherokee Group of early Desmoinesian age in west-central Kansas is a mixed carbonate and siliciclastic succession. The Cherokee sandstones are important oil reservoirs in this region and offer an attractive exploration target for independent oil companies. Some fields have produced over 2.5 million barrels of oil. Individual wells typically produce 50,000 to 80,000 barrels of oil on 40-acre spacing from depths of approximately 4200 ft (1280 m). Most reservoir sands are only 10 to 15 ft (3.0 to 4.6 m) thick, highly variable, and difficult to predict. Although many of the early Cherokee fields were discovered while prospecting for oil in the underlying Mississippian carbonates (Walters and others, 1979), many exploration efforts are now directed specifically to the Cherokee sandstones.

This study is concentrated in eastern Ness County, Kansas, located on the northeastern shelf of the Hugoton Embayment (Fig. 1). It is based on four wells that were cored in the Cherokee Group: NCRA, Pfaff 2; NCRA, Pfaff 3; NCRA, Pfaff 5; and NCRA, Thompson A-2. The total Cherokee section is observable in approximately 170 ft (52 m) of core with overlapping intervals. Open-hole electric logs (dual induction resistivity, compensated neutron-density, and borehole compensated sonic) were run in these wells and complement the core study. A four-township area in eastern Ness County and in the vicinity of the cored wells was selected for regional subsurface analysis. Data from approximately 420 wells were studied. Few wells in this region have cored the Cherokee, and approximately 25 percent of the wells drilled do not have electric logs. Without these data a detailed, high-resolution, stratigraphic analysis is difficult. The typical suite of open-hole electric logs run in most wells is the gamma-ray, guard (resistivity), and neutron porosity. Gamma-ray and neutron porosity logs are presented with the core descriptions and graphic columns for sake of brevity (Figs. 11-14). This presentation is consistent with the more commonly used logging suite and presents characteristic log signatures for the recognizable lithologies.

The Cherokee Group crops out in southeastern Kansas and has been the subject of numerous surface and subsurface studies (a summary of authors is presented in Brenner, 1989). East of the Central Kansas Uplift, the Cherokee consists of interbedded siliciclastic mudrocks, lenticular sandstones, coals, and thin limestones deposited in deltaic complexes (Brenner, 1989), and is considerably different from the geology of the Cherokee Group west of the Uplift. Although much has been published on the geology of the Cherokee Group in eastern Kansas, few publications have addressed the Cherokee in central and western Kansas. Walters and others (1979) described channel sands from the Cherokee Group in Ness County, and Nodine-Zeller (1981) published the results of a core study of the basal Pennsylvanian conglomerate. On a more regional scale, Merriam (1963) briefly discussed the area in an overview of the geology of Kansas, and Rascoe and Adler (1983) presented an overview of accumulations of Permo-Carboniferous hydrocarbons in the Mid-Continent that included paleogeographic maps of the Mid-Continent during the early Desmoinesian.

GEOLOGIC SETTING

Ness County is located on the west flank of the Central Kansas Uplift and the northeastern shelf of the Hugoton Embayment of the Anadarko Basin (Fig. 1). The Central Kansas Uplift is a northwest-southeast trending structural high that separates the Hugoton Embayment on the west from the Salina and Sedgwick basins to the east. It is the largest structurally positive feature in Kansas. The widespread pre-Pennsylvanian unconformity of the Mid-Continent developed during a period of emergence and erosion. It accompanied the cratonic epeirogeny that led to the active movement of the Central Kansas Uplift and other structural features across much of North America at the close of the Mississippian (Rascoe and Adler, 1983). The major influence on sedimentation throughout the Early Pennsylvanian was transgression of the sea on this

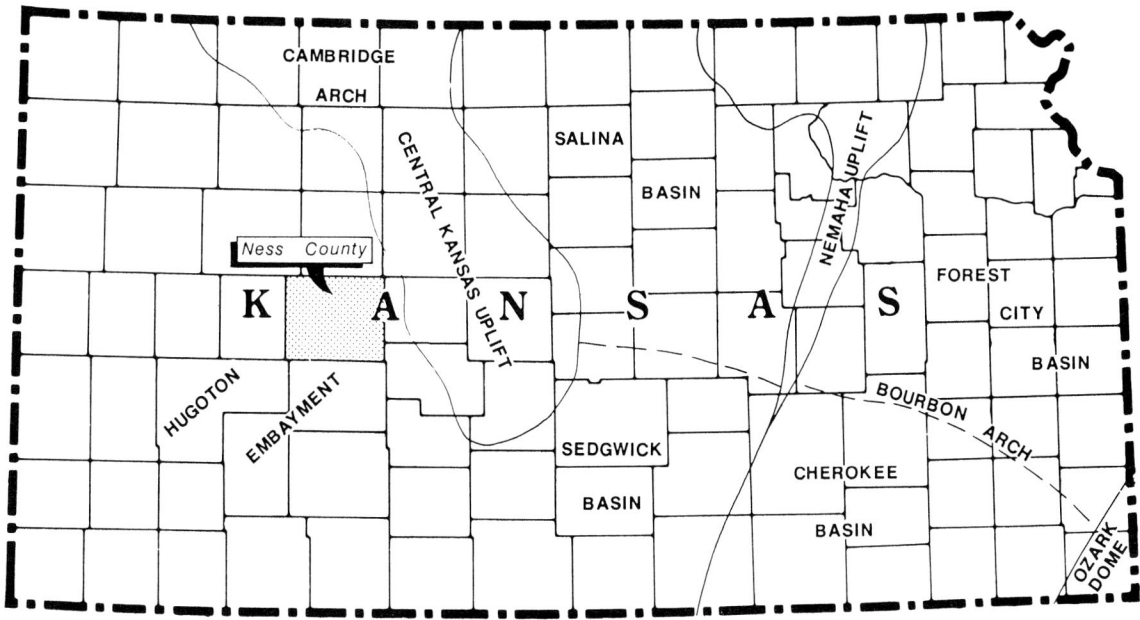

FIG. 1.--Location map, Ness County and major tectonic features of Kansas (adapted from Zeller, 1968).

erosional surface. Sediments were deposited on a shallow, tectonically stable shelf in water depths that probably did not exceed 100 ft (30 m)(McCrone, 1964). On the northeastern shelf of the Hugoton Embayment, sediments of the Desmoinesian Stage rest unconformably on Mississippian carbonates and consist primarily of interbedded carbonate beds and thin shales (Fig. 2). Closer to the Uplift, the fine-grained clastics give way to coarser clastics (sandstones and conglomerates) that parallel the ancient shorelines. By the end of Missourian time, the Central Kansas Uplift was inundated and covered with marine sediments (Rascoe and Adler, 1983).

Regional dip on the Cherokee is exceptionally flat and barely noticeable at 5 ft (1.5 m) per mile to the south-southwest, although local structural relief can be as much as 40 ft (12.2 m) per half mile. This local relief is probably compactional and inherited from structure of the underlying unconformity surface (Figs. 3-6). Farther to the east and in close proximity to the Uplift, the dip rapidly increases to as much as 100 ft (30 m) per mile.

STRATIGRAPHY

The Cherokee Group is a formal stratigraphic unit in Kansas, and the type locality is in Cherokee County (eastern Kansas)(Zeller, 1968). It consists of two formations: Krebs and Cabiness, with distinctive coals, limestones, and sandstones defined as members of these formations (Fig. 2). The Cherokee Group in Ness County rests unconformably on Mississippian carbonates and is

ROCK TYPE		TIME STRAT.		
Holdenville Shale	Marmaton Group	DESMOINESIAN STAGE	MIDDLE PENNSYLVANIAN SERIES	PENNSYLVANIAN SYSTEM
Lanapah Limestone				
Nowata Shale				
Altamont Limestone				
Bendera Shale				
Pawnee Limestone				
Labette Shale				
Fort Scott Limestone				
Cabaniss Formation	Cherokee Group			
Krebs Formation				
Mississippian Undifferentiated		OSAGE STAGE	UPPER MISS. SYS.	MISS. SYSTEM

FIG. 2.--Stratigraphic column for study area (adapted from Zeller, 1968). Formations of the Cherokee Group are not identifiable in the subsurface.

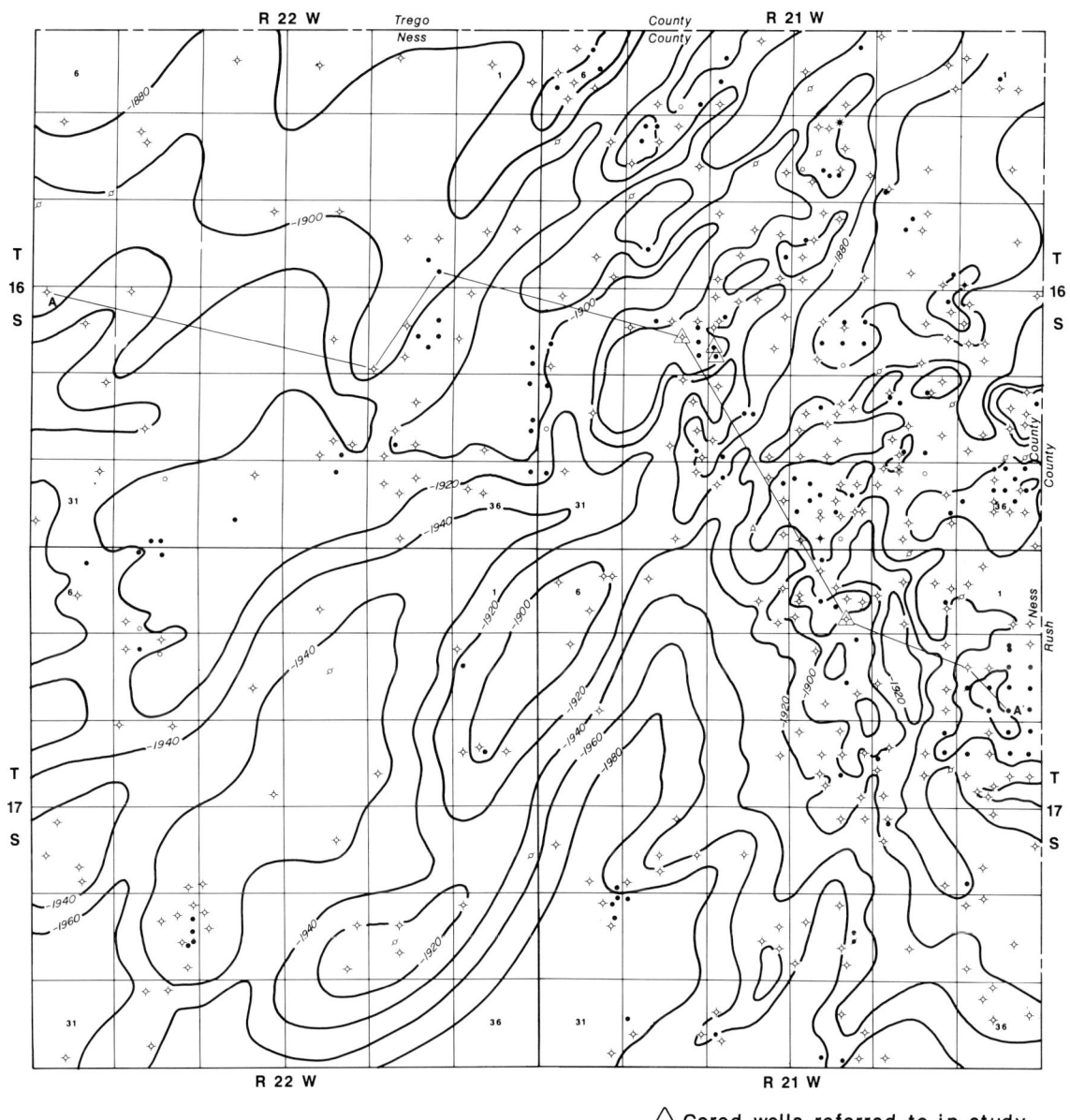

△ Cored wells referred to in study.

FIG. 3.--Structure contour map on top Cherokee Group, contour interval 20 ft. Contours mimic the structural grain of the Mississippian unconformity (compare Fig. 6).

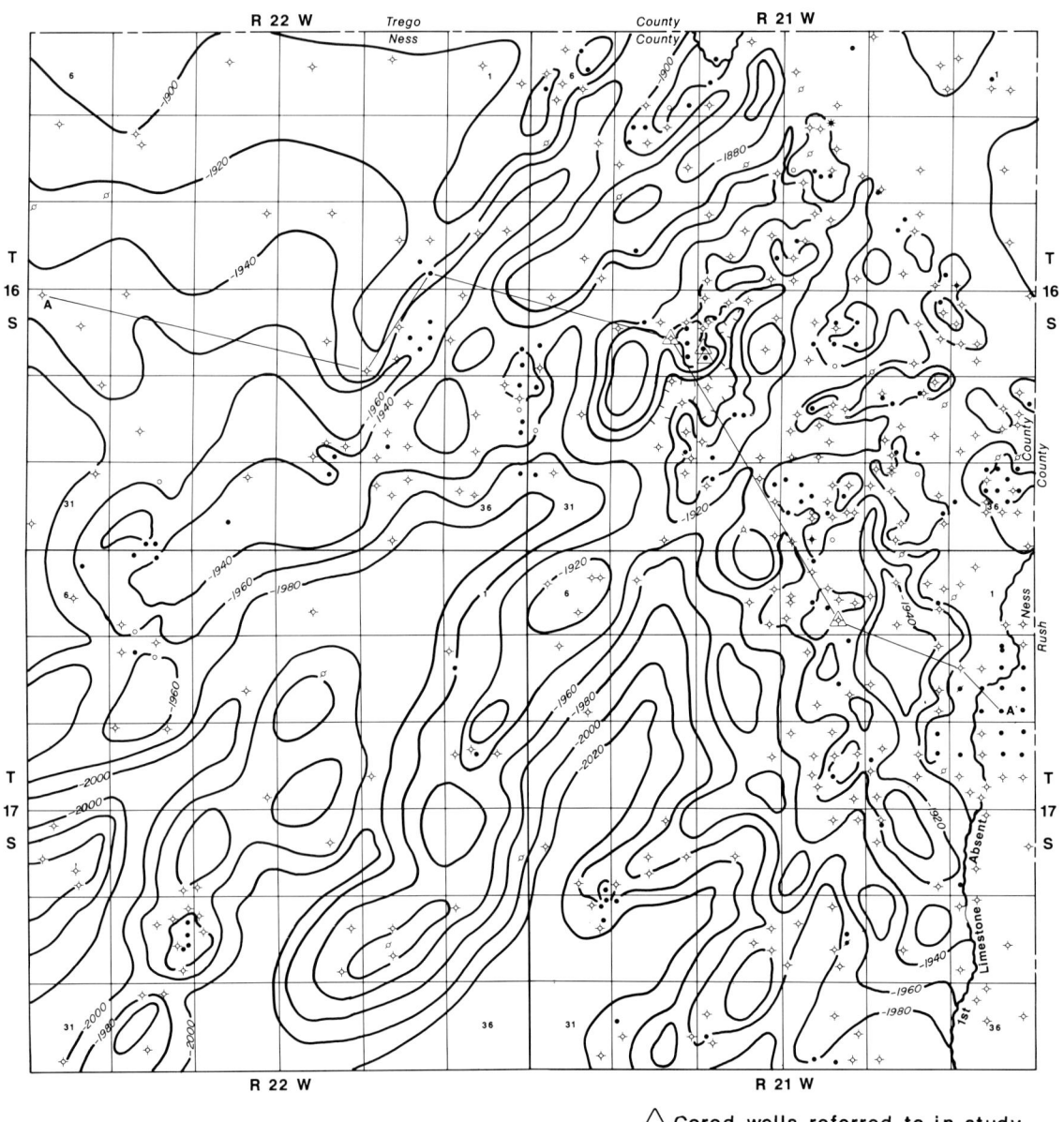

FIG. 4.--Structure contour map on base First Limestone Marker, 20 ft contour interval.

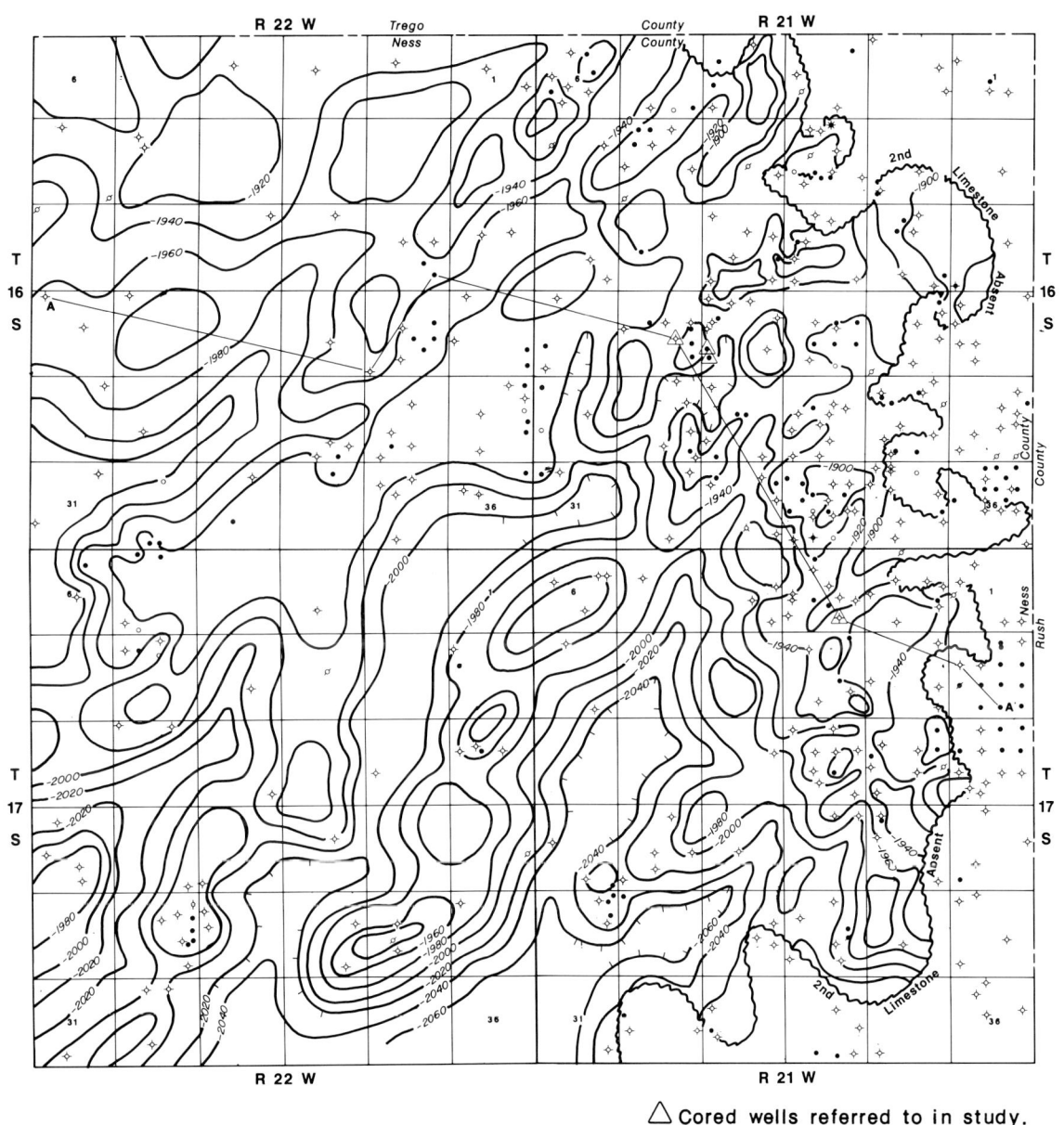

△ Cored wells referred to in study.

FIG. 5.--Structure contour map on base Second Limestone Marker, 20 ft contour interval.

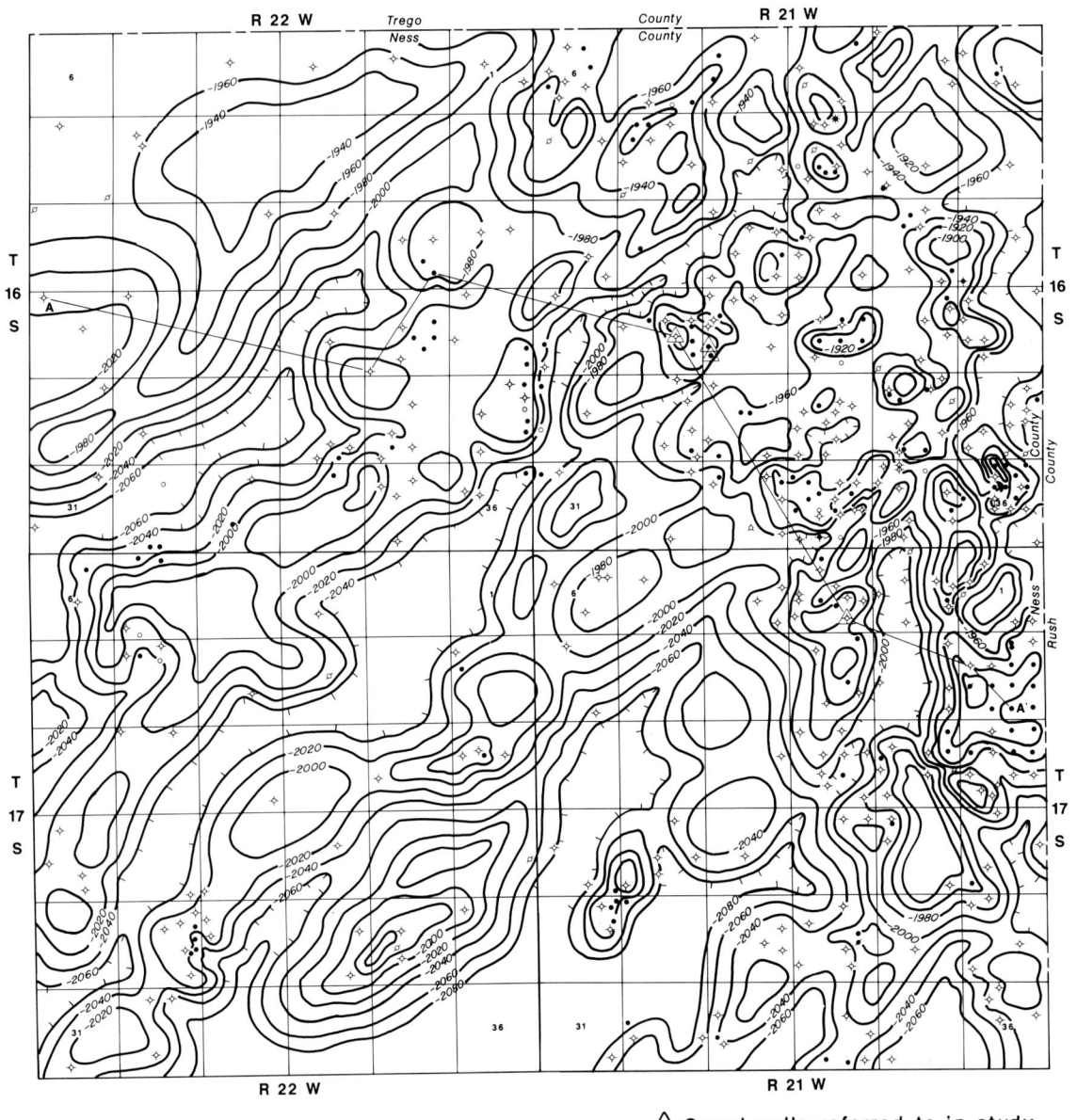

△ Cored wells referred to in study.

FIG. 6.--Structure contour map on top Mississippian unconformity, 20 ft contour interval. The top of the Mississippian is a karstic surface with a system of knobs (highs) and valleys (lows); a northeast-southwest drainage pattern is dominate.

conformably overlain by the Marmaton Group. In the subsurface and west of the Central Kansas Uplift, the Cherokee Group is not divisible into formations (Zeller, 1968). Overlying the Cherokee Group is the Marmaton Group, consisting of eight recognized formations. The Fort Scott Limestone is the lowermost and conformably overlies the Cherokee. It consists of two limestone members and an intervening shale that are not readily distinguishable in the subsurface. The Fort Scott is conformably overlain by the Labette Shale. In the study area, the Labette is overlain by the Anna Shale member of the Pawnee Limestone and is characterized by a distinctive, high gamma-ray log signature, making it a useful stratigraphic datum. Detailed examination of drill cores and regional stratigraphic cross sections indicates the presence of two persistent and widespread limestone units throughout most of Ness County (Figs. 10-14). These form useful stratigraphic markers within the Cherokee and provide a convenient framework for regional mapping and facies analysis. Three divisions are defined based on electric log correlations and drill core in order to facilitate discussion of the Cherokee Group. They are, in ascending order: (1) top of the Mississippian unconformity to the base of the second limestone marker, (2) base of the second limestone marker to the base of the first limestone marker, and (3) base of the first limestone marker to the top of the Cherokee. Farther east and toward the Uplift, however, the limestone beds pinch out and are replaced by interbedded fine and coarse clastics (Fig. 11). In this region where the limestone beds are absent, this subdivision cannot be applied. Eventually the entire Cherokee section thins and pinches out against the Uplift (Fig. 10).

Top of the Mississippian Unconformity to Base Second Limestone

The pre-Pennsylvanian unconformity surface is a karstic, erosional surface developed on Mississippian-age carbonates. A system of knobs (highs) and valleys (lows) is depicted on the structure on top Mississippian unconformity (Fig. 6). Developed on the Mississippian is a latosolic-type paleosol. This paleosol is present in Pfaff 3, Thompson A-2, and Pfaff 2 drill core. The lower portion is a matrix-supported breccia, containing clasts of white chert derived from the underlying Mississippian rocks. The clasts are poorly sorted and vary greatly in size, ranging from tens of centimeters to a few millimeters. They are both highly irregular in shape and extremely angular (Fig. 15A). The enclosing matrix is silt and clay and varies in color from maroon and ocher to pale green. The breccia is overlain by a thin layer that is probably a preserved remnant of the top horizon of the paleosol. In Pfaff 3 this is a sandy siltstone of variegated color, ranging from maroon to green, and includes some isolated chert clasts, abundant slickensides, and possible rhizoliths (Fig. 15B). In the Thompson A-2 and Pfaff 2, the breccia is overlain by a green and maroon mottled shale. Localized isolated green spots are probably reduction spots caused either by burrows or decayed plant roots. This is the basal Pennsylvanian conglomerate that mantles the Mississippian rocks and is similar to the section described by Nodine-Zeller (1981). In drill core it is distinguishable from the superjacent strata but somewhat less distinctive on electric logs. The angularity and poor sorting of the chert clasts indicate that they have undergone very little transport and probably developed in situ from weathering of the underlying rocks. It is likely that some clasts may have been transported from surrounding highs to lower areas and deposited as fanglomerates.

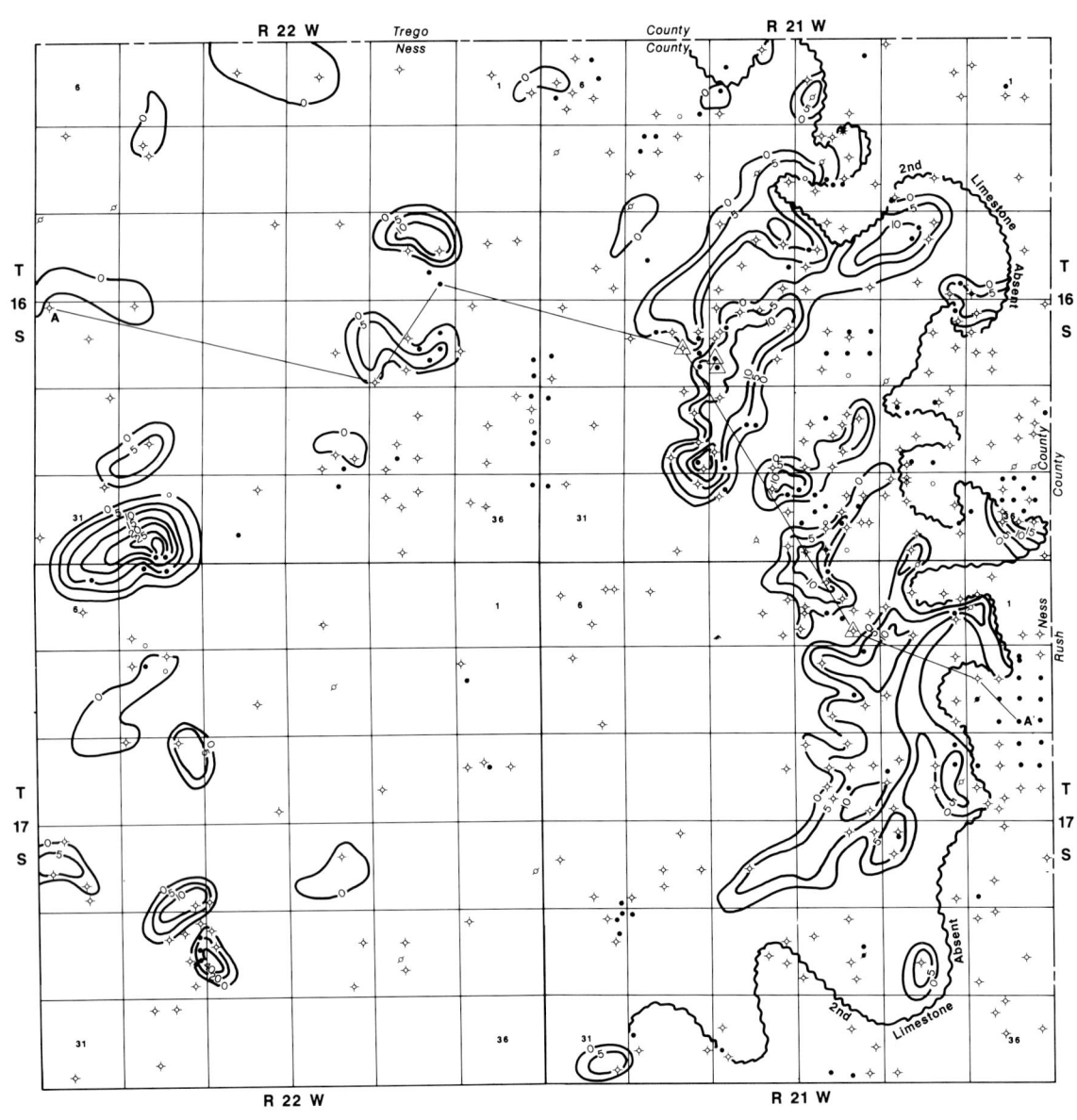

FIG. 7.--Isopach of gross sand in interval from top Mississippian to base of Second Limestone Marker, 5 ft contour interval; sand bodies generally occupy valleys developed on the Mississippian surface.

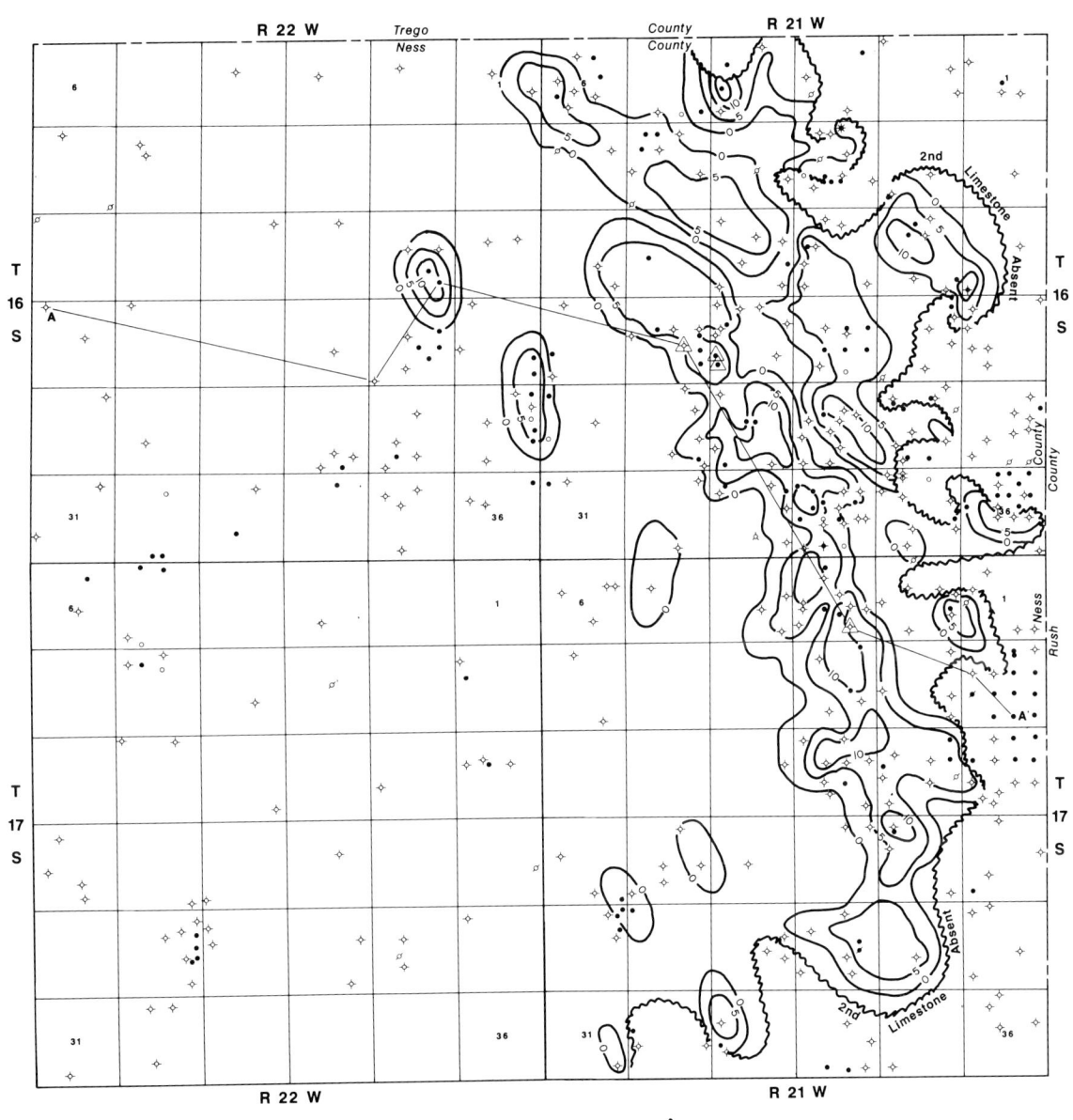

△ Cored wells referred to in study.

FIG. 8.--Isopach of gross sand in interval from base Second Limestone Marker to base First Limestone Marker, 5 ft contour interval. Sand bodies trend generally north-south and parallel the shoreline defined by pinchout of Second Limestone marker, thicker sand occurs on structural highs (compare with structure on base of Second Limestone Marker Fig. 5). The sand diminishes to the west where it is replaced by interbedded fine clastics and limestone.

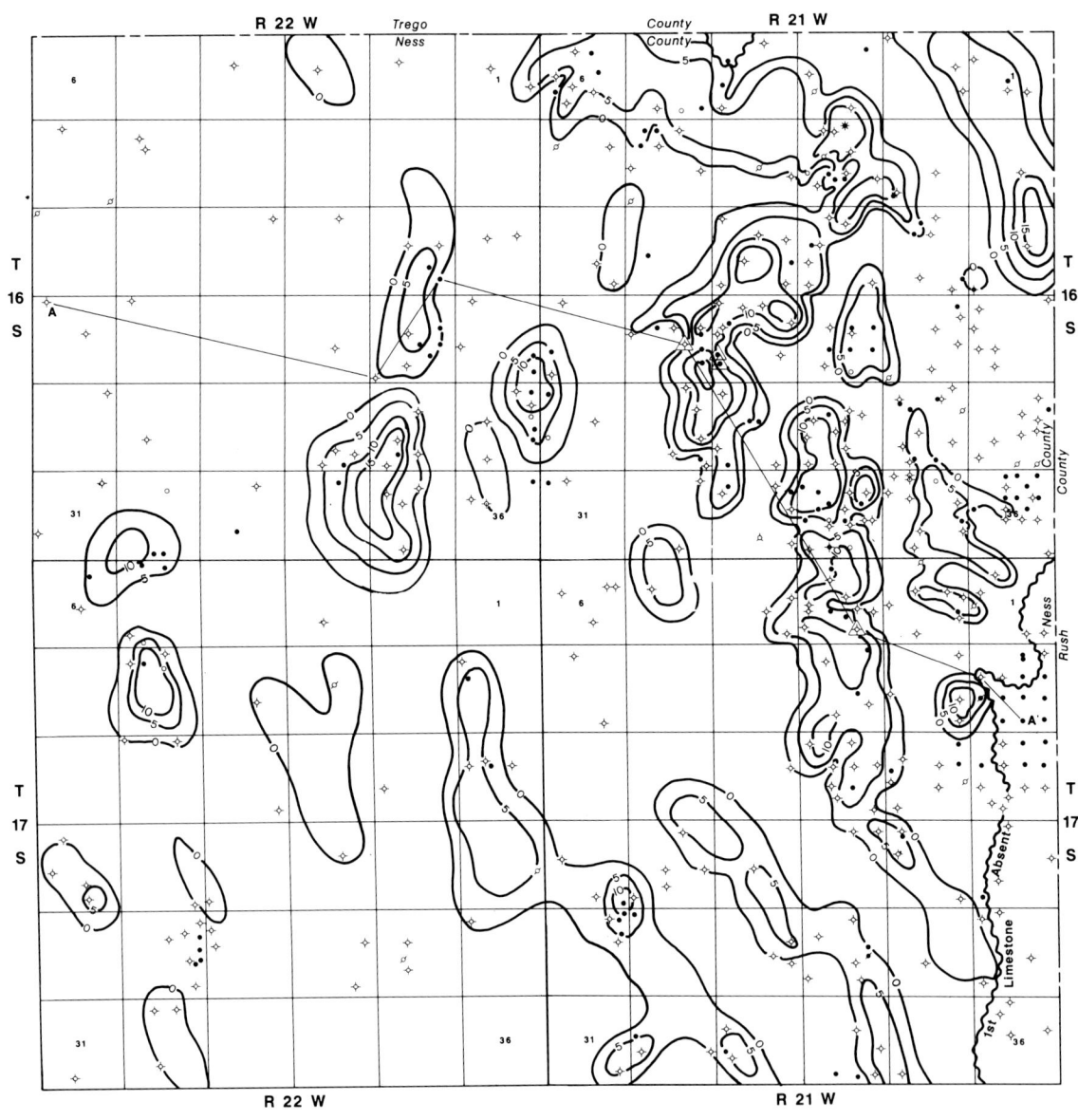

FIG. 9.--Isopach of gross sand in interval from base Second Limestone Marker to First Limestone Marker, 5 ft contour interval. Sand bodies trend generally north-south and with smaller digitate lobes extending toward the shoreline (east); thicker sand is developed over structural highs (compare with structure on base of First Limestone Marker Fig. 4).

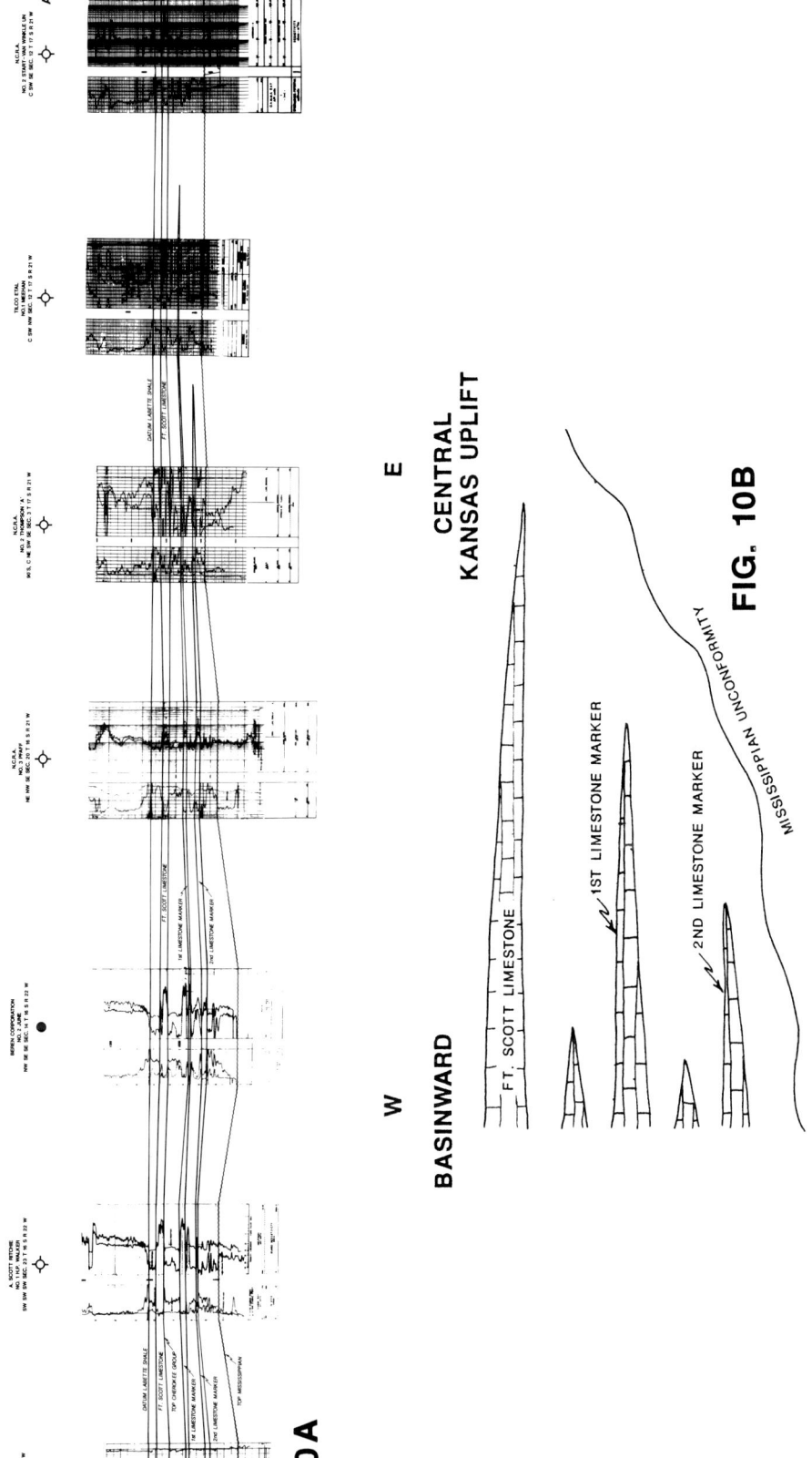

FIG. 10A.--Stratigraphic cross section A-A'. Datum is the top of the Labette Shale or the base of the radioactive Anna Shale. Limestone markers onlap and pinch out to the east towards the Central Kansas Uplift. Where the limestone markers are absent, the proposed correlation scheme cannot be applied.

FIG. 10B.--Schematic cross-section.

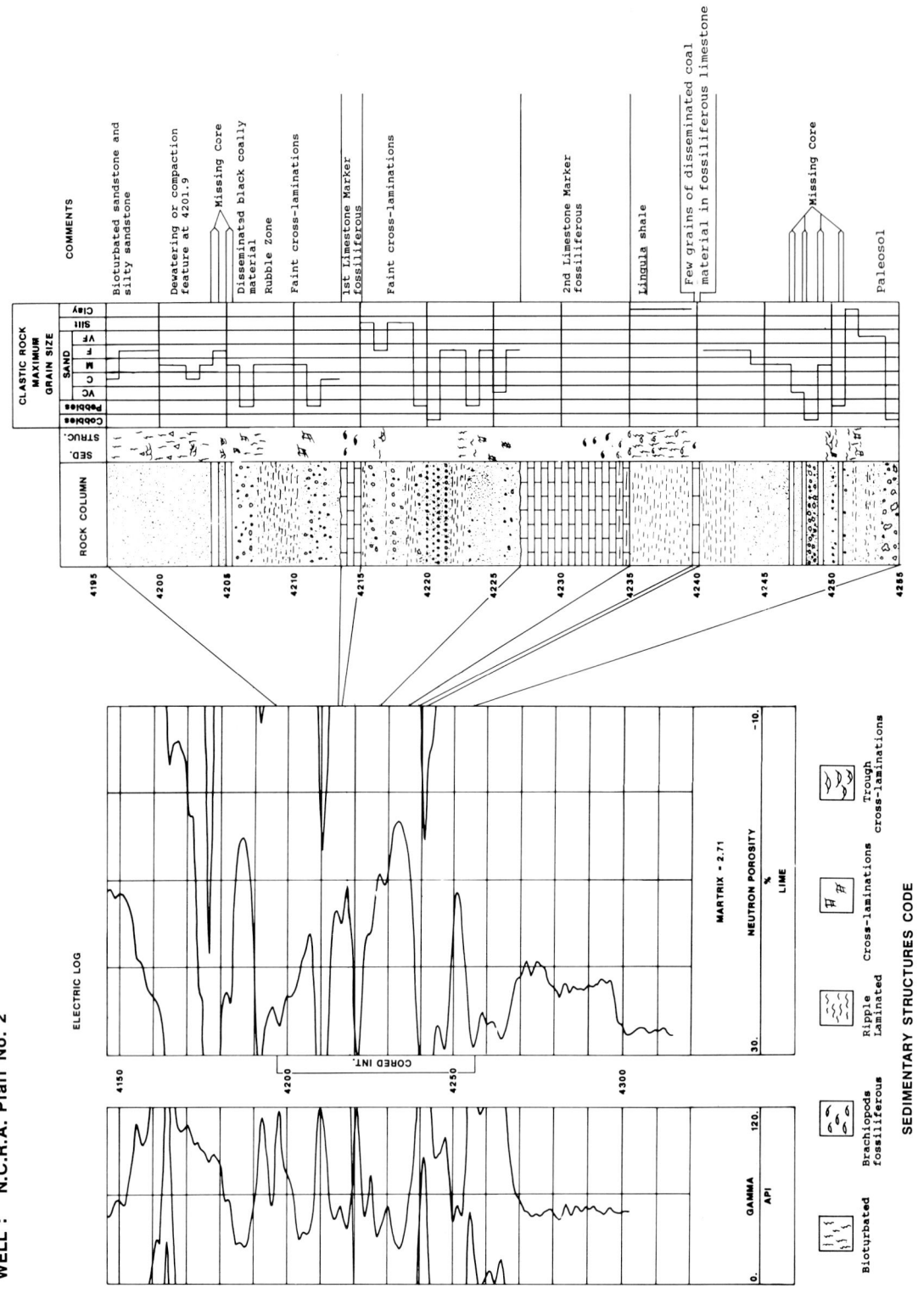

FIG. 11.--Log of core NCRA, Pfaff 2. Core adjusted 1 ft uphole to log depth.

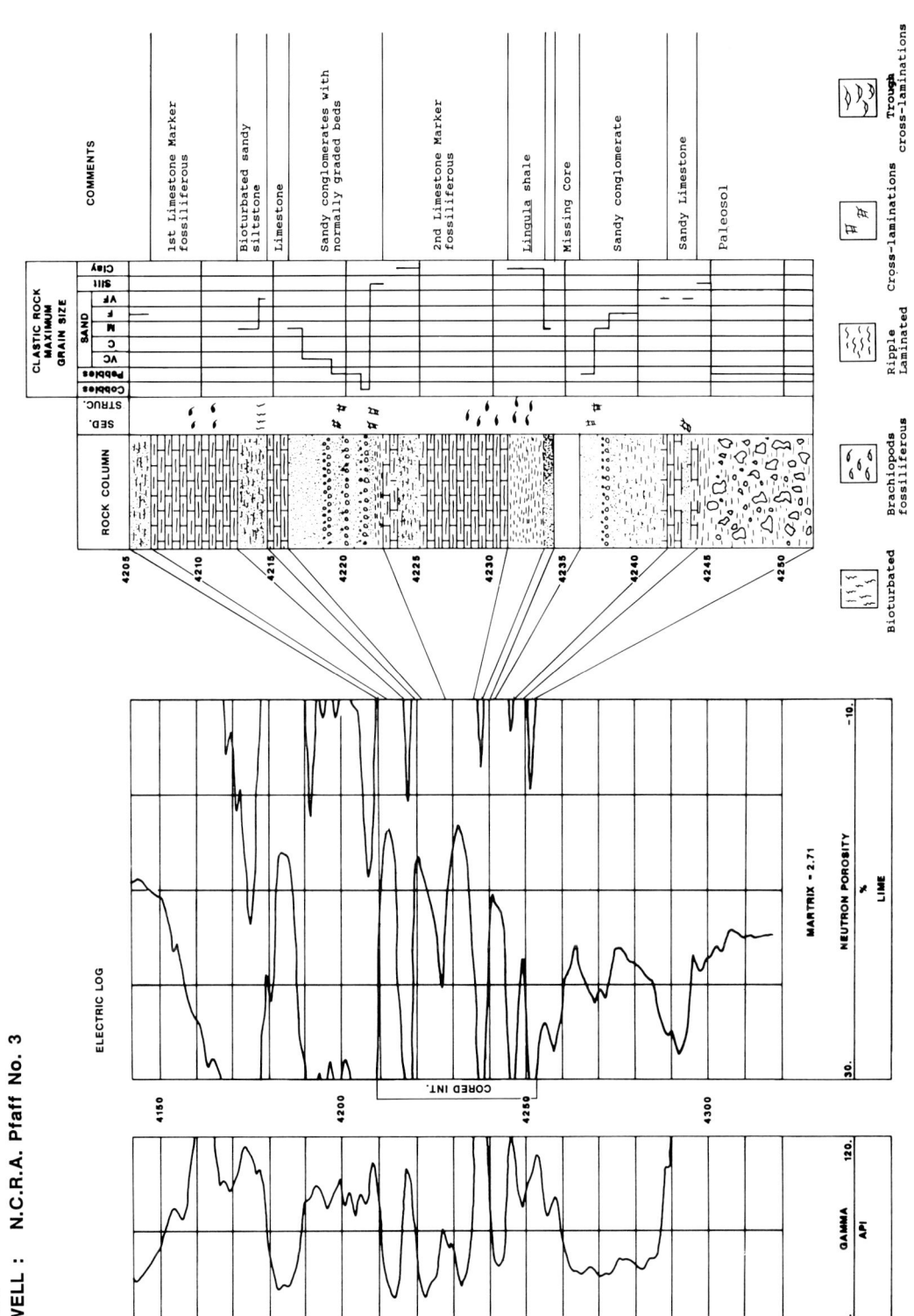

FIG. 12.--Log of core NCRA, Pfaff 3. Core adjusted 5 ft downhole to log depth.

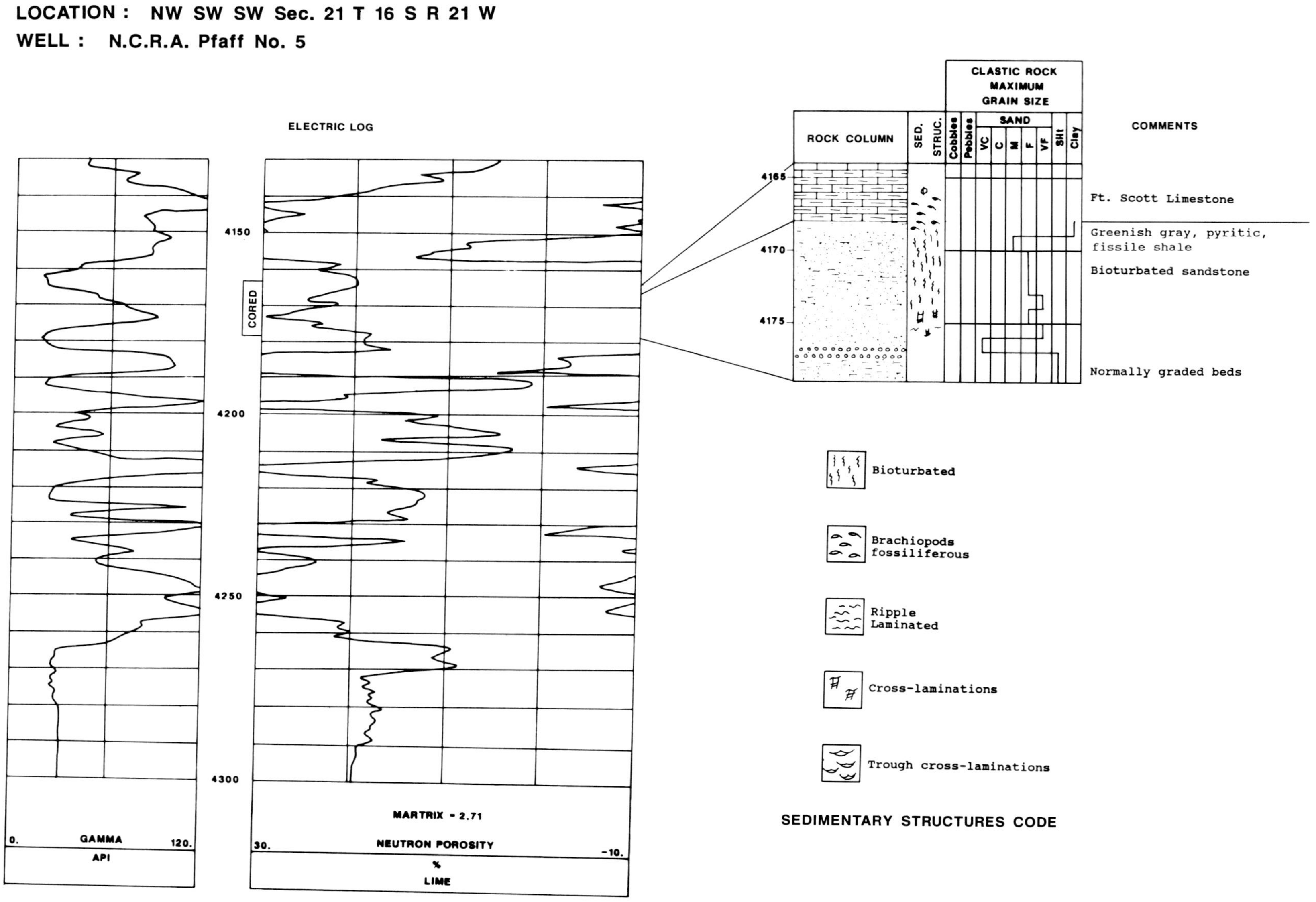

FIG. 13.--Log of core NCRA, Pfaff 5. Core depth equals log depth.

LOCATION : NE SW SE Sec. 3 T 17 S R 21 W
WELL : N.C.R.A. Thompson A-2

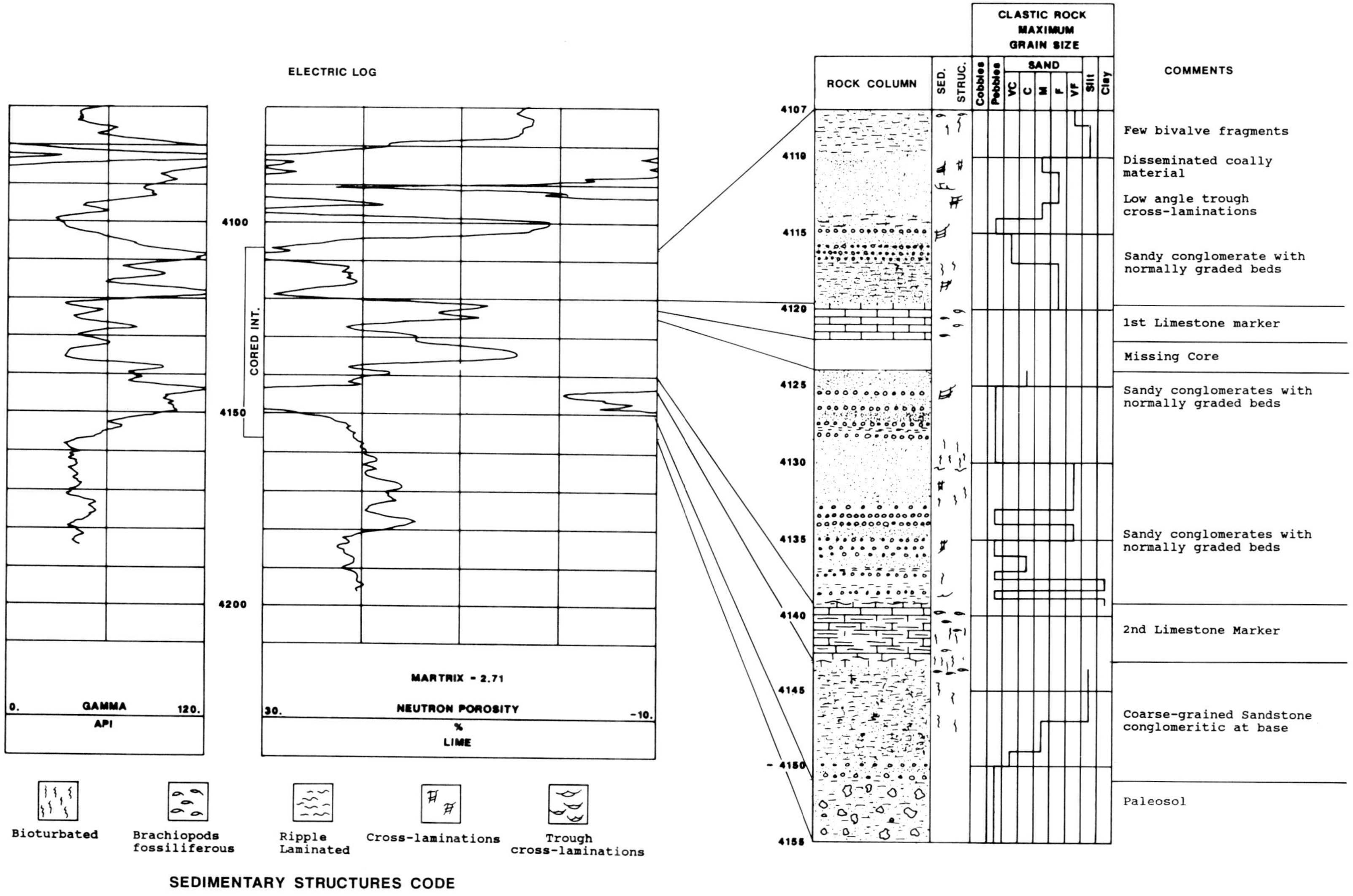

FIG. 14.--Log of core NCRA, Thompson A-2. Core adjusted 1 ft downhole to log depth.

FIG. 15A.--Photograph of slabbed core, NCRA, Thompson A-2, 4149.7 to 4156.0 ft (1264.8 to 1266.7 m). Paleosol (basal Pennsylvanian conglomerate) developed on Mississippian carbonates. Note large chert clasts (noticeably below depth 4151 ft (1265.2 m)) in matrix-supported conglomerate, overlain by thin mottled shale that is a remnant of an upper paleosol horizon (rubble zone below arrow at contact). Paleosol is overlain by fine- to medium-grained sandstone having low-angle cross-laminations and foresets (above arrow at contact); sandstone is interpreted to have been deposited in a fluvial environment.

FIG. 15B.--Photograph of slabbed core, NCRA, Pfaff 3, 4237.6 to 4244.5 ft (1291.6 to 1293.7 m). Upper horizon of paleosol (rubble zone below arrow at contact) is sandy siltstone with abundant slickensides and some isolated chert clasts, overlain by sandy nodular limestone (above arrow at contact). Limestone is overlain by shaly siltstone (rubble zone 4240.0 to 4241.5 ft (1292.3 to 1292.8 m)) that is interbedded with, and eventually grades upward to, sandstone (4239.7 to 4237.5 ft (1292.3 to 1291.6 m)). The sandstone was probably deposited on the upper shoreface or foreshore.

FIG. 16A.--Photograph of slabbed core, NCRA, Pfaff 3, 4229.1 to 4237.6 ft (1289.0 to 1291.6 m). Oil-stained quartz arenite with clasts up to 3 mm in size and normally graded beds abruptly overlain by fissile shale with abundant *Lingula*. Shale is gradationally overlain by Second Limestone Marker; nodules of fossiliferous micrite are enclosed in shale. Nodular texture is in-situ brecciation (micro karst), attributed to weathering.

FIG. 16B.--Photograph of slabbed core, NCRA, Thompson A-2, 4133.4 to 4141.5 ft (1259.9 to 1262.3 m). Second Limestone Marker is overlain by interbedded sandstone and sandy conglomerates. The limestone consists of nodules of fossiliferous micrite encased in shale. The contact of the sandstone with the underlying limestone is sharp and probably erosional (arrow at contact); clasts of limestone are incorporated in the sandstone (above 4139 ft (1261.6 m)). Sandy conglomerates are characterized by normally graded beds 1.1 to 2 in. (3 to 5 cm) thick and coarse chert clasts up to 3.5 mm in size. They may have been deposited by braided streams that prograded across the tidal flat or by storms on the upper shoreface or foreshore. Note scour surface at 4133.8 ft (1260 m)(near left top of photograph) where sandy conglomerate overlies massive bedded sandstone.

FIG. 15B

FIG. 15A

FIG. 16B

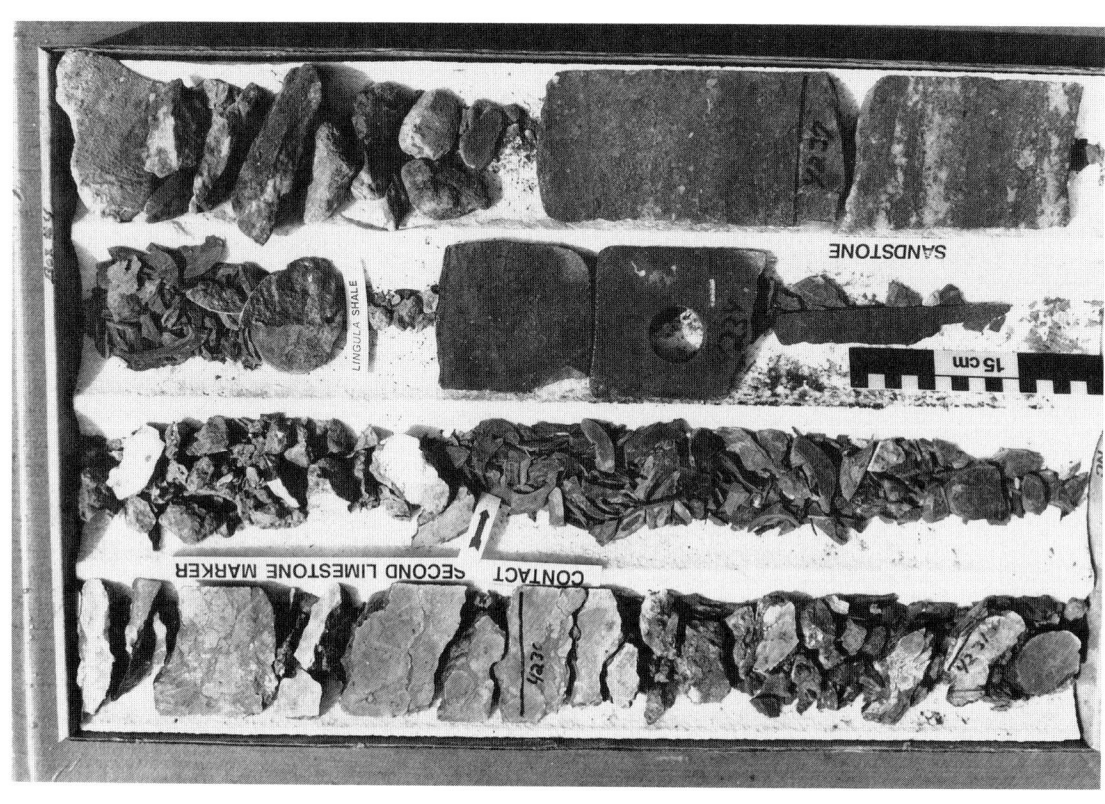

FIG. 16A

Overlying the basal Pennsylvanian conglomerate and paleosol are mixed lithologies of siliciclastics and limestones. In the Thompson A-2, the paleosol is overlain by a fine- to medium-grained, light to medium gray sandstone that coarsens upward into medium- to coarse-grained (Fig. 15A). Low-angle cross-laminae with foresets are made visible by thin layers of green and some red (oxidized) clay. This is overlain by a maroon sandstone that is very fine-grained and occasionally ripple-laminated; it fines upward into a mottled brownish maroon and light green sandy siltstone that caps the unit. The green spots may be caused by burrows or are rhizoliths caused by decayed plant roots. Disseminated throughout the entire unit are larger (up to 2.0 mm), isolated, angular chert clasts.

In the Pfaff 2, the paleosol is overlain by 12 ft (3.7 m) of sandstone. At the base it is a green, fine-grained litharenite that has continuous wavy laminae and is gradationally overlain by a green sandy shale. This in turn grades to an oil-stained quartz arenite having occasional large chert clasts. It varies from clay-rich and moderately sorted to finer-grained beds and little clay. Faint low-angle and trough cross-laminations are also present. These beds are overlain by a coarser-grained, poorly sorted, clay-rich, matrix-supported, oil-stained sandy conglomerate and medium gray quartz arenite. Most of the clasts are rounded quartz (up to 15 mm in diameter), but some clasts are chert and granitic rock fragments that are also enclosed in a matrix of medium-grained, quartz sand and clay. This is in turn overlain by a fine- to medium-grained sandy siltstone with some slickensides. It is mainly maroon to mottled greenish gray in color. These beds grade upward into a thin, dark green shale and finally into a bioturbated, maroon, medium-grained silty sandstone. The burrows are filled with a green clayey sandstone. Abruptly overlying the sandy siltstone is a thin limestone. It is a fossiliferous micrite with thin, horizontal laminae of green shale. Macrofossils present are predominately brachiopods and corals. The thin limestone is abruptly overlain by a dark gray fissile shale that contains abundant *Lingula.*

Unlike the Pfaff 2 and Thompson A-2, the paleosol in the Pfaff 3 is overlain by a nodular sandy limestone interbedded with a calcareous sandstone (Fig. 15B). The limestone nodules are composed of micrite and are enclosed by very fissile, green and maroon shale at the base. A thin bed of green calcareous litharenite, having faint low-angle cross-laminations and horizontal laminations, is interbedded within the limestone. This grades upward to a nodular sandy limestone enclosed in green sandy siltstone. This is overlain and in sharp contact with a mottled green and maroon shaly siltstone interbedded with a thin, two inch-thick bed of greenish gray quartz arenite that is ripple-laminated. This is in turn overlain by alternating thin, pale green, very fine- to fine-grained litharenites that have clay matrix and medium-grained, oil-stained quartz arenites having some coarse grains up to 5 mm in size. Both lithologies have faint horizontal laminations. Overlying these sandstones is a gray to tan and oil-stained, well-cemented, very coarse-grained quartz arenite having some chert clasts up to 5 mm in size and some normally graded beds 1.1 to 2 in. (3 to 5 cm) thick. At the top of these beds is a mottled dark gray and dark green, fissile shale that is moderately bioturbated and contains abundant *Lingula* and black, amorphous organic material (Fig. 16A).

The earliest Cherokee sediments overlying the basal Pennsylvanian conglomerate are fine and coarse clastics. They probably were deposited in fluvial environments in a karstic valley system that developed in the underlying Mississippian carbonates. The interbedded sandstones, sandy conglomerates, and thin shales, coupled with normally graded beds and faint horizontal and low-angle cross-laminations present in the Pfaff 2 and Thompson A-2 core, suggest a fluvial origin. The geometry of these sands is generally linear, and the thicker sand bodies typically occupy valleys developed in the underlying Mississippian carbonates (Figs. 6 and 7). Similar lithologies, sedimentary structures, distribution, and occurrences were reported by Walters and others (1979). Some sand bodies occur on Mississippian structural highs. Coastal processes may have redistributed the fluvial sands during marine transgression and concentrated them on paleohighs. As the shallow sea began transgressing onto the coastal plain, marginal marine sedimentation patterns were initiated. The nodular limestone overlying the paleosol in the Pfaff 3 was deposited in a marine environment with low wave energy and rests on a Mississippian high (Fig. 6). The nodular texture is the result of weathering and erosion. Similar textures were noted in limestones of the Lansing-Kansas City groups by Watney (1980) and were attributed to in-situ brecciation and erosion. The sandstone that overlies the limestone in Pfaff 3 may have been deposited by a storm on the upper shoreface or foreshore. The fluvial clastics in both Pfaff 2 and Thompson A-2 grade upward into finer-grained, bioturbated clastics probably deposited in a tidal flat. Some of the fluvial sediments were probably reworked and redistributed by marine processes. In both the Pfaff 2 and 3 core, the sequence is capped by a dark gray fissile shale containing abundant *Lingula*. This is interpreted as a brackish-water, tidal-flat facies and is common to continental-marine transitional settings (Miall, 1984). The Thompson A-2 is capped by a maroon sandy siltstone probably deposited in a tidal flat that was later subjected to exposure and weathering, as evidenced by the maroon clays, slickensides, and possible rhizoliths. The *Lingula* shale beds in Pfaff 2 and 3 correspond with a high gamma-ray log signature that is recognizable in other wells in the study area and typically occurs below the second limestone marker.

The normally graded beds of coarse chert clasts and the isolated clasts that occur in finer-grained sediment suggest intermittent high-energy processes. These beds may have been deposited by storms or in fluvial environments. The chert is derived from the underlying basal conglomerate, either from a landward source (i.e., toward the Uplift) or locally. The coarse chert clasts in the finer-grained sediments were most likely brought to the depositional site by streams during periods of high runoff and later reworked by coastal processes. The presence of maroon clays suggests periods of subaerial exposure, oxidation, and weathering. Environments transitional from continental to marginal marine and very shallow, fluctuating water depths are recorded by this vertical lithologic succession.

*Base of the Second Limestone Marker
to the Base of the First Limestone Marker*

Overlying the dark gray *Lingula* shale and maroon sandy siltstone is the second limestone marker. In Pfaff 2 and 3 the contact is gradational; whereas, in the Thompson A-2, the contact is sharp. The basal portion of the limestone is interbedded with a gray fissile shale that contains occasional molluscs and some black amorphous material. This lower portion of the second limestone marker is radioactive and may be equivalent to the *Lingula* shale noted in Pfaff 2 and 3. The second limestone marker varies in thickness from 5.0 ft (1.5 m) in the Thompson A-2 to 8.8 ft (2.7 m) thick in Pfaff 2. In general the limestone has a nodular texture of fossiliferous micrite enclosed in a maroon to gray and green shale (Fig. 16A). The limestone and shale contain macrofossils that are brachiopods, crinoids, rugose corals, gastropods, and sponges. Slickensides occur in the shale along breaks, and incipient stylolites are developed occasionally in the limestone.

In general the sequence of beds overlying the limestone marker in the Pfaff 2, Pfaff 3, and Thompson A-2 are in sharp contact and of similar lithologies. There are two lithologic types: a sandstone and a clay-rich, sandy conglomerate. The sandstone is a beige to light gray, fine- to coarse-grained quartz arenite that is either massive bedded and slightly bioturbated, or has faint horizontal or low-angle cross-laminations and an occasional flaser of green shale. Thin laminae of greenish gray fissile shale are occasionally interbedded with the sandstone. These sandstones have minor amounts of pyrite and fine, disseminated black coal material. The conglomerate is either beige or a pale green to maroon, matrix-supported, occasionally clay-rich and sandy conglomerate that has chert clasts up to 3.5 mm in size and repetitive, normally graded beds 1.1 to 2 in. (3 to 5 cm) thick.

Directly overlying the limestone marker in the Pfaff 2 and Thompson A-2 is a fine- to medium-grained quartz arenite. The contact is an angular scour surface. In the Thompson A-2, clasts of limestone are incorporated in the base of the sandstone (Fig. 16B). In Pfaff 3, at the top of the limestone, is a thin, mottled maroon and gray siltstone which is in turn overlain by quartz arenite (Fig. 17A). The sandy conglomerate gradationally overlies the sandstone and, in some cases, is separated by a scour surface (Fig. 16B). Overlying the conglomeritic, normally graded beds in Pfaff 2, Pfaff 3, and Thompson A-2 is a beige to greenish gray and occasionally mottled maroon, fine- to coarse-grained sandstone interbedded with green fissile shale. These beds cap the sequence in Pfaff 2 and Thompson A-2 (Fig. 17B). In the Pfaff 3, the sandstone is overlain and in sharp contact with a thin nodular limestone. The limestone is a fossiliferous micrite enclosed in maroon and greenish gray shale, similar to the underlying second limestone marker. This is in turn overlain by a mottled greenish gray, ocher, and maroon sandy siltstone that is moderately bioturbated with horizontal burrows.

The second limestone marker is a regionally continuous and correlative bed that can be traced for many miles downdip. Toward the Uplift, however, it eventually pinches out (Fig. 5). Structure contours on the base of this limestone mimic the structural grain of the underlying Mississippian unconformity; this is probably due to compaction (Figs. 5 and 6). The limestone was deposited in a subtidal, shallow marine environment with low wave energy and probably more distant offshore than the brackish-water tidal-flat (*Lingula* shale) facies that it overlies. This suggests a regional increase in water depth and inundation accompanied by a decrease in clastic supply, although the clastics supplied from the eroding Uplift were probably not carried out to sea very far. The nodular texture of the limestone and the maroon clay that encloses the nodules is interpreted to be a micro-karst feature and the result of weathering either on an exposed tidal flat or by meteoric diagenesis. Subaerial exposure, weathering, and erosion that followed deposition of the limestone suggest at least a local shallowing of the sea and a return to clastic sedimentation patterns.

The quartz arenite that overlies the limestone and caps the sequence in Pfaff 2 and Thompson A-2 was probably deposited either as a shoreline sand or in a tidal channel (Fig. 16B). Grain-size and small-scale sedimentary structures are similar to the shoreline sands of the

FIG. 17A.--Photograph of slabbed core, NCRA, Pfaff 3, 4221.3 to 4229.1 ft (1286.6 to 1289.0 m). Second Limestone Marker, a nodular limestone encased in green shale at the base that becomes maroon toward the top, is overlain by thin bed of maroon claystone with slickensides (arrow at contact). This may be a remnant of a thin soil horizon developed on top of the limestone. The claystone is abruptly overlain by sandstone with faint ripple laminations (above 4222.1 ft (1286.9 m)).

FIG. 17B.--Photograph of slabbed core, NCRA, Pfaff 2, 4212.9 to 4220.9 ft (1284.1 to 1286.5 m). Interbedded sandstone and sandy conglomerate grade upward to sandstone with thin shale laminae (4215.5 ft (1284.9 m)) that is in turn overlain by the First Limestone Marker (contact at about 4215 ft (1284.7 m)). The First Limestone Marker, like the Second, consists of fossiliferous micrite nodules encased in shale. The nodular texture is considered to be a micro karst feature attributed to weathering. The First Limestone Marker is abruptly overlain by sandy conglomerate separated by a high-angle scour surface (arrow at contact).

FIG. 18A.--Photograph of slabbed core, NCRA, Pfaff 2, 4203.4 to 4212.9 ft (1281.2 to 1284.1 m). Interbedded sandstone and sandy conglomerates. Sandstones have low-angle cross-laminations and abundant fine, black, coaly material, and conglomerates are characterized by coarse chert clasts and normally graded beds. Rubble zone (4207.6 to 4210.5 ft (1282.5 to 1283.4 m)) is sandy siltstone with occasional slickensides.

FIG. 18B.--Photograph of slabbed core, NCRA, Pfaff 2, 4196.0 to 4203.4 ft (1278.9 to 1281.2 m). Sandstone with faint horizontal and low-angle cross-laminations grades upward (above 4199 ft (1279.8 m)) to interbedded, bioturbated, sandy siltstone (rubble zones below 4198 and 4197 ft (1279.5 and 1279.2 m)) and silty sandstone. Dark specks in sandstone (around 4203 ft (1281.1 m)) are pyrite nodules. Dewatering structure or compaction feature is at 4201.8 ft (1280.7 m) and angular mud clasts above depth 4202 ft (1280.7 m).

FIG. 17B

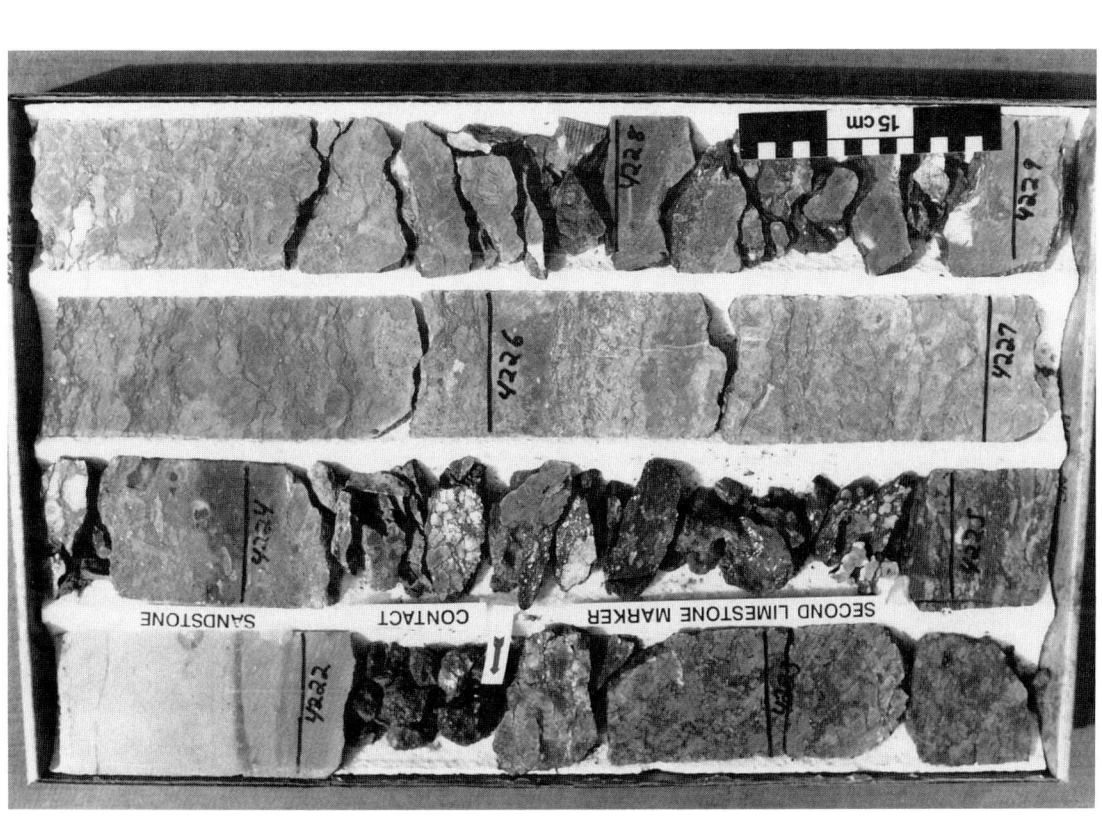

FIG. 17A

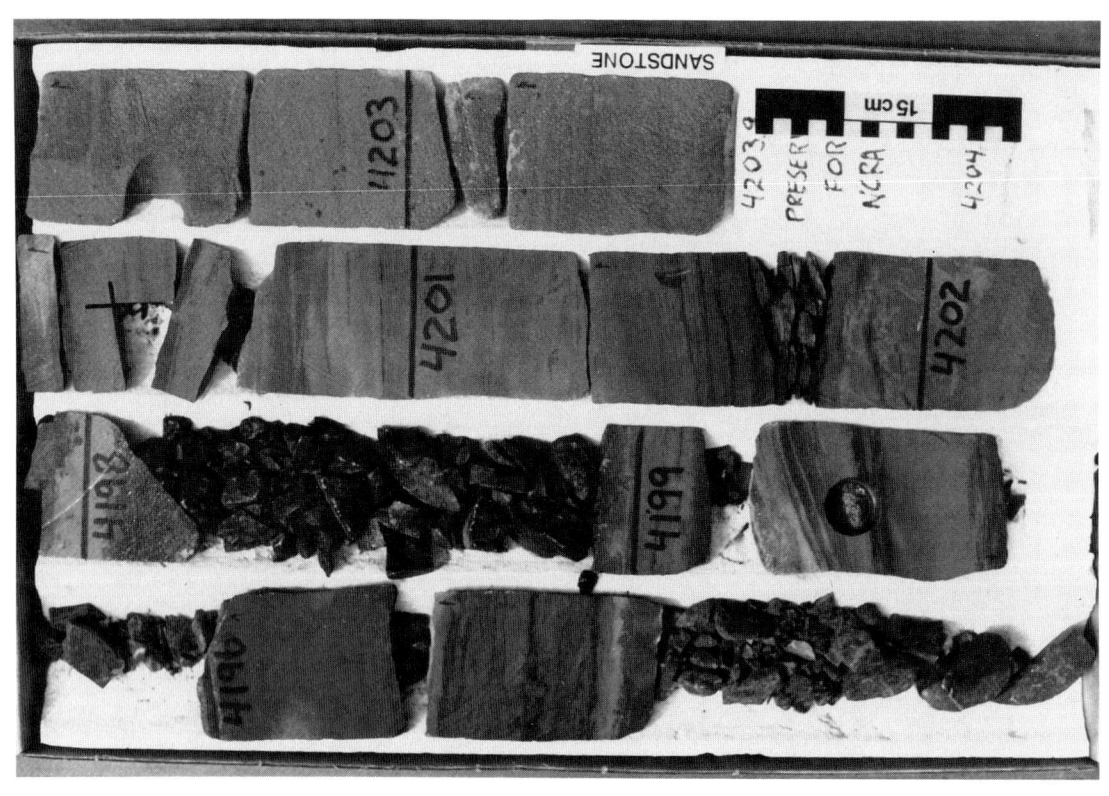

FIG. 18B

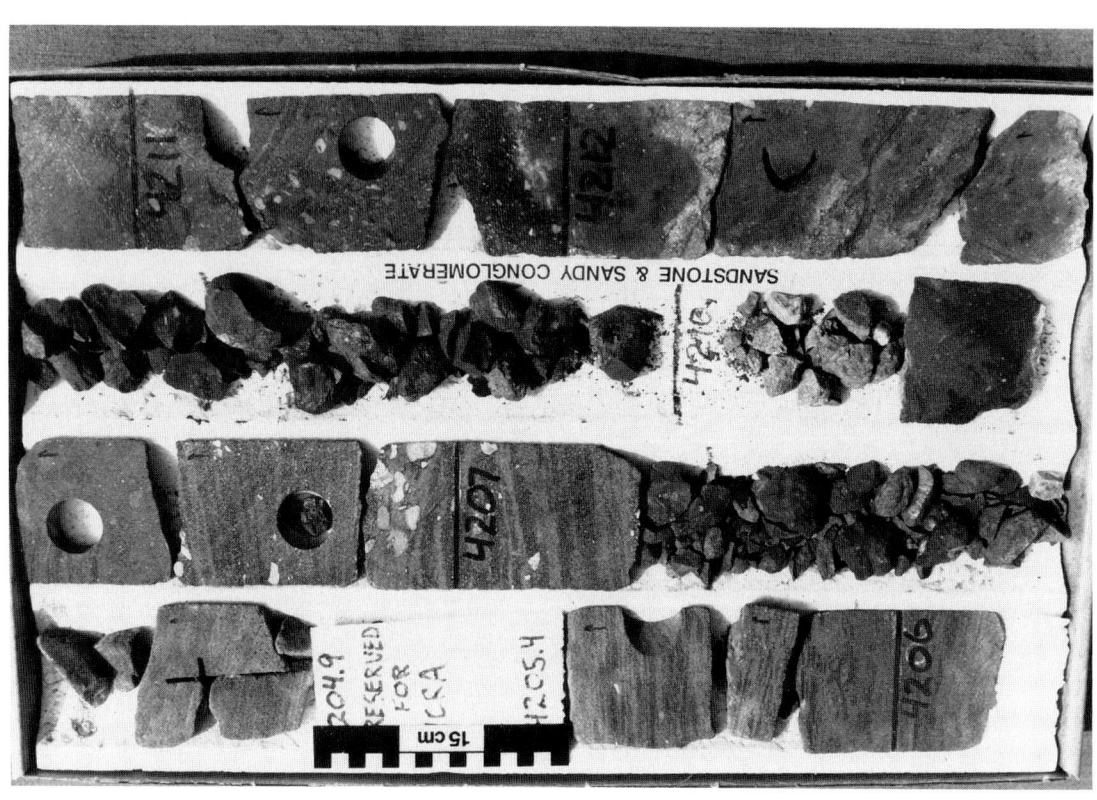

FIG. 18A

Aux Vases Formation (Mississippian of the Illinois Basin) described by Weimer and others (1982). The clay-rich sandy conglomerate, having normally graded beds that overlie the quartz arenite and, in some cases, is interbedded with it, represents episodes of high energy juxtaposed to the overall lower-energy environments recorded by the other strata. The coarse chert clasts were derived from the nearby eroding Uplift. These conglomerates may have been deposited onto the tidal flat as braided-stream deposits or as storm deposits on the foreshore or upper shoreface. The conglomerates cannot be distinguished on electric logs from the interbedded sandstones. A map of gross sand takes into account both sandy lithologies. The geometry and distribution of the sand bodies in this interval parallel the shoreline as defined by the pinchout of the second limestone marker (Fig. 8). Nearshore wave and tidal processes probably reworked the conglomerates, winnowing out the sand and redepositing it in tidal channels and as shoreline sands on the upper shoreface or foreshore. The thicker sand bodies tend to be located on structural highs defined by structure on the base of the second limestone marker (Figs. 5 and 8). The sand diminishes to the west of the study area and tends to be concentrated toward the shoreline, suggesting that coarse clastic sedimentation was very near shore. Sedimentary structures and facies relationships of these sandy conglomerates resemble the conglomerates of the Triassic Ivishak Sandstone of the Sadlerochit Group described by Atkinson and others (1988). They interpreted the sandy conglomerates that are interbedded with finer-grained flood-plain deposits to have been deposited by braided streams in a fluvially dominated coastal-plain environment.

The thin limestone and overlying sandy siltstone in Pfaff 3 suggests a local subtidal accumulation of limestone that was later exposed and capped by tidal-flat, fine-grained, clastic sediments. These strata record a contemporaneous environment, adjacent to the shoreline sand deposits, that probably did not experience the higher-energy processes responsible for the sandstones. Farther downdip, the coarse clastic section thins and is replaced by interbedded limestone and shale. These strata probably record similar low-energy peritidal environments.

Base First Limestone Marker to Top of the Cherokee

The entire sequence from the first limestone marker to the top of the Cherokee is not completely represented in any one drill core but is present with overlapping intervals amongst all four drill cores. The contact of the Cherokee with the overlying Fort Scott Limestone is present only in the Pfaff 5 drill core.

Overlying the sandy siltstone in Pfaff 3 and the quartz arenite in Pfaff 2 and Thompson A-2 is the first limestone marker. It consists of nodules of fossiliferous micrite with incipient stylolites enclosed in green or brownish maroon shale and having occasional slickensides developed in the shale. The faunal assemblage includes crinoids, brachiopods, fusulinids, and bryozoa. The enclosing shale contains some disseminated pyrite, occasional isolated quartz grains, and fossils of *Crurythyris*, sponge spicules, fusulinids, conodonts, *Rhombopora*, and plant fragments. The limestone varies in thickness from 6.0 ft (1.8 m)(Pfaff 3) to 1.5 ft (0.5 m)(Pfaff 2).

Directly overlying the limestone and in sharp contact are interbedded sandstones, sandy siltstones, and sandy conglomerates that grade upward to interbedded sandy siltstones and shales that are moderately to intensely bioturbated. The limestone is overlain by a maroon sandy siltstone in Pfaff 3 and green sandy siltstone in Thompson A-2. The siltstone contains interbedded green and brown, horizontal laminae and some low-angle cross-laminae in Thompson A-2. The first limestone marker in Pfaff 2 is sandy near the top, and the contact with the overlying maroon sandstone is a sharp and angular scour surface (Fig. 17B). In Thompson A-2 and Pfaff 2, these beds are overlain by a greenish gray, fine- to coarse-grained quartz arenite that is either massive bedded or has faint low- to medium-angle cross-laminations and abundant fine, disseminated coal material, and occasional disseminated pyrite (Figs. 18A and B). This sandstone is interbedded with a sandy conglomerate having coarse chert clasts up to 16 mm in size and repetitive, normally graded beds 0.8 to 2 in. (2 to 5 cm) thick (Fig. 18A). In Pfaff 2 near the top of this bed, the sandstone is more laminated. Trough cross-laminations, dewatering or compaction features, and low-angle cross-laminations are common (Fig. 18B). The sandstone fines upward to a mottled greenish gray and maroon, moderately bioturbated sublitharenite that is in turn overlain by a greenish gray, highly bioturbated silty sandstone. The top of the sequence and top of the Cherokee are marked by a mottled dark gray and greenish gray, pyritic, fissile shale (present in Pfaff 5 only)(Fig. 19) that probably correlates with the Excello Shale recognized east of the Uplift (Brenner, 1989). The contact of the shale with the overlying Fort Scott Limestone is sharp and distinctive both in drill core and electric logs.

The depositional environments recorded by these strata are very similar to that of the underlying interval from the base of the second limestone marker to the base of the first limestone marker. The first limestone marker, like the second, can be traced for many miles downdip; however, toward the Uplift, it too eventually pinches out (Fig. 4). The first limestone marker persists farther updip toward the Uplift than the second in this onlapping succession (Figs. 4, 5, and 10B). Structure on the base of the first limestone marker bears a similar structural grain as the second, i.e., mimicking the Mississippian unconformity (Figs. 4 and 6). The first limestone marker was deposited in a shallow, subtidal setting with low wave energy, in a similar environment as the second limestone marker. Later the limestone bed was subjected to subaerial exposure, weathering, and erosion, just as the second. The nodular texture of the limestone is similar to that of the second limestone marker and is interpreted to be a micro karst developed during weathering on an exposed tidal flat or by meteoric diagenesis. An increase in water depth and a decrease in clastics supplied from the eroding Uplift are recorded by the first limestone marker. This is followed by a shallowing of water depths and a return to clastic sedimentation patterns. The overlying sandy siltstones are interpreted to have been deposited in a tidal-flat environment, and the sandstones were probably deposited as shoreline sands. The sandstone with angular mud clasts (evidenced in Pfaff 2) was probably deposited in a tidal channel and the mud clasts derived from erosion of the surrounding tidal flats. The coarser conglomeritic beds were probably deposited in braided streams that prograded over the tidal flats and were later reworked by coastal processes that redistributed the sands into shoreline deposits and in tidal channels. The sandstones and sandy conglomerates are indistinguishable on electric logs. A map of gross sand consists of both sandy lithologies and portrays isolated sand bodies oriented parallel to the shoreline with some lobes extending toward the shoreline; thicker portions are concentrated on structural highs (Figs. 4 and 9). The bioturbated sandy siltstones and thin shale that cap the sequence at the top of the Cherokee are an overall fining-upward lithologic succession that

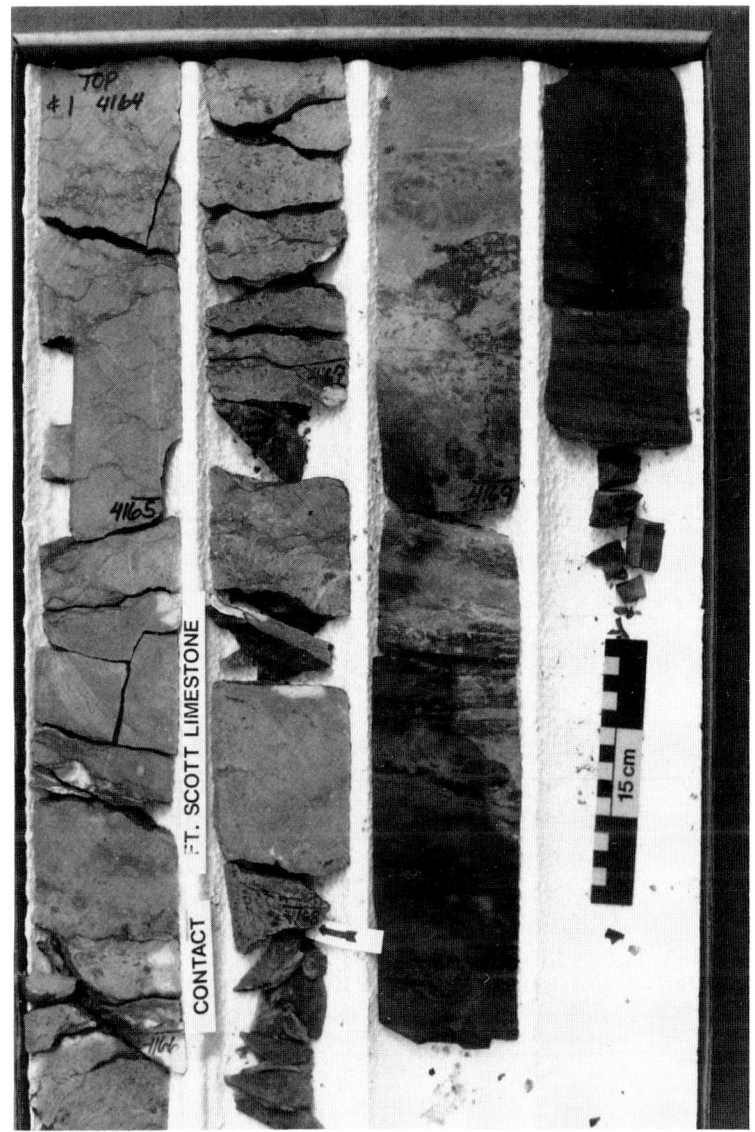

FIG. 19

FIG. 19.--Photograph of slabbed core, NCRA, Pfaff 5, 4164 to 4171.5 ft. Bioturbated sandy siltstones and silty sandstones are overlain by pyritic, fissile shale (arrow at contact) and mark the top of the Cherokee Group. The Ft. Scott Limestone is a bioclastic lime mudstone/wackestone with occasional green shale laminae. In contrast to the limestones of the Cherokee Group, the Ft. Scott does not have the characteristic nodular texture attributed to weathering.

suggests a gradual transgression over the tidal flats and a gradual decrease in clastic supply from the eroding Uplift. The Fort Scott Limestone that overlies the Cherokee records conditions of shallow, quiet marine waters and very little clastic supply. In western Ness County the Fort Scott thickens to more than 20 ft (6.1 m); whereas, toward the Uplift, it thins and eventually pinches out.

DISCUSSION AND IMPLICATIONS

The Cherokee Group in Ness County was deposited in environments that are transitional from continental to marginal marine as the sea transgressed onto the Central Kansas Uplift. Six basic lithofacies are recognized in the cores studied: (1) basal Pennsylvanian conglomerate, (2) fluvial sands, (3) fine-grained tidal-flat deposits, (4) shallow marine limestones and shales, (5) shoreline sands and tidal-channel sands, and (6) braided-stream sandy conglomerates. The basal Pennsylvanian conglomerate and the fluvial sands are confined to the stratigraphic interval below the second limestone marker and typically are associated with valleys developed in the Mississippian carbonates. These facies were deposited in a continental setting and, in some cases, were later modified by coastal processes as the sea began to transgress the region. The shoreline sands and tidal-channel deposits, tidal-flat deposits, braided-stream sandy conglomerates, and shallow marine limestones and shales overlie the terrigenous deposits and are interbedded throughout the unit. These sediments were deposited during transgressive-regressive pulses that occurred within the overall transgressive phase recorded by the Cherokee sediments.

Heckel (1977) described the components of a basic Kansas cyclothem to characterize the cyclic pattern of lithofacies recognized in the Lansing-Kansas City groups of (Missourian) Pennsylvanian age. This concept was further developed by Watney (1980) and was applied by Heckel (1986) to establish sea-level curves for the Pennsylvanian of the Mid-Continent. The basic cycle representing a transgressive-regressive sequence, as presented by Heckel (1977), consists of: (1) an outside (nearshore) shale, (2) middle (transgressive) limestone, (3) core (offshore) shale, (4) upper (regressive) limestone, and (5) outside (nearshore) shale (Fig. 20A). Sediments deposited during transgressive-regressive cycles vary according to the local geologic conditions (e.g., water depth, coastal processes, source area, sediment supply, subsidence, etc.), and the rock record is dependent on the limits of sediment preservation. The components that form the basic Kansas cyclothem with some local variations are present in the stratigraphic succession of the Cherokee Group represented in the cores studied (Fig. 20B).

The middle (transgressive) limestone is deposited in an open-marine environment and below effective wave base (Heckel, 1977). These strata are equivalent to the initial flooding unit described by Watney and others (1989). They can be fossiliferous limestones, but are usually thin beds and oftentimes absent. They are characterized by a sharp basal contact and commonly contain terrestrial organic matter contributed by fresh-water runoff. The thin limestone bed, present in the interval from the Mississippian unconformity to the base of the second limestone marker in Pfaff 2 and 3, is interpreted to be equivalent to the transgressive limestone that records the initial marine inundation.

FIG. 20A.--The four components of an individual transgressive-regressive, Kansas-type cyclothem (from Heckel, 1986).

FIG. 20B.--Generalized lithostratigraphy, equivalent facies, and interpreted transgressive-regressive cycles. Two transgressive-regressive cycles within the overall transgressive phase are recorded by the lithologic succession of the Cherokee Group in Ness County. The first transgression begins with initial flooding that is recorded by a thin limestone bed that overlies the fluvial and terrigenous clastics at the base of the Cherokee. This limestone is oftentimes absent and, when present, is difficult to recognize on wireline logs. This unit is followed by marginal marine clastic sediments that are equivalent to the outside (nearshore) shale. Continued gradual inundation culminates with deposition of the core shale (i.e., *Lingula* shale and lower portion of the Second Limestone Marker). Shoreline retreat and regression rapidly follow and are marked by erosion and scour at the top of the Second Limestone Marker and the return of marginal marine clastic sediments. These beds are equivalent to the upper (regressive) limestone and outside shale. The second transgressive pulse is followed by rapid regression and is recorded by the First Limestone Marker. It is equivalent to the middle (transgressive) limestone in the basal part and the upper (regressive) limestone at the top. Deposition of marginal marine clastics accompanies the regressive phase. A gradual increase in water depth and inundation is marked by a fining-upward succession that culminates with deposition of the Excello shale at the top of the Cherokee.

The black phosphatic shale that accompanies maximum transgression is the main constituent of the core (offshore) shale (Heckel, 1977) and similar to the condensed section described by Loutit and others (1988) and Watney and others (1989). In the study area the core shale is considered equivalent to the radioactive *Lingula* shale that underlies the second limestone marker in Pfaff 2 and 3, the thin shale interbedded with the basal portion of the second limestone marker in the Thompson A-2, and the Excello Shale at the top of the Cherokee. The characteristic high gamma-ray log signature that is associated with these strata is observable in most wells and suggests widespread distribution. The thin, greenish gray shale that occurs at the top of the Cherokee and is directly overlain by the Fort Scott Limestone is correlatable with the Excello Shale, a unit identified with widespread transgression that is eustatic in nature (Brenner, 1989). The deeper-water anoxic conditions conducive to deposition of black phosphatic core shale were never attained during any of the transgressive pulses.

The upper (regressive) limestone is deposited in a similar environment as the middle (transgressive) limestone but with more agitated conditions as water depth shallowed and waves became more effective (Heckel, 1977). The regionally correlative first and second limestone markers were deposited in subtidal conditions. The second limestone marker was probably deposited in more deeper-water conditions than the underlying *Lingula* shale, and it is equivalent, in part, to the core shale in the basal portion and the regressive limestone in the upper portion. The first limestone marker is equivalent to both the transgressive and regressive limestone with no core shale and records a transgressive-regressive pulse. The limestone markers are distinguished by a nodular texture that developed during exposure and weathering, and, in some cases, a thin paleosol rests on the top of the limestone bed. These characteristics represent a

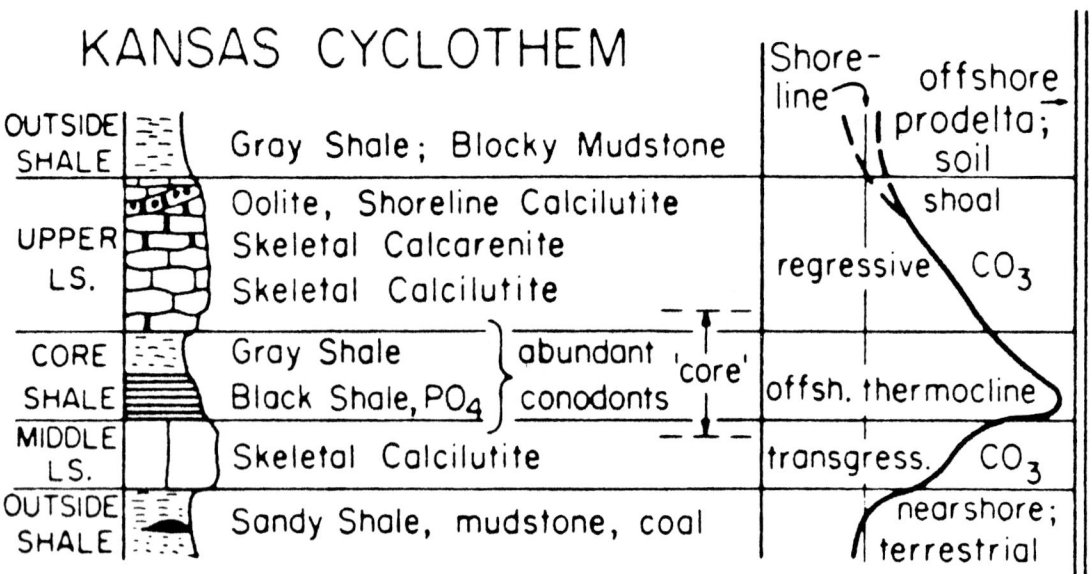

FIG. 20A

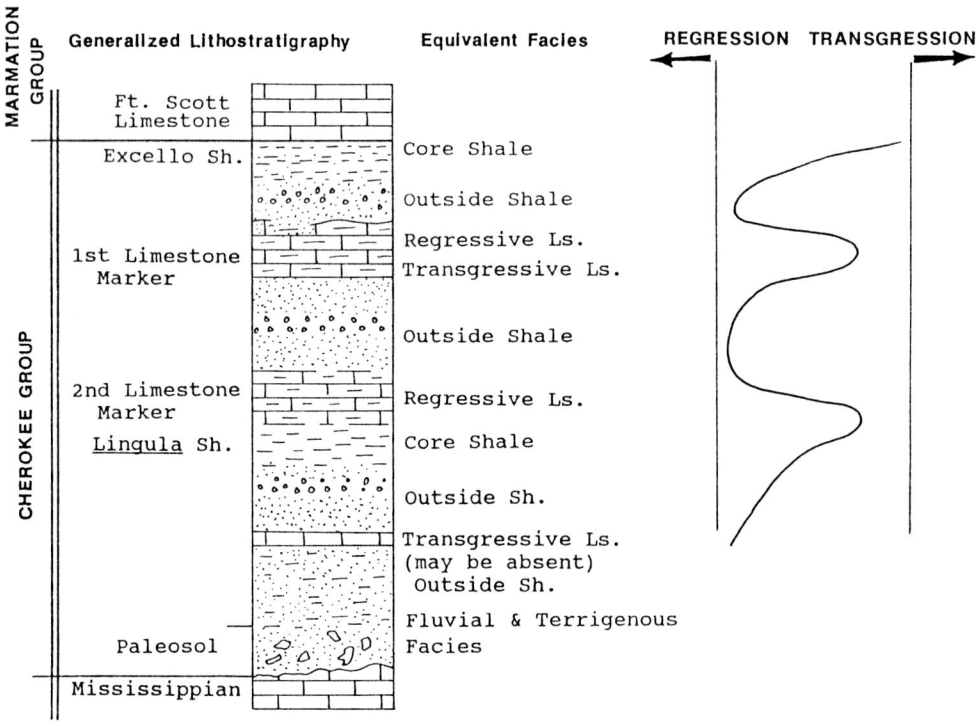

FIG. 20B

shallowing-upward lithologic succession. The shoreline sands that overlie the limestone markers are also considered equivalent to the upper (regressive) limestone, as these sands were deposited inside the effective wave base during shoreline retreat. In some cases these sands rest on scour surfaces at the top of the limestone markers. These characteristics suggest that regression and clastic depositional patterns immediately followed deposition of the first and second limestone markers.

The (outside) nearshore shale is deposited in a variety of complex environments from shallow subtidal marine to nonmarine terrigenous settings (Heckel, 1977). The fine-grained tidal-flat deposits, the shoreline sands and tidal-channel sands, and the braided-stream sandy conglomerates are considered equivalent to the outside (nearshore) shale. These sediments were deposited in very shallow conditions on an intermittently exposed tidal flat and subjected to periods of exposure and weathering.

CONCLUSIONS

The vertical lithologic succession observed in cores of the Cherokee Group is highly variable and records rapidly changing depositional environments that laterally coexist in a transitional, continental to marginal marine setting. It is predominately a siliciclastic-dominated system but farther offshore more carbonate rocks enter the system. The paleotopography of the coastal plain had very little relief once the karstic valleys developed in the Mississippian were filled. Minor changes in sea level of the shallow sea would have caused widespread transgressions and regressions on this relatively featureless coastal plain. Depositional environments and sedimentation patterns quickly responded to these corresponding sea-level changes. A gradual rise in sea level is accompanied by a rise in base level, and the clastic supply would decrease, giving way to deposition of fine clastics and carbonates in a shallow sea. Clastics were deposited in a wide range of flow regimes and were readily available from the nearby eroding Central Kansas Uplift. It is probable that, at least locally, sedimentation rates exceeded the increasing space available for sediment accommodation, causing local shoreline retreat and eventually subaerial exposure. The weathering features noted throughout the Cherokee sequence are related to subaerial exposure but may have formed below the actual exposure surface by meteoric diagenesis while the strata were in the vadose zone (Goldstein and others, 1989).

Oil production from the Cherokee Group in Ness County is from sandstones in combination structural-stratigraphic traps. The sandstones identified in the three stratigraphic subdivisions defined in this study are each productive in their proper setting. The sandstones were deposited in a variety of environments that are contemporaneous and therefore difficult to predict. The correlation scheme defined in this study provides a useful framework for recognizing different sand bodies and corresponding oil habitat. This correlation may be a useful predictive tool for oil exploration and development.

ACKNOWLEDGMENTS

The authors wish to acknowledge the management of National Cooperative Refinery Association, Exploration and Production Division, for their help and support throughout this study and permission to publish this paper. Specifically we thank Dr. W. Lynn Watney of the Kansas Geological Survey for his useful suggestions and discussion throughout the course of this study and for critical review of the manuscript. We express our deepest gratitude to David Buell for drafting and preparation of figures and Marilynn Ellis for clerical and editorial assistance in preparing the manuscript.

REFERENCES CITED

ATKINSON, C. D., TRUMBLY, P. N., AND KREMER, M. C., 1988, Sedimentology and depositional environments of the Ivishak Sandstone, Prudhoe Bay Field, North Slope, Alaska, in Lomando, A. J., and Harris, P. M., eds., Giant Oil and Gas Fields: A Core Workshop: Society of Economic Paleontologists and Mineralogists Core Workshop no. 12, v. 2, p. 561-613.

BRENNER, R. L., 1989, Stratigraphy, petrology, and paleogeography of the upper portion of the Cherokee Group (Middle Pennsylvanian), eastern Kansas and northeastern Oklahoma: Kansas Geological Survey, Geology Series 3, 70 p.

GOLDSTEIN, R. H., CARLSON, R. C., BOWMAN, M. W., AND ANDERSON, J. A., 1989, Diagenetic responses to sea-level change - integration of field, stable-isotope, paleosol, and cement-stratigraphy research to determine history and magnitude of sea-level fluctuation, in Franseen, E. K., and Watney, L. K., eds., Sedimentary Modeling: Computer Simulation of Depositional Sequences: Kansas Geological Survey, Subsurface Geology Series 12, p. 59-60.

HECKEL, P. H., 1977, Origin of phosphatic black shale facies in Pennsylvanian cyclothems of Mid-Continent North America: American Association of Petroleum Geologists Bulletin, v. 61, no. 7, p. 1045-1068.

_____, 1986, Sea-level curve for Pennsylvanian eustatic marine transgressive-regressive depositional cycles along Mid-Continent outcrop belt, North America: Geology, v. 14, p. 330-334.

LOUTIT, T. S., HARDENBOL, J., VAIL, P. R., AND BAUM, G. R., 1989, Condensed sections: the key to age dating and correlation of continental margin sequences, in Wilgus, C. K., Hastings, B. S., Kendall, C. G. St. C., Posamentier, H. W., Ross, C. A., and Van Wagoner, J. C., eds., Sea-Level Changes - An Integrated Approach: Society of Economic Paleontologists and Mineralogists, Special Publication no. 42, p. 183-213.

MCCRONE, A. W., 1964, Water depth and Mid-Continent cyclothems, *in* Merriam, D. F., ed., Symposium on Cyclic Sedimentation: Kansas Geological Survey Bulletin no. 169, v. 1, p. 275-281.

MERRIAM, D. F., 1963, The geologic history of Kansas: Kansas Geological Survey Bulletin 22, 256 p.

MIALL, A. D., 1984, Principles of sedimentary basin analysis: Springer-Verlag, New York, 490 p.

NODINE-ZELLER, D. E., 1981, Karst-derived Early Pennsylvanian conglomerate in Ness County, Kansas: Kansas Geological Survey Bulletin 222, 30 p.

RASCOE, B., AND ADLER, F. J., 1983, Permo-Carboniferous hydrocarbon accumulations, Mid-Continent, U.S.A.: American Association of Petroleum Geologists Bulletin, v. 67, no. 6, p. 979-1001.

WALTERS, R. F., GUTRU, R. J., JAMES III, A., 1979, Channel sandstone oil reservoirs of Pennsylvanian age in Northwestern Ness County, Kansas, *in* Hyne, N. J., ed., Pennsylvanian Sandstones of the Mid-Continent: Tulsa Geological Society Special Publication no. 1, p. 313-326.

WATNEY, W. L., 1980, Cyclic sedimentation of the Lansing-Kansas City groups in northwestern Kansas and southwestern Nebraska: Kansas Geological Survey Bulletin no. 220, 72 p.

_____, FRENCH, J., AND FRANSEEN, E. K., 1989, Sequence stratigraphic interpretations and modeling of cyclothems in the Upper Pennsylvanian (Missourian) Lansing and Kansas City groups in eastern Kansas: Guidebook, Kansas Geological Society 41st Annual Field Trip, 211 p.

WEIMER, R. J., HOWARD, J. D., AND LINDSAY, D. R., 1982, Tidal flats and associated tidal channels, *in* Scholle, P. A., and Spearing, D., eds., Sandstone Depositional Environments: American Association of Petroleum Geologists, Memoir 31, p. 191-245.

ZELLER, D. E., 1968, The stratigraphic succession in Kansas: Kansas Geological Survey Bulletin 189, 81 p.

LITHOSTRATIGRAPHY AND DEPOSITIONAL ENVIRONMENTS OF THE ANCELL GROUP IN CENTRAL ILLINOIS: A MIDDLE ORDOVICIAN CARBONATE-SILICICLASTIC TRANSITION

T. H. SHAW
Department of Geology, Queens College, CUNY,
Flushing, NY 11367

B. C. SCHREIBER
Lamont-Doherty Geological Observatory,
Columbia University, New York, NY 10964

ABSTRACT

The Middle Ordovician Ancell Group is the basal transgressive deposit of the Tippecanoe sequence in the Illinois Basin region. The mixed carbonate and siliciclastic composition of its constituent lithologies is a reflection of depositional conditions related to regional tectonism, the lithic character and nature of erosional processes in adjacent source areas, and changes in relative sea level and paleoclimate. Differentiation of lithofacies, their inferred depositional environments, and their vertical and lateral relationships provide a basis for identification of three major depositional cycles. Each cycle is made up of small-scale, shallowing-upward sequences that tend to reflect increasingly subtidal conditions upsection. The cycles are regionally correlative; consequently, they provide a basis for subdividing the Ancell Group into genetic units and for interpreting regional sedimentological, paleoclimatic, and tectonic processes during early-Middle Ordovician time in the Illinois Basin region.

INTRODUCTION

The rocks discussed in this study make up a mixed carbonate- siliciclastic sequence, the Ancell Group, that was deposited during Middle Ordovician time in the Illinois Basin. The Ancell Group (cycle)(Witzke and Kolata, 1988) is the basal, transgressive subsequence of the Tippecanoe sequence in the North American Midcontinent (Sloss, 1963). The depositional environments and mixed carbonate-siliciclastic lithic succession of the Ancell Group reflect the subtle interplay of variations in relative sea level, rates of basin subsidence, paleoclimate, and the character and rate of erosion in adjacent source areas. Our goals in this study are to (1) briefly summarize the regional stratigraphy and tectonic setting of the Illinois Basin during the early-Middle Ordovician, (2) describe and interpret the lithic succession of the Ancell Group in the subsurface of central Illinois, and (3) discuss the relationship of sedimentation to Middle Ordovician tectonism, relative sea-level variation, and paleoclimate.

GEOGRAPHIC AND STRUCTURAL SETTING

The Illinois Basin is a spoon-shaped, intracratonic basin bounded by positive tectonic features in the interior region of eastern North America (Fig. 1); this structural configuration developed in post-Ordovician time. The apparent limits of the Illinois Basin in the lower-Middle Ordovician (Whiterockian) are in part delineated by the geographic distribution of the Everton Dolomite; the present basin limit is represented by the 500-ft (150 m) structural contour on top of the Middle to Upper Ordovician Galena Group (Trenton equivalent; Collinson and others, 1988)(Fig. 1). During much of the Paleozoic, the proto-Illinois Basin depocenter was associated with the New Madrid Rift Complex, the now-buried Lower Cambrian rift system in southern Illinois, southwestern Indiana, western Kentucky, and southeastern Missouri (Nelson and Lumm, 1987). This paleo-rift and its associated lower-crust intrusives are an isostatically uncompensated mass that, in part, governed subsidence from the Middle Ordovician onward (Braile and others, 1986; Kolata and others, 1990).

Based on isopach maps of the Lower Ordovician, upper-Sauk sequence (see Collinson and others, 1988), it is apparent that, during the upper-Lower Ordovician, a broad, slowly subsiding carbonate ramp was developed throughout the Illinois Basin region. On this ramp the peritidal and shallow subtidal carbonates of the Knox Supergroup were deposited (Droste and Shaver, 1983; Collinson and others, 1988).

In the lower-Middle Ordovician (early Whiterockian)(Fig. 2), a drop in sea level resulted in exposure and erosion of the basin margins, while deposition may have been continuous within the southern depocenter (Shaw, 1988). This hiatus is at the Sauk-Tippecanoe sequence boundary of Sloss (1963). In the middle-Middle Ordovician (late Whiterockian), subsidence rates increased and/or relative sea level rose, and the paleoshoreline transgressed northward over the post-Sauk paleoerosional surface. By the upper-Middle Ordovician (early Mohawkian), the paleoshoreline transgressed as far north as St. Paul, Minnesota, and subtidal carbonate deposition was widespread throughout the proto-Illinois Basin. It was during the initial stages of this transgression that the Ancell Group was deposited.

LITHOSTRATIGRAPHY OF THE ANCELL GROUP

The Ancell Group (Templeton and Willman, 1963) is a disconformity-bounded stratigraphic unit considered to be the basal transgressive-regressive cycle, or subsequence, of the Tippecanoe sequence (Witzke and Kolata, 1988). We informally define the Ancell Group as overlying the regional, Sauk-Tippecanoe unconformity, and underlying the intraregional sub-Platteville unconformity (Fig. 2). This definition expands the concept of the Ancell Group to include the Everton Dolomite, which was excluded from the Ancell by Templeton and Willman (1963). The Everton does not appear to be genetically related to the Knox carbonates of the Sauk sequence, and, based on its stratigraphic position overlying the post-Knox disconformity, it is included in the Ancell Group for the purpose of this discussion.

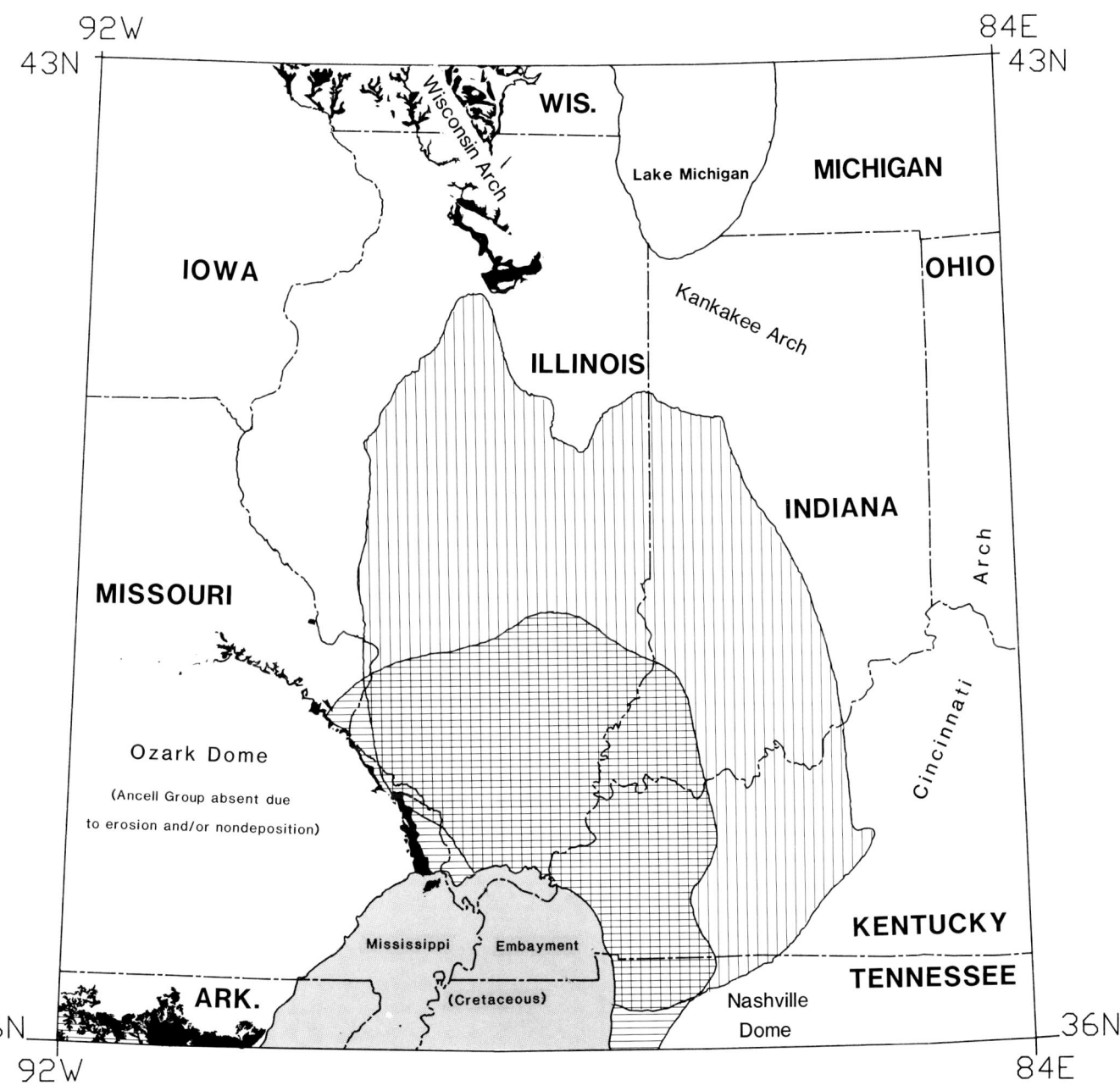

Fig. 1--Map of Illinois Basin region showing (1) outcrop distribution of the Ancell Group (black), (2) major peripheral tectonic features, (3) approximate limit of the early-Middle Ordovician Basin (horizontal pattern), (4) the present limit of the Illinois Basin (vertical pattern), and (5) the present distribution of Cretaceous Mississippi Embayment sediments.

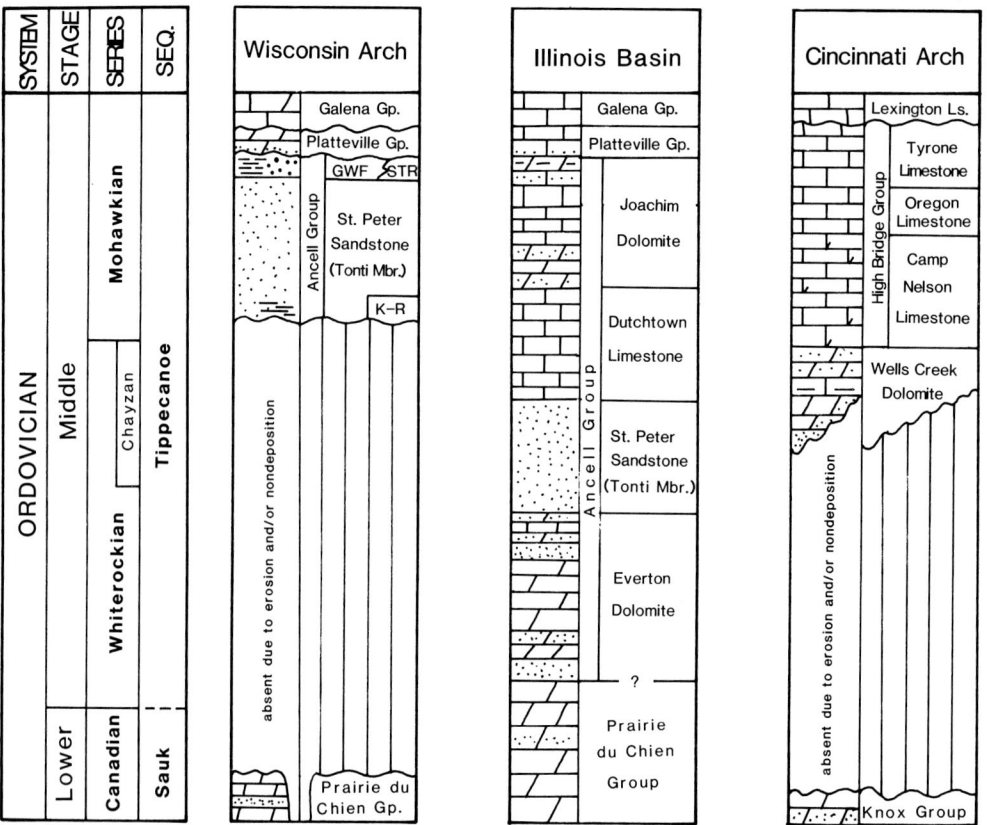

Fig. 2--Upper-Lower and Middle Ordovician stratigraphic nomenclature, age relationships, and gross lithologies in the Illinois Basin region.

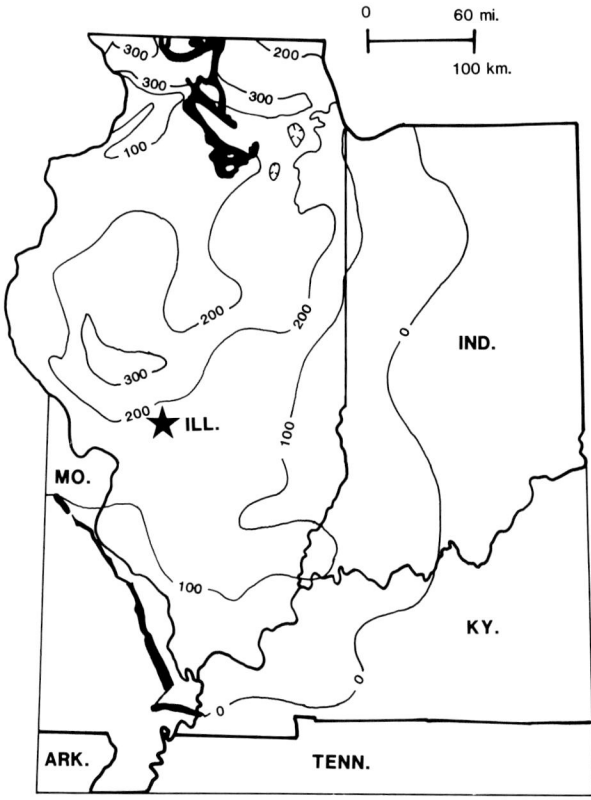

Fig. 3--Thickness map of the St. Peter Sandstone in the Illinois Basin region; contour interval 100 ft, outcrop shown in black (modified from Collinson and others, 1988). Star indicates location of the Illinois Power Company, Truitt #1 well.

The Ancell Group is subdivided into four formations, in ascending order: Everton Dolomite, St. Peter Sandstone, Dutchtown Limestone, and Joachim Dolomite (Fig. 2). The Everton, Dutchtown, and Joachim are absent in northern Illinois and Wisconsin, and the St. Peter Sandstone is conformably overlain by the Glenwood Formation (Fig. 2). Based on the bounding disconformities, the Ancell Group is consistent with the definition of a parasequence set (van Wagoner and others, 1990).

The Middle Ordovician Ancell Group disconformably overlies strata of the Potsdam supergroup (Cambrian), the Prairie du Chien Group (Lower Ordovician), or the Knox Group (Upper Cambrian to Lower Ordovician); however, the stratigraphic sequence may be conformable in the central Illinois Basin. The Ancell is, in part, laterally equivalent to the Wells Creek Dolomite, an argillaceous dolostone with sandstone, shale, and limestone in central Kentucky, western Ohio, and northern Tennessee. Throughout much of the basin, the post-Sauk unconformity is overlain by either the Everton Dolomite or the St. Peter Sandstone; along the eastern and southern margins of the basin, eastern Indiana, Ohio, and central Kentucky, the unconformity is overlain by carbonates of the Wells Creek Dolomite or Dutchtown Limestone (Stith, 1979; Droste and others, 1982). This discussion of the Ancell will be primarily limited to the region west of the St. Peter Sandstone limit (Fig. 3).

DEPOSITIONAL FACIES

Although this discussion will focus on the lithostratigraphy of a 3.5-inch (8.5 cm) core from the Illinois Power Company, Truitt #1 well in Montgomery County, Illinois (Fig. 3), observations from other cores and surface exposures in the basin will be introduced as are pertinent to the discussion. The core was obtained from a depth of 2902 to 3278.5 ft (882 to 997 m); the well reached a total depth of 3320 ft (1110 m). This core is on file at the Illinois State Geological Survey (ISGS #C13443). The section described is from a depth of 3000 to 3278.5 ft (912 to 997 m); the stratigraphic interval represented is the upper Everton Dolomite up through the basal Platteville Group. The lithostratigraphy of the Ancell Group in the core is shown in Figure 4. In the following discussion the lithofacies of each stratigraphic unit will be summarized and its depositional environment inferred.

Everton Dolomite

Only one lithofacies of the upper Everton Dolomite is represented in the Truitt #1 core. This lithofacies, the mottled dolomitic wackestone lithofacies, is predominantly limy micrite and anhedral to subhedral, finely crystalline dolomite with "floating" fine- to medium-grained, well-rounded to subrounded quartz grains (Fig. 5a and b). The dolomitic areas appear as light-colored mottles, 1 to 2 cm in diameter, that have sharp stylolitic boundaries or grade into limy micrite (Fig. 5b). Intercrystalline porosity is developed within the dolomitic mottles, and vugs are present in the micrite. Moderately abundant conodont faunas recovered from this interval are dominated by species assignable to the genera *Leptochirognathus* and *Multioistodus*. These genera are common components of the North American Midcontinent Province and are commonly associated with semi-restricted to restricted marine, inner-shelf environments.

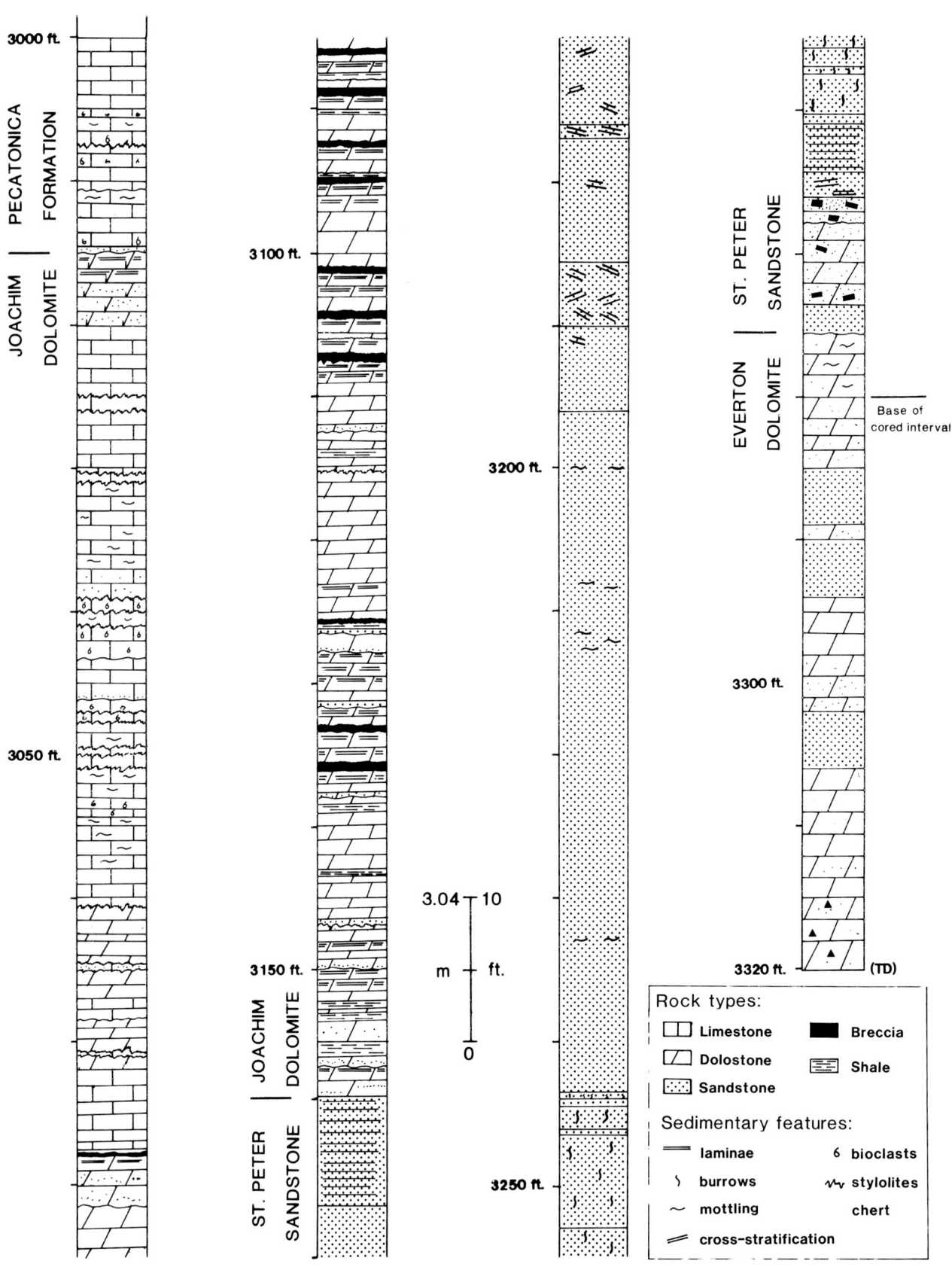

Fig. 4--Lithostratigraphy of the Ancell Group in the subsurface of central Illinois: Illinois Power Company, Truitt #1, Montgomery County, Illinois.

FIG. 5.--Lithofacies of the uppermost Everton Dolomite and the lowermost St. Peter Sandstone: (A) dolomitic wackestone lithofacies (3278.2 ft); (B) photomicrograph showing the contact of limy micrite and anhedral to subhedral, crystalline dolomite, note presence of quartz grains in both lithologies (3278.4 ft)(plane-light); (C) photograph of interbedded dolostone and sandstone lithofacies (3273.5 ft) showing dolostone (medium gray) enclosed in fine-grained sandstone bounded by stylocontacts; (D) pebble-conglomerate lithofacies (3267.1 ft), note overturned mud-chip at arrow. Scales in centimeters for A, B, and C; D is in inches.

FIG. 6.--Core interval photo (3255-3265 ft) showing lithofacies transition from the laminated sandstone lithofacies (arrows C and D) to the *Skolithos* sandstone facies (arrows A and B). Burrow traces are less well developed in medium-grained sandstones (3257-3260 ft).

FIG. 7.--Lithofacies from the main body of the St. Peter Sandstone: (A) planar cross-stratified sandstone (3196 ft), (B) *Skolithos* sandstone (3259 ft), (C) massive sandstone (3218.2 ft), and (D) cross-stratified sandstone (3189-3190 ft). All photos are at the same scale.

FIG. 8.--Clay-bearing sandstones: (A) laminated sandstone lithofacies (3261.4 ft); (B) photomicrograph of clay laminae, dark central area is organic-rich with clay-cemented quartz sand above, and microcrystalline feldspar and clay minerals beneath (3160.8 ft)(cross-nicols); (C) clay-cemented sandstone (3159.9 ft)(cross-nicols); (D) calcite-cemented sandstone with clay parting (3158.5 ft). All scales in centimeters.

FIG. 9.--Photomicrographs showing typical textures and types of cementation in the St. Peter Sandstone: (A) fine-grained, well-sorted, well-rounded quartz sandstone (3186.6 ft)(plane-light); (B) medium-grained, moderately sorted, rounded to subrounded quartz sandstone (3201.5 ft)(plane-light); (C) calcite (dark gray) cemented sandstone, note that most of the grains are cement-supported (3160.9 ft)(cross-nicols); and (D) quartz overgrowth-cemented sandstone, note planar and sutured grain contacts and euhedral quartz grains (3091.5 ft)(cross-nicols). All scales in centimeters.

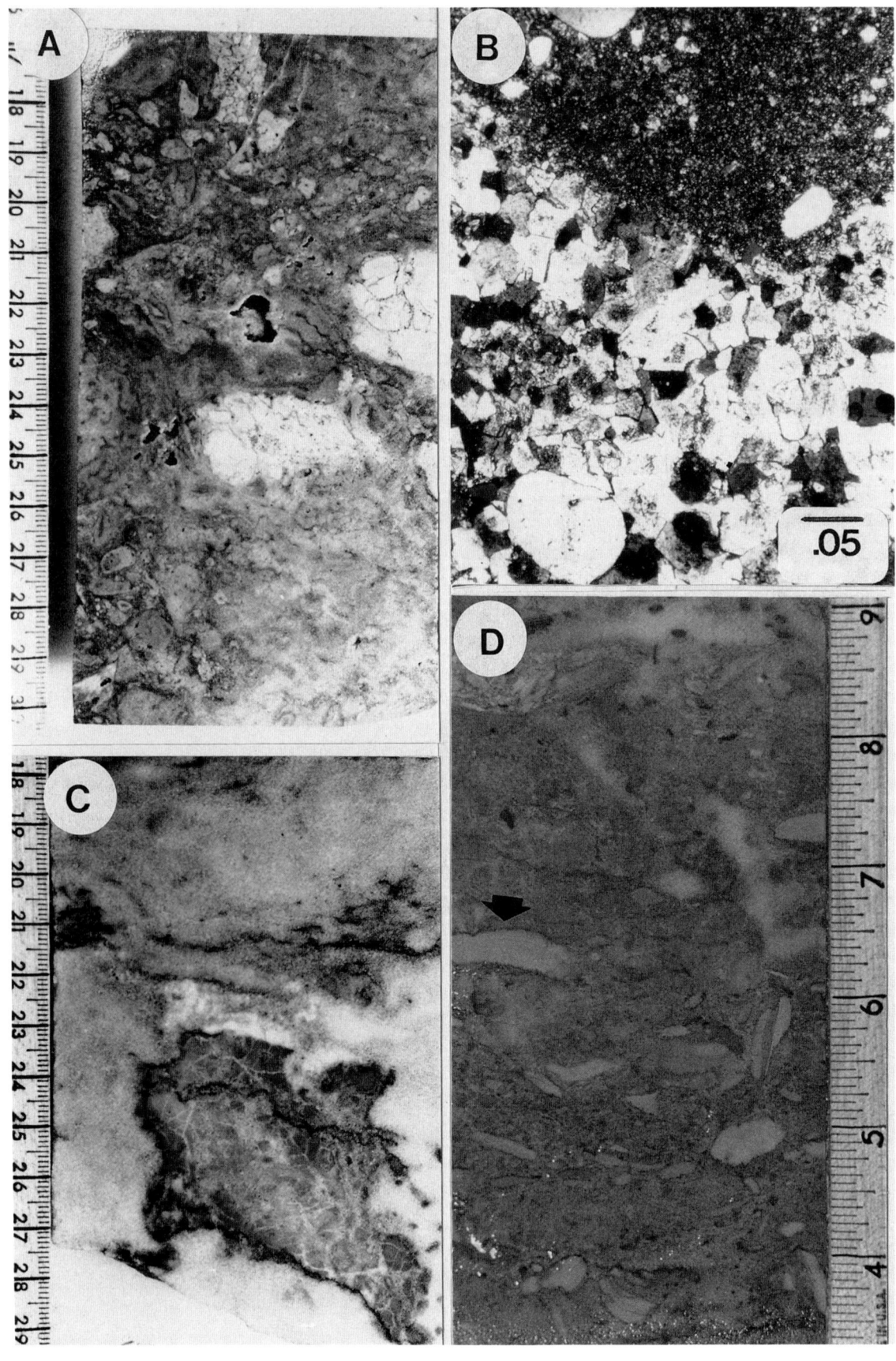

Fig. 5

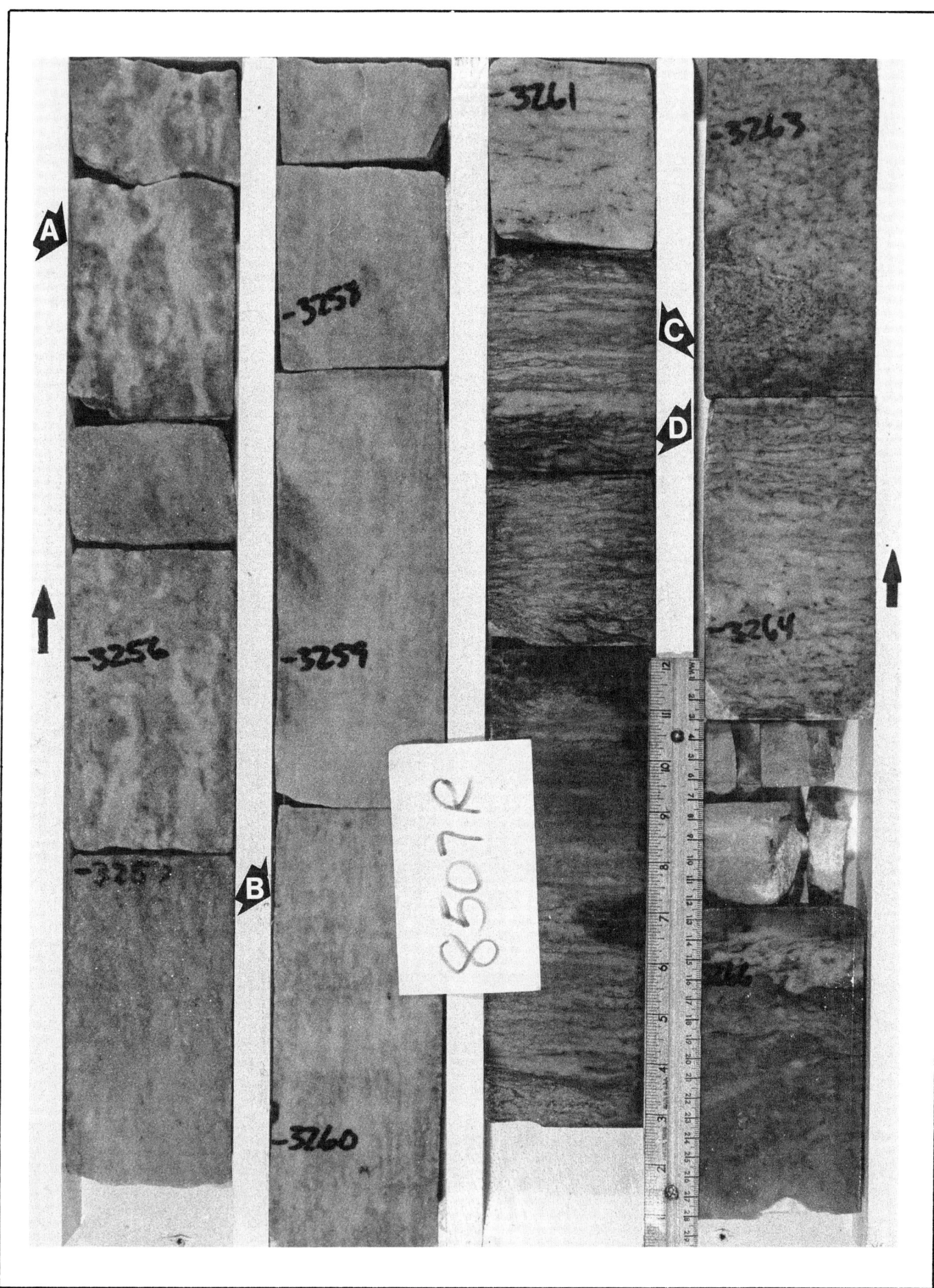

Fig. 6

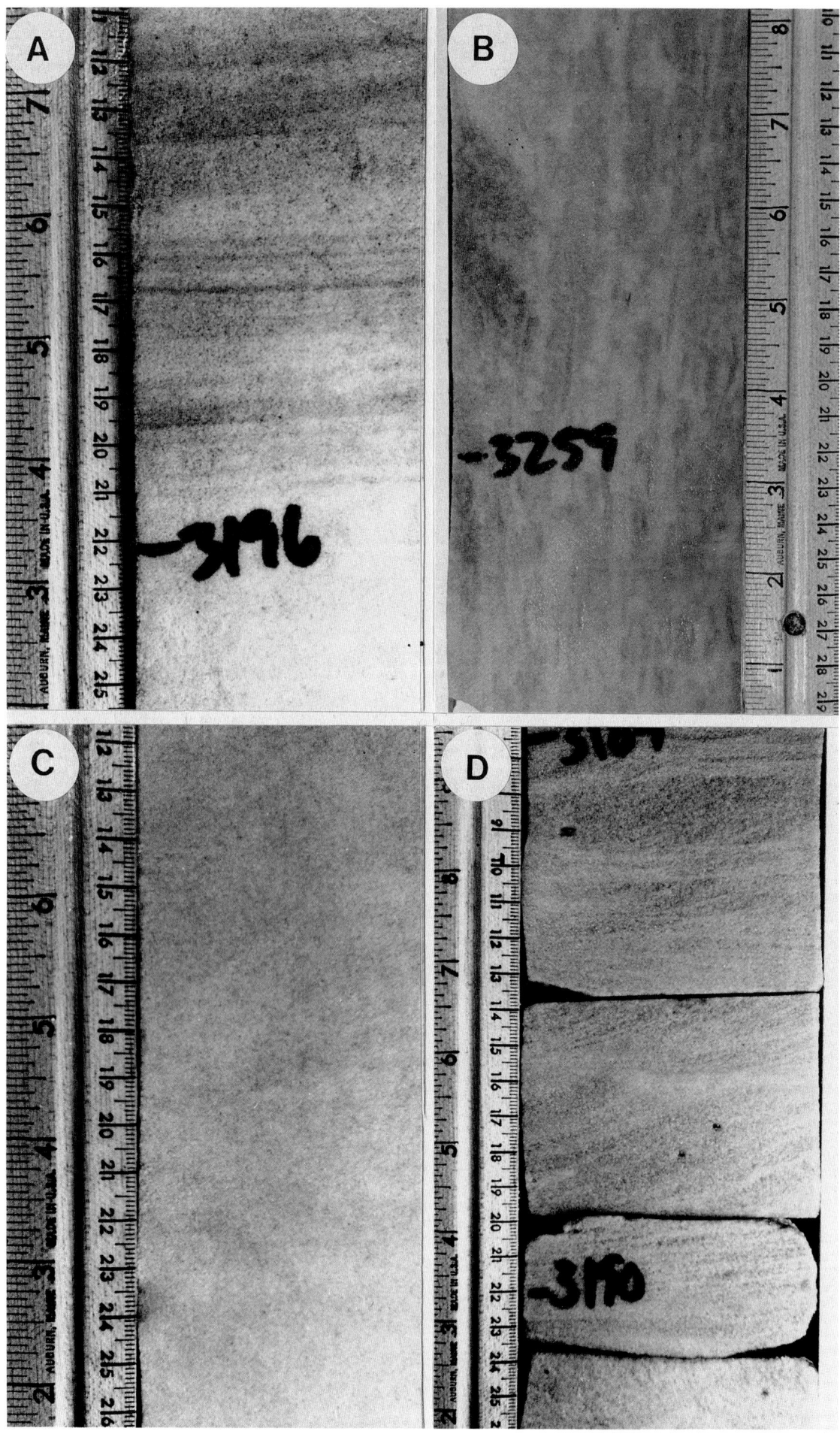

Fig. 7

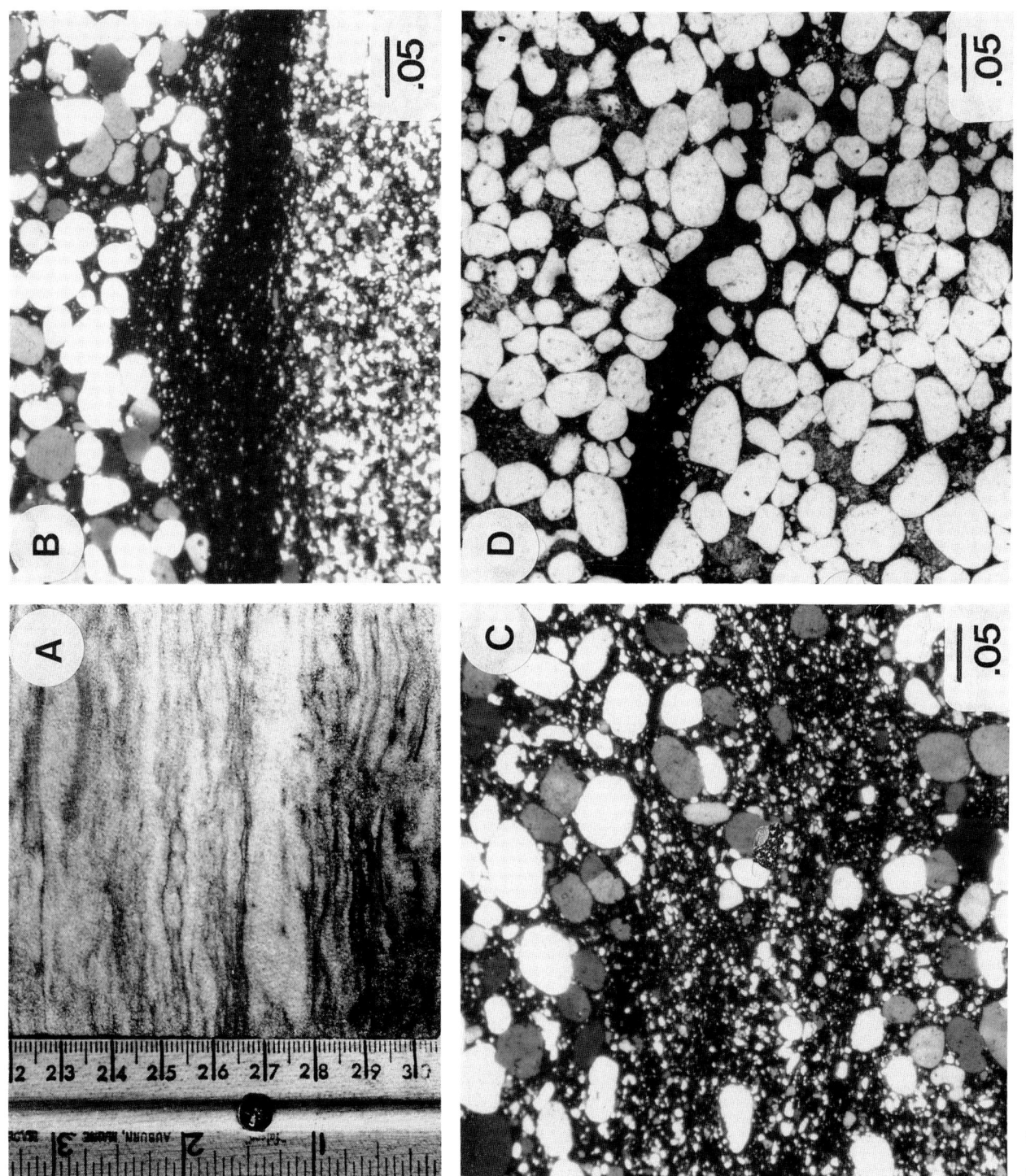

Fig. 8

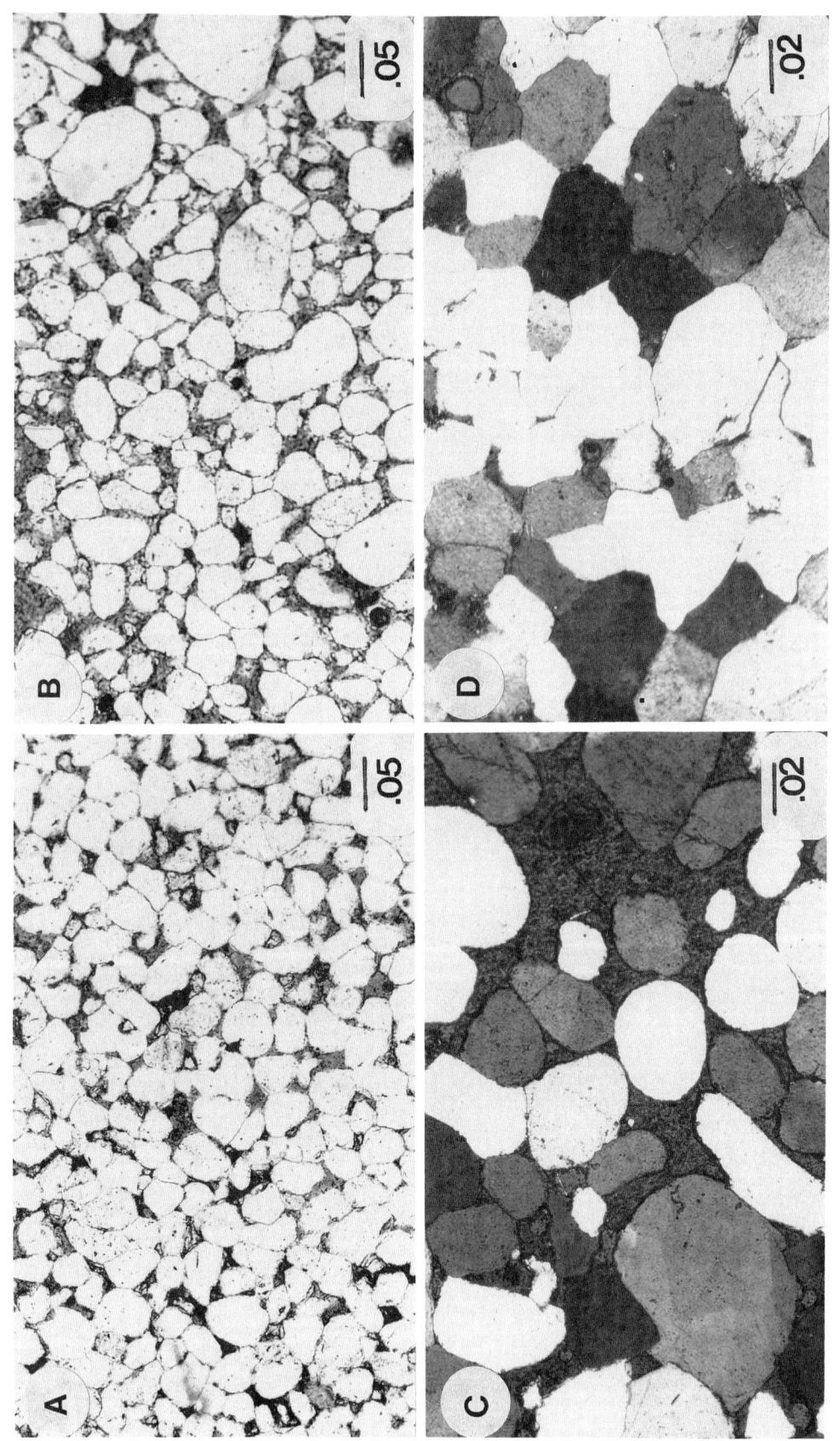

Fig. 9

The mottled dolomitic wackestone lithofacies is interpreted as bioturbated lagoonal mud. The quartz grains were transported into the lagoon area by eolian and storm-generated currents. Preferential dolomitization of bioturbated fabrics has produced the mottled appearance.

St. Peter Sandstone

Based on sedimentary structures, the St. Peter Sandstone can be subdivided into a thin lower unit and a thicker main body. The lower St. Peter contains four lithofacies: (1) pebble conglomerate, (2) interbedded sandstone and dolostone, (3) planar cross-stratified sandstone, and (4) laminated sandstone (Figs. 5c and d, and 6).

The pebble conglomerate lithofacies contains angular to moderately rounded clasts of lime mudstone and dolostone in a fine- to medium-grained sandstone matrix; the pebble conglomerates are poorly sorted and tend to grade upsection to sandstones (Fig. 5d). Locally, convex flat-pebbles of mudstone are present. The clasts commonly exhibit soft-sediment deformation and are similar in composition to underlying lithologies.

The pebble-conglomerate lithofacies is associated with the interbedded sandstone and calcareous dolostone lithofacies; the contact is generally sharp and irregular, but may be stylolitic locally (Fig. 5c). The sandstone interbeds are fine- to medium-grained, well-rounded quartz sandstone that tends to fine upsection. Calcareous dolostones are composed of finely crystalline, anhedral to subhedral dolomite with some micrite; well-rounded, fine- to medium-grained quartz sand is present locally. The contact between interbeds is commonly stylolitic.

The pebble conglomerate lithofacies is in turn overlain by the planar cross-stratified sandstone lithofacies, a fine-grained, well-sorted, well-rounded sandstone with planar-stratified laminae (Figs. 6 and 7a). The sandstone is well-indurated with a quartz overgrowth cement. Laminae are the product of minor, but uniform, variations in grain size from very fine- to fine-grained. Finer-grained laminae are more extensively cemented than the coarser-grained laminae.

This lithofacies grades upsection into the laminated sandstone lithofacies (Figs. 6 and 8a). The sandstone is fine- to medium-grained, well-sorted, well-rounded quartz sand with thin, planar clay-laminae. The sands display a moderate amount of quartz overgrowth and are well indurated. Quartz grains in contact with the clay laminae do not possess overgrowths. The clay laminae, predominantly composed of illite and some authigenic potassium feldspar, contain organic partings (Fig. 8b). Upsection the laminar layers are laterally discontinuous.

The lithofacies of the lower St. Peter are interpreted to have been deposited in shallow lagoonal and wash-over fan environments that grade upsection into algal-controlled, laminated sandstones that were deposited under intertidal and supratidal conditions. "Floating" sand grains in the mottled-wackestone lithofacies of the upper Everton, the increasing abundance of sandstone, and coarsening of the pebble-conglomerates upsection suggest increasing proximity to dune-foreshore environments. This transition is interpreted as the cratonward migration of the clastic shoreface during a rise in sea level in medial Whiterockian time. The first appearance upsection of the transgressive sandstones and pebble conglomerates are considered to mark the base of the

St. Peter Sandstone, and are overlain by the laminated sandstone lithofacies. The laminated sandstone lithofacies was deposited in the peritidal environments of the beach ridge, where algal mats locally bound sediments. With more open-marine conditions, bioturbation disrupted the algal mats. The algal mats, preserved as organic laminae, became accumulation sites for detrital and/or authigenic clays, and resulted in some of the observed clay laminae. The laminated sandstone lithofacies grades into the overlying bioturbation-dominated sandstones that make up the main body of the St. Peter Sandstone.

Differentiation of lithofacies within the main body of the St. Peter Sandstone is based on the presence or absence of cross-stratification and bioturbation structures (Fig. 6); three lithofacies can be identified: (1) *Skolithos* sandstone lithofacies (Fig. 7b), (2) massive sandstone lithofacies (Fig. 7c), and (3) cross-stratified sandstone lithofacies (Fig. 7d).

The *Skolithos* sandstone lithofacies is fining-upward, coarse- to fine-grained, well-rounded to subrounded quartz sand with vertical *Skolithos* burrows. These burrows are 1 to 15 cm in length, 1 to 5 mm in diameter, and are commonly infilled with medium- to coarse-grained sand; similar burrow structures are observed in the outcrop area of southern Wisconsin (Dott and others, 1986). The coarse-grained sandstone is friable with poorly developed quartz overgrowth cements; it grades upsection into fine-grained sand that is clay cemented and moderately indurated (Fig. 8c). The clay cements in sandstones of the St. Peter are commonly associated with authigenic potassium feldspar; these intervals are commonly identifiable in gamma-ray geophysical logs (Gendron and others, 1988). The potassium feldspar associated with the clay cements of the St. Peter may be related to regional, potassic diagenetic events due to migration of basinal brines (Hay and others, 1988). The number of burrows and presence of coarser-grained fractions decrease upsection, where it grades into the massive sandstone lithofacies.

The massive sandstone lithofacies is fine- to medium-grained, well-sorted, well-rounded, moderately to well-indurated quartz sandstone with both calcite and silica cements (Figs. 7c and 9). Internal sedimentary structures are absent, with the possible exception of vaguely defined, horizontal burrow traces. Where calcite cement is present, the grains are either cement-supported or have point contacts. Where well-indurated sandstone is present, euhedral quartz overgrowths are common and grain contacts tend to be sutured or planar. Disseminated pyrite is common; pyrite nodules are present where this lithofacies immediately underlies the Joachim Dolomite.

This lithofacies in turn is overlain by the cross-stratified sandstone lithofacies, a fine- to coarse-grained, well-rounded, well-indurated quartz sandstone with low-angle cosets (Fig. 7d). The relationship of the cosets to the bedding planes and the geometry of the cross-stratification in the core are unclear. The development of cross-stratification changes upsection from planar to low-angle to ripple cross-stratification; in addition, the grain size decreases. The basal contact of the cross-stratified lithofacies with the massive sandstone is sharp and planar; the upper contact is gradational.

The lithofacies in the main body of the St. Peter Sandstone represent three phases of deposition in nearshore subtidal environments. The *Skolithos* sandstone lithofacies reflects high-energy depositional conditions, most commonly associated with foreshore conditions in epicontinental seas (Frey and others, 1990; Droser and Bottjer, 1990). The massive sandstone lithofacies was deposited in shallow subtidal conditions, but has been extensively reworked by biogenic activity destroying most of the primary depositional structures.

The cross-stratified sandstone lithofacies reflects an increase in current energy. The relatively sharp contact of this lithofacies with the underlying massive sandstone lithofacies at 3196 ft (974 m) suggests that the contact represents a significant change in current energy; locally this contact is sharp and may be an erosional surface or a surface of nondeposition. This horizon can be identified in the subsurface throughout much of central and southern Illinois and suggests that the increase in current energy is due to a regional event, either a relative sea-level fall or a major storm event, and is not due to localized channelization. The upper gradational contact suggests a simple decrease in current activity or more extensive exposure to biogenic reworking.

The massive sandstone lithofacies in the upper St. Peter Sandstone grades into the clay-cemented, laminated sandstone, suggesting shoaling upward into peritidal conditions (Figs. 8c and d, and 10). In the upper lithofacies of the St. Peter, calcite cement is more common, although its distribution is highly variable. Pyrite nodules in the upper St. Peter may be related to penecontemporaneous downward movement of anoxic waters into the St. Peter Sandstone during the deposition of algal marsh sediments represented by the shale lithofacies in the Joachim Dolomite (Fig. 10)(Pye and others, 1990; Kirchner, 1989).

Joachim Dolomite

The Joachim Dolomite is a highly variable carbonate unit containing 11 lithofacies that are present in discrete shallowing-upward sequences. The vertical stacking of shallowing-upward sequences reflects an upsection transition from predominantly peritidal to subtidal environments, and a return to peritidal conditions prior to the relative fall of sea level at the end of Ancell deposition.

The large-scale, shallowing-upward sequences grossly correspond to stratigraphic subdivisions, as proposed for the Joachim by Templeton and Willman (1963). This relationship suggests that regional correlation of these shallowing-upward cycles may be possible.

The contact of the Joachim Dolomite and the St. Peter Sandstone at 3159 ft (963 m) is gradational, as it is throughout much of central and southern Illinois, but a significant change in lithology from mostly sandstone to mostly carbonate occurs within a relatively limited vertical interval (Figs. 10 and 11). Locally, dissolution breccias and intraformational conglomerate are present within this interval; however, these features are related to early diagenetic and/or penecontemporaneous processes. The profound decrease in siliciclastic content suggests that the source area had been greatly reduced, the shoreline had prograded to a point where basinward transport of siliciclastic sediment was minimal, and/or sediment is bypassing this portion of the carbonate ramp.

FIG. 10.--Uppermost St. Peter Sandstone (3158.5-3169.6 ft); contact with Joachim Dolomite at 3159 ft (arrow A). Note shale (arrow B), laminated sandstone (arrow C), and pyrite nodules in massive sandstone (arrow D).

FIG. 11.--Photograph of the lower Joachim Dolomite (3149.5-3159.4 ft) showing interrelationship of (A) laminated dolostone, (B) shale, and (C) patterned mudstone lithofacies.

FIG. 12.--Photomicrographs of dolomudstone lithofacies: (A) very fine-grained quartz-sand stringers in dolomudstone (3159.4 ft); (B) dolomudstone matrix at greater magnification showing very finely crystalline, subhedral dolomite; (C) calcite pseudomorphs after gypsum or anhydrite laths (3131.8 ft); and (D) calcite pseudomorph after gypsum or anhydrite lath (3125.6 ft). All photomicrographs in plane-light; scales in centimeters.

FIG. 13.--Lithofacies of the lower Joachim Dolomite: (A) laminated dolostone lithofacies (3090 ft), (B) patterned dolomudstone lithofacies interbedded with shale lithofacies (3155.05-3155.15 ft), (C) patterned mudstone lithofacies showing variable distribution of intercalations (3121 ft), and (D) finely laminated dolomudstone grading upsection to patterned dolomudstone (3150 ft). All photos are at the same scale.

FIG. 14.--Photo of lower Joachim Dolomite (3082.6-3091.5 ft) showing interbedded laminated dolostone, patterned mudstone, shale, and breccia lithofacies. Note possible calcite pseudomorphs after gypsum or anhydrite at arrows A and B.

FIG. 15.--Photomicrographs of shale and fine-grained sandstone lithofacies: (A) organic-rich shale (black) with silt-sized quartz stringer (3146.9 ft); (B) shale lithofacies overlying dolomudstone (contact at arrow), note irregular quartz infillings of voids in dolomudstone (3091.5 ft); (C) fine-grained sandstone with shale intraclasts (black)(3152.9 ft); and (D) fine-grained sandstone overlying shale laminae (3160.9 ft). All photomicrographs in plane-light; scales in centimeters.

FIG. 16.--Core interval (3061-3070 ft) predominated by the patterned mudstone lithofacies; note the wide variation in intercalations and the fluid-escape structure at the arrow.

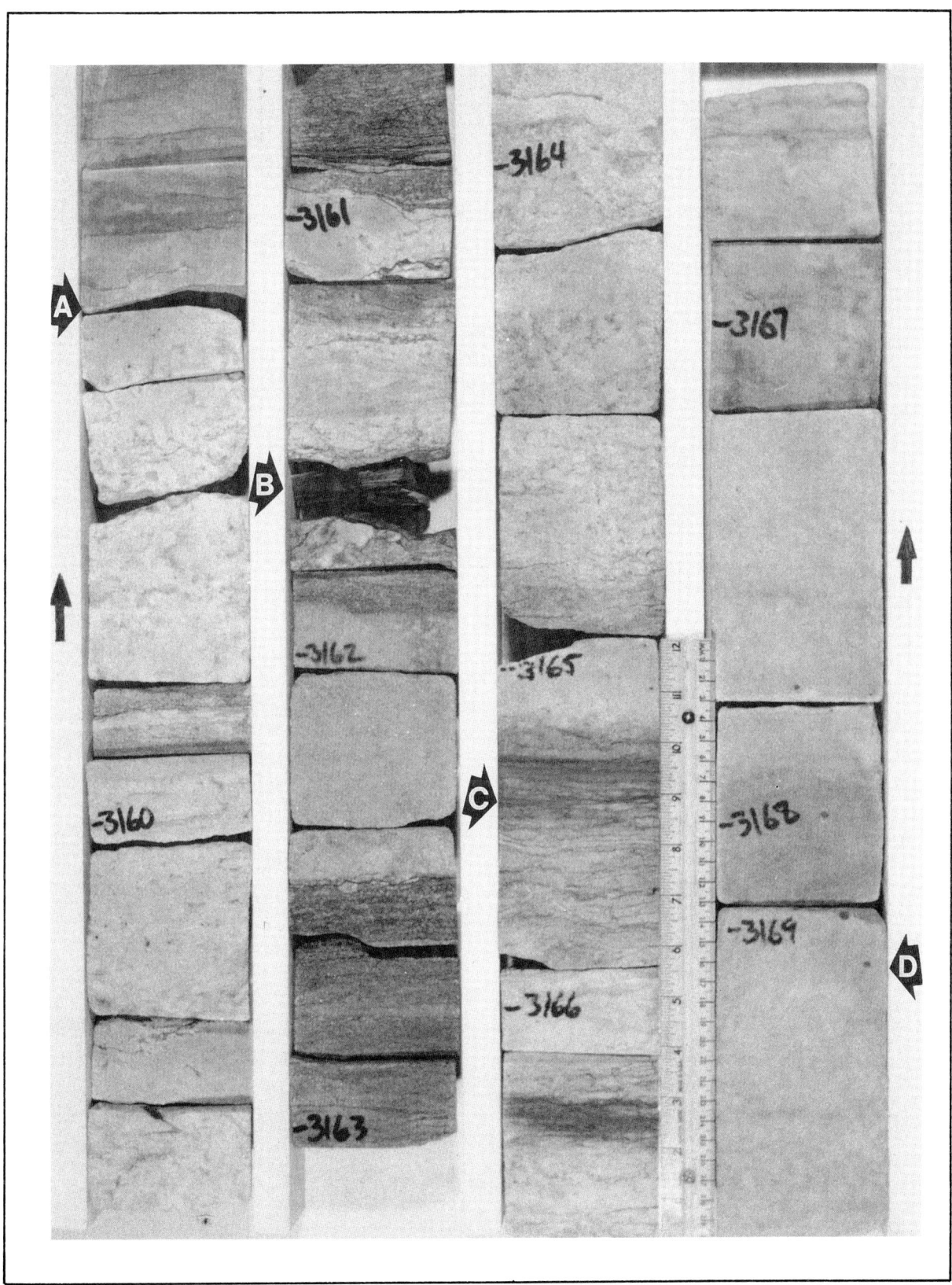

Fig. 10

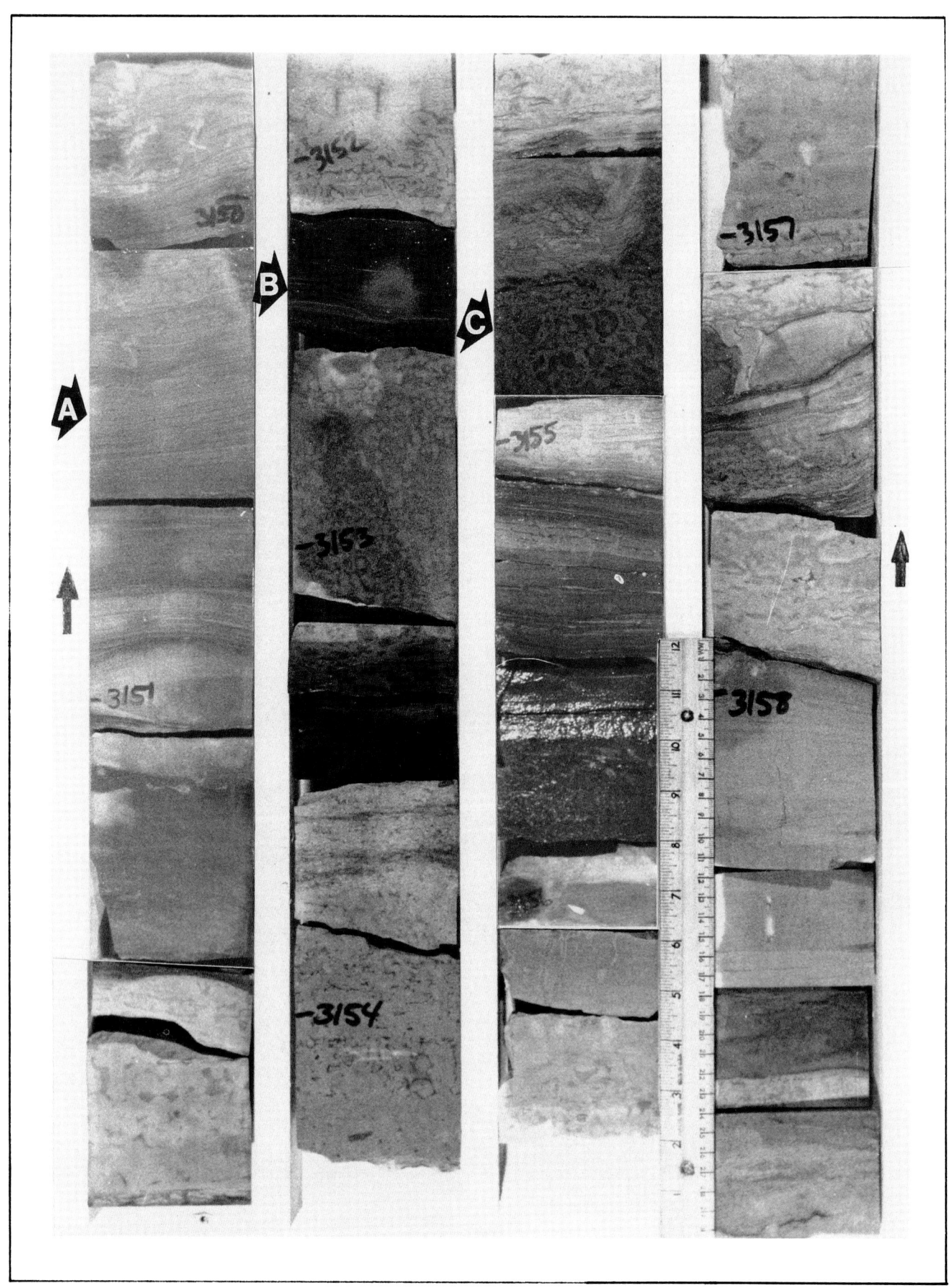

Fig. 11

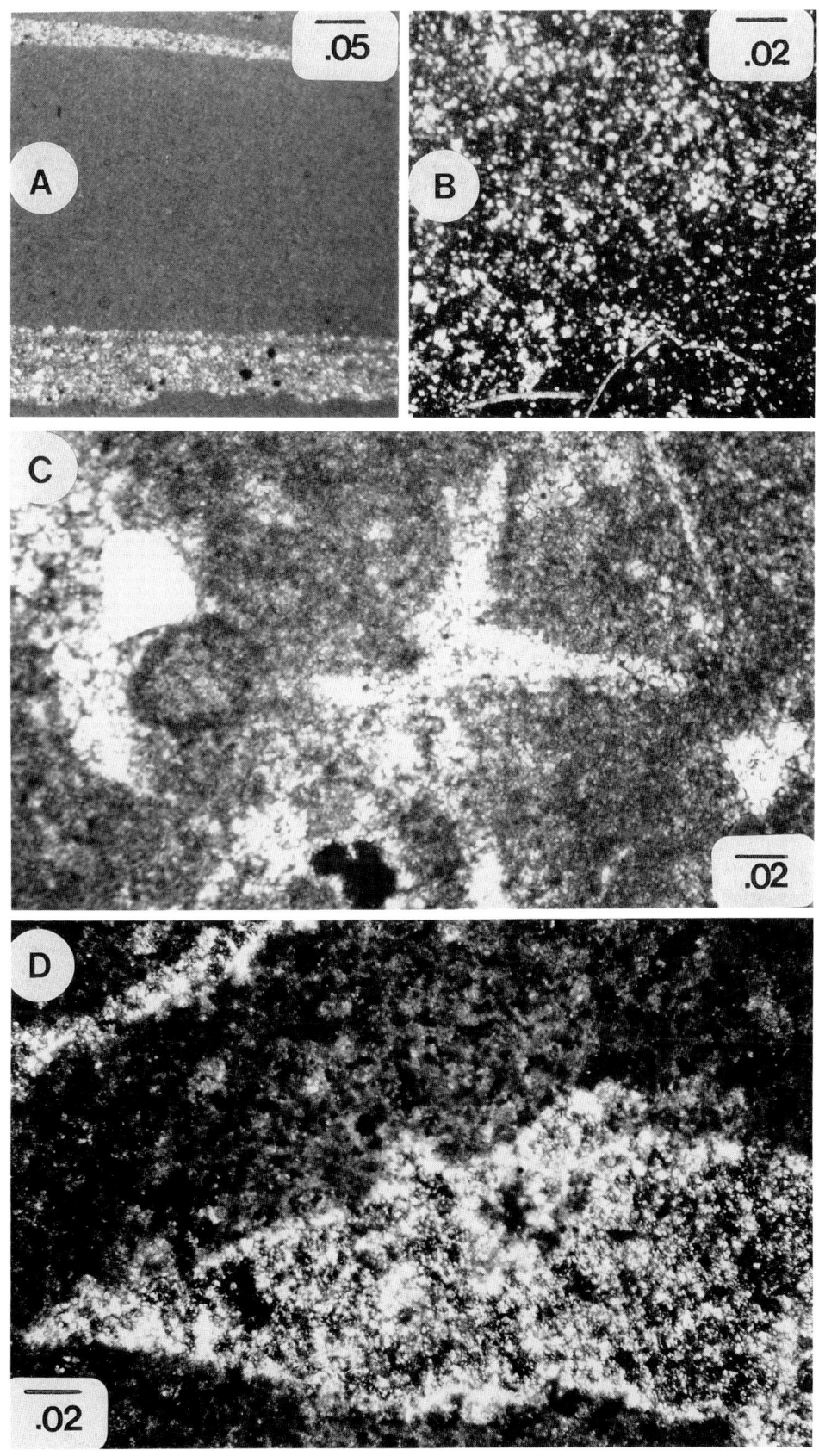

Fig. 12

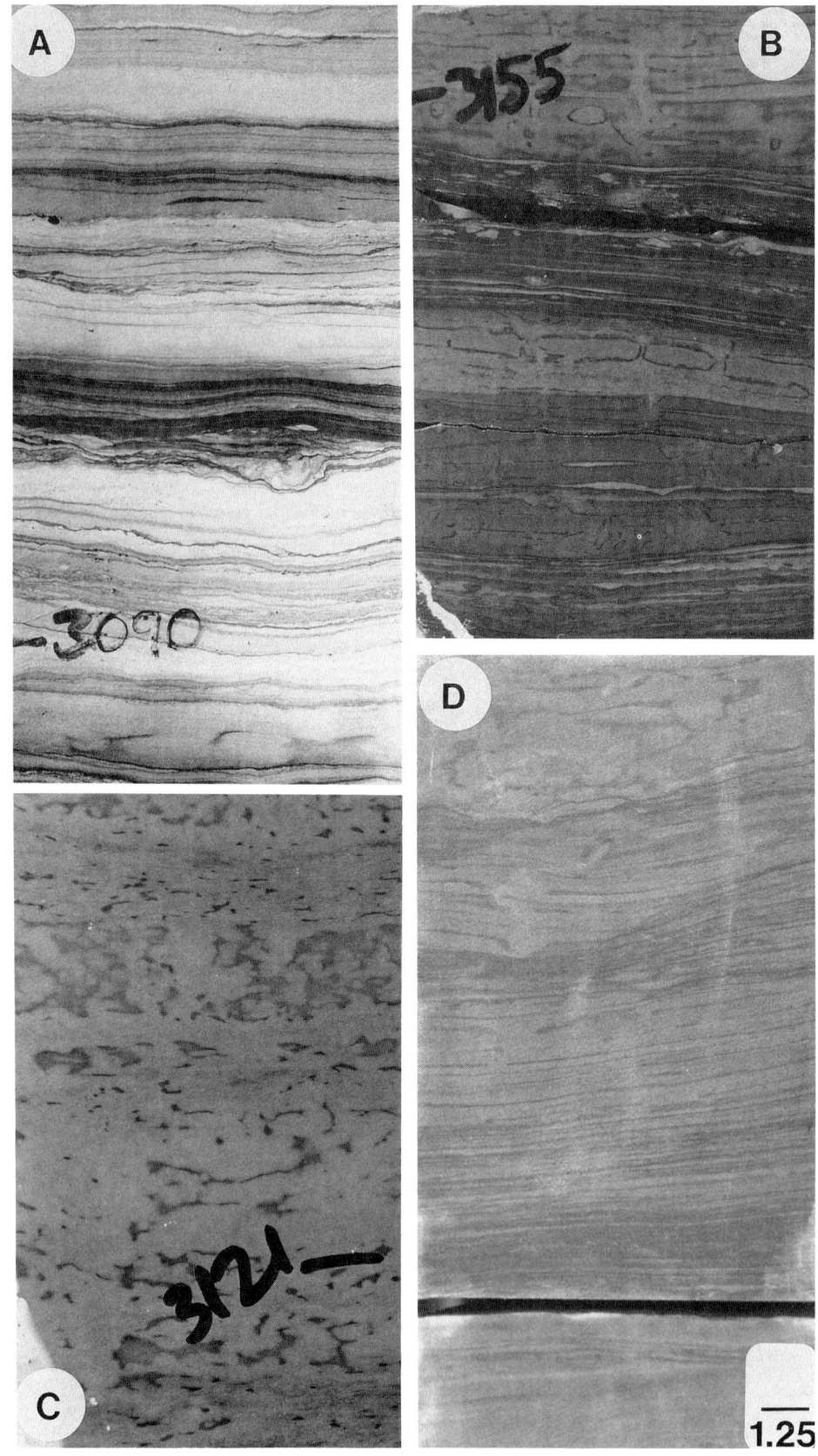

Fig. 13

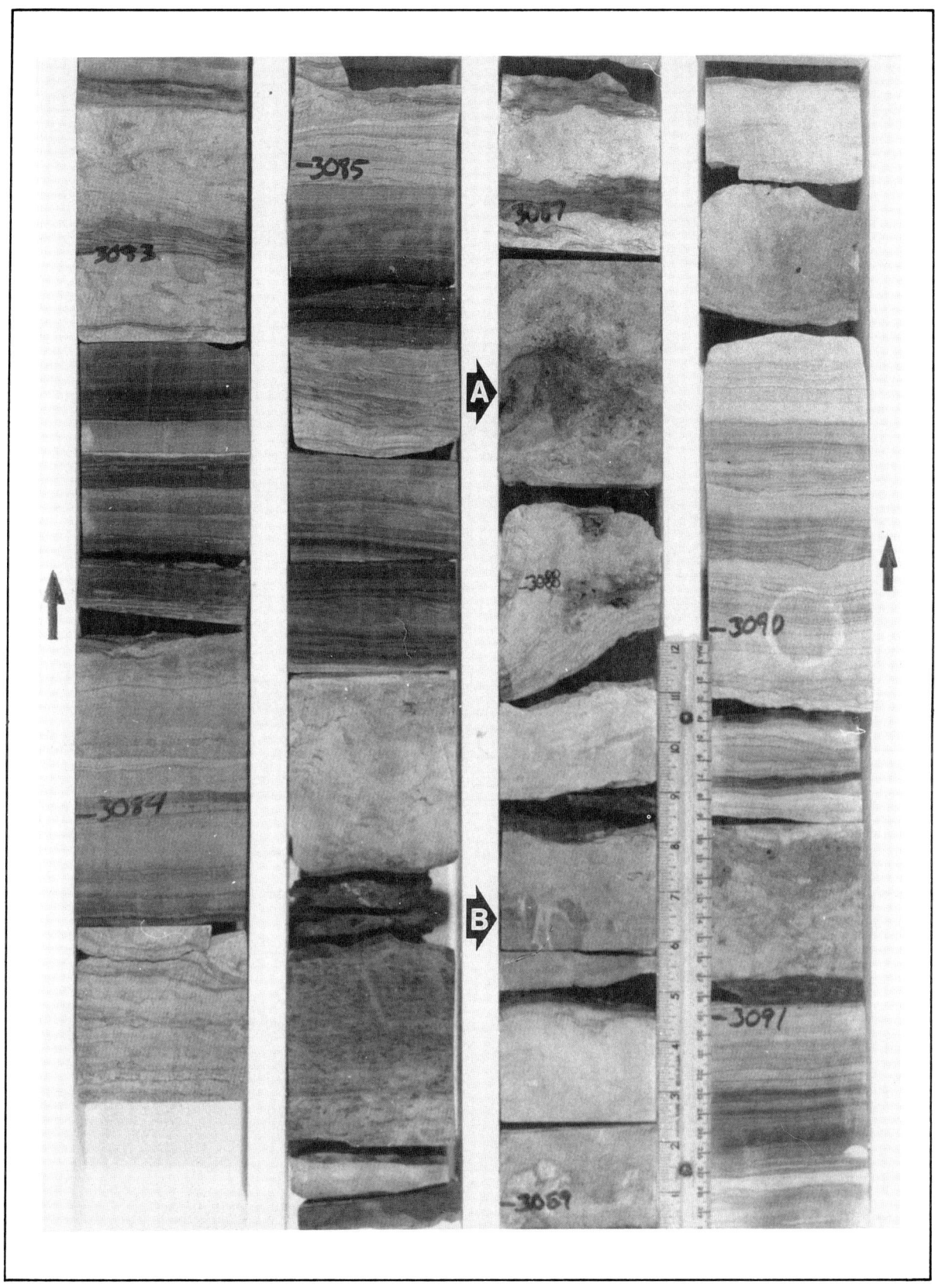

Fig. 14

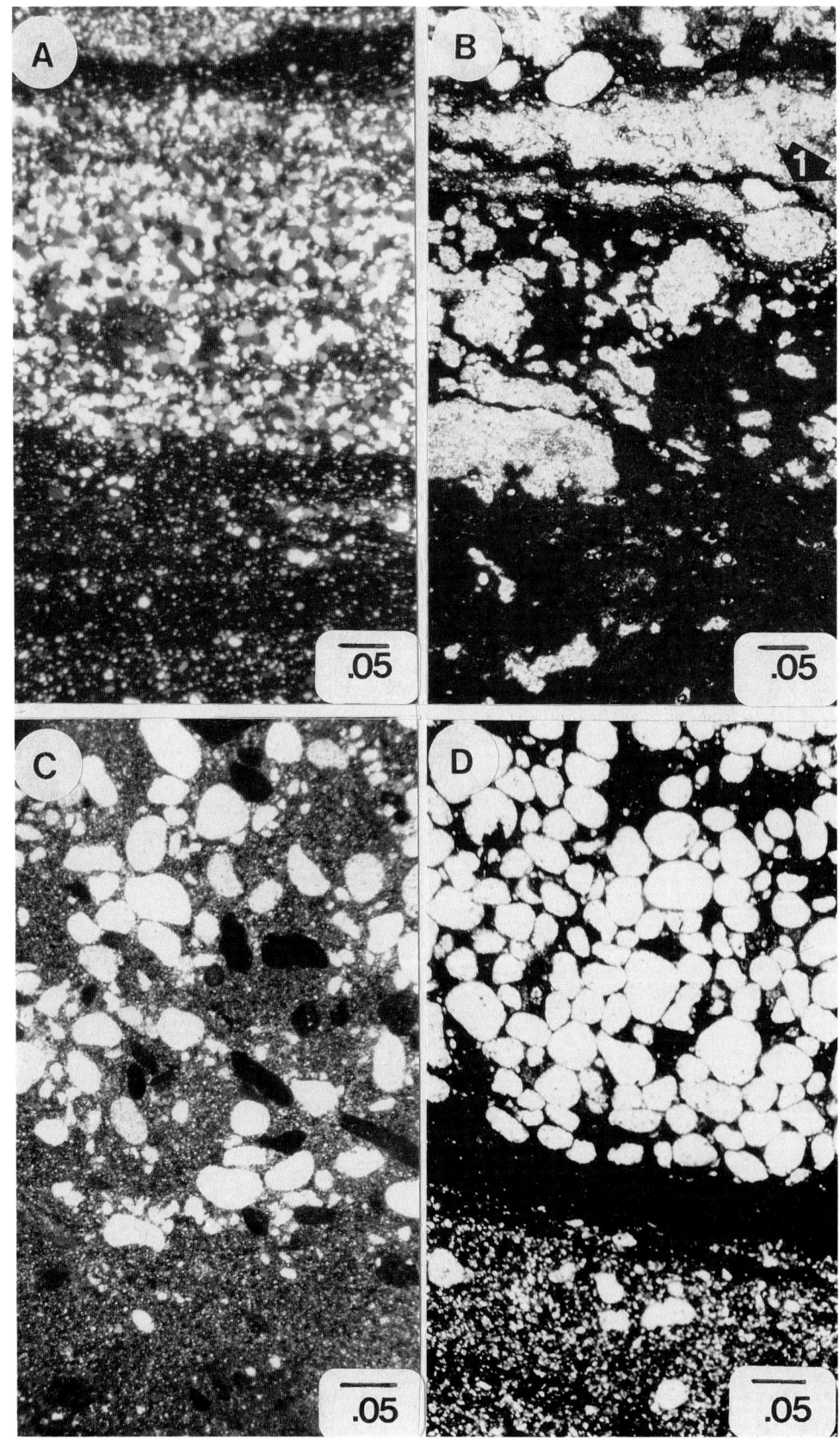

Fig. 15

Fig. 16

A common lithofacies in the lower Joachim is silty dolomudstone, a tan, chalky-textured, commonly dolomitized mudstone with abundant quartz silt (Fig. 12a and b). In most samples dolomitization has masked the primary sedimentological features, and it appears to be featureless both mega- and microscopically. Where unaltered, it is predominantly limy micrite with rare gypsum or anhydrite laths replaced by calcite (Fig. 12c and d). The mudstone commonly possesses relatively high intercrystalline porosity and has a low density. In the absence of primary sedimentary features, the depositional environment is inferred to be intertidal to supratidal based on associations with overlying and underlying lithologies.

The silty dolomudstone lithofacies is associated with laminated dolomudstone lithofacies that differs primarily in the presence of laterally continuous, brown, organic-rich laminae (Fig. 13a). Thickness of the laminae ranges from 1 to 3 mm and varies from planar to highly crinkled. Bedding varies from laminated to thin-bedded within any one-foot section of core. The micritic matrix tends to be darker in laminated intervals. This lithofacies is interpreted as intertidal where algal mats have bound the sediment, and sporadic desiccation or specialized types of algal growth may have produced the crinkle laminae. Mudcracks are rarely observed in the core, but are abundant in the rock in outcrop.

Within intervals where laminated dolomite is common, the dark brown silty shale lithofacies is present (Figs. 11 and 14). The lowermost appearance of this facies is in the uppermost St. Peter Sandstone. In addition to organic-rich shale, this lithofacies contains plentiful intralaminae of quartz (Fig. 15a). These sand laminae are silt-sized to fine-grained, well-rounded, and well-sorted, and exhibit graded bedding locally. Void-filling quartz cements are present locally in overlying and underlying carbonates (Fig. 15b).

Based on whole-rock pyrolysis data (pers. comm., M. Chou and J. Crockett, Illinois State Geological Survey) for the shales in the basal Joachim, in an adjacent well (Illinois Power Company, Morrell #2), and palynomorph studies, the shales have three notable organic characteristics: (1) they have relatively high total organic carbon content (1 to 4.3 percent TOC); (2) the organics possess relatively high hydrogen-index values (greater than 300), indicative of algally derived organic matter and commonly associated with Type I kerogen (a possible precursor of hydrocarbons); and (3) the shales contain acritarchs and dispersed organic matter associated with normal marine conditions. T_{max} values, an indicator of the thermal maturation, fall between 400 and 430, indicating that these shales are thermally immature (Peters, 1986). These data are in agreement with the observed conodont (CAI = 1) and palynomorph (TAI = 1) color-alteration indices.

These shales are interpreted to have been deposited in algal ponds formed in the supratidal zone by storm events, when large volumes of normal-marine water inundated the tidal flat. Wind storms intermittently transported sand into the ponds, producing sand laminae. With time, the ponds infilled with organic detritus or tidal-flat carbonate sediments.

Anomalously low hydrogen-index (83) and total organic content (less than 0.06), and high oxygen-index (150) values, suggestive of a nonmarine organic detritus, were obtained from the basal shale in the Joachim Dolomite in the Truitt #1 core. Organic extracts analyzed for biomarkers and carbon-isotope data for each of the organic fractions suggest that post-Devonian

hydrocarbons with a terrestrial source component have flushed the lowermost shale (pers. comm., M. Helman, Imperial College). This fluid migration may also be related to the authigenic cements found in the St. Peter Sandstone.

Fine-grained sandstone is present in the Joachim as laminae (less than 1 cm), stringers (1 to 2 cm), and as beds up to 18 cm thick. It is primarily composed of very well-sorted, fine to medium, very well-rounded, frosted quartz grains (Fig. 15d). If the basal contact of the sandstone bed is sharp or irregular, intraclasts of the underlying lithology are commonly incorporated into the sandstone (Fig. 15c). Grain contacts are mostly sutured to planar, and quartz overgrowth cements predominate. This lithofacies reflects the incursion of eolian dunes onto the tidal flat and infilling of small depressions or deflation surfaces where residual material of the underlying lithology was present.

Patterned mudstone, a distinctively color-mottled carbonate which locally may be dolomitized, is common throughout the lower Joachim and present within limited intervals in the upper Joachim (Figs. 13 and 16). The distinctive intercalations of green-gray and light gray-brown "matrix" are not reflected in mega- or microscopic sedimentary features. Pyrite is also locally associated with the green-gray intercalations. Although the origin of this patterned feature remains enigmatic, it is probably related to the distribution and oxidation state of clay minerals. This lithology is commonly considered to be deposited in the intertidal zone where the color mottling is produced by the intermixing supratidal brines and marine waters.

Breccias are present within restricted horizons in the core (Figs. 14 and 17). Most clasts are poorly sorted, angular to subrounded, composed of peloidal and/or fenestral mudstone or dolomudstone, and range from fine-grained sand to cobble-sized clasts. The matrix is locally micritic, but it is mostly a coarse calcite-spar. The clasts are mostly grain-supported, but may also be mud-supported. Locally, problematic chevron- to lath-shaped clast morphology suggests possible replacement after gypsum or anhydrite (Fig. 14). Diagenetic overprint, however, has obscured the petrographic character of these clasts, precluding a more definitive interpretation of these features. This lithofacies is interpreted as collapse breccia formed by the dissolution or replacement of gypsum and/or anhydrite. A similar lithofacies was described from the Joachim in the outcrop area of southeastern Missouri (Okharavi and Carozzi, 1983).

The upper Joachim Dolomite mostly contains limestone, with rare to abundant, complete and incomplete bioclasts that are intergradational in character (Fig. 18). These can be subdivided into five representative lithofacies: (1) peloidal mudstone, (2) bioclastic wackestone, (3) intraclastic packstone/grainstone, (4) stylolaminated wackestone/packstone, and (5) cryptolaminated mudstone (Figs. 18-20).

Peloidal mudstone is a common component of the upper Joachim. It is mostly micritic with regions of diffuse peloidal features. Contacts between peloids are vague and locally discontinuous; where complete, the peloids are cemented with rims of fine, sparry calcite. Peloidal mudstones are common components of modern-marine, subtidal carbonate environments. Preservation of peloids is thought to be the result of early cementation in the oxygenated phreatic zone.

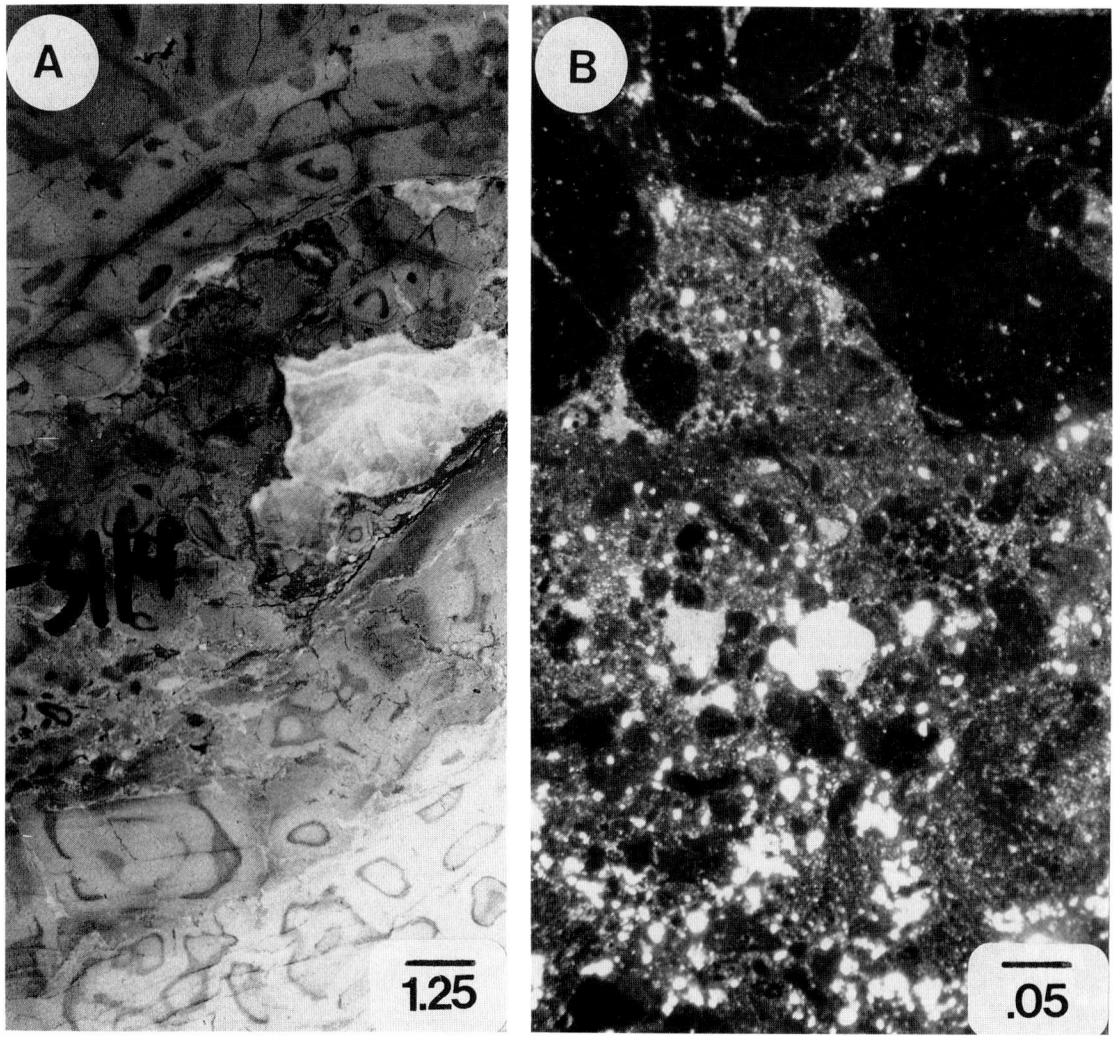

Fig. 17--Breccia lithofacies: (A) breccia with void-filling calcite developed in patterned mudstone, note that contacts of the breccia are highly irregular and well defined (3114 ft); (B) photomicrograph of breccia showing poorly sorted, pebble-sized to coarse-grained, angular clasts with micritic matrix containing quartz sand (3085.2 ft) (plane-light). Scales are in centimeters.

FIG. 18.--Cored interval (3044.5-3053.5 ft) from the upper Joachim Dolomite showing the four major lithic components: end members of the intraclastic packstone/grainstone lithofacies, (A) bioclastic grainstone and (B) intraclastic packstone; (C) peloidal mudstone lithofacies interbedded with bioclastic wackestone, and (D) stylolaminated wackestone/packstone lithofacies.

FIG. 19.--Lithofacies of the upper Joachim Dolomite: (A) bioclastic/intraclastic grainstone (3046 ft); (B) stylolaminated wackestone/packstone (3050.5 ft); (C) peloidal mudstone interbedded with bioclastic wackestone (3048.5 ft); and (D) intraclastic grainstone, note imbrication of grains. All photos are at the same scale.

FIG. 20.--Photomicrographs of lithofacies in the upper Joachim Dolomite: (A) bioclastic/intraclastic grainstone lithofacies (3044.1 ft); (B) stylolaminated wackestone and packstone lithofacies (3046.5 ft), note stylolite-bounded lens of bioclastic packstone near base of photo; (C) intraclastic grainstone (3112.3 ft); and (D) cryptolaminated mudstone lithofacies, base of laminae are indicated by arrows (3113.6 ft). All photomicrographs in plane-light; scales are in centimeters.

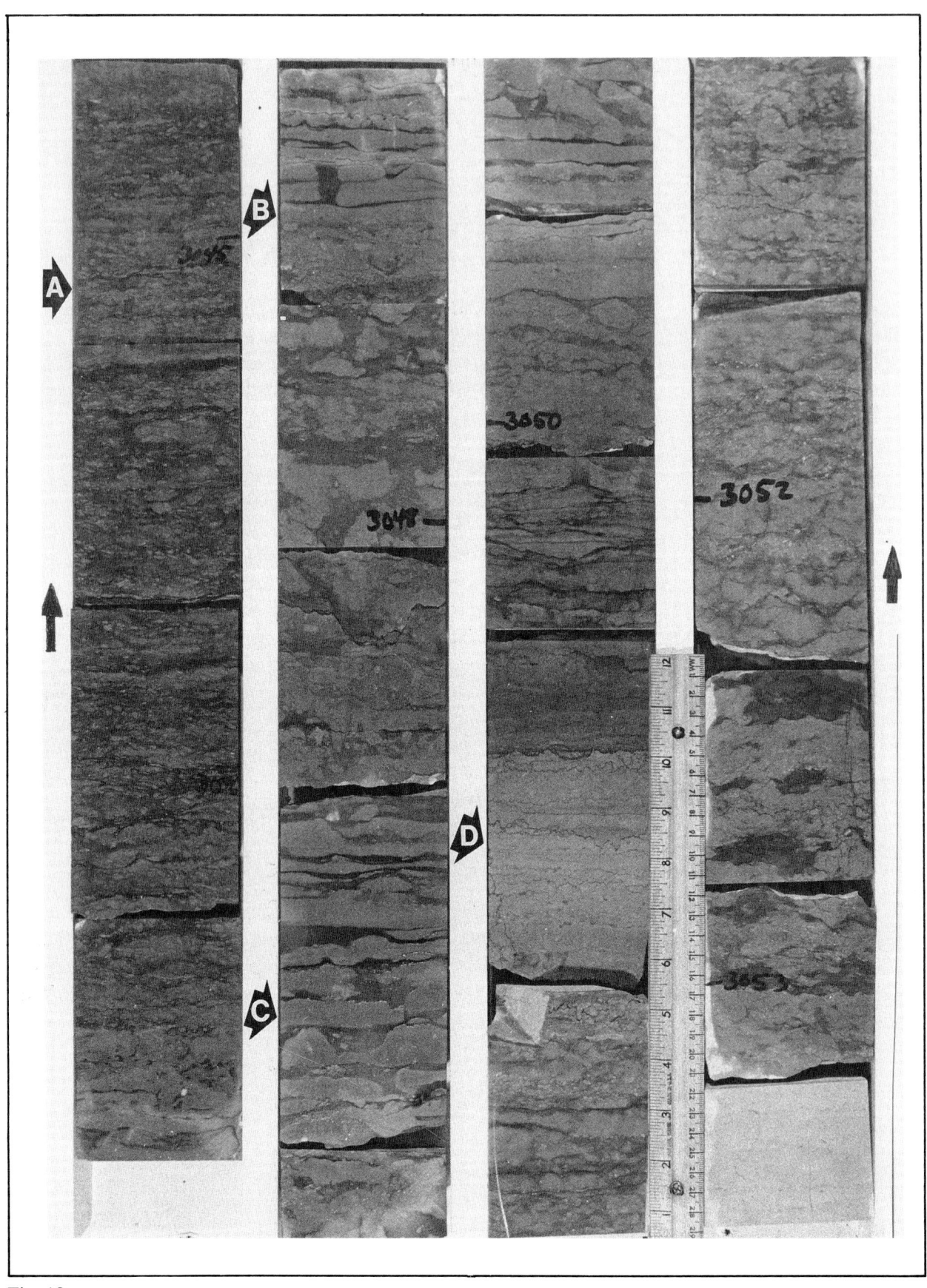

Fig. 18

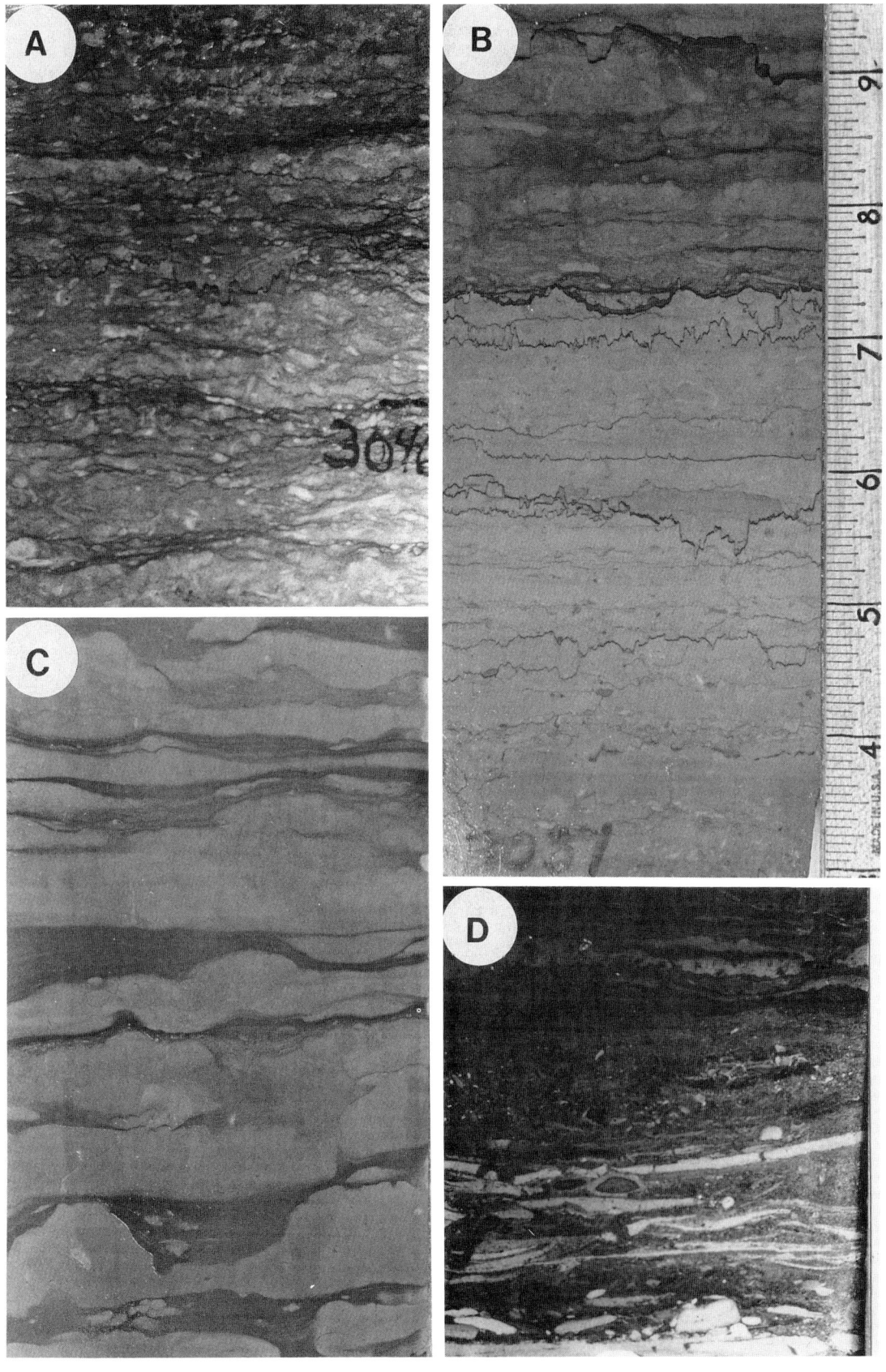

Fig. 19

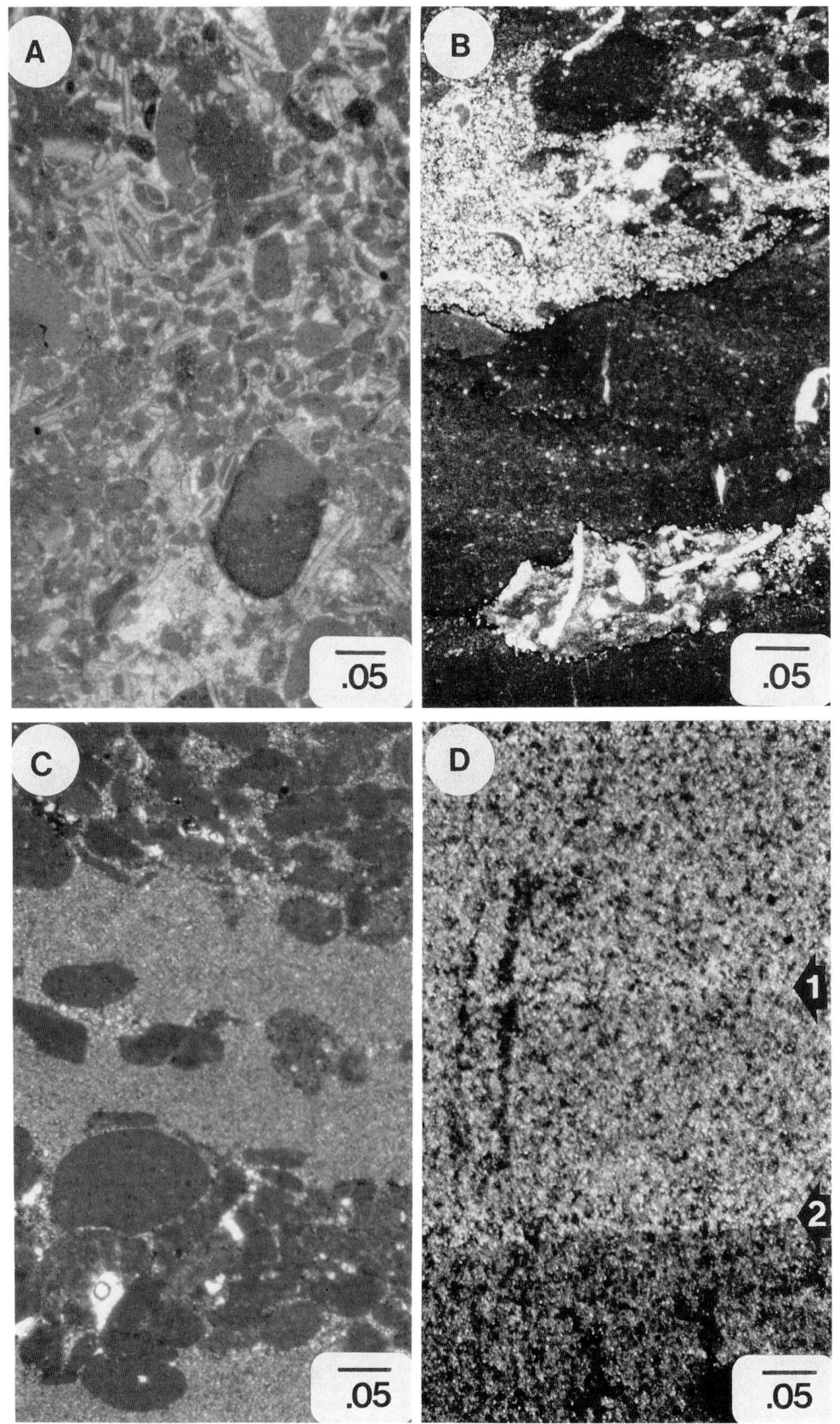

Fig. 20

The peloidal mudstone lithofacies is commonly interbedded with the bioclastic wackestone facies (Fig. 19). This facies commonly contains complete ostracods, sparse brachiopods, and other indeterminate bivalves. The bioclasts are mostly complete and oriented in life position with localized stringers of bioclastic packstone; bioclasts in the stringers show no preferred orientation. The matrix is limy micrite with sparse spar-filled fenestrae. This facies is interpreted as shallow, subtidal lagoonal muds. The predominance of ostracods suggests somewhat restricted marine conditions. Storm currents sporadically reworked the sediments, winnowed out the fine-grained components, and resulted in the observed bioclastic packstone stringers.

Intraclastic packstone/grainstone contains moderately sorted bioclasts, intraclasts, peloids, and rare ooids (Figs. 19 and 20). Grains are mostly medium to coarse, subrounded to subangular, with calcite-spar cements. Identifiable bioclasts include trilobites, brachiopods, ostracods, and gastropods; complete fossils are uncommon. Intraclasts are micritic and rarely possess any internal structure. The intraclastic/bioclastic grainstone lithofacies is commonly gradational with the bioclastic wackestone lithofacies.

The bioclastic wackestone lithofacies is similar in character but contains fewer intraclasts. The bioclasts are generally complete, and the grains are completely mud-supported. These two textures are commonly intermixed and reflect subtidal depositional conditions. The local abundance of both fossil debris and intraclasts is due to winnowing of finer-grained muds by wave and current action.

The stylolaminated wackestone/packstone lithofacies is similar to the previously discussed bioclastic packstone lithofacies. Microfacies of bioclastic wackestone, with sparse complete bioclasts of brachiopods and ostracods, are stylolaminated with bioclastic packstone containing incomplete bioclasts, and have a calcite-spar cement (Figs. 19 and 20). The large number of complete bioclasts implies that this lithofacies was deposited under subtidal conditions.

Abundant organic laminae of variable lateral continuity are present throughout this lithofacies. The preservation of these organic laminae conflicts with the evidence for an oxygenated subtidal environment, as indicated by the number of complete bioclasts. The abundance of organic laminae in this lithofacies suggests that the generation of organic acids by diagenetic alteration of organic matter has resulted in dissolution of carbonate along the boundaries of the differentially cemented lithologies, and also concentration of organic matter locally. This interpretation is further supported by the observed intergradational relationship of the laminae and the stylolites.

The cryptolaminated mudstone lithofacies, which is locally dolomitic, is characterized by internal layering of microscopic-scale, graded laminae (Fig. 20d). The basal contacts of the laminae are sharp and commonly overlain by quartz silt that grades upsection into micrite. Disseminated pyrite is present locally. This lithofacies was probably deposited offshore, under subtidal anoxic conditions where both storm-transported micrite and quartz silt were deposited, and the graded laminae preserved, in the absence of biogenic activity.

THE RELATIONSHIP OF FACIES DISTRIBUTION TO STRATIGRAPHY

The inferred depositional environments of the lithofacies and their position relative to sea level (i.e., subtidal versus intertidal versus supratidal), and the vertical stacking of lithofacies, record the progressive changes in sea level through time. The predominantly subtidal environments of the St. Peter Sandstone are overlain by predominantly peritidal environments of the lower Joachim. These are, in turn, overlain by almost entirely subtidal environments of the upper Joachim (Fig. 4). In the uppermost Joachim, the recurring supratidal and intertidal environments mark the relative drop in sea level at the end of Ancell deposition.

The dominance of siliciclastics in the St. Peter implies that the sediment supply to the basin increased substantially during the time when it was being deposited. The lack of a sharp change in depositional environment between the uppermost St. Peter Sandstone and the Joachim Dolomite suggests that the decrease in the supply of siliciclastics is the result of inundation of the source area, rather than a major change in depositional conditions.

Conodont biostratigraphic data from subsurface reference sections along a north-south cross section (Fig. 21; sections 2, 3, 5, 6, and 9) and previously published reports of conodonts from several localities in the Illinois Basin region provide a basis for regional correlation of the constituent formations of the Ancell Group. Although the paucity of conodonts in much of the section, the difficulties in recovering conodonts from the St. Peter Sandstone, and the taxonomic uncertainties concerning common components of recovered faunas limit the biostratigraphic resolution, it does provide a basis for a chronostratigraphic framework.

Based on the lithofacies succession within this chronostratigraphic framework, three major depositional intervals can be identified and related to the present lithostratigraphic classification.

1. Everton Dolomite Interval: Lower to Upper-Middle Whiterockian

The Sauk-Tippecanoe regression was initiated in the North American Midcontinent during latest Canadian to early Whiterockian time. The absence of core and the limited number of deep tests in the central Illinois Basin inhibit the understanding of this interval. However, a few observations can be made.

First, the paleoshoreline apparently regressed into southernmost Illinois, western Kentucky, and southeastern Missouri during the initial sea-level fall in the early Whiterockian. Persistent subsidence resulted in a south- to southwestern-opening marine embayment (present coordinates), somewhat analogous in distribution to the Cretaceous Mississippi Embayment (Fig. 1). Evaporites, siltstones, and finely crystalline dolostones present in the lower Everton suggest sabkha to restricted marine, subtidal conditions. Sandstone is present locally, but shows only rare lateral persistence. The limited number of siliciclastic components suggests that the source areas exposed on the margins were mostly carbonate terrains in which karst was locally developed; this karst is represented in the rock record by the Readstown and Kress members of the St. Peter Sandstone. By the middle Whiterockian a relative sea-level rise resulted in the transgression of

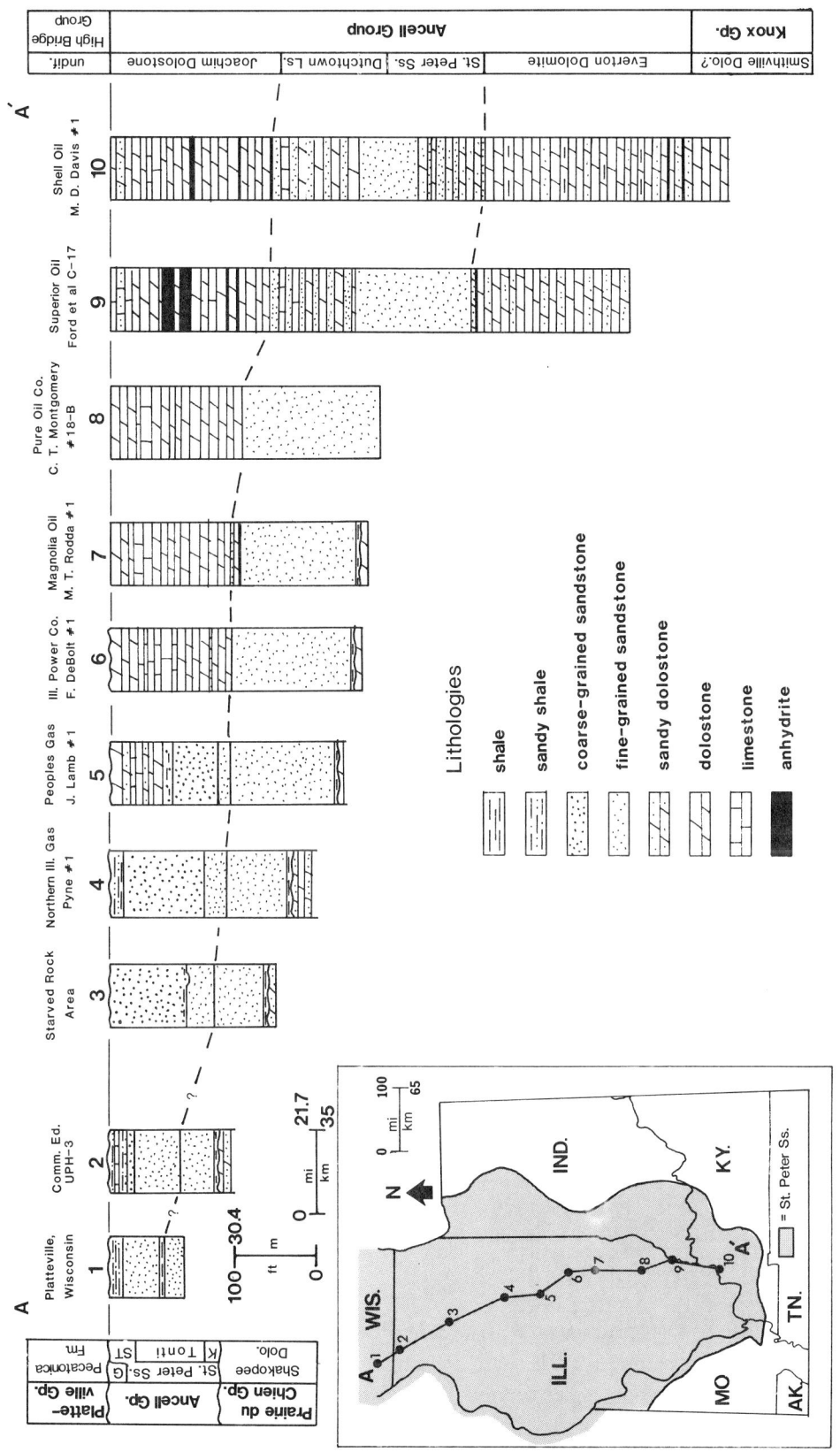

Fig. 21--Cross section from Platteville, Wisconsin, south to Crittenden County, Kentucky, showing lines of correlation for the St. Peter Sandstone-Dutchtown Limestone interval. Datum is top of the Ancell Group (K = Kress Member, G = Glenwood Formation).

the Everton embayment over the post-Sauk erosional surface to approximately its present distribution. Although composed of predominantly peritidal environments, subtidal marine conditions were established locally, as evidenced by fossiliferous carbonates (open-marine conditions) in the upper Everton. The gradual increase in siliciclastic input resulted in the episodic progradation of eolian sands over the carbonate peritidal environments and the local development of subtidal sheet-sands and bars (Suhm, 1976).

2. St. Peter Sandstone-Dutchtown Limestone Interval: Late Whiterockian (Chayzan) to Early Mohawkian

Exposure of Cambrian siliciclastics on the Wisconsin Arch resulted in a massive influx of reworked quartz sand into the proto-Illinois Basin; these sands were redeposited after south-southwestern transport by streams, prevailing winds, and longshore currents. The St. Peter Sandstone was deposited in fluvial, eolian, and subtidal environments in northern Illinois and southern Wisconsin. The influx of sand into the basin, and the rapid transgression of the Tippecanoe Sea over the relatively flat post-Sauk paleo-erosional surface, resulted in the rapid migration of the shoreline cratonward, reworking of eolian shoreline dunes, and transgression of siliciclastic offshore bars over the subtidal and peritidal carbonate environments, as represented by the upper Everton. Locally, disconformable surfaces developed on top of the Everton carbonates, where the tidal-inlet channels eroded the underlying strata (Suhm, 1976). Relief developed in this manner resulted in the local, apparently disconformable relationship of the St. Peter Sandstone with the Everton Dolomite. Siliciclastic sabkha, intertidal, tidal-inlet channel, foreshore, and shallow subtidal environments were developed within the St. Peter Sandstone. Fluvial and sinkhole deposits were locally preserved in paleotopographic lows. Shallow subtidal conditions prevailed to the south, where the carbonates of the Dutchtown Limestone were deposited.

3. Joachim Dolomite-Starved Rock Sandstone-Glenwood Formation Interval: Early Mohawkian

A regional drop in sea level resulted in the progradation of the Joachim shoreline basinward during latest Whiterockian or earliest Mohawkian time and established sabkha to shallow subtidal environments throughout much of the basin. Dolomitic mudstone, bedded anhydrite, and algal-laminated mudstone predominate in this interval. Reestablishment of an eolian environment during the deposition of the upper Tonti Member of the St. Peter Sandstone, which interfingers with the Joachim in central Illinois, implies the progradation of nonmarine eolian environments basinward.

Later, renewed transgression is suggested by the increased abundance of subtidal environments basinward and upsection, as represented by bioclastic packstone and grainstones in the middle and upper Joachim Dolomite. Bimodal, quartz-sand stringers in this interval in central Illinois suggest it may be coeval with the Starved Rock Sandstone Member of the St. Peter, which is a clastic, offshore bar complex that prograded across northern Illinois from the northeast (Fraser, 1976; Nunn, 1986), and is coeval with the subtidal lagoonal and tidal-flat environments of the

Glenwood Formation (Fraser, 1976) to the north. The offshore bar complex may have impounded the finer siliciclastics, as evidenced by the Glenwood Shale, and reduced the supply of siliciclastics basinward; as a result, the Joachim carbonates are relatively pure limestones. Alternatively southwest-trending paleocurrents, which prevailed over much of the midcontinent, may have resulted in longshore transport of the available clastic sediments, thus bypassing the southeastern Illinois depocenter (Dapples, 1955; James and Oaks, 1976; Witzke, 1980; Mai and Dott, 1985). The carbonates of the upper Joachim are similar in character to the Dutchtown Limestone, suggesting that these units may have a facies relationship and/or the carbonates represent a repetition of the Dutchtown environment. Silty dolostones of the upper Joachim Dolomite, with abundant mudcracks and scour surfaces, reflect the shoaling-upward conditions prior to the development of the Ancell-Platteville disconformity.

The Post-Sauk Disconformity

The key to the understanding of sedimentation during Ancell time is the development and nature of the post-Sauk unconformity. The origin of this unconformity may be tectonic, resulting from the migration of a peripheral uplift cratonward in response to loading of the plate margin during the Taconic Orogeny (Jacobi, 1981; Lash, 1989), and/or a worldwide eustatic sea-level fall (Vail and others, 1977; Worsley, 1984). Where the rate of basin subsidence exceeded the rate of relative sea-level fall, the duration of the post-Sauk hiatus decreased or deposition was uninterrupted. Depocenters at which deposition was continuous or nearly continuous during the Sauk-Tippecanoe regression have been identified in the Champlain Valley, New York; eastern Pennsylvania; and the Black Warrior and Michigan basins (Repetski and Harris, *in* Fisher and Barratt, 1985; Repetski and Harris, 1986; Shaw, 1988).

The base of the Ancell Group becomes older basinward from early Mohawkian in the Upper Mississippi Valley (Sweet, 1984, 1987), to late Whiterockian or early Mohawkian in north-central Illinois (Shaw, 1990), to early Whiterockian (lower Middle Ordovician) in southeastern Illinois (Norby and others, 1986). In southernmost Illinois and western Kentucky, an additional 200 to 400 ft (60 to 120 m) of Everton may be present in the subsurface, suggesting that a more complete early Whiterockian section is present in the central portion of the basin, and that deposition may have been uninterrupted during the early Whiterockian (Collinson and others, 1988).

The Ancell Group disconformably overlies progressively younger strata basinward. In southern Wisconsin and northern Illinois, the Ancell overlies the Lower Ordovician carbonates of the Prairie du Chien Group, the Upper Cambrian to Lower Ordovician Eminence and Potosi Dolomites, and the Cambrian siliciclastics of the Potsdam supergroup (Willman and Buschbach, 1975; Mai and Dott, 1985). In northern and eastern Indiana the Ancell overlies the Oneota Dolomite (Droste and Patton, 1985; Ethington and others, 1986). Elsewhere in the Illinois Basin region, the Ancell overlies the Shakopee Dolomite and its lateral equivalents. The age of the uppermost Shakopee decreases basinward from early-late Canadian in northern Illinois (Shaw, 1987) to latest Canadian in southwestern Illinois (Shaw and others, 1988). Well-preserved, Lower Ordovician conodonts recovered from the basal St. Peter Sandstone (Readstown and Kress members) indicate that younger strata of the Shakopee Dolomite were removed by dissolution prior to deposition of the St. Peter Sandstone (Grether and Clark, 1982; Shaw, 1987). Erosion of these underlying strata provided a source for siliciclastic sediments during Ancell deposition (Mai and Dott, 1985).

The decreasing magnitude of the unconformity, as indicated by the decreasing age of the disconformably underlying strata and the increasing age of the basal Ancell Group, implies that the rate of subsidence in the central proto-Illinois Basin exceeded, or closely approximated, the rate of eustatic sea-level fall during the early Whiterockian.

The change in the magnitude of the unconformity basinward suggests that the duration of exposure of the paleoerosional surface decreases basinward; this decrease has two important implications.

First, it can be inferred that the duration of exposure should, in part, control the magnitude of denudation of the post-Sauk erosional surface, and it would be expected that paleotopographic relief should increase from north to south. Paleotopographic relief on the disconformity is greatest on the basin margins and progressively decreases basinward. Relief decreases from a maximum of 1700 ft (518 m), with a local average of 300 to 400 ft (92 to 122 m) in southern Wisconsin (Mai and Dott, 1985), to a maximum of 400 ft (122 m) with an average of 200 ft (61 m) in northern Illinois (Buschbach, 1961, 1964), to an average of less than 100 ft (30.5 m) in central Illinois. In eastern Indiana and northern Kentucky, the relief on the paleotopographic surface may exceed 200 ft (61 m), but averages 50 to 100 ft (15 to 30.5 m)(Schmidt and Warner, 1964; Wolcott and others, 1972; Keller and Abdulkareem, 1980). Up to 500 ft (152 m) of relief is present on the post-Sauk erosional surface in north-central Tennessee (Stearns and Reesman, 1986). Paleotopography on the basin margins was primarily karstic, with southward-developed stream-cut drainage (Buschbach, 1961, 1964; Mai and Dott, 1985; Dott and others, 1986; Keller and Abdulkareem, 1980).

Second, the lithologies exposed on the paleoerosional surface should control the volume and character of the siliciclastic sediments during Ancell deposition. Clearly the volume of siliciclastic sediment derived from erosion of the Prairie du Chien Group carbonates, 15 to 40 percent insoluble content (Stevenson and others, 1975), is less than that derived from the sandstones of the Potsdam supergroup (greater than 90 percent siliciclastics). Based on these two observations, the siliciclastic input should increase as erosion progressively downcut the post-Sauk paleoerosional surface, and the greatest influx of siliciclastics should occur when the Cambrian sandstones were eroded. Conversely it can be expected that, as the Tippecanoe transgression progressed, the volume of siliciclastic sediment within the basin should rapidly decrease as the siliciclastic-source area diminished due to flooding of the paleoerosional surface, and the sediment accommodation space within the basin would increase due to the relative rise in sea level.

Three additional parameters may have influenced the rate of erosion on the post-Sauk erosional surface: (1) relative uplift of the source area, (2) surface gradient, and (3) paleoclimate.

Although regional early-Middle Ordovician uplift of the basin-bounding Wisconsin, Kankakee, and Cincinnati Arches and the Nashville and Ozark Domes has been proposed by Tikrity (1968), Atherton (1971), and Quinlan and Beaumont (1984), there is little evidence of sub-Ancell structure in these areas. Witzke and Kolata (1988), and Wolcott and others (1972), based on isopach mapping of the Platteville and High Bridge groups, respectively, suggested that the uplift of these features did not occur until the late-Middle Ordovician. Watso's (1988) subsidence analysis of a deep well on the flank of the Wisconsin Arch did not recognize any significant uplift of the arch during Ancell deposition. Isostatic modeling suggests that the Nashville Dome was tectonically stable during Ancell deposition (Reesman and Stearns, 1990).

The apparent "basin and arches" configuration developed during Ancell deposition may have been the result of differential regional subsidence in which the tectonically positive margins subsided at a rate lower than the relative sea-level fall, and the "basin" subsided at a greater rate than relative sea-level fall. Consequently the marginal "arches" and "domes" would undergo erosion, and flanking strata would dip gently toward the subsiding depocenter. Peripheral uplift due to flexural loading, as a result of continued subsidence and sedimentation within the depocenter, may have accentuated the relative uplift of the basin margins (Watso and Klein, 1989).

Even with minor uplift of the basin periphery, the topographic gradient on the paleoerosional surface was still very low. The present regional gradient of the post-Sauk surface is 1 to 4 ft/mile (less than 1 m/km) in central Kentucky (Wolcott and others, 1972) and 18 to 25 ft/mile (5 to 8 m/1 km) in southern Wisconsin (Mai and Dott, 1985). If corrected for post-Ancell subsidence, the gradient of the paleoerosional surface would be further reduced. The low regional gradient, and consequently low stream gradient on the paleoerosional surface, suggests that stream downcutting was subordinate to marked karst dissolution in reducing the paleosurface (Desrochers and James, 1987).

The subordinate role of fluvial systems to dissolution as the predominant erosive force on the post-Sauk paleosurface is further supported by the limited paleoclimatological data available. During the Middle Ordovician the proto-Illinois Basin was located within an arid trade-wind belt 10° to 20° south of the equator that was located to the northwest, and trended northeast of the proto-Illinois Basin; trade winds prevailed from the northeast (present coordinates)(Witzke, 1980). The limited development of fluvial lithofacies within the basal Ancell Group, the identification of well-developed eolian lithofacies within the St. Peter Sandstone, and the presence of evaporites within the Everton Dolomite and Joachim Dolomite support this interpretation. The limited lateral and vertical distribution of evaporites in the stratigraphic section, the establishment of a storm-influenced shelf by the late-Middle Ordovician (Bakush and Carozzi, 1986), and evidence for moisture-associated eolian features in the St. Peter Sandstone (Dott and others, 1986) suggest that humid conditions, possibly analogous to a monsoonal climate pattern, may have prevailed seasonally. As a consequence, fluvial transport of sediments was probably limited to a few major, persistent drainage networks fed by intermittent tributaries, and the paleosurface was primarily reduced by karst dissolution and eolian abrasion.

In summary, the post-Sauk erosional surface, which is represented in the stratigraphic record by the Sauk-Tippecanoe unconformity, was a relatively low-gradient, predominantly carbonate terrain in an arid to semi-arid climate. In the absence of well-developed land plants, karst dissolution, eolian abrasion, and locally persistent fluvial networks were the predominant erosive agents; time of exposure was the primary control in the rate of erosion by these processes. The nature of the sediments produced by the denudation of the paleoerosional surface was controlled by the lithic character of the exposed bedrock. These sediments were transported into the proto-Illinois Basin by eolian and fluvial processes. The rate and volume of siliciclastic input into the proto-Illinois depocenter was the primary control of the distribution of siliciclastics within the region.

CONCLUSIONS

Careful analysis of vertical and lateral lithofacies distributions within a chronostratigraphic framework indicates that the Ancell Group can be subdivided into three major depositional cycles. These cycles possess sedimentological characteristics that reflect changes in basinal processes, relative sea-level and/or basin subsidence, and sedimentological processes in the adjacent source areas controlled by the paleoclimatological conditions and bedrock geology of the post-Sauk paleoerosional surface. Basin subsidence and sea-level changes influenced the available accommodation space within the basin and areal extent of the source area; paleoclimate and paleo-bedrock geology of post-Sauk erosional surface controlled the nature and volume of the siliciclastic input into the basin. The interplay of these factors controlled the distribution of siliciclastics within the Ancell Group during early-Middle Ordovician time.

ACKNOWLEDGMENTS

The authors are grateful to J. Morgan, Illinois Power Company, who made the core and geophysical logs available for study; and to J. Crockett and M. Chou, Illinois State Geological Survey, and Dr. M. Helman, Imperial College, who provided the organic-geochemical data. We have greatly profited from discussions with Dr. D. Barnes, Western Michigan University, and Drs. D. R. Kolata, R. D. Norby, and M. L. Sargent of the Illinois State Geological Survey; Drs. D. Barnes, D. Kolata, and M. Sargent also provided critical review of this manuscript. Portions of this study were conducted at the Illinois State Geological Survey and funded, in part, by a research gift from the Atlantic Richfield Corporation.

REFERENCES

ATHERTON, E., 1971, Tectonic development of the Eastern Interior Region of the United States, *in* Bond, D. C., ed., Symposium on Future Petroleum Potential of NPC Region No. 9: Illinois State Geological Survey Petroleum 96, p. 29-43.

BAKUSH, S. H., AND CAROZZI, A. V., 1986, Subtidal storm-influenced carbonate ramp model: Galena Group (Middle Ordovician) along Mississippi River (Iowa, Wisconsin, Illinois, and Missouri), U.S.A.: Archives des Sciences, v. 39, no. 2, p. 10-183.

BRAILE, L. W., HINZE, W. J., KELLER, G. R., LIDIAK, E. G., AND SEXTON, J. L., 1986, Tectonic development of the New Madrid Rift Complex, Mississippian Embayment, North America: Tectonophysics, v. 128, p. 1-21.

BUSCHBACH, T. C., 1961, Morphology of the sub-St. Peter surface of northeastern Illinois: Illinois Academy of Science, v. 54, p. 83-89.

_____, 1964, Cambrian and Ordovician strata of northeastern Illinois: Illinois State Geological Survey Report of Investigations 218, 90 p.

COLLINSON, C., SARGENT, M. L., AND JENNINGS, J. R., 1988, Illinois Basin region, in Sloss, L. L., ed., Sedimentary Cover - North America: Geological Society of America, The Geology of North America, v. D-2, p. 383-426.

DAPPLES, E. C., 1955, General lithofacies relationship of the St. Peter Sandstone and Simpsom Group: American Association of Petroleum Geologists Bulletin, v. 39, p. 444-467.

DESROCHERS, A., AND JAMES, N. P., 1987, Early Paleozoic surface and subsurface paleokarst: Middle Ordovician carbonates, Mingan Islands, Quebec, in James, N. P., and Choquette, P. W., eds., Paleokarst: Springer-Verlag, New York, p. 183-210.

DOTT, R. H., JR., BYERS, C. W., FIELDER, G. W., STENZEL, S. R., AND WINFREE, K. E., 1986, Aeolian to marine transition in Cambro-Ordovician cratonic sheet sandstones of the northern Mississippi Valley, U.S.A.: Sedimentology, v. 33, p. 345-367.

DROSER, M. L., AND BOTTJER, D. J., 1990, Ichnofabric of sandstone deposited in high-energy nearshore environments: measurement and utilization: Palaios, v. 4, p. 598-604.

DROSTE, J. B., ABDULKAREEM, T. F., AND PATTON, J. B., 1982, Stratigraphy of the Ancell and Black River Groups (Ordovician) in Indiana: Indiana Geological Survey Occasional Paper 36, 15 p.

_____, AND PATTON, J. B., 1985, Lithostratigraphy of the Sauk Sequence in Indiana: Indiana Geological Survey Occasional Paper 47, 24 p.

_____, AND SHAVER, R. H., 1983, Atlas of Early and Middle Paleozoic paleogeography of the southern Great Lakes region: Indiana Geological Survey Special Report 32, 32 p.

ETHINGTON, R. L., DROSTE, J. B., AND REXROAD, C. B., 1986, Conodonts from subsurface Champlainian (Ordovician) rocks of eastern Indiana: Indiana Geological Survey Special Paper 37, 32 p.

FISHER, J. H., AND BARRATT, M. W., 1985, Exploration in Ordovician of central Michigan Basin: American Association of Petroleum Geologists Bulletin, v. 12, p. 2065-2076.

FRASER, G. S., 1976, Sedimentology of a Middle Ordovician quartz arenite-carbonate transition in the Upper Mississippi Valley: Geological Society of American Bulletin, v. 86, p. 833-845.

FREY, R. W., PEMBERTON, G., AND SAUNDERS, T. D., 1990, Ichnofacies and bathymetry: a passive relation: Journal of Paleontology, v. 64, p. 155-158.

GENDRON, C. R., CAHILL, R. A., AND GILKENSON, R. H., 1988, Comparison of spectral gamma-ray (SGR) well logging data with instrumental neutron activation analysis (INAA) data for rock types in northern Illinois: The Log Analyst, v. 29, p. 345-357.

GRETHER, W. J., AND CLARK, D. L., 1980, Conodonts and stratigraphic relationships of the Readstown Member of the St. Peter Sandstone in Wisconsin: Geoscience Wisconsin, v. 4, p. 1-29.

HAY, R. L., LEE, M., KOLATA, D. R., MATTHEWS, J. C., AND MORTON, J. P., 1988, Episodic potassic diagenesis of Ordovician tuffs in the Mississippi Valley area: Geology, v. 16, p. 743-747.

JACOBI, R. D., 1981, Peripheral bulge - a causal mechanism for the Lower/Middle Ordovician unconformity along the western margin of the Northern Appalachians: Earth and Planetary Sciences, v. 56, p. 245-251.

JAMES, W. C., AND OAKS, R. Q., JR., 1977, Petrology of the Kinnikinic quartzite (Middle Ordovician), east-central Idaho: Journal of Sedimentary Petrology, v. 47, p. 1491-1511.

KELLER, S. J., AND ABDULKAREEM, T. F., 1980, Post-Knox unconformity - significance at Unionport gas-storage project and relationship to petroleum exploration in Indiana: Indiana Geological Survey Occasional Paper 31, 19 p.

KIRCHNER, J. G., 1989, Origin of pyrite nodules in the St. Peter Sandstone, Buffalo Rock State Park, Illinois: Illinois Academy of Sciences, v. 82, p. 47-56.

KOLATA, D. R., TREGWORGY, J. D., SARGENT, M. L., AND NELSON, W. J., 1990, Subsidence history of the Illinois Basin: Geological Society of America, Abstracts with Programs, v. 22, no. 5, p. 37.

LASH, G., 1989, Middle and Late Ordovician shelf activation and foredeep evolution, central Appalachian Orogen, *in* Keith, B., ed., The Trenton Group (Upper Ordovician Series) of Eastern North America: Deposition, Diagenesis, and Petroleum Occurrence: American Association of Petroleum Geologists, Studies in Geology no. 29, p. 37-53.

MAI, H., AND DOTT, R. H., JR, 1985, A subsurface study of the St. Peter Sandstone in southern and eastern Wisconsin: Wisconsin Geological and Natural History Survey, Information Circular 47, 26 p.

NELSON, W. J., AND LUMM, D. K., 1987, Structural geology of southeastern Illinois and vicinity: Illinois State Geological Circular 538, 70 p.

NORBY, R. D., SARGENT, M. L., AND ETHINGTON, R. L., 1986, Conodonts from the Everton Dolomite (Middle Ordovician) in southern Illinois: Geological Society of America, Abstracts with Programs, v. 18, no. 3, p. 258.

NUNN, J. R., 1986, The petrology and paleogeography of the Starved Rock Sandstone in southeastern Iowa, western Illinois, and northeastern Missouri: M.S. Thesis, University of Iowa, Iowa City, 144 p.

OKHARAVI, R., AND CAROZZI, A. V., 1983, The Joachim Dolomite: a Middle Ordovician sabkha of southeastern Missouri and southern Illinois: Archives des Sciences, v. 36, p. 373-424.

PETERS, K. E., 1986, Guidelines for evaluating petroleum source rock using programmed pyrolysis: American Association of Petroleum Geologists Bulletin, v. 70, p. 318-329.

PYE, K., DICKSON, A. D., SCHIAVON, N., COLEMAN, M. L., AND COX, M., 1990, Formation of siderite-Mg-calcite-iron sulphide concretions in intertidal marsh and sandflat sediments, north Norfolk, England: Sedimentology, v. 37, p. 325-343.

QUINLAN, G. M., AND BEAUMONT, C., 1984, Appalachian thrusting, lithospheric flexure, and the Paleozoic stratigraphy of the eastern interior of North America: Canadian Journal of Earth Sciences, v. 21, p. 973-996.

REESMAN, A. L., AND STEARNS, R. G., 1990, The Nashville Dome - an isostatically induced erosional structure - and the Cumberland Plateau - an isostatically suppressed late Paleozoic extension of the Jassamine Dome: Southeastern Geology, v. 20, p. 147-173.

REPETSKI, J. R., AND HARRIS, A. G., 1986, Conodont biostratigraphy and biofacies of upper Knox/Beekmantown Groups and overlying Ordovician rocks in Appalachian Basin, New York to Alabama: American Association of Petroleum Geologists Bulletin, v. 70, no. 5, p. 637.

SCHMIDT, R. G., AND WARNER, R. A., 1964, Knox as the target for northern Kentucky: Oil and Gas Journal, v. 62, no. 16, p. 178-180.

SHAW, T. H., 1987, Lower Ordovician conodonts from the subsurface of north-central Illinois: Geological Society of America, Abstracts with Programs, v. 19, no. 4, p. 244.

SHAW, T. H., 1988, The conodont biostratigraphy of the Prairie du Chien Group in its type area: implications for petroleum exploration in the Illinois and Michigan Basins, *in* Barnes, D. A., and Harrison, W. B., III, eds., Symposium on the Lower Paleozoic of the Michigan Basin: Abstracts, Western Michigan University, Kalamazoo, Michigan, p. 15.

_____, 1990, A conodont fauna from the base of the St. Peter Sandstone (Ordovician), north-central Illinois: Geological Society of America, Abstracts with Programs, v. 22, no. 5, p. 44.

_____, NORBY, R. D., AND SARGENT, M. L., 1988, Lower Ordovician conodonts from the Shakopee Dolomite (Upper Prairie du Chien Group) in southwestern Illinois: Geological Society of America, Abstracts with Programs, v. 20, no. 5, p. 388.

SLOSS, L. L., 1963, Sequences in the cratonic interior of North America: Geological Society of America Bulletin, v. 74, p. 93-114.

STEARNS, R. G., AND REESMAN, A. L., 1986, Cambrian to Holocene structural and burial history of Nashville Dome: American Association of Petroleum Geologists Bulletin, v. 70, p. 143-154.

STEVENSON, D. L., CHAMBERLIN, T. L., AND BUSCHBACH, T. C., 1975, Insoluble residues of the Sauk Sequence (Cambrian and Lower Ordovician) rocks of the Fairfield Basin, Illinois: an aid in correlation and petroleum exploration: Illinois State Geological Survey Petroleum 106, 12 p.

STITH, D. A., 1979, Chemical composition, stratigraphy, and depositional environments of the Black River Group (Middle Ordovician), southwestern Ohio: Ohio Division of Geological Survey, Report of Investigations 113, 36 p.

SUHM, R. W., 1974, Stratigraphy of Everton Formation (Early Medial Ordovician), northern Arkansas: American Association of Petroleum Geologists Bulletin, v. 58, p. 685-707.

SWEET, W. C., 1984. Graphic correlation of upper Middle and Upper Ordovician rocks, North American Midcontinent Province, *in* Bruton, D. L., ed., Aspects of the Ordovician System: Palaeontological Contributions from the University of Oslo, no. 295, p. 23-35.

_____, 1987, Distribution and significance of conodonts in Middle and Upper Ordovician strata of the upper Mississippi Valley region, *in* Sloan, R. E., ed., Middle and Late Ordovician Lithostratigraphy and Biostratigraphy of the Upper Mississippi Valley: Minnesota Geological Survey, Report of Investigations 35, p. 167-172.

TEMPLETON, J. S., AND WILLMAN, H. B., 1963, Champlainian Series (Middle Ordovician) in Illinois: Illinois Geological Survey Bulletin 89, 260 p.

TIKRITY, S. S., 1968, Tectonic genesis of the Ozark Uplift: Ph.D. dissertation, Washington University, St. Louis, Missouri, 196 p.

VAIL, P. R., MITCHUM, R. M., JR., AND THOMPSON, S., III, 1977, Seismic stratigraphy and global changes in sea level, part 4: global cycles of relative sea level, *in* Seismic Stratigraphy Applications to Hydrocarbon Exploration: American Association of Petroleum Geologist Memoir 26, p. 83-97.

VAN WAGONER, J. C., MITCHUM, R. M., JR., CAMPION, K. M., AND RAHMANIAN, V. D., 1990, Siliciclastic sequence stratigraphy in well logs, cores, and outcrops: American Association of Petroleum Geologists, Methods in Exploration Series no. 7, 55 p.

WATSO, D. C., 1988, The effect of tectonic subsidence on sedimentation processes during deposition of four Late Cambrian formations, upper middle west: M.S. Thesis, University of Illinois at Urbana-Champaign, Urbana, Illinois, 260 p.

_____, AND KLEIN, G. deV., 1989, Origin of the Cambro-Ordovician sedimentary cycles of Wisconsin using tectonic subsidence analysis: Geology, v. 17, p. 879-881.

WILLMAN, H. B., AND BUSCHBACH, T. C., 1975, The Ordovician System, *in* Willman, H. B. and others, eds., Handbook of Illinois Stratigraphy: Illinois State Geological Survey Bulletin 95, p. 47-86.

_____, AND KOLATA, D. R., 1978, The Platteville and Galena groups in northern Illinois: Illinois State Geological Survey Circular 502, 75 p.

WITZKE, B. J., 1980, Middle and Upper Ordovician paleogeography of the region bordering the Transcontinental Arch, *in* Fouch, T. D., and Magathan, E. R., eds., Paleozoic Paleogeography of the Western United States: Proceedings of the Rocky Mountain Section Meeting of the Society of Economic Paleontologists and Mineralogists, Denver, Colorado, p. 1-18.

_____, AND KOLATA, D. R., 1988, Changing structural and depositional patterns, Ordovician Champlainian and Cincinnatian Series of Iowa and Illinois, *in* Ludvigson, G. A., and Bunker, B. J., eds., New Perspectives on the Paleozoic History of the Upper Mississippi Valley, An Examination of the Plum River Fault Zone: Guidebook for the 18th Annual Field Conference of the Great Lakes Section, Society of Economic Paleontologist and Mineralogists, Guidebook no. 8, p. 55-77.

WOLCOTT, D. E., CRESSMAN, E. R., AND CONNOR, J. J., 1972, Trend surface analysis of the thickness of the High Bridge Group (Middle Ordovician) of central Kentucky and its bearing on the nature of the post-Knox unconformity: U.S. Geological Survey Professional Paper 800-B, p. B25-B33.

WORSLEY, T. R., NANCE, R. D., AND MOODY, J. B., 1984, Tectonic cycles and the history of the earth's biochemical and paleooceanographic record: Paleoceanography, v. 1, p. 233-263.

SEDIMENTARY HISTORY OF THE MOYVOUGHLY AREA, COUNTY WESTMEATH, IRELAND: EVIDENCE FOR SYNSEDIMENTARY FAULT MOVEMENTS IN A MIXED CARBONATE-SILICICLASTIC SYSTEM OF COURCEYAN AGE

GILL M. HARWOOD
School of Environmental Sciences
University of East Anglia, Norwich NR4 7TJ, U.K.
MICHAEL SULLIVAN
c/o Ennex International plc
Piggott Street, Loughrea, County Galway, Ireland

ABSTRACT

The sediments of the Moyvoughly Beds within the Moyvoughly-Moate area of the Irish Midlands comprise ooid grainstones, skeletal ooid grainstones, and marine sandstones, punctuated by rarer calcareous mudstones. These sediments were deposited in agitated environments on a gently sloping south- or southeast-facing ramp. Core investigations and construction of isopach maps have shown changes of thicknesses within individual sedimentary units across known faults. These are interpreted as the result of synsedimentary faulting during sediment deposition. Changes in sediment lithologies can be related both to relative changes in sea level plus faulting activity, with the sandstones being sourced from erosion of local, active, fault-bounded highs. Relative sea-level highs are characterized by calcareous mudstones, with periods of lower sea level being characterized by grainstones or sandstones dependent on local siliciclastic sourcing. No major sequence breaks have been recognized with the Moyvoughly Beds. However, a regional change in sedimentation occurs at the base of the Moyvoughly Beds, where corals colonized the lithified surface of the underlying Micrite Unit. These lower sediments were deposited in a shallow lagoon, with periodic marsh progradation and subsequent exposure with paleosol formation. The Micrite Unit can be traced across the Irish Midlands and, although no regional detailed studies have yet been carried out, appears to be of similar lithologies throughout. In contrast, units within the Moyvoughly Beds, although correlatable within the area studied, cannot be traced to other areas, and their characteristics are more controlled by local tectonic movements.

INTRODUCTION

Within the Irish Midlands many shallow (less than 660 ft (200 m)) cores have been drilled in Mississippian carbonates in the search for base metal sulphide deposits. These cores are commonly closely spaced around potential prospects and give a great deal of information about the otherwise poorly exposed sediments. Major lead-zinc ore deposits within the Mississippian (Lower Carboniferous) of Ireland occur within two major lithotypes: (a) Waulsortian mounds or their off-mound equivalents, as host the deposits at Tynagh (Boast and others, 1979; Clifford and others, 1986; Akhurst, 1989), Ballinalack (Jones and Brand, 1986) and at Silvermines

(Samson and Russell, 1983; Andrew, 1986b); and (b) shallow-water, high-energy carbonate-siliciclastic facies of the Navan Group, as at Navan (Andrew and Ashton, 1985; Ashton and others, 1986), Tatestown (Andrew and Poustie, 1986) and the Moyvoughly-Moate area (Poustie and Kucha, 1986; Kucha, 1988). This study concentrates on cores taken by Ennex International plc near the village of Moyvoughly, about 3.7 miles (6 km) to the northeast of the town of Moate, County Westmeath, situated in the center of the Irish Midlands (Fig. 1a and b). The sequence studied, informally known as the Moyvoughly Beds, is of Courceyan age and forms part of the Navan Group (Fig. 2)(Andrew, 1986a). It is the lateral equivalent of the formation that hosts the ore at Navan, the Pale Beds. Shallow cores used in this study were drilled by Ennex International plc while prospecting for base metal sulphides within the area. The shows of mineralized core drilled in the initial target area defined a deposit of about 125,000 tonnes at a grade of 8 weight percent combined lead and zinc (Poustie and Kucha, 1986; Kucha, 1988). Despite further drilling during 1984 and 1985 to the southeast of the original prospect area, the area proved subeconomic and prospecting terminated.

Within the Moyvoughly-Moate area, the Pale Beds of the Navan Group (Philcox, 1984) can be subdivided into three units. The Moyvoughly Beds, an informal name used by Ennex International plc (Andrew, 1986a), comprise an alternating sequence of ooid grainstones and sandstones with minor siltstones, mudrocks, wackestones, and mudstones (Fig. 2) that average some 131 to 148 ft (40 to 45 m) in thickness. These units can be traced throughout the area but cannot be correlated with cores some tens of kilometers distant. They lie above a thick (75 to 108 ft (23 to 33 m)) series of algal wackestones and packstones, the Micrite Unit (Philcox, 1984), and below a 82 to 98 ft (25 to 30 m) series of calcareous sandstones with skeletal debris and minor siltstones, the Middle Sandy Unit (informal name from Ennex International plc)(Andrew, 1986a)(Fig. 2). All these formations lie some 660 ft (200 m) lower in the succession than the Waulsortian mounds.

On a regional scale these sediments were deposited on the open marine shelf, which gradually deepened southward into the South Munster Basin (Fig. 3)(Phillips and Sevastopulo, 1986). North of Moyvoughly the proportions of sandstones and mudrocks within the succession increase, although carbonates dominate until some 18.6 miles (30 km) further to the northwest. There conditions during the Courceyan were restricted, with the development of evaporites interbedded with sandstones and dolostones (Phillips and Sevastopulo, 1986)(Fig. 3). Philcox (1984) places Moate near the junction of two major provinces within the Irish Midlands: the Limerick Province to the south, where sandstones formed only a minor part of the succession throughout much of the Mississippian; and the North Midlands Province further north, where siliciclastics become more dominant. At the junction between these two provinces, detailed lateral correlation of the Pale Beds and the equivalent Middle Ballysteen Limestone to the south is unsure (Philcox, 1984). It is possible that many lateral facies transitions throughout the Midlands were the result of synsedimentary fault movement, with extensional tectonics controlling sedimentation patterns. Cores taken through the carbonate-siliciclastic facies within the Moyvoughly area were used to study one part of this transition between the Limerick and North Midland provinces. As mineralization within the Moyvoughly Beds is concentrated along faults, some meters of sedimentary units are missing in many of these cores, although this is not the case with the core studied in the most detail (core 1117-15). Cores are also commonly drilled at an angle to the vertical; core 1117-15 was drilled at 60°.

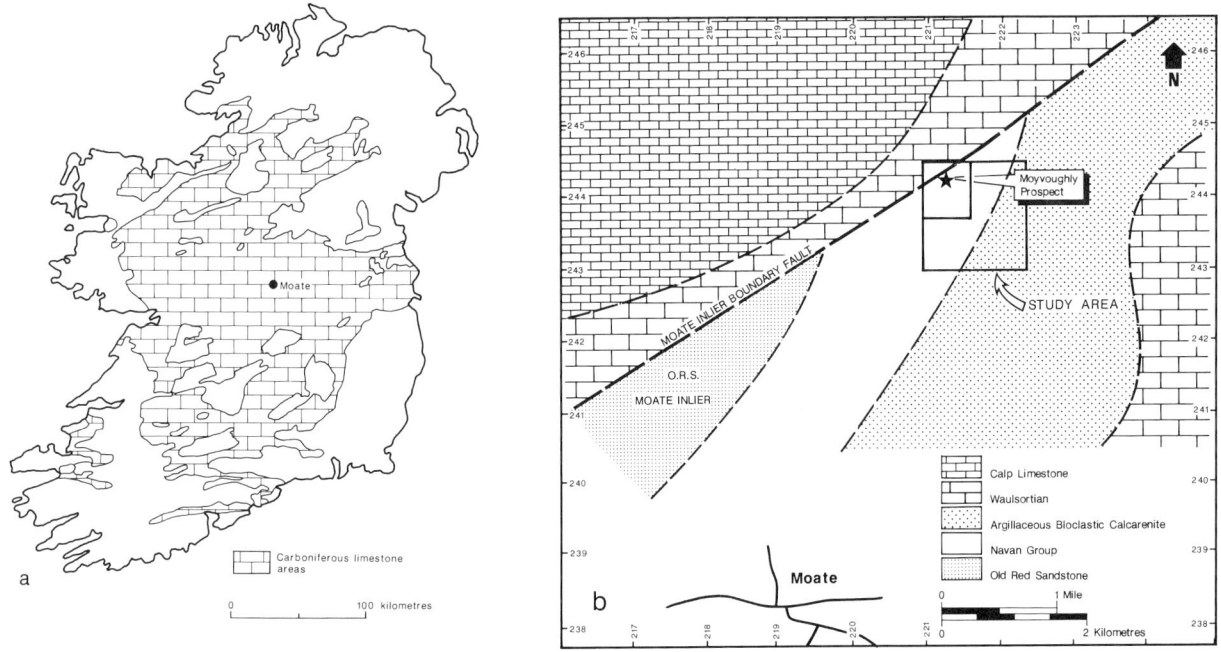

FIG. 1.--Location maps of the area studied; a.--generalized geologic map of Ireland, showing the distribution of Mississippian sediments (Carboniferous Limestone) and the location of Moate; b.--location of the Moyvoughly Prosect, Co. Westmeath, with the major geological formations (sub-glacial deposits outcrop) (modified from Poustie and Kucha, 1986).

Techniques Employed in This Study

Within the area three cores (1117-15, 1117-21, and 1117-32)(Figs. 4 and 5) were logged in detail and sampled. The resultant sedimentological logs were compared with those previously compiled by Ennex International plc, and these were used with a further 42 logs (Fig. 6) to compile isopach maps for the subunits within the Moyvoughly Beds. Samples from the three cores were slabbed and stained peels prepared; thick and polished thin sections were made from samples from core 1117-15. Thin sections were used for petrographic (transmitted light, blue light, reflected light) and cathodoluminescence studies.

LITHOFACIES RELATIONSHIPS WITHIN THE MOYVOUGHLY AREA

Twelve subunits within the Moyvoughly Beds were found to be traceable over much of the area studied (Table 1). Erosion, prior to deposition of the glacial deposits, resulted in insufficient information to define thickness variations within the three uppermost subunits of ooid skeletal grainstones and skeletal grainstones/packstones, and upper mudrocks. These subunits were absent, due to faulting or erosion, in the cores studied; research therefore concentrated on the lower subunits and the upper parts of the Micrite Unit. Generalized lithological logs of the three

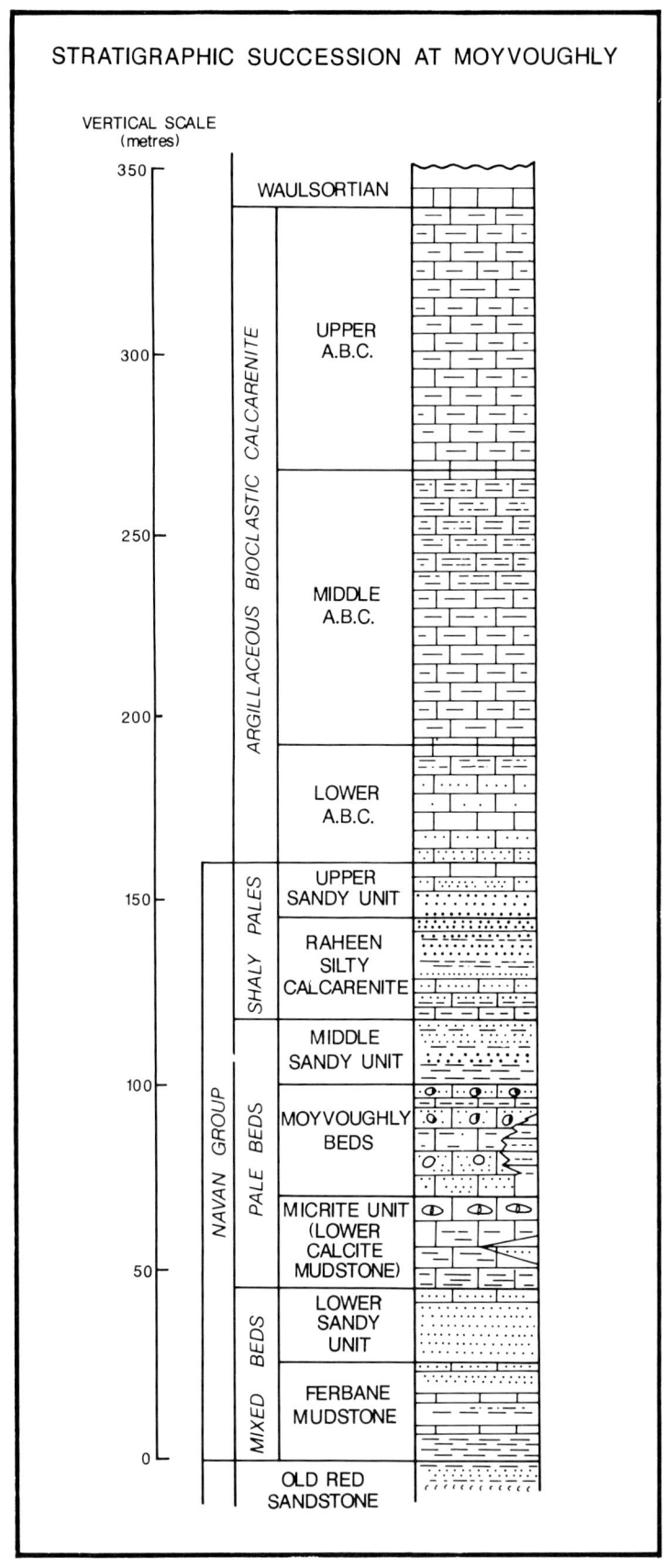

FIG. 2.--Courceyan stratigraphy of the Moyvoughly area (modified from Andrew, 1986a)

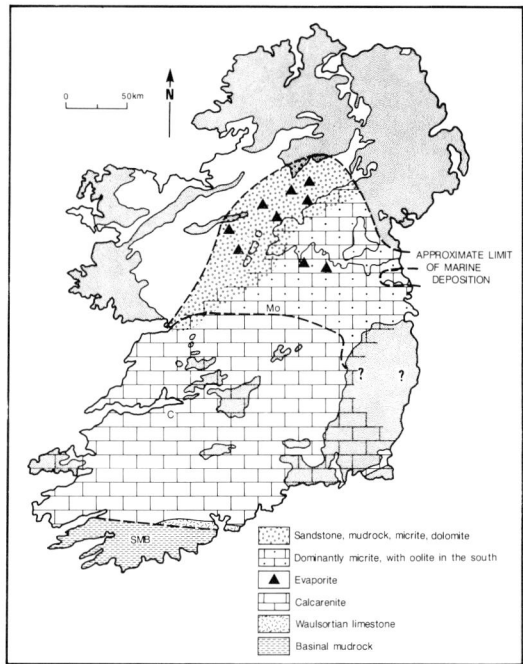

FIG. 3.--Generalized paleogeography of Ireland during the deposition of the Pale Beds of the Navan Group and the Ballysteen Limestone (southern equivalent) (modified from Sevastopulo and Phillips, 1986)

cores (Figs. 4 and 5) demonstrate the lateral continuity of these subunits; correlation between these cores is shown in Figure 5. The composition, sedimentological structures, and lateral variations within these subunits are presented below.

Upper Ooid Grainstone

The upper ooid grainstone is not present in core 1117-15. Elsewhere it comprises a medium gray, quartz sand-rich, massive, cross-stratified grainstone containing coarse ooids (greater than 500 µm) with minor skeletal debris (commonly echinoid fragments with thick-walled brachiopod valves being more common in the center of the subunit). Coarse skeletal beds form the base of cross sets. The quartz sand content is variable, commonly between 5 to 10 percent. Quartz grains have authigenic overgrowths, with resultant euhedral/subhedral forms. Fresh feldspar grains also occur. In both logged cores (1117-21 and 1117-32, Fig. 5) where this subunit is present, the upper ooid grainstone is faulted against the overlying unit, the Middle Sandy Unit, and fractures therein are filled with calcite. Within the area few of the cores contain the full thickness of the upper ooid grainstone; it commonly has a faulted margin or has been eroded. What core control there is demonstrates a thickening of the subunit toward the southeast, from less than 6.6 ft (2 m) in the northwest to over 33 ft (10 m) in the southeast (Fig. 7). Tight control around borehole 1117-21 demonstrates a thickening of the grainstone across the present fault, perhaps indicative of synsedimentary fault movement in this area. The data set is not, unfortunately, complete enough for full justification of this, nor for detailed investigation of thickness changes elsewhere in the area.

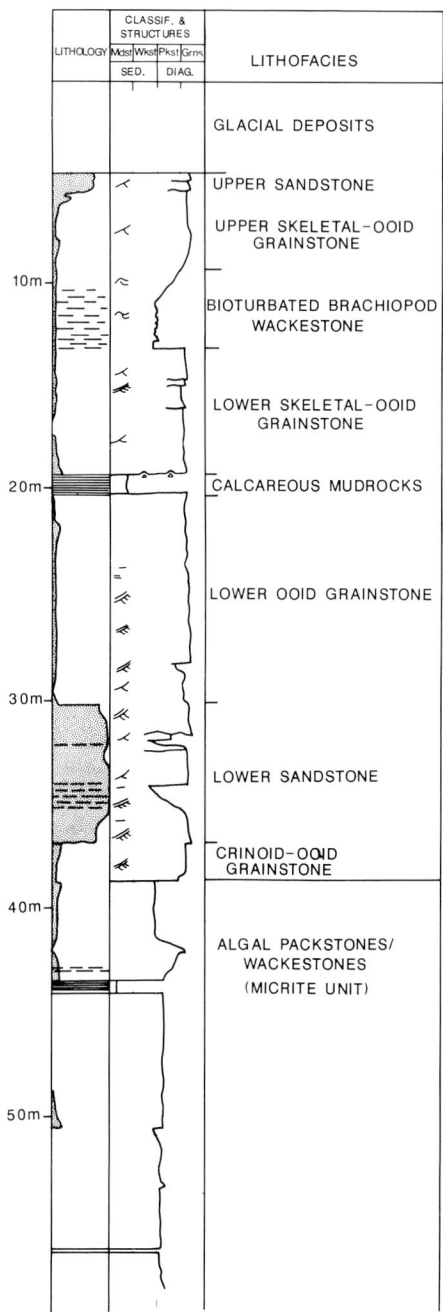

FIG. 4.--Sedimentological log through the Moyvoughly Beds and upper Micrite Unit, core 1117-15. Core log for drilling angle initially 60°.

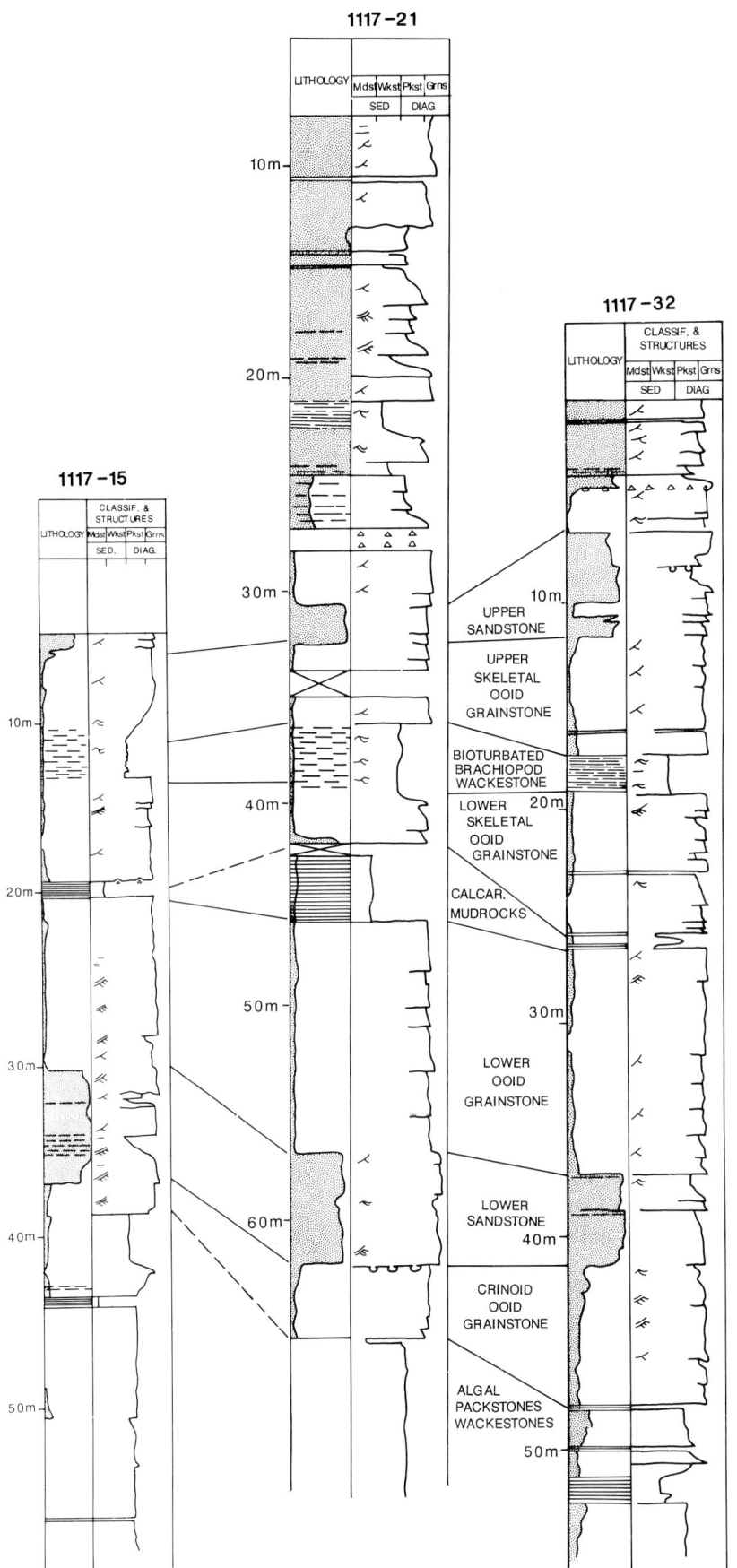

FIG. 5.--Correlation of subunits within the Moyvoughly Beds between cores 1117-15, 1117-21 and 1117-32. Core 1117-15 has been rescaled to show true bed thickness.

FIG. 6.--Structural trends and borehole coverage within the Moyvoughly area. The major inlier-bounding fault is to the northwest of the area; minor faults with variable throws and hades cross the area on a conjugate NW-SE trend. Location of logged cores are indicated. Map compiled from data supplied by Ennex International plc.

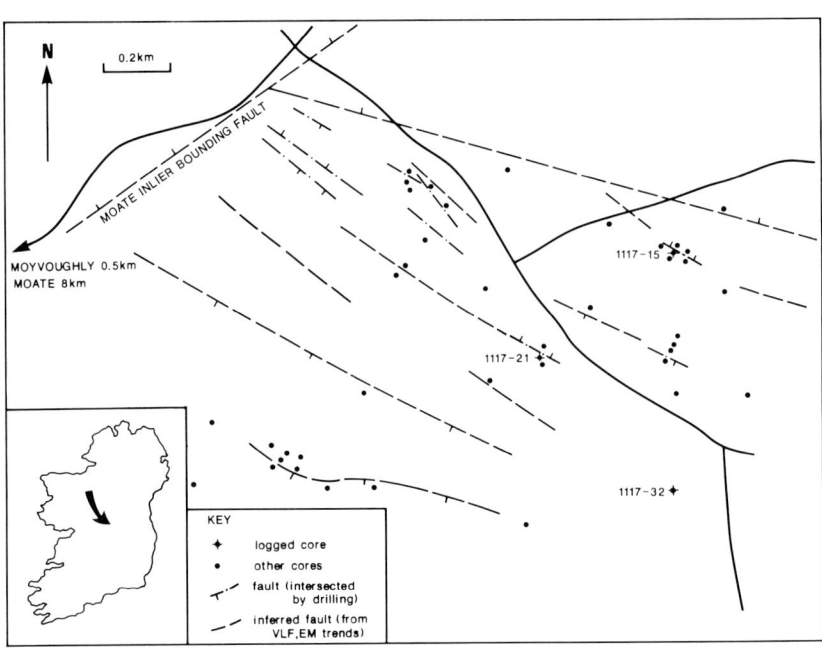

TABLE 1.--Lithologic subunits within the Moyvoughly Beds and uppermost Micrite Unit in the Moyvoughly - Moate area

Subunit	Average thickness within area (m)
ooid-skeletal grainstone	3
skeletal packstone/grainstone	1.7
upper mudrock	2
upper ooid grainstone	2
upper sandstone	5
upper skeletal-ooid grainstone	2
biotubated brachiopod wackestone	3.2
lower skeletal-ooid grainstone	4.8
calcareous mudrock	4.8
lower ooid grainstone	9.8
lower sandstone	7.9
crinoid-ooid grainstone	3.2
algal packstones/wackestones (Micrite Unit)	34.7

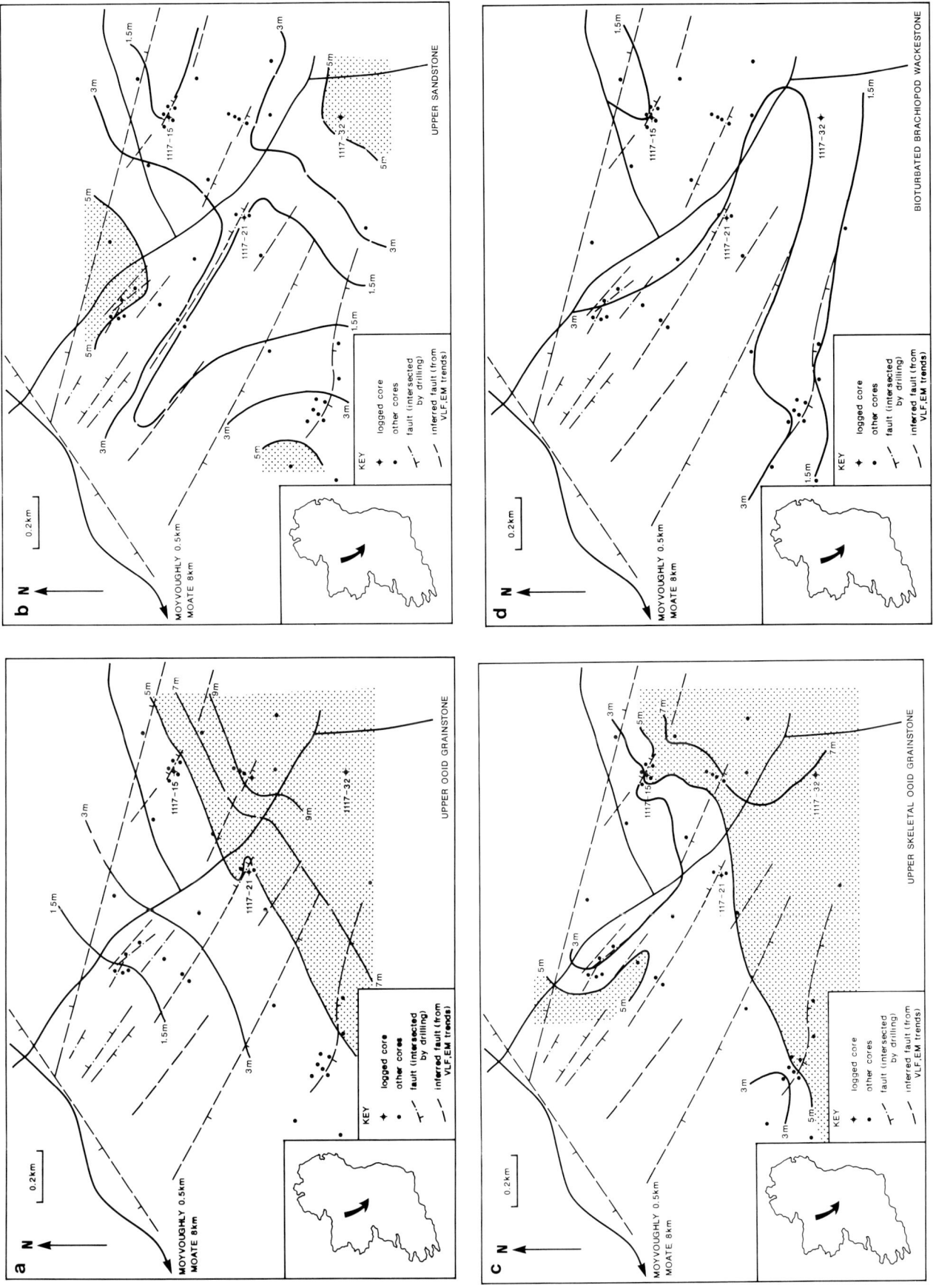

FIG. 7.--Isopach maps for the four upper subunits in the Moyvoughly area; a.--upper ooid grainstone; b.--upper sandstone; c.--upper skeletal ooid grainstone; d.--bioturbated brachiopod wackestone.

Upper Sandstone

The upper sandstone subunit is both calcareous and micaceous, and contains evidence of minor bioturbation and cross- and horizontal stratification (Figs. 4 and 8), with rarer, more clay- and micaceous-rich layers where wispy bedding and bioturbation are present. Carbonate-rich laminae commonly alternate with siliciclastic laminae (Fig. 8a). Within the subunit quartz sand grains average 150 µm in the sand-rich beds but range in size (100 to 250 µm) with, on average, larger grains in mixed carbonate-siliciclastic layers; grains are angular/subangular, although this is in part a diagenetic modification as many have overgrowth cements. Quartz grains are commonly monocrystalline, in places with slight undulose extinction, showing evidence of some strain; rare polycrystalline grains are present. Many grains, more than is readily evident by core examination, are of fresh feldspar, a mixture of plagioclase with lesser amounts of orthoclase and microcline. These have a similar morphology and size to the quartz sand grains and also have overgrowth cements (Fig. 8b and c); the feldspars are readily recognized under cathodoluminescence by a distinctive blue/purple luminescence (Fig. 8c) contrasting with the nonluminescent quartz. Feldspars may comprise almost 50 percent of the siliciclastic grains within this 'sandstone'. Micas (chiefly muscovite) are also fresh.

The most common skeletal components within this subunit are echinoid fragments (Fig. 8), although ooids, peloids, brachiopods, goniatites, bryozoans, foraminifers, and algal fragments also occur. The echinoid fragments in many cases have micritized margins, evidencing algal and/or fungal boring after death and break up of the organism. Some 'fresh', nonmicritized echinoid fragments are present and develop overgrowth cements into available pore spaces (Fig. 8). No early marine cements were recognized on the other skeletal fragments.

The upper sandstone ranges from near 3.3 ft (1 m) to near 16 ft (5 m) in thickness (Fig. 7b). The thicker sandstone in the north and west of the area may represent sources of siliciclastic input, with open marine conditions more prevalent toward the southeast, as suggested by the thickening of the overlying upper ooid grainstone in that direction. The thicker sandstones in the southeast contain interdigitations of skeletal-ooid grainstones (core 1117-32, Fig. 5), indicating active carbonate and siliciclastic sourcing. In parts of the area, isopachs are subparallel to fault trends, again suggesting some synsedimentary movement, although thickening trends are both on the hangingwall (near 1117-21) and footwall (near 1117-15) margins (Fig. 7b). The thickening trend near 1117-21 is consistent with that postulated for the upper ooid grainstone (Fig. 7a).

Upper Skeletal-Ooid Grainstone

The underlying skeletal grainstone contains large (700 to 800 µm) ooids - in places with well-developed radial fabric - coated skeletal grains, and skeletal fragments, of which echinoids are the most common. The proportions of ooids and skeletal fragments vary throughout the subunit, as does the coated grain/superficial ooid content. In places ooids are micritized and their radial structure is barely evident; elsewhere, they are fresh with excellent textural preservation. Similarly echinoid fragments are predominantly fresh, with little or no surface coating or micritization in some areas, but mostly coated and/or micritized in others. From the samples available it does appear that ooid micritization is more common where echinoid fragments have

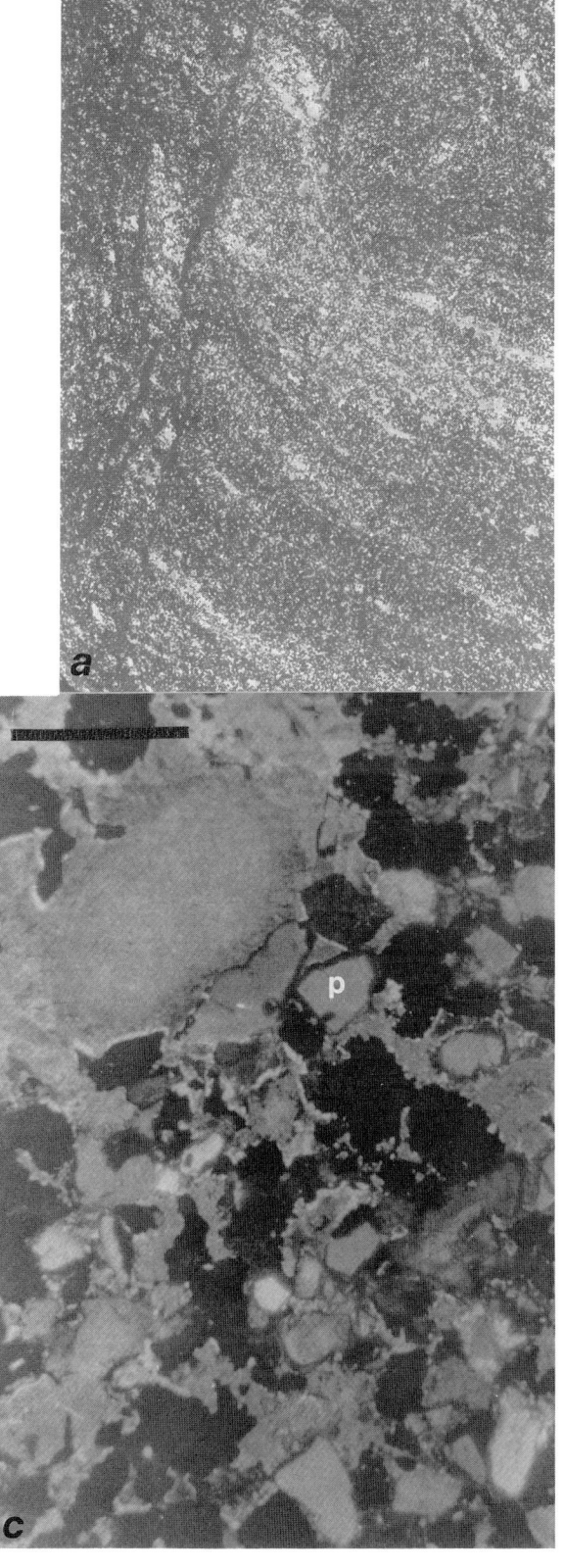

FIG. 8.--Upper sandstone; a.--negative print of stained peel, showing faulted, siliciclastic-carbonate laminae with some bioturbation (scale 2x natural); b.--transmitted light photomicrograph showing large fresh echinoid grain, other carbonate grains plus angular siliciclastic grains; c.--cathodoluminescence view of b. Plagioclase (p) and orthoclase grains have overgrowth cements; quartz, also with overgrowth cements, is non-luminescent. Echinoid grain is surrounded by syntaxial carbonate overgrowth cement (scale bar 500um).

micritized margins, or are coated, but the sampling density was not sufficient to quantitatively demonstrate such a trend within this grainstone. Within the grainstone a small, but variable, percentage of quartz is present (Figs. 4, 5, and 9), originally probably quartz sand but now obscured by quartz overgrowth cements. Fresh plagioclase, orthoclase, and microcline feldspars with overgrowth cements are as abundant as the quartz grains.

At least one hardground with an irregular bored surface is present near the top of this subunit (Fig. 9b). Ooids with early isopachous, fine columnar circumgranular cements were truncated during hardground formation (Fig. 9b). Ooids, peloids, and coated intraclasts with fine columnar circumgranular cements are present immediately beneath the hardground (Fig. 9c), but, 0.8 in. (2 cm) further below, many ooids have spalled cements and outer margins, indicating compaction against the upper rigid hardground surface, and thus showing that cementation was concentrated, in this instance, at the sediment-water interface. Ooids and skeletal grains that overlie the hardground are, on average, larger with less well-developed circumgranular cements (although this may be a function of replacement during diagenesis).

This grainstone varies in thickness between less than 9.8 ft (3 m) and over 19.7 ft (6 m), thickening toward the southeast (Fig. 7c). Synsedimentary fault control is evident near borehole 1117-15 (consistent with the thickening in the upper sandstone, Fig. 7b) and associated with the cluster of boreholes approximately 0.6 miles (1 km) to the west-northwest. No other faults seem to have controlled sedimentation at this time (Fig. 7c).

Bioturbated Brachiopod Wackestone

The transition to more muddy facies with bioturbation and common wispy bedding is abrupt in some cores (1117-21, 1117-32, Fig. 5) but gradational in others (1117-15, Fig. 4). This wackestone contains isolated solitary corals together with whole brachiopods, plus single, complete, thin-walled brachiopod valves; crinoids are less common and ooids only present, with other fragmented skeletal debris, in storm horizons. The occurrence of abundant brachiopods at the top of this subunit was used as a marker horizon by the mining company.

The bioturbated brachiopod wackestone does not greatly vary in thickness throughout the area, and there were no apparent structural controls on its sedimentation (Fig. 7d). It is thickest to the northwest and central parts of the area, thinning to less than 5.3 ft (1.6 m) to the northeast and southwest (Fig. 7d).

Lower Skeletal-Ooid Grainstone

The junction between the bioturbated brachiopod wackestone and the underlying lower skeletal-ooid grainstone is abrupt (Figs. 4 and 5). The grainstone is cross-stratified and, more rarely, ripple-laminated (Figs. 4 and 5); cross-stratified beds contain coarse (greater than 0.4 in. (1 cm) in places) skeletal debris, presumably storm-derived. Fragments include echinoids, echinoderm spines, goniatites, corals, coated algal flakes, and thick-walled brachiopods. One brachiopod valve was encrusted by both algae (probably blue-green) and a foraminifer before redeposition

FIG. 9.--Upper ooid skeletal grainstone; a.--negative print of stained peel, showing well-developed ooids, with crinoid ossicles and other (rarer) skeletal grains, some of which are coated (scale 2x natural); b.--hardground development, with eroded and bored surface, both ooids and circumgranular cements being truncated (scale bar 500um); c.--ooids with isopachous, inclusion-rich circumgranular columnar cements, 1cm below hardgroud surface (scale bar 250um).

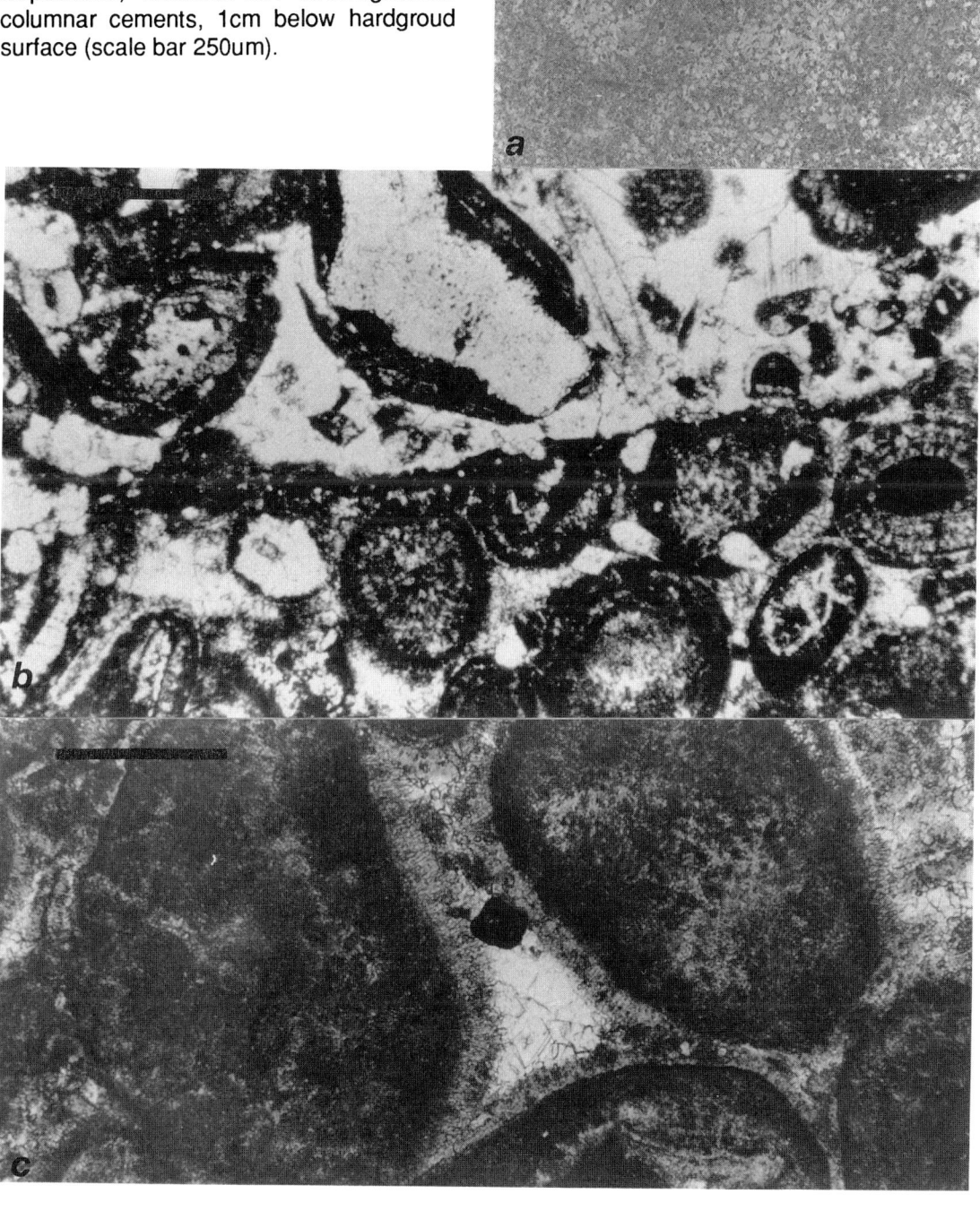

in this grainstone. Large (700 to 800 μm) and smaller (average 500 μm) ooids are present, both exhibiting well-preserved radial fabrics; bimodal sorting of ooids occurs. Skeletal fragments, excepting molluscs, are also well preserved. A small, but variable, percentage of quartz and fresh feldspar is present throughout. In core 1117-15 (Fig. 4) the base of this subunit is brecciated, perhaps a result of preferential cementation and subsequent loading.

The lower skeletal-ooid grainstones change considerably in thickness across the area (Fig. 10a), representing active shoaling at this time. Shoal morphology is variable, with some shoals apparently subparallel to fault trends, whereas others transect known faults. There is no regional thickening trend within this grainstone, although its thinnest occurrence (less than 9.8 ft (3 m)) is to the northeast and extends into the center of the area (Fig. 10a). Shoal thicks within this subunit are located over footwall crests. In places these may be single shoals (as by 1117-15), but elsewhere composite stacked shoals may develop (as west of 1117-15, along fault from 1117-21).

Calcareous Mudrocks

The lower skeletal-ooid grainstone commonly rests abruptly on a series of dark, pyritic, calcareous mudrocks. These sediments contain some fossil skeletons, commonly whole thin-walled brachiopods, although corals and echinoid fragments also occur. Samples of this subunit gave vitrinite reflectance values between 5 and 6 percent (J. M. Jones, personal communication, 1987), demonstrating that the area has been prone to a high heat flow during its diagenetic history.

This subunit thins to the southeast of the area and is not present in some cores. It is thickest (greater than 9.2 ft (2.8 m)) in the northwest of the area. Thicknesses vary considerably across some faults, particularly the fault zone midway between boreholes 1117-15 and 1117-21 (Fig. 10b), a fault where little or no variation has taken place in the overlying subunits. The calcareous mudrocks here thin on the footwall margin of the fault. Possible thickness changes around borehole 1117-15 may also be associated with synsedimentary fault movement. Thickness variations within a short distance in mudrocks are difficult to explain, unless the mudrocks drape some preexisting feature, such as an ooid shoal, or unless synsedimentary tectonism is involved. Here no shoals are evident in the underlying lower ooid grainstone, but there are variations in subunit extent in the overlying lower skeletal-ooid grainstone, which could indicate erosion prior to grainstone deposition.

Lower Ooid Grainstone

The calcareous mudrocks have an abrupt contact with the underlying lower ooid grainstone, the thickest subunit within the Moyvoughly Beds. It contains varied skeletal components and some coated grains (Fig. 11a and b); it is dominated by smaller (400 to 500 μm) ooids with excellently preserved radial fabrics (Fig. 11c), although larger (600 to 800 μm) ooids are present at some levels. Intraclasts of ooid grainstone, usually containing small (less than 500 μm) radial ooids with isopachous, fine columnar circumgranular cements, are also present (Fig. 12a). Cylindrical

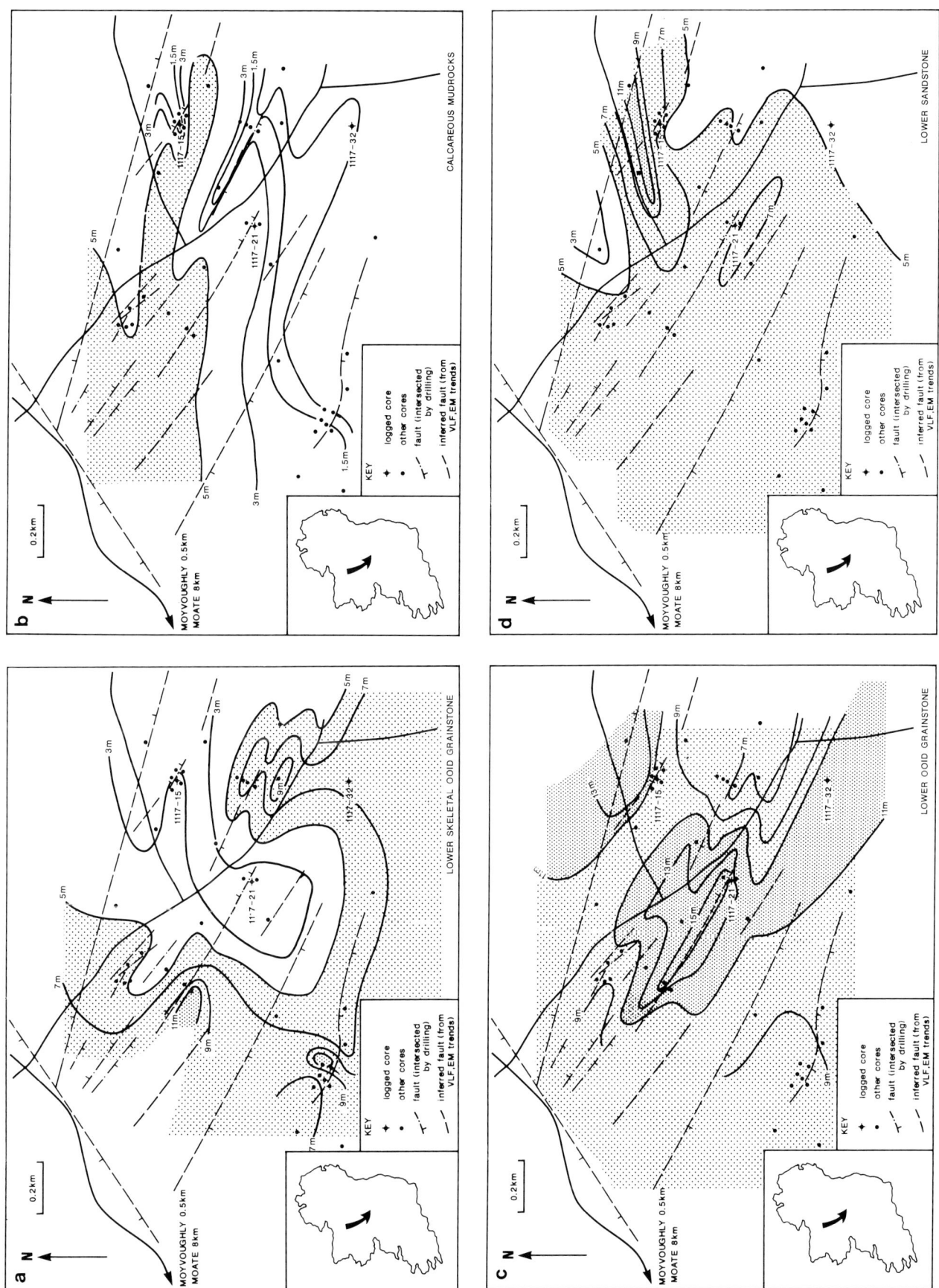

FIG. 10.--Isopach maps of subunits within the Moyvoughly Beds; a.--lower skeletal ooid grainstone; b.--calcreous mudrocks; c.--lower ooid grainstone; d.--lower sandstone.

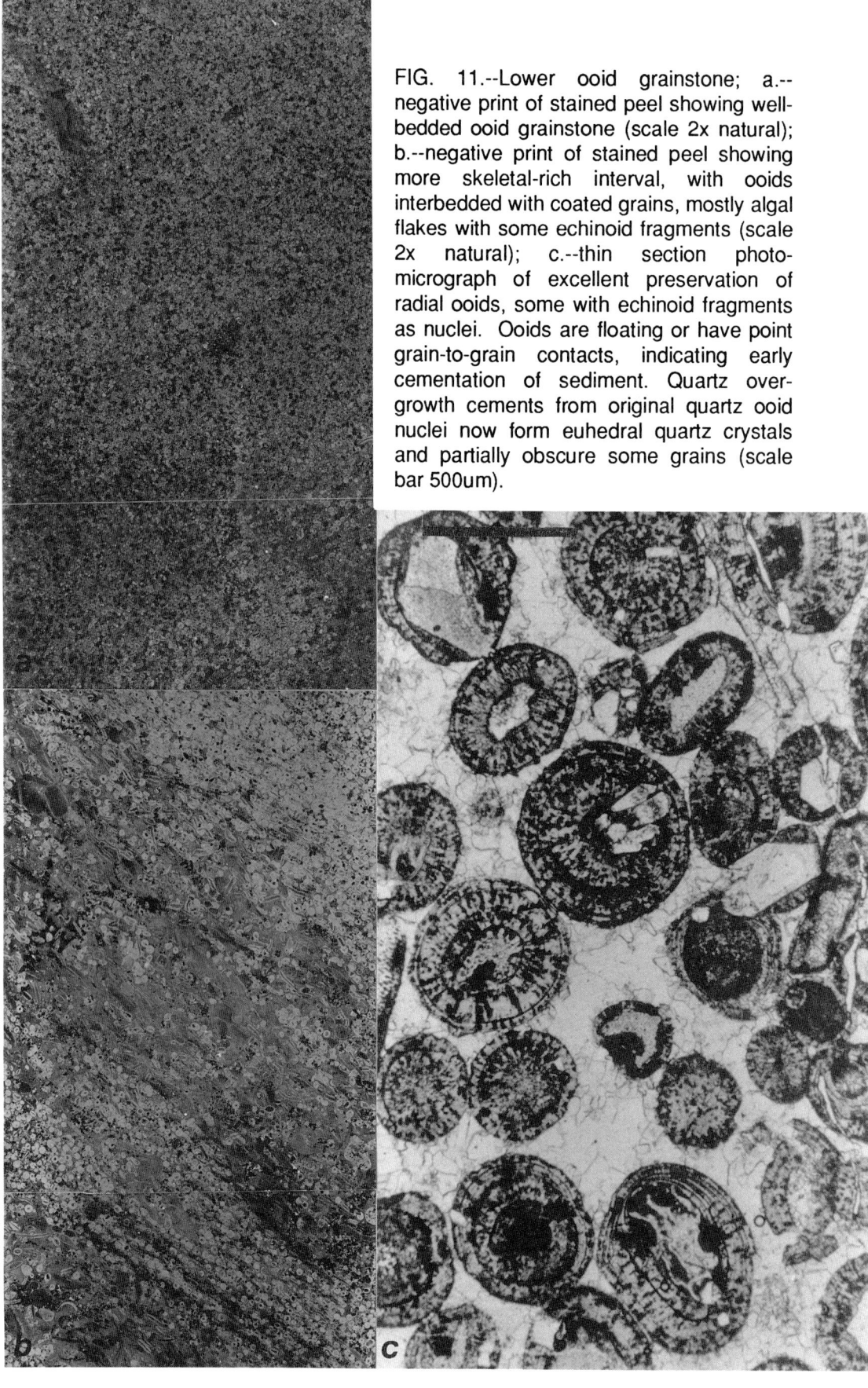

FIG. 11.--Lower ooid grainstone; a.--negative print of stained peel showing well-bedded ooid grainstone (scale 2x natural); b.--negative print of stained peel showing more skeletal-rich interval, with ooids interbedded with coated grains, mostly algal flakes with some echinoid fragments (scale 2x natural); c.--thin section photomicrograph of excellent preservation of radial ooids, some with echinoid fragments as nuclei. Ooids are floating or have point grain-to-grain contacts, indicating early cementation of sediment. Quartz overgrowth cements from original quartz ooid nuclei now form euhedral quartz crystals and partially obscure some grains (scale bar 500um).

FIG. 12.--Lower ooid granstone; a.--intraclast of ooid granstone containing ooids with isopachous circumgranular columnar cements and pore-filling equant cements. Fracture through intraclast filled by overgrowth cement from adjacent echinoid grain (scale bar 500um); b.--well-preserved radial ooids, echinoid fragments and coated grain. One echinoid fragment has micritized margin, whereas another, directly below, has no micritization and syntaxial overgrowth cement (scale bar 500um); c.--blue light view of b, emphasizing fabric preservation.

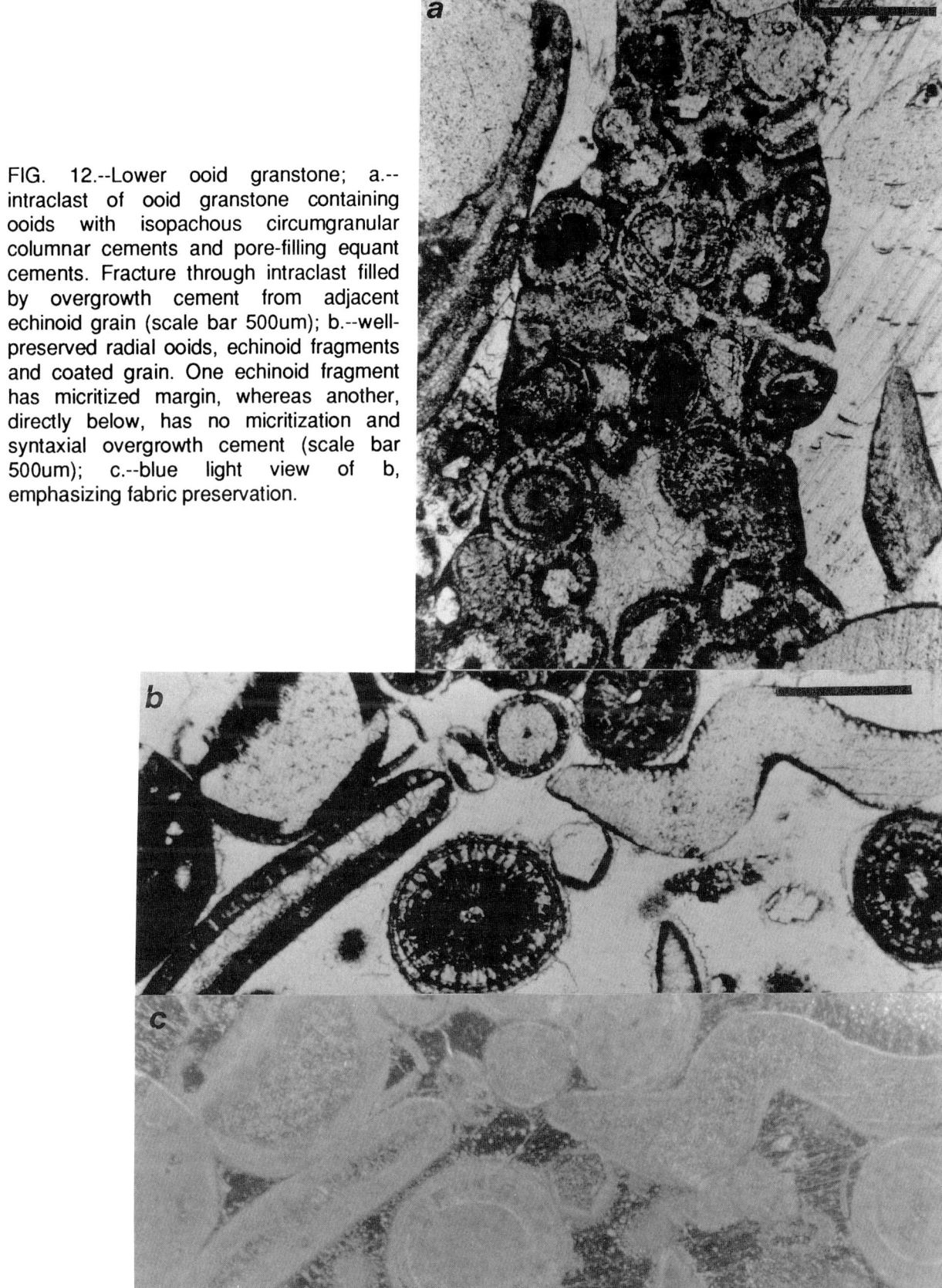

coated grains with nuclei of different skeletal fragments (echinoids, echinoderm spines, fractured brachiopod valves, algal flakes) are common in this interval (Fig. 11b) and in places occur in an imbricate texture within laminae. The ooids and skeletal fragments commonly float within the intergranular cements (Figs. 11c, 12b and c), indicating very little compaction prior to cementation (although detailed petrographic studies demonstrate that this cement may be, in part, replacive). In some areas within this grainstone, however, the grains are overly packed with a resultant fitted texture; circumgrain fractures have emphasized this texture. Quartz and feldspar grains are mostly euhedral with large overgrowth cements (Figs. 11 and 12).

Cross-stratification and ripple lamination are common throughout the subunit (Figs. 4 and 5). Internal variations are from ooid grainstones through coated grain ooid grainstones to skeletal ooid grainstones, although ooid grainstones predominate. Beds are commonly massive, on a meter-plus scale. Isopachs demonstrate subunit thicknesses in excess of 39.4 ft (12 m) in the center of the area, ranging to less than 19.7 ft (6 m) in the southeast (Fig. 10c). Thickness variations again are associated with fault trends, the thicker grainstones being on the hangingwalls of the faults (Fig. 10c). Thicks parallel the structural trend of the area.

Lower Sandstone

There is a sharp boundary between the lower ooid grainstone and the underlying lower sandstone (Figs. 4 and 5). This sandstone is calcareous near its margins but, in the center, contains a higher clay and silt content than is common elsewhere in the Moyvoughly Beds. Clayey partings and clay-rich burrows occur, and load structures are common. These structures become less common where the sandstone is thinner, but clay-rich partings occur in the center of the sandstone in all cores logged. Cross-stratification and ripple lamination are present near the top and base of the subunit, where the sandstone is cleaner (Figs. 4 and 5) and, in places, more calcareous (Fig. 13). Also, near the base, large skeletal clasts and carbonate lithoclasts show the proximity of carbonate environments. Grain sizes and compositions are similar to the upper sandstone, with feldspar forming a large (up to 40 percent) proportion of the siliciclastic grains. Most quartz grains are monocrystalline, some with slight undulose extinction. Fresh muscovite micas are also common.

Where the sandstone is calcareous, skeletal grains include brachiopods, echinoids, algae, foraminifers, and rare solitary corals; ooids and peloids are few or absent. Echinoid fragments rarely have micritized margins; some brachiopod fragments have a circumgrain columnar cement.

This sandstone averages 16.4 to 19.7 ft (5 to 6 m) thick over much of the area (Fig. 10d). The greatest thickness variations are seen in the northeast of the area, where ranges from less than 9.8 ft (3 m) to nearly 33 ft (10 m) occur. These thickness variations extend across fault trends with no apparent offset, indicating little or no synsedimentary movement at this time.

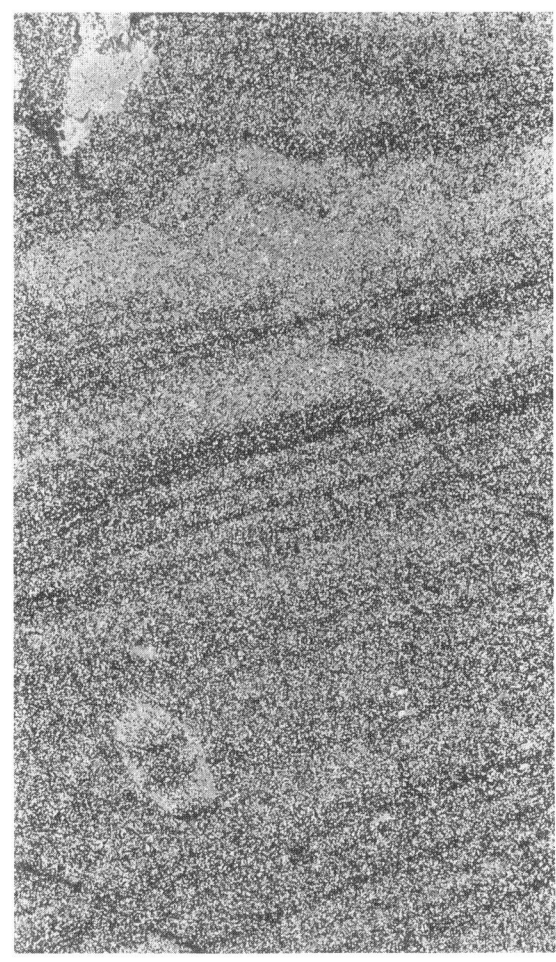

Fig. 13.--Negative print of stained peel of lower sandstone, showing laminae of alternating carbonates and siliciclastics, bioturbation and minor faulting (scale 2x natural).

Crinoid-Ooid Grainstone

The underlying crinoid-ooid grainstone (or Micrite Transition Unit) has a variable contact with the lower sandstone. In core 1117-15 the sandstone loads into the grainstone (Fig. 4); whereas, in the other two cores, there is a decrease in siliciclastic content into the grainstone (Fig. 5). In addition to a variable, but small, quartz/feldspar content, the grainstone contains numerous thin bands of skeletal debris, mostly crinoidal (average 250 µm), characteristic of storm beds. Echinoid and brachiopod spines, brachiopods (almost complete), corals, algal flakes, peloids, goniatites, and foraminifers are also present. Ooids are small (average 200 µm); some retain a radial fabric, whereas others are partially and, in the storm beds, almost completely micritized.

The crinoid-ooid grainstone thickens to the southeast of the area (Fig. 14a); thickness variations run across fault trends, indicating no synsedimentary fault movement, as for the lower sandstone.

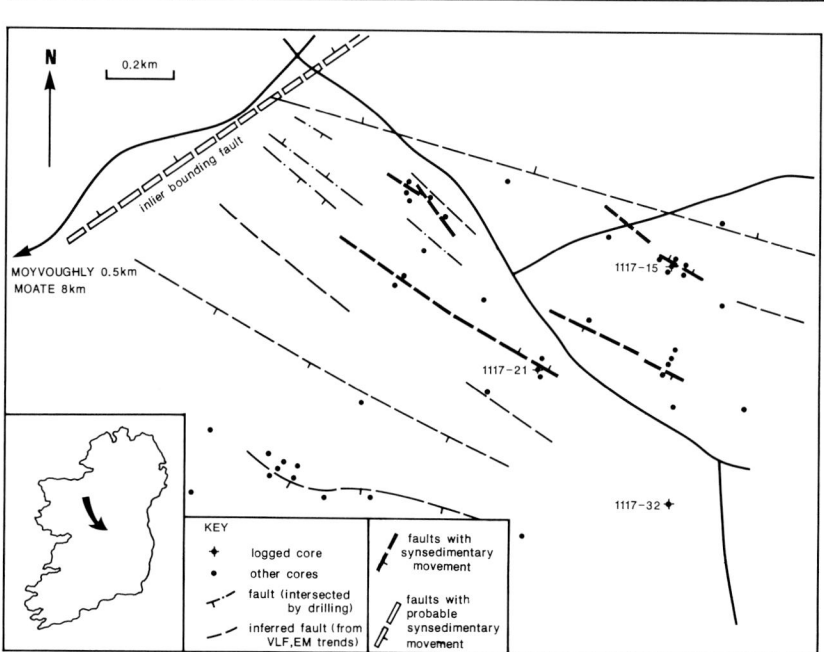

FIG. 14.--Isopach map of crinoid ooid grainstone (a) and (b) summary map showing faults where synsedimentary movement is interpretted to have occurred during deposition of the Moyvoughly Beds.

Algal Packstones/Wackestones

There is an abrupt transition at the base of the crinoid-ooid grainstone, marked in some cores by colonization of corals on the lithified surface of the underlying algal packstones and wackestones of the Micrite Unit. These algal packstones/wackestones were deposited in very different conditions to those of the overlying Moyvoughly Beds. The unit is carbonate mud-rich and dominated by abundant millimeter-scale fenestrae. It contains a variety of algae: blue-green algae that coat oncoids; common clusters of green algae throughout the upper part of the unit, together with red algae (Fig. 15a); plus peloids, oncoids, numerous brachiopod spines, and thin-walled brachiopod valves, micritized crinoid fragments, ostracods, and quartz/feldspar silt.

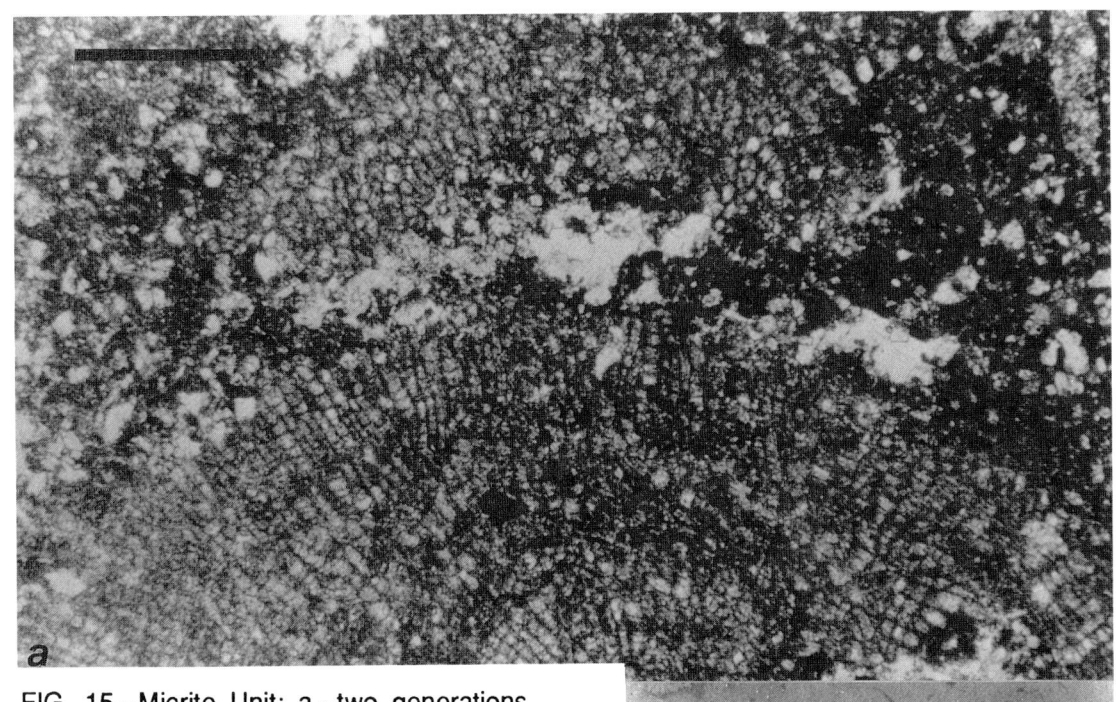

FIG. 15.--Micrite Unit; a.--two generations of red algae encrusting sediment surface; smaller, upper algal head encrusts the larger, lower head (scale bar 500um); b.--negative print of stained peel showing large, irregular vugs with multiple generations of cement fills including early peloidal and columnar cements (scale 2x natural); c.--sketch of varied vug morphology within Micrite Unit. Vertically-oriented vugs appear to be after plant roots; high water tables at other periods resulted in roots decaying to form horizontally-oriented vugs.

Typical of these sediments are vugs larger than the fenestrae - some laminar, some circular, and some irregular in shape - commonly floored by geopetal carbonate muds (Fig. 15b and c). Internal margins of some voids have an algal coating; others contain thin peloidal cements, both factors indicating the voids were open to the passage of circulating marine waters, presumably shortly after sedimentation. A further generation of circumvoid, radial, inclusion-rich cements, formerly high magnesian calcite, supports this contention. The margins of the voids must have been rigid at this time, showing rapid cementation (Fig. 15b and c). Other intervals of core display textures similar to those developed within paleosols.

The algal packstones/wackestones are present across the area, with an average thickness of over 131 ft (40 m)(Kucha, 1988). Most cores, however, terminate within these sediments, and thus no regional variation in thickness can be determined. The so-called 'Micrite Unit' is by no means composed of monotonous carbonate mud and merits a more detailed study.

CONTROLS ON SEDIMENTATION

Sedimentation within the Moyvoughly Beds in this area is controlled by the interplay of several factors: the agitated aspect of the marine shelf, the tectonic setting and synsedimentary tectonic movements, relative sea-level changes, and the location of siliciclastic and carbonate source inputs. The carbonate grainstones, sandstones, and more mud-rich lithologies were all deposited within a marine environment, evidenced by the occurrence of marine organisms and bioturbation throughout the succession. The present-day tectonic setting of the area is on the southeastward-dipping slope directly adjacent to the major fault that bounds the Moat Inlier (Figs. 1b and 6). The area is dissected by numerous, more minor northwest-southeast trending faults (Fig. 6 and following figures), some of which have influenced local sedimentary patterns throughout deposition of much of the Moyvoughly Beds (Fig. 14b). The age of the earliest movement on the Moate Inlier bounding fault is unknown; syndepositional movement of the minor faults, however, indicates that some movement of the bounding fault may well have occurred at this time. This hypothesis is further supported by overall thickening trends of the upper ooid grainstone, upper skeletal ooid grainstone, and crinoid ooid grainstone toward the southeast (Figs. 7a, c, and 14b), indicating somewhat deeper water in this direction at the time. This southeastward dip of the depositional surface, albeit then at a lower angle than the structural dip today, together with the character of the sedimentary sequence, demonstrates that the Moyvoughly Beds were deposited on a localized ramp within the Irish Midlands. Changes in relative sea level combined with these other factors to influence the development of different sedimentary subunits, as described below.

Common to the siliciclastic component of all sediments within the Moyvoughly Beds is the high feldspar:quartz ratio plus the fresh nature of contained feldspar and mica grains. This argues for a common siliciclastic source throughout, possibly from the braid-plains of the Old Red Sandstone, deposited under arid conditions and now unroofed on the Moate Inlier (Fig. 1b). The combination of plagioclase, orthoclase, and muscovite, together with the monocrystalline and unstrained/slightly strained nature of the quartz grains, indicates ultimate provenance from a granitic source of location at present unknown. Phillips and Sevastopulo (1986, Fig. 14) show that there are no known, or inferred, Caledonian granites, perhaps the most probable source of these minerals, within 31 miles (50 km) of the Moyvoughly area, although more distant granites may have contributed to local Old Red Sandstone sediments.

DEPOSITIONAL HISTORY OF THE UPPER MICRITE UNIT AND MOYVOUGHLY BEDS

The depositional environments of the uppermost Micrite Unit and Moyvoughly Beds are summarized below, with emphasis on the interplay between the different controlling factors. The Micrite Unit is traceable over the North Midlands Province where it rests on a major sequence boundary. Philcox (1983) indicates it to form shoreward of an ooid shoal facies and equivalent to the Middle Ballysteen Limestones further south. This hypothesis is supported by observations within the cores studied. The large percentage of carbonate mud within these algal packstones and wackestones argues for somewhat sheltered conditions, especially when contrasted with those during deposition of the overlying Moyvoughly Beds. The abundant fenestrae demonstrate sedimentation in extremely shallow water, with little current action and perhaps with brief periodic exposure, plus very rapid lithification of the sediments. Blue-green algal coatings, present on some grains, do, however, evidence moderate current agitation, and it may be that much of the mud was generated from algal fragmentation. These sediments were thus deposited in shallow, moderate-energy environments, where wave and current action was not sufficiently strong to winnow the carbonate muds but constantly turned both allochems and skeletal grains, resulting in oncoid growth. Soils formed during longer periods of exposure. The wide, shallow (less than 6.6 ft (2 m)) lagoon over which these conditions developed stretched across the North Midlands Province, with little or no structural interruption.

Further evidence for the depositional environment for these sediments can be gained by study of the vugs therein. These voids range to over 0.4 in. (1 cm) in diameter and are irregular in shape, interconnecting in three dimensions. The peloidal cements and algal coatings, geopetal fills, plus the subsequent inclusion-rich radial cements, indicate that these voids were open to circulating marine fluids and that the host sediments must have been at least partially lithified to support the void network. Moreover, changes occur in void morphology, in places repeated over several meters of core, with a transition from laminar through vertical to equidimensional voids (Fig. 15c). This morphological transition is reminiscent of root patterns, and it is probable that each of these 'cycles' represents build-up to base level, marsh plant colonization, and subsequent destruction within extremely shallow marine pools. Partial lithification would have taken place during these periods. Some paleosols developed during these periods of marsh conditions, with perhaps better paleosol development as the area became more silted up and the marshes deteriorated. Exposure, followed by a return to higher relative water levels, with consequent lagoon, resulted in marine water flooding the root systems with multi-phase void fill before muddy conditions could be completely reestablished. Within the lagoon colonization by a varied algal fauna, together with brachiopods and other marine organisms, competed with fine siliciclastic detritus, derived from a landmass to the north (Fig. 16). Storm beds dominated by small ooids lower in this unit testify to the continued existence of a grainstone barrier, either as submarine shoal or, perhaps more likely, as a string of barrier islands.

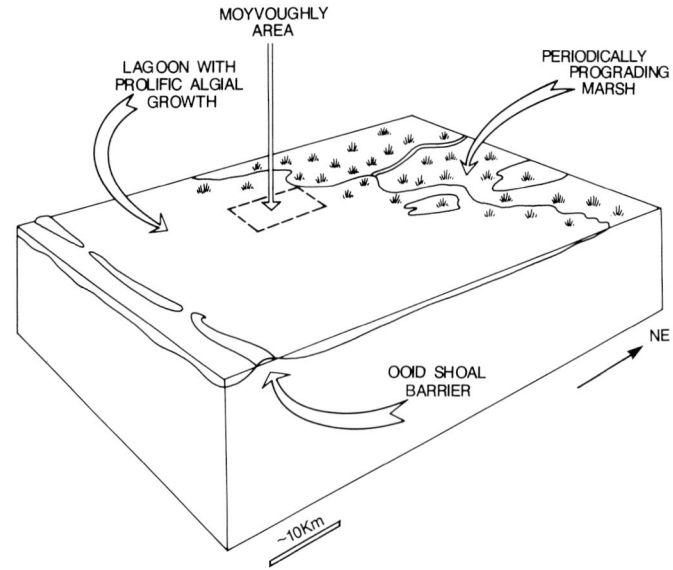

FIG. 16.--Depositional environments within the upper Micrite Unit. An extensive shallow lagoon extended over much of the northern Irish Midlands, with periodic sediment progradation to sea level, marsh plant growth and subsequent exposure with paleosol development. An ooid grainstone barrier to the south protected the area from all but major storms. Algal growth within the lagoon was abundant, sourcing the carbonate muds. Lagoon depths were at maximum a few meters, as fenestrae are common throughout much of the sequence.

An abrupt change from these relatively sheltered conditions took place at the top of the Micrite Unit in the Moyvoughly area. Colonization of the upper surface of the Micrite Unit by corals, themselves in places partially eroded before deposition of the overlying crinoid-ooid grainstone, marks the largest depositional break within this sequence. The crinoid-ooid grainstones were deposited as carbonate sands migrated across the area. This lithofacies marks the breakdown of the grainstone barrier with the Moyvoughly area becoming exposed to fluctuating, agitated, open-marine conditions, an aspect characteristic of much of the Moyvoughly Beds. The adjacent, and presumably southward, open-marine shelf is evidenced by the common crinoid debris, together with, in storm beds, echinoderm spines and plates, corals, brachiopods, and foraminifers (Fig. 17a). Limitation on ooid growth may have been due to rapid sedimentation rates, or to conditions being somewhat deeper, perhaps just above active fairweather wave base, so that ooid generation only occurred at certain intervals. Periodic cessation of ooid development is further indicated by common occurrences of micritized ooids, particularly in storm beds.

The causes for this change in depositional environment were multiple. Increased water depth caused breaching of the grainstone barrier. This may have been enhanced by slight movement along the inlier bounding fault, thus tilting the Moyvoughly Block to the full input of southerly derived oceanic storms and currents, and marking the onset of ramp sedimentation. This is substantiated by the thickening of this unit toward the southeast, although, by the time of deposition of the overlying lower sandstone, this dip had been blanketed by sediment fill.

A major sourcing of siliciclastic sediment curtailed grainstone development, with a thick tongue of sediment being deposited in the northeast (Fig. 10d), a possible splay channel prograding onto the carbonate shelf (Fig. 17b). This progradation direction is evidence that sourcing was not from any erosion of Old Red Sandstone from any palaeocrest of the Moate Inlier, further to the southwest, but from the north or, in this instance, the northeast. Siliciclastic sedimentation was intimately associated with carbonate sedimentation near the base of the lower sandstone, with an upward decrease in carbonate components from the base of the unit. Moreover, the switch over in conditions was rapid with lack of cementation of the underlying grainstones, evidenced by load structures in the base of the sandstone (core 1117-21, Fig. 5). Two major sandstone sequences have been recorded from the three cores logged (Figs. 4 and 5), the lowest defining a fining-upward cycle on the margins of the sandstone thick (Figs. 4 and 10d), further evidence of a splay channel. This sandstone is therefore the marine counterpart of a bay-fill sand, remobilized by the agitated conditions on the shelf.

Abrupt abandonment of the channel led to a sharp decrease in siliciclastic input and the establishment of an active grainstone shoal complex within the Moyvoughly area (Fig. 17c). The thickness of the lower ooid grainstone shows continual agitation and ooid development over a considerable period of time. The small average size of the ooids may represent more active ooid nucleation than in the overlying shoal developments, coupled with a higher rate of deposition (and consequently higher subsidence?), curtailing development to larger ooids. A rise in sea level must have accompanied deposition of this grainstone, keeping the sediment surface within the upper meter or so of the mean sea level. The cemented intraclasts (Fig. 12a) may derive from exposure with beachrock formation or, alternatively, from reworking of bored hardground surfaces, although no hardgrounds were observed in this grainstone. Early cementation, subsequent tight packing, and exposure with constant wetting/drying are further evidenced by the beds of overly packed grains with circumgrain cracks. Early cementation, a process characteristic of shallow marine grainstones, was common throughout. The common occurrence of marine skeletal fragments and coated grains within the grainstone indicates adjacent, less turbulent conditions, most likely to the south or southeast, with debris introduced by northward-trending storms and currents, evidence for the reestablishment of a southward- or southeastward-facing ramp at this time (Fig. 17c).

The establishment of this ramp suggests movement along the inlier bounding fault during this time. Within the lower ooid grainstone, there is also evidence of synsedimentary fault movement on the conjugate (northwest-southeast) trend. Shoals initiated on footwall crests commonly shed sediment onto hangingwall flanks (Figs. 10c and 17c).

A sudden increase in relative sea level was responsible for deposition of the overlying calcareous mudrocks. The presence of solitary corals, thin-walled brachiopods, and echinoid fragments indicates marine sedimentation rather than sedimentation shoreward of a grainstone barrier. Conditions were much less agitated with clay introduction, possibly from a distal fluvial source to the north (Fig. 17d). These mudrocks may have draped shoal crests in the underlying grainstones; their absence in the south of the area and northward thickening are due to erosion prior to deposition of the overlying lower skeletal-ooid grainstone. Synsedimentary movement occurred along some faults during deposition of these mudrocks, with thins over footwall crests (Fig. 10b).

During deposition of the lower skeletal-ooid grainstone, conditions in the area became agitated once more due to initial relative sea-level fall, and active shoal formation recommenced. The increased skeletal content, when compared with the lower ooid grainstone, indicates that the area was closer to areas of faunal colonization, perhaps slightly further offshore. Storm beds had a probable southerly origin. Depositional conditions varied within this unit; bimodal sorting of

FIG. 17.--Depositional environments within the Moyvoughly Beds:

a.--Crinoid ooid grainstone. Southeasterly facing ramp developed, probably due to movement on inlier bounding fault. Ooid grainstone barrier to the south was breached, exposing ramp to major oceanic and tidal currents.

b.--Lower sandstone. Major channelized input of siliciclastic sands from northeast, with some marine reworking of sands, led to switch from carbonate deposition; carbonate grains were washed in from adjacent areas.

c.--Lower ooid grainstone. Minor relative sea-level rise, with abandonment of (or deposition over) siliciclastic source, led to reestablishment of carbonate deposition. Ooid shoals develop over area, which must have remained at or near sea level throughout deposition of this subunit. Minor exposure of shoal crests likely. Evidence of synsedimentary fault movements.

d.--Calcareous mudrock. Rapid rise in relative sea level, with drowning of ramp below the zone of ooid generation and, probably, below wave base. Low sedimentation rate relative to that of the ooid grainstones. Active synsedimentary faulting led to thinning of mudrock over faulted highs.

e.--Lower skeletal ooid grainstone. Initial relative fall of relative sea level to bring ramp, once more, into the zone of ooid generation and shoal formation. Abundant skeletal debris washed in from deeper on ramp. Sea level kept pace with sedimentation. Minor synsedimentary faulting.

f.--Bioturbated brachiopod wackestone. Further relative sea-level rise flooded ramp. Low sedimentation rate with minor synsedimentary fault movements.

g.--Upper skeletal ooid grainstone. Relative rapid sea-level fall, with reestablishment of ooid shoal development across ramp. Sedimentation was periodic, with migration of shoals across the area. Hardgrounds formed when active shoals migrated out of area. Minor synsedimentary fault movement.

h.--Upper sandstone. Relative sea-level stillstand, but active faulting generating siliciclastic inputs (from northeast and southwest). Active feldspar-quartz shoals, with carbonate skeletal debris, formed in shallow marine environment.

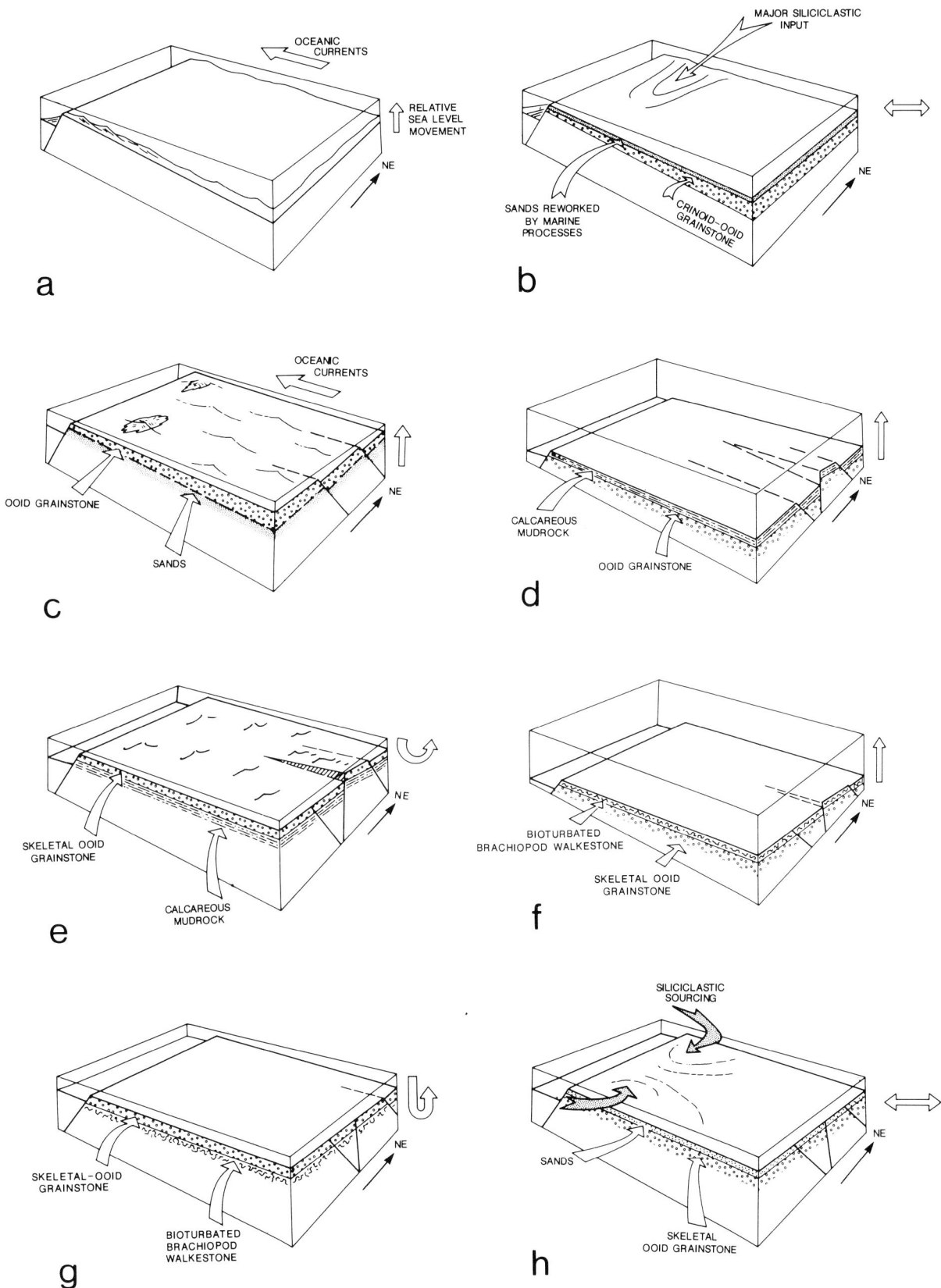

ooids demonstrates different degrees of shoal development within the area. Stillstands of sedimentation are evidenced by encrustation of brachiopod valves, although the majority of skeletal debris, and ooids, are well preserved. It is possible that open ramp conditions were established once more during this time, with reactivation of the inlier bounding fault. Thickening of grainstones occurs on the footwall crests in some areas, indicating synsedimentary movement along these trends. Relative sea-level rise, keeping pace with sedimentation, must have taken place during deposition of this unit (Fig. 17e).

Corals and abundant brachiopods within the bioturbated brachiopod wackestone indicate open-marine conditions in an environment with greatly reduced agitation, also probably indicative of rapid relative sea-level rise (or, equally, relative subsidence increase)(Fig. 17f). Sedimentation rates were low at this time, allowing abundant colonization of the sea floor by brachiopods. Stable conditions are also indicated by the isopach map (Fig. 7d), with no fault movement at this time.

Dominantly quiescent structural conditions persisted into the period of deposition of the upper skeletal-ooid grainstone, although the depositional conditions changed radically, with an agitated, southeast-facing ramp reestablished (Fig. 17g). This transition was gradational in one core (1117-15, Fig. 4), indicating gradual shallowing. In other parts of the area, the transition is more abrupt, with skeletal-ooid shoals migrating over the brachiopod-rich wackestone surface. Hardgrounds developed in areas between the active shoals, when ooid production had shifted out of the immediate area. This periodic production of ooids is further substantiated by occurrence of micritized ooids and coated/micritized skeletal grains at certain horizons, contrasting with well-preserved ooids together with fresh skeletal grains in other beds. The fine preservation of the fabric of many of the ooids, together with the presence of acicular cements in the one hardground, argues against subaerial exposure, as does the bored nature of the hardground.

The upper sandstone differs from the lower sandstone in that thickness changes are less variable across the area (Fig. 7b), with no definite source inputs. This sandstone is also interbedded with the underlying grainstone (core 1117-32, Fig. 5). This upper sandstone developed adjacent to sources of crinoid debris, both from areas of active growth and from where boring of skeletal grains had occurred. The further inclusion of peloids, with brachiopod and goniatite fragments, indicates a marine environment with rapid sand input, curtailing the earlier carbonate development. Cross-stratification, plus the general lack of clay and silt contents, shows the existence of constant current action, perhaps typical of a siliciclastic offshore bar or shoal, adjacent to a land-derived sand supply (Fig. 17h). The similarity of many of the grains with those of the underlying grainstone indicates that water depths and the marine conditions changed very little between the upper sandstone and upper skeletal ooid grainstone, except in that ooid formation was curtailed and a siliciclastic source introduced. There was little (or no?) structural movement during upper sandstone deposition. This sandstone differs from the lower sandstone, therefore, in that it was a true marine sand, constantly reworked with interspersed skeletal material; whereas, the lower sandstone was proximal to a major land-sourced siliciclastic supply channel, with few or no marine skeletal fragments present in some areas.

Switching of the siliciclastic source led to revived ooid formation with deposition of the upper ooid grainstone, representative of active sedimentation on a local ramp, particularly in the down-dip areas. Coarse ooids, plus a robust fauna, indicate prolonged development of ooid shoals in the area at this period. Evidence on the upper ooid grainstone is scant, so no further conclusions can be drawn here.

CONCLUSIONS

The majority of evidence points to a change in depositional conditions from the Micrite Unit into the Moyvoughly Beds, from a near-uniform shallow lagoon, occasionally filled to base level and colonized by marsh plants, to a ramp environment, facing constant agitation from the open ocean to the south. No evidence of synsedimentary tectonism was found within the Micrite Unit, although studies were not so detailed as those of the Moyvoughly Beds. The change in depositional style between the Micrite Unit and Moyvoughly Beds was probably the result of synsedimentary faulting, with block tilting and the formation of a gently dipping ramp. This resulted in breaching the lagoon barrier to the south and the ramp being subjected to constant wave and current agitation. This agitated environment persisted during sedimentation of the Moyvoughly Beds, apart from when two rapid relative rises in sea level brought the area to below fairweather wave base. Deposition was on a south- or southeast-facing ramp, with periodic incursions of siliciclastic sediments resulting from active fault movements. Relative sea-level rises and falls are summarized against the stratigraphy in Figure 18. Sedimentation rates also varied greatly during deposition of the Moyvoughly Beds, from very low, during deposition of the bioturbated brachiopod wackestone, to high, during deposition of the crinoid-ooid grainstone (Fig. 18). Interlaced with these changes were structural movements, with synsedimentary faulting active upward from the lower ooid grainstone (Fig. 18) across the ramp and probable constant reactivation of the present inlier bounding fault. This complex scenario of relative sea-level changes combined with synsedimentary fault movements makes it impossible to determine, within this area, which sea-level changes had a more regional extent and which were the result of local tectonics. At the end of Moyvoughly Beds times, a major siliciclastic unit prograded southward across the area.

The existence of synsedimentary fault movements in this area is, in part, one reason why correlation between the North Midlands and Limerick provinces is disputed. Synsedimentary slide units have been recognized in equivalent-aged strata at Navan (Andrew and Ashton, 1985; Ashton and others, 1986), perhaps an indication that synsedimentary tectonic movements occurred outside of the Moyvoughly area during the same period of time. If this were so, continued small movements across the region would make correlations very difficult, particularly where attempts were made to correlate units across major tectonic lineaments, such as the inlier bounding fault. The depositional history of these sediments does, however, demonstrate the continuous occurrence of synsedimentary tectonics within this part of the Irish Midlands and emphasizes the effects of the regional extensional regime during the Courceyan.

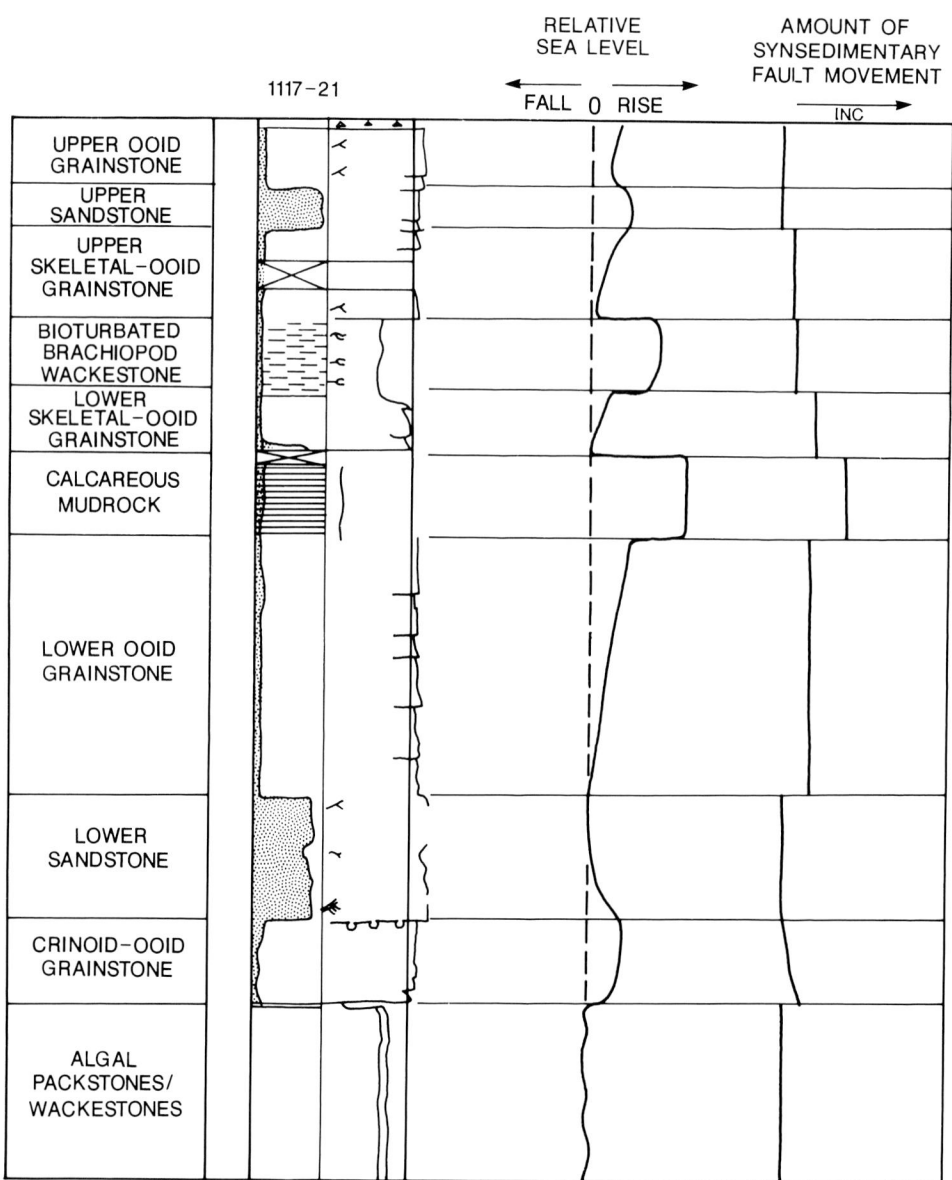

FIG. 18.--Summary diagram, showing relative sea level changes plus relative amounts of syn-sedimentary tectonism compared with the sedimentary sequence within the Moyvoughly Beds. Log of core 1117-21 used for subunit thickness.

ACKNOWLEDGMENTS

This study was encouraged by Getty (now Texaco) Research in Houston. Texaco financed core shipment to the U.S.A. Emily Stoudt, Susan Longacre, Corinne Danielli, and Alan Kendall are thanked for their time and interest. Pat Ryan compiled some of the core logs on which the isopach maps are based. Ennex International plc allowed use of these original descriptive logs plus core examination and sampling, with much help from John Clifford. Mick Jones (University of Newcastle upon Tyne) carried out vitrinite reflectance determinations. This study was financed by NERC (GR3/5496) and aided by the concurrent research of Maxine Akhurst (NERC GT4/84/GS/77) on the diagenesis and mineralization of Mississippian carbonates at Tynagh, County Galway.

REFERENCES

AKHURST, M. C., 1989, Deposition and diagenesis of Waulsortian carbonates from Tynagh, County Galway, Eire; their relationship with base metal mineralization: unpublished Ph.D thesis, University of Newcastle upon Tyne.

ANDREW, C. J., 1986a, A diagrammatic representation of the Courceyan stratigraphy of the Irish Midlands, in Andrew, C. J., Crowe, R. W. A., Finlay, S., Pennell, W. M., and Pyne, J. F., eds., Geology and Genesis of Mineral Deposits in Ireland: Special Publication, Irish Association for Economic Geology, p. 239-242.

_____, 1986b, The tectono-stratigraphic controls to mineralization in the Silvermines area, County Tipperary, Ireland, in Andrew, C. J., Crowe, R. W. A., Finlay, S., Pennell, W. M., and Pyne, J. F., eds., Geology and genesis of mineral deposits in Ireland: Special Publication, Irish Association for Economic Geology, p. 377-418.

_____, AND ASHTON, J. H., 1985, The regional setting, geology and metal distribution patterns of the Navan orebody, Ireland: Transactions of the Institution of Mining and Metallurgy, v. 94, p. B66-93.

_____, AND POUSTIE, A., 1986, Syndiagenetic or epigenetic mineralization - the evidence from the Tatestown prospect, County Meath, in Andrew, C. J., Crowe, R. W. A., Finlay, S., Pennell, W. M., and Pyne, J. F., eds., Geology and Genesis of Mineral Deposits in Ireland: Special Publication, Irish Association for Economic Geology, p. 281-296.

ASHTON, J. H., DOWNING, D. T., AND FINLAY, S., 1986, The geology of the Navan Zn-Pb orebody, in Andrew, C. J., Crowe, R. W. A., Finlay, S., Pennell, W. M., and Pyne, J. F., eds., Geology and Genesis of Mineral Deposits in Ireland: Special Publication, Irish Association for Economic Geology, p. 243-280.

BOAST, A. M., COLEMAN, M. L., AND HALLS, C., 1981, Textural and stable isotopic evidence for the genesis of the Tynagh base metal deposit, Ireland: Economic Geology, v. 76, p. 27-55.

BOYCE, A. J., ANDERTON, R., AND RUSSELL, M. J., 1983, Rapid subsidence and early Carboniferous base-metal mineralization in Ireland: Transactions of the Institution of Mining and Metallurgy, v. 92, p. B55-B66.

CLIFFORD, J. A., RYAN, P., AND KUCHA, H., 1986, A review of the geological setting of the Tynagh orebody, County Galway, *in* Andrew, C. J., Crowe, R. W. A., Finlay, S., Pennell, W. M., and Pyne, J. F., eds., Geology and Genesis of Mineral Deposits in Ireland: Special Publication, Irish Association for Economic Geology, p. 419-440.

JONES, G. V., AND BRAND, S. F., 1986, The setting, styles of mineralization and mode of origin of the Ballinalack Zn-Pb deposit, *in* Andrew, C. J., Crowe, R. W. A., Finlay, S., Pennell, W. M., and Pyne, J. F., eds., Geology and Genesis of Mineral Deposits in Ireland: Special Publication, Irish Association for Economic Geology, p. 355-375.

KUCHA, H., 1988, Zn-Pb sulphides as rim cements, filling cements and replacements of carbonate sediments, Moyvoughly, Ireland - a product of two convective cells: Transactions of the Institution of Mining and Metallurgy, v. 97, p. B64-76.

PHILCOX, M. E., 1984, Lower Carboniferous lithostratigraphy of the Irish Midlands: Special Publication, Irish Association for Economic Geology.

PHILLIPS, W. E. A., AND SEVASTOPULO, G. D., 1986, The stratigraphic and structural setting of Irish mineral deposits, *in* Andrew, C. J., Crowe, R. W. A., Finlay, S., Pennell, W. M., and Pyne, J. F., eds., Geology and Genesis of Mineral Deposits in Ireland: Special Publication, Irish Association for Economic Geology, p. 1-30.

POUSTIE, A., AND KUCHA, H., 1986, The geological setting, style and petrology of zinc-lead mineralization in the Moyvoughly area, County Westmeath, *in* Andrew, C. J., Crowe, R. W. A., Finlay, S., Pennell, W. M., and Pyne, J. F., eds., Geology and Genesis of Mineral Deposits in Ireland: Special Publication, Irish Association for Economic Geology, p. 305-317.

CYCLICITY IN THE PERMIAN QUEEN FORMATION - U.S.M. QUEEN FIELD, PECOS COUNTY, TEXAS

K. E. TUCKER AND R. G. CHALCRAFT
Chevron Oil Field Research Company
P. O. Box 446, La Habra, CA 90633-0446

ABSTRACT

The Queen Formation, as observed in the U.S.M. Queen Field, Pecos County, Texas, consists of alternating siliciclastic and carbonate sequences deposited as shelf sediments during the Middle Permian. Eight of eleven lithofacies present are of siliciclastic origin, and their deposition can be related to sedimentation during periods of sea-level lowstand when detritus was probably transported across the shelf by fluvial and eolian processes. However, subsequent reworking by currents and infauna during sea-level rise destroyed the original sedimentary structures, making the actual processes of transport difficult to interpret directly from these rocks. Carbonate sequences, which are composed of three lithofacies, were deposited during sea-level highstand periods and characteristically formed as initial deepening followed by shallowing-upward sequences.

Thin-section, core, and log studies have focused on the clastic deposits, especially the Gray Sandstones, which are the principal reservoir rocks. Typical Queen sandstones have variable amounts of silt, clay, and cement, which result in relatively low porosities (average 16.9 percent) and very low permeabilities (average 0.39 md). Siltstones often contain enough very fine-grained sand to be considered reservoir rock; however, they have a lower average porosity (12.1 percent) and permeability (0.08 md) than the sandstones due to more abundant clay and cement.

INTRODUCTION

The Guadalupian Queen Formation of the West Texas Permian Basin Region is composed of a complex mixture of lithologies deposited in a shelf environment characterized by extensive sea-level fluctuation and/or variation in siliciclastic input. Alternating cycles of carbonate and siliciclastic sedimentation have produced approximately 350 feet (107 m) of sediment accumulation at the U.S.M. Queen Field. The environment of deposition for Queen Formation siliciclastic lithofacies has been a focal point for discussion. Proposed environments of deposition include shallow subtidal (Franseen, 1985; Wheeler, 1989), coastal mudflat (Spencer and Warren, 1986), sabkha (Siegal, 1989; Malicse, 1988; Newsom, 1989; Harper, 1990), eolian (Mazzullo and others, 1984), fluvial (Mazzullo and others, 1990), and alluvial plain (Holley and Mazzullo, 1988). The objective of this study has been to utilize available core, petrographic studies, and core analysis data to generate a geologic model that defines depositional environments and accounts for the distribution of reservoir lithofacies.

The U.S.M. Queen Field, discovered in 1964, is located approximately two miles (3.2 km) north of Fort Stockton, Texas, in the north-central part of Pecos County, which corresponds to the southwestern end of the Central Basin Platform during the middle Guadalupian (Figs. 1 and 2). The productive area of the field covers nearly 3000 acres and has been developed on 40-acre proration units. Fifteen wells were cored, seven of which have been studied for this project (Fig. 3).

Coarse siltstones and very fine-grained sandstones are the primary reservoir rocks. The reservoir section consists of four separate pay zones: the Upper Queen and three Lower Queen pays designated as 1, 2, and 3 (Fig. 4). Wells on the eastern side of the field do not contain an Upper Queen pay zone. Petrophysical parameters of each pay zone vary widely throughout the field. Porosity ranges from less than 1 to 23 percent and air permeability values range from less than 1 to 170 md.

GENERAL GEOLOGY

Stratigraphic nomenclature is complex for Guadalupian Series sediments that were deposited as time-equivalent basinal clastics, shelf margin limestones, and shelf carbonates, siliciclastics, and evaporites (Silver and Todd, 1969). The Queen Formation is a shelfal equivalent of the Goat Seep shelf margin dolomitized carbonates and the basinal siliciclastics of the Cherry Canyon Formation. At the U.S.M. Queen Field, the Queen Formation consists of siltstones and very fine-grained sandstones that cyclically alternate with dolomites. Evaporitic shelf lithologies of the Seven Rivers Formation overlie the Queen, which in turn overlies dolomitized limestones of the Grayburg Formation (Fig. 2).

Low-amplitude, high-frequency eustatic sea-level changes are believed to be responsible for the cyclicity of the carbonate and clastic deposits of the Yates Formation during the late Guadalupian (Borer and Harris, 1989). The depositional setting of the Queen Formation is similar to the mixed carbonate-siliciclastic Yates Formation described by Borer and Harris (1989). Queen sediments were deposited across a gently sloping surface in subtidal to nonmarine environments behind the Goat Seep shelf margin. Cyclicity within the Queen is illustrated by Figures 5, 6, and 7, a series of stratigraphic cross sections through the U.S.M. Queen Field. Alternating carbonate and siliciclastic lithologies define the reservoir zones.

PREVIOUS WORK

A type section for the Queen Formation was established by Moran in 1962. The type section is located in Dark Canyon, Guadalupe Mountains, Eddy County, New Mexico. It consists of 421 feet (128 m) of interbedded sandstones, sandy dolomites, and dolomites. The uppermost 100 feet (30.5 m) were identified as the Shattuck Member, a distinctive sandstone unit that is often the only part of the Queen Formation described in the literature.

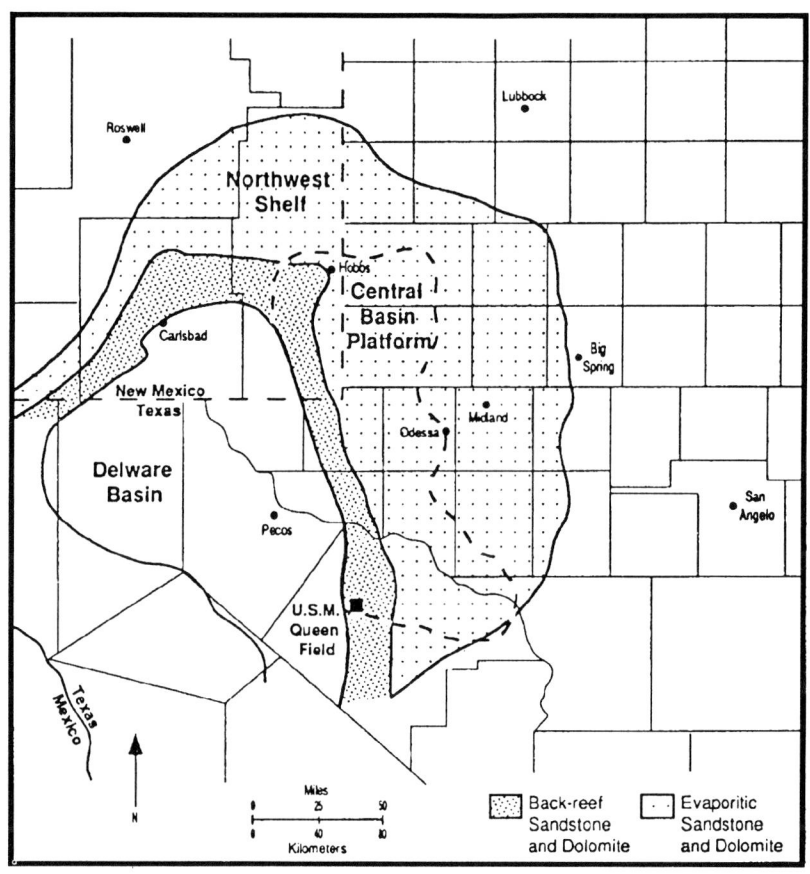

Figure 1
General location map for the U.S.M. Queen Field showing the major tectonic elements of the Permian Basin (modified from Ward et al., 1986)

System	Series	Delaware Basin		Central Basin Platform		Eastern Shelf	Sloss Sequences
Permian	Ochoa	Dewey Lake		Dewey Lake		Dewey Lake	
		Rustler		Rustler		Rustler	
		Salado		Salado		Salado	
		Castile					
	Guadalupe	Delaware Mt.	Bell Canyon	Tansill	Artesia Gr.	Artesia Group	Absaroka
			Cherry Canyon	Yates			
				Seven Rivers			
			Brushy Canyon	Queen			
				Grayburg			
		Victoria Peak		San Andres		San Andres	
				Glorieta Ss.		San Angelo	
	Leonard	Bone Spring Limestone		Clear Fork	Yeso	Clear Fork	
				Wichita-Aba		Wichita	
	Wolfcamp	Wolfcamp		Wolfcamp		Wolfcamp	

Figure 2
Principal stratigraphic units of the Permian Basin Region (Modified after Frenzel et al., 1989).

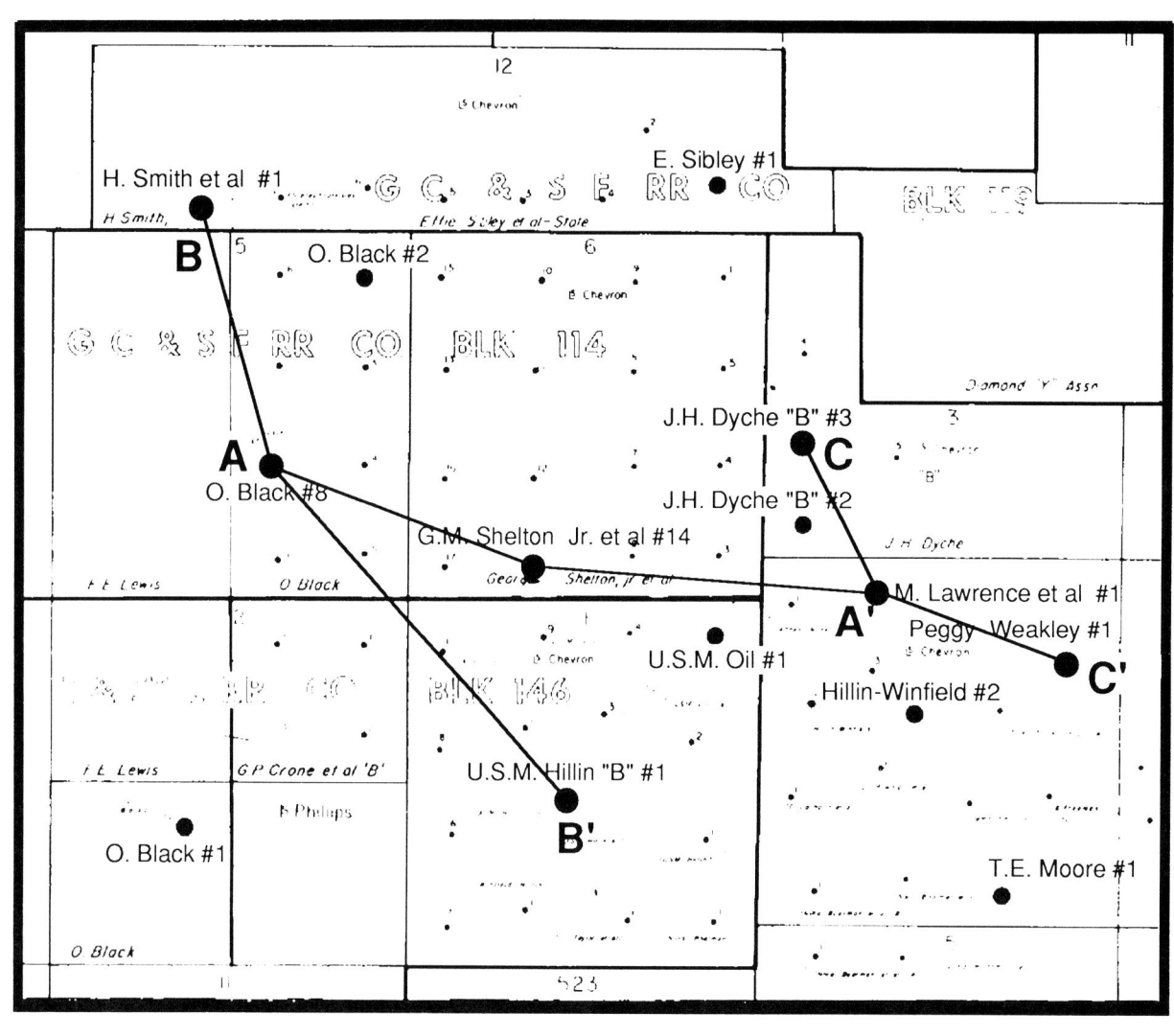

Figure 3
U.S.M. Queen Field map with cored wells highlighted.
(● = cored wells, ● = cored wells used in this study)
Lines of sections correspond to figures 5, 6, and 7.

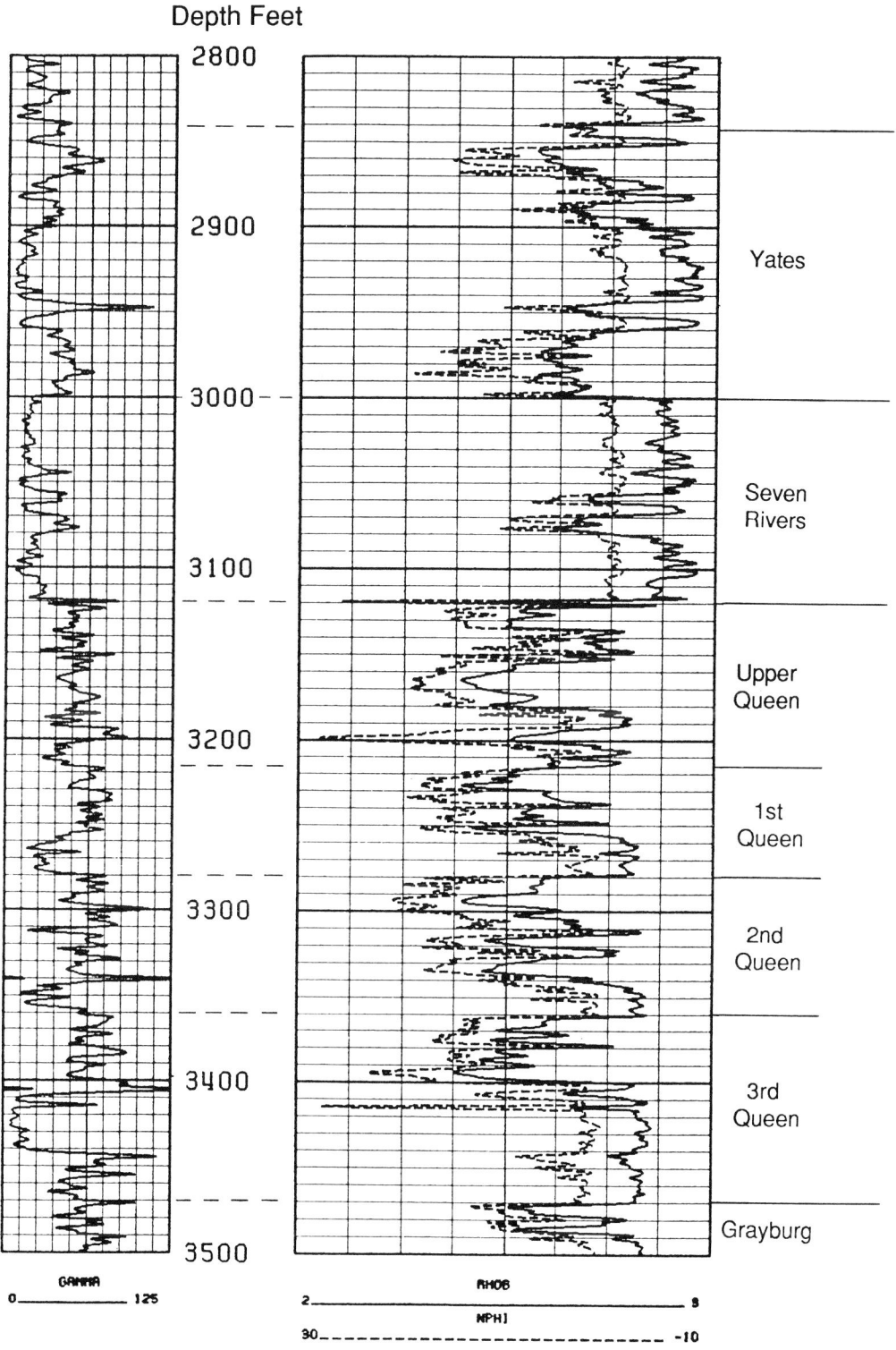

Figure 4
Type log showing pay zones for the Queen Formation, U.S.M. Queen Field, Pecos County, Texas.

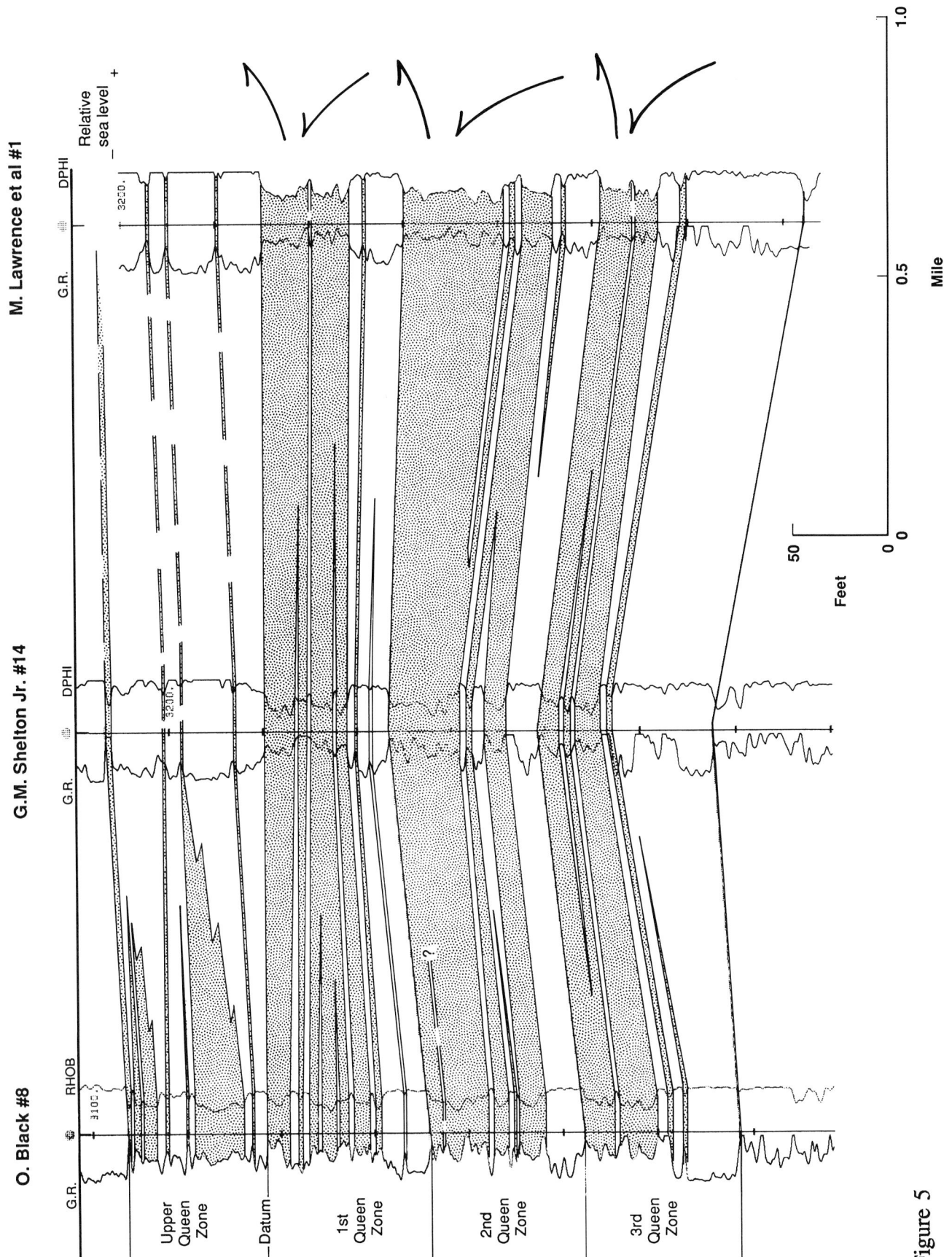

Figure 5

Stratigraphic cross section A-A' oriented west to east (dip direction). Reservoir zones (stippled) are separated by thick, continuous carbonate cycles. Thin carbonate subcycles vary in lateral continuity and are interbedded with reservoir sandstones, creating a complex reservoir stratigraphy. Note the abrupt pinch-out (less than 1 mile; 1.6 km) of the Upper Queen reservoir sandstones.

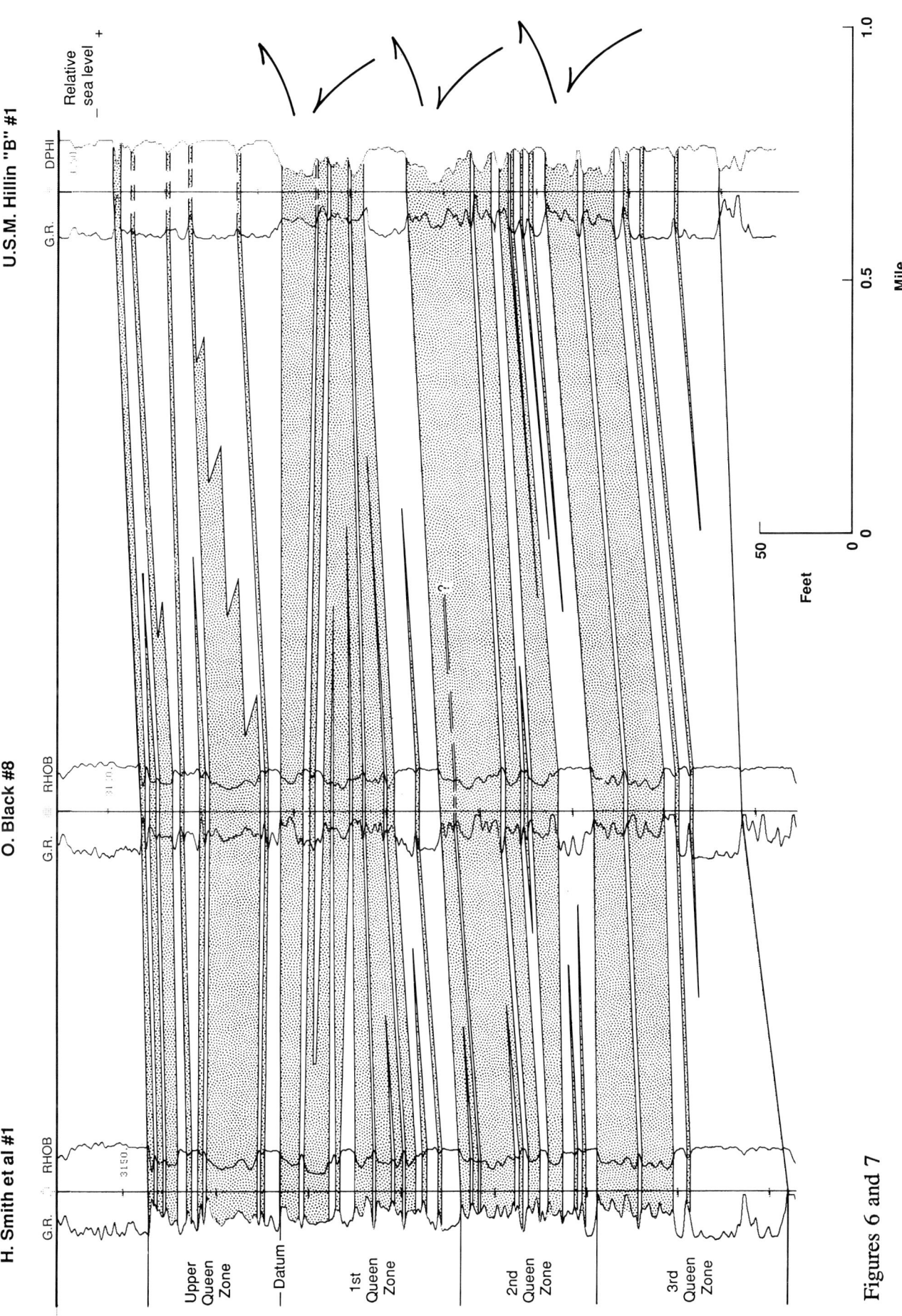

Figures 6 and 7

Stratigraphic cross sections B-B' and C-C' oriented north to south (strike direction). Cross section B-B' (Fig. 6) is located west of cross section C-C' (Fig. 7) and has more thin carbonate subcycles interbedded with the reservoir sandstones. The thin carbonate subcycles represent low-amplitude rises in sea level, which are recorded in the more basinward cross section B-B', but not in the more landward (approximately 2 miles; 3.2 km) cross section C-C'. Note again the abrupt pinch-out of the Upper Queen reservoir sandstones on cross section B-B' and their absence on cross section C-C'.

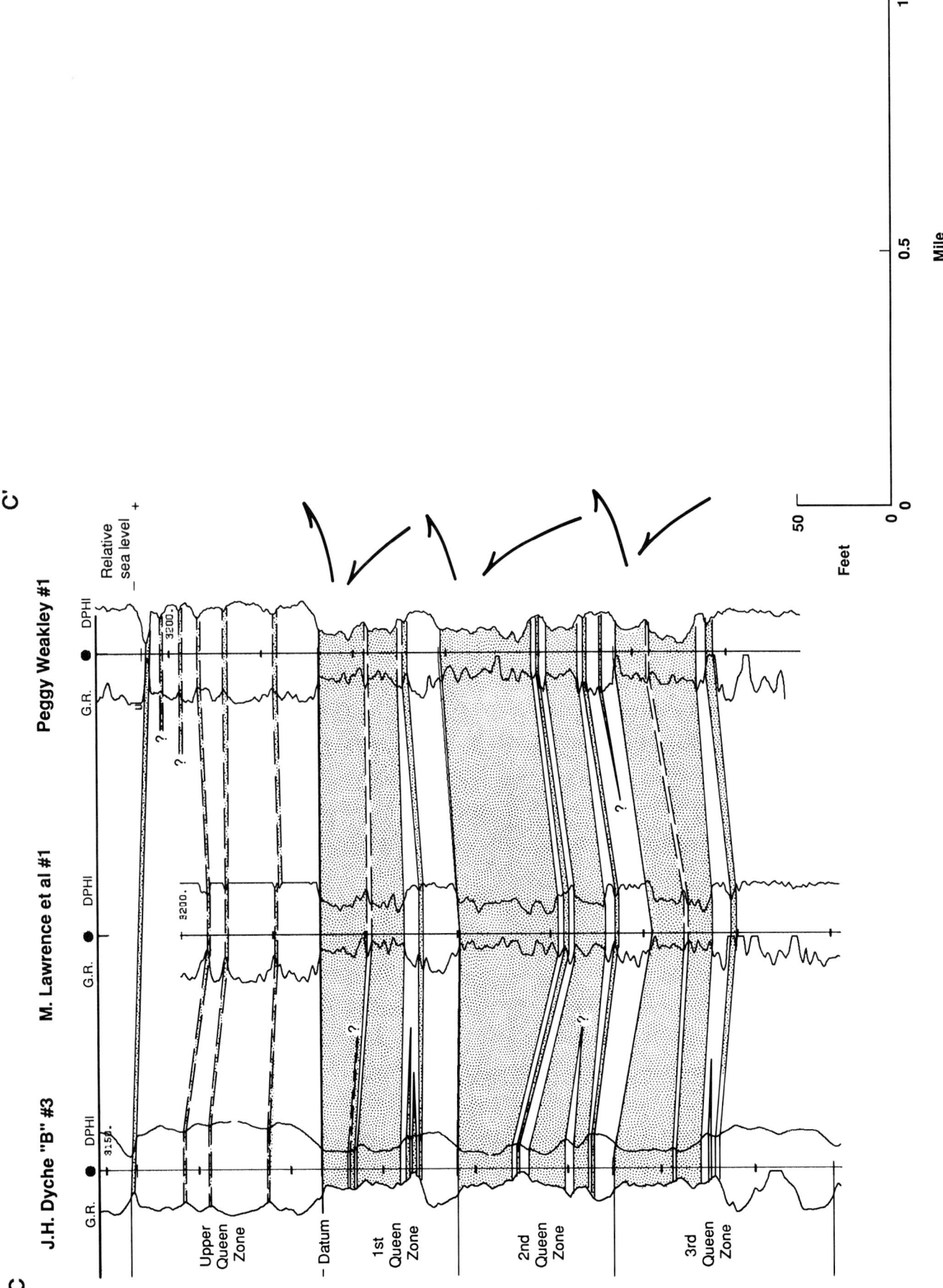

Figure 7

Moran (ibid) gave no indication of the environment of deposition at the type section locality; however, on a recent trip to the area, K. E. Tucker quickly walked the lower part of the section with P. M. Harris, J. M. Borer, and C. W. Grant. We tentatively concluded that the sandstones and dolomites were deposited in shallow marine environments; however, this section needs a more detailed sedimentological description. Faint wavy to planar laminations disrupted by horizontal and vertical burrows suggest a low-energy tidal flat(?) environment for the sandstones. Dolopackstones to dolomudstones, with no clear evidence for subaerial exposure, suggest a low-energy subtidal to intertidal environment for the dolomites.

Franseen (1985) studied the Grayburg and Queen formations on the Western Escarpment of the Guadalupe Mountains. He described the lower 275 feet (84 m) of the outermost 1.2 miles (2 km) of the Queen Formation as shallow water to emergent, peritidal, dolomite facies mixed with massive, burrowed, and cross-bedded sandstones that pinch out at the shelf margin. The cross-bedded sandstones were interpreted as migrating submarine dunes.

A similar outcrop study by Wheeler (1989) describes the sedimentology of the Shattuck Member of the Queen Formation (upper 98-131 ft (30-40 m)) within 1.5 miles (2.5 km) of the Goat Seep shelf margin. He defined three facies belts that roughly paralleled the shelf margin as outer shelf, shelf crest, and inner shelf. The outer shelf was characterized as an open marine, shallow subtidal, high-energy shoreface environment, while tidal flats formed on the shelf crest, and a restricted, low-energy lagoon developed in the inner shelf. He found some evidence of exposure on the shelf crest, but only rare indications of emergence in the outer and inner shelf facies.

J. Mazzullo and students (pers. comm.) have completed subsurface studies of the Shattuck Member of the Queen Formation (16-64 ft (4.9-19.5 m) in thickness) on the Northwest Shelf and the Upper Queen Formation (38-50 ft (11.6-15.2 m) in thickness) on the eastern edge of the Central Basin Platform. The sandstones they describe are dominantly nonmarine in origin and interbedded with evaporites and dolomudstones that are sabkha and playa deposits. The location of the fields studied suggests that these rocks were deposited 24 to 48 miles (40-80 km) behind the shelf margin.

CORE STUDY

Seven cores from the U.S.M. Queen Field, totalling 1250 feet (381 m), were described; 11 lithofacies were identified based on fabric, detrital grain size, and rock color. The lithofacies may be grouped as follows:

Reservoir

1. Gray Sandstone (GSS)

Possible Reservoir

2. Gray Siltstone (GSL)
3. Red Sandstone (RSS)
4. Dolomite (DL)

Nonreservoir

5. Gray and Red Sandstone (RGSS)
6. Gray and Red Siltstone (RGSL)
7. Anhydrite (A)
8. Gray Shale (GSH)
9. Red Siltstone (RSL)
10. Red Shale (RSH)
11. Anhydritic Dolomite (ADL)

Gray Sandstone

Gray, brownish gray, or greenish gray sandstones are the primary reservoir rock at the U.S.M. Queen Field (Fig. 8A and B). The sandstones are typically composed of medium- to very fine-grained sand and are often silty and/or argillaceous. Distinct bedding is absent in the Gray Sandstones, possibly due to bioturbation (Fig. 8D); however, faint, discontinuous, wavy laminations are common, and planar or ripple laminations are present. Pyrite nodules are present in some Gray Sandstones (Fig. 8C).

Gray Siltstone

Gray, green-gray, or brown-gray siltstones (Fig. 9A) are the most common lithofacies observed at the U.S.M. Queen Field. Gray Siltstones contain variable amounts of sand, clay, and dolomite, which makes classification of the rock difficult. Lamination in the Gray Siltstones is typically wavy, although planar and lenticular laminations are present. Convoluted and bioturbated beds are relatively common. Dolomite tightly cements most of the Gray Siltstones, which frequently grade into the Dolomite Lithofacies. Small pyrite nodules are common.

Red Sandstone

The Red Sandstone Lithofacies is composed of dark to light red sandstones (Fig. 9B) that are frequently mottled. Red sandstones are medium- to very fine-grained with variable amounts of silt and clay. Wavy and ripple lamination are common when bedding is not disrupted by bioturbation (Fig. 9C). Variable amounts of dolomite cement are present, which cause the Red Sandstones to be poorly to well consolidated.

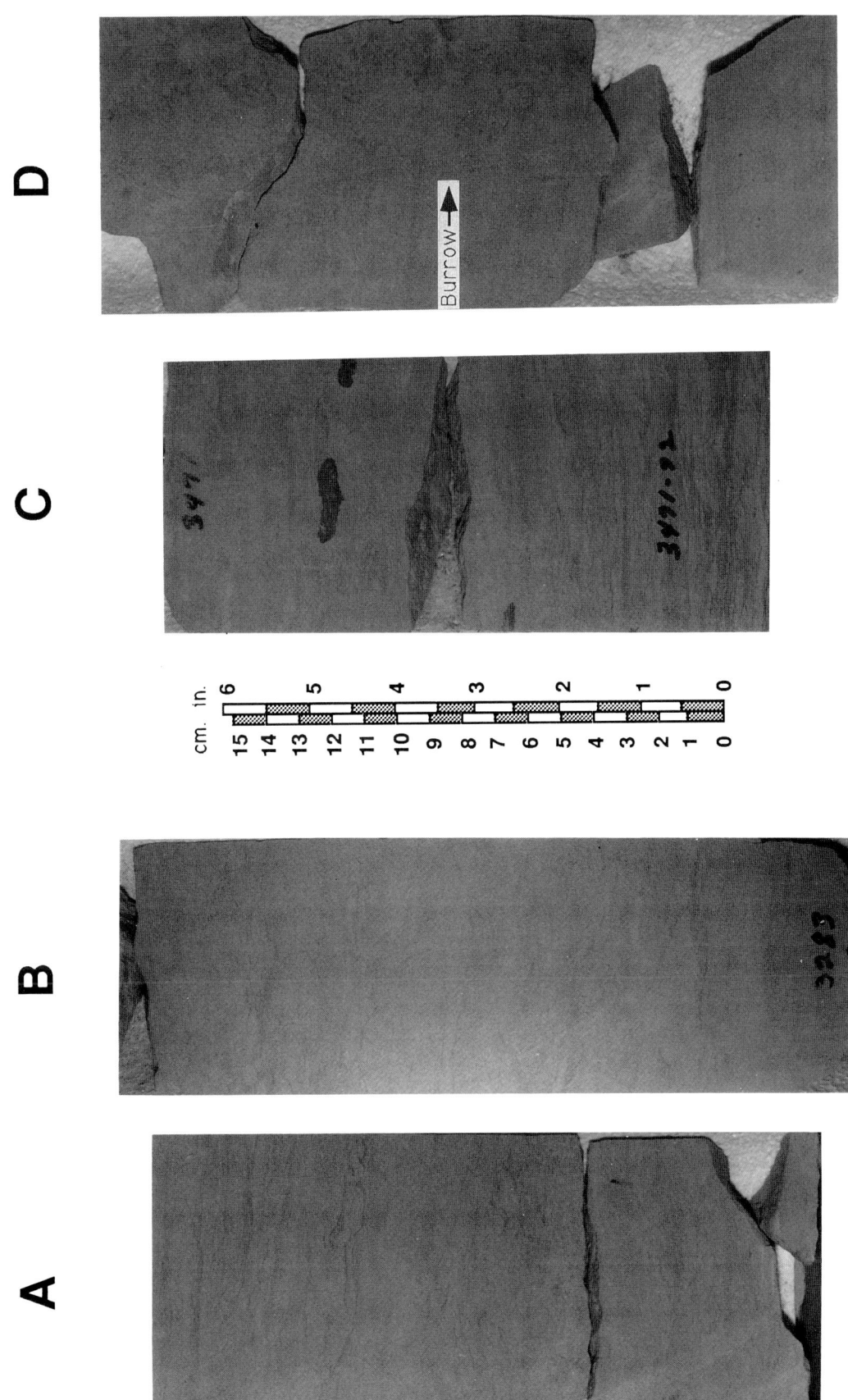

Figure 8

The Gray Sandstone is the primary reservoir in the U.S.M. Queen Field. (A) Discontinuous, wavy laminations are the most common physical structure in the Gray Sandstone. Small, horizontal burrows are also common. Sample: H. Smith et al. #1, 3201.3'-3202.0'. (B) Gray Sandstone with discontinuous, wavy laminations and possible burrows. Sample: J. H. Dyche "B" #3, 3282.2'-3283.0'. (C) Pyrite nodules in a wavy- to ripple-laminated Gray Sandstone. Sample: J. H. Dyche "B" #3, 3471.0'-3471.6'. (D) Gray Sandstone with large vertical burrow, possibly a *Conichnus*. Sample: H. Smith et al. #1, 3204.1'-3204.9'.

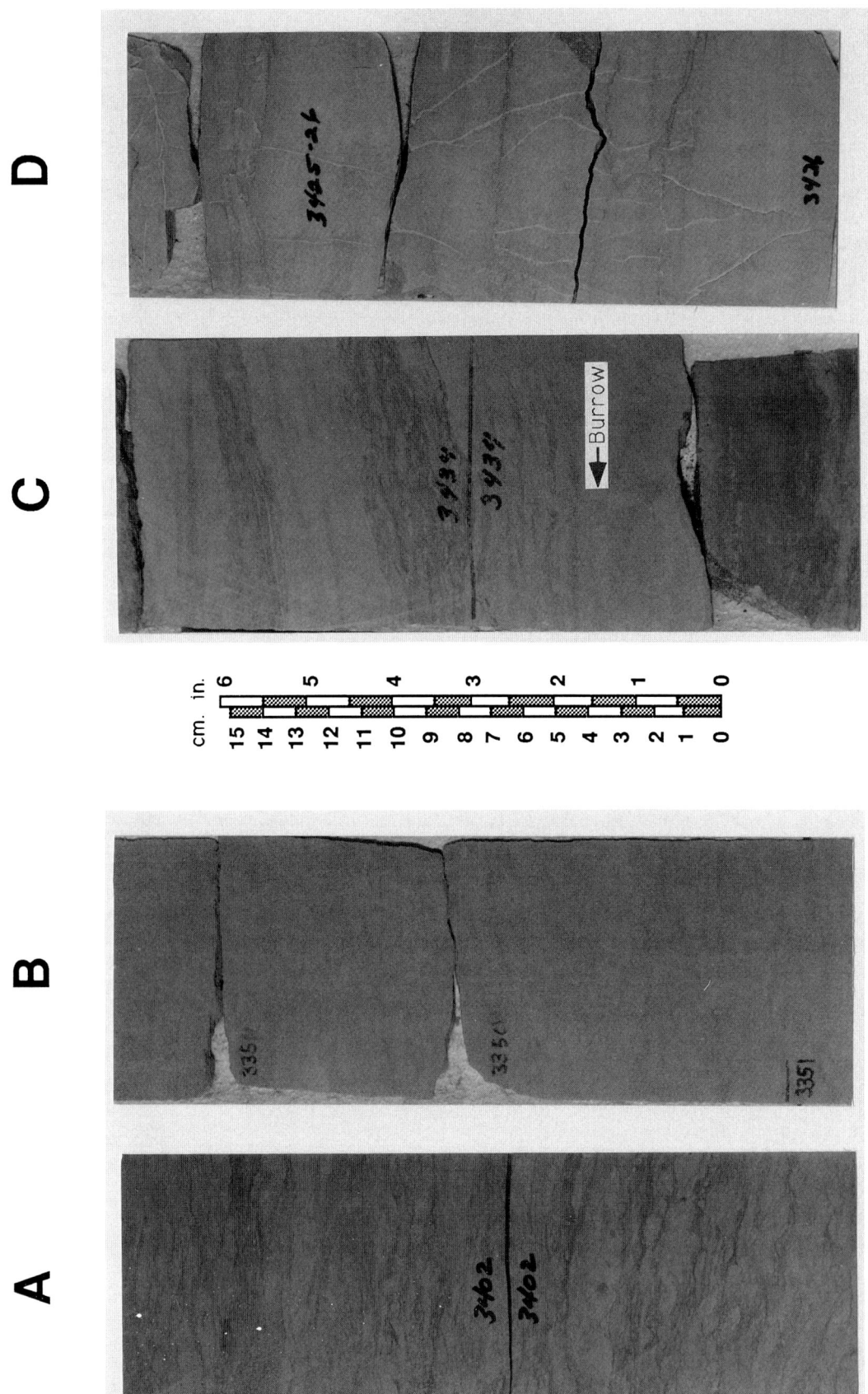

Figure 9

The Gray Siltstone, Red Sandstone, and Dolomite are possible reservoir targets in the U.S.M. Queen Field. (A) Discontinuous, wavy to lenticular laminations are common physical structures in the Gray Siltstone. Small, horizontal burrows are also common. Sample: G. M. Shelton, Jr. et al. #14, 3401.6'-3402.4'. (B) Red Sandstone with faint discontinuous, wavy laminations. Sample: H. Smith et al. #1, 3350.3'-3351.1'. (C) Small vertical burrow through ripple-laminated Red Sandstone. Sample: J. H. Dyche "B" #3, 3433.7'-3434.2'. (D) Massive dolomudstone with stylolites and fractures. Sample: J. H. Dyche "B" #3, 3425.1'-3426.0'.

Dolomite

Queen Formation dolomites are mudstones (Fig. 9D) to packstones (Fig. 10A) that are gray, green-gray, brown, or tan in color. Red to pink coloration is caused by iron-oxide staining of light brown to cream-colored dolomite. Dolomites often contain significant amounts of sand, silt, or clay, and thin, organic-rich layers. Algal-laminated dolomites (Fig. 10B) and intraclast dolomites (Fig. 10C and D) are common. Wavy lamination, bioturbation, and extensive stylolitization are present locally. Moldic and vuggy porosity are common, and vertical fractures are frequently oil-stained (Fig. 11A).

Gray and Red Sandstone

Mottled sandstones colored red and gray (Fig. 11B) by oxidation and reduction reactions are common. The composition, grain size, bedding, and sedimentary structures are the same as the Gray or Red Sandstone lithofacies.

Gray and Red Siltstone

Discontinuous and abrupt color changes from red to gray, due to postdepositional diagenetic processes, are also common in siltstones (Fig. 11C). Mixed-color siltstones have similar sedimentary structures, grain size, and composition to the Gray or Red Siltstone lithofacies.

Anhydrite

Anhydrite occurs as cement that fills intercrystalline or interparticle pores, molds, and fractures. Coarsely crystalline anhydrite also occurs as distinct beds or laminae associated with anhydritic dolomite. Anhydrite is typically light gray or light blue in color and is frequently associated with brecciated dolomite, creating "chicken-wire" anhydrite texture (Fig. 11D).

Gray Shale

Gray Shale occurs in well-laminated (Fig. 12A), fissile thin beds or laminae. The shale coloration ranges from light gray to dark gray, often displaying a green cast. Mottling is present in green-gray shales, which yields a characteristic spotted red appearance. Most Gray Shales contain variable amounts of silt and/or sand.

Red Siltstone

Dark to light red siltstones (Fig. 12B) are one of the most common lithofacies of the Queen Formation. Red Siltstones contain variable amounts of sand, clay, and dolomite. Well-developed wavy lamination, contorted bedding, and bioturbation are present in most Red Siltstones.

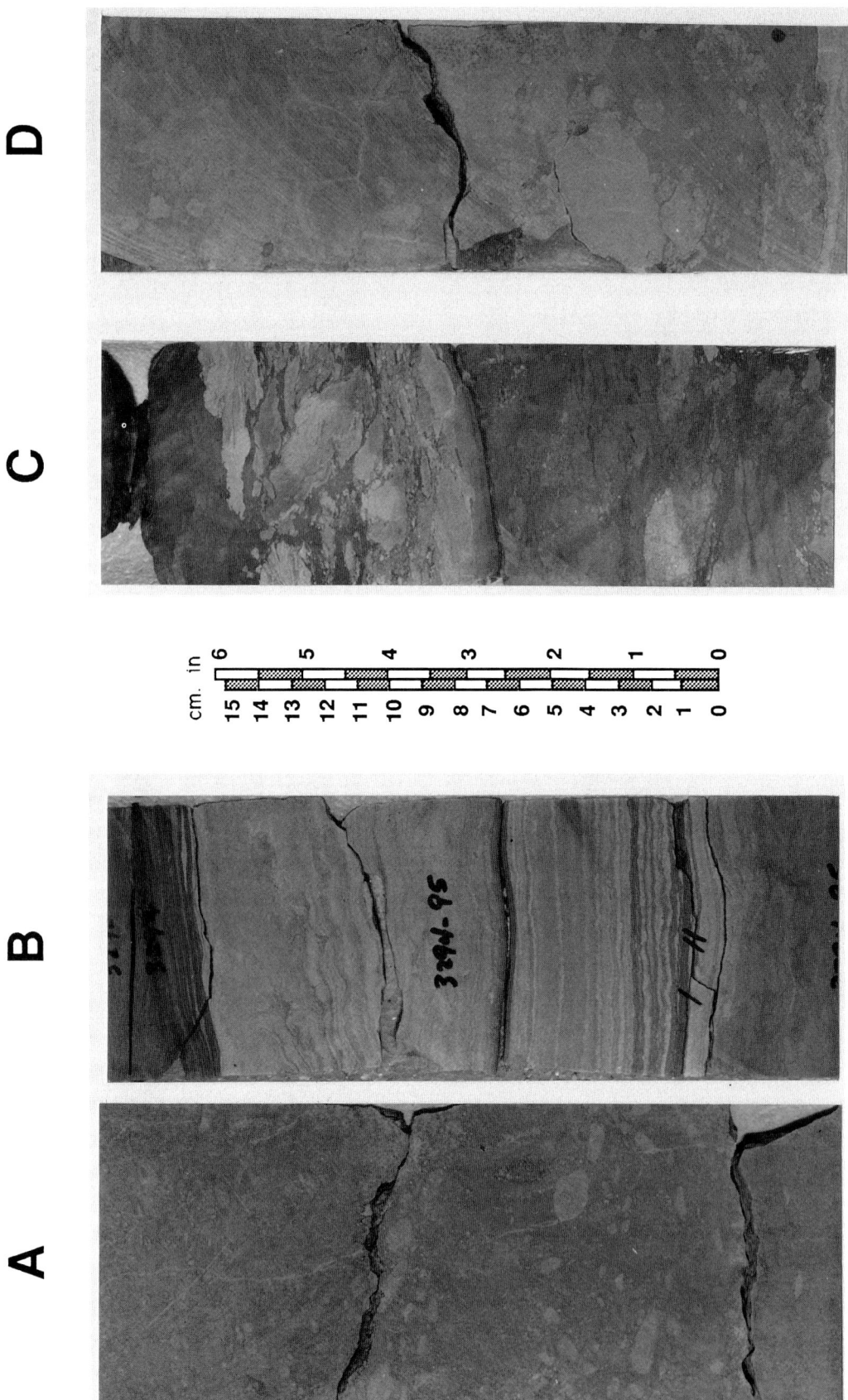

Figure 10

Common textures in the Dolomite Lithofacies. (A) Oncolite dolopackstone. The light gray grains are oncolites. They are irregularly shaped and sized (less than 1 cm; see also Plates 9D and 14A). Sample: H. Smith et al. #1, 3321.0'-3321.7'. (B) Algal-laminated dolowackestone. Crinkly to wavy laminations interbedded with more massive dolowackestone. Sample: G. M. Shelton, Jr. et al. #14, 3294.0'-3294.7'. (C) Dolointraclast breccia. Light gray to pink, angular clasts in a red and gray siltstone matrix. Sample: O. Black #8, 3303.0'-3304.0'. (D) Dolointraclast breccia. Light gray, angular clasts in a gray sandstone matrix. Sample: O. Black #8, 3376.3'-3377.2'.

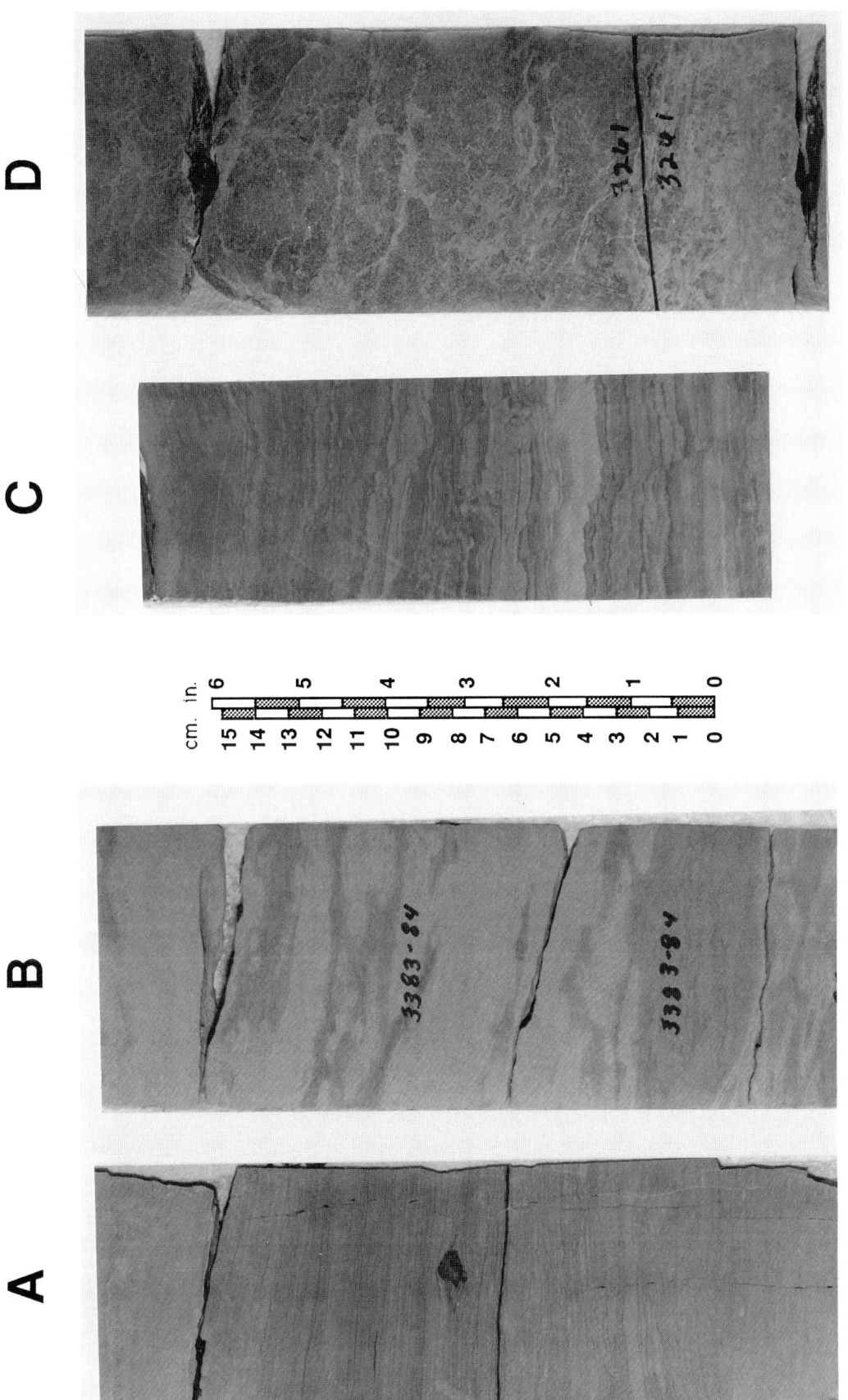

Figure 11

Gray and Red Sandstone, Gray and Red Siltstone, and Anhydrite appear to be nonreservoir lithofacies in the U.S.M. Queen Field. (A) Parallel- to wavy-laminated dolomudstone with pyrite nodule and oil-stained fracture. Laminations may be algal in origin. Sample: O. Blank #8, 3207.8'-3208.7'. (B) Mottled Gray and Red Sandstone with discontinuous, wavy laminations. Sample: J. H. Dyche "B" #3, 3383.1'-3383.9'. (C) Wavy- to lenticular-laminated Gray and Red Siltstone with soft-sediment deformation features (vertical microfaults?). Sample: O. Black #8, 3172.1'-3172.9'. (D) Blue-gray chickenwire anhydrite with dolomite. Sample: J. H. Dyche "B" #3, 3260.4'-3261.2'.

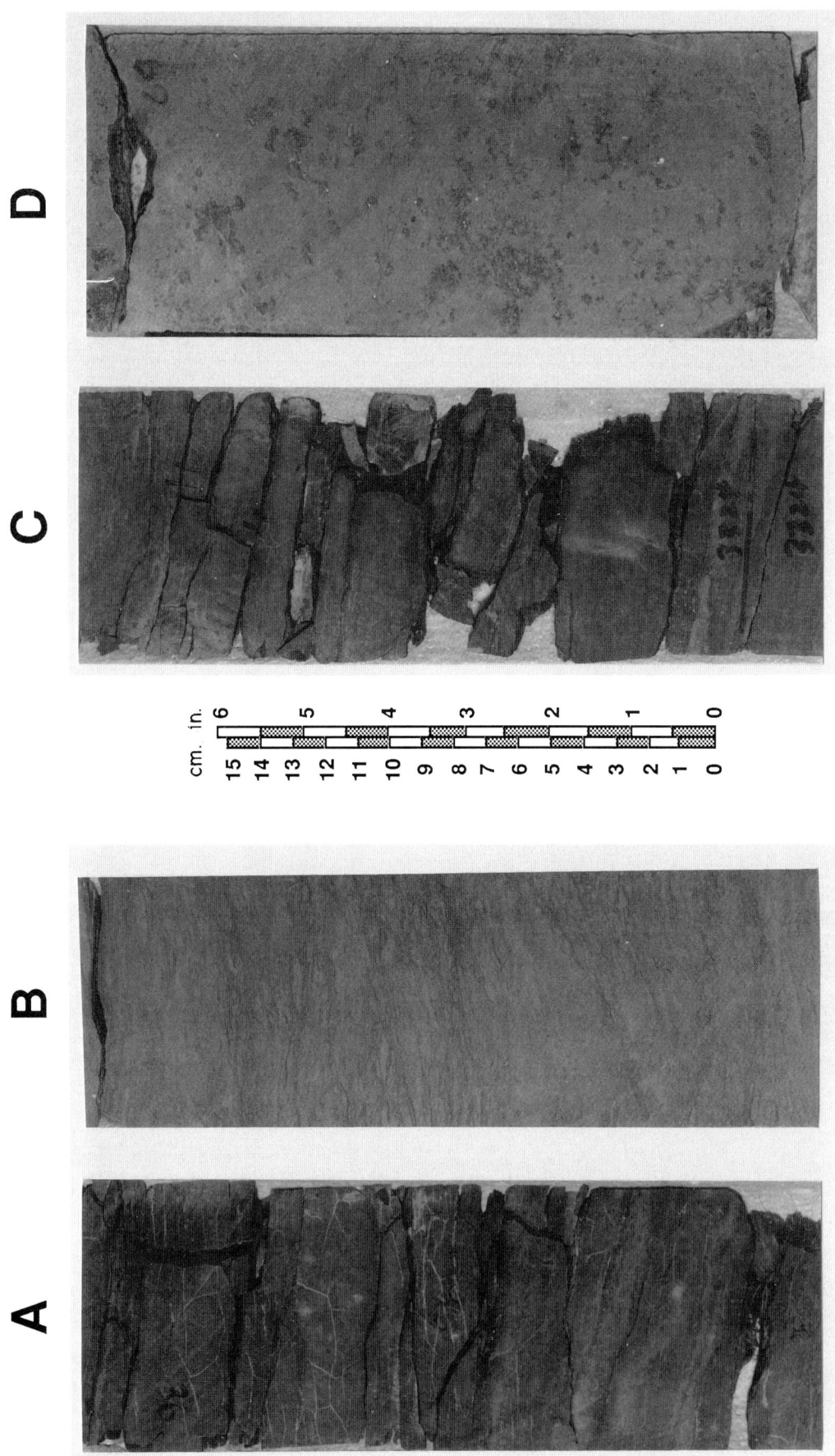

Figure 12

Red Siltstone, Gray Shale, and Red Shale are nonreservoir lithofacies in the U.S.M. Queen Field. (A) Fissile, dark Gray Shale. Sample: H. Smith et al. #1, 3198.0'-3198.9'. (B) Wavy- to lenticular-laminated Red Siltstone with small, horizontal burrows. Sample: G. M. Shelton, Jr. et al. #14, 3371.1'-3371.9'. (C) The Red Shale is fissile and dark red. Sample: G. M. Shelton, Jr. et al. #14, 3323.1'-3323.9'. (D) Anhydrite-filled vugs in dolowackestone/packstone. Sample: H. Smith et al. #1, 3164.3'-3165.1'.

Red Shale

Brick red, fissile shales (Fig. 12C) are commonly associated with the Red Siltstone Lithofacies. The shales are typically well laminated and are often dolomitic.

Anhydritic Dolomite

Gray, brown-gray, cream, or tan mudstone to wackestone with anhydrite cement (Fig. 12D) characterizes this lithofacies. Light gray to faint blue-gray anhydrite occurs as pore- and fracture-filling cement and as small nodules in wavy- or algal-laminated dolomites.

SIGNIFICANCE OF CORE STUDY

Examination of the distribution of lithofacies in each core reveals the existence of cyclical patterns in carbonate and siliciclastic sedimentation. Carbonate cycles and subcycles frequently display deepening followed by shallowing-upward trends, and siliciclastic cycles and subcycles usually coarsen, then fine upward.

For example, the carbonate cycle at the base of the First Queen zone (Figs. 13 and 14A) grades from an anhydritic, silty and argillaceous dolomudstone with abundant dolomudstone intraclasts at the base to an algal-laminated dolopackstone to dolowackestone; both are suggestive of intertidal to supratidal deposition. A bioturbated dolopackstone to dolowackestone overlies this unit and is subtidal in origin. Alternating dolowackestones and dolomudstones punctuated by thin, organic-rich(?) shales record a trend of decreasing fossils and increasing micrite indicative of increasing stress and decreasing water depth on the environment. This unit grades into alternating algal-laminated dolomudstones to dolowackestones and burrowed, argillaceous siltstones, which represent the first pulses of siliciclastic sediment brought on by shallowing and/or increased siliciclastic input. The uppermost unit contains abundant dolomudstone intraclasts in an argillaceous siltstone matrix, which suggests a return of siliciclastic deposition following exposure(?) of the shelf. Carbonate subcycles follow this general pattern of deepening then shallowing, but are thinner (1-5 ft; 0.3-1.5 m).

An example of a typical siliciclastic subcycle from the Second Queen Zone is shown in Figure 14B. A burrowed argillaceous siltstone at the base grades upward into a burrowed silty sandstone. This unit is overlain by alternating bioturbated silty sandstones and wavy- to ripple-laminated silty sandstones. An algal-laminated dolowackestone (less than 1 ft (0.3 m) thick) punctuates this unit and may represent a lull in sediment input and/or a slight sea-level rise(?). The alternating unit is overlain by a burrowed argillaceous and silty sandstone that grades upward into an argillaceous siltstone. This represents waning of sediment input and/or a subsequent gradual rise in sea level. Siliciclastic cycles consist of several of these subcycles stacked together, usually with carbonate subcycles in between (Fig. 13).

J.H. Dyche "B" #3 Core Summary

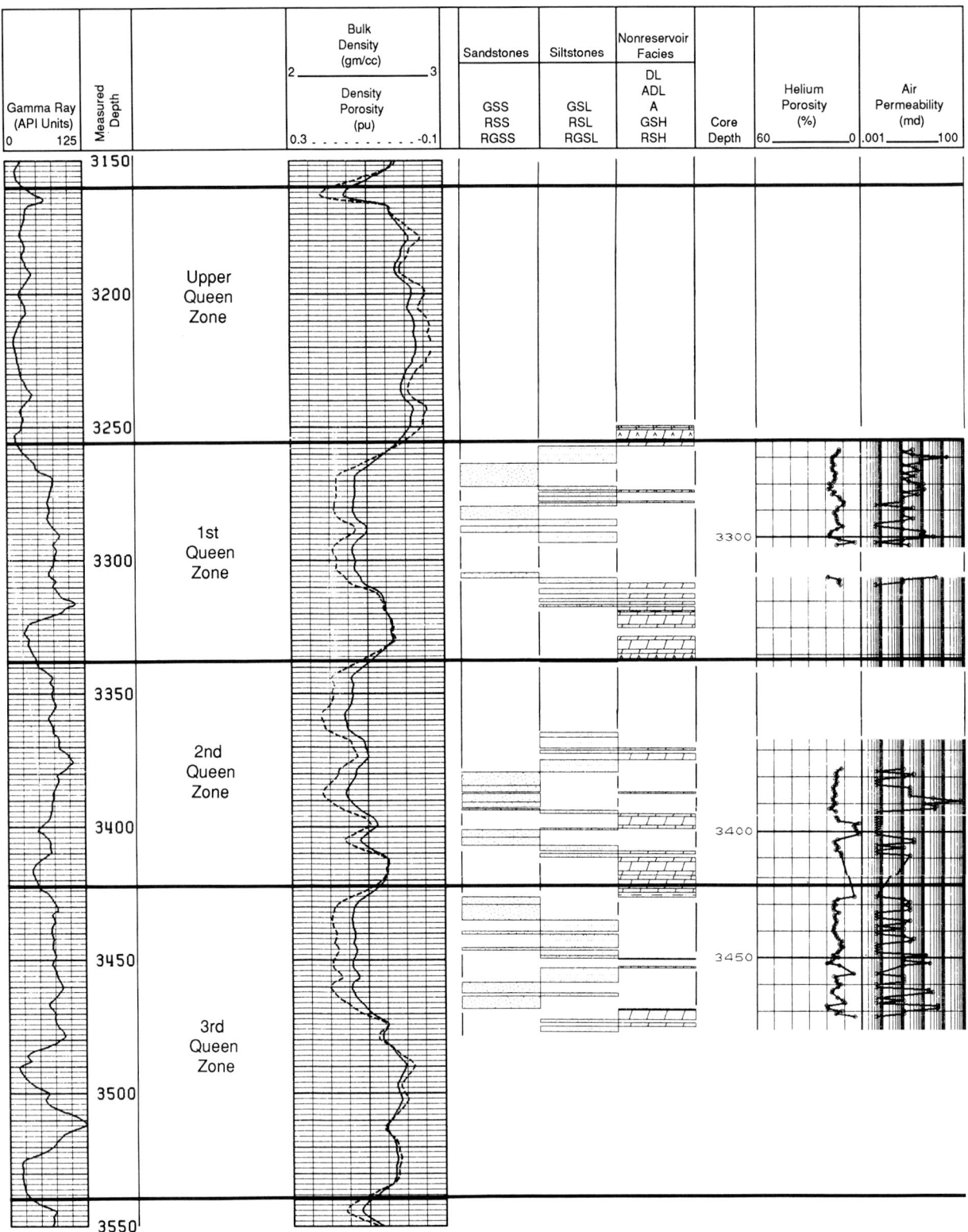

Figure 13
Log and core data summary by zone for J.H. Dyche "B" #3 well.

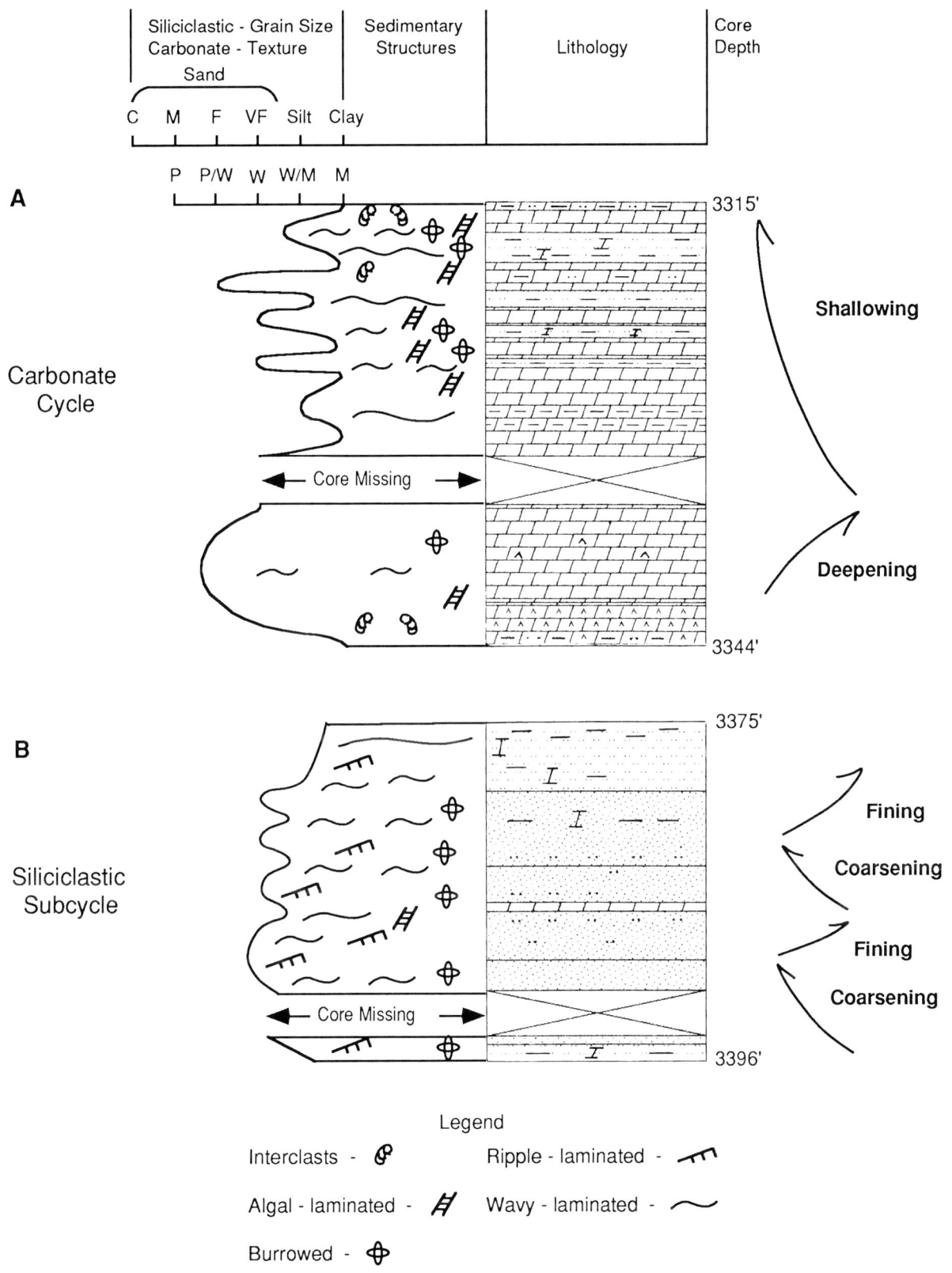

Figure 14
Common cycles in Queen Formation from J.H. Dyche "B" #3 core from U.S.M. Queen Field. Carbonate cycles are typically 20 to 30 feet thick. Siliciclastic subcycles are 5 to 20 feet thick.

PETROGRAPHY

A total of 128 thin sections from 6 wells were studied to determine detrital mineralogy, matrix and cement content and distribution, porosity types and distribution, rock texture, and diagenesis. Rock composition was estimated rather than determined by point-count techniques, and the average gross composition by lithofacies is given below.

	All Sandstone	GSS	RSS	All Siltstone	GSL	RSL	ADL,DL,RSH
Det. grains	74.8	74.4	75.1	68.4	68.3	68.4	28.3
Matrix	6.0	4.2	8.3	12.3	8.2	16.4	62.2
Cement	12.1	12.0	12.3	18.2	22.2	14.2	7.6
Porosity	6.2	8.4	3.4	0.9	1.0	0.8	1.9

Textural parameters (grain size, sorting, and roundness) were determined qualitatively. The Wentworth scale was used for grain-size determination.

Ten SEM samples were analyzed to better identify clay and cement (overgrowths) types, their distribution, and the paragenetic sequence.

Rock Types

The sandstones and siltstones were classified petrographically by their detrital composition. Figure 15 shows that the sandstones and siltstones are dominantly arkosic arenites. However, some sandstones and siltstones classify as arkosic wackes because they have more matrix.

Sandstones and Siltstones

Texture.--The red and gray lithofacies are texturally very similar. The sandstones are dominantly very fine-grained, but grain size ranges from fine-grained sand to silt. The siltstones are primarily silt with variable amounts of very fine-grained sand. In general, the sandstones are very well- to well-sorted. Sandstones with a significant amount of silt and the siltstones are moderately well- to poorly sorted. Grains are subangular to subrounded.

Figure 16 illustrates some additional textural attributes of the sandstones and siltstones. Laminations defined by alternating grain size and grain composition (Fig. 16A) and normally graded laminations (Fig. 16B) are common. Local variations in sorting and heavy mineral concentration in a sample may be the result of bioturbation (Figs. 16C and 17A). Samples with abundant dolomitic intraclasts are poorly sorted (Fig. 16D).

Detrital grains.--The average detrital grain composition is very similar for the gray and red lithofacies (Fig. 15). Quartz is the most abundant, averaging 47 percent. Potassium feldspar and plagioclase feldspar together average 30 percent. Rock fragments consisting of mostly chert

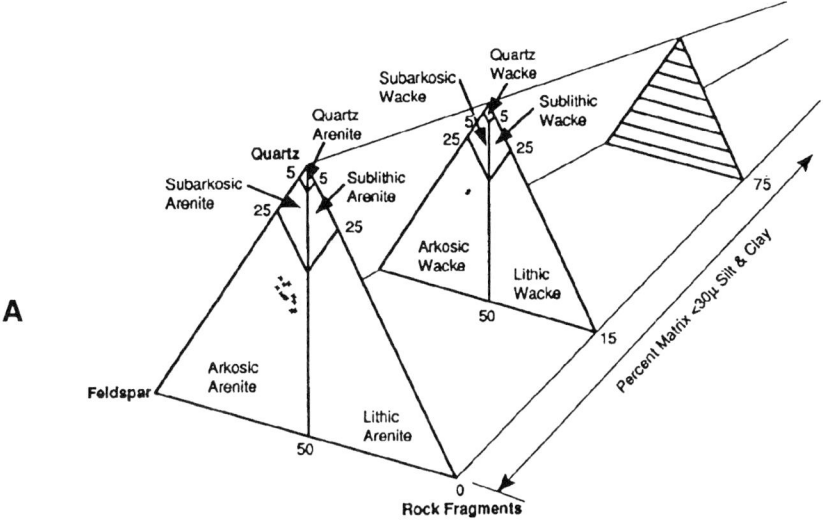

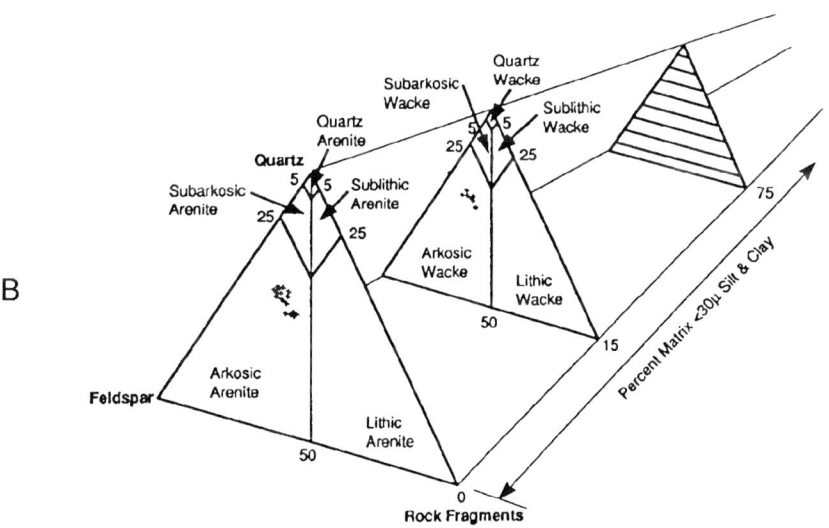

Modified from:
Pettijohn, F.J., P.E. Potter, and R. Siever (1972) Sand and Sandstone, Springer - Verlag, New York, p. 158.

Figure 15
Petrographic classification of sandstones (A) and siltstones (B) from wells
J.H. Dyche #3, USM. Hillin "B" #1, Marie Lawrence #1, and G.M. Shelton #14.

Figure 16

The sandstone and siltstone lithofacies have similar textures. (A) Laminations are commonly defined by alternating grain size and have heavy mineral trains or lags. Sample: U.S.M. Hillin "B" #1, 3266.1'. Plane polarized light, x7.7. (B) Normal grading is common in laminations, suggesting deposition by currents. Sample: G. M. Shelton, Jr. #14, 3428.4'. Plane polarized light, x79. (C) Burrow-fill is often cleaner and better sorted than the host sediment. Note the concentration of heavy minerals along the burrow wall. Sample: G. M. Shelton, Jr. et al. #14, 3374.6'. Plane polarized light, x79. (D) Dolomite intraclasts are often deformed into pseudomatrix due to compaction. Sample: J. H. Dyche "B" #3, 3367.1'. Plane polarized light, x31.

Figure 17

The Gray Siltstone lithofacies are typically tightly cemented arkosic arenites, while the Red Siltstone lithofacies are clay-rich arkosic wackes. (A) The Gray Siltstones are commonly laminated; however, the laminations may be partially to totally disrupted by burrows. Sample: Marie Lawrence et al. #1, 3265.1'. Plane polarized light, x7.7. (B) Similarly, the red siltstones are commonly laminated, although the laminations may be disrupted by burrows. Sample: Marie Lawrence et al. #1, 3310.5'. Plane polarized light, x7.7. (C) Fine dolomite crystals and minor pyrite tightly cement this Gray Siltstone. Sample: Marie Lawrence et al. #1, 265.1'. Plane polarized light, x200. (D) Fine dolomite crystals tightly cement this argillaceous Red Siltstone. Dolomitic intraclasts (center of photo) are common in some samples. Sample: Marie Lawrence et al. #1, 3310.5'. Plane polarized light, x200.

Figure 18

The Red and Gray Sandstone lithofacies are typically arkosic arenites with variable amounts of intergranular porosity, clay matrix, and cement. (A) At low magnification the Gray Sandstones commonly appear massive to mottled and often vary from gray to red in the same thin section. Sample: G. M. Shelton, Jr. et al. #14, 3291.6'. Plane polarized light, x7.7. (B) Similarly, the Red Sandstones often appear massive to mottled at low magnification. Sample: U.S.M. Hillin "B" #1, 3330.0'. Plane polarized light, x7.7. (C) Most Gray Sandstones have intergranular porosity and some dissolution porosity. Thin clay rims are very common, but only recognizable in fairly porous samples. Dolomite cement is present in most samples as fine (less than 10 µm) to coarse (pore-filling) rhombic crystals. Sample: G. M. Shelton, Jr. et al. #14, 3291.6'. Plane polarized light, x200. (D) Good intergranular porosity and possibly some dissolution porosity are present in this Red Sandstone. The thin clay rims in this sample appear to be stained by oil and/or iron-oxide. Fine to coarse dolomite crystals are also present. Sample: U.S.M. Hillin "B" #1, 3330.0'. Plane polarized light, x200.

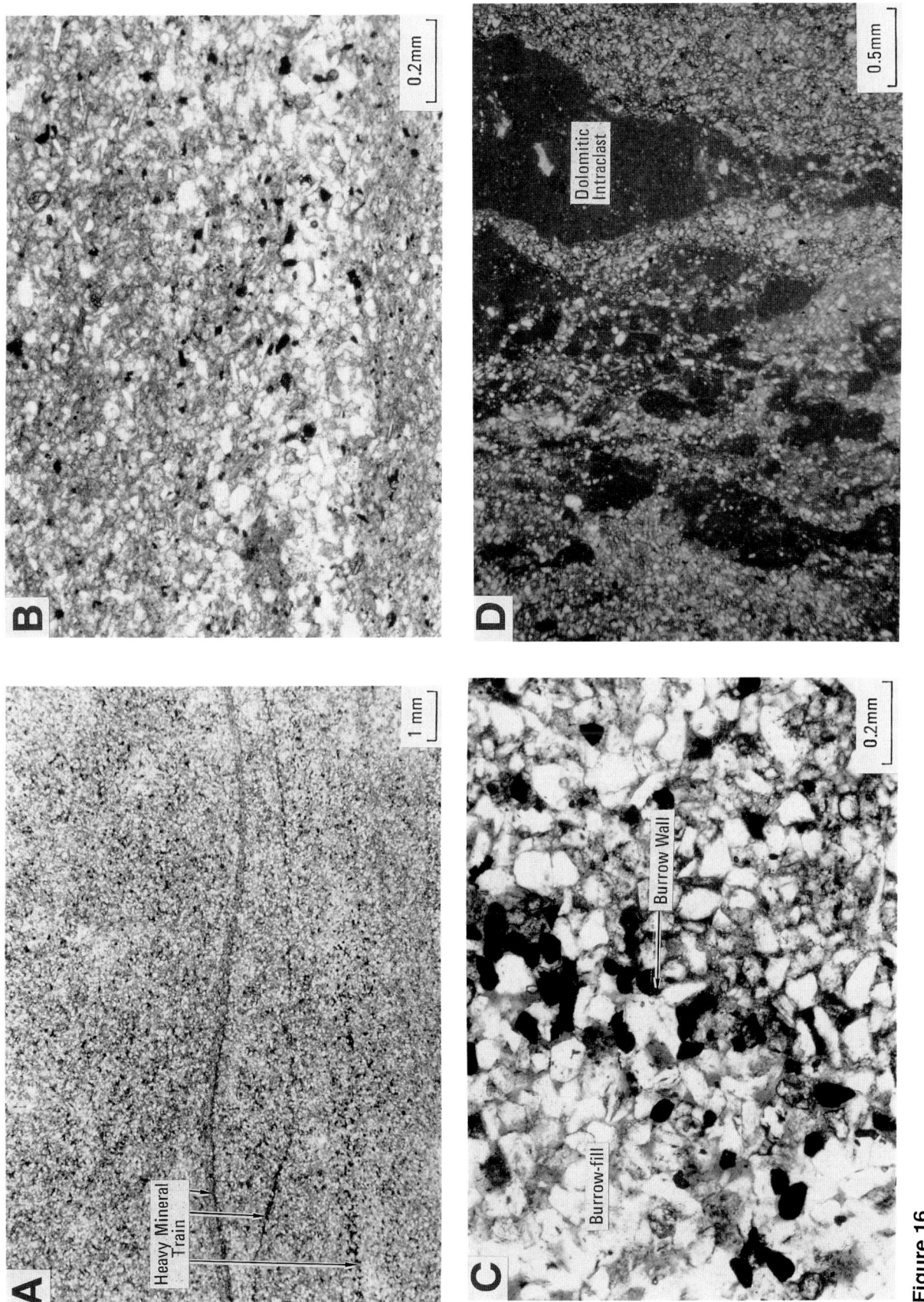

Figure 16

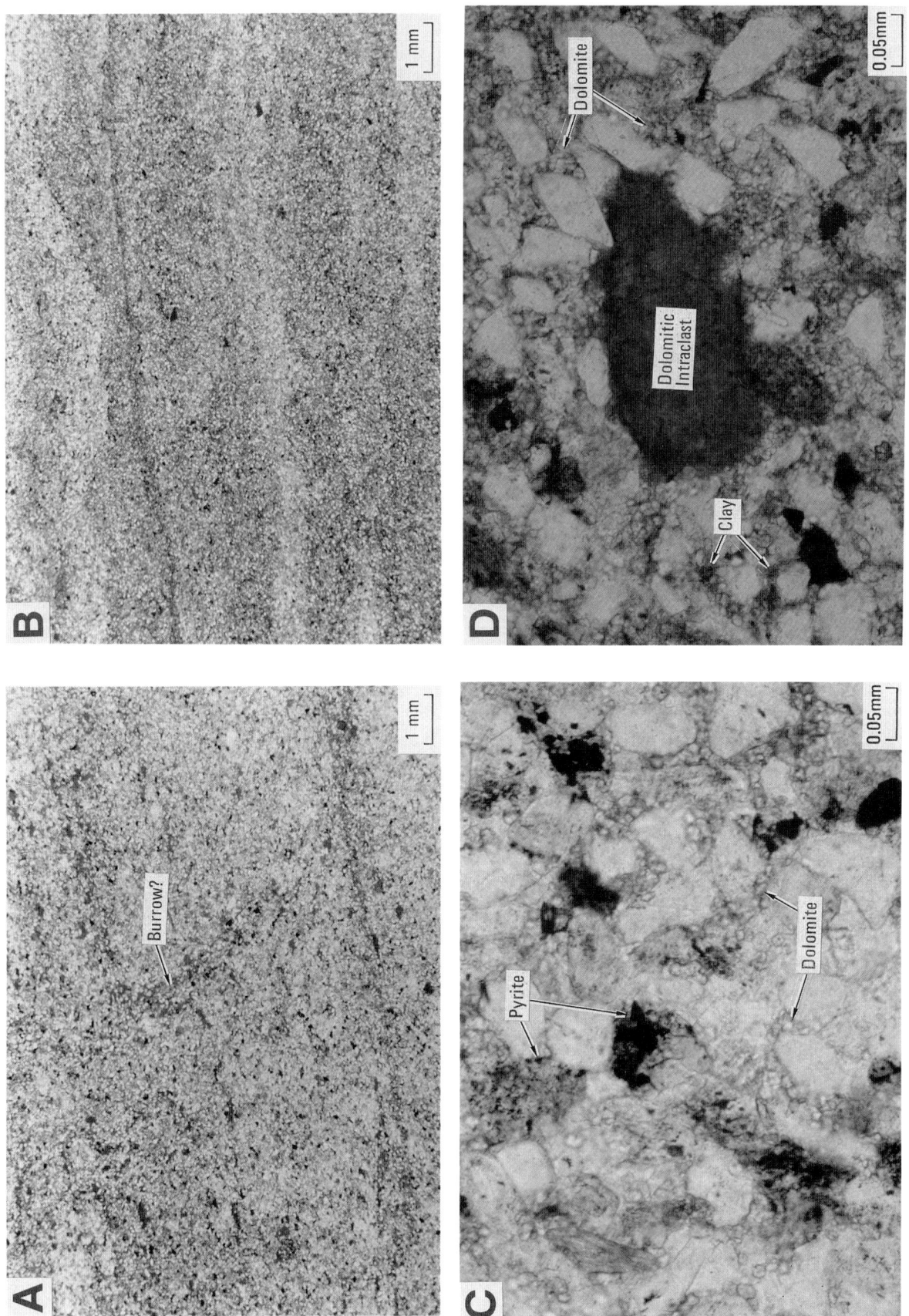

Figure 17

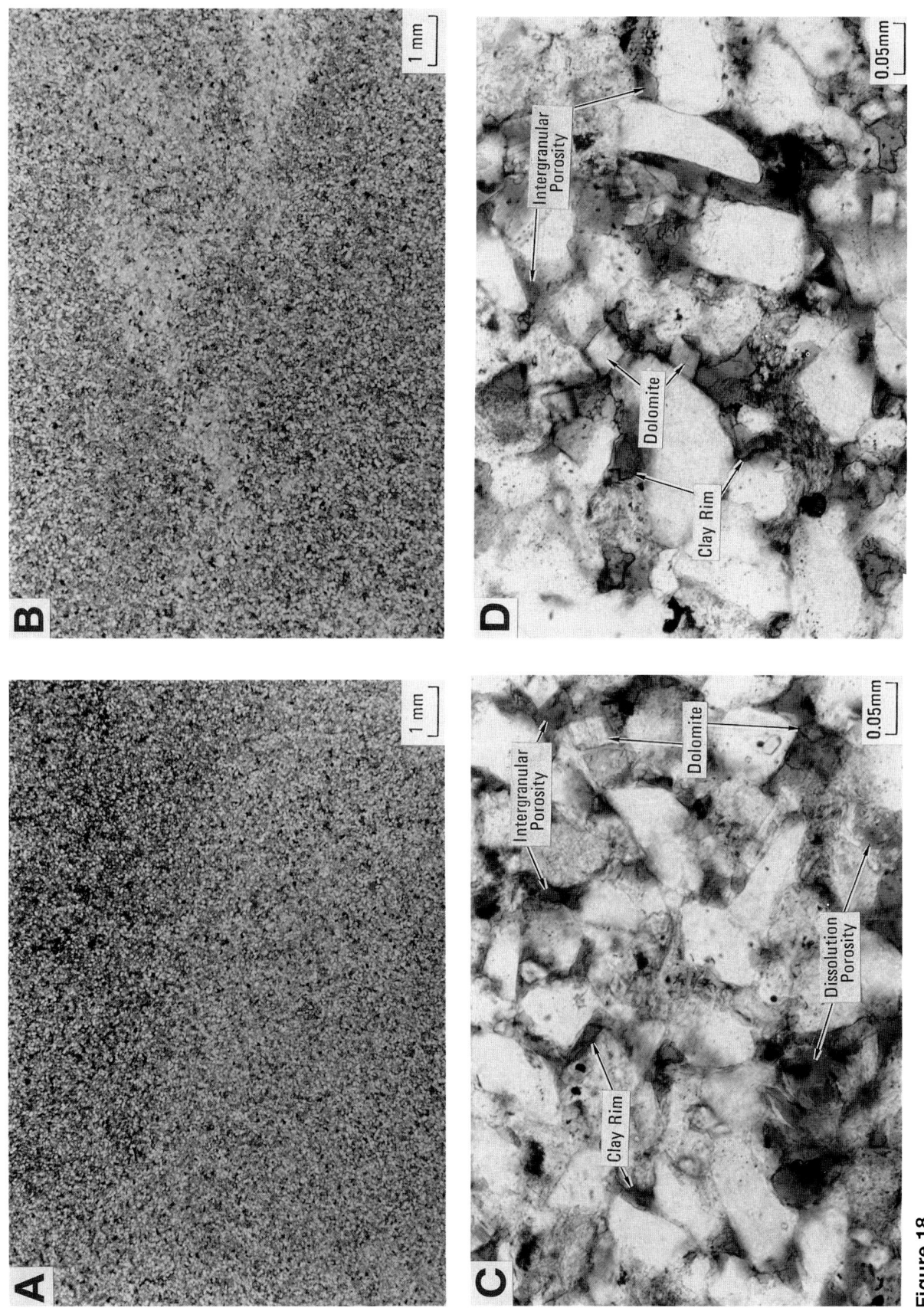

Figure 18

and metasedimentary grains(?) average 15 percent. Muscovite, biotite, and chlorite make up the mica fraction, which averages 6 percent. A variety of heavy minerals, including ilmenite, zircon, tourmaline, and epidote, average 2 percent combined.

Matrix.--Clay matrix occurs as laminations, lenses (Fig. 17A and B), and disseminated pore-fill (Figs. 16B and C, and 17D). Distinct grain rims (Figs. 18C and D, and 20C) were considered authigenic clay because of their light color and delicate pore-lining habit; however, this clay might be recrystallized detrital clay matrix. The clay (whether matrix or authigenic) is probably corrensite (Fig. 19).

In general, the siltstones are more argillaceous than the sandstones. The average amount of clay matrix in the siltstones is about twice that in the sandstones. In addition, the red lithofacies are more argillaceous than the gray lithofacies.

Cement.--Authigenic cements are present in all samples. Dolomite is the most common, averaging 12.3 percent for sandstones and siltstones combined. It occurs as microcrystalline rhombs 1 to 10 µm in diameter (Figs. 17C and D, 20B, and 21B and C) and coarser pore-filling crystals (Figs. 18C and D, and 21A). Calcite cement is most common in the gray lithofacies. In some samples, large poikilotopic crystals fill all intergranular space; whereas, in other samples, single crystals cement patches of grains (Fig. 20A).

Minor cements are silica, feldspar, pyrite, gypsum, and anhydrite. Silica cement is most abundant in the sandstones, especially the Red Sandstones. In thin section, silica cement occurs as quartz overgrowths or pore-filling cement with no crystal faces (Fig. 20C and D). Quartz overgrowths of various sizes are present in all SEM samples (Fig. 21), but are volumetrically minor. Feldspar overgrowths (mostly K-spar) were rarely identified in thin section, but were commonly observed in SEM (Fig. 21). Like silica cement, feldspar overgrowths are volumetrically minor. Pyrite is most common in the gray lithofacies, especially the Gray Sandstones. It occurs as small crystals and crystal aggregates commonly mixed with clay (Fig. 17C). Gypsum and anhydrite are also most common in the gray lithofacies. They occur as large poikilotopic crystals that cement patches of grains (Fig. 20B).

The siltstones are slightly more cemented than the sandstones. The Gray Siltstones are more cemented than the Red Siltstones, but the Gray and Red Sandstones are equally cemented.

Porosity types and distribution.--Primary intergranular porosity is the most common porosity type (Fig. 18C and D). Minor amounts of secondary dissolution porosity are associated with dissolved plagioclase grains and rock fragments (Fig. 18C). In addition, some primary intergranular porosity may actually be secondary dissolution porosity associated with the dissolution of calcite and/or anhydrite cement. No conclusive evidence was found, but a few samples with patchy calcite or anhydrite are suspicious because anhedral patches of cement adjacent to areas with open porosity are in optical continuity. Minor amounts of ineffective microporosity are associated with the clay matrix and microcrystalline dolomite cement.

Figure 19

XES (X-ray Energy-dispersive Spectra) analysis of the clay from all samples typically yields a spectrum with major Si, Mg, Al, and Fe peaks. The smectite-like morphology, relatively high Mg and Fe peaks, and typical grain-rimming habit suggest this clay is corrensite. Sample: Marie Lawrence et al. #1, 3415.7'. SEM photomicrograph, x3000.

Figure 20

Calcite, anhydrite, and silica cements are minor diagenetic minerals in this sample set. (A) Coarse poikilotopic calcite has a patchy distribution. It occurs primarily in Gray Sandstones and Siltstones. Sample: Marie Lawrence et al. #1, 3343.5'. Cross polarized light, x79. (B) Patchy anhydrite cement is present in some samples of both sandstone and siltstone lithofacies. Sample: Marie Lawrence et al. #1, 3265.1'. Cross polarized light, x79. (C) Patchy silica cement is observed in only a few sandstone samples; however, small to large quartz overgrowths were observed in most of the SEM samples. Sample: U.S.M. Hillin "B" #1, 3330.0'. Plane polarized light, x200. (D) Same as "C" with cross polarized light.

Figure 21

Diagenesis is very similar for the sandstone and siltstone lithofacies. (A) Dolomite (rhombic crystals) and corrensite (clay coating on most grains) have not filled all pore spaces in most sandstones. Sample: J. H. Dyche "B" #3, 3391.2'. SEM photomicrograph, x400. (B) The siltstones have little or no porosity. Dolomite cement (rhombic crystals) and corrensite (clay coating on most grains) occlude most of the original porosity. Quartz and potassium feldspar overgrowths also reduce porosity. Sample: G. M. Shelton, Jr. et al. #14, 3374.4'. SEM photomicrograph, x300. (C) Potassium feldspar and quartz overgrowths also reduce porosity in some sandstones. Sample: G. M. Shelton, Jr. et al. #14, 3426.4'. SEM photomicrograph, x1000. (D) Corrensite apparently crystallized first on grain surfaces. Dolomite followed or was contemporaneous with the end of corrensite crystallization. Quartz overgrowths formed after corrensite and dolomite. Sample: J. H. Dyche "B" #3, 3299.6'. SEM photomicrograph, x1200.

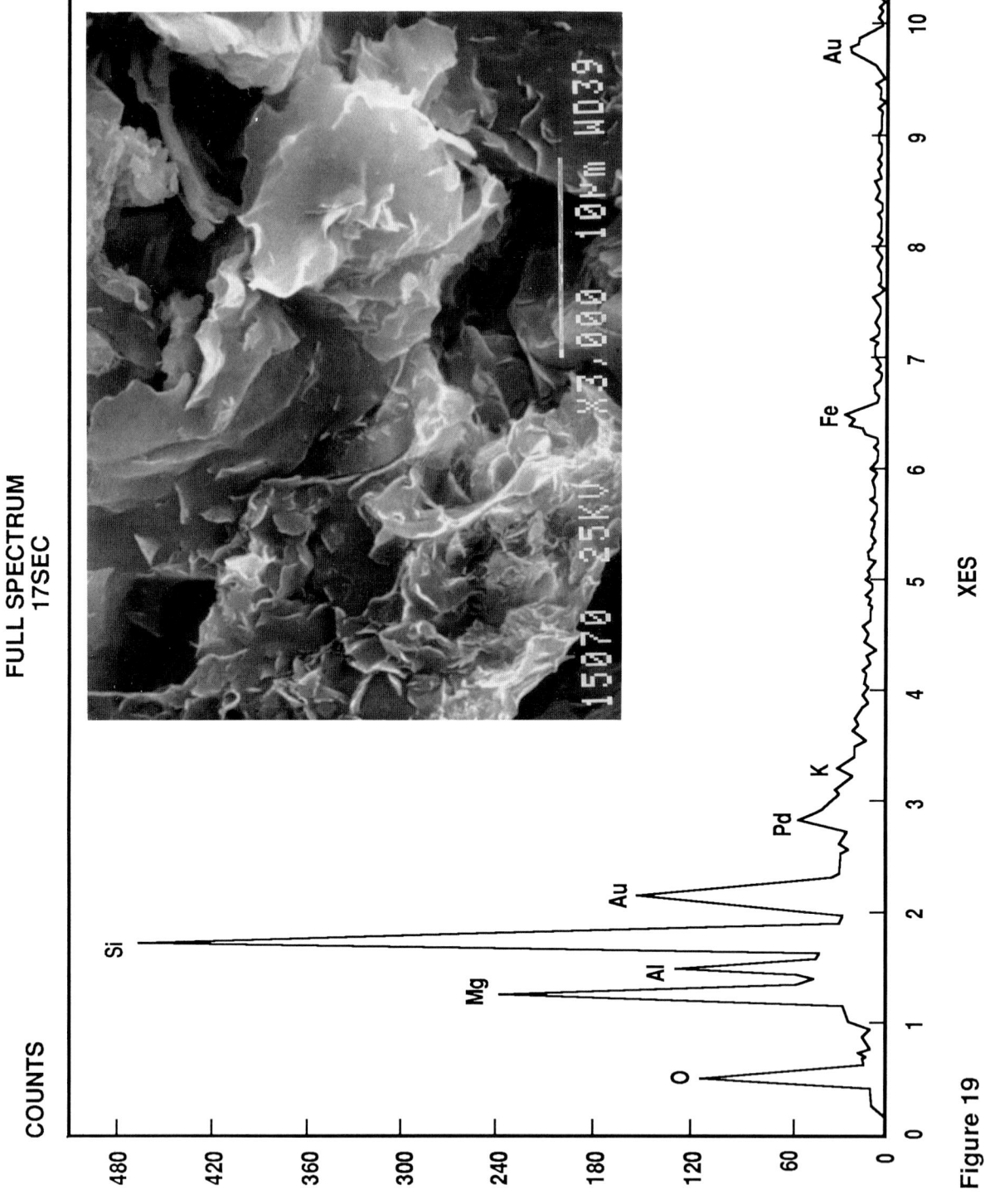

Figure 19

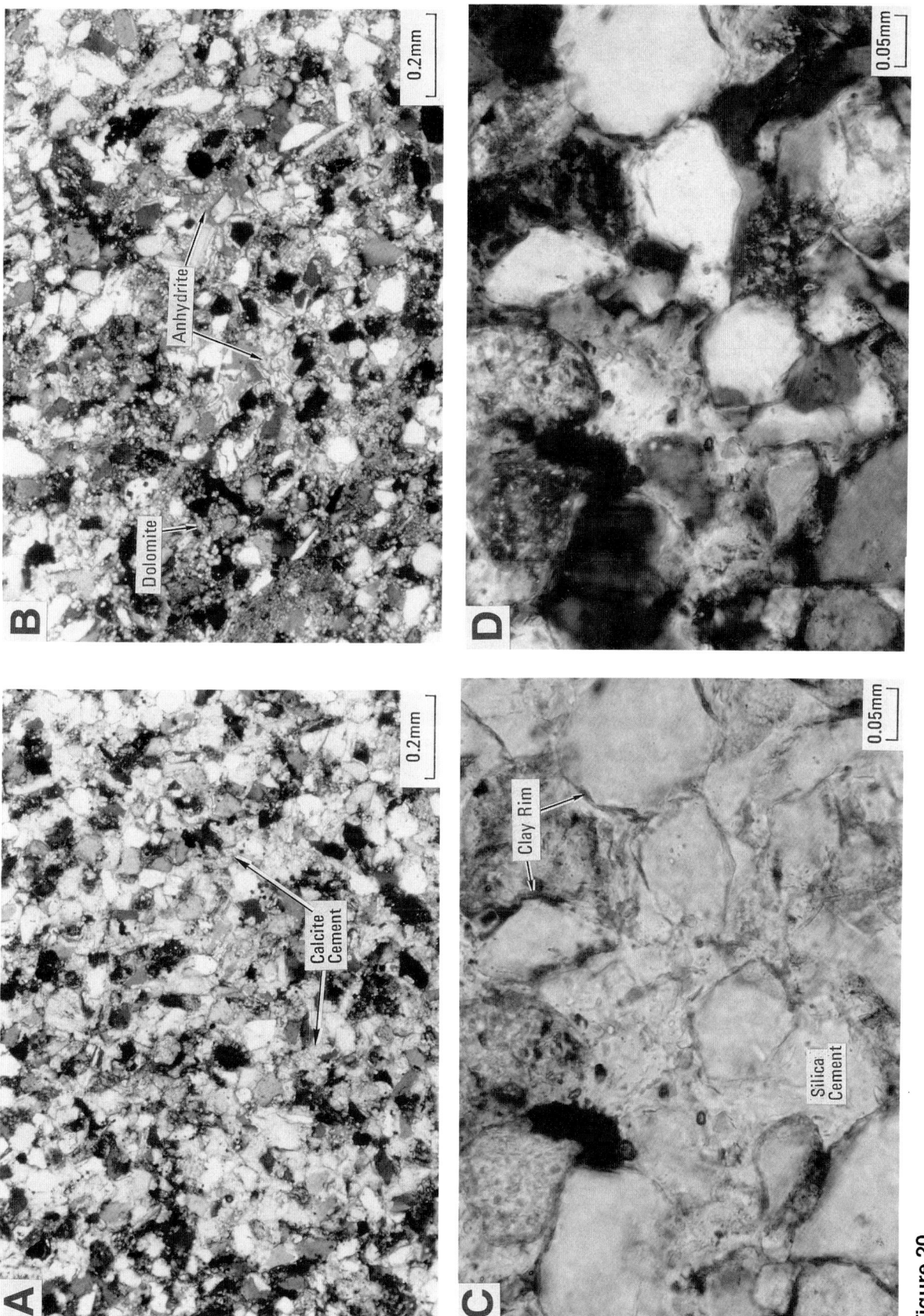

Figure 20

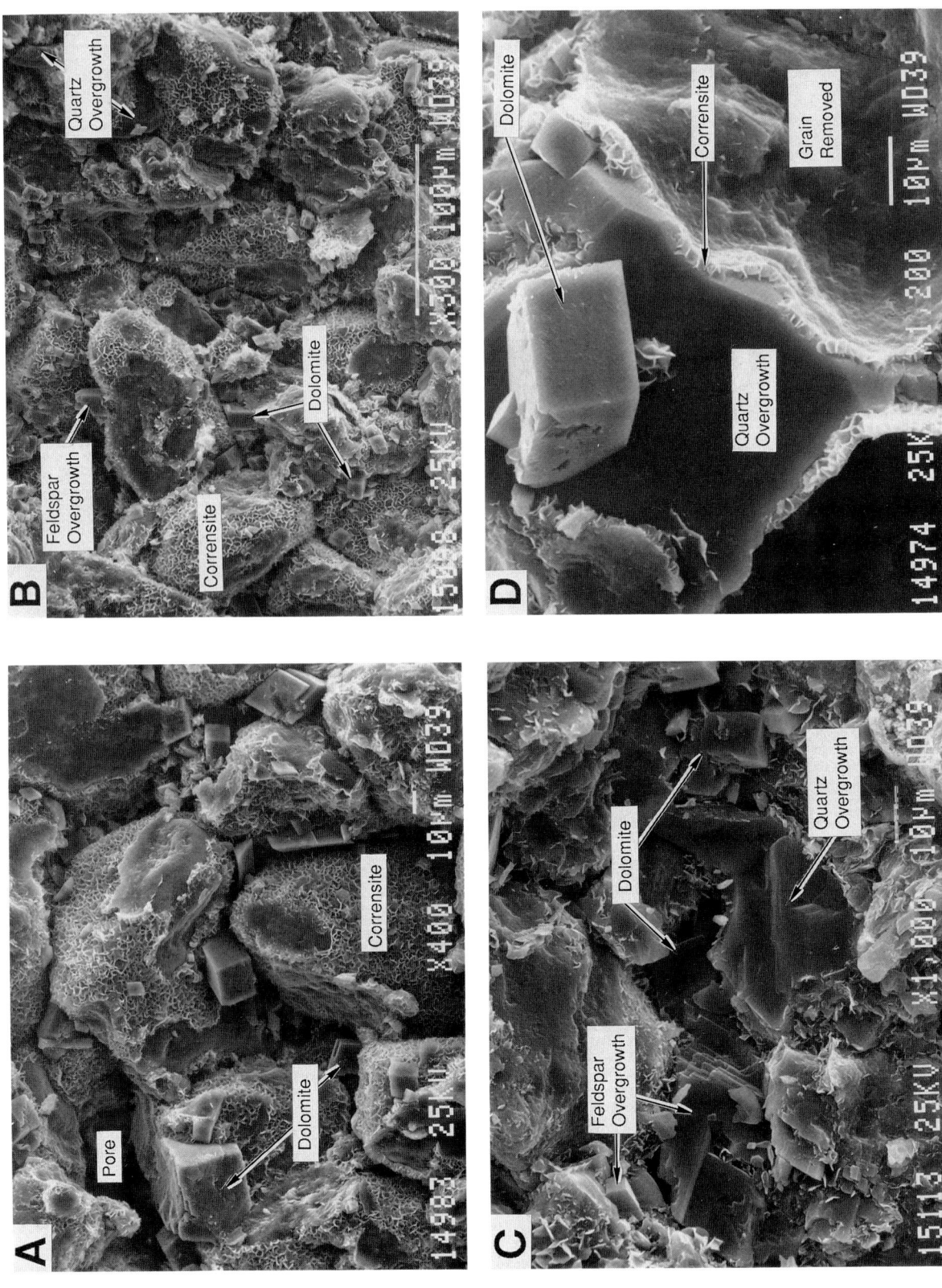

Figure 21

The sandstones have more visual porosity than the siltstones. Visually estimated porosity does not include microporosity, which is probably more abundant in the siltstones since they are more clay-rich. The Gray Sandstones have the most visual porosity (average 8.4 percent) of all the lithofacies.

Diagenesis of Sandstones and Siltstones

Calcite cement, dolomite cement, and corrensite clay have the greatest impact on reservoir quality. Visual relationships indicating relative timing of these and other minor diagenetic events are ambiguous or absent in most cases.

The earliest diagenetic events were probably the infiltration of detrital clays into exposed sands and silts, and the subsequent alteration/oxidation of iron-bearing minerals, which resulted in the iron staining of the red lithofacies. Calcite and anhydrite cements were emplaced early before much compaction. These cements are inferred to be relatively early because they cement grains with relatively loose packing. Since most samples have pervasive dolomite cement, its absence in calcite-cemented samples suggests that calcite is earlier than dolomite and prevented dolomite crystallization (Fig. 20A). Dolomite cementation and corrensite crystallization are probably contemporaneous, but may overlap with anhydrite and calcite cementation. Dolomite crystals often appear to be intergrown with corrensite, although the larger pore-filling dolomite crystals may have been emplaced following corrensite crystallization (Fig. 21). Quartz and feldspar overgrowths appear to be contemporaneous but, for the most part, follow dolomite and corrensite precipitation. The timing of quartz overgrowth crystallization is documented in Figure 21D, where a quartz overgrowth clearly engulfs a dolomite crystal.

Grain and cement dissolution are inferred to be later because none of the above phases were seen in secondary pores. In fact, the majority of the total porosity may be a result of cement dissolution (Siegal, 1989; Malicse, 1988). They (ibid) also suggest that the acidic pore fluids necessary for dissolving the cements reduce the red ferric oxide staining the clays, bleaching the Red Sandstones and Red Siltstones to a pale yellowish brown.

Dolomites

Texture.--Although dolomitization can effectively destroy original depositional fabrics, a variety of depositional textures, from dolomudstones to dolopackstones, are discernible in these samples (Figs. 22 and 23). Structureless, mottled, and a variety of laminated fabrics are present, of which crinkly algal laminations are the most common (Figs. 22B and 23A and B). In general, the dolomites are composed of 50 percent or greater dolomicrite. No grainstones were identified.

Grain types.--Grain-rich textures are the most common in the thin sections studied; however, grain types are often difficult to identify due to dolomitization. The most common grain types are peloids and skeletal fragments, and the most common skeletal grains are

Figure 22

The Dolomite lithofacies exhibit a variety of depositional textures, grain types, and porosity types. (A) This dolomudstone has a mottled texture that suggests it was originally a pellet (dark patches) packstone. Dolomitization has effectively hidden original texture. Porosity types include intercrystalline and microvuggy. Dead oil is present in some of the pores. Sample: U.S.M. Hillin "B" #1, 3222.6'. Plane polarized light, x7.7. (B) The crinkly laminations in this dolowackestone are probably algal in origin. The porosity is intercrystalline and microvuggy. Sample: U.S.M. Hillin "B" #1, 3224.7'. Plane polarized light, x7.7. (C) This skeletal, peloid dolopackstone has moldic and vuggy porosity. The large skeletal grains have been dissolved and the small preserved skeletal grains are calcispheres and benthic foraminifera. Sample: U.S.M. Hillin "B" #1, 3346.2'. Plane polarized light, x7.7. (D) The irregularly shaped and variably sized grains are oncolites. The algal coatings encase mud, calcispheres, and forams. Porosity types in this oncolite dolopackstone are microvuggy and intercrystalline. Sample: U.S.M. Hillin "B" #1, 3348.6'. Plane polarized light, x7.7.

Figure 23

These stromatolitic and oncolitic features suggest shallow marine deposition, probably intertidal to supratidal. (A) This stromatolite, oncolite dolopackstone consists of dark stromatolite fragments encasing mud and calcispheres on the left (top of sample), faintly algal-laminated dolowackestone in the center, and oncolite dolopackstone on the right (bottom of sample). Sample: U.S.M. Hillin "B" #1, 3345.9'. Plane polarized light, x7.7. (B) This sample is very similar to 'A', except for the graded storm(?) lamination in the center (top of sample to the left). Sample: G. M. Shelton, Jr. et al. #14, 3437.7'. Plane polarized light, x7.7. (C) A close-up from the top of 'B' shows stromatolite fragments with authigenic pyrite and geopetal structures (oil-stained sediment in the bottom of vugs). Sample: same as 'B', x31. (D) A close-up from the bottom of 'B' shows a poorly sorted mixture of flattened mud clasts and rounded peloids. The stylolites are oil-stained. Sample: same as 'B', x31.

Figure 24

The most common grain types in the Dolomite Lithofacies are calcispheres, peloids, and foraminifera. Thin pelecypod shells are less common. (A) Abundant calcispheres and a pelecypod fragment may be bound together by algae into oncolite grains (darker patches). Sample: U.S.M. Hillin "B" #1, 3345.9'. Plane polarized light, x31. (B) Calcispheres, quartz silt to very fine sand, and peloids are common in this dolopackstone. Sample: Marie Lawrence et al. #1, 3312.5'. Plane polarized light, x79. (C) Miliolid foraminifera and pelecypod shells are common in this dolopackstone. Sample: J. H. Dyche "B" #3, 3477.3'. Plane polarized light, x79. (D) Small benthic foraminifera(?) are also present in this sample. Sample: same as 'C'.

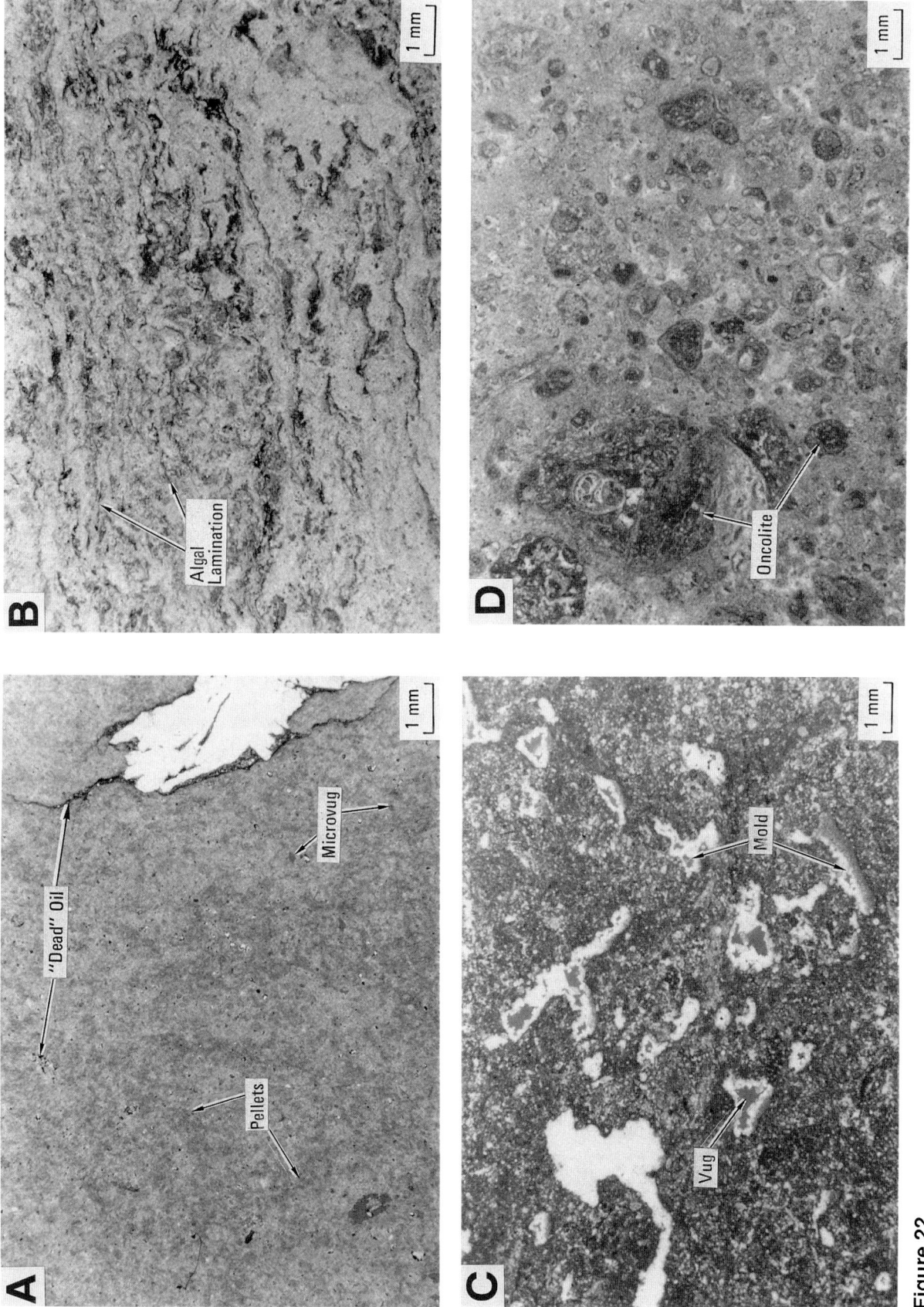

Figure 22

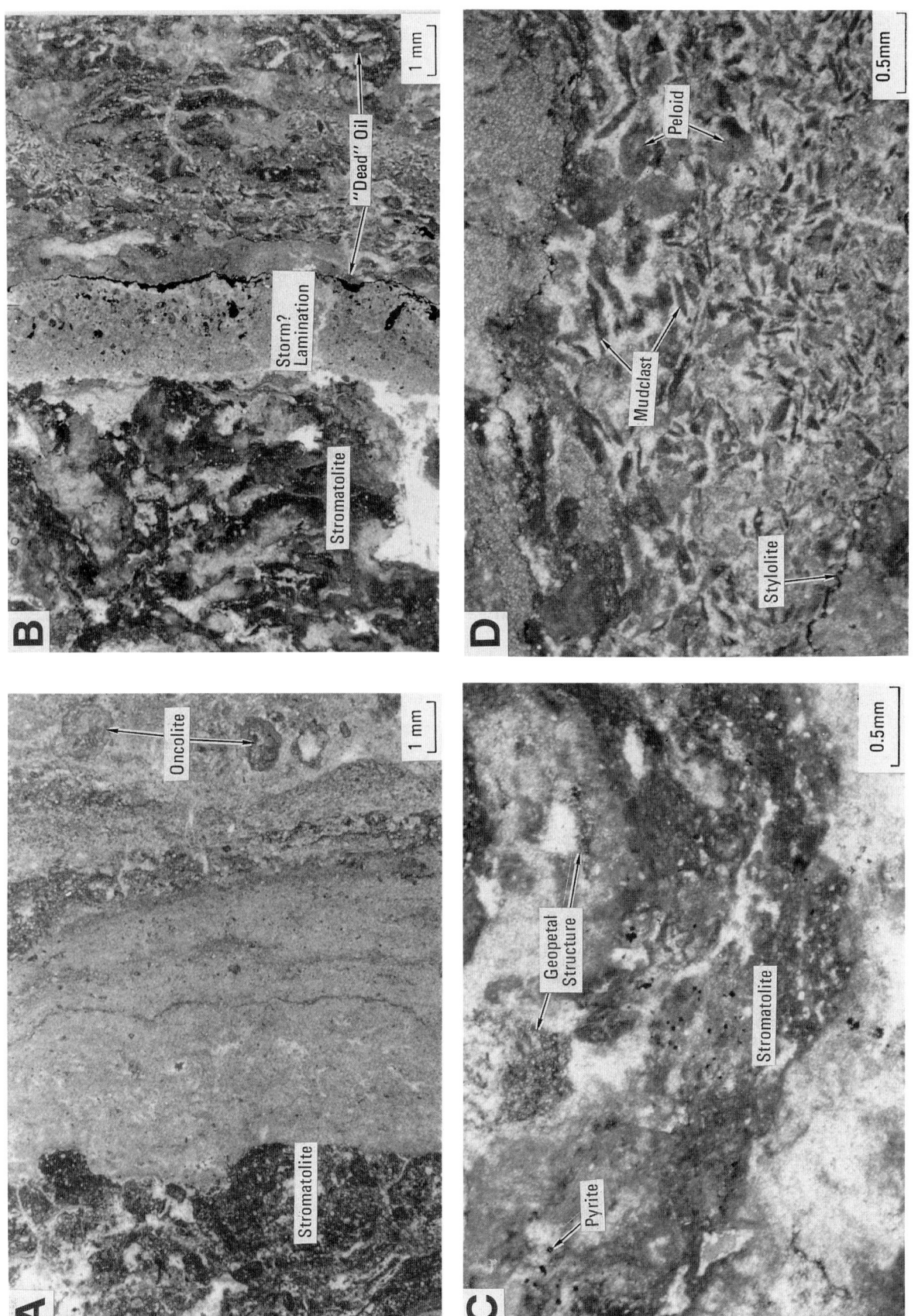

Figure 23

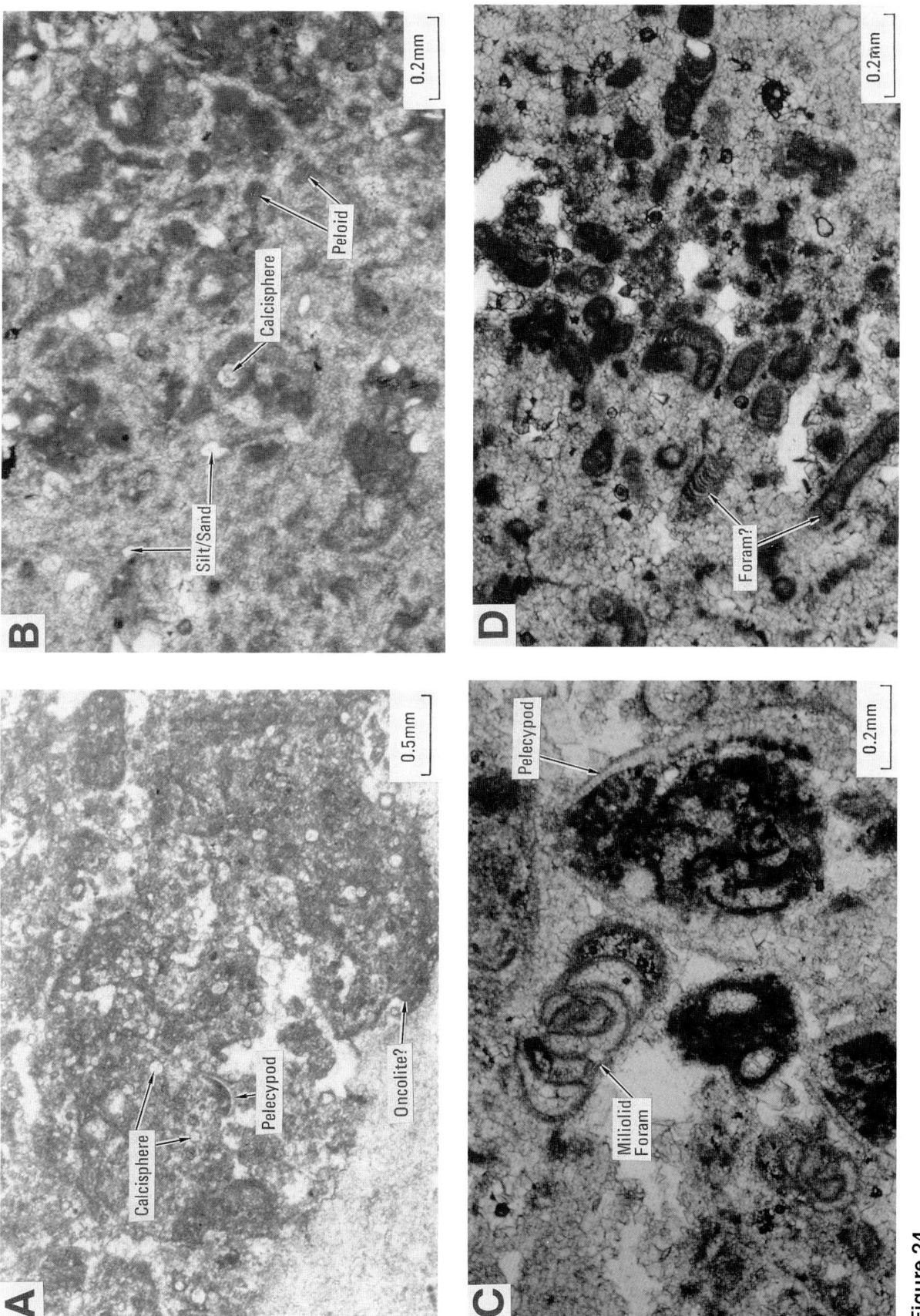

Figure 24

calcispheres and foraminifera (Fig. 24). Oncolites and stromatolite fragments are locally abundant (Fig. 23A-C). Intraclasts, usually dolomicrite, and siliciclastic silt and sand are also locally abundant (Figs. 23D and 24B).

Cements.--The most common cements are dolomite, calcite, and anhydrite. Coarsely crystalline calcite and anhydrite fill fractures, molds, and vugs (Fig. 22A and C). Dolomite cement is fine to medium crystalline and distinguished from the dolomicrite by larger and clearer crystals (Fig. 23D, between mudclasts). Pyrite cement occurs as nodules or fine crystals disseminated throughout the sample. Total cement in the dolomites is rarely more than 10 percent, and usually only trace to 2 percent.

Porosity types.--Samples with moldic, vuggy, and/or fracture porosity (Fig. 22C) have the highest visual porosities (up to 10 percent). Intercrystalline and/or microvuggy pores are common in most samples (Fig. 22A-C), but are difficult to estimate visually (trace to 1 percent).

Significance of Petrography

The Gray Sandstones have the best reservoir quality. Although variable amounts of intergranular material are present, they have the lowest average amount of clay matrix and cement and the highest average amount of porosity. The Red Sandstones, Gray Siltstones, and Dolomites have poorer reservoir quality overall. However, select samples from all of these lithofacies may have good reservoir quality. The Red Siltstones and Red Shales are nonreservoir lithofacies.

In general, the textures and variety of grain types observed in the dolomite lithofacies are indicative of subtidal to intertidal deposition; however, a few samples have textures that may be representative of supratidal environments.

PETROPHYSICAL DATA

Core analysis for porosity, air permeability, hydrocarbon saturation, and water saturation was completed on 793 samples taken from the cored wells examined for this study. Petrophysical data tabulated by lithofacies and lithofacies groups are compiled for all cores in this study and presented below.

	All Sandstone	GSS	RSS	All Siltstone	GSL	RSL	ADL,DL,RSH
Porosity (%)	16.9	19.2	13.6	12.1	12.3	11.3	9.4
Air Perm. (md)	0.39	0.87	0.11	0.08	0.08	0.07	0.03
Oil Sat. (%)	4.4	5.5	1.6	2.0	2.2	1.4	6.3
Water Sat. (%)	63.7	63.4	65.0	72.8	74.9	69.6	65.4

Although whole-core analyses were done for two of the wells and plug analysis for the rest, all the data were averaged together.

U.S.M. Queen Field oil production is principally from the Gray Sandstones and Siltstones. Measured hydrocarbon saturations for the Dolomites are often greater than values determined for the siliciclastic reservoir rocks. This anomalous occurrence may be the result of flushing of hydrocarbons out of the more porous and permeable sandstones and siltstones during the core-cutting process.

A crossplot of porosity vs permeability for both Gray and Red Sandstones indicates that no reliable correlation exists between the parameters (Fig. 25A). Scatter in the crossplot is in part attributed to the degree of sorting. Poorly sorted sandstones and siltstones contain variable amounts of clay and microcrystalline dolomite cement. These samples typically have considerable microporosity, but low permeability. The crossplot also clearly indicates that the Gray Sandstones have higher porosities and permeabilities than the Red Sandstones.

Siltstones (red and gray), when crossplotted for porosity and permeability, also yield a scatter plot (Fig. 25B) with no reliable correlation between the parameters. Substantially lower porosity and permeability values were measured for siltstones than corresponding sandstones because of their finer grain size and greater amounts of clay and microcrystalline dolomite cement. As in the sandstones, Gray Siltstones generally have greater porosity and permeability than the Red Siltstones.

DISCUSSION

The southern end of the Central Basin Platform was located approximately 5° to 15° from the equator during the Guadalupian Epoch, and at this latitude dry easterly winds originating in the subtropical high-pressure belt, or Horse Latitudes, created the warm and arid climate favorable to formation of carbonates, evaporites, and wind-modified alluvial clastic sediments (Mazzullo and others, 1990).

The most striking stratigraphic change in the U.S.M. Queen Field occurs within the Upper Queen reservoir zone. Upper Queen siliciclastic lithologies are restricted to the western half of the field. This pinchout occurs along a north-south trending line located in the western half of section 1, the G. M. Shelton lease of the G.C. and S.F.RR. Survey Block 114 (Figs. 3, 5, and 6). The carbonate- and evaporite-dominated Upper Queen may represent the encroachment of sabkha facies from the east prior to deposition of the Seven Rivers Formation.

The dominant sedimentological feature of the Queen Formation at the U.S.M. Queen Field is the development of alternating sequences of carbonate and siliciclastic lithofacies. Eustatic sea-level changes and/or variable siliciclastic sediment input are believed to be responsible for this mixed carbonate-siliciclastic deposition. Most carbonate deposition occurred during periods of sea-level highstand, while siliciclastic sedimentation is related to increased detrital input during sea-level lowstand. Figure 26 is an interpreted sea-level curve that relates sea-level fluctuations to deposition of carbonate and siliciclastic lithofacies sequences. Perturbations of sea-level cyclicity and/or siliciclastic sediment input result in the formation of complexly interbedded sequences.

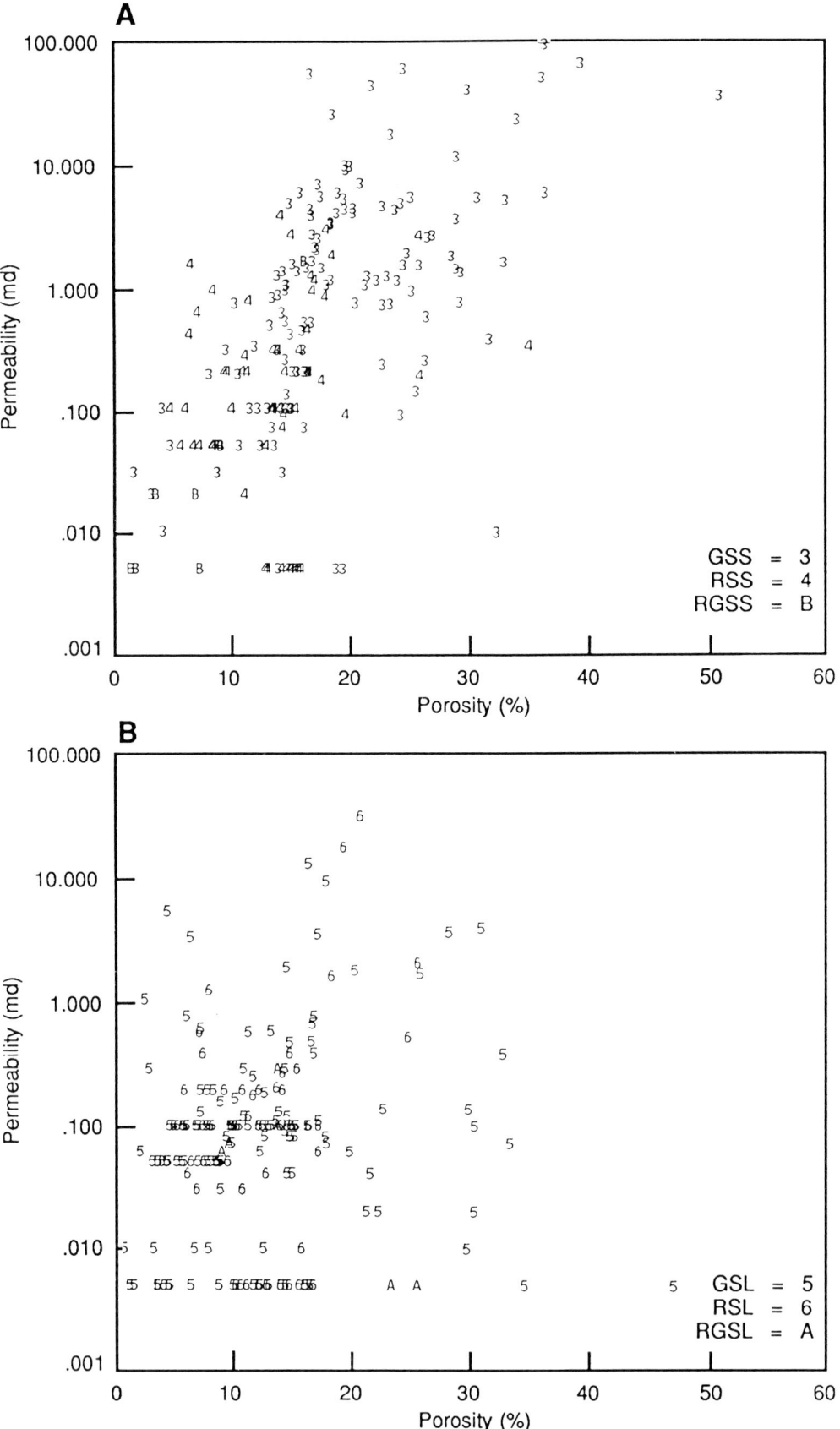

Figure 25
Crossplots of porosity vs. permeability for the sandstones (A) and siltstones (B). Crossplot A shows no reliable correlation between porosity and permeability for the sandstones, although the gray sandstones (GSS) clearly have the highest porosities and permeabilities. Crossplot B shows no reliable correlation between porosity and permeability for the siltstones.

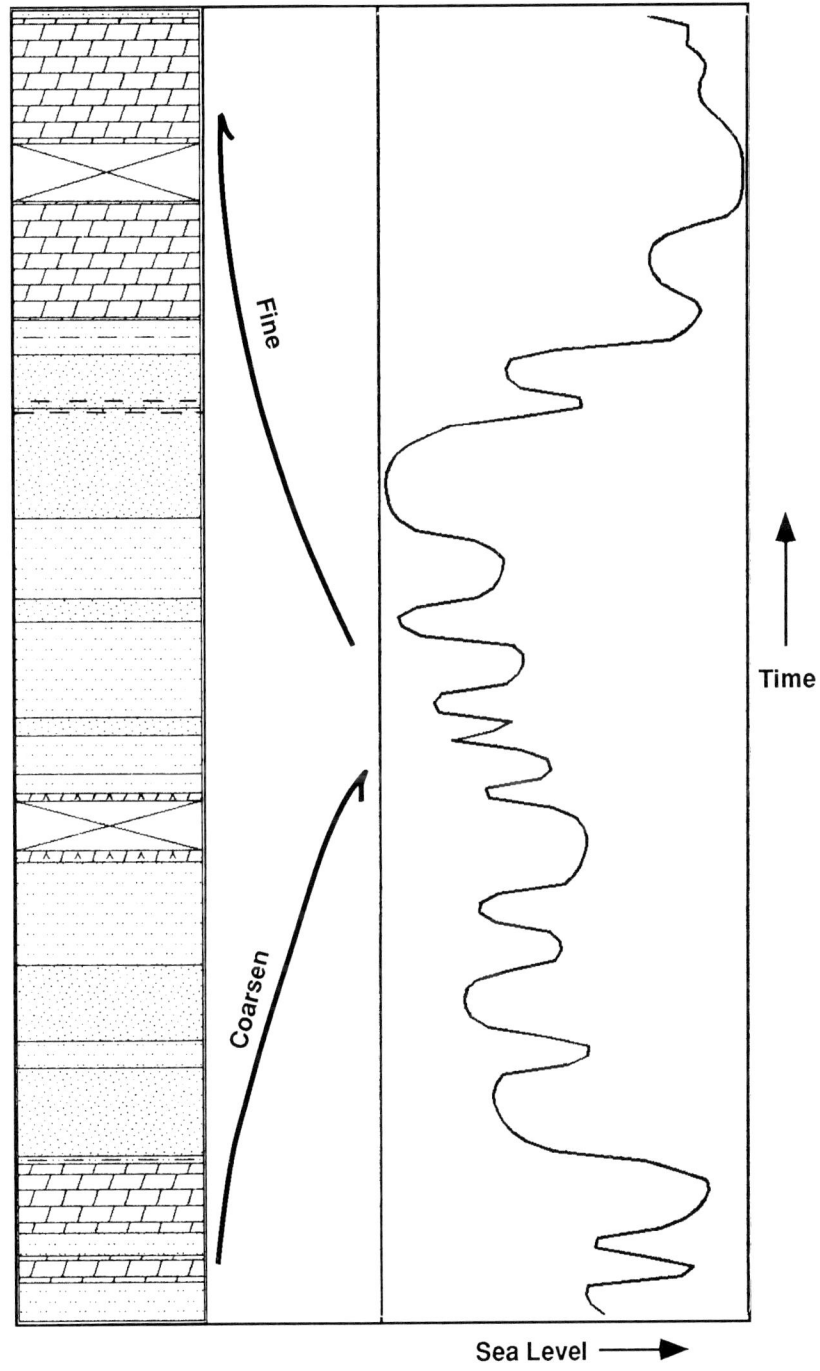

Figure 26
Depositional sequence related to sea level cyclicity for the Third Queen Zone (3412-3480 ft.), J.H. Dyche "B" #3 well.

Original limestone lithofacies, now thoroughly dolomitized, were deposited during marine transgressions (Fig. 27B). Subtidally deposited skeletal packstones form the deepest component in the middle of the carbonate cycle. The percentage of allochems generally decreases upward and downward in the cycle, resulting in the formation of more wackestone and mudstone. Oncolite packstones and wackestones are associated with algal-laminated wackestones and mudstones, which are most common near the top and bottom of the cycle. Algal laminations, oncolites, and reduced faunal abundance and diversity are indicators of increased environmental stress related to development of penesaline conditions representative of the intertidal environment.

A hypersaline environment existed during deposition of some carbonate sequences, resulting in the formation of anhydritic dolomites and, in some instances, thin anhydrite beds. Deposition in a supratidal setting is suggested by the occurrence of evaporites. The presence of reddish coloration near the top of carbonate cycles is attributed to wind-blown, fine clastic detritus transported from a nearby alluvial plain where it was oxidized. In addition, brecciated dolomite clasts near the top of dolomite cycles are attributed to desiccation of carbonate muds during periods of subaerial exposure.

Progradation of clastic sediments occurred during sea-level fall and lowstand periods when detritus, originating from the Central Basin Platform and the infilled Midland Basin, was probably transported by wind and ephemeral streams south and southwestward toward the Delaware Basin. A significant amount of sediment probably bypassed the shelf during maximum lowstand. Sands and silts were transported across the shelf and out into the Delaware Basin, forming the siltstones and fine-grained sandstones of the Cherry Canyon Formation.

The processes that transported the sands and silts across the shelf are not clear, and subsequent reworking of the sediments trapped on the shelf during the next sea-level rise may have obliterated their original depositional texture. Sedimentary structures in the cores studied are faint and ambiguous in most cases, but tend to favor deposition by currents (ripple, wavy, and normally graded laminations) over eolian processes (planar and inversely graded laminations). Bioturbation may be responsible for disrupting sedimentary structures and homogenizing the sediment, which would also suggest subtidal to intertidal deposition. Small horizontal burrows (*Planolites*) are common and represent grazing traces, which are common in tidal flat and low-energy subtidal environments. Less common, large vertical burrows (*Conichnus?*) represent dwelling traces, which are also common in tidal flat and subtidal settings.

The wide, low-gradient Central Basin Platform was probably mostly flooded during sea-level highstands (Fig. 27B), creating a wide tidal flat. This interpretation is supported by the dolomites from the U.S.M. Queen Field, which were deposited under dominantly shallow, restricted water conditions with only rare open marine indicators and no clear exposure features.

The lagoon (Fig. 27A) may or may not have totally evaporated during sea-level lowstands. If siliciclastic sediments bypassed the shelf at this time, sedimentary processes documenting the processes of transport are not preserved in the sandstones and siltstones from the U.S.M. Queen Field. The siliciclastic sediments must have been reworked by currents and infauna during the subsequent rise in sea level, since they are dominated by current structures and ambiguous structures disrupted by bioturbation.

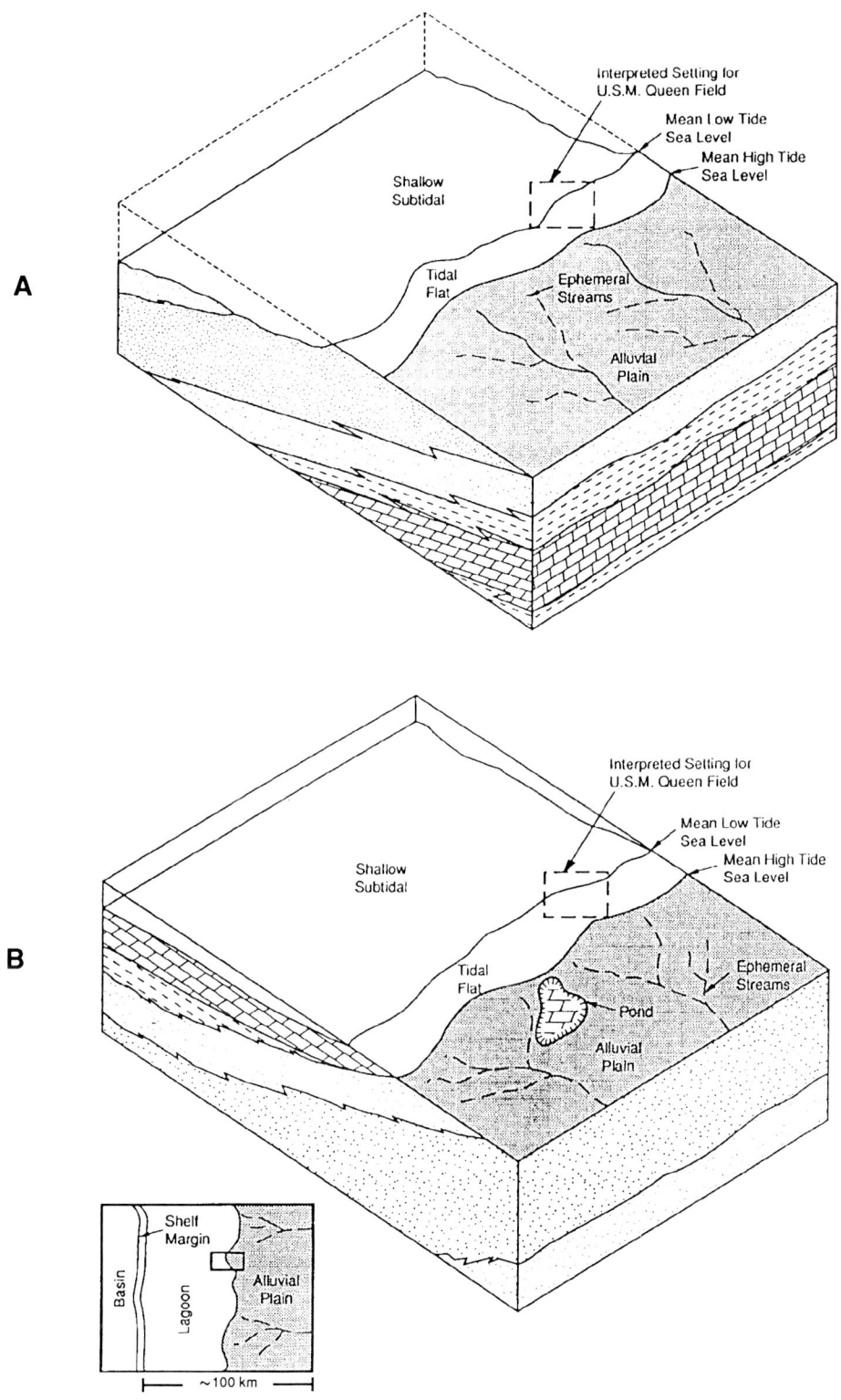

Figure 27
Schematic block diagram of lowstand depositional environments (A) and highstand depositional environments (B) interpreted from cores at U.S.M. Queen Field.

Evidence for nonmarine deposition given by Mazzullo and others (1990); Holley and Mazzullo (1988); Newsom (1989); and Harper (1990) is absent at this location, suggesting it is much closer to the shelf edge. The lithofacies and interpreted environment of deposition for the reservoir rocks of the U.S.M. Queen Field correlate better with the study by Spencer (1987) at the Yates Field, the localities described by Franseen (1985) and Wheeler (1989), and the Queen type section (Moran, 1962). However, the exact location of the Goat Seep shelf margin on the southern end of the Central Basin Platform is not well defined. Considering the similarities and differences between the Franseen and Wheeler studies and our study, the U.S.M. Queen Field is probably located 1.2 to 6.2 miles (2-10 km) behind the shelf margin.

ACKNOWLEDGMENTS

We wish to thank Chevron U.S.A. Inc. for cores, logs, and petrophysical data, and permission to publish this paper. In addition, we would like to thank Chevron Oil Field Research Company for drafting and word processing support.

REFERENCES

BALL, S. M., ROBERTS, J. W., NORTON, J. A., AND POLLARD, W. D., 1971, Queen Formation (Guadalupian, Permian) outcrops of Eddy County, New Mexico, and their bearing on recently proposed depositional models: American Association of Petroleum Geologists Bulletin, v. 55, p. 1348-1355.

BORER, J. M., AND HARRIS, P. M., 1989, Depositional facies and cycles in Yates Formation outcrops, Guadalupe Mountains, New Mexico, *in* Harris, P. M. and Grover, G. A., eds., Subsurface and Outcrop Examination of the Capitan Shelf Margin, Northern Delaware Basin: Society of Economic Paleontologists and Mineralogists Core Workshop No. 13, p. 305-318.

FRANSEEN, E. K., 1985, Sedimentology of the Grayburg and Queen formations (Guadalupian) and the shelf margin erosion surface, Western Escarpment, Guadalupe Mountains, West Texas: Master's Thesis, University of Wisconsin-Madison, 189 p.

FRENZEL, H. N., BLOOMER, R. R., CLINE, R. B., CYS, J. M., GALLEY, J. E., GIBSON, W. R., HILLS, J. M., KING, W. E., SEAGER, W. R., KOTTLOWSKI, S. THOMPSON III, LUFF, G. C., PEARSON, B. T., AND VAN SICLEN, D. C., 1988, The Permian Basin region, *in* Sloss, L. L., ed., Sedimentary Cover - North American Craton: U.S.: Geological Society of America, Geology of North America, v. D-2, p. 261-306.

HARPER, J. B., 1990, The lithology, environment of deposition, and reservoir evaluation of sandstones in the Upper Queen Formation (Guadalupian, Permian) at Concho Bluff North and Jennifer fields, Upton and Ector counties, Texas: Master's Thesis, Texas A&M University, 149 p.

HOLLEY, C., AND MAZZULLO, J., 1988, The lithology, depositional environments, and reservoir properties of sandstones in the Queen Formation, Magutex North, McFarland North, and McFarland fields, Andrews County, Texas, *in* Permian and Pennsylvanian Stratigraphy, Midland Basin, West Texas: Study to Aid Hydrocarbon Exploration: Society of Economic Paleontologists and Mineralogists Special Publication 88-28, p. 55-63.

MALICSE, J. A. E., 1988, The environment of deposition and diagenesis of the Shattuck Member of the Queen Formation (Guadalupian, Permian) at Caprock Queen Field, Chavez County, New Mexico: Master's Thesis, Texas A&M University, 160 p.

MAZZULLO, J., MALICSE, J. A. E., AND SIEGAL, J. D., 1990, Depositional environments of the Shattuck Member of the Artesia Group, Northwest Shelf, Permian Basin: Journal of Sedimentary Petrology preprint.

_____, WILLIAMS, M., AND MAZZULLO, S. J., 1984, The Queen Formation of Millard Field, Pecos County, Texas, its lithologic characteristics, environments of deposition, and reservoir petrophysics: Transcripts of the Southwest Section of the American Association of Petroleum Geologists, p. 103-109.

MORAN, W. R., 1962, Surface type localities of the Queen and Grayburg formations in the Guadalupe Mountains, Eddy County, New Mexico: Permian of the Central Guadalupe Mountains, Eddy County, New Mexico, Guidebook, West Texas, Roswell and Hobbs Geological Societies, p. 76-86.

NEWSOM, D. F., 1989, The lithology, environment of deposition, and reservoir properties of sandstones in the Upper Queen Formation (Guadalupian, Permian) at Concho Bluff Queen Field, Crane County, Texas: Master's Thesis, Texas A&M University, 187 p.

SIEGAL, J. D., 1989, The lithology, environment of deposition, and diagenesis of the Shattuck Member of the Queen Formation (Guadalupian, Permian) at Central Corbin Queen Field, Lea County, New Mexico: Master's Thesis, Texas A&M University, 159 p.

SILVER, B. A., AND TODD, R. G., 1969, Permian cyclic strata, Northern Midland and Delaware basins, West Texas and southeastern New Mexico: American Association of Petroleum Geologists Bulletin, v. 53, No. 11, p. 2223-2251.

SPENCER, A. W., 1987, Evaporite facies related to reservoir geology, Seven Rivers Formation (Permian), Yates Field, Texas: Master's Thesis, University of Texas at Austin, 125 p.

_____, AND WARREN, J. K., 1986, Depositional styles in the Queen and Seven Rivers formations - Yates Field, Pecos County, Texas, *in* Bebout, D. G. and Harris, P. M., eds., Hydrocarbon Reservoir Studies, San Andres/Grayburg Formations, Permian Basin: Permian Basin Section, Society of Economic Paleontologists and Mineralogists Publication 86-26, p. 135-137.

WARD, R. F., KENDALL, C. G. S. C., AND HARRIS, P. M., 1986, Upper Permian (Guadalupian) facies and their association with hydrocarbons - Permian Basin, West Texas and New Mexico: American Association of Petroleum Geologists Bulletin, v. 70, No. 3, p. 239-262.

WHEELER, C., 1989, Stratigraphy and sedimentology of the Shattuck Member (Queen Formation) and lowermost Seven Rivers Formation (Guadalupian), North McKittrick and Dog canyons, Guadalupe Mountains, New Mexico and West Texas, *in* Harris, P. M. and Grover, G. A., eds., Subsurface and Outcrop Examination of the Capitan Shelf Margin, Northern Delaware Basin: Society of Economic Paleontologists and Mineralogists Core Workshop No. 13, p. 353-364.

INTERRELATIONSHIP OF PLATFORM DOLOSTONE AND SILICIC SILTSTONE FACIES - GRAYBURG FORMATION (PERMIAN), CENTRAL BASIN PLATFORM AND OZONA ARCH, PERMIAN BASIN, WEST TEXAS

DON G. BEBOUT
Bureau of Economic Geology,
University of Texas at Austin, Austin, TX 78713

ABSTRACT

Early Guadalupian Grayburg dolostones and siltstones accumulated on shallow-water to exposed platforms that encircled the Midland Basin, West Texas. The Grayburg Formation is the uppermost of several upward-shoaling sequences of the lower Guadalupian and is transitional between the dolostone strata of the underlying San Andres Formation and the siltstone and anhydrite of the overlying Queen Formation. The siltstone and silty dolostone units within the Grayburg mark the base of a number of thinner, upward-shoaling cycles.

In Dune Field, Central Basin Platform, siltstones and silty dolostones are associated with subtidal fusulinid wackestones to supratidal pisolite wackestone to grainstone; in some areas a single marker is associated with this entire range of depositional environments. Most of the siltstone cores exhibit structures interpreted to have developed under subaqueous conditions. In Farmer Field, Ozona Arch, siltstones mark the base of many thin cycles within the reservoir section.

INTRODUCTION

Grayburg carbonates and siltstones of Permian early Guadalupian age were deposited on shallow-water to exposed platforms that encircled the Midland Basin of West Texas (Fig. 1). These sediments are the reservoirs for the several Grayburg fields located along the Central Basin Platform and Ozona Arch, on the west and south sides of the Midland Basin respectively. Cores from Dune Field, on the Central Basin Platform, and Farmer Field, on the Ozona Arch, form the basis for this paper (Fig. 2).

The lower Guadalupian section on the Central Basin Platform is composed of several 300- to 400-ft-thick (91- to 122-m), upward-shoaling sequences. Each sequence is initiated at the base with a thin transgressive grainstone to packstone over the supratidal facies and exposure surfaces of the previous sequence. This transgressive facies is followed by subtidal, normal-marine, fusulinid wackestones and packstones, which in turn grade upward into subtidal to intertidal grainstones, and finally supratidal algal mudstones, pisolite packstones, anhydrite, and minor amounts of siltstone.

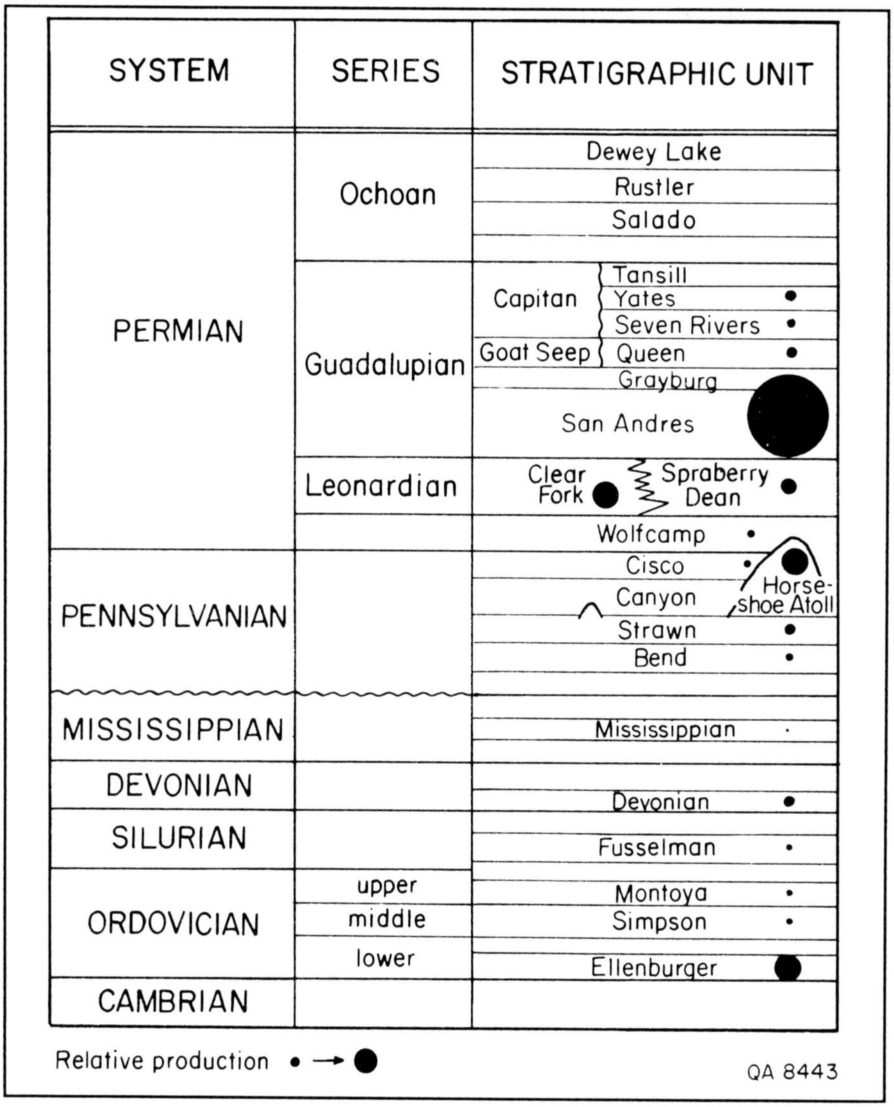

SYSTEM	SERIES	STRATIGRAPHIC UNIT	
PERMIAN	Ochoan	Dewey Lake	
		Rustler	
		Salado	
	Guadalupian	Capitan { Tansill	•
		Capitan { Yates	•
		Capitan { Seven Rivers	•
		Goat Seep { Queen	•
		Grayburg	●
		San Andres	●
	Leonardian	Clear Fork ● ≋ Spraberry ≋ Dean	•
		Wolfcamp	•
PENNSYLVANIAN		Cisco	•
		Canyon	Horse-shoe Atoll
		Strawn	•
		Bend	•
MISSISSIPPIAN		Mississippian	·
DEVONIAN		Devonian	•
SILURIAN		Fusselman	•
ORDOVICIAN	upper	Montoya	•
	middle	Simpson	•
	lower	Ellenburger	●
CAMBRIAN			

Relative production • → ●

QA 8443

FIG. 1.—Stratigraphic section of the Paleozoic strata of the Central Basin Platform/Ozona Arch. Also shown is the relative stratigraphic distribution of produced oil. From Galloway and others (1983).

The Grayburg Formation represents the uppermost of these sequences and occurs between the underlying dolostone strata of the San Andres Formation and the overlying siltstone and anhydrite of the Queen Formation. In this transitional position the Grayburg contains increasing amounts of siltstone, particularly toward the top of the formation. The Grayburg sequence is divisible into a number of thinner, less well-defined, upward-shoaling cycles that are commonly marked by the presence of a siltstone or silty dolostone at the base.

The siltstone and silty dolostone units in the Grayburg on the Central Basin Platform and Ozona Arch serve as excellent local correlation markers because of their higher radioactivity response on the gamma-ray log. The amount of gamma-ray deflection generally depends upon the amount of silt in the marker and can be used as an indicator of the silt content in the dolostone facies.

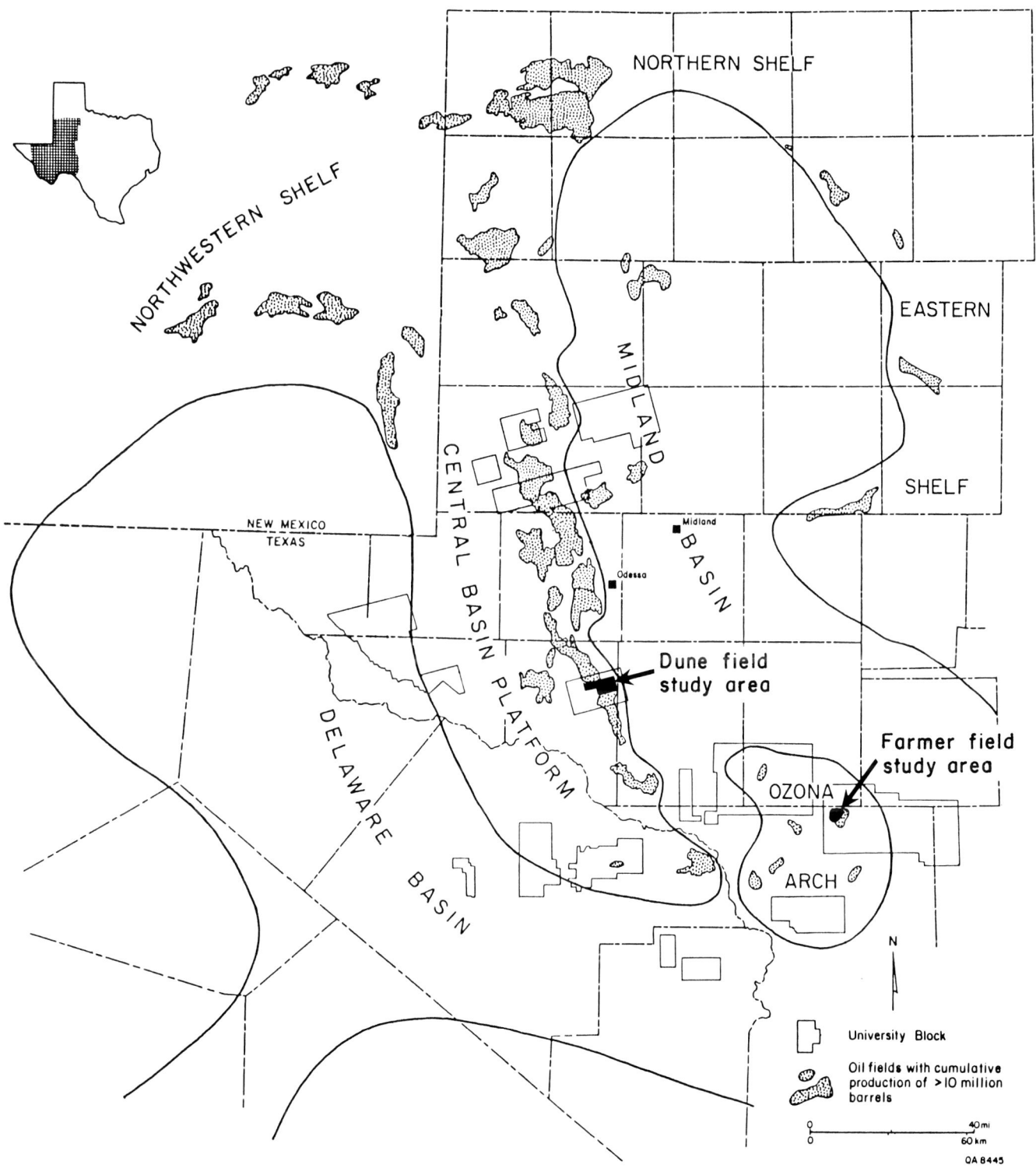

FIG. 2.—Index map of the Permian Basin showing the location of Dune and Farmer fields (black). Also shown are the San Andres and Grayburg fields (stippled) that have produced more than 10 million barrels of oil. From Bebout and others (1987).

DESCRIPTION AND DISTRIBUTION OF SILICICLASTICS AND ASSOCIATED DOLOSTONES

The siliciclastics occur as siltstones and very fine-grained sandstones with dolomite or anhydrite cement to dolostone units with less than ten percent silt and sand grains. The siliciclastic units are most commonly burrowed, but locally laminations and cross laminations occur. Burrows are small, generally less than 3 mm in diameter. Vertical water-escape structures occur locally. The siliciclastic grains are angular and range in size from 20 to 250 µm. Quartz is the dominant mineral, but feldspar grains are present. Moldic porosity in the siltstones is the result of leaching of the feldspar grains.

Dolostone facies associated with the siltstone range from open-shelf fusulinid wackestone to arid tidal-flat mudstone and pisolite and bird's-eye packstone and grainstone. Cross laminations are common in the siltstone where associated with ooid facies. Contacts between siltstone and dolostone vary from sharp to gradational; in some cores the two sediments are mixed by burrowing.

GRAYBURG FIELDS

Dune Field, Crane County

The Grayburg Formation in Dune Field has been divided into three units (Figs. 3 and 4). The Lower Unit, which rests unconformably on the tidal-flat dolostone and siltstone at the top of the San Andres Formation, consists solely of fusulinid wackestone and records maximum flooding of the Grayburg Formation. The Middle Unit consists of two dolostone facies: vertically structured wackestone (landward) and crinoid packstone/grainstone (basinward). These lower two units are below the first occurrence of siltstone in the Grayburg and are not discussed further here. The Upper Unit, which extends from the A siltstone marker upward to the top of the Grayburg, consists of a number of upward-shoaling, progradational cycles. In contrast to the lower two units, the dolostone of the Upper Unit is interrupted by several thin siltstone to silty dolostone sections. The siltstone sections are thinner in the lower part and thicker and more numerous in the top part of the Upper Unit.

The dolostone vertical sequence in the Upper Unit consists of fusulinid wackestone at the base; pellet and ooid grainstone near the top; and pisolite wackestone to grainstone, mudstone, and anhydrite at the top. The siltstone zones have been used to subdivide this unit into several parts, each of which demonstrates the change from the pisolite wackestone to grainstone facies to the west (landward) to fusulinid wackestone facies to the east (basinward). The siltstone zones are correlatable throughout the area and are associated with tidal-flat to subtidal environments. These siltstones do not appear to be characteristic of a particular depositional environment but rather assume the depositional environment of the associated dolostone. In the Mobil 1560 core (Figs. 4-7), the siltstones in the lower subtidal section are thin but cleaner (less dolomite) than those in the upper tidal-flat section, which contain 50 percent or more dolomite.

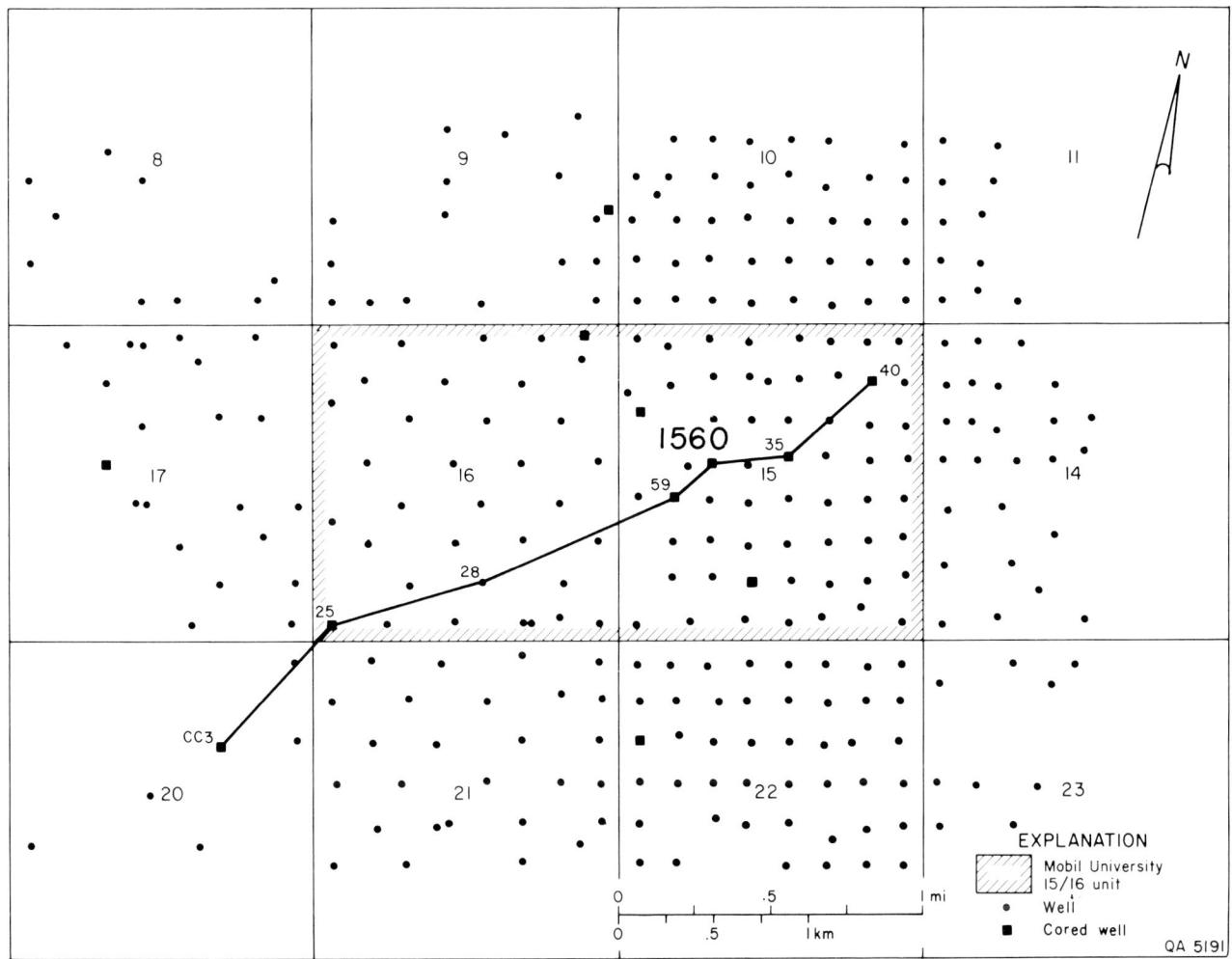

FIG. 3.—Map showing location of the facies cross section (Fig. 4) in Mobil University Unit 15/16 located in Dune field, Sections 15 and 16, University Block 30, Crane County, Texas. From Bebout and others (1987).

Siltstone markers (Fig. 4) were assumed to represent local time lines for delineating mapping units. These markers are readily identified on the gamma-ray logs by the high-radioactivity response (Fig. 8) and were used to correlate all wells in the study area. Correlations of the siltstone units displayed in the 1560 core are shown on the facies cross section (Fig. 4), which crosses Dune Field in a depositional dip direction (Fig. 3). Siltstones A and B (Figs. 4 and 8) are associated with subtidal fusulinid wackestone facies in the 1560 core and updip on the cross section, with tidal-flat pisolite facies. The siltstones in the upper part of the 1560 core (D, Y, Z, and C; Figs. 4 and 8) are associated with intertidal and supratidal grainstone and pisolite facies; updip equivalent beds contain more siltstone and anhydrite facies, as indicated by the gamma-ray and sonic log responses.

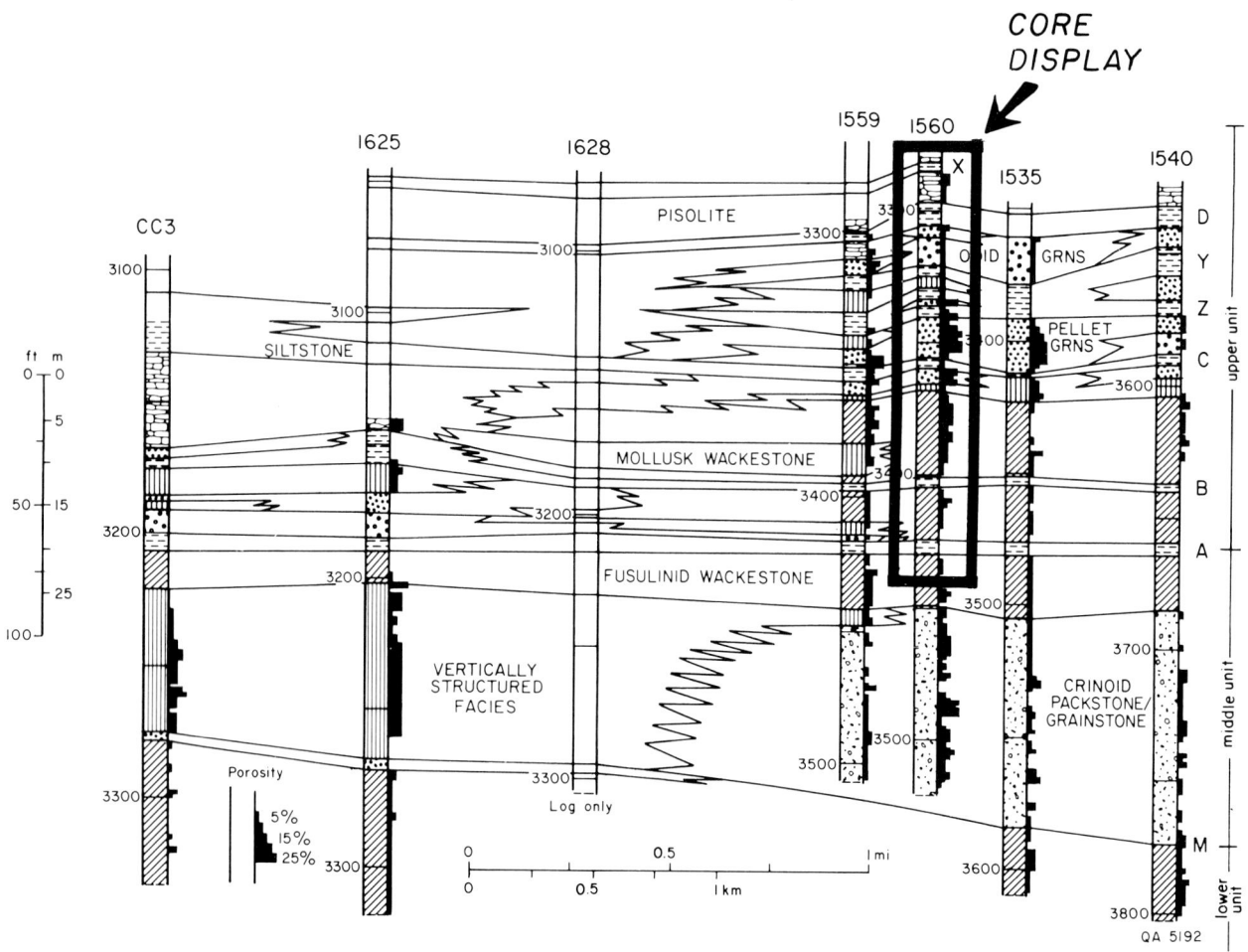

FIG. 4.—Facies dip section across the Mobil University Unit 15/16, Dune field. Location of the section is shown in Figure 3. Location of the displayed core is outlined on the facies column for well 1560. From Bebout and others (1987).

Farmer Field, Crockett and Reagan Counties

The Grayburg sequence at Farmer Field is composed of 14 thin siltstone/dolostone cycles within the reservoir section (Figs. 9-13). These major cycles are readily defined by prominent siltstone markers. The cycles vary in thickness from 12 to 62 ft (3.6 to 18.9 m), and some contain a number of even thinner, less distinct cycles. The lower cycles are generally thicker than the higher cycles.

A complete cycle includes siltstone, silty dolostone, and skeletal wackestone at the base; skeletal and pelletal packstone and fine-grained grainstone in the center; and coarse-grained grainstone at the top. Many of the cycles are incomplete. The gamma-ray log is effective in defining each cycle because of the presence of silt at the base and decrease in silt content upward. The coarse-grained grainstones at the top of each cycle are heavily cemented with gypsum and lack siliciclastics; consequently, the top of each cycle is identified by high resistivity (greater than 1000 ohm-m) and low gamma-ray readings (less than 30 API units)(Fig. 14).

The major siltstone markers used to identify the 14 cycles have been correlated throughout the area (Fig. 9). These correlative markers occur as siltstone in some cores to silty dolostone with silt content ranging as low as ten percent in others; in all cases the high gamma value provides the signature for the correlation, although those of the siltstones are much higher and more distinctive than those of the silty dolostones in which siltstone makes up a small percentage of the rock.

To the west, grainstone units are thicker and more abundant than those to the east, where the section is dominantly wackestone and packstone. The siltstone units are thicker to the east (64 ft (19.5 m) in the DD10 core) than to the west (49 ft (14.9 m) in the P5 core)(Fig. 9). Because the overall section is thinner to the east, the siltstone makes up more of the total (19 percent) than to the west (10 percent).

DEPOSITIONAL ENVIRONMENTS

Because the Central Basin Platform and Ozona Arch are remote from major sources of terrigenous sediments to the east and northwest, it is assumed that the primary method of transport of the silt-sized siliciclastics was wind. The very fine grain size and excellent sorting support the assumption that they arrived at the general location by eolian means. The siliciclastics were not transported continuously, as evidenced by the presence of thick sections of dolostone without silt grains. Thus, there must have been episodic periods of intense or constant wind activity, perhaps during times of lower sea level, when vast platform areas were exposed that resulted in transportation and deposition of large quantities of siliciclastics on the Central Basin Platform and Ozona Arch.

Some siltstone units that are correlatable across this field are associated with subtidal fusulinid wackestone downdip and subtidal to intertidal pellet and ooid grainstone and supratidal pisolite facies updip. However, the sedimentary structures (dominantly small burrows) preserved in most siltstone units, even those associated with the supratidal carbonates, indicate final deposition and accumulation under subaqueous conditions.

REFERENCES

BEBOUT, D. G., LUCIA, F. J., HOCOTT, C. R., FOGG, G. E., AND VANDER STOEP, G. W., 1987, Characterization of the Grayburg reservoir, University Lands Dune Field, Crane County, Texas: The University of Texas at Austin, Bureau of Economic Geology Report of Investigations No. 168, 98 p.

GALLOWAY, W. E., EWING, T. E., GARRETT, C. M., TYLER, NOEL, AND BEBOUT, D. G., 1983, Atlas of major Texas oil reservoirs: The University of Texas at Austin, Bureau of Economic Geology, 139 p.

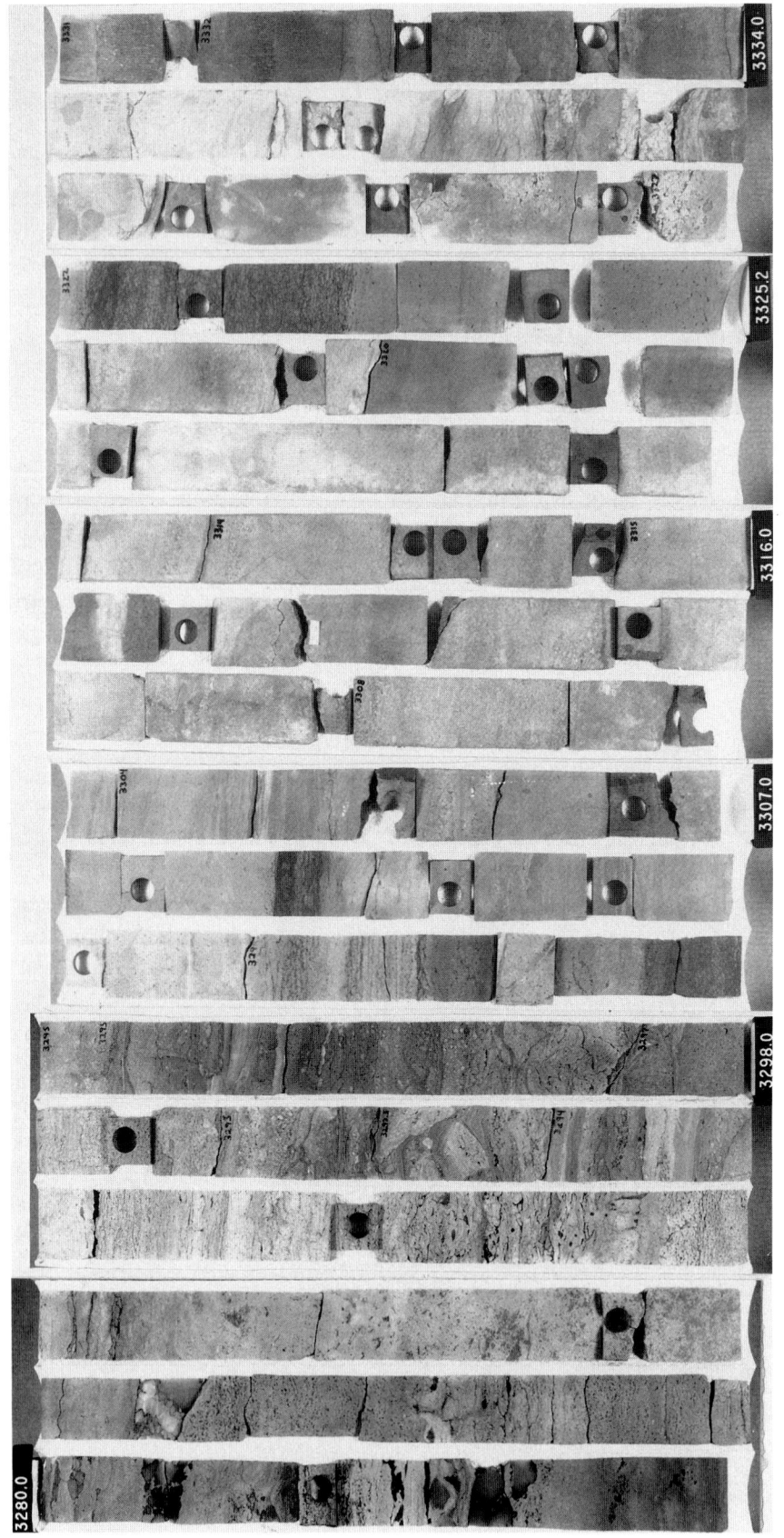

FIG. 5.—Continuous core from the Mobil 1560 well. Each core segment is three feet long.

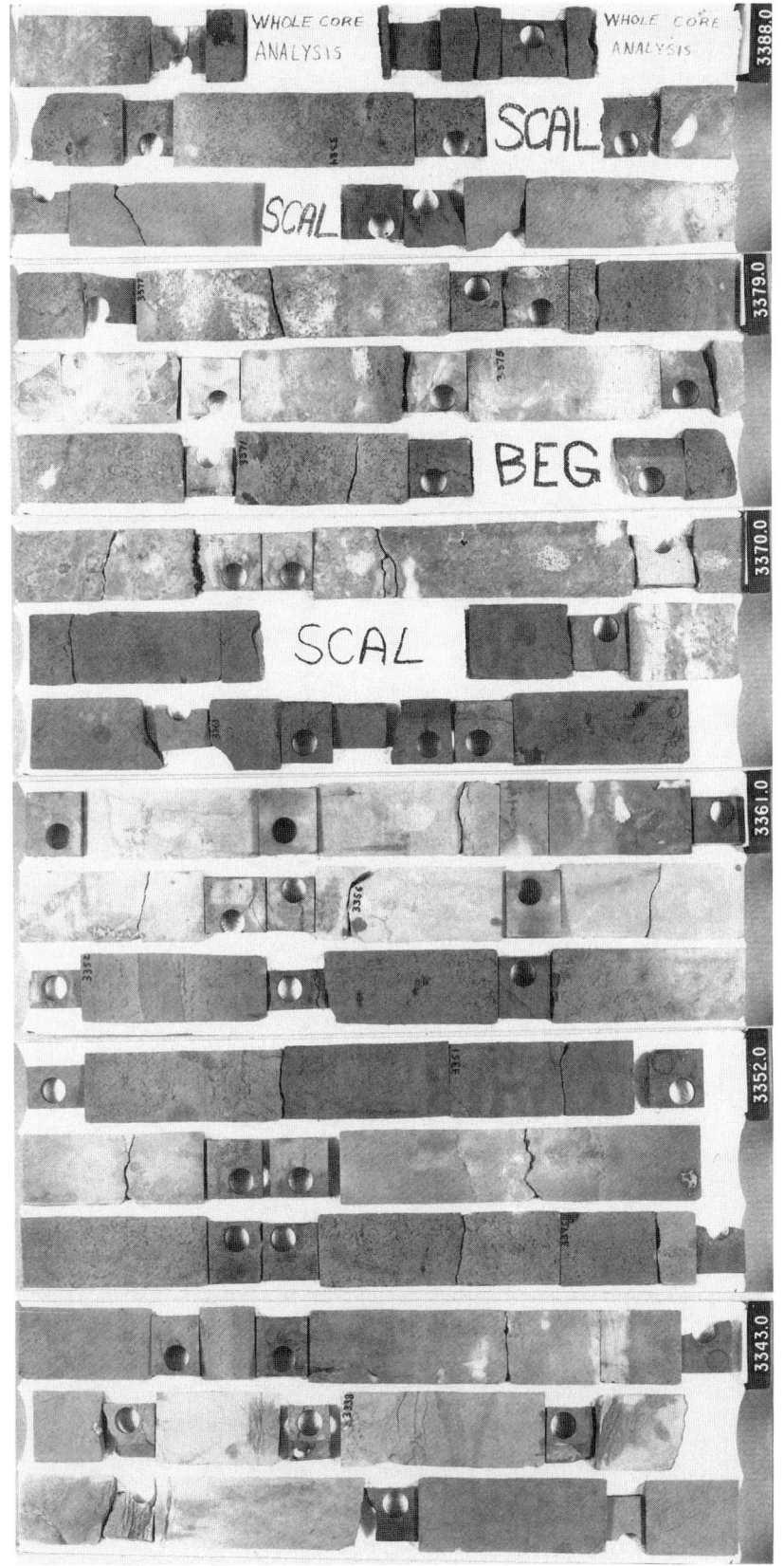

FIG. 6.—Continuous core from the Mobil 1560 well. Each core segment is three feet long.

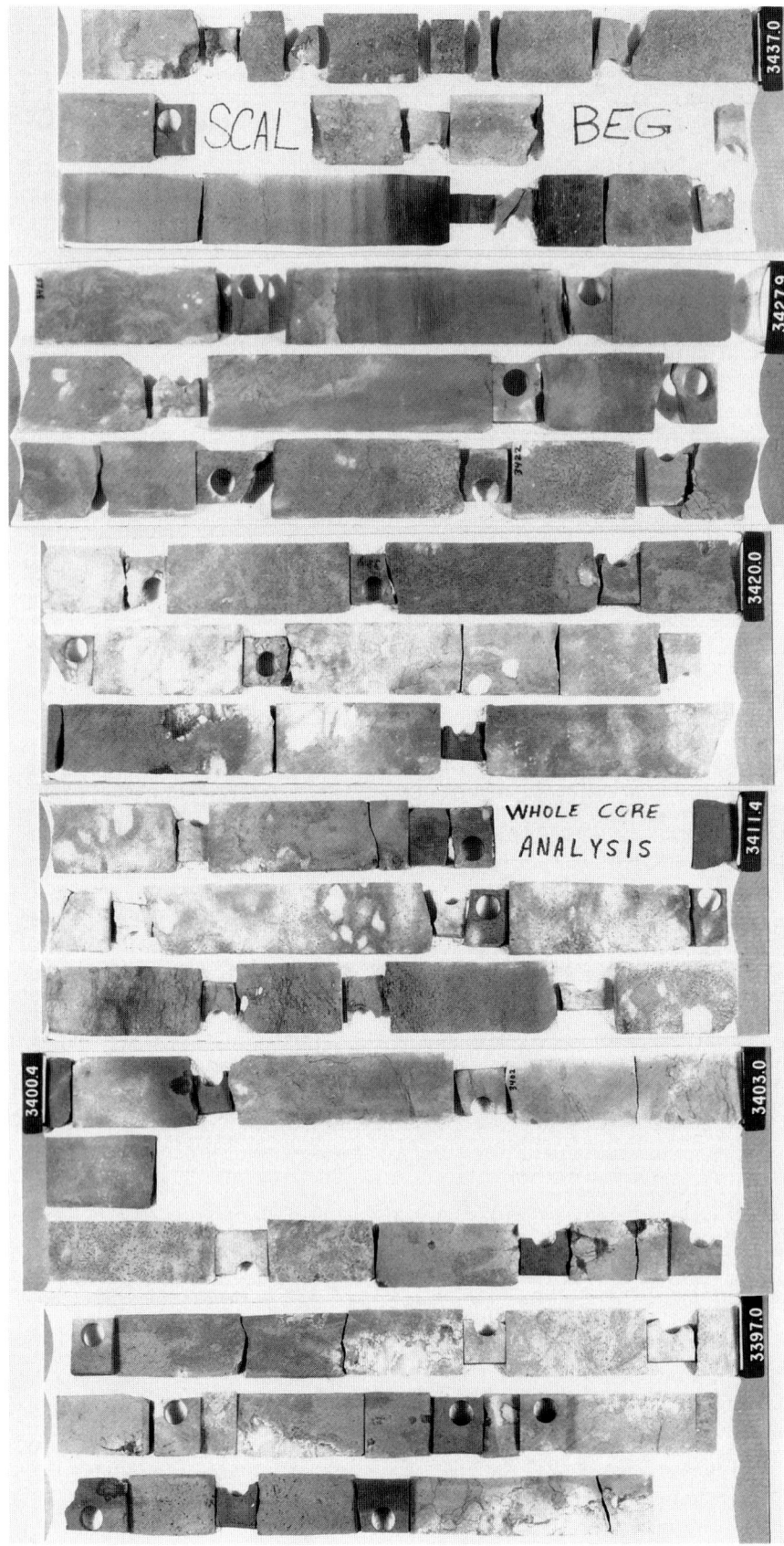

FIG. 7.—Continuous core from the Mobil 1560 well. Each core segment is three feet long.

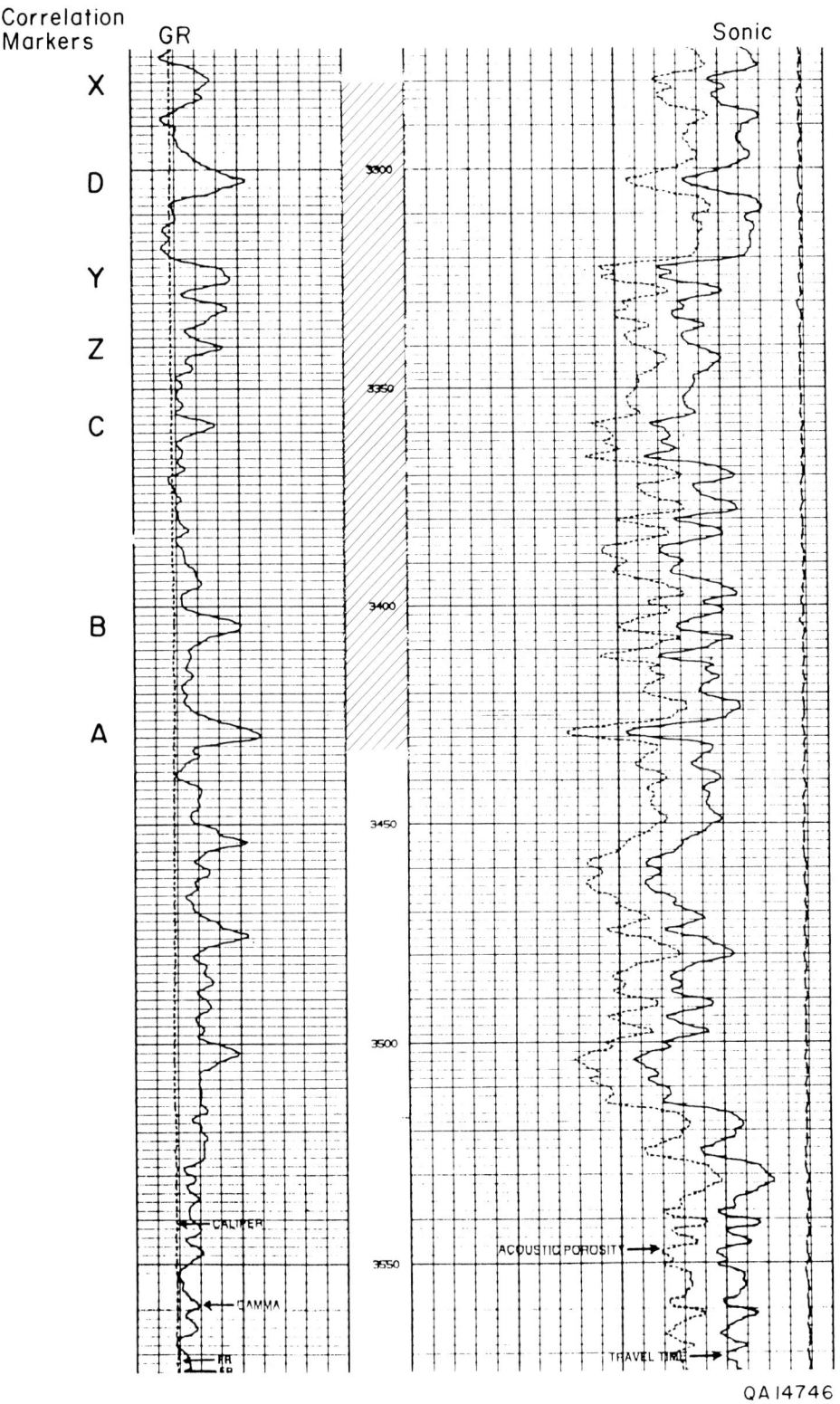

FIG. 8.—Gamma-ray and sonic logs from the Mobil 1560 well. Location of the displayed core is shown.

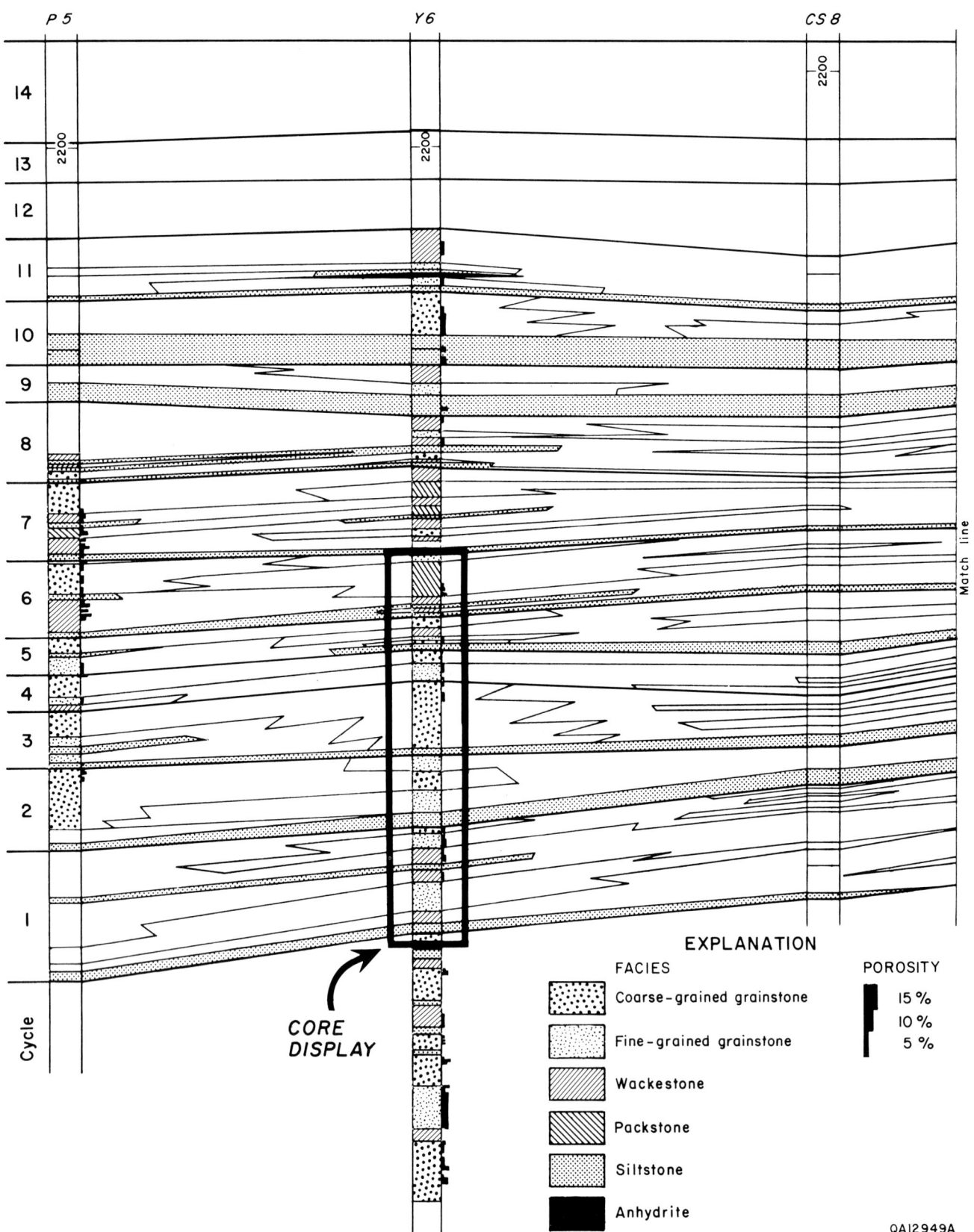

FIG. 9.—Facies dip section across a portion of Farmer field, Sections 8 and 9, University Block 50, and Section 10, University Block 47, Crockett County, Texas. Location of the displayed core is outlined on the facies column for well Y6.

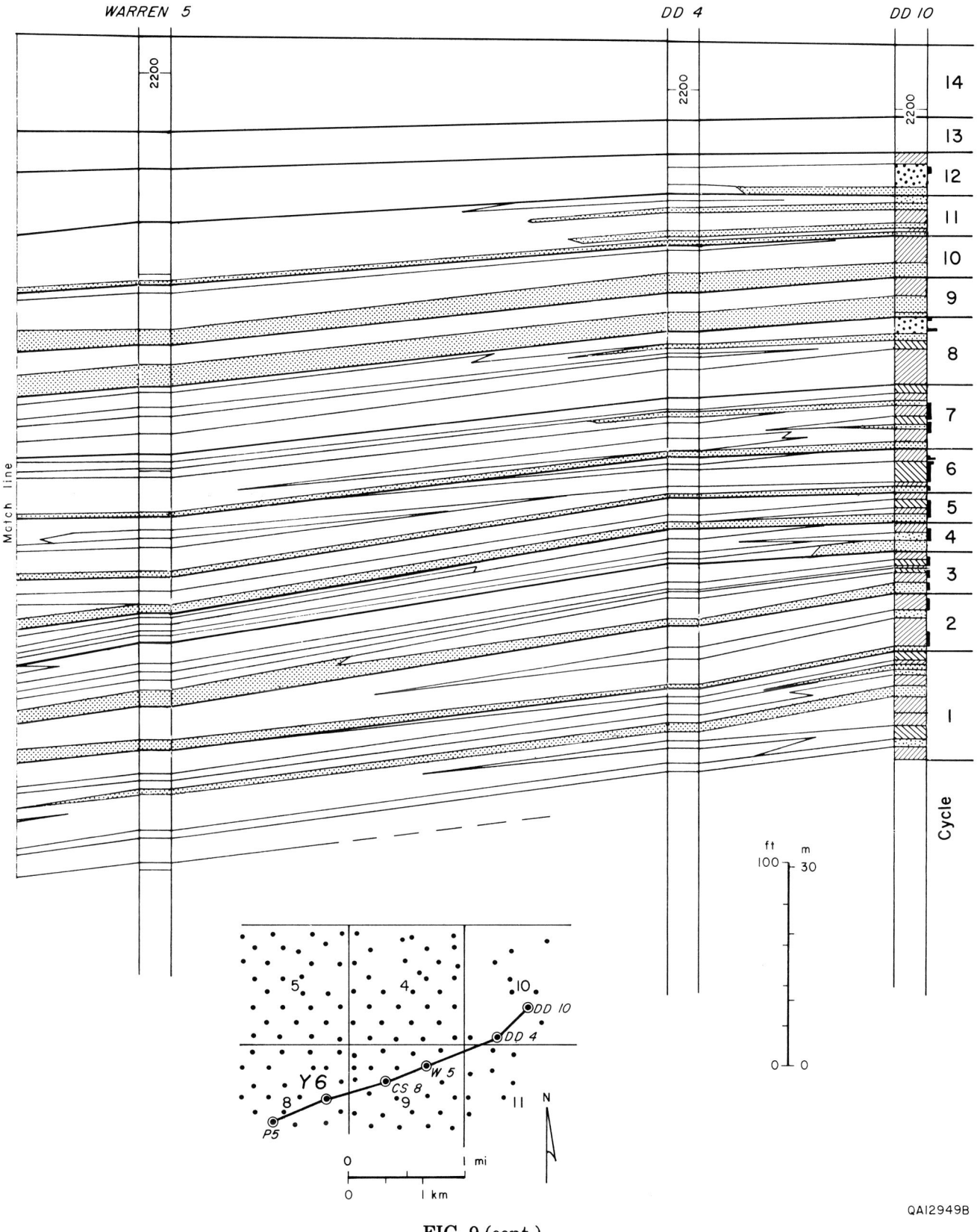

FIG. 9 (cont.)

FIG. 10.—Continuous core from the Marathon Y6 well. Each core segment is 3 ft long.

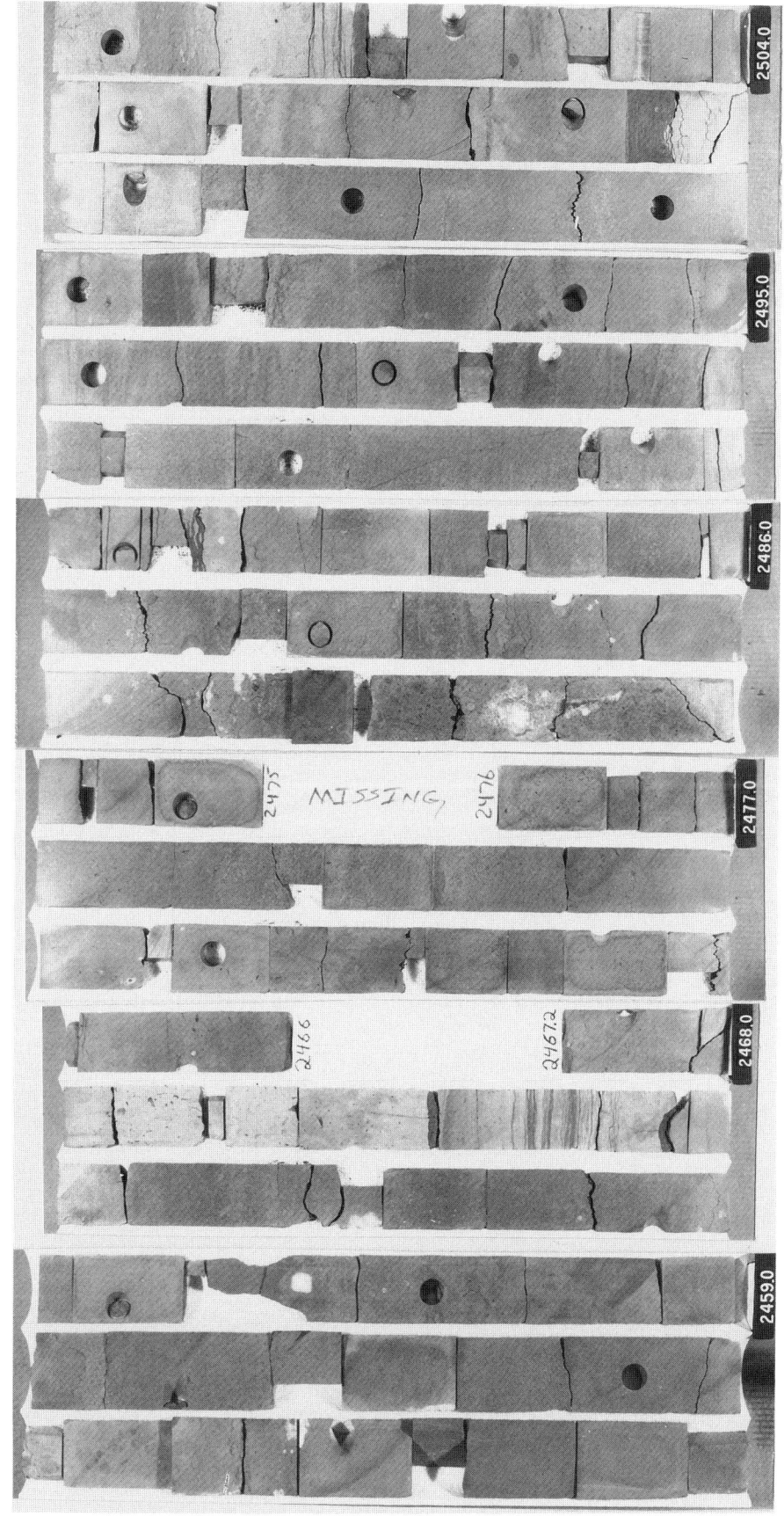

FIG. 11.—Continuous core from the Marathon Y6 well. Each core segment is 3 ft long.

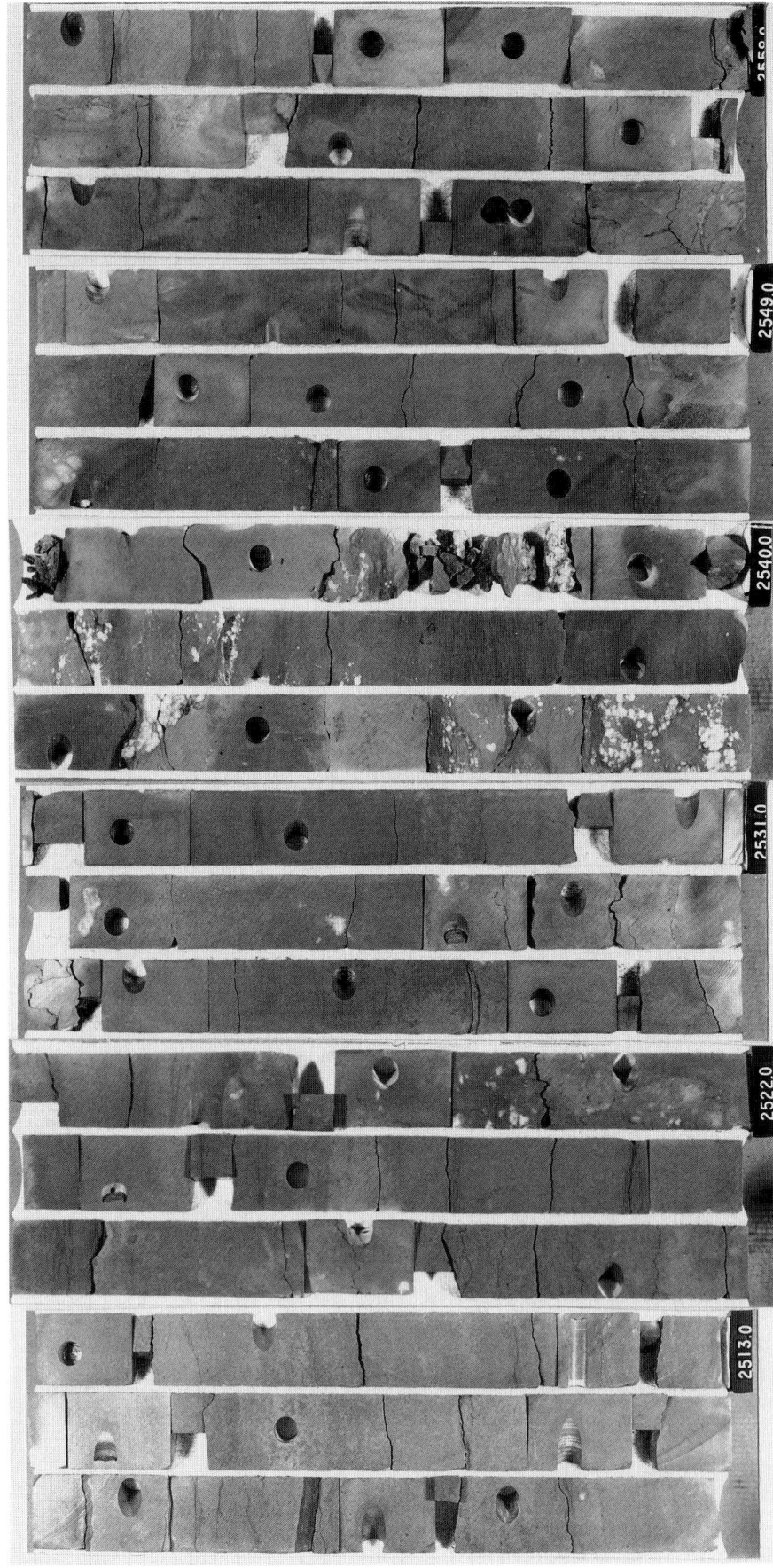

FIG. 12.—Continuous core from the Marathon Y6 well. Each core segment is 3 ft long.

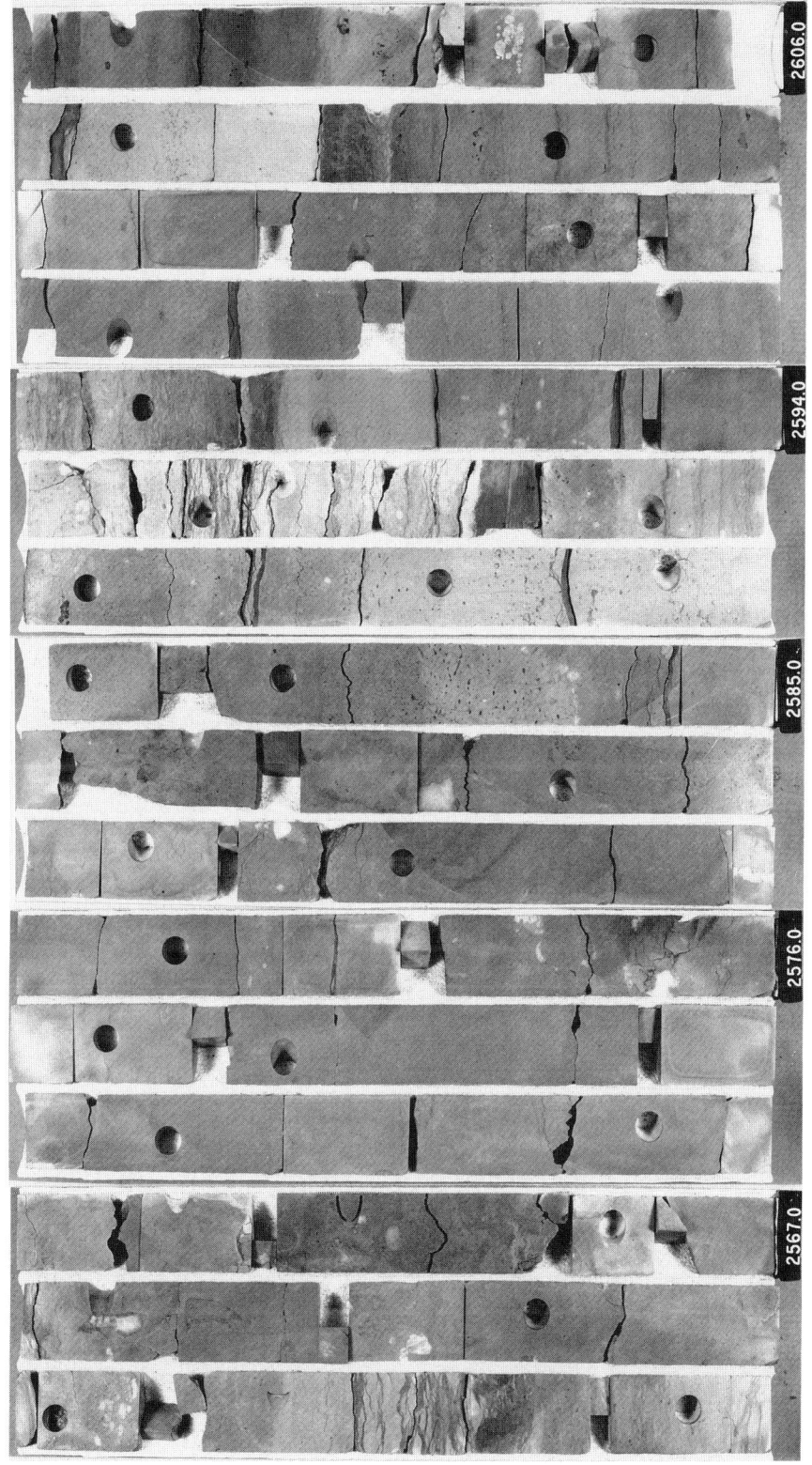

FIG. 13.—Continuous core from the Marathon Y6 well. Each core segment is 3 ft long.

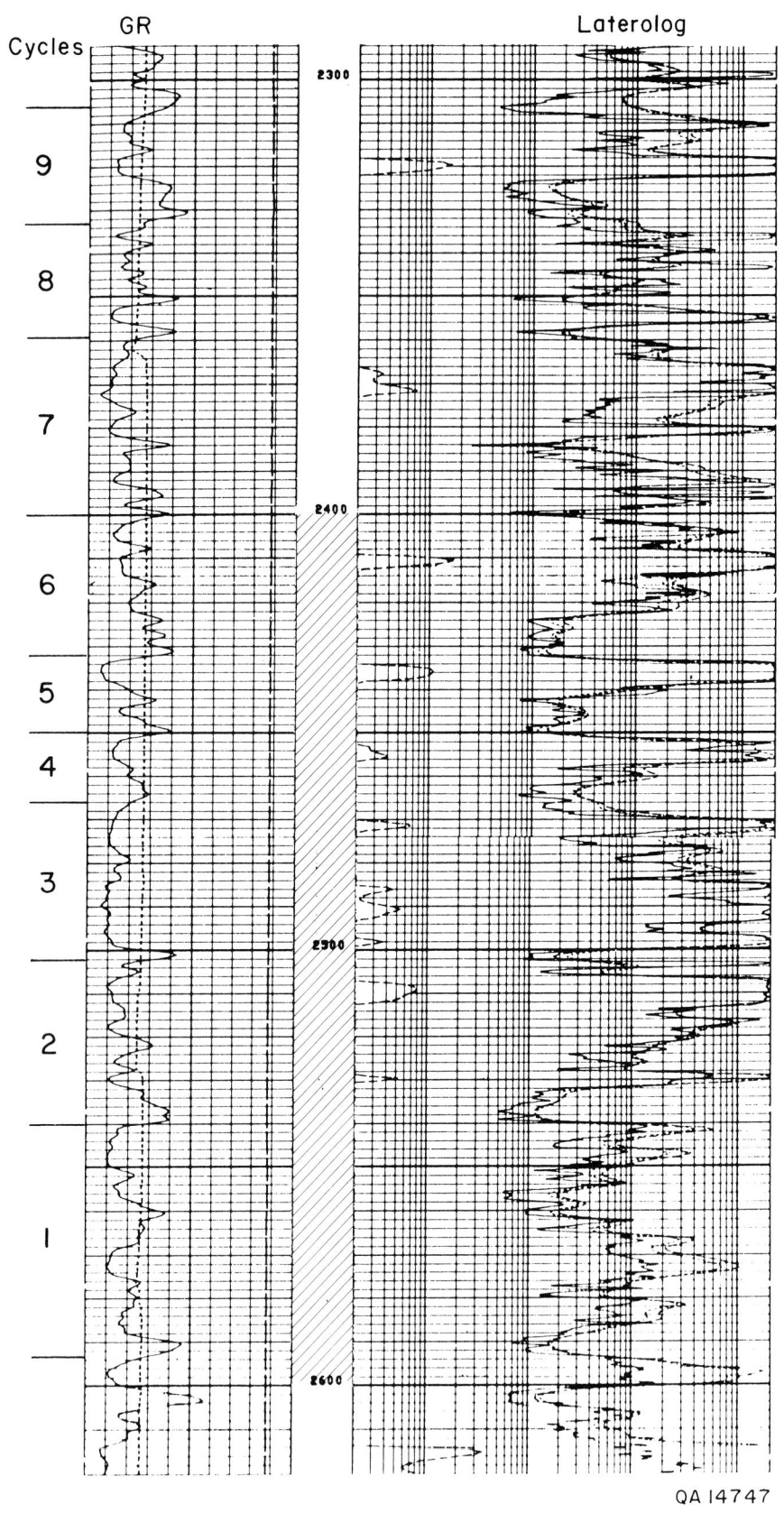

FIG. 14.—Gamma-ray and laterolog logs from the Marathon Y6 well. Location of the displayed core is shown.

MIXED CARBONATE/SILICICLASTIC SEDIMENTATION: NORTHERN INSULAR SHELF OF PUERTO RICO

DAVID M. BUSH
Duke University Department of Geology
Durham, North Carolina 27708

ABSTRACT

The northern shelf of Puerto Rico is a mixed-sediment (carbonate/siliciclastic), high-energy, storm-dominated, steep, narrow shelf. Muddy, allochthonous siliciclastic sediments are introduced directly onto the shelf from small rivers during floods. Clean, autochthonous carbonate sediments accumulate away from river mouths. Shelf-sediment cover is in equilibrium with shelf processes compositionally and texturally. Sedimentation is controlled by two processes acting at different time scales: storms and river-mouth migration.

Storms introduce, rework, transport, and mix sediment on the shelf. Degree of river influence (not always simple distance from river mouth) and water depth are the most important controls on storm stratification in the study area. Water depth controls flux of wave energy to the sea bottom (potential for physical reworking of sediments). Degree of river influence controls distribution of cohesive versus noncohesive sediments.

River-mouth location controls which portion of the shelf is actively receiving siliciclastic sediment at a given time. Changes in location influence mixing of sediment between river systems and between shelf- and river-sediment sources. River-mouth location switching may also help explain the great number of submarine canyons found indenting the shelf.

INTRODUCTION

The growing emphasis over the past 20 years in storm stratification has opened a new study realm in sedimentology. More and more deposits are now being interpreted, or reinterpreted, to have formed from intense "catastrophic" events. Another important concept that has evolved more or less concurrently has been the idea of "proximality." That is, if deposition of sediments or intensity of reworking is a function of water depth or some other predictable and measurable variable or variables, the resultant deposits will change in character in a recognizable pattern. Such patterns are called proximality trends (Aigner and Reineck, 1982). The northern shelf of Puerto Rico is a good location in which to study both concepts.

The study area is a high-energy, storm wave-dominated, mixed carbonate/siliciclastic shelf. The shelf-sediment cover is in equilibrium with present-day physical processes in both composition and texture. Allochthonous, river-derived siliciclastic sediments dominate the shelf in front of river mouths. Autochthonous, shelf-derived, nonreefal carbonates cover remaining portions of the shelf. Sedimentation on the northern shelf of Puerto Rico is controlled by two processes acting at different time scales. They are storms and river-mouth migration. Individual storms last from a few hours to a few days. Storms introduce new siliciclastic sediment, rework sediment, transport sediment, and mix sediment on the shelf. They vary in intensity, duration, and direction of travel. Storms form event deposits of varying thickness, areal distribution, and initial sedimentary structures. Storms also rework previously deposited material and control the frequency, degree, and type of reworking, whether it is largely physical (many storms) or biological (few storms).

River mouths may migrate every few to several hundred years. Location of river mouths controls which portion of the shelf is actively receiving sediment at any given time, thereby influencing mixing of sediment between river systems. This causes stacking of siliciclastic and carbonate clastics, spreads river-derived sediment over a large shelf area, and explains lenses of river-derived material on the shelf not presently connected to their fluvial source. In addition, river-mouth location switching may help to explain the great number of submarine canyons found indenting the shelf.

The northern shelf of Puerto Rico is an ideal location to study storm sedimentation and the concept of proximality. Frequency of storms to rework sediment and introduce new sediment (or both) and relatively high fairweather wave energy mean sediments reach their "equilibrium stance" relatively quickly (weeks or months). There are many factors, however, even in this "ideal" setting, that complicate what should be simple cross-shelf proximality relationships or trends. This paper attempts to describe some of the complex relationships and develop a model of sedimentation for a steep, narrow, high-wave energy, mixed-sediment equilibrium shelf.

STUDY AREA

Puerto Rico is the easternmost and smallest of the Greater Antilles Islands (Fig. 1). It is roughly rectangularly shaped, approximately 99 miles (160 km) long (east-west) and 31 to 37 miles (50 to 60 km) wide (north-south). The insular shelf surrounding Puerto Rico ranges from less than 1.2 miles (2 km) on the northern shelf to over 15.5 miles (25 km) on the western shelf, and the shelf break occurs at about 262 ft (80 m)(Fig. 2). The northern shelf ranges from 0.6 to 1.9 miles (1 to 3 km) in width. The shelf slope is steep, ranging from about 1:15 to 1:50. The insular slope off the northern coast of Puerto Rico is the landward slope of the Puerto Rico trench (Heezan and others, 1959). This study deals with only a portion of the western part of the northern shelf. The "northern shelf study area" extends across the shelf and along the coast for about 50 miles (80 km), from the Rio Guajataca in the west to the Rio de la Plata in the east (Fig. 2).

Rainfall in Puerto Rico is orographically controlled. The east-west trending Cordillera Central divides the island into a humid northern two-thirds and a much more arid southern one-third (Ehlmann, 1968). Since most of the land area drains to the north, the northern shelf receives by far the bulk of the terrigenous sediment eroded from the island (Ehlmann, 1968; Calversbert, 1970). Study area rivers drain a variety of rock types: Moderate to basic Cretaceous volcanics, Late Cretaceous and Early Tertiary granitic intrusives, and thick Tertiary limestones (Fig. 2).

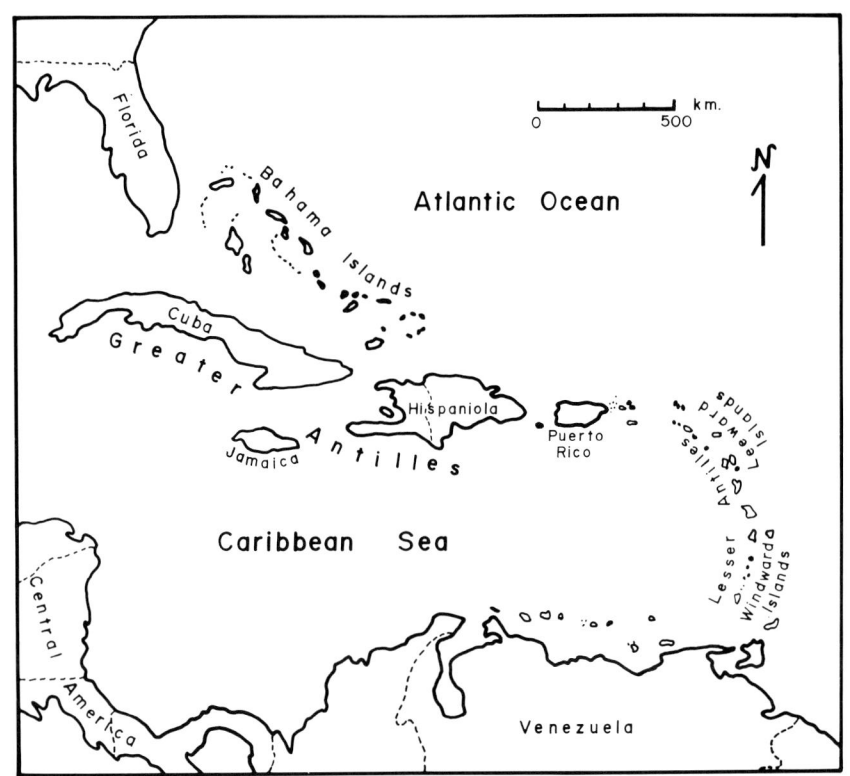

FIG. 1.--Regional location of Puerto Rico.

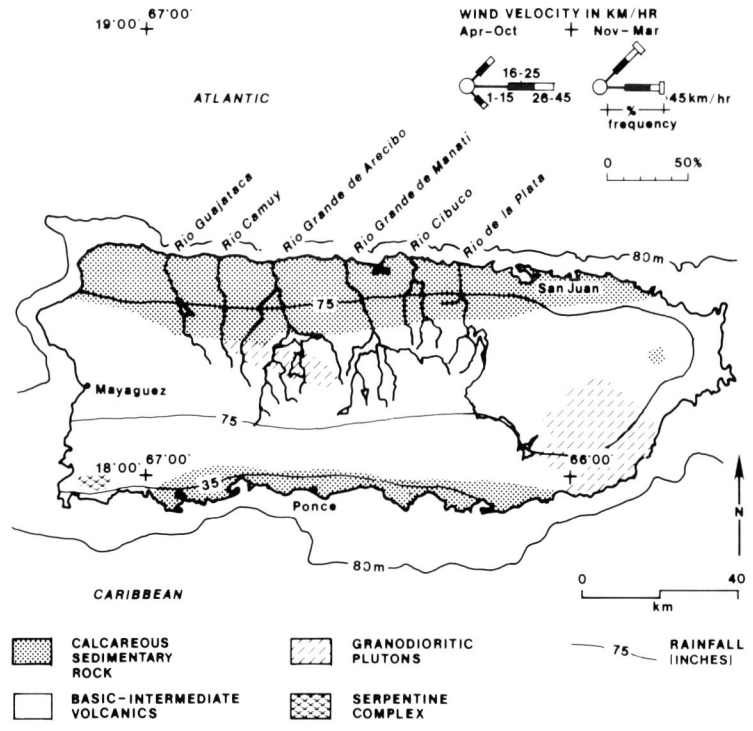

FIG. 2.--Study area rivers; wind velocity data for San Juan; generalized geologic subdivisions, rainfall distribution and location of shelf edge (80m isobath) for Puerto Rico (after Ehlmann, 1968; Morelock, 1984).

Rivers of Puerto Rico discharge up to 70 percent of their annual sediment load during a few days of torrential rains occurring through the year (Lugo and others, 1980). Muddy sediments are introduced directly onto the shelf largely in response to floods caused by major storms. The northern shelf is subjected to wave systems formed by the northeast trade winds, to tropical cyclones (including the entire spectrum from tropical waves, tropical depressions, tropical storms, and hurricanes), and to swell from distant North Atlantic winter storms. Tropical storms cause most of the river flooding and sediment introduction. North Atlantic winter storms are responsible for waves greater than 6 m in the study area (Fields and Jordan, 1972; Morelock and Taggart, 1988). Winter storm waves rework the shelf sediment without accompanying rainfall, thus without introducing new sediment.

On an annual basis dominant winds are from the east and east-northeast, reflecting the strong effect of the northeast trade winds (see wind data on Fig. 2). Extreme significant wave heights of greater than 5 m occur in January, August, and November (Coastal Dynamics Incorporated, 1986). Dominant current direction on the northern shelf is to the west (Wood and others, 1975a, 1975b; Morelock and others, 1985). Tidal range is only 0.3 m.

Pleistocene eolianite armors much of the coastline in the study area, giving it an irregular shape. The lithified dunes occur in several rows or ridges (Kaye, 1959). Some of the ridges are offshore, either emergent as small islands or submerged and forming linear, rocky shoals. Presence of eolianite leads to complex nearshore bathymetry (especially to the 20-m isobath), complicating nearshore sediment transport. Rocky coastal headlands compartmentalize the shoreline, blocking alongshore sediment transport forming "pure" siliciclastic or "pure" carbonate pocket beaches. On coastal stretches without rocky headlands, mixing of siliciclastics and carbonates commonly forms a mixed-sediment facies (Fig. 3).

Surficial sediment distribution has been well studied (Schneidermann and others, 1976; Pilkey and others, 1987; Pilkey and others, 1988). The sediment cover on the northern shelf of Puerto Rico is patchy and diverse, with little lateral continuity and with sharp boundaries between sediment types (Fig. 3). Anomalous muddy patches off the river mouths in the high-energy environment of the northern shelf lead Schneidermann and others (1976) to conclude that the sediment cover in front of river mouths is in textural equilibrium with present-day processes.

High-resolution seismic (Puerto Rico Water Resources Authority (PRWRA), 1975) reveals variations in thickness of unconsolidated sediment on the shelf (Fig. 4). Sediment thickness ranges from greater than 131 ft (40 m) off the Arecibo River mouth to essentially zero on portions of the shelf away from river influence. This leads to the conclusion that the unconsolidated sediment seen on the seismic was largely of river origin. The three-dimensional geometric form of the river-derived sediment deposits on the shelf is best described as a "lens," using the seismic stratigraphic terminology of Mitchum and others (1977).

River-derived sediment is introduced via river-mouth bypassing from point sources (the river mouths) and is then distributed largely cross-shelf with some along-shelf movement (Pilkey and others, 1978; Pilkey and Lincoln, 1984). Typically the entire shelf directly in front of a river mouth is covered with river-derived material. Away from river mouths, only a thin veneer of calcareous sediment exists overlying submerged Pleistocene or Tertiary rock. Siliciclastic and

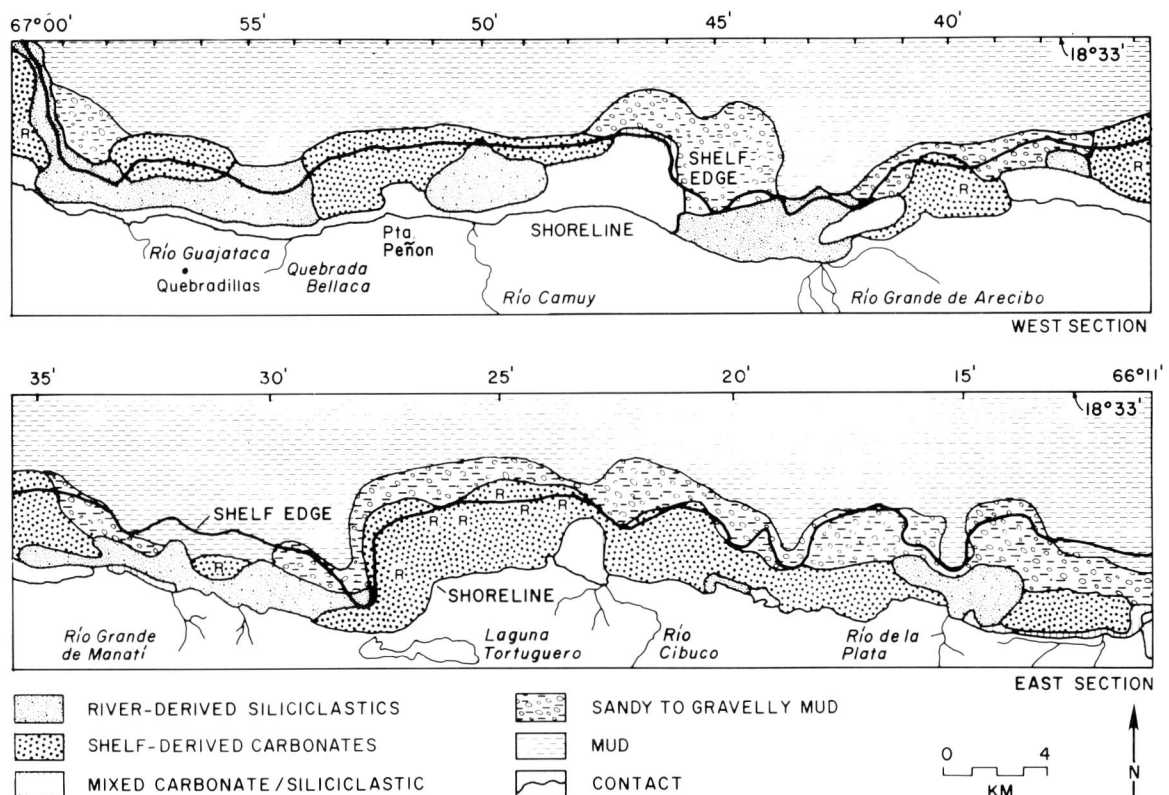

FIG. 3.—Sediment type distribution. "R" denotes areas where rhodoliths are locally abundant.

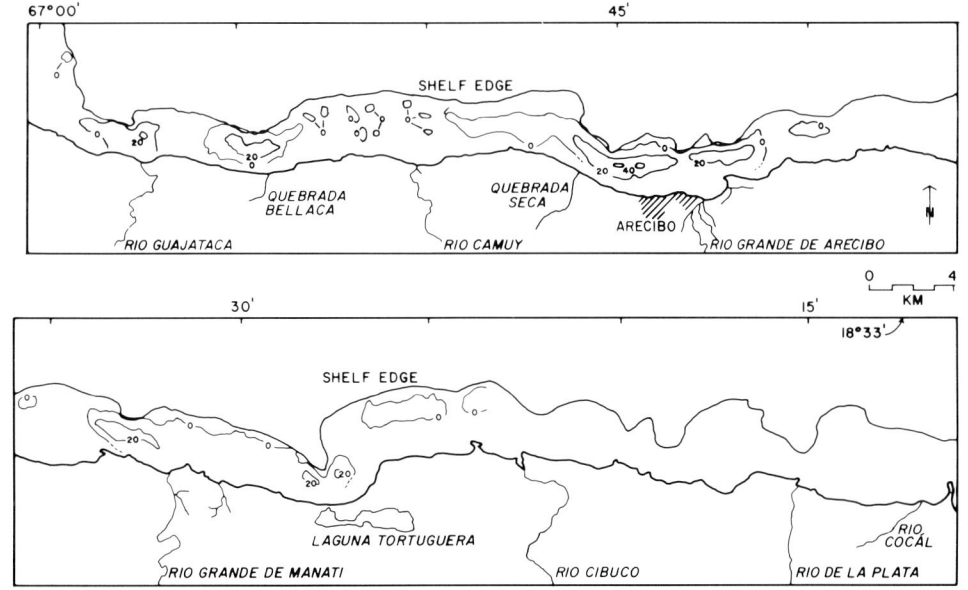

FIG. 4.—Isopachs (in meters) of unconsolidated sediment (after PRWRA, 1975). Seismic coverage ends at Rio Cibuco.

carbonate sediments combine in several places to form a mixed-sediment facies (Fig. 3) that is between the other two in grain size and calcium carbonate content, and is typically devoid of mud.

METHODS

The primary data base for this study consists of 66 Ewing piston cores and 60 Alpine vibracores collected February 1984 and January 1985 from R/V Cape Hatteras, plus 21 diver-operated jackhammer vibracores taken from shallow water off R/V Jean A during several cruises from 1984 to 1986. A total of 147 cores are utilized in this study: 82 from the Arecibo shelf, 19 from Tortuguero, 28 from Manati, and 18 from la Plata. Cores were split, described, and photographed. Relief casts (resin peels) were made to enhance sedimentary structures for description. Subsamples were taken as necessary to adequately define and describe textures and sediment types.

Standard textural analyses were performed on the core subsamples. Median grain size of the sand fraction was determined using a 15 cm-diameter settling tube. Percent calcium carbonate was calculated by digestion with 10 percent Hcl. The heavy mineral suite of the fine sand fraction of selected samples was "fingerprinted" using X-ray diffraction to help trace movement of sediment on the shelf, utilizing the technique described in Grossman (1978). Detailed identification of sediment constituents is discussed by several investigators (see, for example, Schneidermann and others, 1976; Bush, 1977; Pilkey and others, 1978; Pilkey and others, 1979; Brackett, 1985; and Benes, 1988). No additional constituent identification was done for this study. A detailed high-resolution (Uniboom and Sparker) seismic survey is available for most of the study area (PRWRA, 1975).

STORM SEDIMENTATION AND PROXIMALITY TRENDS

The dominant role of storms in shelf sedimentation has become widely accepted. Most information has come from the study of ancient systems, not modern, owing largely to good exposure in outcrops versus poor exposure in cores and to obvious problems involved in sampling during storms. A concise summary of the history of the concept and study of storm sedimentation is given by Walker (1985).

Hummocky cross-stratification (HCS) is often cited as evidence of storm-induced sedimentation in ancient systems (for example, Hamlin and Walker, 1979; Bourgeois, 1980; Harms and others, 1982). Paleogeographic distribution of HCS is discussed in Marsaglia and Klein (1983) and Duke (1985). HCS alone is not absolute proof of storm sedimentation, as its origin remains controversial (see, for example, Dott and Bourgeois, 1982; Swift and others, 1983; Walker and others, 1983; Greenwood and Sherman, 1986; Nøttvedt and Kreisa, 1987). HCS is probably polygenetic in origin (Southard and others, 1990). Storm deposits are described from many different modern shelf settings: the North Sea (Reineck and Singh, 1972; Aigner and Reineck, 1982), off the east coast of the U.S. (Kumar and Sanders, 1972), the Bering Sea (Nelson, 1982), and the Great Barrier Reef shelf (Gagan and others, 1988).

Storm deposits in both modern and ancient systems display an "ideal" vertical sequence of sedimentary structures (Aigner, 1982). The idealized tempestite sequence shares many similarities with the Bouma sequence for turbidites (Nelson, 1982; Aigner, 1982). A major contrast is presence of wave-ripple laminations in tempestites, evidence of an oscillatory component to depositional currents (Aigner, 1982). Tempestite units typically display a scoured base with or without a basal lag, evidence of peak storm-erosion activity. The remainder of the tempestite is evidence of waning storm flow (Aigner, 1982) and of the transition from higher to lower flow regimes (Seilacher, 1982). Above the basal unit internal laminations of tempestites grade from parallel laminations to HCS to wave-ripple laminations. Tops of storm beds typically grade into a muddy unit or a post-storm flood cap.

The most important effects of storms on bodies of shallow water are (1) the onshore movement of water, creating a near-surface current and a positive set up (storm tide); and (2) an offshore bottom current formed by pressure gradient return flow (Allen, 1982). The result is suspension of a great deal of sediment, winnowing, offshore transport of large volumes of sediment, and deposition as sharp-based graded units. Mechanisms of storm-sediment transport are still controversial and probably vary. Five types of storm-driven currents (rip currents, incremental storm flows, storm-surge ebb currents, wind-forced currents, and turbidity currents) are discussed by Walker (1985, after Walker and others, 1983; and Swift and others, 1983).

The concept of proximality is based on the fact that major controls of storm sedimentation are decreasing effects of storms (1) away from the coast and (2) with increasing water depth (Aigner, 1985). Dott and Bourgeois (1982) also include storm intensity as an important factor. Owing to these controls, storm deposits often show qualitative and quantitative changes in sedimentological parameters with increasing water depth and increasing distance from the sediment source (Aigner, 1985). Such patterns, or proximality trends, aid in paleogeographic reconstructions of the basin of deposition (Aigner and Reineck, 1982). The proximality concept is applied to shelf-storm sedimentation by Aigner and Reineck (1982) and is adapted from Walker's (1967, 1970) proximality concept for turbidites.

Applying the Concepts to Puerto Rico

Water depth and distance from river mouth are the most important controls on event stratification in the Puerto Rico study area. Water depth controls flux of wave energy to the sea bottom at a given location (potential for physical reworking of sediments). Distance from river mouth is distance from sediment source in this system. New sediment is introduced to the study area shelf by river-mouth bypassing (terminology of Swift, 1976). In a setting where shoreface bypassing predominates, distance from shore is equivalent to distance from sediment source. Distance from river mouth, then, controls input of muddy, fine, river-derived siliciclastic sediments to a given shelf area. Water depth and distance from river mouth combine in varying degrees to control the type of sediment that accumulates on any given portion of the shelf.

In the high-wave energy, geologically compressed system of the northern shelf of Puerto Rico, input and reworking of sediment occur quickly and predictably. In addition, the shelf is steep and narrow, meaning water depth changes rapidly and effect of water depth on sedimentary attributes should be recognizable in relatively closely spaced cores. Changes in various sedimentary attributes should reveal simple proximality trends.

The picture is more complex. Even though some proximality trends are obvious, most are not straightforward. The mixed-sediment environment may be the most important factor complicating the picture in Puerto Rico. The autochthonous carbonate sands and gravels are largely noncohesive; the allochthonous, river-derived siliciclastic sediment is muddy, fine grained, and is cohesive. The majority of the volume of the mixed-sediment facies is also noncohesive. Cohesive sediments respond differently from noncohesive sediments to the same fluid forces, and modern muddy shelf sediments make better analogs to ancient settings (Kreisa, 1981). Mud influences response of sediments to physical forces by compacting to form scour-resistant beds, increasing the critical shear stress needed to move sediment (De Raaf and others, 1977; Collinson and Thompson, 1982). Rapid burial of storm layers by mud can also reduce subsequent reworking (Aigner and Reineck, 1982).

The first study of individual storm units on the northern shelf of Puerto Rico was the correlation of a single storm sequence between 32 cores on the Arecibo River shelf (Brackett, 1985; Brackett and Bush, 1986). The thickness of the storm unit (termed the "SS1" for Storm Sequence #1) varied from 12 to 19.7 in. (30 to 50 cm) on the mid-shelf to 0.4 to 4 in. (1 to 10 cm) on the outer shelf. On the inner shelf the SS1 sequence is truncated, amalgamated, or reworked by wave action. The SS1 is described as a tempestite/inundite couplet. The SS1 sequence is characterized by a basal lag, overlain by a graded storm-sand layer, capped by a muddy, post-storm flood layer. Basal lag material is an admixture of skeletal carbonate gravel and river-derived rock-fragment gravel near river mouths and coarse skeletal carbonate gravel away from river mouths. Graded storm-sand layers display parallel and wave-ripple laminations, but primary physical sedimentary structures are often disrupted by bioturbation. In nearshore, shallow-water cores, primary physical sedimentary structures dominate.

The thin, muddy, uppermost portion of the SS1 is an inundite, not part of the tempestite. Increasing percent of calcium carbonate away from the river, the thinning of the unit away from the river mouth, and the sharp erosional contact (not gradational) suggest Arecibo River flooding is the source of the fine-grained sediment. The inundite/tempestite boundary is often diffuse from burrowing and hard to recognize without detailed textural analyses.

Location of mud on a shelf is a function of the balance between supply and shelf physical processes working to remove the muds to deeper water (McCave, 1972). Physical processes overwhelm mud supply on the northern shelf of Puerto Rico except directly off river mouths, in areas of high degrees of river influence. A muddy layer similar to the SS1 inundite was observed in two studies off the Rio de la Plata. After Tropical Storm Eloise in 1975 (Pilkey and others, 1978), and after Hurricanes David and Frederick in 1979 (Grove and others, 1982), the shelf floor was covered with muddy layers 0.4 and 4 in. (1 and 10 cm) thick respectively. In both cases the muddy layer had been removed to deeper water when sample locations were reoccupied 4 to 6 months later.

Proximality Trends

This report discusses average characteristics of storm units in 147 cores across the entire study area. Characteristics of storm units do not seem to be as simple and straightforward as the SS1 unit on the Arecibo shelf. Figures 5 and 6 show portions of select cores from the study area arranged by water depth. Cores in Figure 5 are in water depths up to 102 ft (31 m). Cores in Figure 6 are in water depths of 128 ft (39 m) and greater. These figures illustrate several types of sedimentary attributes that will be referred to in the following sections. Labeling codes for Figures 5 and 6 are given on Table 1.

TABLE 1. LABELING CODES FOR FIGURES 5 AND 6

Core-number prefix:

First letter denotes shelf area:

A=Arecibo   T=Tortuguero Lagoon
M=Manati

Second letter denotes core type:

V=Alpine Vibracore
D=Diver-operated jackhammer core
P=Piston core

Sedimentary attributes:

S=Sedimentary structures

Sp=parallel planar laminations
Sl=Low-angle planar laminations
Sh=high-angle planar laminations
Sw=wave ripple laminations
Ss=massive sand
Sg=massive gravel
Sm=massive sand and gravel
Sd=deformed by coring
Sf=faintly laminated bioturbated sand

L=Lag type

Ls=coarse sand   Lg=organics
Lo=old-looking shells Lm=mixed
Lf=fresh-looking shells Lh=rhodoliths

TABLE 1. (contd.)

B=Bioturbation

    B0=no bioturbation
    B1=low degree of bioturbation
    B2=moderate degree of bioturbation
    B3=high degree of bioturbation
    Bb=distinct burrows

C=Nature of the basal contact

    Cs=scoured   Cp=planar
    Cg=gradational   Cb=bioturbated

A=Amalgamation

    Ah=high degree of amalgamation
    Al=low degree of amalgamation

RM=River-mouth location switch resulting in sharp change in sediment type

Distribution of units.--Storm layers do not always show characteristic evidence of a "catastrophic origin" (Kreisa, 1981). There are large areas of the shelf where cores reveal no obvious storm units (scoured bases, basal lags, or megascopic grading). Often cores with evidence of distinct event stratification are found immediately adjacent to cores without distinct units (Fig. 7). Characteristics as subtle as burrow terminations (rarely observed) and slight (though abrupt) changes in grain size were used to identify bases of events. Additional X-radiography or textural analyses may help identify a greater number of units. Examples of amalgamated units are shown in Figure 5, cores AD-2, AD-1, and TD-14.

Units per meter of core.--Units per meter of core can be thought of as number of storms per meter of core. Obviously these must be thought of as minimum number of storms, since there is no way of knowing how many times the same sediment has been reworked. The number of units per meter of core tends to be higher nearshore, with several exceptions (Fig. 7).

Basal lag composition.--Many storm units on the northern shelf of Puerto Rico do have easily identifiable, coarse basal lags. Lags are typically composed of skeletal material away from river mouths, river gravel very near river mouths, and mixed composition in between. The lack of an ubiquitous basal lag means that a shell pavement discussed by Kreisa and Bambach (1982) and Aigner (1982) does not exist in the study area. Incorporation of calcareous shelf-skeletal material into the thick sediment wedges of muddy, river-derived sediment tends to protect the shells and prevent physical, chemical, or biological degradation (Perkins and Halsey, 1971;

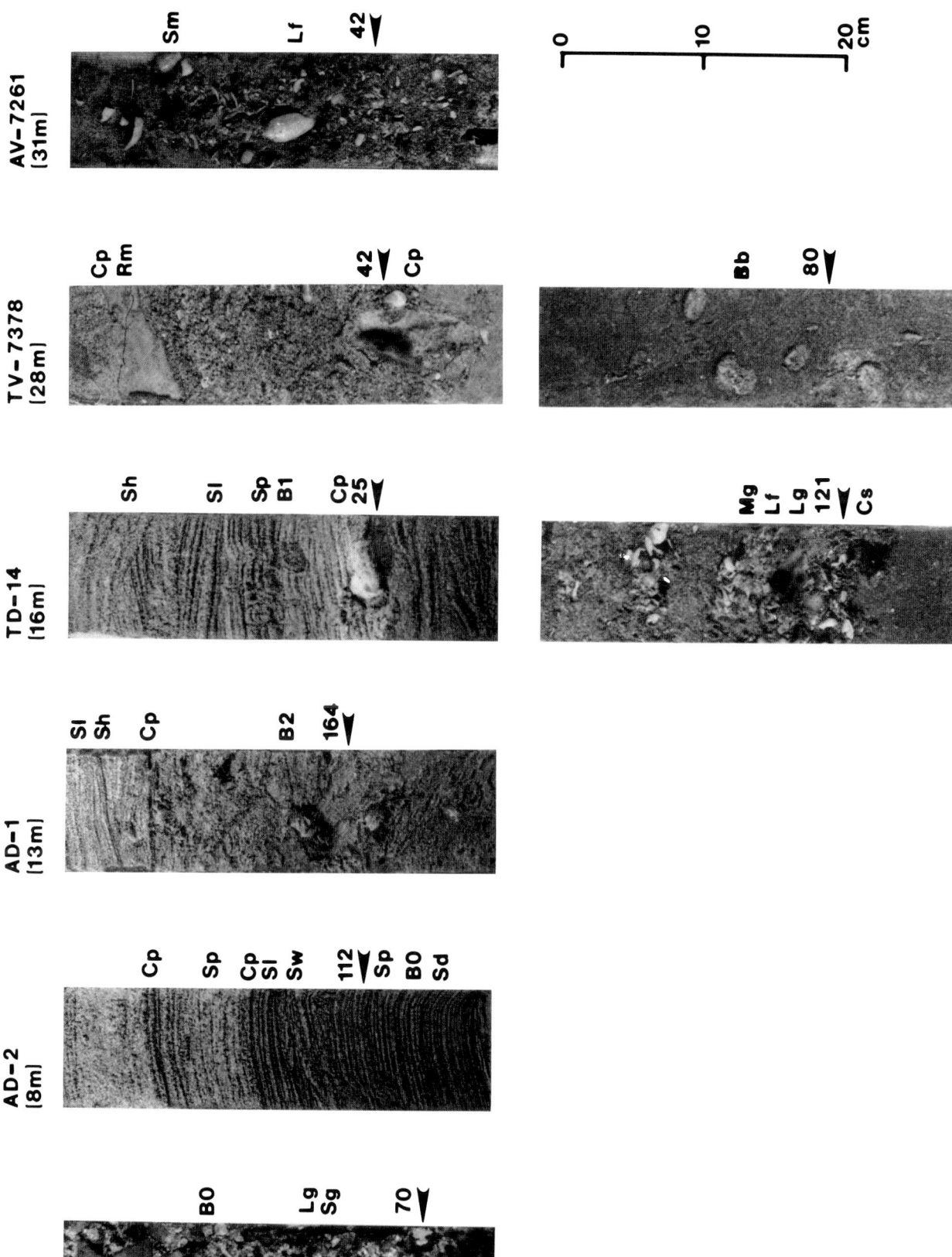

FIG. 5.—Portions of representative cores (photographs of peels) from water depths to 31 meters. Labeling codes given in Table 1.

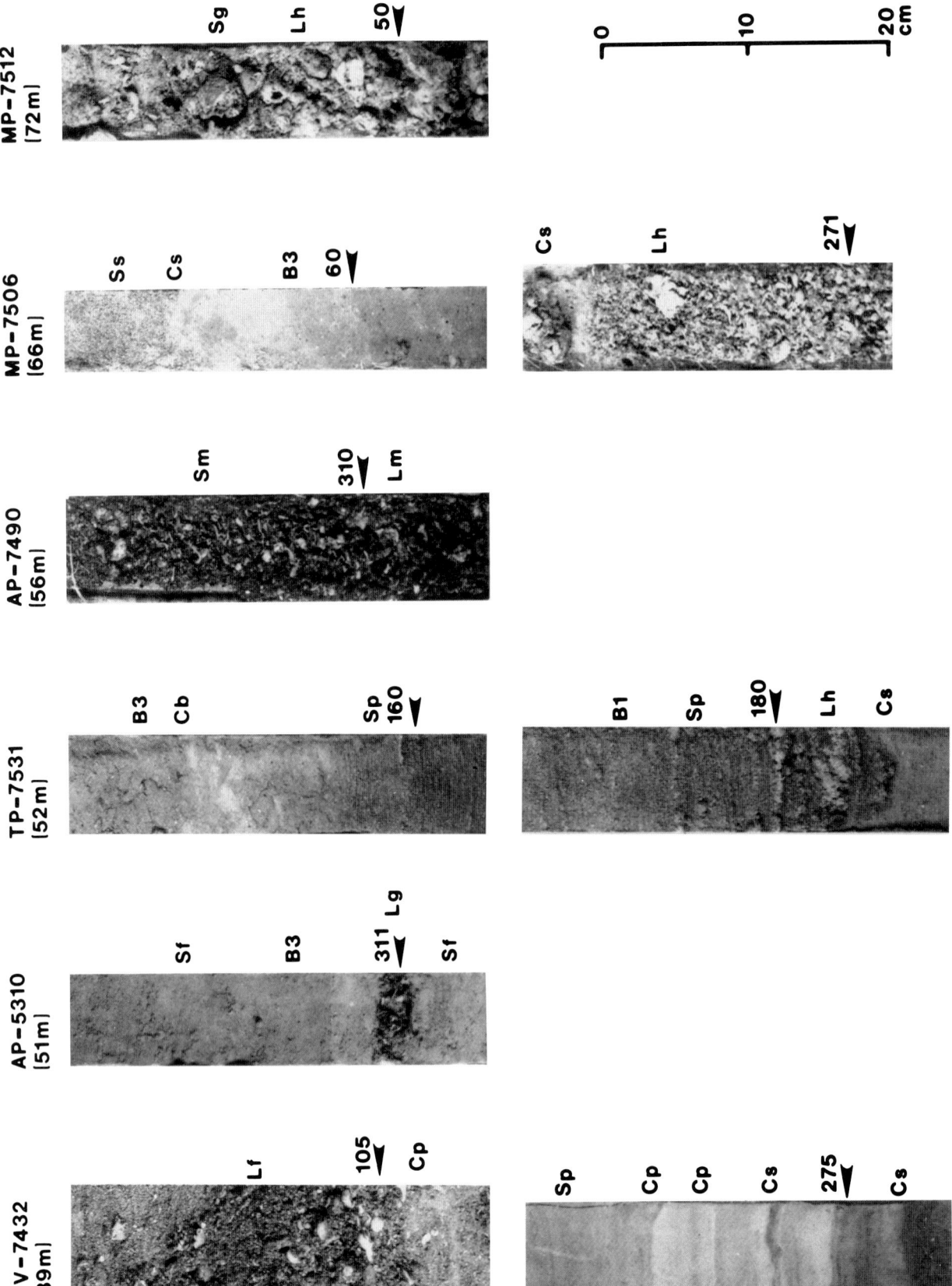

FIG. 6.—Portions of representative cores (photographs of peels) from water depths greater than 39 meters. Labeling codes given in Table 1.

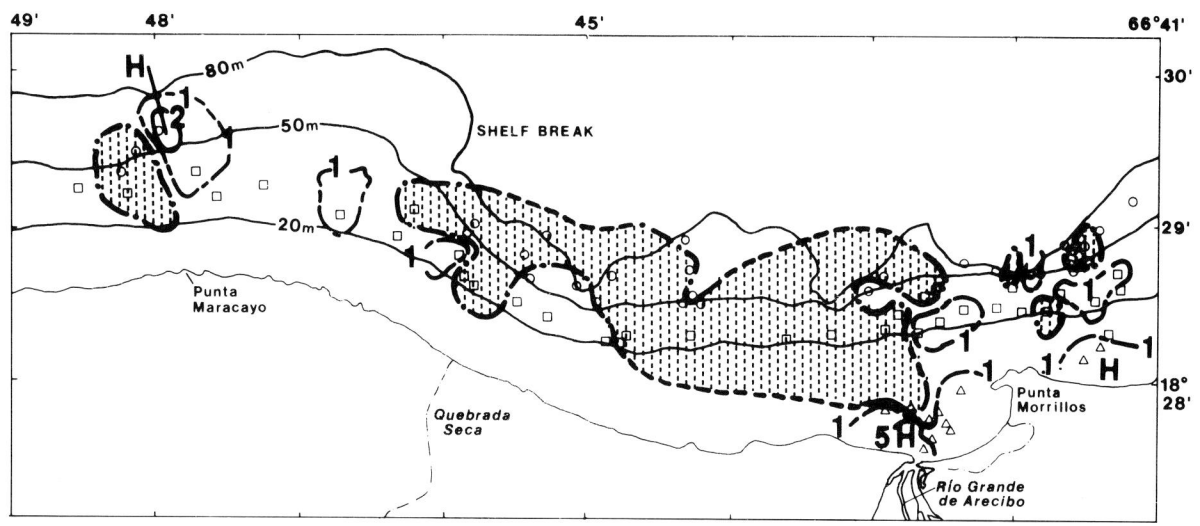
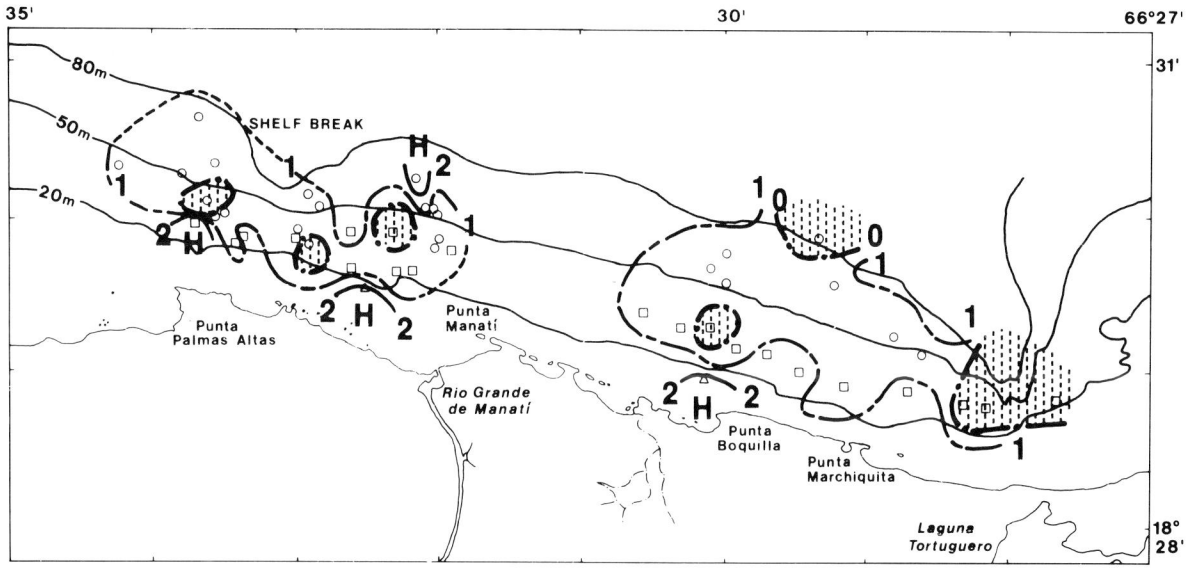
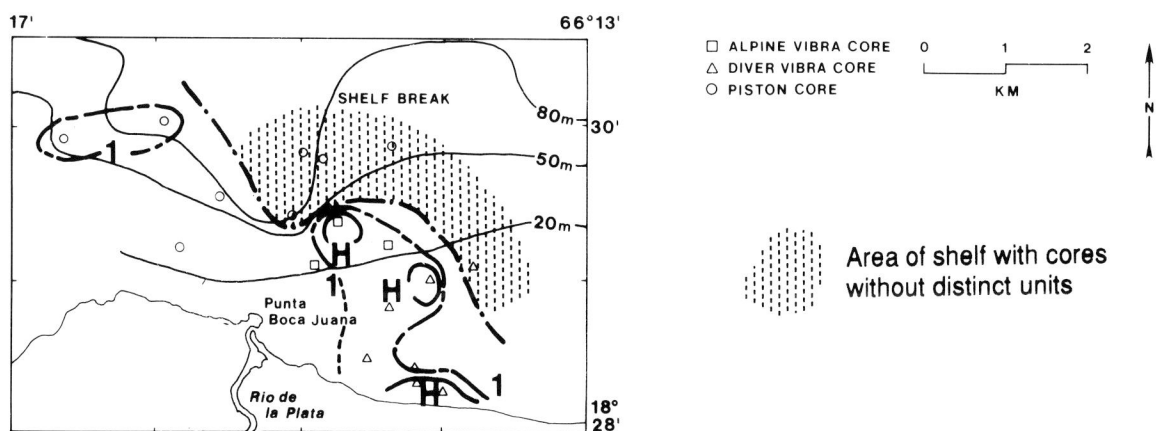

FIG. 7.—Distribution of storm units on the shelf. Contours are number of units per meter of core. Area within zero contour is portion of shelf without easily identifiable units. "H" denotes areas of >1 units per meter of core.

Perkins and Tsentas, 1976; Pilkey and others, 1978; Pilkey and others, 1979; Budd and Perkins, 1980). Benes (1988) showed that, almost without exception, shells are fresh-looking when found incorporated into the muddy, river-derived sediment wedges. Lags made up of fresh-looking shells are found largely in cores from the river-derived sediment wedges (Fig. 5, cores TD-14 and AV-7261; Fig. 6, core TV-7432).

Basal lags composed of "old"-looking shells typically are located far away from river mouths, on portions of the shelf away from river influence, where they are left exposed to degradational processes. Many of the old shell lags are composed predominantly of rhodoliths, a major constituent on portions of the shelf farthest from river mouths (Fig. 6, cores AP-7490, MP-7506 and MP-7512; Fig. 8).

Rock fragment gravel is found only in the very nearshore, off river mouths (present or ancient position), and are unquestionably river-derived (Fig. 5, cores AD-7 and TD-14).

Not all cores with "distinct" storm stratification have gravelly basal lags. Basal lags are often composed of medium to very coarse sand (Fig. 5, core TV-7378, above muddy unit near top of core) if no gravel-size material is present. These lags are termed simple "sand" lags and were identified as bases of storm units because of a sharp basal contact and an easily identifiable change in grain size and/or mud content across that boundary. Sandy lags directly off river mouths are, almost without exception, siliciclastic. The distribution suggests major transport paths of the coarsest fraction of river-derived material.

Basal lag thickness.--Of the 147 cores analyzed in this study, only 98 had obvious (that is, easily discernible) units. Those 98 cores contained 309 identified units of which only 135 had coarse sand and/or gravel basal lags. Of the 135 coarse basal lags, 88 were gravelly lags and 47 were sandy lags. Average thickness of basal lags (regardless of composition of the lag) is shown in Figure 8. The thinnest lags are found on the mid-shelf, with thicker lags found both landward and seaward. Rhodolith massive sands and gravels form the thickest lags and are found mostly on the outer shelf, far from river influence. Lags in very nearshore cores also tend to be very thick and are often composed of very coarse sand and/or river gravel. Very thick nearshore lags are often found in cores from within Pleistocene river channels or pockets eroded into or between ridges of the submerged Pleistocene eolianite.

Nature of the basal contact.--Most basal contacts are planar or scoured, as would be expected. Several contacts are visible on Figures 5 and 6. An occasional gradational basal contact was noted, but, in such cases, diffusion of the contact by bioturbation could not be ruled out.

Amalgamation of storm units.--Amalgamation of storm layers is one trend that is rather straightforward. Nearshore units are usually amalgamated (Fig. 5). Meanwhile, in deeper water, it is common to find complete units or units separated by a bioturbated muddy cap. The degree

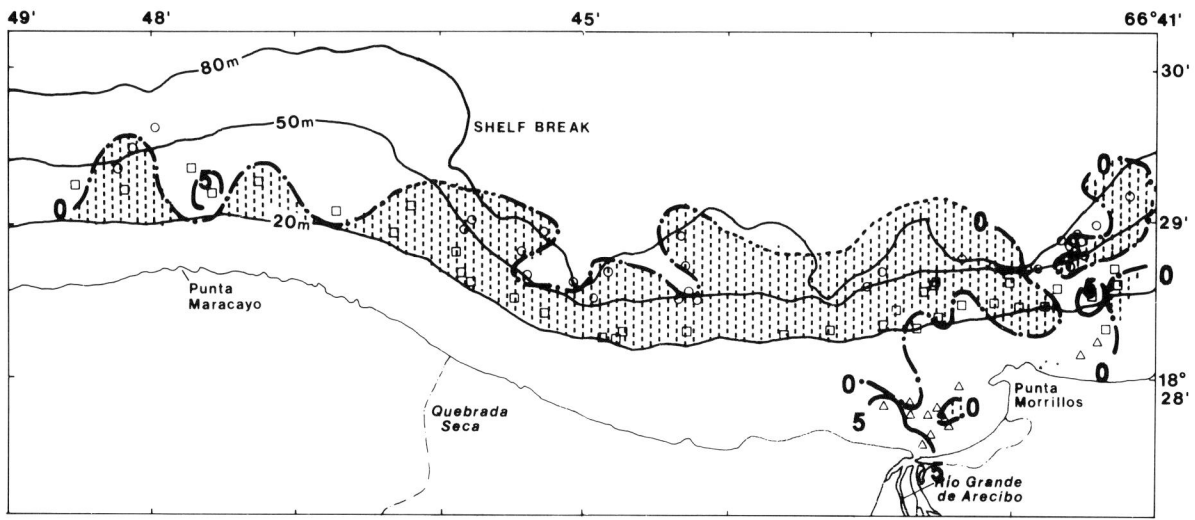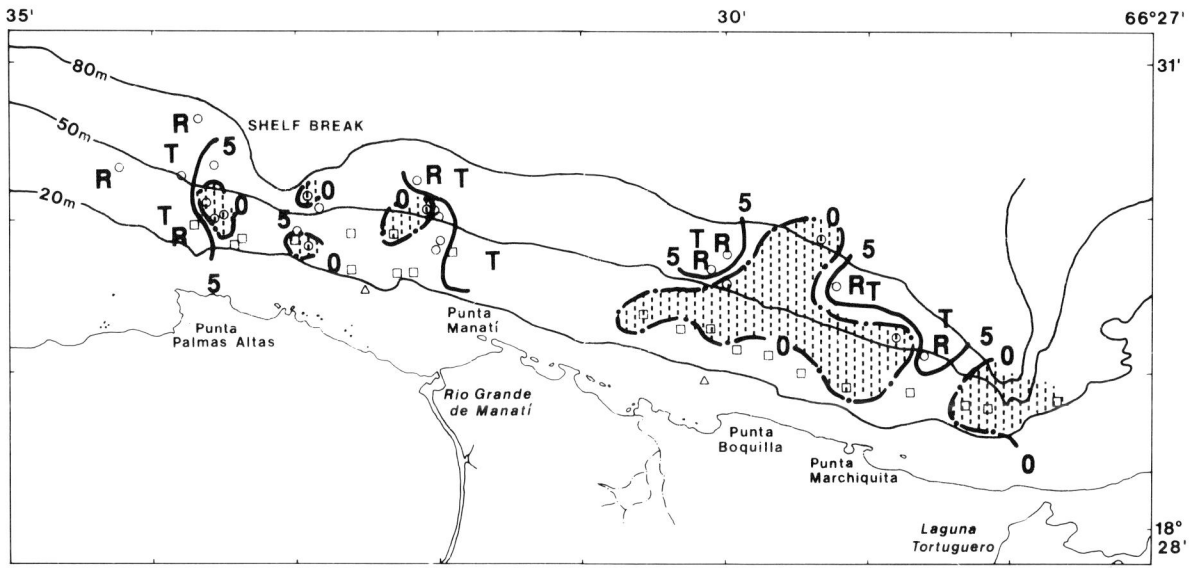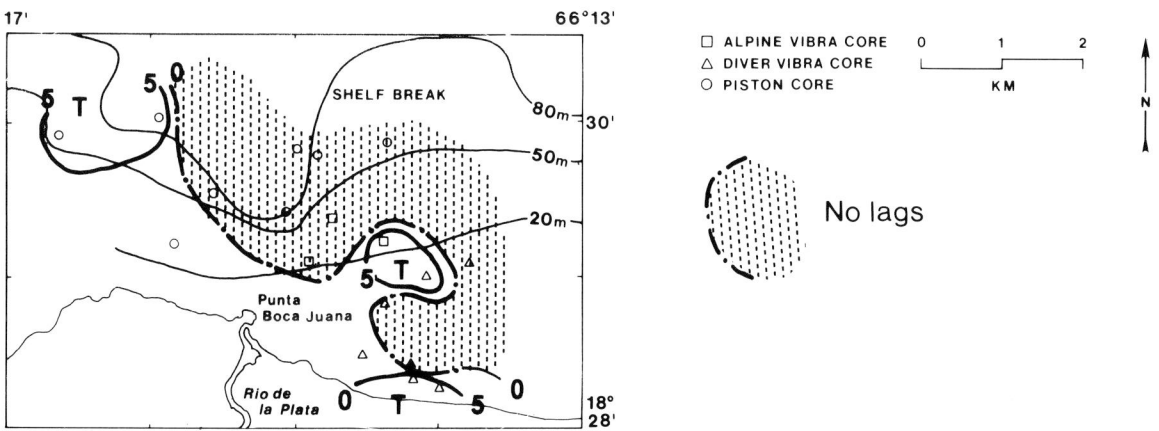

FIG. 8.—Average thickness (cm) of basal lags. Zero contour encloses both cores without units and cores with units, but without coarse basal lags. "T" denotes areas with basal lags >5cm thick. "R" denotes areas where rhodoliths are dominant constituent of lag.

and type of amalgamation of the idealized unit are a function of local rate of influx of river-derived sediment, relative mix of sand and mud, and frequency, duration, and magnitude of storms (Dott and Bourgeois, 1982).

Average unit thickness.--Units are thinnest on the inner shelf, thickest on the mid-shelf, and thin on the outer shelf, with a great deal of variability (Fig. 9). The thickest units are composed of rhodolith gravel lags overlain by a graded rhodolith/*Halimeda* unit (Fig. 6, cores MP-7506 and MP-7512). The graded nature indicates that the material was in suspension, testimony to the intense reworking capabilities of storms affecting the northern shelf study area. Rhodolith lags were probably formed by extremely powerful storms (1000-year storms?) and are left unreworked by smaller storms. Several nearshore cores have thick units much as they had thick basal lags. Those cores are also located in old river channels or in pockets of eroded, submerged Pleistocene eolianite.

Units are thin on the inner portion of the sediment wedges because shallower water depth means greater number of storms and storm waves that can affect the sea bottom, causing increased reworking and degree of amalgamation. Thin units on the outer shelf are a function of less flux of wave energy to the sea bottom and less suspension of sediment to form units. Thick units found on the outer portions of the shelf may represent units deposited by major storms that have not been reworked. Thick units also may be artifacts caused by nonrecognition of unit boundaries either because of lack of a distinct basal lag or because bioturbation has diffused unit boundaries.

Sediment Texture

Cross-shelf or along-shelf textural trends in the study area are not always straightforward. Textural trends are a function of sediment type distribution. Sediment-type boundaries are sharp and do not necessarily follow any bathymetric contour or distance from river-mouth relationship other than in the most general sense (Fig. 3).

Coarse siliciclastic sediments are found directly off river mouths. Coarse calcareous sediments are more abundant and are found across the entire shelf, but are found concentrated in basal lags. The coarsest calcareous sediments are rhodolith gravels. What follows is a distinct correlation between calcium carbonate content and grain size, mud content, sand content, and gravel content. Distance alone does not determine or control input of river-derived sediments to a particular location of the shelf. Portions of the shelf in the shadow of a rocky headland may be very close to a river mouth but still receive little, if any, siliciclastic sediment. As such, sediment texture, as with sediment facies, is a function of degree of river influence, not strictly distance from river mouth or water depth. River influence is controlled by size (discharge) of river, distance from river, interruptions of river-derived sediment transport paths, and, to some degree, water depth.

Median grain size.--Median grain-size values are for the sand fraction only. River-derived

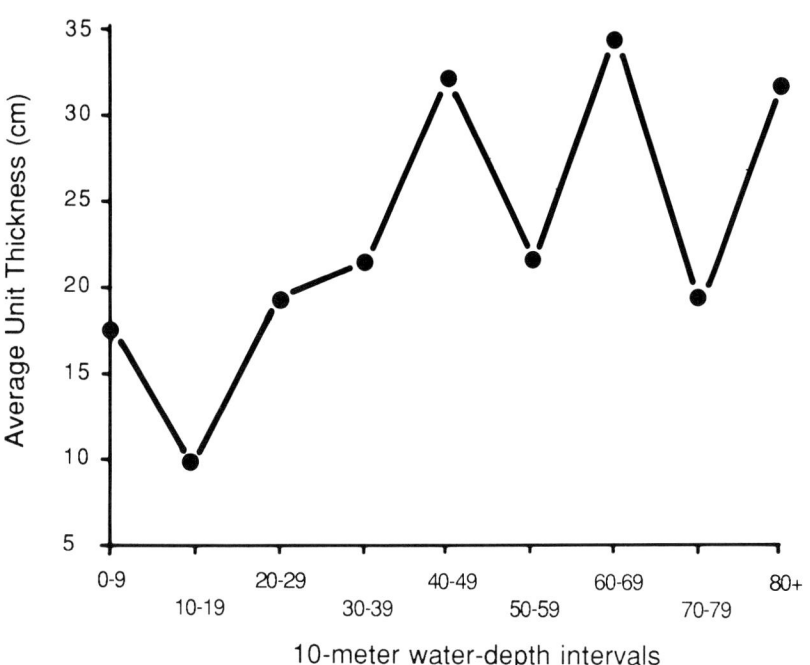

FIG. 9.--Average thickness (cm) of all storm units by 10-m water depth intervals.

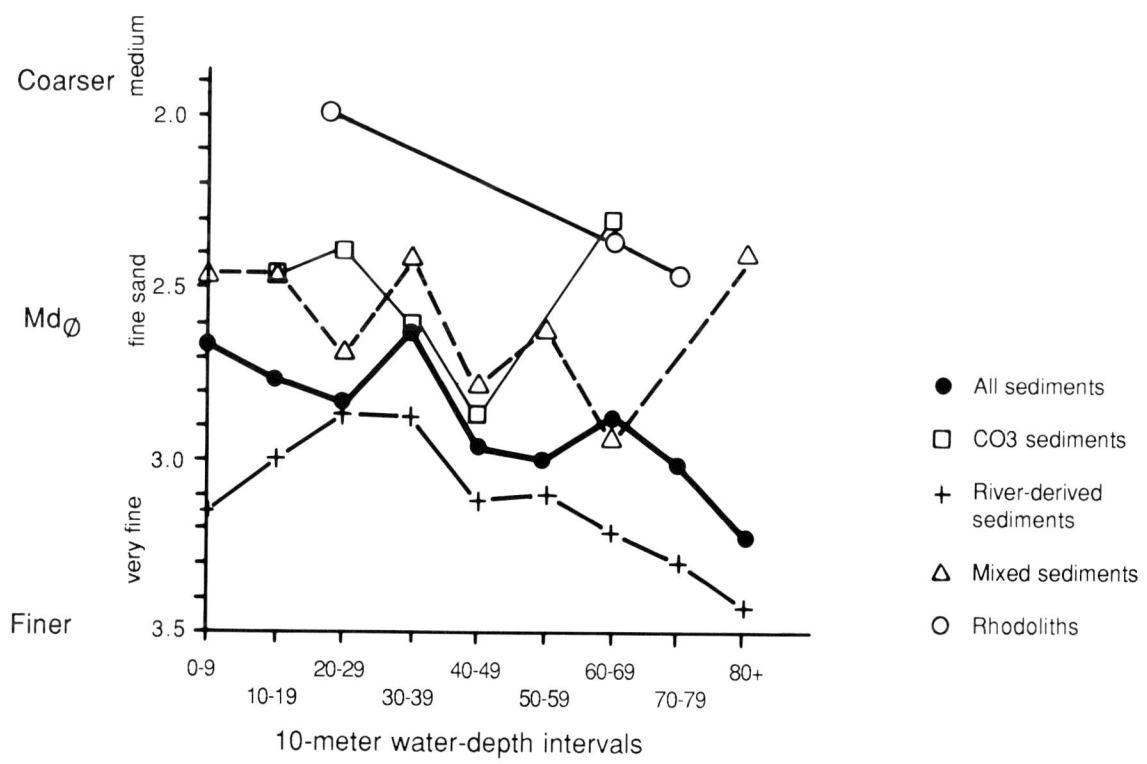

FIG. 10.--Average median grain size (phi units) of the sand fraction for all sediments by 10-m water depth intervals.

Median grain size.--Median grain-size values are for the sand fraction only. River-derived sediments show a fining-seaward sequence, expected on an equilibrium shelf (Fig. 10). Other sediment types show a more scattered distribution of grain size. Grain size versus distance from river mouth (distance from sediment source for river-derived sediments) shows a coarsening trend away from river mouths, indicating an overall decrease of fine siliciclastic sand and an increase in coarse carbonate sediment (Fig. 11).

Percent gravel, sand, and mud.--Gravel content has a bimodal distribution (Fig. 12) reflecting two sources of coarse material on the shelf. In deep water (50+ m) several cores have almost 30 percent gravel. Nearshore gravels are typically rock fragments introduced by rivers, although occasionally mollusk shells are the components. In deeper-water cores, the gravels are almost exclusively rhodoliths. Sand content decreases with increasing water depth (Fig. 12). Mud content of the cores increases with water depth (Fig. 12). As river-derived mud is moved seaward into deeper water, larger storms are needed to resuspend it. The process by which fine-grained sediment is removed to deeper water in a series of jumps driven by storms has been termed Mud Hopping (Pilkey and others, 1978; Grove and others, 1982).

Calcium Carbonate Content

Calcium carbonate content is lowest directly off the river mouths. Autochthonous portions of the shelf, where skeletal carbonates and rhodoliths are accumulating, typically have the highest calcium carbonate values. Portions of the shelf with the mixed-sediment facies have intermediate values (Fig. 13).

Physical Sedimentary Structures

Sedimentary structures recognized in cores from the northern shelf of Puerto Rico are parallel planar laminations; low-angle planar laminations; high-angle planar laminations; wave-ripple laminations (small-scale trough cross-stratification); massive, poorly sorted sands and gravels; and faintly laminated, bioturbated sand and fine sand. Most of the sediment column is composed of faintly laminated, bioturbated sands and fine sands. Examples of sedimentary structures are shown on Figures 5 and 6.

Planar laminations.--Parallel planar lamination or flat beds are produced in a wide range of current velocities and can represent low flow velocities or high flow velocities (Harms and others, 1982). Thick accumulations of parallel planar laminations and association with trough cross-stratified sands suggest upper plane-bed conditions (Harms and others, 1982). Parallel planar laminations are much more common in the study area than low- or high-angle planar laminations. All are indicative of relatively high flow velocities (Harms and others, 1982). All types of planar laminations decrease rapidly with increasing water depth (Fig. 14). Upper flow

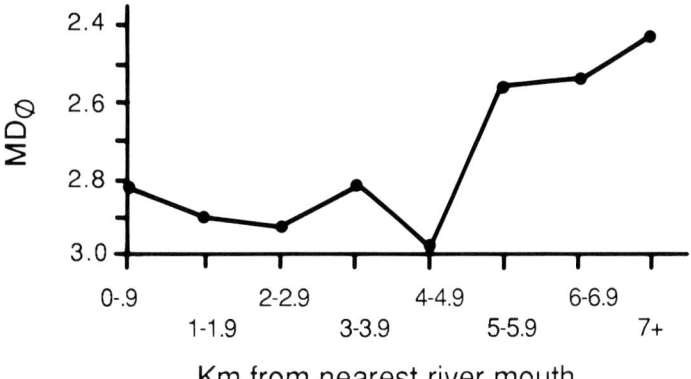

FIG. 11.--Average median grain size (phi units) of the sand fraction for all sediments by 1-km distance from river mouth intervals.

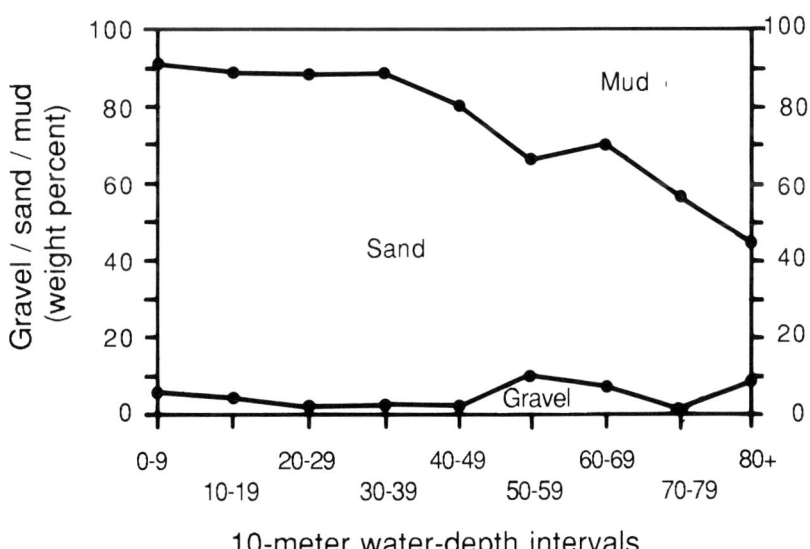

FIG. 12.--Average percent gravel, sand and mud for all sediment types by 10-m water depth intervals.

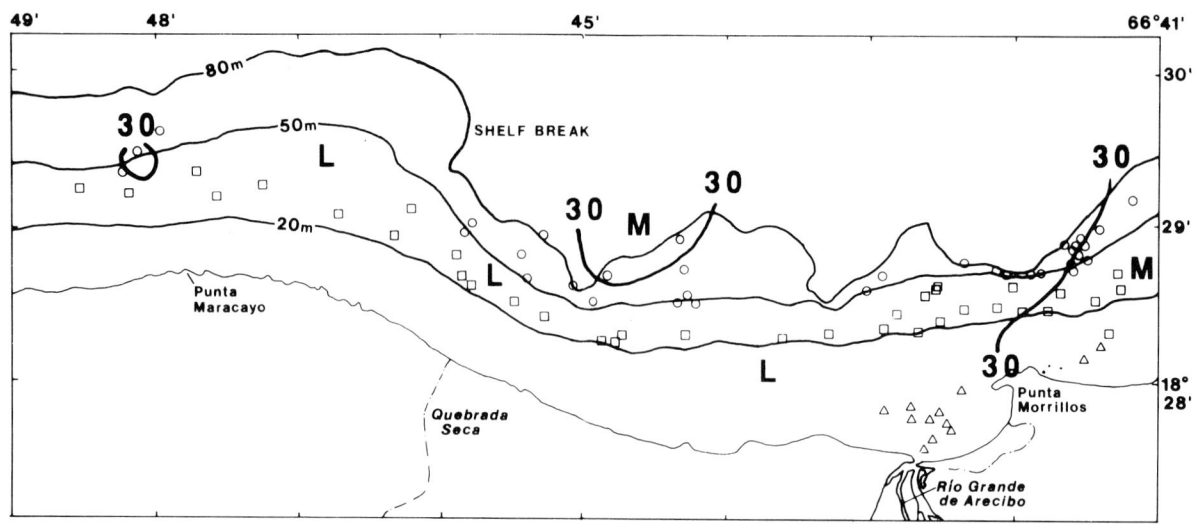
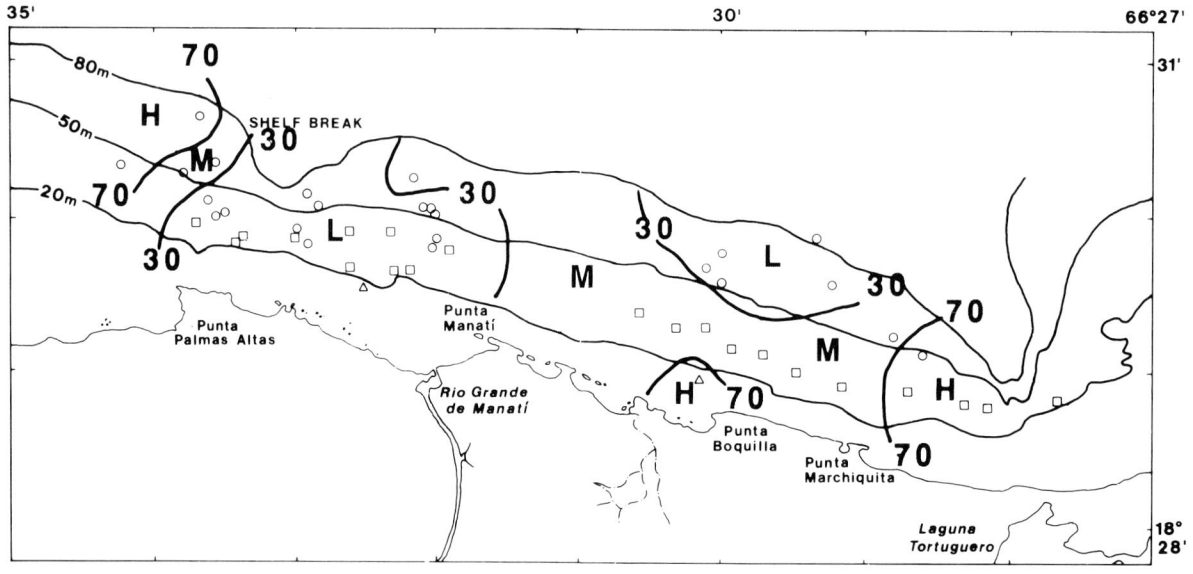
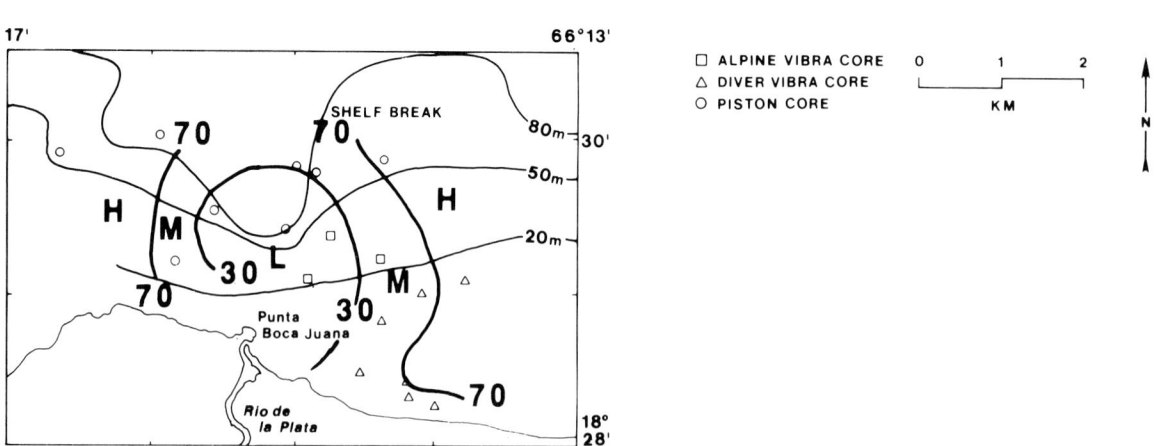

FIG. 13.—Average percent calcium carbonate distribution. L = 0-30%, M = 30-70%, H = >70%.

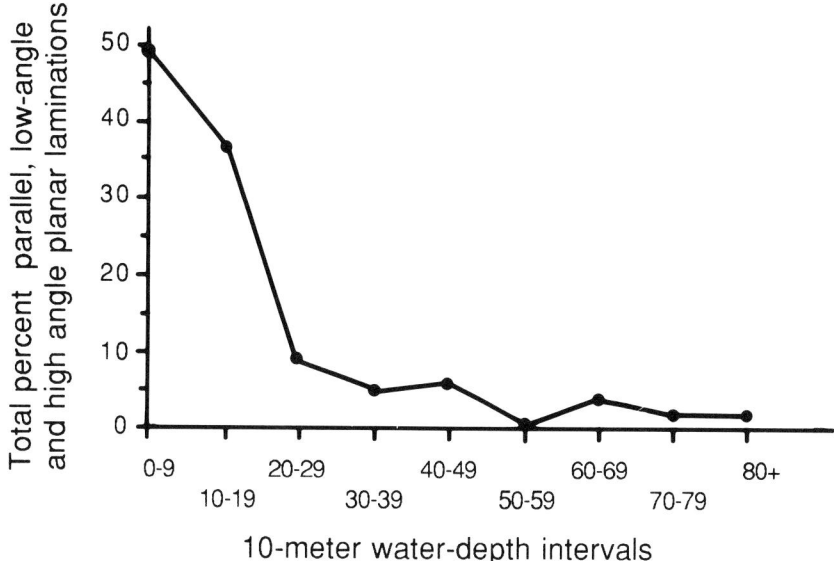

FIG. 14.--Average total percent of all parallel, low-angle and high-angle planar laminations in all sediments by 10-m water depth intervals.

regime planar laminations should decrease with increasing water depth because wave energy impacting the sea bottom decreases, and because increased bioturbation destroys primary physical structures. Some of the low-angle laminations may suggest HCS.

Wave-ripple laminations.--Wave-ripple cross-lamination is an indicator of the presence of oscillatory currents during the deposition of sediments (see, for example, De Raaf and others, 1977; Allen, 1982). Wave-ripple laminations should be rare in nearshore locations where wave energy is too high and upper flow regime planar beds result (Aigner and Reineck, 1982; Allen, 1982; Collinson and Thompson, 1982), or where the wave-rippled part of the sequence is reworked by subsequent events (Aigner and Reineck, 1982). Wave-ripple laminations then increase in the mid-shelf in an "optimum zone," and decrease again in deeper water, where fewer waves with sufficient intensity to form wave ripples impact the sea bottom and where increasing levels of bioturbation destroy primary physical sedimentary structures. In this study wave-ripple laminations decrease with increasing water depth, with much variation, and overall the sedimentary structure is not common (Fig. 15; compare to Aigner and Reineck, 1982). Possible explanations are that nearshore, parallel laminations dominate because of high wave energy, and, offshore, bioturbation dominates, destroying primary physical sedimentary structures. Also high mud content offshore could preclude formation of wave ripples.

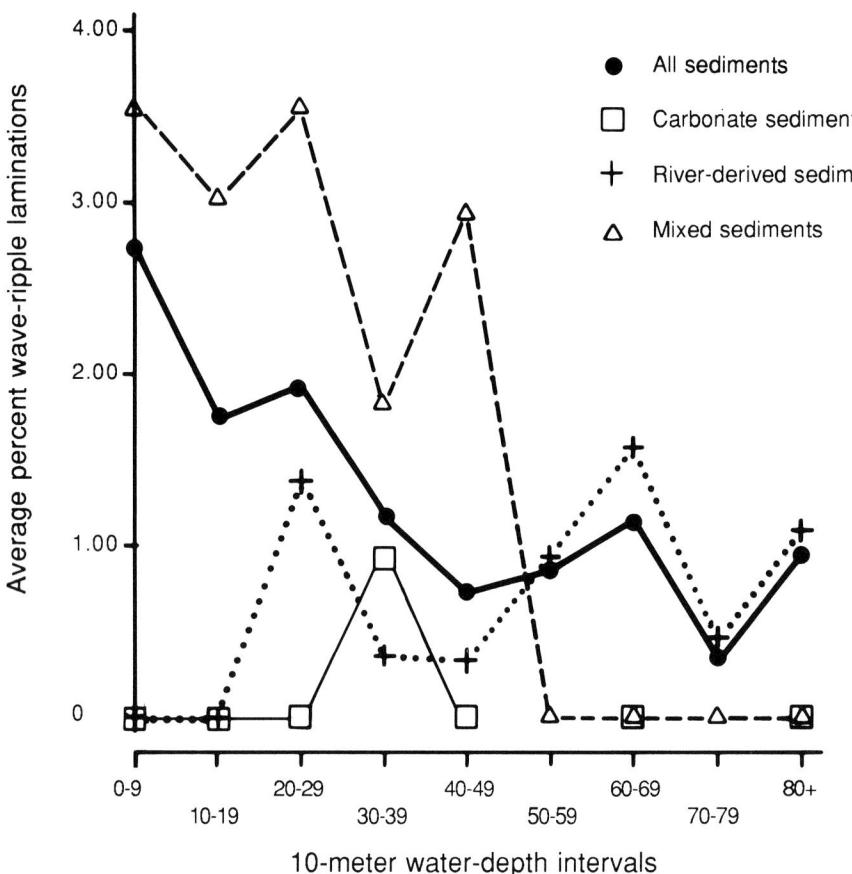

FIG. 15.—Average percent wave-ripple laminations by individual sediment types and for all sediments by 10-m water depth intervals.

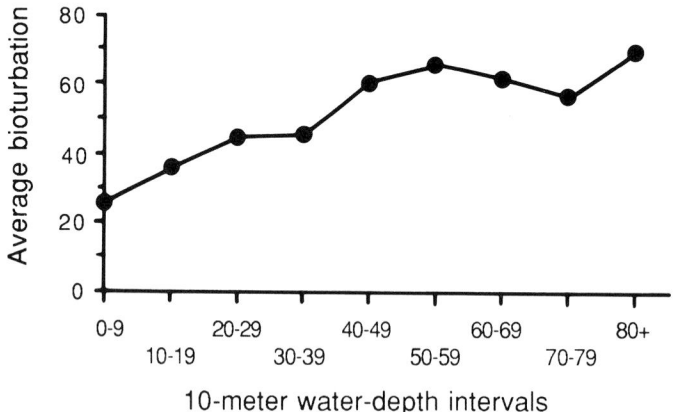

FIG. 16.—Average bioturbation by 10-m water depth intervals.

Bioturbation

Degrees of bioturbation were described using scales modified after those of Howard and Reineck (1972). In that study, off Sapelo Island, Georgia, seven scales of biogenic reworking are defined: 0, trace, less than 30 percent, 30 to 60 percent, 60 to 90 percent, 90 to 99 percent, and 100 percent. In the present study four scales are defined: 0, low (less than 30 percent), moderate (30 to 70 percent), and high (greater than 70 percent) degree of bioturbation (see Figs. 5 and 6 for examples). A shorthand notation is used, naming each scale of bioturbation 0biot, 1biot, 2biot, or 3biot for no, low, moderate, or high degree of bioturbation respectively. Percent of cores with each scale of bioturbation was determined. An attempt was then made to derive a single number that would describe whole-core "average" bioturbation. Percent of each scale of bioturbation was weighted by multiplying the percent by a number 0 through 3 corresponding to no bioturbation through a high degree of bioturbation. Then a whole-core average bioturbation number was calculated.

"Average" Bioturbation = ((Percent 0biot x 0)+(Percent 1biot x 1)
  + (Percent 2biot x 2)
  + (Percent 3biot x 3)) / 3

As an example, core 5292, on the Arecibo shelf, is 698 cm long, all of it slightly bioturbated. The core, then, has 0biot = 0 percent, 1biot = 100 percent, 2biot = 0 percent, and 3biot = 0 percent. Its "Average" bioturbation is calculated as follows:

((0*0) + (100*1) + (0*2) + (0*3)) / 3 = (0 + 100 + 0 + 0) / 3
  = 100 / 3 = 33 "percent"

Therefore, the figure that best "represents" the average bioturbation of the core in this example is 33 "percent."

There are several scenarios that can be envisioned where cores with the same "average" bioturbation have very different actual bioturbation. For example, a core with 0biot = 30 percent, 1biot = 45 percent, 2biot = 20 percent, and 3biot = 5 percent would have an "average" bioturbation of 33, the same as core 5292, which had 100 percent 1biot. Overall, however, both cores can be properly described as being slightly to moderately bioturbated.

Average bioturbation generally increases with increasing water depth (Fig. 16), but there are many instances of bioturbation decreasing on the outer shelf/upper slope (Fig. 17). Bioturbation increases with water depth until approximately 164 ft (50 m) water depth, then decreases slightly until the shelf break, then increases in slope cores. Bioturbation also increases with increasing distance from river mouth up to a point, then decreases with some exceptions.

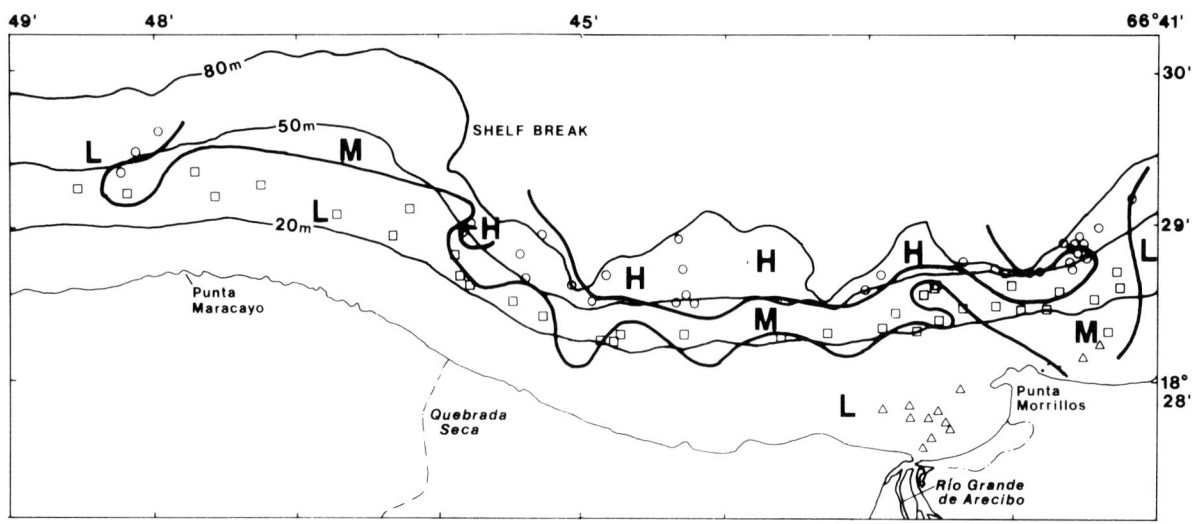
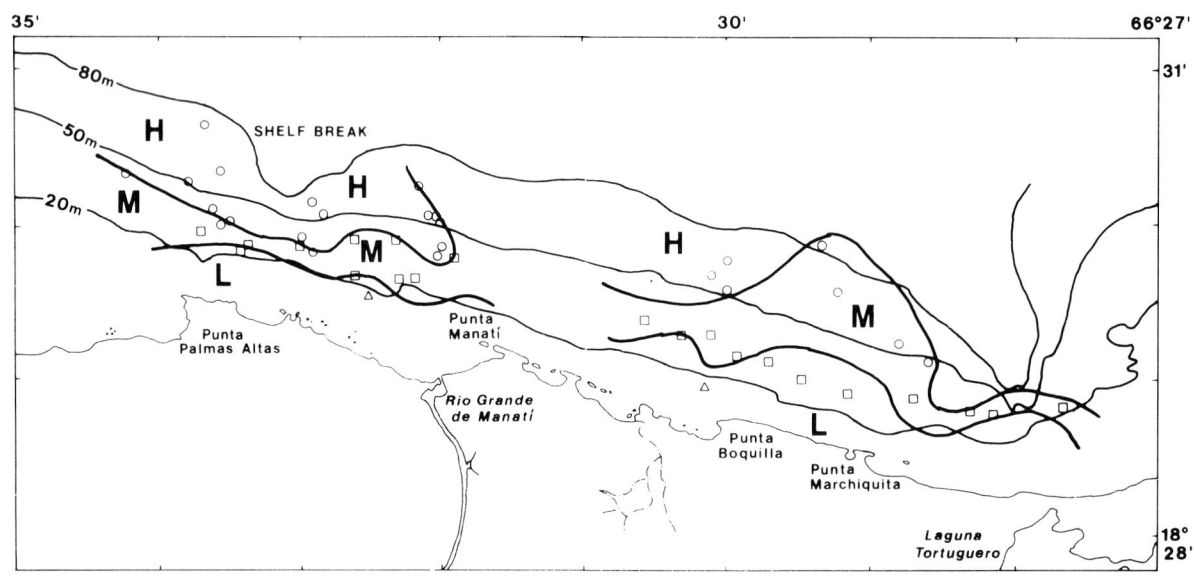
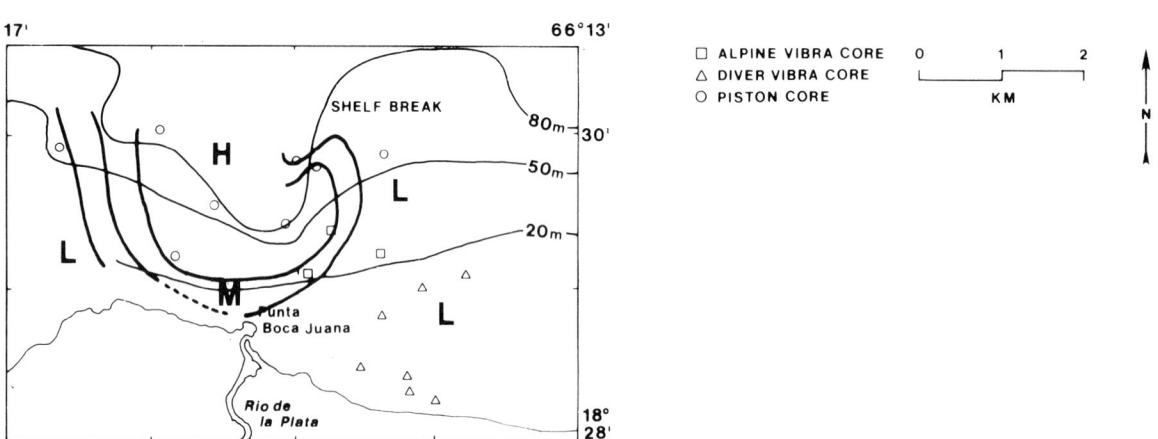

FIG. 17.--Whole core average bioturbation. L = none to slight, M = moderate, H = intense.

Storm Stratification and Proximality Trends - Summary

Using the shelf strata classification scheme of Einsele and Seilacher (1982), the deposits of the northern shelf of Puerto Rico study area should be properly termed "event strata" - differentiated from "periodites." Periodites are formed by cyclic events (for example, tidal cycles, climatic cycles). Event strata are formed by "random" events (for example, turbidity currents, storms, floods).

Based on Brackett (1985), Brackett and Bush (1986), and the present study, two "end-member"-type storm units are identified on the northern shelf based on the type of sediment in which they are formed (see Figs. 5 and 6). Storm units in cohesive sediments tend to be thick, have basal lags of "fresh-looking" skeletal carbonates (or river gravel or coarse sand near river mouths), have high mud content, high degrees of bioturbation, and show less degrees of amalgamation.

Storm units in noncohesive sediments typically have lags composed of skeletal carbonates and coralline algae (or river gravel near rivers), and amalgamation is common. Bioturbation can range from none to intense. Carbonate grain physical condition is typically much less degraded in cohesive sediments, as the mud tends to offer protection from degradation. Storm units on the outer shelf in noncohesive sediments often consist of thick rhodolith basal lags grading to a massive rhodolith/*Halimeda* coarse sand layer that is rarely amalgamated and rarely contains more than a few percent mud.

In summary, water depth and distance from river mouth (or degree of river influence) are the most important controls on event stratification in the study area. Water depth controls flux of wave energy to the sea bottom at a given location (the potential for physical reworking of sediments). Distance from river mouth is actually distance from sediment source in this system. Distance from river mouth controls the input of muddy, fine, river-derived siliciclastic sediments to a given area of the shelf. Water depth and distance from river mouth combine in varying degrees to control the type of sediment accumulating on a given portion of the shelf. These factors work in concert leading to strict, though not always straightforward or obvious, trends in event stratigraphy. The important lesson to be learned here is that even in this "ideal" situation to study storm sedimentation, there appear so many complicating factors as to make it difficult to explain (or predict) all relationships or associations with simple cross-shelf models.

IMPACT OF RIVER-MOUTH MIGRATION

The migration of river mouths from one location to another is a central factor of sedimentation in this dynamic, storm-dominated environment, and has two major effects. First, it results in vertical stacking of the various sediment facies found on the shelf. For example, a river mouth can migrate and supply river-derived siliciclastic sediments to a portion of the shelf that had been previously accumulating only shelf-derived carbonate sediments. Second, it spreads sediment from the many small rivers over a large area on the shelf. Dominant sediment movement pathways in the study area have been shown by previous investigations to be largely cross-shelf, and rivers typically influence only the small portion of the shelf directly off their mouths (Pilkey and others, 1978; Pilkey and Lincoln, 1984; Pilkey and others, 1987). By moving the location of a river mouth along the coast, thus changing the focus of sedimentation, a relatively small amount of river-derived sediment can be spread out to cover an uncharacteristically large portion of the shelf. Additionally, migration of river mouths may help to explain the large number of submarine canyon tributaries that indent the shelf. Two main lines of evidence - geomorphic and sedimentologic - indicate that river-mouth switching is an important process in the study area.

Geomorphic Evidence for River-Mouth Switching

The flat-lying northern coastal lowlands are replete with marshes, swamps, and lagoons that offer extreme potential, over a large area, for river-channel and river-mouth migration. In addition, abandoned river mouths, meander scars, oxbows, and oxbow lakes indicate where rivers and river mouths have been. Figure 18 is a summary of geomorphic evidence for river-mouth migration in the study area. The area from Punta Salinas west to Arecibo contains the most potential for, and evidence of, river-channel and river-mouth migration. Seaward of Cienaga Tiburones, high nearshore elevations (not depicted on Fig. 18) would seem to preclude river-mouth migration. Here elevations are commonly in excess of 33 ft (10 m). Behind this 0.6 to 0.9 mile (1 to 1.5 km) wide strip, however, is extremely flat and low-lying land.

Sedimentologic Evidence for River-Mouth Switching

Three main kinds of sedimentologic evidence indicate river-mouth switching in the northern shelf study area: (1) relation of river-mouth location to location of the shelf-sediment lens; (2) shelf-sediment provenance; and (3) core stratigraphy.

First, lenses of Holocene, river-derived sediment exist on the shelf that are not presently tied to an active river system. Also present are river mouths with no sediment lens located offshore (Fig. 4).

Second, heavy mineral suite "fingerprinting" of river and shelf-sediment samples helps to define the transport paths of river-derived sediments on the shelf. A diagnostic mineral, Flour-Tremolite, has a 110-face peak found at approximately 10 degrees 2-theta on X-ray diffractograms (Grossman, 1978). Fortuitously Flour-Tremolite is present in the Cibuco River and absent from the Manati. It has thus been possible to establish that the Cibuco River previously flowed through the present site of Laguna Tortuguero, and that the active mouth of the Manati River was at one time located a few kilometers to the east of the present mouth (Fig. 19).

Third, core stratigraphy reveals stacking of river- and shelf-derived sediment, indicating a change in the focus of sedimentation of river-derived sediment. This is illustrated in Figures 20 and 21 in 6 cores off the Manati River mouth, the location of which are shown in Figure 19. Cores 7506, 7339, and 7433 (Fig. 20) are from off the presently active river; cores 7529, 7377, and 7378 (Fig. 21) are from off the "Old" Manati River. The most obvious feature is an inverse

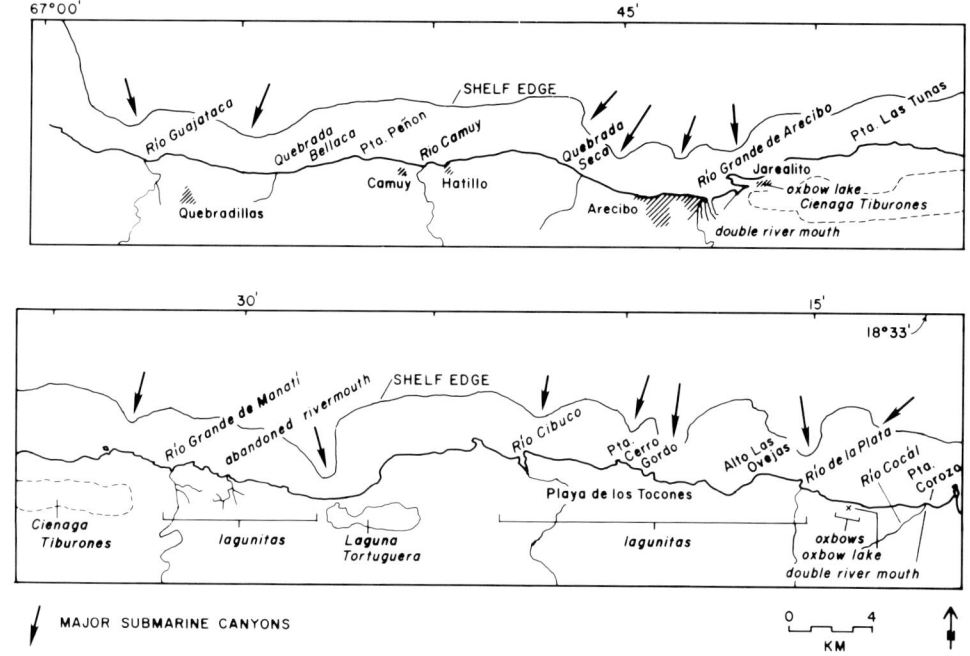

FIG. 18.--Geomorphic evidence of river-mouth switching. Arrows point to major submarine canyons.

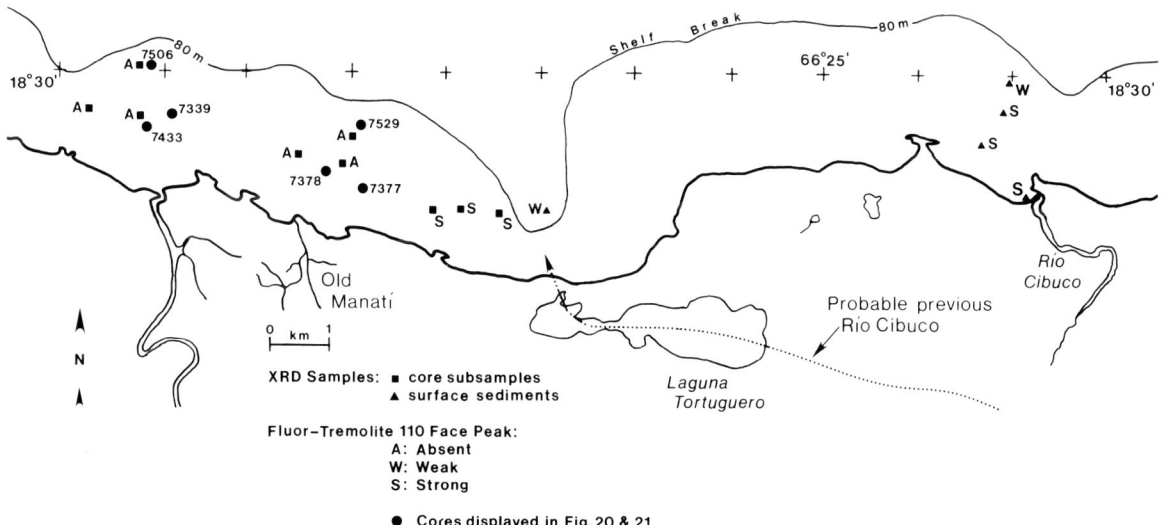

FIG. 19.--Distribution of samples "fingerprinted" for presence/absence and strength of Fluor-Tremolite 110-face peak. Cibuco River sediments contain the mineral, Manati River sediments do not. Also shown are locations of Manati shelf display cores, shown in Figures 20 and 21.

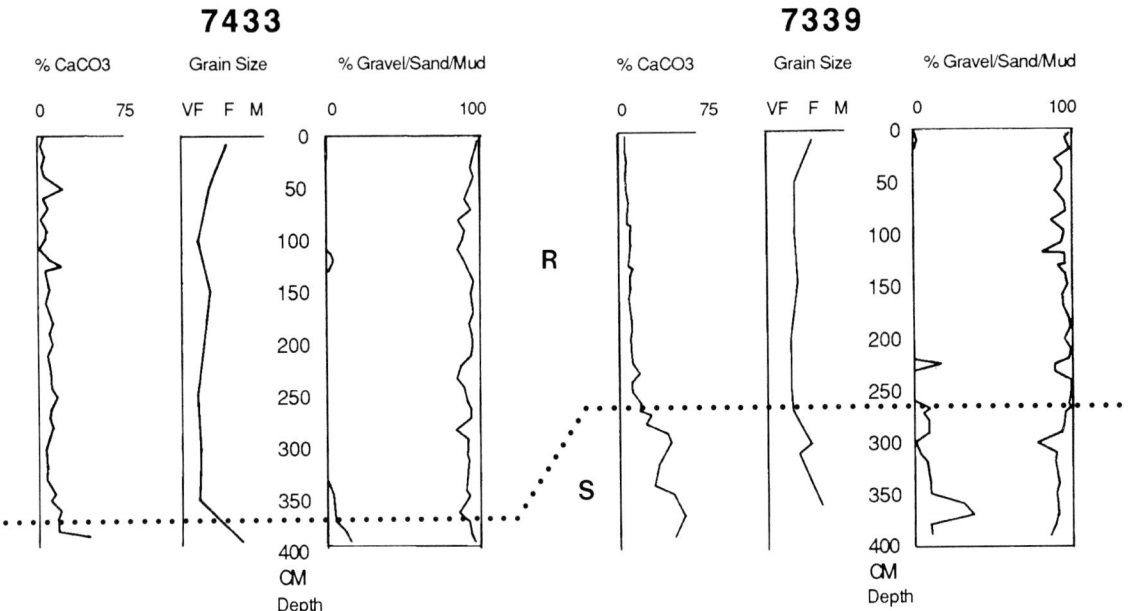
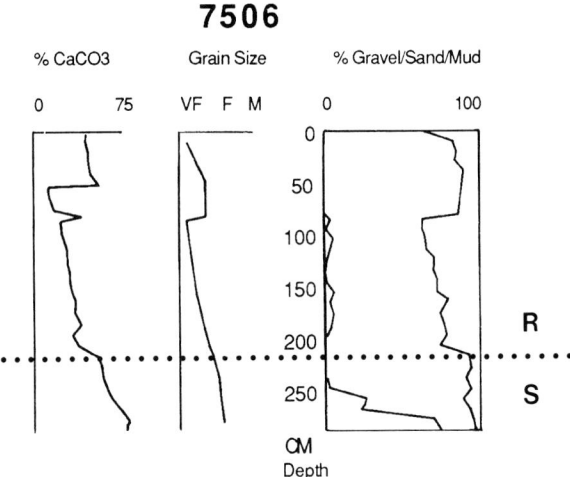

FIG. 20.—Textural data from representative cores off the active Manati River. Grain size is median of sand fraction (very fine, fine or medium). Dashed line depicts change in sediment type indicating river-mouth switch. R=river-derived siliciclastic sediment, S=shelf-derived carbonate sediment.

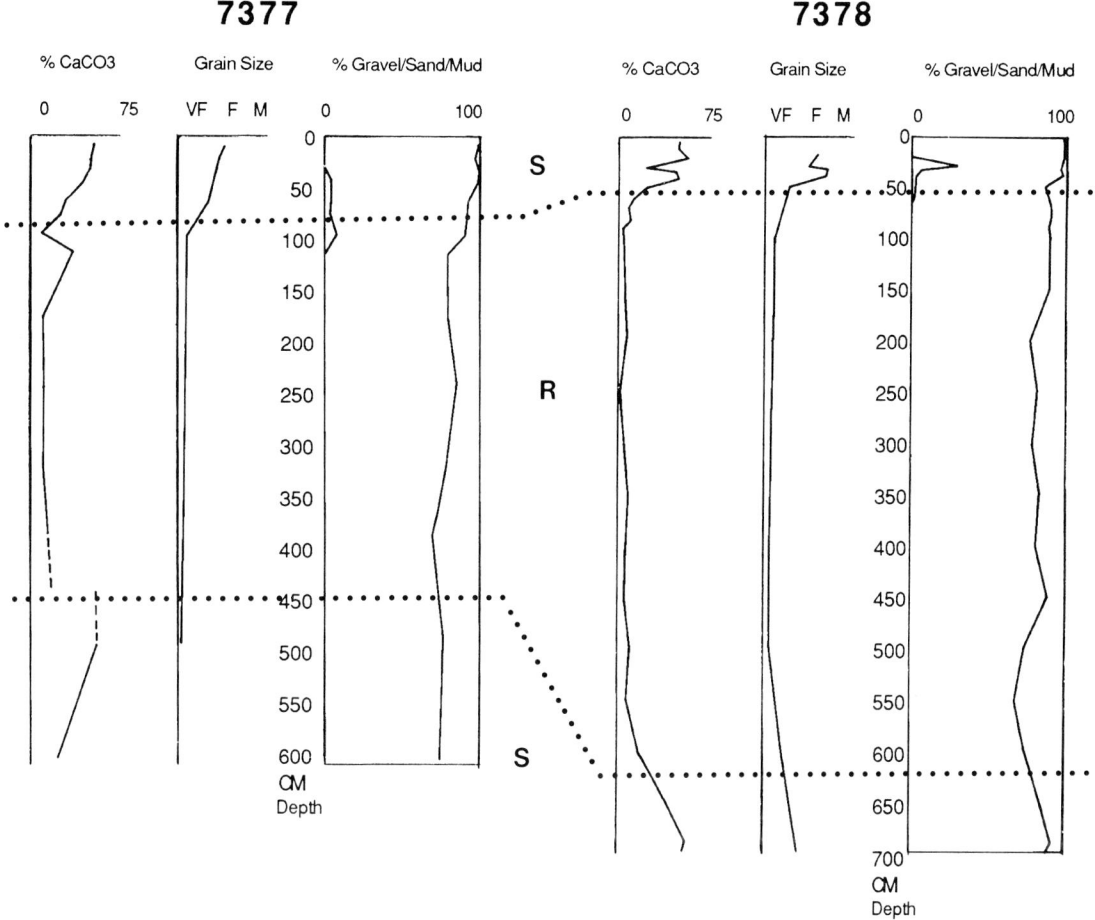
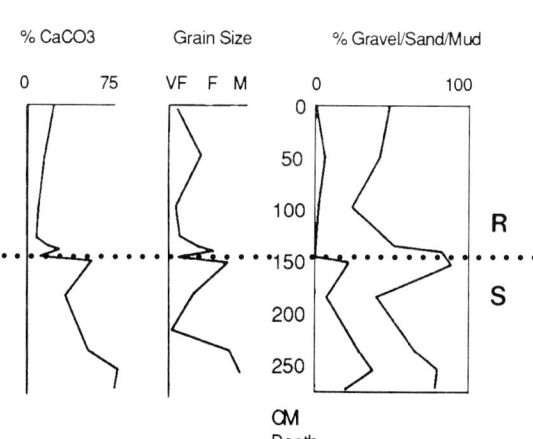

FIG. 21.—Textural data from representative cores off the abandoned Manati River mouth. Grain size is median of sand fraction (very fine, fine or medium). Dashed line depicts change in sediment type indicating river-mouth switch. R=river-derived siliciclastic sediment, S=shelf-derived carbonate sediment.

relationship in stratigraphy between cores from the two portions of the shelf. The two main facies shown in these cores are river-derived siliciclastic sand (R) and shelf-derived carbonate sands and gravels (S).

SEISMIC EXPRESSION OF THE HOLOCENE SECTION

The Puerto Rico northern shelf study area represents only part of a single sea-level rise event, not even an entire "cycle" of sea-level rise and fall described by Vail and others (1977). High-resolution seismic (PRWRA, 1975) has given insight into sediment types and processes in the vertical dimension (Brackett, 1985; Brackett and Bush, 1986; Pilkey and others, 1984; and Pilkey and others, 1988) but has not been incorporated in great detail.

The seismic data are the basis for defining wedges or, more precisely, "lenses" of unconsolidated sediment on the shelf off the mouths of most of the rivers (Fig. 4). Reflectors within the wedges exhibit prograding configurations similar to those described from other modern shelf settings (for example, the Amazon shelf, Kuehl and others, 1982; California shelf off San Diego County, Darigo and Osborne, 1986; South Africa shelf, Martin and Flemming, 1986). Seismic further reveals earlier (Pleistocene?) deposits that have been planed off and partially eroded.

The shelf-sediment lens as revealed by seismic data certainly was deposited entirely during the Holocene sea-level rise. On such a narrow shelf, sea level must have been below the shelf break during the last sea-level minimum (see Lighty and others, 1982, for a regional sea-level curve) and previously deposited highstand deposits were eroded.

SUMMARY AND CONCLUSIONS

Storms are the dominant force driving sedimentation on the northern shelf of Puerto Rico. Episodic switching of river-mouth location complicates sediment distribution and facies relationships. Storms (including floods) introduce new sediment and rework sediment. Changing location of river mouths controls loci of deposition of siliciclastic sediment and controls vertical stratigraphic changes.

The northern shelf of Puerto Rico is a case study in how complex even a small mixed-carbonate/siliciclastic environment can be. Factors contributing to the complexity of the northern shelf of Puerto Rico relative to other modern shelves are (1) two sources of sediments - river and shelf - giving a mixed-sediment environment, (2) two major types of sediments - cohesive and noncohesive, (3) irregularly shaped coastline, (4) antecedent topography of several submarine canyons indenting the shelf, (5) several different types of storms affecting the area, and (6) fairly high fairweather wave energy.

Important aspects of sediment accumulation history in the study area are depicted in Figure 22. They are (1) changing foci of deposition of river-derived sediment; (2) differences in rates of deposition of sediment between the river- and shelf-derived material; (3) winnowing and reworking of sediment that is ongoing, but can result in erosion after a river mouth moves away,

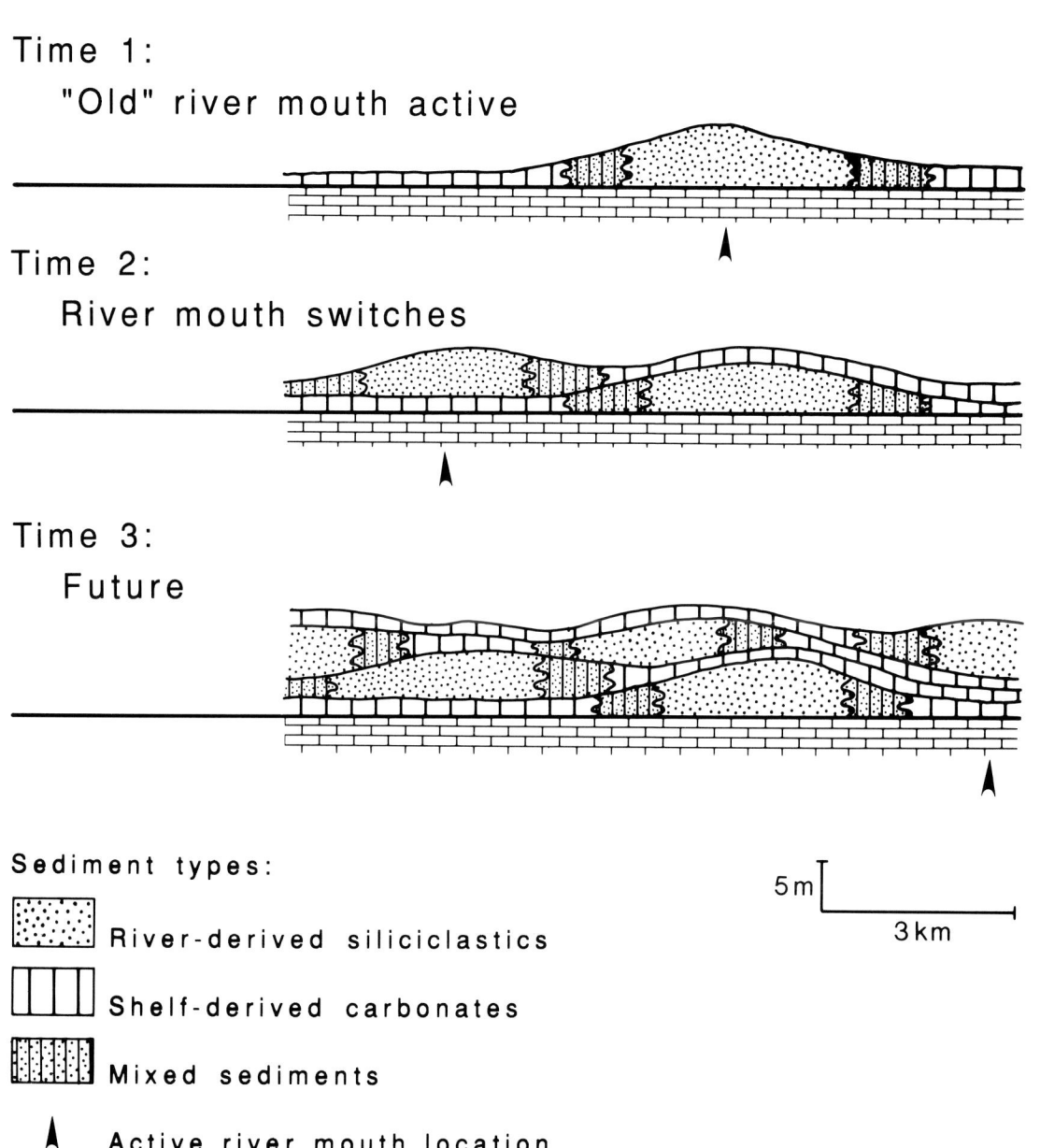

FIG. 22.—Conceptual model of sediment accumulation on the northern shelf of Puerto Rico. See text for discussion. Note the extreme vertical exaggeration.

cutting off the supply of new sediment from a portion of the shelf; (4) differential compaction of muddy sediments versus clean sediments; and (5) preferential siting of an active "lobe" of river-derived sediment accumulation in between previously deposited lobes. This is only hypothesized for the study area, but would be similar to what is found in deltas and deep-sea fan lobes, only on a much smaller scale. The number of individual lobes of allochthonous material that are deposited is controlled by discharge of river and frequency of river-mouth switching, and by frequency of flooding events.

This study has provided a modern analog for ancient storm-dominated shelves. Examples of modern shelf study areas for comparison with ancient shelf sequences are lacking because the majority of modern continental shelves are "drowned" and out of equilibrium with modern-day processes. Relict, noncohesive material dominates most modern shelves, while estuaries trap most modern cohesive sediments (Meade, 1972). Storm response of muddy and sandy sediments can be substantially different (Kreisa, 1981). The abundance of mud in ancient open-shelf environments and lack of mud in many modern shelves do not allow application of most modern storm-sedimentation models to ancient open-shelf settings. Thus, the muddy equilibrium northern shelf of Puerto Rico is a perfect natural laboratory for continued research of storm-sedimentation principles.

ACKNOWLEDGMENTS

This paper contains highlights of the author's PhD dissertation research. Separate papers dealing with each of the main facets of sedimentation on the northern shelf of Puerto Rico in much greater detail are forthcoming. Research was carried out under the supervision of the author's major professor, Orrin H. Pilkey, Jr., funded largely by NSF grant #333-0721 to Orrin Pilkey. The author acknowledges the generous support of the U.S. Geological Survey and particularly Willie Rodriguez and Juan Trias of the USGS Marine Geology Project Office in San Juan. Special thanks to captain and crew of Research Vessels *Cape Hatteras* and *Jean A*. Amber Taylor did a ton of last-minute drafting and managed to keep her sense of humor.

REFERENCES

AIGNER, T., 1982, Calcareous tempestites: storm-dominated stratification in upper Muschelkalk limestones (Middle Trias, S.W. Germany), *in* Einsele, G., and Seilacher, A., eds., Cyclic and Event Stratification: Berlin, Springer-Verlag, p. 180-198.

_____, 1985, Storm depositional systems, dynamic stratigraphy in modern and ancient shallow-marine sequences: Lecture Notes in Earth Science, v. 3, Berlin, Springer-Verlag, 174 p.

_____, AND REINECK, H. E., 1982, Proximality trends in modern storm sands from the Helgoland Bight (North Sea) and their implications for basin analysis: Senckenbergiana Maritima, Vol. 14, p. 183-215.

ALLEN, J. R. L., 1982, Sedimentary structures, their character and physical basis: Developments in Sedimentology, 30A and B, Amsterdam: Elsevier, A: 593 p., B: 663 p.

BENES, P. S., 1988, Relationship between physical condition of the carbonate fraction and sediment environments: northern shelf of Puerto Rico: unpublished M.S. thesis, Duke University, Department of Geology, 178 p.

BOURGEOIS, J., 1980, A transgressive shelf sequence exhibiting hummocky stratification: the Cape Sebastion Sandstone (Upper Cretaceous), southwestern Oregon: Journal of Sedimentary Petrology, v. 50, p. 681-702.

BRACKETT, R. S., 1985, Characteristics of a modern storm-generated sedimentary sequence: north insular shelf of Puerto Rico: unpublished M.S. thesis, Duke University, Department of Geology, 140 p.

_____, AND BUSH, D. M., 1986, Sedimentary characteristics of modern storm-generated sequences: north insular shelf, Puerto Rico, in Moslow, T. F., and Rhodes, E. G., eds., Modern and Ancient Shelf Clastics: A Core Workshop: Society of Economic Paleontologists and Mineralogists Core Workshop No. 9, p. 125-168.

BUDD, D. A., AND PERKINS, R. D., 1980, Bathymetric zonation and paleoecological significance of microborings in Puerto Rican shelf and slope sediments: Journal of Sedimentary Petrology, v. 50, no. 3, p. 881-904.

BUSH, D. M., 1977, Equilibrium sedimentation: north insular shelf of Puerto Rico: unpublished M.S. thesis, Duke University, Department of Geology, 120 p.

CALVERSBERT, R. G., 1970, Climate of Puerto Rico and U.S. Virgin Islands: Climatography of the United States, no. 60-52, U.S. Department of Commerce.

COASTAL DYNAMICS INCORPORATED, 1986, Coastal erosion and protection evaluation, Condado-Ocen Park to Isla Verde: Draft final report, prepared for Puerto Rico Department of Natural Resources, 107 p.

COLLINSON, J. D., AND THOMPSON, D. B., 1982, Sedimentary structures: London, George Allen & Unwin, 194 p.

DARIGO, N. J., AND OSBORNE, R. H., 1986, Quaternary stratigraphy and sedimentation of the inner continental shelf, San Diego County, California, in Knight, R. J., and McLean, J. R., Shelf Sands and Sandstones: Canadian Society of Petroleum Geologists Memoir 11, p. 73-98.

DE RAAF, J. F. M., BOERSMA, J. R., AND VAN GELDER, A., 1977, Wave-generated structures and sequences from a shallow marine succession, Lower Carboniferous, County Cork, Ireland: Sedimentology, v. 24, p. 451-483.

DOTT, R. H., AND BOURGEOIS, J., 1982, Hummocky stratification: significance of its variable bedding sequences: Geological Society of America Bulletin, v. 93, p. 663-680.

DUKE, W. L., 1985, Hummocky cross-stratification, tropical hurricanes, and intense winter storms: Sedimentology, v. 32, p. 167-194.

EHLMANN, A. J., 1968, Clay mineralogy of weathered products and of river sediment, Puerto Rico: Journal of Sedimentary Petrology, v. 38, p. 885-889.

EINSELE, G., AND SEILACHER, A., eds., 1982, Cyclic and event stratification, Berlin: Springer-Verlag, 536 p.

FIELDS, F. K., AND JORDAN, D. G., 1972, Storm-wave swash along the north coast of Puerto Rico: U.S. Geological Survey Hydrologic Investigations: Atlas H.A. 432, in two sheets.

GAGAN, M. K., JOHNSON, D. P., AND CARTER, R. M., 1988, The Cyclone Winifred storm bed, central Great Barrier Reef shelf, Australia: Journal of Sedimentary Petrology, v. 58, p. 845-856.

GREENWOOD, B., AND SHERMAN, D. J., 1986, Hummocky cross-stratification in the surf zone: flow parameters and bedding genesis: Sedimentology, v. 33, p. 33-46.

GROSSMAN, Z. N., 1978, Distribution and dispersal of Manati River sediments: Puerto Rico insular shelf: unpublished M.S. thesis, Duke University, Department of Geology, 72 p.

GROVE, K. A., PILKEY, O. H., AND TRUMBULL, J. V. A., 1982, Mud transportation on a steep shelf, Rio de la Plata shelf, Puerto Rico: Geo-Marine Letters, v. 2, p. 71-75.

HAMLIN, A. P., AND WALKER, R. G., 1979, Storm-dominated shallow-marine deposits: The Fernie-Kootenay (Jurassic) transition, southern Rocky Mountains: Canadian Journal of Earth Science, v. 16, p. 1673-1690.

HARMS, J. C., SOUTHARD, J. B., AND WALKER, R. G., 1982, Structures and sequences in clastic rocks: Society of Economic Paleontologists and Mineralogists Short Course Notes No. 9, 249 p.

HEEZAN, B. C., THARP, M., AND EWING, M., 1959, The floors of the oceans: I. The North Atlantic: Geological Society of America Special Paper 65, 137 p.

HOWARD, J. D., AND REINECK, H.-E., 1972, Georgia coastal region, Sapelo Island, U.S.A., sedimentology and biology: IV. Physical and biogenic sedimentary structures of the nearshore shelf: Senckenbergiana Maritima, v. 4, p. 81-123.

KAYE, C. A., 1959, Shoreline features and Quaternary shoreline changes, Puerto Rico: U.S. Geological Survey Professional Paper 317-B, p. 49-140.

KREISA, R. D., 1981, Storm-generated sedimentary structures in subtidal marine facies with examples from the Middle and Upper Ordovician of southwestern Virginia: Journal of Sedimentary Petrology, v. 51, p. 823-848.

_____, AND BAMBACH, R. K., 1982, The role of storm processes in generating shell beds in paleozoic shelf environments, in Einsele, G., and Seilacher, A., eds., Cyclic and Event Stratification: Berlin, Springer-Verlag, p. 200-207.

KUEHL, S. A., NITTROUER, C. A., AND DEMASTER, D. J., 1982, Modern sediment accumulation and strata formation on the Amazon continental shelf: Marine Geology, v. 49, p. 279-300.

KUMAR, N., AND SANDERS, J. E., 1976, Characteristics of shoreface deposits: modern and ancient: Journal of Sedimentary Petrology, v. 46, p. 145-162.

LIGHTY, R. G., MACINTYRE, I. G., AND STUCKENRATH, R., 1982, *Acropora palmata* reef framework: a reliable indicator of sea level in the western Atlantic for the past 10,000 years: Coral Reefs, v. 1, p. 125-130.

LUGO, A. E., QUINONES MARQUES, F., AND WEAVER, P. L., 1980, La erosion y sedimentacion en Puerto Rico: Caribbean Journal of Science, v. 16 (1-4), p. 143-165.

MCCAVE, I. N., 1972, Transport and escape of fine-grained sediment from shelf areas, in Swift, D. J. P., Duane, D., and Pilkey, O. H., eds., Shelf Sediment Transport: Process and Pattern: Stroudsburg, PA: Douden Hutchinson and Ross, p. 225-248.

MARSAGLIA, K. M., AND KLEIN, G. D., 1983, The paleogeography of Paleozoic and Mesozoic depositional systems: Journal of Geology, v. 91, p. 117-142.

MARTIN, A. K., AND FLEMMING, B. W., 1986, The Holocene shelf sediment wedge off the south and east coasts of South Africa, in Knight, R. J., and McLean, J. R., Shelf Sands and Sandstones: Canadian Society of Petroleum Geologists Memoir 11, p. 27-44.

MEADE, R. H., 1972, Sources and sinks of suspended matter on continental shelves, in Stanley, D. J., and Swift, D. J. P., eds., Marine Sediment Transport and Environmental Management: New York, John Wiley, p. 249-262.

MITCHUM, R. M., JR., VAIL, P. R., AND SANGREE, J. B., 1977, Seismic stratigraphy and global changes of sea level: Part 6: stratigraphic interpretation of seismic reflection patterns in depositional sequences, in Payton, C. E., ed., Seismic Stratigraphy - Applications to Hydrocarbon Exploration: American Association of Petroleum Geologists Memoir 26, p. 117-134.

MORELOCK, J., 1984, Coastal erosion in Puerto Rico: Shore and Beach, January 1984, p. 18-27.

MORELOCK, J., SCHWARTZ, M. L., HERNANDEZ-AVILA, M., AND HATFIELD, D. M., 1985, Net shore-drift on the north coast of Puerto Rico: Shore and Beach, October 1985, p. 16-21.

_____, AND TAGGART, B., 1988, U.S.A.-Puerto Rico, *in* Walker, H. J., ed., Artificial Structures and Shorelines: Kluwer Academic Publishers, p. 649-658.

NELSON, C. H., 1982, Modern shallow-water graded sand layers from storm surges, Bering shelf: a mimic of Bouma sequences and turbidite systems: Journal of Sedimentary Petrology, v. 52, p. 537-545.

NOTTVEDT, A., AND KREISA, R. D., 1987, Model for the combined-flow origin of hummocky cross-stratification: Geology, v. 15, p. 357-361.

PERKINS, R. D., AND HALSEY, S. D., 1971, Geologic significance of microboring fungi and algae in Carolina shelf sediments: Journal of Sedimentary Petrology, v. 41, p. 843-853.

_____, AND TSENTAS, C. I., 1976, Microbial infestation of carbonate substrates planted on the St. Croix shelf, West Indies: Geological Society of America Bulletin, v. 87, p. 1615-1628.

PILKEY, O. H., BUSH, D. M., AND RODRIGUEZ, R. W., 1987, Bottom sediment types of the northern insular shelf of Puerto Rico: Punta Penon to Punta Salinas: U.S. Geological Survey Miscellaneous Investigations Map I-1861.

PILKEY, O. H., BUSH, D. M., AND RODRIGUEZ, R. W., 1988, Carbonate-terrigenous sedimentation on the north shelf of Puerto Rico, *in* Doyle, L. J., and Roberts, H. H., eds., Carbonate-Clastic Transitions: Developments in Sedimentology 42, Amsterdam: Elsevier, p. 231-250.

_____, FIERMAN, E. I., AND TRUMBULL, J. V. A., 1979, Relationship between physical condition of the carbonate fraction and sediment environments: northern Puerto Rico shelf: Sedimentary Geology, v. 24, p. 283-290.

_____, AND LINCOLN, R. B., 1984, Insular shelf heavy mineral partitioning: northern Puerto Rico: Journal of Marine Mining, v. 4, p. 403-414.

_____, TRUMBULL, J. V. A., AND BUSH, D. M., 1978, Equilibrium shelf sedimentation: Rio de la Plata shelf, Puerto Rico: Journal of Sedimentary Petrology, v. 48, p. 389-400.

PRWRA (Puerto Rico Water Resources Authority), 1975, Shallow water bathymetric, sonar and seismic investigations for siting the north-coast nuclear plant (NORCO-NP-1) on Puerto Rico, Appendix 2.5R, Amendment 27.

REINECK, H.-E., AND SINGH, I. B., 1972, Genesis of laminated sand and rhythmites in storm-sand layers of shelf mud: Sedimentology, v. 18, p. 123-128.

SCHNEIDERMANN, N., PILKEY, O. H., AND SAUNDERS, C., 1976, Sedimentation on the Puerto Rico insular shelf: Journal of Sedimentary Petrology, v. 46, p. 167-173.

SEILACHER, A., 1982, Distinctive features of sandy tempestites, in Einsele, G. and Seilacher, A., eds., Cyclic and Event Stratification: Berlin, Springer-Verlag, p. 333-349.

SOUTHARD, J. B., LAMBIE, J. M., FEDERICO, D. C., PILE, H. T., AND WEIDMAN, C. R., 1990, Experiments on bed configurations in fine sands under bidirectional, purely oscillatory flow, and the origin of hummocky cross-stratification: Journal of Sedimentary Petrology, v. 60, p. 1-17.

SWIFT, D. J. P., 1976, Continental shelf sedimentation, in Stanley, D. J., and Swift, D. J. P., eds., Marine Sediment Transport and Environmental Management, New York: John Wiley, p. 311-350.

_____, FIGUEIREDO, A. G., JR., FREELAND, G. L., AND OERTEL, G. F., 1983, Hummocky cross-stratification and megaripples: a geological double standard?: Journal of Sedimentary Petrology, v. 53, p. 1295-1317.

VAIL, P. R., MITCHUM, R. M., AND THOMPSON, S., III, 1977, Seismic stratigraphy and global changes of sea level: Part 4: global cycles of relative changes of sea level, in Payton, C. E., ed., Seismic Stratigraphy - Applications to Hydrocarbon Exploration: American Association of Petroleum Geologists Memoir 26, p. 83-98.

WALKER, R. G., 1967, Turbidite sedimentary structures and their relationship to proximal and distal depositional systems: Journal of Sedimentary Petrology, v. 37, p. 25-43.

_____, 1970, Review of the geometry and facies organization and turbidite-bearing basins, in Lajoie, L., ed., Flysch Sedimentology in North America: Geological Association of Canada Special Paper 7, p. 219-251.

_____, 1985, Geological evidence for storm transportation and deposition on ancient shelves, in Tillman, R. W., Swift, D. J. P., and Walker, R. G., eds., Shelf Sands and Sandstone Reservoirs: SEPM Short Course Notes No. 13, p. 243-301.

_____, DUKE, W. L., AND LECKIE, D. A., 1983, Hummocky stratification: significance of its variable bedding sequences: discussion: Bulletin of the Geological Society of America, v. 94, p. 1245-1249.

WOOD, E. D., YOUNGBLUTH, M. J., NUTT, M. E., YOSHIOKA, P., AND CANOY, M. J., 1975a, Tortuguero Bay environmental studies: Puerto Rico Nuclear Center Publication #181, 227 p.

WOOD, E. D., YOUNGBLUTH, M. J., NUTT, M. E., YEAMAN, M. N., YOSHIOKU, P., AND CANOY, M. J., 1975b, Punta Manati environmental studies: Puerto Rico Nuclear Center Publication #182, 225 p.

ORDOVICIAN LIMESTONE AND SHALE IN THE CENTRAL APPALACHIAN BASIN: EARLY SEDIMENTARY RESPONSE TO PLATE COLLISION

RICHARD SMOSNA
Department of Geology & Geography
West Virginia University, Morgantown, WV 26505
DOUGLAS G. PATCHEN
West Virginia Geological and Economic Survey,
Morgantown, WV 26507

ABSTRACT

The Trenton Limestone of West Virginia was deposited on a gentle carbonate ramp that sloped eastward into a deep foreland basin. Unlike many limestones this unit formed during an early stage of orogeny, is transgressive in nature, and accumulated during a large influx of shale. Initial uplift of an eastern fold-thrust belt, created as the North American plate collided with a volcanic-arc system, led to lithospheric flexure and downwarping of the foreland basin under the load of an accreted terrane. Basin subsidence was rapid, being fast enough to outpace sedimentation but not so fast as to suppress carbonate sedimentation under deep, anaerobic conditions. Thus, limestone deposition continued, although with a deepening-upward sequence. Introduction of terrigenous mud was discontinuous through time, which generated a periodicity to the limestone and shale interbedding. These muds did temporarily downgrade or eliminate skeletal lime production, especially to the east where shale influxes were greater. With each return to clear water, however, carbonate deposition resumed. Eventually plate convergence and uplift of the source area proceeded to the point where clastic deposition overwhelmed the carbonate ramp. Trenton sedimentation then gave way to the flysch wedge of the overlying Martinsburg Formation. Thus, the relationship in space and time between subsiding basin and rising orogenic landmass exerted a major influence on the Trenton's internal stratigraphy and facies development.

INTRODUCTION

The Upper Ordovician Trenton Limestone in eastern North America was deposited under rather unusual conditions. These carbonate rocks accumulated during a period of major tectonic activity (in contrast, limestones often reflect tectonic quiescence), during a period of large-scale transgression (carbonate stratigraphic sequences are characteristically regressive), and during a period of significant terrigenous influx (introduction of terrigenous clays frequently suppresses the biochemical production of lime sediment). Because an exception may prove the rule, knowledge of the Trenton's depositional framework should broaden our understanding of sedimentary and tectonic controls over limestones in general. In addition, this tectonic/depositional framework can serve as a model for other exceptional rock units: synorogenic, transgressive, mixed carbonate-siliciclastic formations.

The unit has economic value as well as its academic importance, being a minor petroleum reservoir in the central Appalachians. In extreme western Virginia, for example, the fractured Trenton Limestone produces oil from two fields. Between 1981 and 1983 about 40 wells in Lee County, Virginia, produced over 128,000 barrels of oil (Bartlett, 1988). Since the 1960s Trenton exploration in north-central New York has enjoyed mixed success. Production comes from dolomitized fractures, and recoverable reserves are estimated at approximately 0.25 million cubic feet of natural gas per acre (Drazan, 1988). In West Virginia only 28 wells penetrate the deep Trenton Limestone, but 7 had shows of natural gas, enough to continue industry's interest. The limestones have minor intergranular porosity because of abundant mud matrix and cementation; the only notable porosity is contributed by fractures. Wells with gas shows are epitomized by high initial potentials (as high as 1.3 to 3.0 MMCFPD) that decrease rapidly.

GEOLOGICAL SETTING

A great volume of carbonate sediment accumulated on the passive margin of North America during Cambro-Ordovician time, before onset of the Taconic orogeny. This accumulation, informally referred to as the Great American Bank, covered the area from southern Ontario to central Alabama and from western Maryland to the Central Interior province (Colton, 1970). In the central Appalachians subsidence was greatest in a Pennsylvania depocenter (maximum thickness exceeds 6562 ft (2000 m)), but everywhere sedimentation kept pace with subsidence and water depth remained shallow (Read, 1980).

At this time the eastern margin of North America (Fig. 1) had a zigzag outline (Thomas, 1977). The continental margin consisted of several promontories (convex oceanward), including those centered in New York and Virginia, and reentrants (concave oceanward), including the depocenter of Pennsylvania. The irregular outline first became established in the late Precambrian, and Thomas (1977) related its origin to transform faulting associated with a period of rifting and breakup of a continental landmass.

Beginning in the Late Ordovician (Pennsylvania reentrant), collision of the North American plate with a volcanic-arc system caused the margin of the carbonate bank to deepen rapidly. The mechanism proposed for this collision involves a subduction zone of reversed polarity between continent and arc (Jacobi, 1981). With a reversed sense of polarity, the continental plate (including a small segment of adjacent oceanic crust along the leading edge) was undergoing subduction beneath the proto-Atlantic, or Iapetus, ocean basin. As the North American plate drifted into the subduction zone, its margin experienced significant downwarping by southeast underthrusting beneath the oceanic lithosphere (Church and Stevens, 1971). Crustal convergence with a volcanic arc is also recorded by numerous and widespread bentonites over much of the eastern United States (Rodgers, 1971). Volcanic activity eventually ceased when the relatively buoyant continental lithosphere jammed the subduction zone.

As the collision progressed, the outer-arc allochthons and related deep-water sediments were thrust westward, uplifted, and accreted as suspect terranes onto the leading edge of the continent (Williams and Hatcher, 1983). Underthrusting beneath the arc thickened the crust, and uplift resulted from isostatic compensation. Vertical tectonic loading by thrust sheets of the

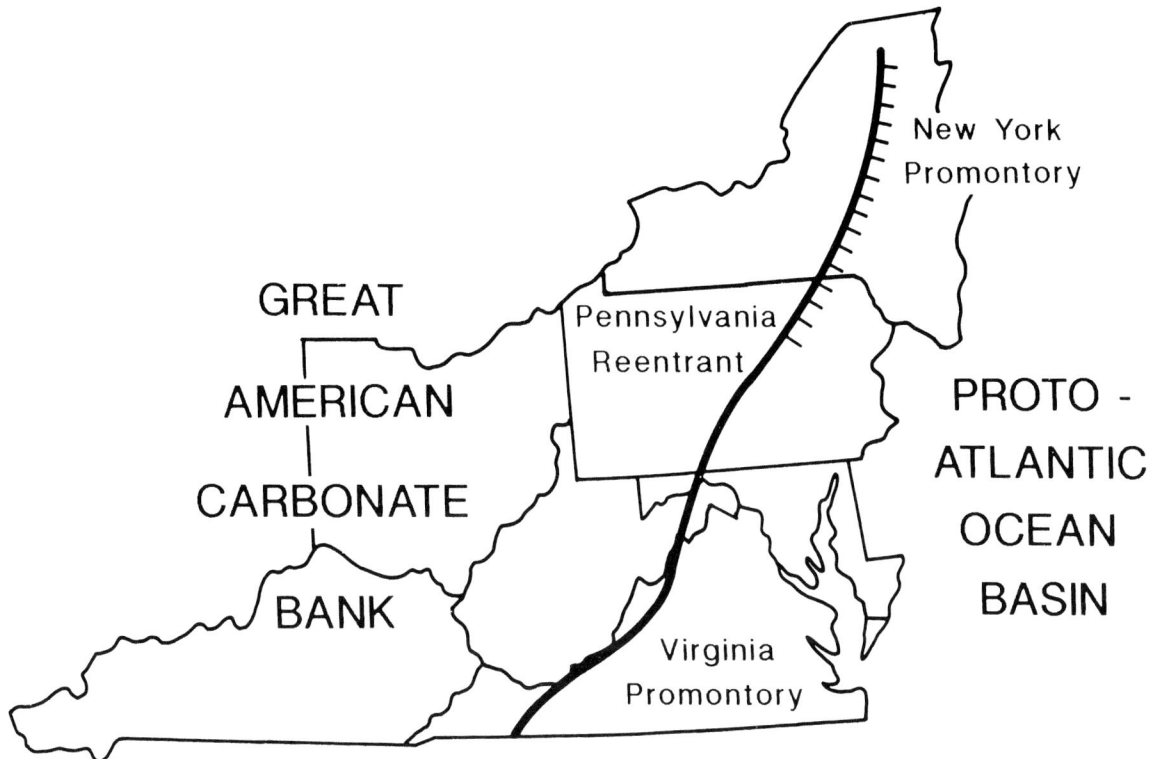

Fig. 1.--Margin of the Ordovician continent (after Thomas, 1976; Read, 1980; Mehrtens, 1988; Carter and others, 1988). Hatching depicts a steep slope in the north, whereas the slope was a more gentle ramp to the south.

accretionary prism in turn produced a foreland basin behind the welded, convergent margin (Quinlan and Beaumont, 1984). Large-scale tectonic movements generated significant relief in the orogenic fold-thrust belt, and a flood of terrigenous material was then shed westward to the adjacent foreland basin.

In the central Appalachians, earliest Taconic collision (Middle Ordovician) may have occurred along the New York and Virginia promontories, which jutted out into the proto-Atlantic Ocean (Lash, 1988). With initial convergence these headlands experienced uplift, deformation, and erosion, while carbonate deposition persisted in the intervening Pennsylvania reentrant. Flysch wedges, however, finally overwhelmed carbonate sedimentation everywhere.

Figure 2 summarizes the profound influence of Taconic plate convergence within the Late Ordovician Pennsylvania reentrant. In particular, three stages of basinal development can be defined as they relate to paleogeograhic reconstruction, tectonic events, and sedimentation patterns. First there was a stable carbonate shelf on the passive continental margin. Shallow-water carbonates of the Black River Group (or Limestone in the western subsurface where it is not divided) accumulated atop the Great American Bank. The shelf edge passed close to the present-day Appalachian Front between Plateau and Valley-and-Ridge provinces. The second

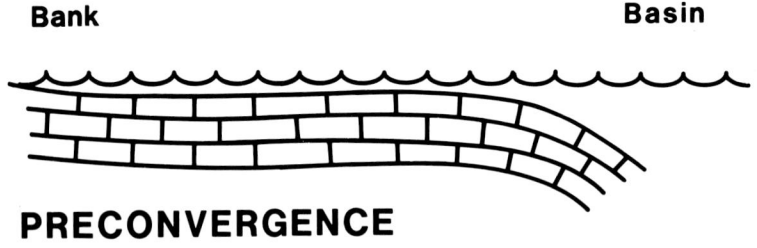
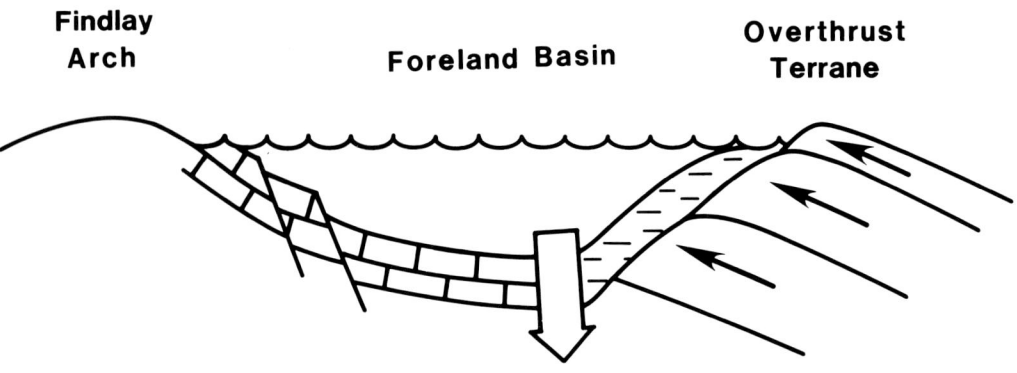
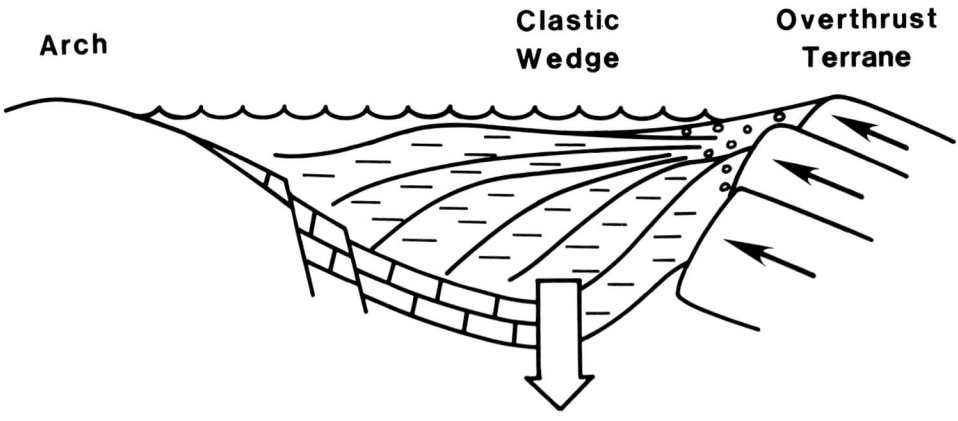

Fig. 2.--Three stages of basinal development related to plate convergence. Top: shallow-water Black River limestones of the stable Great American Bank. Middle: rapid subsidence of a carbonate ramp and deposition of the shallow-to-deep Trenton Limestone. Bottom: clastic flysch wedge of the Martinsburg Formation overwhelms carbonate sedimentation.

stage was marked by rapid subsidence of the basin and formation of a carbonate ramp. At this time of early plate collision, the shallow-to-deep Trenton Limestone (this paper) was deposited across the area. The third stage was that of major collision. Uplift of accreted terranes led to rapid sedimentation in the nearby foreland basin. Flysch deposits of the Martinsburg Formation, consisting of turbidites and hemipelagic mudstones, filled the basin, initially from adjacent promontories toward the reentrant and later from east to west (McBride, 1962). As the flysch wedge progressed up the eastward-sloping carbonate ramp, deep-water turbidites gave way to shallow-water tempestites (Kreisa, 1981).

RAMP FACIES

Not only was the Ordovician continental margin irregular in plan view, it was irregular in profile as well (Fig. 1). Around the New York promontory, the margin had a steep slope. Shelf carbonates with diverse benthonic communities change suddenly eastward to dark, pyritic shales with planktonic faunas: the shallow carbonate bank dropped precipitously into a deep basin. The presence of carbonate turbidites, breccias, and slumps testifies to its steepness (Mehrtens, 1988). Similarly, in Tennessee, the slope separating shelf from basin was very steep, calculated at 2.5° (144 ft/0.6 mile (44 m/km)), and the deep basin filled with turbidites and contourites (Shanmugam and Walker, 1980). In contrast, regional paleogeographic reconstructions by Read (1980) and Carter and others (1988) showed that Middle-Upper Ordovician limestones of the Virginias and Maryland were deposited on a carbonate ramp. Ramps have relatively gentle slopes, as low as 3.3 ft/0.6 mile (1.0 m/km) or lower, and characteristically there is no break in slope.

The Trenton Limestone over the Findlay arch in northwestern Ohio, where it has been a major producer of oil and natural gas since the late 1800s, has been extensively studied. Although controversy still surrounds some aspects of diagenesis and dolomitization, most workers are in agreement on the carbonate lithofacies and a shallow-ramp environment (Wickstrom and Gray, 1988). In particular, sedimentary structures and faunal communities reflect shallow-subtidal, relatively high-energy, normal-marine conditions of the upper ramp (Fig. 3). In the central Appalachians the ramp sloped gently eastward from the shallows of Ohio, across West Virginia, and into the deep foreland basin of Virginia.

The formation consists of thin limestones, often less than 1 ft (30 cm) thick, interbedded with shales and bentonites. On the basis of core analysis, outcrop descriptions, and petrographic study of 269 thin sections, 8 depositional facies have previously been identified in the Trenton of West Virginia (Smosna, 1985; Carter and others, 1988). The Trenton core was taken from the Sandhill well, Wood County (state permit number Wood 351), drilled by Hope Natural Gas Company (Woodward, 1959), and the outcrop is a composite section near Riverton, Pendleton County, in the Valley-and-Ridge province (Fig. 3). These eight facies, the first seven of which are interpreted as having occupied successively deeper positions on the carbonate ramp (Fig. 4), are summarized below.

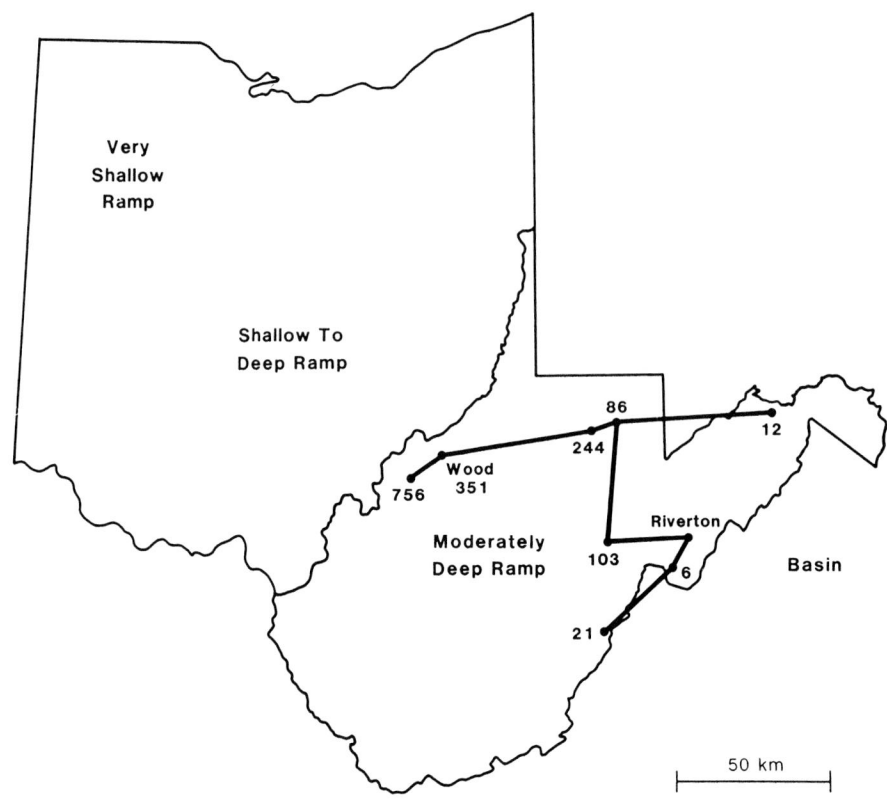

Fig. 3.—Trenton paleogeography across Ohio and West Virginia. Detailed petrologic studies have been carried out on the Trenton Limestone from the Wood-351 core and the Riverton outcrop. Stratigraphic cross-sections are shown in Figure 9.

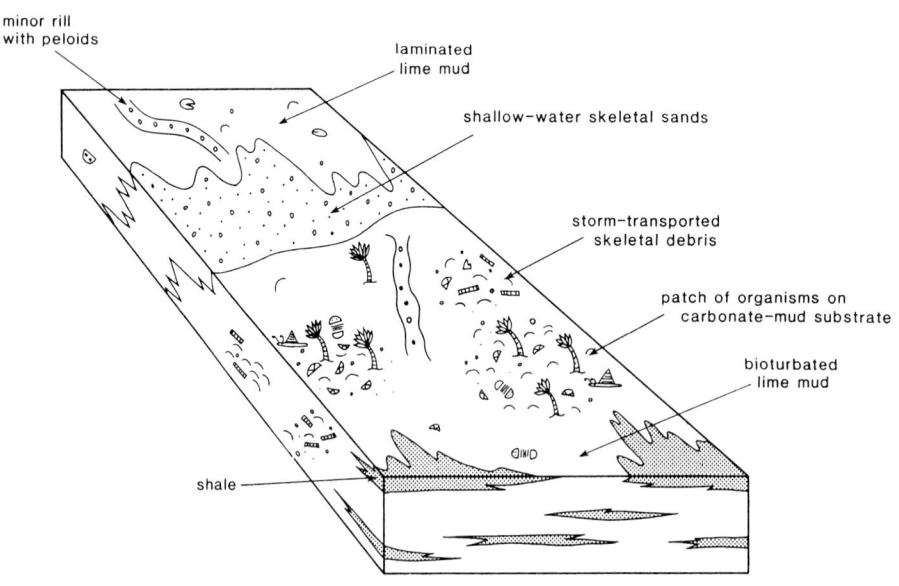

Fig. 4.—Overview of the Ordovician ramp facies. See text for explanation.

Facies 1, Laminated Lime Mudstone

Lime mud generally makes up from 90 to 100 percent of this rock, and carbonate and noncarbonate grains are rare (Fig. 5A). As is typical throughout the Trenton, much of the micrite has undergone recrystallization to microspar. Very fine dolomite is locally abundant. The dominant fossil is leperditiid ostracodes, tolerant of a wide range of environmental factors, specifically subaerial exposure and elevated salinity, but total fossil density and diversity are usually low. These mudstones are well laminated, and several undulatory, discontinuous laminae hint at the presence of blue-green algal mats. Burrows are numerous (though not abundant enough to destroy bedding), especially large dwelling burrows and smaller feeding burrows. Infrequent hardgrounds testify to periods of interrupted sedimentation coupled with marine lithification and scouring or corrosion. These rocks represent an intermittently exposed tidal flat.

Facies 2, Rounded-Skeletal Grainstone

The faunal community of this facies (Fig. 5B) displays a high diversity, consisting of echinoderms, inarticulate and articulate brachiopods, cryptostome and trepostome bryozoans, gastropods, pelecypods, corals, trilobites, ostracodes, conodonts, and nautiloid cephalopods. Skeletal algae and algal structures are also numerous, including solenopora nodules, *Girvanella* encrustations, oncolites, and dasycladaceans. Obviously the environment was a most favorable habitat for marine organisms. Many of the fossils have been abraded and rounded. Plane beds and a general lack of biogenic structures suggest moderate to strong waves and a mobile substrate. Additionally, it seems clear that this facies was deposited in shallow, turbulent water, as evidenced by erosional intraclasts, shale galls, good sorting, washing of the matrix, coated grains, scattered grapestones, and micritized skeletal grains. The facies is interpreted as a skeletal-sand shoal of the upper ramp. The lower depth limit was marked by normal wave base, where the amount of matrix increased (grading into packstones) and sorting became poorer.

Facies 3, Graded-Skeletal Grainstone

Taxa in this facies are the same as in other facies; the fossils, however, are not in growth position but have been exhumed, displaced, and redeposited. Gobbets of lime mud adhering to the shells, shells with internal geopetal structure that is not horizontal, and the jumbled, edgewise, or current-aligned orientation of many shells attest to their movement and redeposition. Intraclasts and shale balls formed by the erosion of adjacent semilithified sediments. Sorting is generally poor, and samples with abundant peloids display a bimodal grain-size distribution. The grainstones occur in lenses that cut down into underlying beds or directly overlie muddy sediments with a sharp contact. The similarity between fossil content of these rocks and stratigraphically subjacent rocks illustrates that the skeletal grains were derived from nearby communities.

A distinctive structure of this facies is graded bedding (Fig. 5C). Beds grade upward from grainstone to packstone to lime mudstone to shale. Shells become smaller and fewer in number, peloids become smaller, and the mud matrix is more abundant. In many places the micrite

Fig. 5.—Core photographs. Length of white labels is 1.5 cm. Sample number is the subsurface depth in feet. A. Laminated lime mudstone with calcite-filled burrows and possible mud crack (break in bedding at center). B. Rounded-skeletal grainstone with large intraclasts (dark) and fossil debris (white). Width of core is 5.2 cm. C. Graded-skeletal grainstone showing graded bedding: skeletal grainstone immediately above label grades upward through skeletal packstone, lime mudstone, and shale (at top). Core is fractured in middle (horizontal black band). D. Peloidal grainstone with horizontal laminae (below) and scour-and-fill structure (left center).

matrix displays an infiltration fabric, and calcite cement occupies the position of shelter porosity. Collectively these features are indicative of deposition by storm waves of fairly high energy but of little persistence. Deposition probably took place at a depth below normal wave base; otherwise, the distinctive textures and structures would have been destroyed by the daily reworking of waves and currents.

Facies 4, Peloidal Grainstone

In several samples silt-sized peloids are the only common carbonate grain. Fossils are present but not abundant, but of these, ostracodes predominate. For the most part shells show some degree of transportation, such as breakage, rounding, and gobbets of attached mud. Quartz silt is ubiquitous. A complete range of textures is present: some samples have good sorting, some are poorly washed, and some contain an abundant matrix (actually packstones). Thus, rocks of this facies reflect a wide range of hydrodynamic energy, but the overall energy level is thought to have been low, as illustrated by the rarity of sand-sized particles. Sedimentary structures (Fig. 5D) include horizontal laminae, scour-and-fill, cross-laminae, graded bedding, and an alternation of peloidal grainstone with lime mudstone. Often the peloidal sediments are confined to small lenses, just a few centimeters deep and with scoured bases. These rocks are generally interbedded with shallow-water and storm deposits of Facies 1, 2, and 3. This association argues that the peloid grainstones themselves are shallow-water deposits, transported and deposited by weak or intermittent currents that flowed through minute rills on the sea floor. Currents responsible for their sedimentation may have been generated by storms.

Facies 5, Skeletal Packstone-Wackestone

The diverse assemblage of fauna in this facies was dominated by suspension-feeders: crinoids, brachiopods, and bryozoans (Fig. 6D). Bivalves, gastropods, barnacles, edrioasteroids, cystoids, trilobites, corals, and ostracodes complete the list of preserved benthonic fauna. Algal remains include solenoporaceans, *Girvanella*, and oncolites. Tentaculitiids, cephalopods, and conodonts swam freely above the sea floor.

The packstones and wackestones occur as thin to very thin, discontinuous beds interlayered with shales and lime mudstones, and they are interpreted as small scattered patches of organisms that inhabited an otherwise muddy sea floor. Patchiness of invertebrate communities is complex and not fully understood, but one important control is substrate consistency. Trenton muds were soupy when first deposited (as indicated by numerous burrows with indistinct outlines), and, for this reason, many epifauna were initially excluded. A firm substrate was eventually provided by flat, thin-shelled brachiopods that pioneered the soft bottom. Other organisms then moved in, and development of the low-density benthonic community began. The sea floor was below normal wave base but above storm wave base; additionally, the small number of algae suggests a depth near the base of the photic zone.

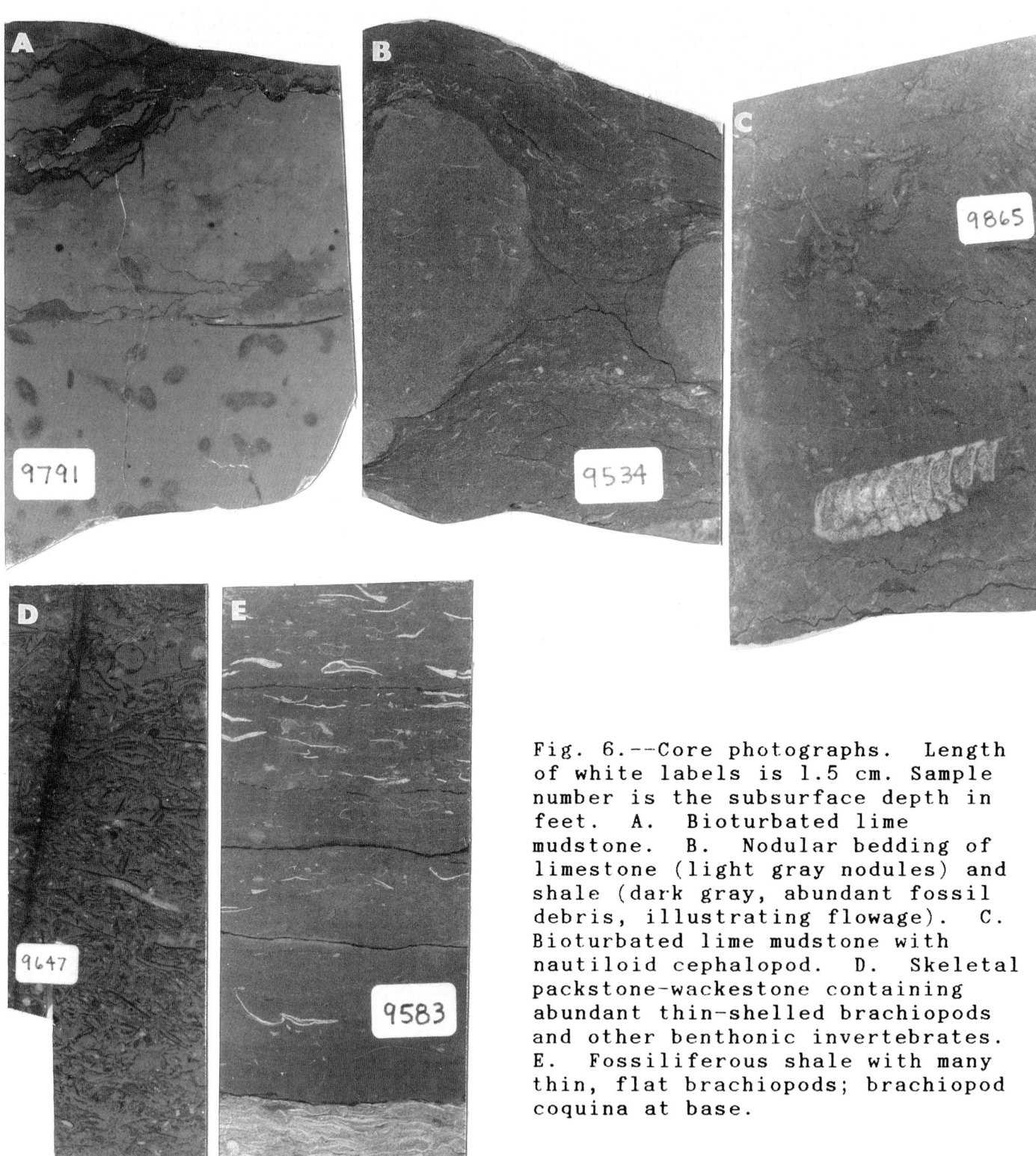

Fig. 6.--Core photographs. Length of white labels is 1.5 cm. Sample number is the subsurface depth in feet. A. Bioturbated lime mudstone. B. Nodular bedding of limestone (light gray nodules) and shale (dark gray, abundant fossil debris, illustrating flowage). C. Bioturbated lime mudstone with nautiloid cephalopod. D. Skeletal packstone-wackestone containing abundant thin-shelled brachiopods and other benthonic invertebrates. E. Fossiliferous shale with many thin, flat brachiopods; brachiopod coquina at base.

Micritic matrix of the packstones in the Sandhill well has been partly washed by waves and currents (replaced by small volumes of intergranular calcite cement); whereas, those of the Riverton outcrop are not washed at all. Sandhill packstones also exhibit a slightly greater fossil diversity. These regional differences support a deepening of the ramp to the east. The Riverton facies was situated in considerably deeper water where waves and currents had little effect on bottom sediments, and fewer organism lived in this more stressful environment.

Facies 6, Bioturbated Lime Mudstone

Deep-water lime mudstones have been extensively bioturbated (Figs. 6A and C), indicating a slow rate of sedimentation. Fossils (trilobites, brachiopods, bryozoans, ostracodes, and molluscs), although scarce, locally constitute up to 10 percent of the rock. Abundant mud, unabraded allochems, and undisturbed clay layers attest to quiet water. The low fossil diversity and density are attributed to extreme environmental stress, in particular the soupy, inhospitable nature of the muddy bottom, rapid influxes of terrigenous clays, and perhaps a lowered oxygen level due to restricted circulation at depth.

Facies 7, Shale

The clay minerals illite and kaolinite (identified by X-ray diffraction) dominate Trenton shales, but also present are quartz, 14-angstrom clays, muscovite, pyrite, and feldspar. Nodular bedding is a characteristic structure of the interbedded shales and limestones (Fig. 6B). Brittle limestone layers have segregated into nodules around which the ductile shales flowed, a feature attributed to a combination of differential compaction and pressure solution. Shales at the Riverton outcrop are abundant and unfossiliferous, ascribed in part to a very fast sedimentation rate; whereas, those in the Sandhill core, deposited far away from the eastern source area, are fewer in number and contain many fossils (Fig. 6E). Such fossil concentrations may be deep-water storm deposits or short-lived benthonic communities on the muddy sea floor.

Not only does the shale content increase toward the east, the number and thickness of shale interbeds dramatically increase upward in the stratigraphic section. Because the Trenton was deposited under transgressive conditions, deep-water shales are more abundant closer to the top. Shales, however, are present in the lower half of the formation, and they do interbed with upper- and mid-ramp facies. Obviously terrigenous clays were introduced into the foreland basin throughout the entire length of Trenton deposition. Currents carried these clays in suspension across the basin and up the carbonate ramp, but accumulation always remained greatest in deep water.

Facies 8, Bentonite

Bentonites or beds of altered volcanic ash are quite common in the stratigraphic section, and they extend from the Appalachian basin to the Midcontinent region. Their areal distribution, number, and thickness indicate a volcanic source along the eastern seaboard of North America. Rodgers (1971) suggested a site in the Carolinas for the volcanic arc of the central Appalachians, but its exact geographic location remains only vaguely known.

Fifteen thin (1 to 16 in. (3 to 40 cm)) bentonites have been identified in the outcrop belt of eastern West Virginia, composed of potassium-bearing, mixed-layer, randomly interstratified illite-montmorillonite clays (Perry, 1972). In the nearby subsurface geophysical logs prove useful for their identification and correlation. The high gamma-ray values are distinctive, presumably due to their contained radioactive potassium. Individual bentonites or complexes of bentonites can be traced laterally from well to well over large areas, each representing a synchronous ash fall that was distributed over much of West Virginia and beyond.

A lower bentonite package is present in the Black River and lowermost Trenton limestones to the southwest, and an upper package occurs in the upper Trenton (Dolly Ridge Member) to the east and north (Fig. 7). This distribution hints that the site of maximum volcanic activity shifted through time. Perhaps the southern bentonites reflect an early plate collision between North America and island-arc system at the Virginia promontory, which projected out into the proto-Atlantic Ocean. As the collision proceeded, tectonic deformation and volcanic activity may have migrated into the Pennsylvania reentrant - indicated by the northern bentonites - where the impact took place at a later time.

DISCUSSION

Ramp Geometry

The Trenton carbonate ramp sloped gently eastward across West Virginia toward the basinal axis in Virginia. Facies transgressed westward through time, and a spectrum of ramp lithologies is present in both the Sandhill core and Riverton outcrop. Although all facies are present in the western core, shallow-water facies dominate (Fig. 8). Tidal-flat lime mudstones account for 18 percent of the limestones (excluding shale and bentonite), and rounded-skeletal grainstone another 14 percent. This grainstone facies was deposited on mobile sand shoals in shallow water, and, in comparison with modern coastal seas (Dietz, 1963), normal wave base in the Trenton sea may have been as shallow as 33 ft (10 m). Storm deposits (graded-skeletal grainstone and peloidal grainstone) make up 31 percent. Thus, in the Sandhill well, almost two-thirds of the limestones consist of sediments deposited above storm wave base, at depths ranging from zero to a few tens of meters. Muddy skeletal patches and bioturbated lime mudstone of the deep sea constitute only one-third of the limestone core. The rarity of algae in these rocks indicates a depth near the base of the photic zone, taken to be about 330 ft (100 m)(Riding, 1975) in the

Fig. 7.--Stratigraphic range and areal distribution of Upper Ordovician bentonite packages, including the widespread S-3 bentonite. Three wells with gamma-ray log (left side of column) and neutron log (right) exemplify the bentonite character. Subsurface depths are in feet.

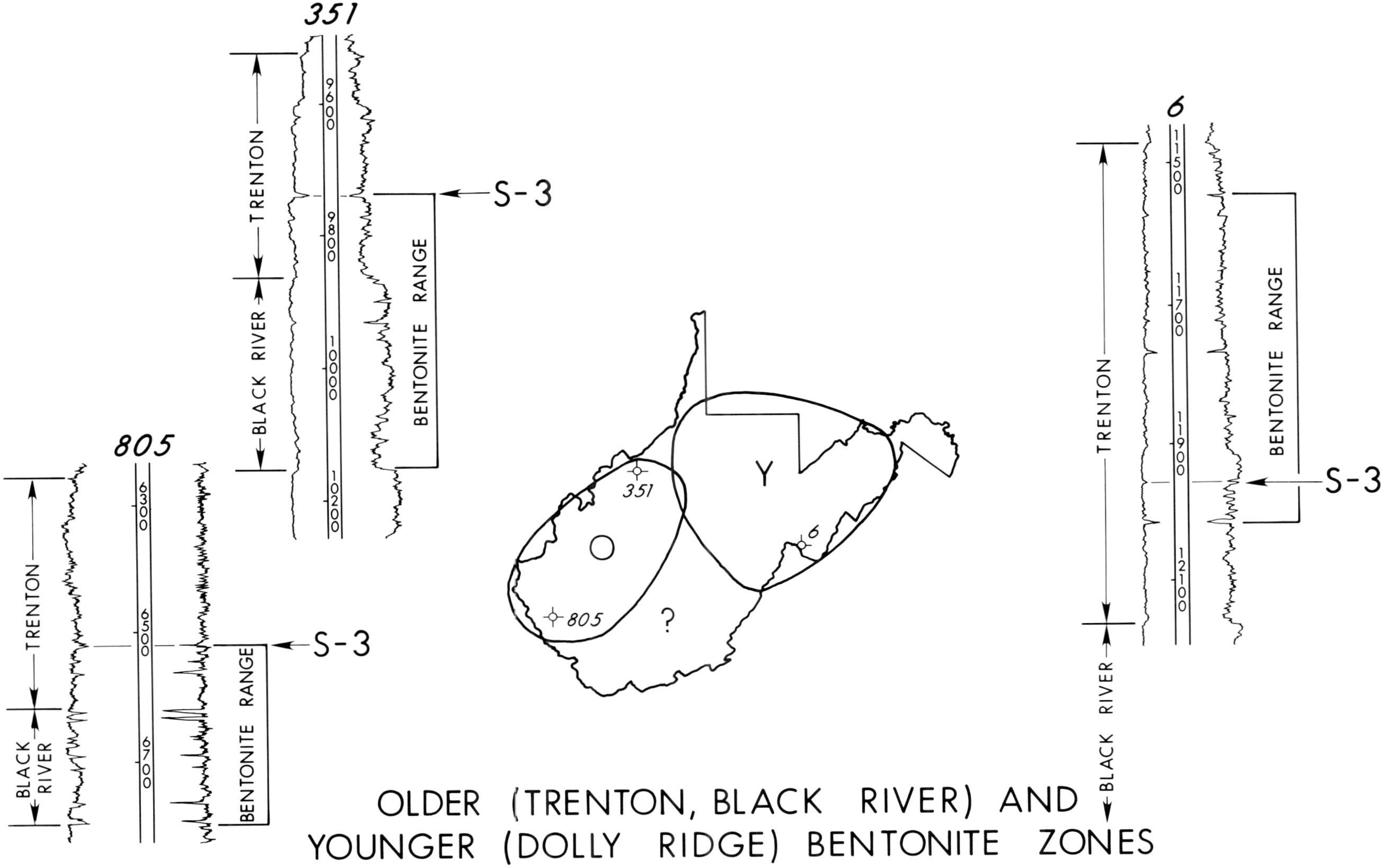

OLDER (TRENTON, BLACK RIVER) AND YOUNGER (DOLLY RIDGE) BENTONITE ZONES

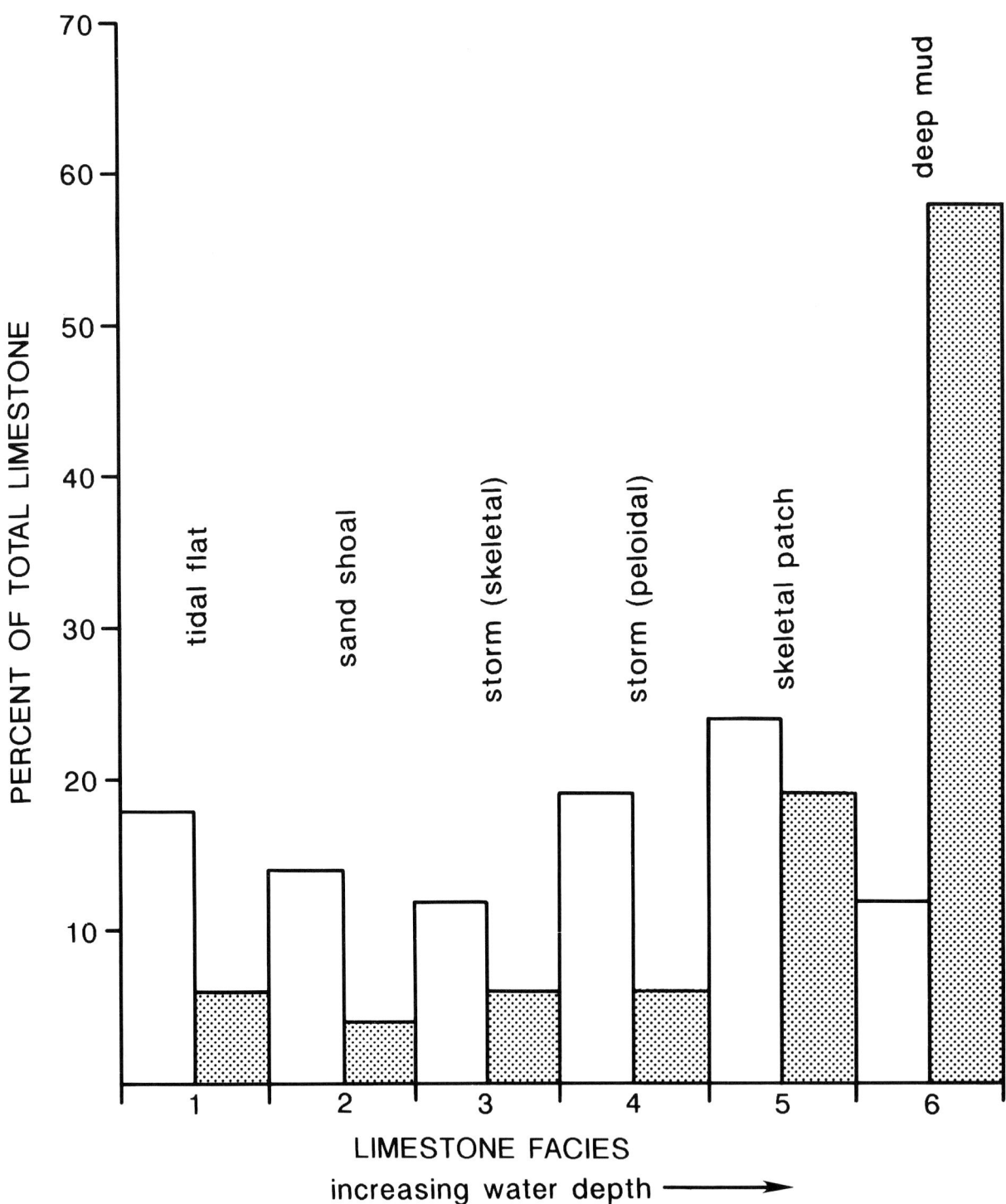

Fig. 8.—Abundance histogram of the six limestone facies at Sandhill (white column, left) and Riverton (black column, right). In general, water depth increased from facies 1 through 6.

turbid water of the Trenton sea. In contrast, deep-water lime mudstone and skeletal packstone-wackestone make up 77 percent of the limestone section at Riverton (Fig. 8). Storm deposits are much less common (12 percent), as are grainstones of the sand shoals (4 percent) and tidal-flat limestones (6 percent).

A subsurface stratigraphic framework has been established on the basis of 33 geophysical logs: 28 in West Virginia (including the Sandhill well), 2 in Ohio, 2 in Kentucky, and 1 in Virginia, plus the Riverton outcrop. Correlations by geophysical logs (gamma ray, density, and neutron), especially correlations of the bentonites, allow construction of four regional cross sections in a loop-and-tie network. These correlations demonstrate that a package of skeletal-shoal grainstones in the Sandhill well, Wood 351, approximately 100 ft (30 m) below the top of the Trenton Limestone, is equivalent to a package of deep-water lime mudstones at Riverton, approximately 330 ft (100 m) below the formation top (bentonite S-4, Fig. 9). Using water depths stated above, the slope angle between these two points (112 miles (180 km) distant on a palinspastic map) is calculated to be 1.6 ft/0.6 mile (0.5 m/km). Although a gentle slope, this value greatly exceeds the slope angle across modern carbonate shelves or postulated for ancient epeiric seas (Coogan, 1972). The Trenton sea floor had a slope comparable to that of the Texas Gulf Coast or the Qatar peninsula in the Persian Gulf.

Transgressive Carbonate Sequence

In the Sandhill well the Trenton constitutes a transgressive or deepening-upward stratigraphic sequence (Fig. 10). The lowest 84 ft (26 m, from 9790 to 9874 ft (2983.9 to 3009.6 m) below the surface) consists mostly of laminated lime mudstone, peloidal grainstone, and shale, deposited at or near sea level on the upper ramp. The overlying 125 ft (38 m, from 9665 to 9790 ft (2945.9 to 2984 m)) was not cored. Above the missing interval is 25 ft (8 m) of rounded-skeletal grainstone, deposited on sand shoals in water less than 33 ft (10 m) deep, followed by 28 ft (8 m) of mostly graded-skeletal grainstone that accumulated below normal wave base and above storm wave base. The uppermost 84 ft (26 m, from 9528 to 9612 ft (2904 to 2929.7 m)) is dominated by bioturbated lime mudstone, shale, and skeletal packstone-wackestone of the deep ramp, laid down on the sea floor below storm wave base and below the photic zone (assumed to be near 330 ft (100 m)). Thus, while the 345 ft (105 m) thick Trenton Limestone was being deposited on the western ramp, sea level rose on the order of 330 ft (100 m).

At Riverton the Trenton likewise displays a deepening-upward sequence. The underlying stratigraphic unit, the Black River Group (including the McGraw and McGlone limestones), represents sedimentation on a shallow shelf (Fig. 2 top), as evidenced by numerous laminated lime mudstones (tidal flat), graded-skeletal grainstones, and peloidal grainstones (storm deposits), and, to a lesser degree, rounded-skeletal grainstones (water depth less than 33 ft (10 m)). Although bioturbated lime mudstone is most abundant in the overlying Nealmont Member of the Trenton (minimal thickness of 200 ft (60 m), lowermost part is covered; Fig. 9), storm and shoal deposits are common, particularly in the upper half. The stratigraphically higher Dolly Ridge Member (414 ft (126 m)) consists almost exclusively of deep-ramp lithologies: bioturbated lime mudstone, skeletal packstone-wackestone, and unfossiliferous shale. An absence of algae indicates that deposition was below the photic zone. Again, while the Trenton limestones

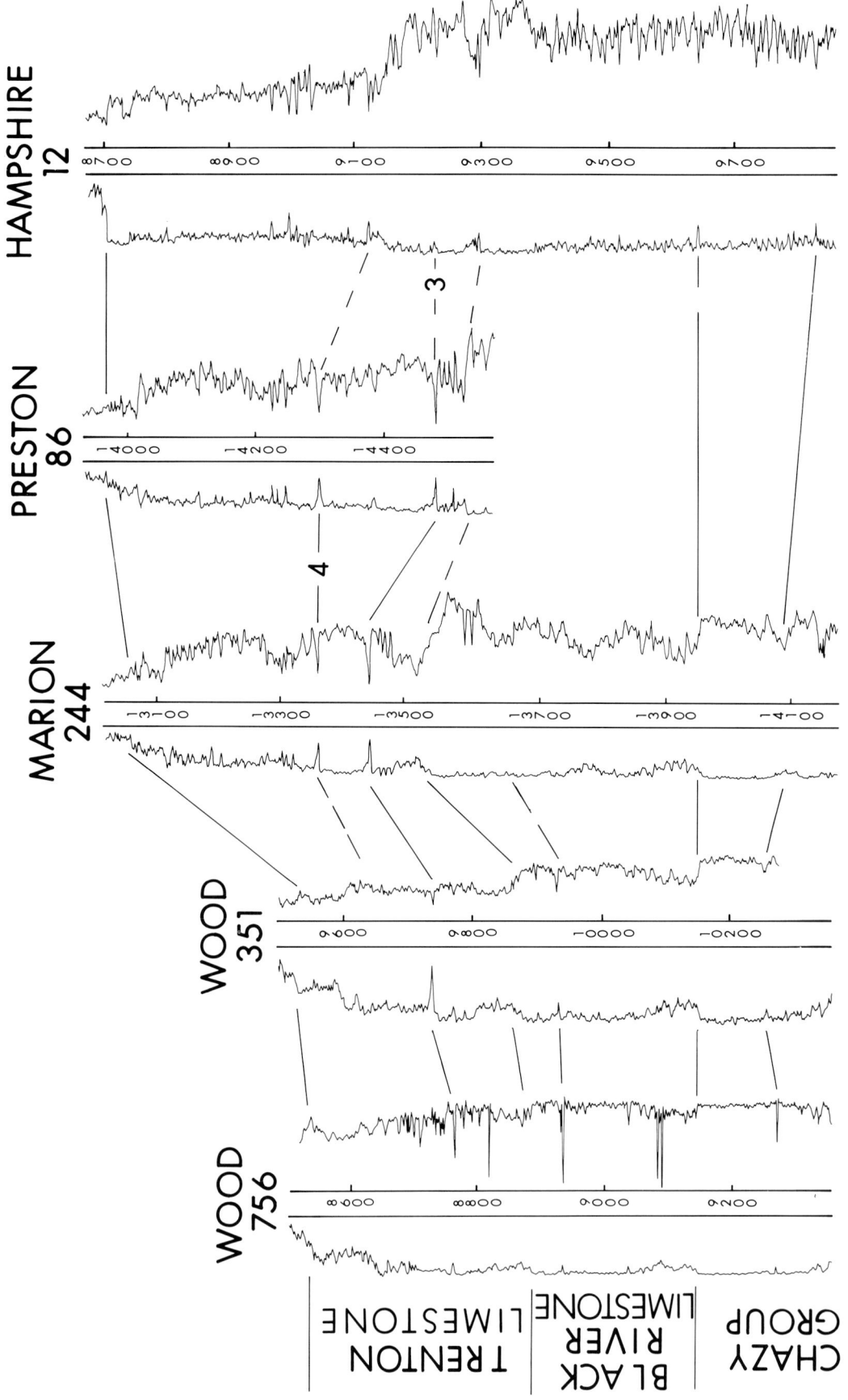

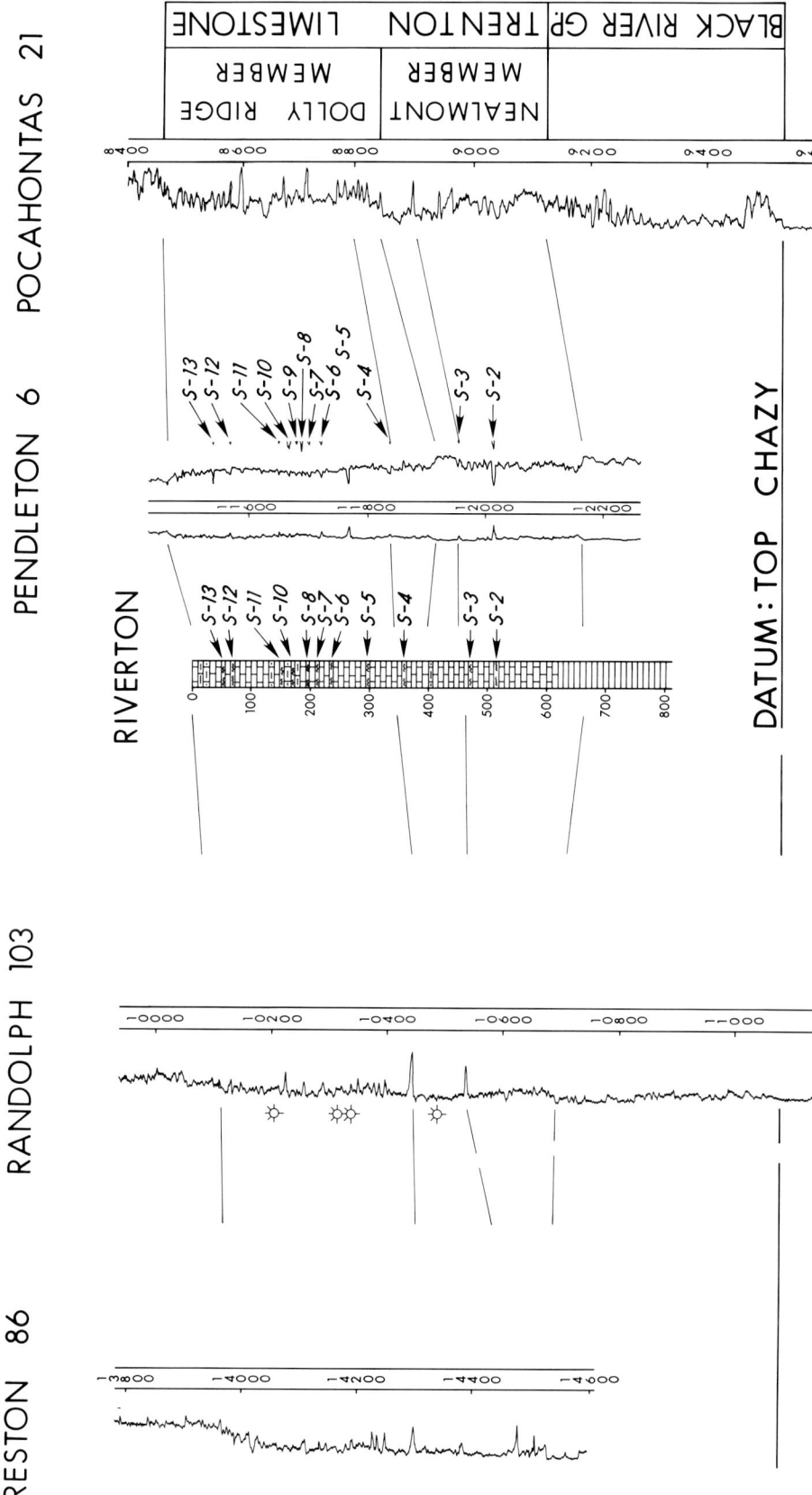

Fig. 9.--Regional cross-section in two parts. Correlations are based on bentonites, including the widespread S-4, identified by gamma-ray (left side of each column) and neutron (right side) logs. Subsurface depths are in feet. Lines of section are shown in Figure 3.

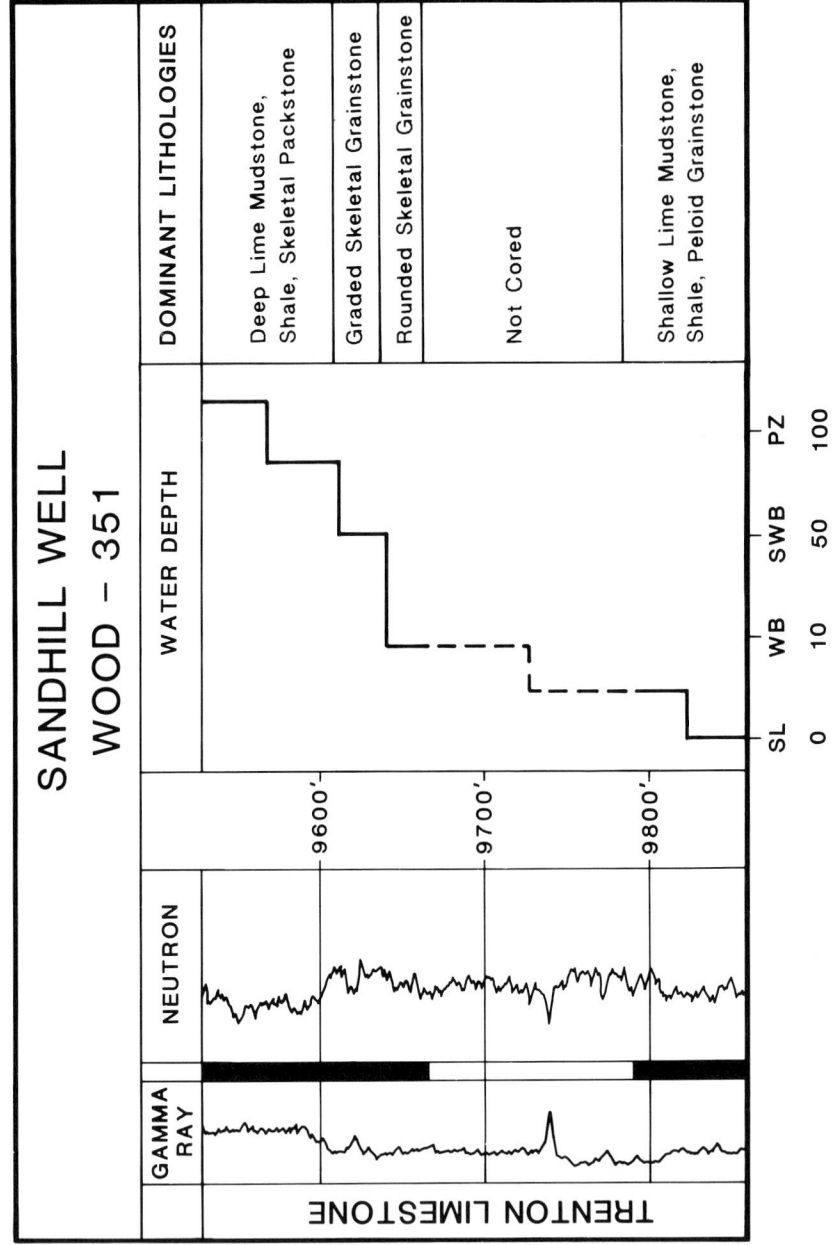

Fig. 10.—Deepening-upward lithologies of Trenton Limestone at Sandhill. Interpreted water depth (in meters) is related to sea level (SL), normal wave base (WB), storm wave base (SWB), and photic zone (PZ). Cored interval indicated in black in column between geophysical logs. Subsurface depths in feet.

accumulated on the eastern ramp, sea level rose by more than 330 ft (100 m). It is interesting to note that Quinlan and Beaumont (1984), in modeling Taconic thrusting and related stratigraphic responses, also postulated a major eustatic sea-level rise at this time, as did Hay and Cisne (1988).

Subsidence

The Trenton stratigraphic unit constitutes a wedge-shaped mass that thickens eastward into the foreland basin. The thickness ranges from 295 ft (90 m) near the western border of West Virginia to near 660 ft (200 m) along the eastern border of the state (Fig. 11). Overall thickening eastward is simply a function of greater basinal downwarping toward the orogenic fold-thrust belt, and this differential subsidence is readily observed on the cross section of Figure 9. However, in the west-central and northwestern parts of the state, isopach contours are slightly discordant. There they outline an area of relatively thick Trenton (492 to 590 ft (150 to 180 m)), a linear trend that parallels the Ordovician continental margin and lies above the Rome trough.

Vertical stresses produced by thrust loading of the crust during Taconic plate collision reactivated basement faults beneath the Appalachian foreland basin (Fig. 2 middle; Jacobi, 1981). This process may have reactivated normal faults of the Rome trough, a Cambrian graben or failed rift basin related to opening of the proto-Atlantic Ocean (Donaldson and Shumaker, 1981). Trenton subsidence of the foreland basin in Pennsylvania has been attributed to displacement along normal faults (Lash, 1988). In presumedly related movements, synsedimentary faulting fragmented the Trenton shelf of New York into a series of horsts and grabens (Mehrtens, 1988). And, in West Virginia, normal faulting associated with flexure of the continental lithosphere apparently led to renewed downwarping over the Rome trough.

Assuming that Trenton deposition lasted for five million years (middle portion of the Caradocian Stage), the average subsidence rate can be calculated. In the Sandhill well 345 ft (105 m) of limestone and shale accumulated (ignoring compaction), while simultaneously water depth increased perhaps 330 ft (100 m). Thus, absolute subsidence was 673 ft (205 m), and the average subsidence rate was (at least) 1.6 in. (4 cm) per 1000 years. Similarly, at Riverton, a minimum of 610 ft (186 m) of rock accumulated while water depth rose 330 ft (100 m), indicating an absolute subsidence of 938 ft (286 m) and a subsidence rate of 2.4 in. (6 cm)/1000 years.

Water depth in the foreland basin of Tennessee was bathyal, estimated at 2300 ft (700 m) by Shanmugam and Walker (1980). The basin was initially starved but later filled with hemipelagic shales and then turbidites. The basin floor in New York was likewise quite deep and dysaerobic to anaerobic (Hay and Cisne, 1988). Sediments there also consisted of hemipelagic shales and carbonate turbidites. In-situ production of skeletal sediments was minimal and the benthonic community impoverished because of the low oxygen concentration. By contrast, water depth was not as great (maximum interpreted to have been not much more than 330 ft (100 m)), nor the slope angle as steep within the Pennsylvania reentrant. In West Virginia basin subsidence was

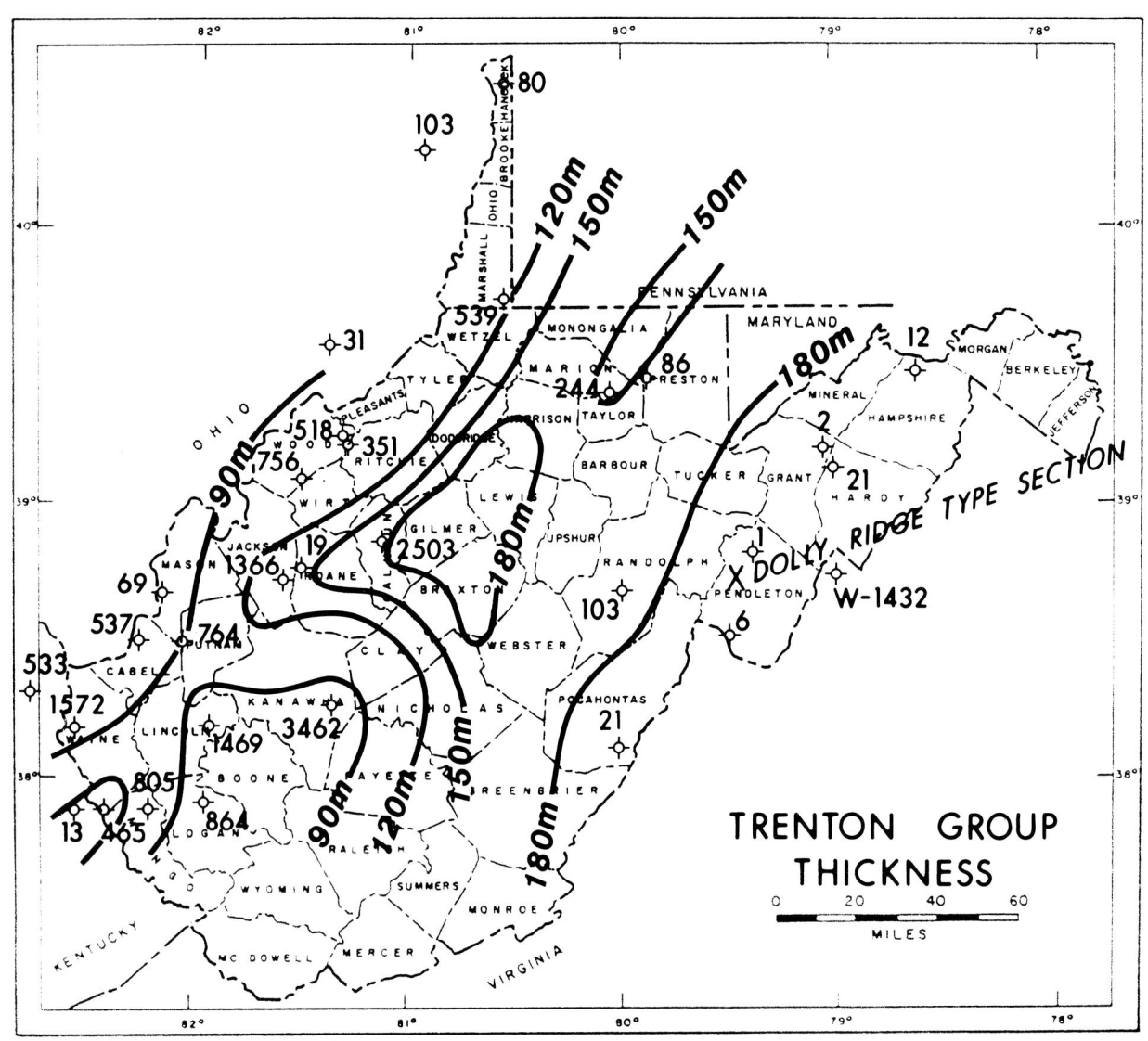

Fig. 11.--Isopach map of the Trenton Limestone across West Virginia. The subsurface stratigraphic framework is established on the basis of geophysical logs from these 33 wells plus the Riverton outcrop. Riverton (X) is the type section for the upper Dolly Ridge Limestone Member.

rapid, being fast enough to outpace sedimentation, but not so fast as to suppress carbonate sedimentation under deep, anaerobic conditions. Limestone deposition therefore continued, but with a deepening-upward sequence.

Interbedded Shale

Shale is a major, but secondary, lithology that constitutes approximately one-third of the stratigraphic section at the Riverton outcrop. Although Trenton limestones contain low-density, low-diversity communities, shales tend to be unfossiliferous. Moreover, the least fossiliferous limestones, the deep-water bioturbated lime mudstones (with which shales are frequently associated), contain the greatest amount of insoluble residue, as much as 20 percent and more. Introduction of shale into the Trenton carbonate facies typically downgraded the community to one of lower diversity (or even abiotic) and of greater dominance by hardy pioneering animals. In addition, bedding contacts between barren shales and fossiliferous limestones are sharp, confirming that the environmental changes were rapid.

Clearly a negative relationship existed between fine terrigenous sediments and fossil content and diversity. Mud alone was not a controlling factor of community development: fossil diversity is actually higher in pure clastic sequences, such as the overlying Martinsburg or Reedsville formations (Bretsky, 1969). Rather, environmental perturbations, particularly the repeated incursions of clays, disrupted the benthonic communities at Riverton. And absence of stenotropic pelmatozoans in the shales argues for extremely turbid water and/or a high sedimentation rate (Fell, 1966). Thus, the adverse relationship is attributed to a combination of factors: (1) the high turbidity in the water column, which would have decreased light intensity for algae and clogged the feeding mechanisms of invertebrates; (2) stressful environmental instability brought about by frequent fluctuations of the clastic input; (3) a high rate of shale sedimentation that would bury benthonic organisms; and (4) the soupy or fluid consistency of a muddy substrate that would not support most epifauna (Carter and others, 1988).

On the other hand, shales of the western Sandhill core, constituting only one-fifth of the formation, are generally quite fossiliferous. In fact, several with more than 50 percent fossils grade into argillaceous limestones. Fossils include bryozoans, brachiopods, echinoderms, ostracodes, trilobites, molluscs, sponges, and conodonts (Smosna, 1985). The difference between barren shales in the east and fossiliferous shales in the west is explained by geographic position. Riverton was situated almost 124 miles (200 km) nearer to the Taconic fold-thrust belt. In this position terrigenous sediment shed from the growing tectonic source area was more common, the water column more turbid, the sedimentation rate more rapid, and the substrate more fluid. Overall, shales had progressively greater influence eastward down the Trenton ramp.

CONCLUSIONS

The Trenton Limestone consists of thin limestones interbedded with shales and bentonites. These rocks were deposited on a gently sloping carbonate ramp. Wave-agitated grainstones and protected lime mudstones were present in the nearshore/shoreline environments; that is, lime-sand banks and shoals and associated mud flats. They passed gradually downslope into deep-water muddy sands, skeletal packstones, and wackestones, which passed farther downslope into lime mudstones. For the most part the rocks accumulated under the influence of day-to-day depositional processes, but fair-weather sedimentation was frequently interrupted by storms, influxes of terrestrial material, and volcanic fall-out.

These rocks formed during a major tectonic event, the early stage of the Middle-Late Ordovician Taconic orogeny. Active tectonism produced very rapid subsidence of the foreland basin and the transformation of a shallow carbonate shelf into a gently sloping, but generally deep-water ramp. The slope angle across West Virginia averaged 1.6 ft/0.6 mile (0.5 m/km). Rapid but differential subsidence (greater downwarping toward the outer-arc complex) led to a wedge-shaped geometry for the Trenton body of rock and a doubling of its thickness from west to east. In addition, the probable reactivation of basement normal faults was triggered by tectonic uplift of the suture zone, generating a local thickening above the Rome trough. Approach of the volcanic arc resulted in numerous bentonites within the formation, interbedded with limestones and shales over the entire ramp.

The stratigraphic sequence is transgressive since sedimentation could not keep pace with basin subsidence. Average sedimentation rate for the Trenton ranged between 0.8 to 1.6 in. (2 to 4 cm)/1000 years (330 to 660 ft (100 to 200 m) of section in 5 million years). This rate compares with typical sedimentation rates calculated for ancient carbonate sequences, which rarely exceeded 1.6 in. (4 cm)/1000 years (Wilson, 1975). The basin subsided, on the other hand, by 1.6 to 2.4 in. (4 to 6 cm)/1000 years. Therefore, relative sea level rose in the basin approximately 330 ft (100 m) during Trenton deposition, and this change is reflected in the deepening-upward sequence of facies.

Trenton limestones accumulated despite significant terrigenous influxes because these were discontinuous through time. Along the eastern edge of the carbonate ramp, close to the source area, introduction of clays did temporarily downgrade or eliminate the skeletal communities. With each return of clear water, however, skeletal sedimentation of limestone resumed. Higher up the ramp at the Sandhill well, introduction of clays - though still in discrete pulses - had a less detrimental effect on limestone sedimentation.

The periodicity of limestone-shale deposition appears to follow dilution rhythms, as described by Einsele (1982). At Riverton a comparatively high rate of terrigenous-mud deposition is indicated by the presence of unfossiliferous shales and ecological downgrading of the benthonic communities. Presumably, too, the rate of skeletal limestone production was low, especially in the relatively deep, inhospitable ramp facies. Episodic flooding of clays could have been a function of (1) discontinuous thrusting and uplift in the growing orogenic source area; (2) variation in the rate of river supply or point of discharge, related to climatic modification or channel switching; (3) changing sea level and its effect on base level and erosion; and (4) intermittent marine transport by storm currents or sediment gravity flow. A combination of quick shale rhythms, therefore, superimposed on slow, steady-state carbonate deposition produced the distinctive interbedding of lithologies.

At Sandhill the rate of mud deposition may have been somewhat high. During these periods, however, carbonate sedimentation continued - as evidenced by fossils, peloids, and micrite/microspar in the shales. Farther from the source area an equilibrium was established between the rate of clay introduction and the production of skeletal lime sediments. Here, slow shale rhythms, coupled with steady-state carbonate deposition, also produced an interbedding of lithologies, although slightly different from that of the east.

Eventually the continent-arc collision and resulting uplift of the fold-thrust belt proceeded to the point where the rate of terrigenous supply soared (Fig. 2 bottom). Carbonate sediments were then overwhelmed by clastics, and the pre-flysch limestones and shales of the Trenton gave way to the flysch wedge of the Martinsburg Formation (McBride, 1962; Kreisa, 1981). This overlying unit represents a stage of basin filling (up to 2953 ft (900 m) thick) when sediment supply exceeded basin subsidence. The rate of sedimentation was about 3.9 in. (10 cm)/1000 years (Kay, 1951), nearly four times greater than the Trenton rate, and the basin filled with hemipelagic mudstones, turbidite sandstones, and tempestite limestones.

REFERENCES

BARTLETT, C. S., 1988, Trenton Limestone fracture reservoirs in Lee County, southwestern Virginia, in Keith, B. D., ed., The Trenton Group (Upper Ordovician Series) of Eastern North America: American Association of Petroleum Geologists Studies in Geology #29, p. 27-35.

BRETSKY, P. W., 1969, Central Appalachian Late Ordovician communities: Geological Society of America Bulletin, v. 80, p. 193-212.

CARTER, B., MILLER, P., AND SMOSNA, R., 1988, Environmental aspects of Middle Ordovician limestones in the central Appalachians: Sedimentary Geology, v. 58, p. 23-36.

CHURCH, W. R., AND STEVENS, R. K., 1971, Early Paleozoic ophiolite complexes of Newfoundland Appalachians as mantle-oceanic crust sequences: Journal of Geophysical Research, v. 76, p. 1460-1466.

COLTON, G. W., 1970, The Appalachian basin - its depositional sequences and their relationships, in Fisher, G. W., Pettijohn, F. J., Reed, J. C., and Weaver, K. N., eds., Studies of Appalachian Geology: Central and Southern: Interscience, New York, p. 5-47.

COOGAN, A. H., 1972, Recent and ancient carbonate cyclic sequences, in Elam, J. C., and Chuber, S., eds., Cyclic Sedimentation in the Permian Basin: West Texas Geological Society, Midland, TX, 2nd ed., p. 5-16.

DIETZ, R. S., 1963, Wave base, marine profile of equilibrium, and wave-built terraces: a critical appraisal: Geological Society of America Bulletin, v. 74, p. 971-990.

DONALDSON, A. C., AND SHUMAKER, R. C., 1981, Late Paleozoic molasse of central Appalachians, *in* Miall, A. D., ed., Sedimentation and Tectonics in Alluvial Basins: Geological Association of Canada Special Paper 23, p. 99-124.

DRAZAN, D. J., 1988, Overview of Trenton exploration and development in New York State, *in* Keith, B. D., ed., The Trenton Group (Upper Ordovician Series) of Eastern North America: American Association of Petroleum Geologists Studies in Geology #29, p. 135-138.

EINSELE, G., 1982, Limestone-marl cycles (periodites): diagenesis, significance, causes - a review, *in* Einsele, G., and Seilacher, A., eds., Cyclic and Event Stratigraphy: Springer-Verlag, Berlin, p. 8-53.

FELL, H. B., 1966, Ecology of crinoids, *in* Boolootian, R. A., ed., Physiology of the Echinodermata: Interscience, New York, p. 49-62.

HAY, B. J., AND CISNE, J. L., 1988, Deposition in the oxygen-deficient Taconic foreland basin, Late Ordovician, *in* Keith, B. D., ed., The Trenton Group (Upper Ordovician Series) of Eastern North America: American Association of Petroleum Geologists Studies in Geology #29, p. 113-134.

JACOBI, R. D., 1981, Peripheral bulge - a causal mechanism for the Lower/Middle Ordovician unconformity along the western margin of the Northern Appalachians: Earth Planetary Science Letters, v. 56, p. 245-251.

KAY, M., 1951, North American geosynclines: Geological Society of America Memoir 48, 143 p.

KREISA, R. D., 1981, Storm-generated structures in subtidal marine facies with examples from the Middle and Upper Ordovician of southwestern Virginia: Journal of Sedimentary Petrology, v. 51, p. 823-848.

LASH, G. G., 1988, Middle and Late Ordovician shelf activation and foredeep evolution, central Appalachian orogen, *in* Keith, B. D., ed., The Trenton Group (Upper Ordovician Series) of Eastern North America: American Association of Petroleum Geologists Studies in Geology #29, p. 37-53.

MCBRIDE, E. F., 1962, Flysch and associated beds of the Martinsburg Formation (Ordovician), central Appalachians: Journal of Sedimentary Petrology, v. 32, p. 39-91.

MEHRTENS, C. J., 1988, Bioclastic turbidites in the Trenton Limestone: significance and criteria for recognition, *in* Keith, B. D., ed., The Trenton Group (Upper Ordovician Series) of Eastern North America: American Association of Petroleum Geologists Studies in Geology #29, p. 87-112.

PERRY, W. J., 1972, The Trenton Group of Nittany Anticlinorium, eastern West Virginia: West Virginia Geological Survey Circular 13, 30 p.

QUINLAN, G. M., AND BEAUMONT, C., 1984, Appalachian thrusting, lithospheric flexure, and the Paleozoic stratigraphy of the Eastern Interior of North America: Canadian Journal of Earth Sciences, v. 21, p. 973-996.

READ, J. F., 1980, Carbonate ramp-to-basin transitions and foreland basin evolution, Middle Ordovician, Virginia Appalachians: American Association of Petroleum Geologists Bulletin, v. 64, p. 1575-1612.

RIDING, R., 1975, *Girvanella* and other algae as depth indicators: Lethaia, v. 8, p. 173-179.

RODGERS, J., 1971, The Taconic orogeny: Geological Society of America Bulletin, v. 82, p. 1141-1178.

SHANMUGAM, G., AND WALKER, K. R., 1980, Sedimentation, subsidence, and evolution of a foredeep basin in Middle Ordovician, southern Appalachians: American Journal of Science, v. 280, p. 479-496.

SMOSNA, R., 1985, Day-to-day sedimentation on an Ordovician carbonate ramp punctuated by storms and clastic influxes: Northeastern Geology, v. 7, p. 167-177.

THOMAS, W. A., 1977, Evolution of Appalachian-Ouachita salients and recesses from reentrants and promontories in the continental margin: American Journal of Science, v. 277, p. 1233-1278.

WICKSTROM, L. H., AND GRAY, J. D., 1988, Geology of the Trenton Limestone in northwestern Ohio, *in* Keith, B. D., ed., The Trenton Group (Upper Ordovician Series) of Eastern North America: American Association of Petroleum Geologists Studies in Geology #29, p. 159-172.

WILLIAMS, H., AND HATCHER, R. D., 1983, Appalachian suspect terranes, *in* Hatcher, R. D., Williams, H., and Zietz, I., eds., Contributions to the Tectonics and Geophysics of Mountain Chains: Geological Society of America Memoir 158, p. 33-53.

WILSON, J. L., 1975, Carbonate facies in geologic history: Springer-Verlag, New York, 471 p.

WOODWARD, H. P., ed., 1959, A symposium on the Sandhill deep well, Wood County, West Virginia: West Virginia Geological Survey Report Investigations No. 18, 182 p.

SLOPE AND BASINAL CARBONATE DEPOSITION IN THE NOLICHUCKY SHALE (UPPER CAMBRIAN), EAST TENNESSEE: EFFECT OF CARBONATE SUPPRESSION BY SILICICLASTIC DEPOSITION ON BASIN-MARGIN MORPHOLOGY

J. L. FOREMAN, K. R. WALKER,
L. J. WEBER, AND S. G. DRIESE
Department of Geological Sciences,
University of Tennessee, Knoxville, TN 37996
R. B. DREIER[1]
Oak Ridge National Laboratory,
Environmental Sciences Division, Oak Ridge, TN 37831

ABSTRACT

Intraclastic limestone, thin fossiliferous limestone, and oolitic limestone interbedded with fissile shale are characteristic of the Nolichucky Shale (Upper Cambrian) in the Whiteoak Mountain thrust sheet, eastern Tennessee. These lithologies are interpreted as slope and basinal units deposited in water depths that may have exceeded 250 to 300 m. Intraclastic limestone debris flow deposits and skeletal and oolitic calcarenite turbidites indicate that nearly all carbonate is allochthonous, derived from up-slope and platform shallow-water environments. Storms were a dominant factor in initializing down-slope transport from shallower-water settings.

Polymictic limestone conglomerate and shallow-water-derived calcarenite, composed of ooids, peloids, and fossils, indicate that depositional slopes were steep enough at times to permit considerable transport (in excess of 20 km) from up-slope areas. During times of increased siliciclastic sedimentation, carbonate production was suppressed and the basin margin had a gentle ramp morphology. Algal and oolitic shoals built up close to sea level when siliciclastic input was reduced, resulting in the progradation of carbonate facies into the basin. Depositional slopes at the basin margin were steepened to the point that gravity sediment flows were effective in transporting shallow-water carbonate lithologies into the basin.

INTRODUCTION

Study Area

Continuous 2.5-in. (6.4 cm) core was obtained from shallow bore holes drilled to 1000 ft (305 m) below the surface on the U.S. Department of Energy Oak Ridge Reservation (ORR, Fig. 1). The present study is based on detailed examination of core from four boreholes (GW134, GW139,

---

[1]Research supported by the Office of Defense Waste and Transportation Management, U.S. Department of Energy, under contract DE-AC05-84OR21400 with Martin Marietta Energy Systems, Inc.

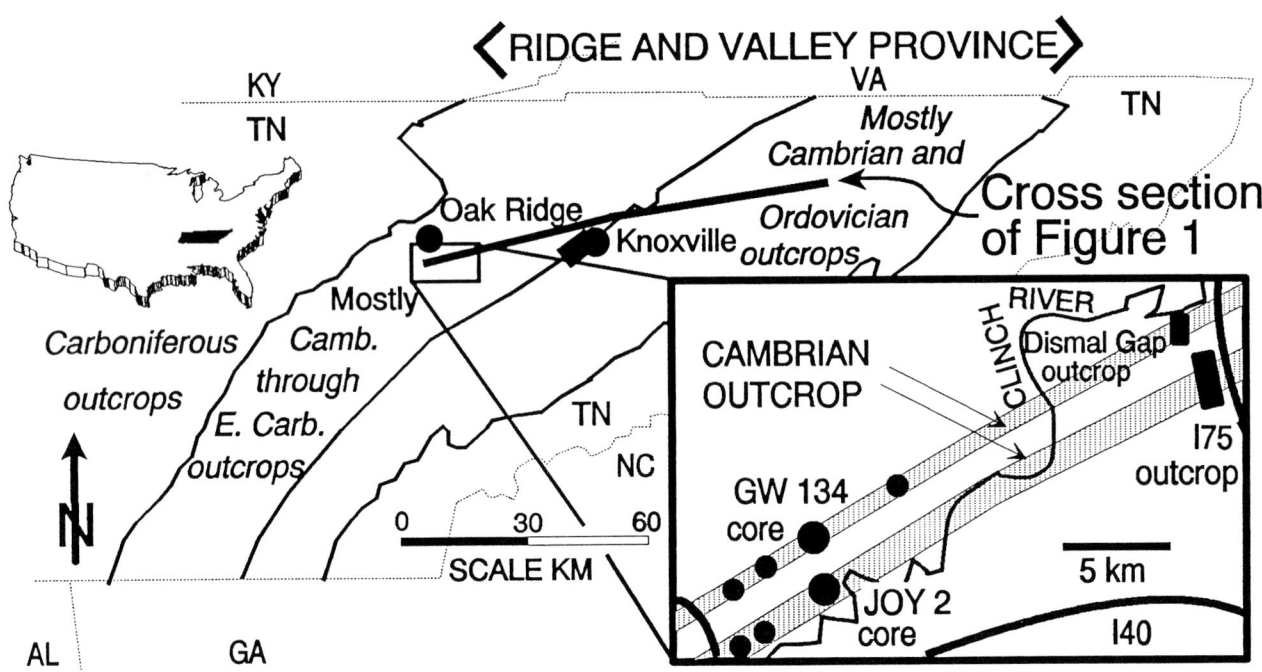

Figure 1. Map of outcrops (filled rectangles) and cored boreholes (filled circles) examined. Northwest and southeast stripped bands in inset figure correspond to Conasauga Group outcrops in the WOM and Copper Creek thrust sheets, respectively. From SW to NE, the boreholes are GW137, GW117, GW134, and GW130.

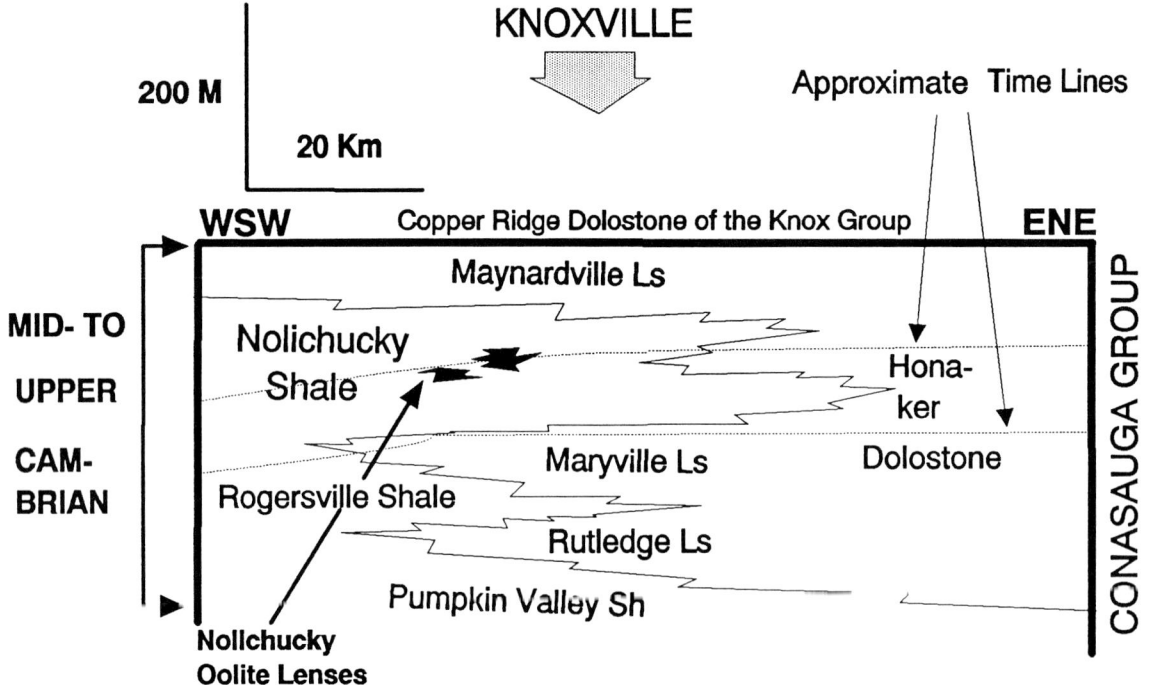

Figure 2. Cross-section oriented nearly perpendicular to depositional strike for Conasauga Group members in the vicinity of Knoxville, TN. Approximate locations of shallow-water oolite from the Copper Creek thrust sheet indicated. Modified from NW-SE cross-section of Rodgers (1953).

GW140, and Joy-2), cursory study of several other cores, and detailed outcrop examination (Fig. 1). The ORR contains a portion of the Appalachian foreland Whiteoak Mountain (WOM) and Copper Creek thrust sheets. Strata dip to the southeast at approximately 40° (King and Haase, 1987) and range in age from Lower Cambrian (Rome Formation) to Middle Ordovician (Chickamauga Group). In the Oak Ridge area, there is a minimum displacement on the Copper Creek thrust of 7.5 miles (14 km) (Lemiszki and Hatcher, 1990)' therefore, the minimum palinspastic separation between boreholes in the WOM sheet (GW134, GW139, and GW140) and the Copper Creek sheet (Joy-2) is approximately 12.5 miles (20 km). A similar palinspastic separation is expected between the I-75 and Dismal Gap outcrops (Fig. 1). In the vicinity of Oak Ridge, structural strike and depositional strike are nearly parallel (Weber, 1988). Consequently, significant facies changes can be observed between the WOM and Copper Creek thrust sheets.

Previous Studies

Rodgers (1953) recognized a northwestern, central, and southeastern phase for the Conasauga Group (Middle to Upper Cambrian), characterized by shale and siltstone, shale and limestone, and dominantly limestone and dolostone respectively. The study area lies in the central Conasauga phase (Rodgers, 1953; Weber, 1988). Here, southeastward-thickening carbonate units of the Rutledge Limestone, Maryville Limestone, and Maynardville Formation interfinger with northwestward-thickening siliciclastic units of the Pumpkin Valley Shale, Rogersville Shale, and Nolichucky Shale (Fig. 2). The Nolichucky Shale (Dresbachian) is approximately 500 to 575 ft (150-175 m) thick in the WOM and Copper Creek thrust sheets respectively (Weber, 1988). The paleogeography and distribution of Conasauga Group lithofacies in eastern Tennessee have been summarized in very general terms by Hasson and Haase (1988).

In southwest Virginia and northeast Tennessee, the Nolichucky has been described in detail by Markello (1979), and Markello and Read (1981; 1982). In eastern Tennessee, Weber (1988) examined the Nolichucky Shale and Maynardville Limestone, with particular emphasis on outcrop exposures and drill core from the Copper Creek thrust sheet. Similarities in the Nolichucky Shale are recognized in both Virginia and Tennessee. These include (1) deposition in an intracratonic basin, (2) an east-west trend from shallow peritidal limestone to deeper-water limestone respectively, and (3) an abundance of storm-generated features in the deeper-water limestone (Markello and Read, 1981, 1982; Weber, 1988). However, significant differences also exist in the sedimentology of the Nolichucky Shale in southwestern Virginia and eastern Tennessee. This study focuses on the most basinward lithologies of the Nolichucky Shale that contrast significantly with the up-slope, storm-dominated units of the Nolichucky in the Copper Creek thrust sheet described by Weber (1988).

LITHOFACIES DESCRIPTION AND INTERPRETATION

Gross Vertical Trends

The vertical trends described here are based on observations in core from borehole GW134 in the WOM thrust sheet (Fig. 1). Similar trends are also observed in core from other boreholes in the same thrust sheet. In general, intraclastic limestone interbedded with shale and calcareous siltstone are the dominant lithologies in the lower Nolichucky Shale (Figs. 3 and 4). Allochthonous oolitic and skeletal packstone and grainstone interbedded with shale are the dominant lithologies in the middle Nolichucky (Figs. 5 and 6). Laminated peloidal packstone, lime mudstone, and shale are the most abundant lithologies toward the top of the Nolichucky Shale (Fig. 7). Thrombolitic limestone overlying oolitic limestone and lime mudstone occurs in the uppermost Nolichucky (Fig. 7). On a formation scale there is an upward progression toward shallower-water lithofacies, although numerous shallowing-upward (less shale upward) and deepening-upward (more shale upward) sequences are superimposed on this overall trend (Fig. 8). These shallowing and deepening intervals can be correlated between wells in the WOM sheet (Bear Creek Valley, Fig. 9) with confidence; correlations between wells in the Copper Creek sheet are more difficult, as are correlations between the two thrust sheets.

Carbonate Lithologies and Interpretation

Intraclastic limestone conglomerate: description.--Intraclastic limestone (flat-pebble) conglomerate is one of the most common Nolichucky limestone lithologies in the GW134 core, especially in the lower third of the unit (709-575 ft (216-175 m), Fig. 8). Intraclastic limestone beds are typically less than 1.0 ft (0.30 m) thick and have sharp lower and upper contacts. Some intraclastic limestone beds can be correlated between boreholes over a distance of at least 4 miles (6 km) along strike. Across-strike correlations between the Joy-2 and GW134 boreholes are tenuous; however, several intraclastic limestone intervals in the lower Nolichucky may be correlative in both wells. Clast size sorting in individual beds is generally lacking, and the fabric can be either matrix-supported (Figs. 10a and 11a) or clast-supported (Fig. 10b and d). In a single bed the intraclastic limestone may be clast-supported at the base and matrix-supported at the top of the bed (Fig. 10c). Shale, fine skeletal debris, peloids, and more rarely ooids are common constituents of the matrix (Figs. 10a and 12a). Clasts are tabular-shaped and clast orientations range from nearly parallel to bedding (Fig. 10a and d) to nearly perpendicular to bedding (Fig. 10b). It is not uncommon to see different clast orientations in the same bed over a short distance. Clasts may project into overlying shale (Fig. 11b). Imbricated clasts (long axis parallel to transport direction and dipping against the transport direction; Walker, 1979) are observed in some intraclastic limestone (Figs. 10a-b and 11b); however, in core, it is commonly difficult to be sure of the maximum clast axis orientation.

Intraclastic limestone may be either monomictic or polymictic. Laminated peloidal packstone, trilobite and echinoderm packstone, oolitic packstone and grainstone, and lime mudstone are the most common clast varieties. Monomictic clasts are commonly similar in nature to underlying interbeds of limestone or calcareous siltstone, suggesting a local origin (Fig. 10c). Polymictic limestone conglomerate is most abundant at the base of the Nolichucky Shale, with clasts

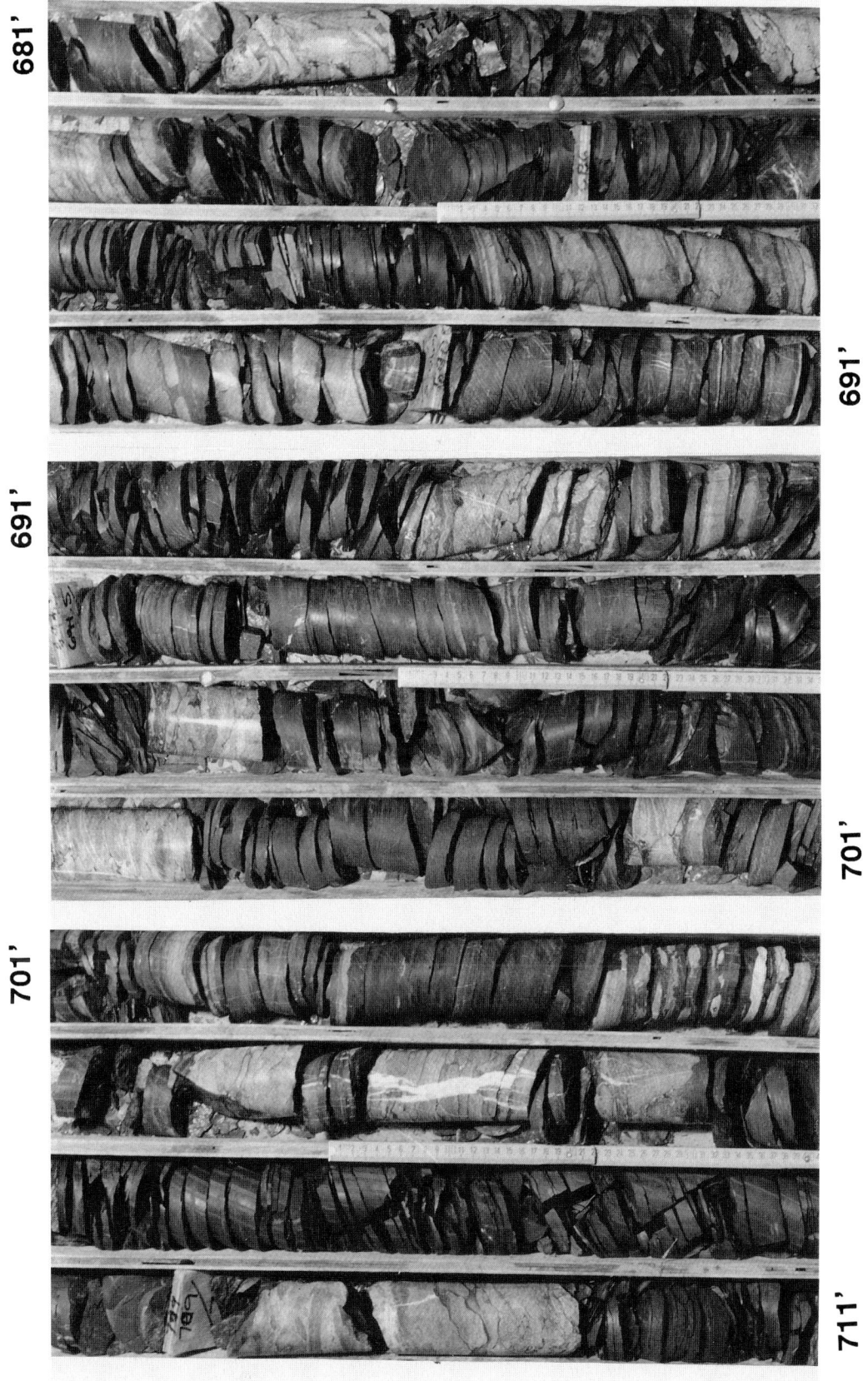

Figure 3. Lowermost 30 ft (9 m) of the Nolichucky Shale in 2.5 in. (6.4 cm) diameter core from borehole GW134. Intraclastic limestone conglomerate is dominant carbonate lithology; Calcareous siltstone and very fine-grained skeletal calcarenite occur in 690-688 ft interval.

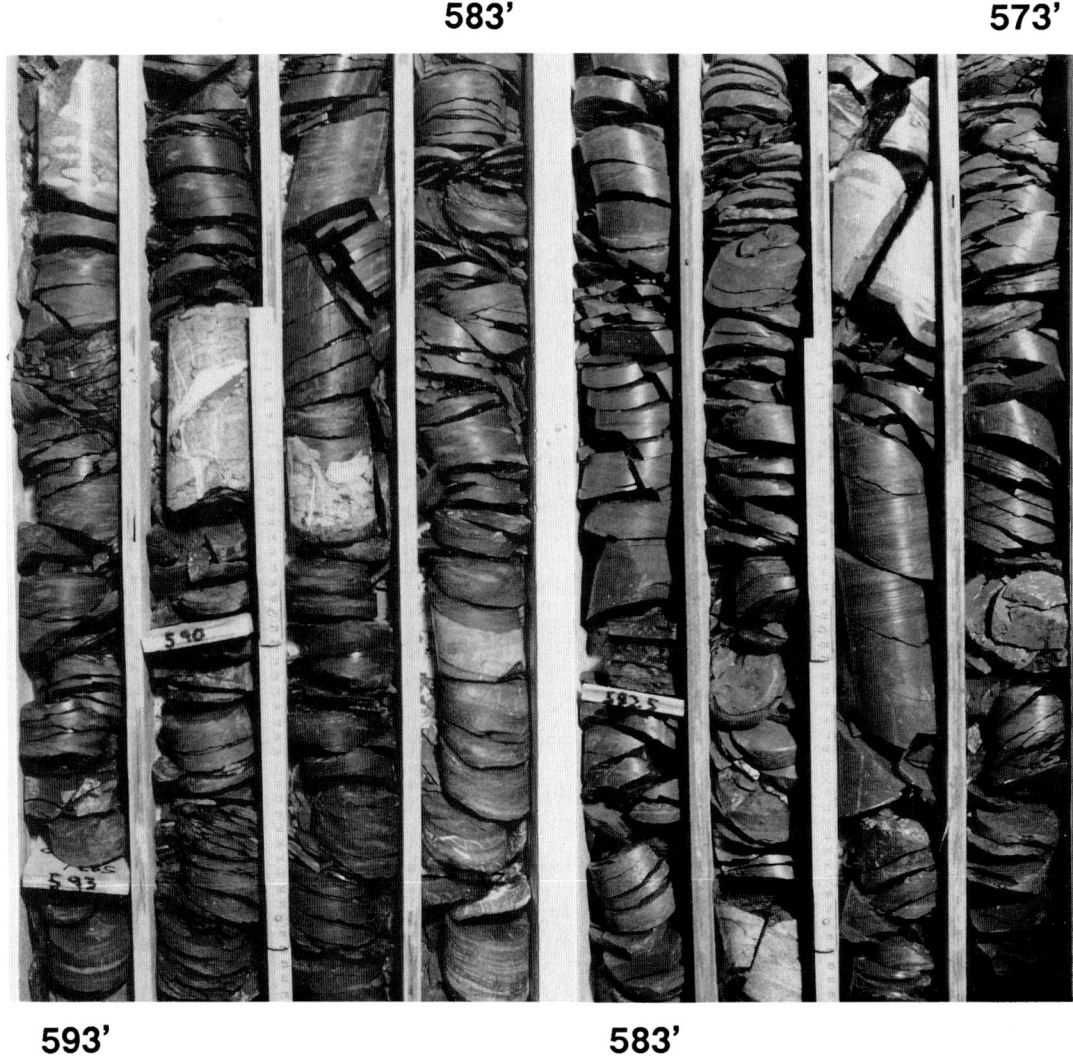

Figure 4. Intraclastic limestone debris flow deposits and black shale. Fine-grained calcarenite at 584.5 ft with sharp top and base. Glauconite abundant in skeletal limestone calcarenite at 575.5 ft. GW134, 593-573 ft.

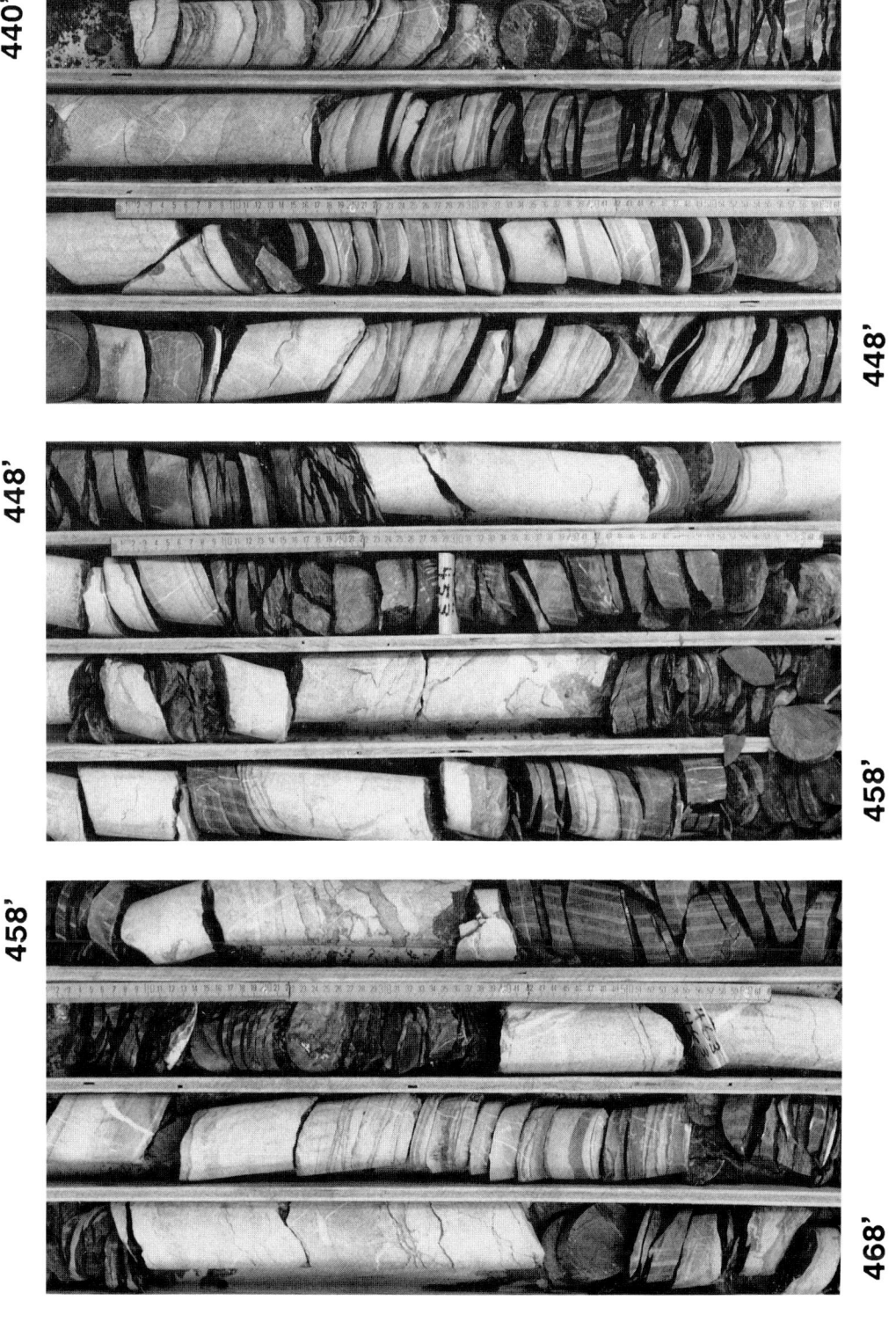

Figure 5. Middle Nolichucky Shale characterized by fining-upward sequences of thick oolitic and skeletal limestone overlying calcareous siltstone and skeletal calcarenite. Well developed fining-upward sequences at 464-462 ft. Intraclastic limestone at 458.5 and 454 ft. GW134, 468-440 ft.

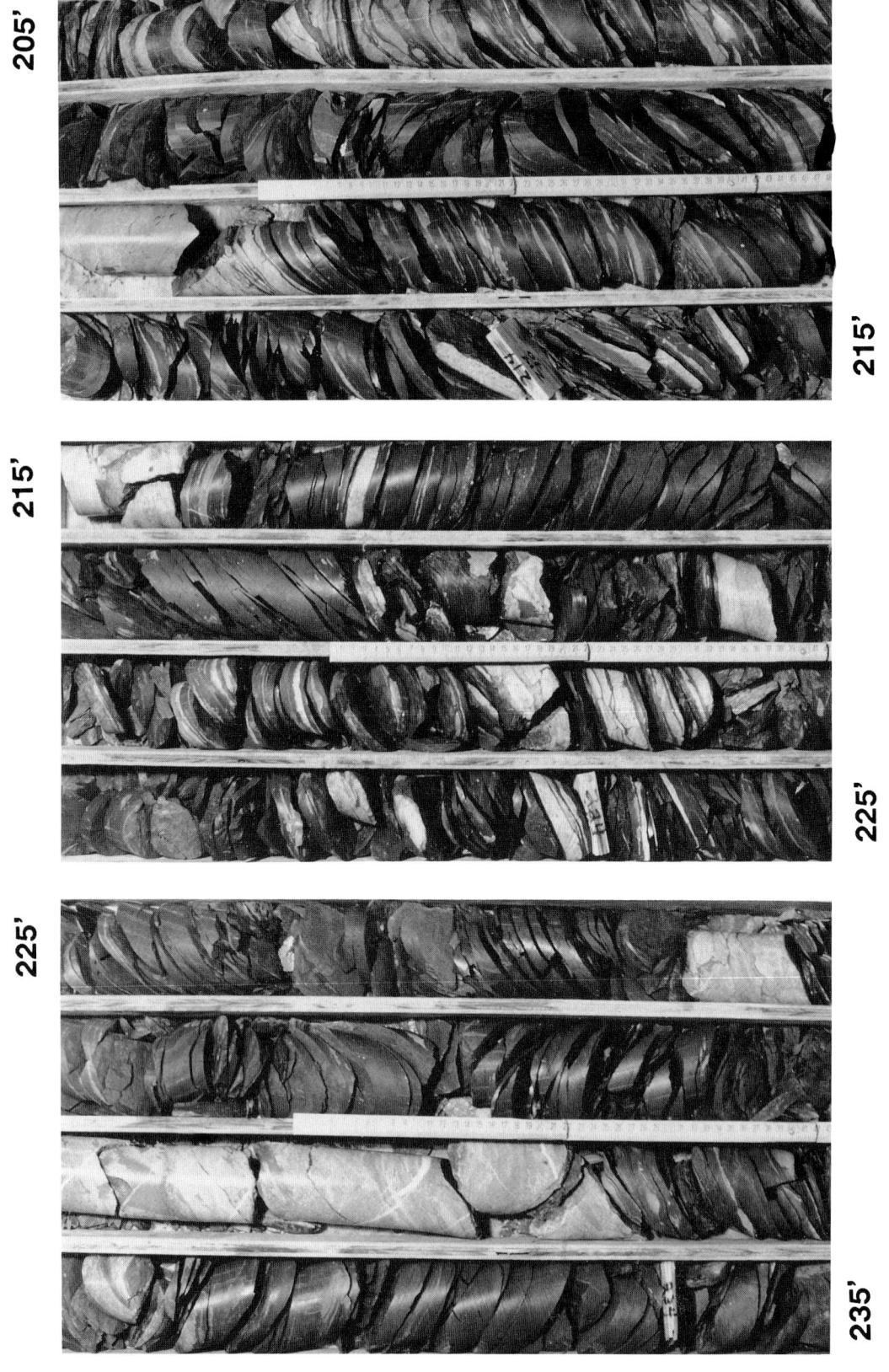

Figure 6. Dark brown-black shale with thicker intraclastic limestone and thin skeletal calcarenite beds. Intraclastic limestone clasts include lime mudstone and laminated peloidal limestone. GW134, 235-205 ft.

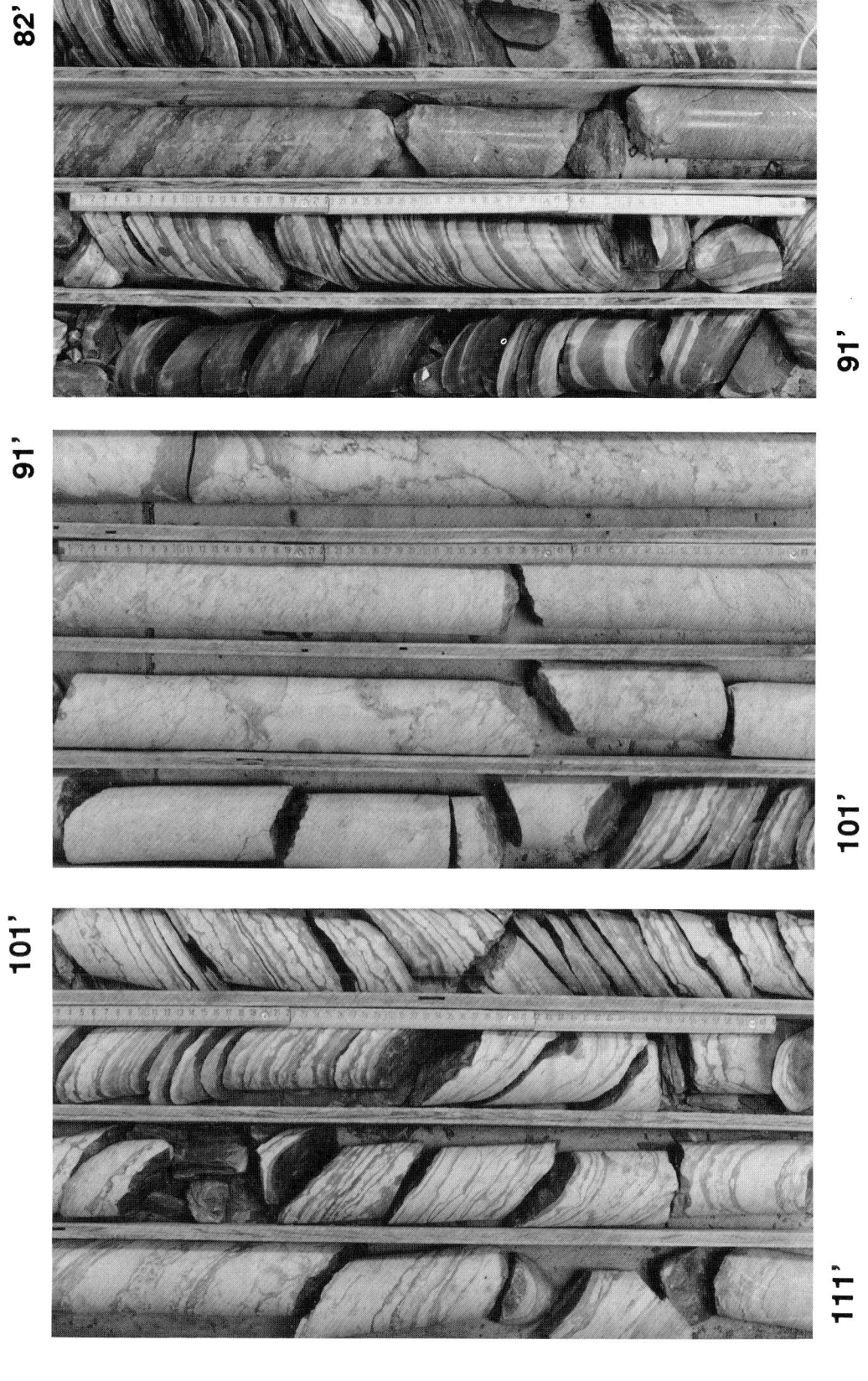

Figure 7. Wavy-bedded lime mudstone("ribbon-rock") grading up into thrombolitic limestone (100-91 ft). Oncolitic limestone from 87.5-84 ft. Bedding-parallel lime mud-filled burrows common from 107-101.5 ft. GW134, 111-101 ft.

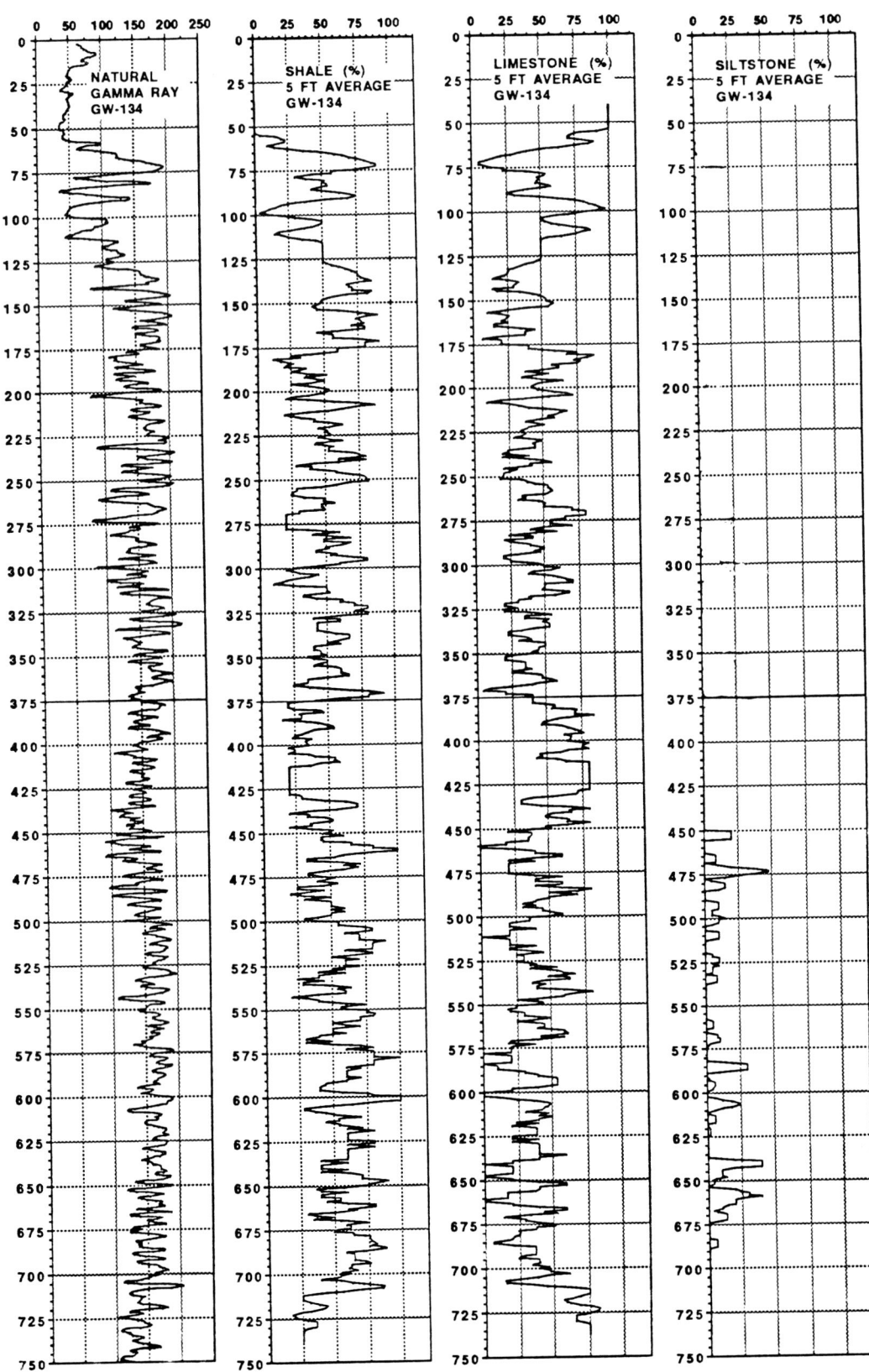

Figure 8. Natural gamma ray log for GW134 with corresponding shale, limestone, and siltstone relative abundance from detailed core descriptions. Lithologic logs were produced from semi-quantitative descriptions of core at 1 ft sample intervals. Upper Nolichucky contact at 55 ft; lower contact at 710 ft. Shallowing-upward sequences seen in gamma ray log at 500-450 ft, 375-300 ft, and 175-50 ft.

NOLICHUCKY SHALE
BEAR CREEK VALLEY

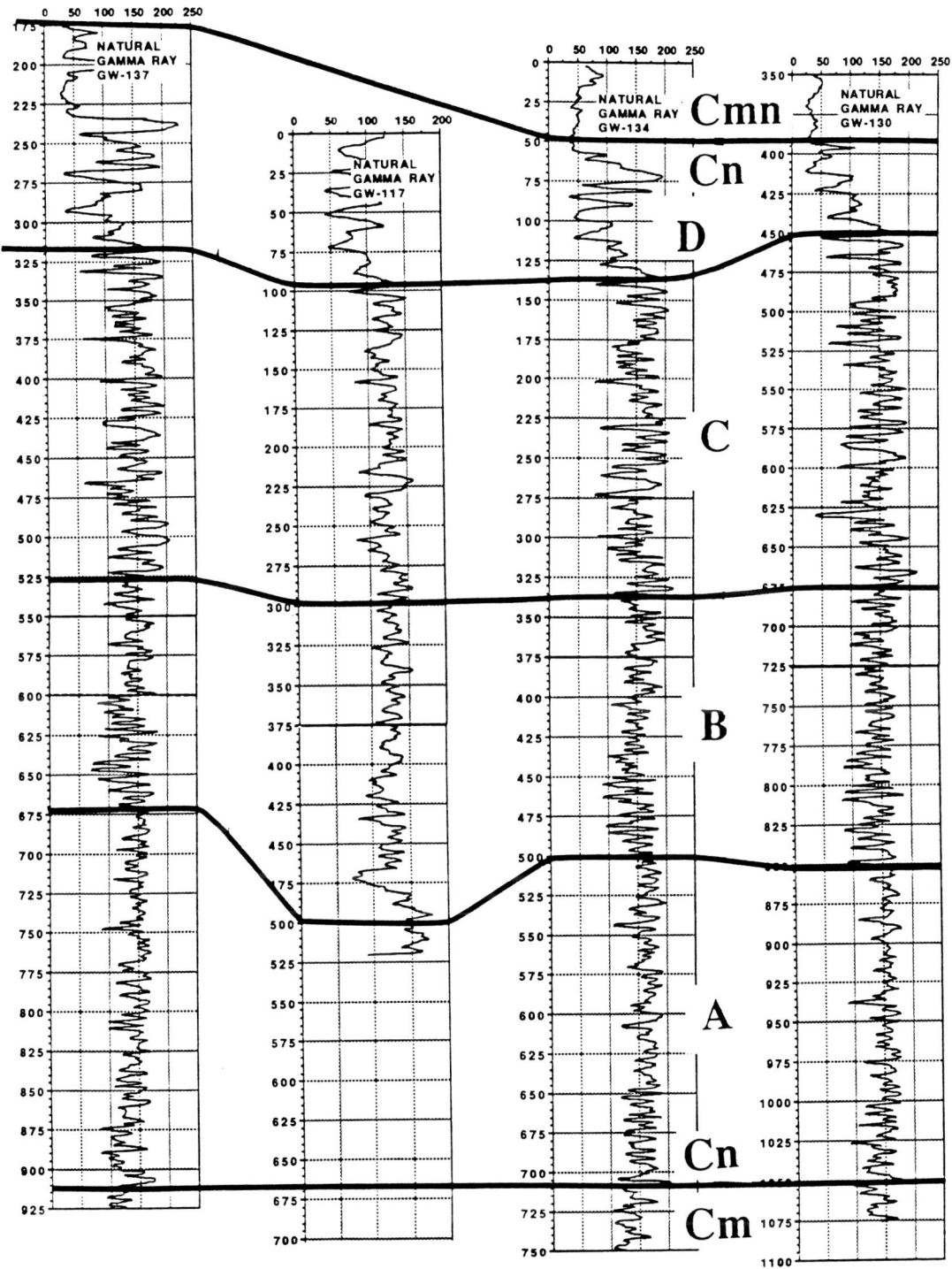

Figure 9. Natural gamma ray log correlations in the Nolichucky for boreholes in WOM thrust sheet. A,B,C, and D interval correlations based on baseline gamma ray shifts and correlation of single beds. Southwest (GW137) to northeast (GW130) transect is parallel to strike. Borehole locations in Figure 1.

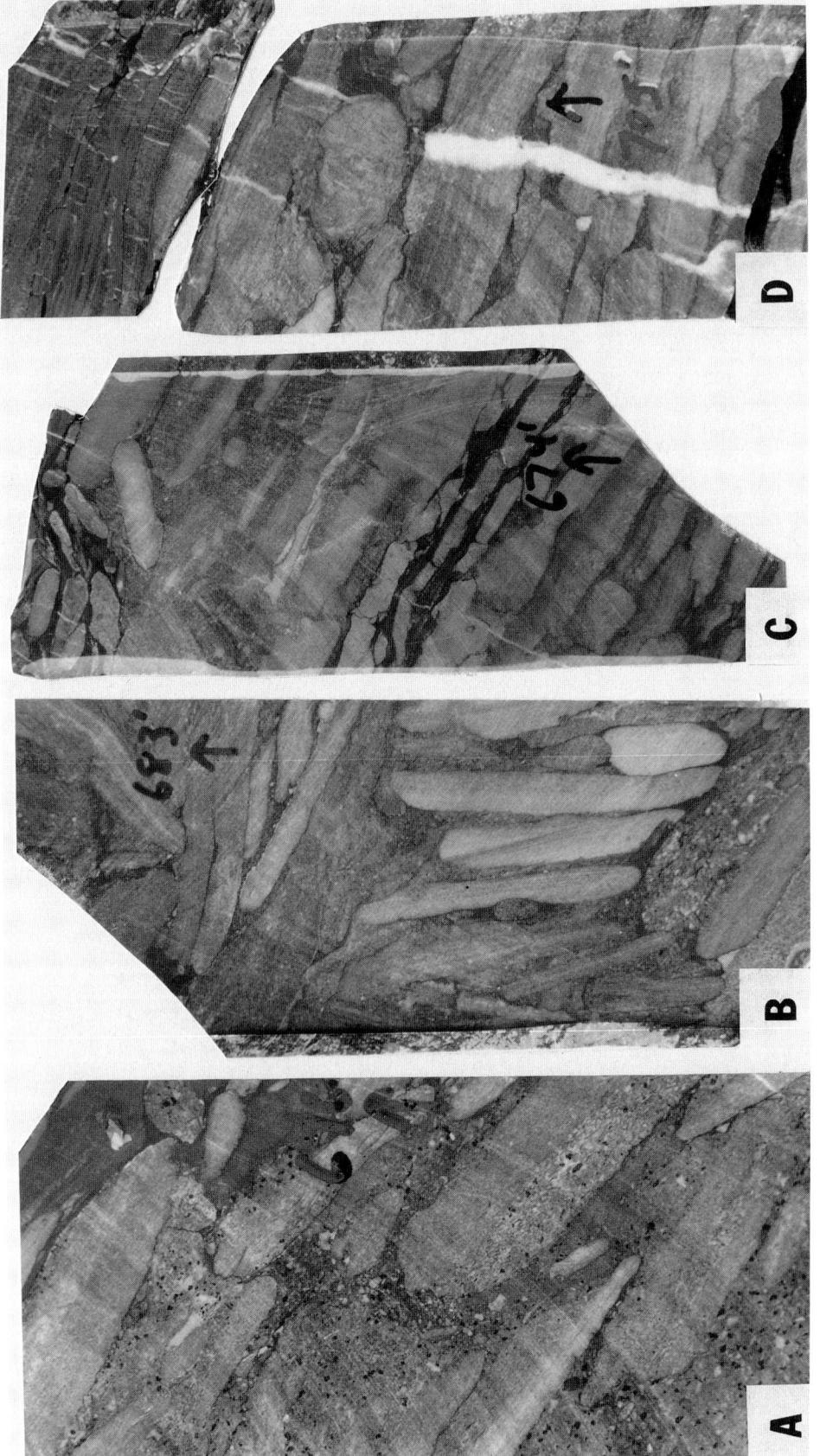

Figure 10. Intraclastic limestone in slabbed core (GW134). A. Matrix-supported conglomerate with peloidal and skeletal limestone clasts. Dark specks are glauconite. Width of photo-1.75 in., 676 ft. B. Edgewise limestone conglomerate with clasts of lime mudstone (light-colored) and laminated peloidal siltstone and laminated peloidal limestone. Width of photo-2.5 in., 683 ft. C. Pseudo-breccia at base consisting of laminated peloidal limestone interbedded with shale, overlain by matrix-supported intraclastic limestone with lime mudstone, peloidal limestone, and calcareous siltstone clasts. width of photo-2.5 in., 674 ft. D. Clast-supported intraclastic limestone consisting of laminated peloidal limestone clasts, overlain by shale with stringers of glauconite. width of photo-1.75 in., 705 ft. Labels on core are positioned at base of core in this and subsequent figures.

Figure 11. Outcrop photos from Dismal Gap, TN (see Fig. 1 for exact location). A. Matrix-supported polymicitic intraclastic limestone near base of Nolichucky. Clasts include calcareous siltstone, peloidal limestone, and lime mudstone. Dime coin for scale in all photos. B. Intraclastic limestone bed with "fanned clasts" (near coin). Most clasts are calcareous siltstone. C. Micro-hummocky cross-stratified siltstone with climbing ripple above coin. D. Slabbed bed of micro-hummocky cross-stratified siltstone showing scour of underlying parallel-laminated siltstone. Bed is 5 cm thick.

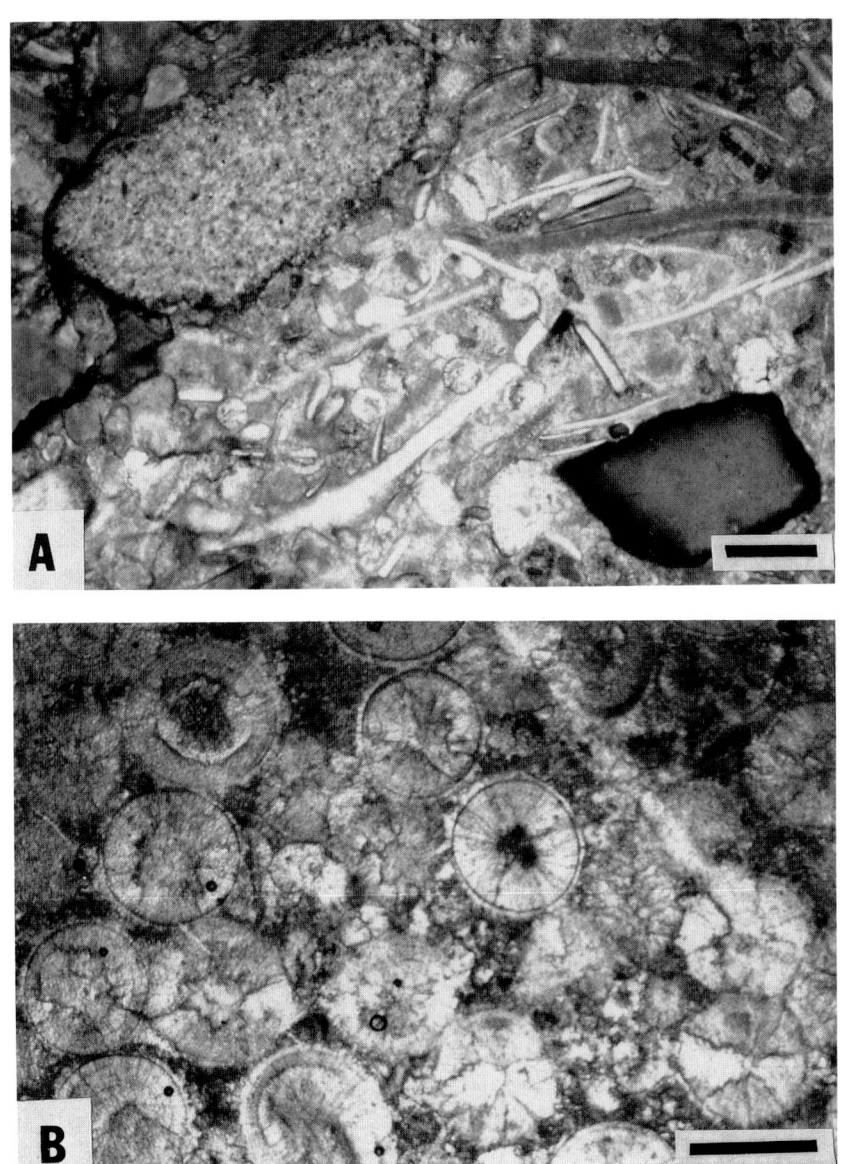

Figure 12. Photomicrographs of core from WOM sheet. A. Skeletal packstone containing trilobites, ooids, echinoderms, and clasts of peloidal ls (upper left) and algal? lime mudstone (lower right), 385 ft, GW134. Scale-1 mm. B. Oolitic packstone showing radial-fibrous ooids and peloids in matrix, 1195 ft, GW140. Scale-1 mm.

consisting of lime mudstone, fossiliferous and peloidal packstone, and calcareous siltstone. From the base of the Nolichucky (709 ft (218 m), GW134) up to approximately 575 ft (175 m)(GW134), monomictic conglomerate becomes more abundant at the expense of the polymictic conglomerate. Clasts in the monomictic conglomerate are predominantly laminated calcareous siltstone, similar in appearance to underlying siltstone that is interbedded with shale. Lime mudstone intraclasts and fossiliferous and peloidal intraclasts become more common above 575 ft (175 m)(GW 134). The lithologies making up these clasts also become more abundant above 575 ft (175 m). In the lower part of the Nolichucky, intraclastic limestone is commonly underlain and overlain by shale (Figs. 3 and 4). Above this interval, intraclastic limestone usually occurs at the top of coarsening-upward sequences (Fig. 5). Across-strike trends in the nature of intraclastic limestone conglomerate, from the WOM sheet to the Copper Creek sheet (west-east), include (1) beds thicker to the east, (2) lime mudstone and skeletal and peloidal limestone clasts more abundant to the east, and (3) calcareous siltstone clasts almost completely disappearing to the east.

Intraclastic limestone: interpretation.--The intraclastic limestone conglomerate has many characteristics of debris flows. These include (1) chaotic internal arrangement of clasts, (2) clasts that are commonly matrix-supported, (3) upper surface containing clasts that project into overlying lithologies, and (4) sheet-like to channel morphology. Numerous examples of intraclastic limestone conglomerate have been attributed in the literature to sediment gravity flows (e.g., Middleton and Hampton, 1973; Cook and Taylor, 1977; Hopkins, 1977; Kozar and others, 1986). Intraclastic limestone is a diagnostic indicator of carbonate slope and basin deposition (McIlreath and James, 1979; Cook and Mullins, 1983; Mullins, 1986). The association of intraclastic limestone with shale, absence of evidence for shallow-water origin (mudcracked lithologies, fenestral limestone, algal limestone), and similarities to published descriptions of intraclastic limestone favor a slope-to-basin setting for the limestone conglomerate in the Nolichucky Shale.

Oolitic limestone: description.--Oolitic packstone is restricted to the upper two-thirds of the Nolichucky (above 500 ft (152 m) in GW134, Fig. 8). Individual beds rarely exceed 1.0 ft (0.3 m) in thickness. This lithology is usually part of coarsening-upward packages and is the coarsest limestone in these packages with the exception of intraclastic limestone. Grading in individual oolitic beds is commonly poorly developed (Fig. 13a); however, oolitic limestone may fine upward from intraclastic limestone (Fig. 13b) to ripple-laminated wackestone caps (Fig. 13c-d). In thin section, oolitic packstone consists of radial-fibrous to radial-concentric ooids in a peloidal matrix (Fig. 12b). Ooid nuclei are typically peloids and micritized skeletal debris. Trilobite and echinoderm fragments are also typical ooid nuclei. Ooids with detrital quartz nuclei do not occur. This indicates that detrital quartz was absent in shallow-water, ooid-forming environments.

Oolitic packstone and grainstone is more abundant, and individual beds are thicker in the Copper Creek thrust sheet (I-75 outcrop) than in cores from the WOM sheet. Two oolitic limestone bodies approximately 20 ft (6 m) thick have been recognized from the middle

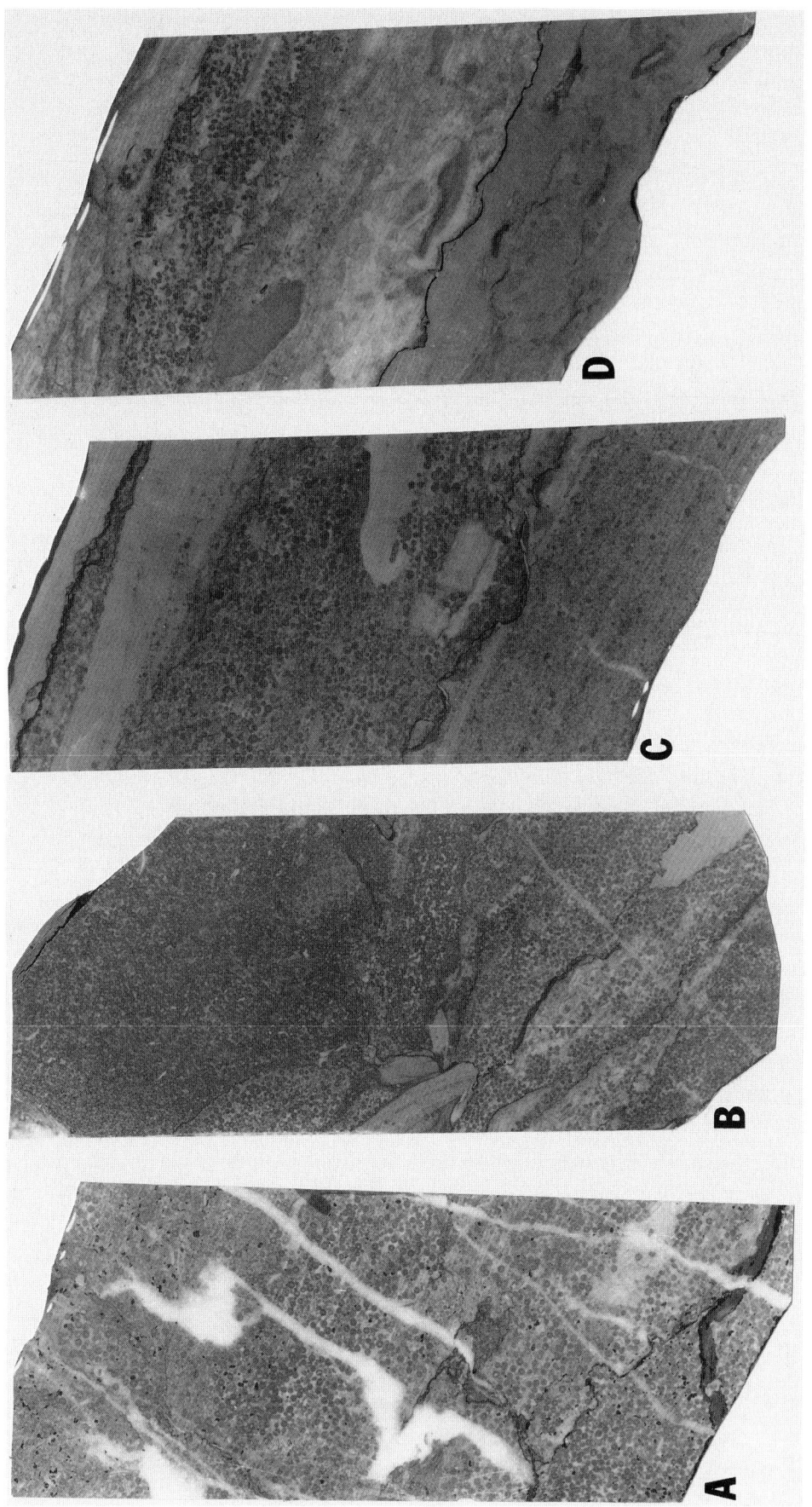

Figure 13. Oolitic limestone in slabbed core from WOM sheet. A. Oolitic packstone composed of oolitic limestone clasts and individual ooids. Bed-normal veins cut from lower left to upper right. width of photo-2 in., 440 ft, GW134. B. Oolitic packstone with lime mudstone and oolitic limestone clasts fining-upward into ooid calcarenite. width of photo-2 in., 183 ft, GW139. C. Ooid calcarenite with clasts at base fining-up into ripple-laminated? lime mudstone. Calcarenite scours into underlying laminated ooid-peloidal packstone. width of photo-1.75 in, 1026 ft, GW140. D. Skeletal limestone with intraclasts grading up into ooid packstone and lime mudstone (top). width of photo-1.75 in., 1111 ft, GW140.

Nolichucky in this outcrop (Weber, 1988). Herringbone cross-stratification is clearly visible in these bodies. Interparticle porosity is nearly completely occluded by fibrous calcite cement of marine origin (Foreman and Weber, 1989).

Oolitic limestone: interpretation.--Oolitic limestone lithologies observed in the WOM thrust sheet are interpreted to be allochthonous for the following reasons: (1) beds are less than 1.0 ft (0.3 m) thick and have sheet-like morphologies; (2) oolitic limestone is associated with slope and basin lithologies (intraclastic limestone and shale); (3) bidirectional cross-stratification, typical of shallow-water, ooid-forming environments, is absent in the WOM cores; and (4) the oolitic limestone shows internal fining-upward fabric. Oolitic limestone is interpreted to be graded calcarenite, the carbonate-equivalent of a siliciclastic turbidite. Carbonate turbidite lithologies are characterized by (1) well-defined bedding (Fig. 5a and b), (2) sharp bases that can be parallel to underlying beds or scour them (Fig. 13b and c), (3) absence of sole marks and load structures, and (4) presence of all or part of the Bouma sequence (Fig. 13b-d; McIlreath and James, 1979). Oolitic calcarenite beds commonly exhibit division A (intraclasts and ooids grading up into ooids) and division B (horizontally laminated ooid packstone) of the Bouma sequence; occasionally division C (ripple-laminated ooid wackestone and lime mudstone) cap the calcarenite. Shale overlies the calcarenite and is equivalent to division E of the Bouma sequence (fine-grained sediment deposited by waning turbidity currents and hemipelagic sedimentation). There is a nearly continuous gradation between graded oolitic calcarenite and oolitic debris-flow and grain-flow deposits (Kozar and others, 1986). This gradation is not unusual because many debris flows can be traced down-slope into turbidite lithologies (Walker, 1979; McIlreath and James, 1979). Oolitic calcarenite is interpreted to have formed by turbidity currents that transported up-slope, ooid sand-sheet deposits into the basin. Ooids in the up-slope sand sheets were swept off the shallow-water oolite shoals by storm currents (Weber, 1988).

Fossiliferous-peloidal limestone: description.--Beds of fossiliferous and peloidal packstone and wackestone, present throughout the Nolichucky Shale, range in thickness from 2.5 in. (6 cm) to less than 0.5 in. (1 cm). In general, thicker beds are characterized by sharp tops and bases (Figs. 5, 14a and c); whereas, thinner beds have sharp bases and diffuse tops (Figs. 5, 12b and d). Internally grading may be absent or fining-upward (Figs. 13d and 14a-d). Thicker beds in the lower Nolichucky commonly have abundant glauconite. In thin section, glauconite is seen to replace peloids and, less commonly, fossils. Thinner beds are laterally discontinuous in core and have lenticular morphologies; inclined current-ripple laminations can be recognized in many of these beds (Figs. 5 and 14b). Fossils include echinoderms, trilobites, and rare phosphatic brachiopods.

Horizontally laminated peloidal packstone with minor fossil debris is an important constituent in the Nolichucky Shale and comprises one of the most common types of clasts as well. The laminae collectively form many fining-upward sequences in individual beds. In Copper Creek thrust sheet outcrops, this lithology commonly exhibits micro-hummocky cross-stratification (MHCS)(Weber, 1988).

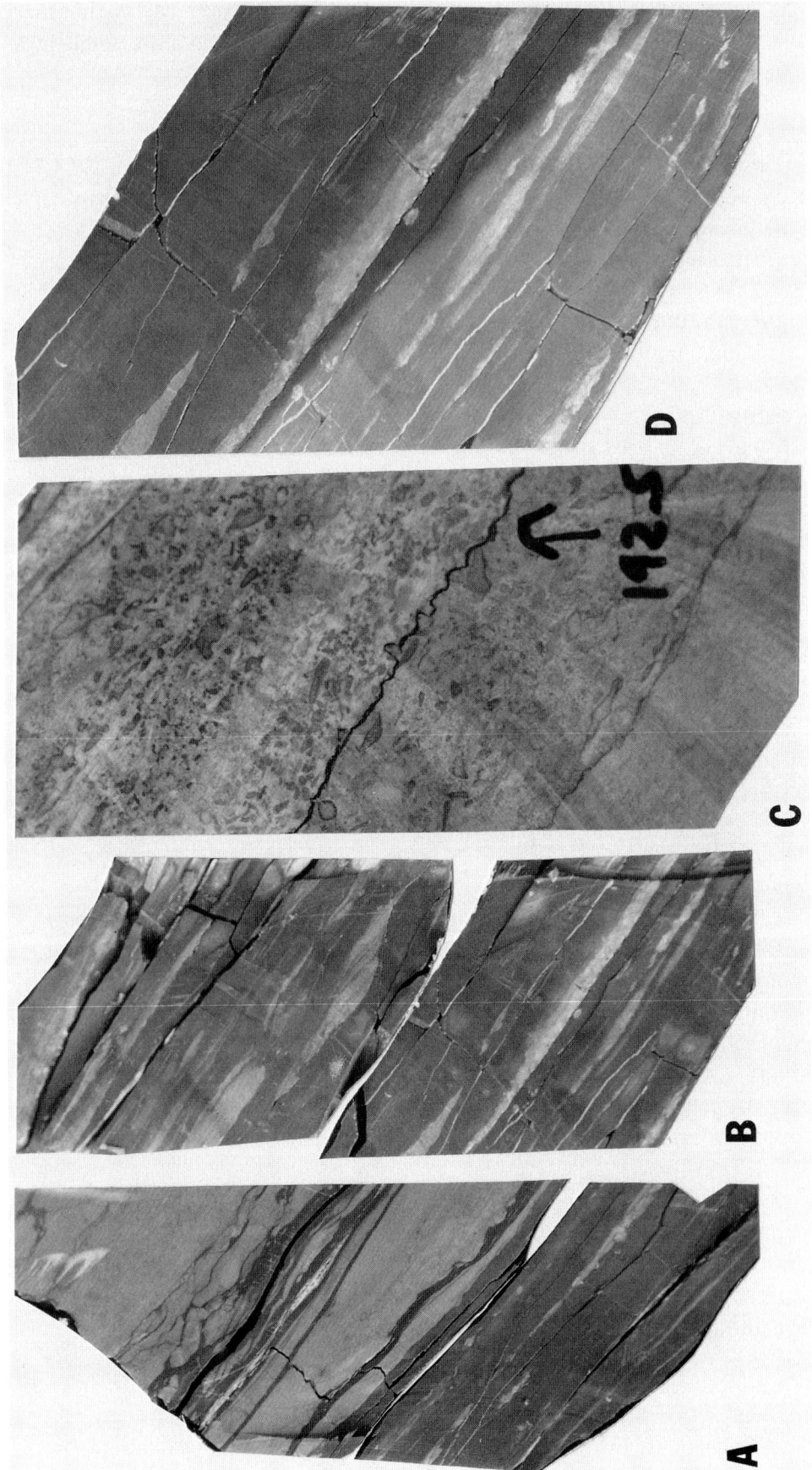

Figure 14. Fining-upward skeletal calcarenite beds in shale from slabbed core (GW134). A. Skeletal packstone grading up into lime mudstone. width of photo-2.0 in., 210.5 ft. B. Skeletal packstone grading up into lime mudstone with diffuse tops. Ripple-laminated beds in upper half of photo may be starved ripples. width of photo-2.0 in., 211 ft. C. Horizontally-laminated skeletal packstone calcarenite with lime mud clasts overlying skeletal wackestone (bottom of photo). width of photo-1.75 in., 192.5 ft. D. enlarged area from lower half of Fig. 14b showing diffuse bed tops and horizontal burrow filled with lime mud in shale (center of photo, just under middle skeletal bed). width of photo-1.75 in.

Fossiliferous-peloidal limestone: interpretation.--Thicker fossiliferous and peloidal limestone beds have many similarities to the oolitic calcarenite interpreted as carbonate turbidite deposits. The horizontally laminated peloidal limestone is interpreted to be a calcarenite with only division B of the Bouma sequence represented. Some fossiliferous grainstone beds, consisting of chaotic, elongate trilobite and echinoderm fragments, were probably deposited as debris or grain flows. Thin beds of fossiliferous limestone grade upward from a fossil lag into ripple-laminated wackestone and mudstone. This lithology is commonly laterally discontinuous and resembles starved ripples. Similarities between the oolitic calcarenite and fossiliferous-peloidal calcarenite favor a slope environment for the latter lithology. Graded calcarenite is a common slope lithology; the nature of the constituent grains depends on the type of sediment available in up-slope settings (McIlreath and James, 1979).

Lime mudstone: description.--Light to medium gray lime mudstone interbedded with shale is most common in the upper Nolichucky (Fig. 7). Individual beds are less than 1 in. (2.5 cm) thick and have irregular tops and bases. Lime mudstone with interbedded shale exhibits a "ribbon-rock" appearance. The proportion of lime mudstone to shale increases toward the top of the Nolichucky Shale and increases across strike from west to east. Lime mudstone grades up into thrombolitic limestone, indicating shallow-water deposition at the top of the Nolichucky (Fig. 7). In thin section, lime mudstone consists of micrite and faint outlines resembling peloids. Laminations are rarely observed. Evidence for bioturbation includes burrow mottles and micrite and fine fossil debris filling bed-parallel burrows (Fig. 7). Lime mudstone occurs as discrete beds or as fining-upward caps on fossiliferous peloidal limestone and oolitic limestone (Figs. 10c and d, 14a and b). Soft-sediment deformation features are common in the lime mudstone beds, exemplified by wavy and contorted bedding, and pseudo-breccias formed by injection of shale into semi-rigid lime mudstone layers. These disrupted mudstone layers are likely sources of mudstone clasts seen in some of the intraclastic limestone beds.

Lime mudstone: interpretation.--A deep-water origin for intervals of lime mudstone interbedded with shale is favored over a shallow-water lagoonal setting, for the following reasons: (1) absence of features typically associated with lagoonal lime mudstone such as fenestrae and desiccation cracks, and (2) interfingering of this lithology with deep-water lithologies such as shale and fossiliferous-peloidal limestone turbidite beds. An exception to the second line of evidence is present at the top of the Nolichucky, where lime mudstone grades into thrombolitic limestone; the thrombolitic limestone is one of the most shallow-water lithologies observed in the Nolichucky Shale (Weber, 1988).

The lime mudstone interbedded with shale probably represents suspension fallout of fine-grained carbonate derived from shallower waters. Storms were a likely mechanism for putting fine-grained carbonate in suspension and spreading it over a large area (Weber, 1988).

Siliciclastics

Shale: description.--In the WOM thrust sheet, shale is the dominant lithology in the Nolichucky (approximate shale/limestone = 1.75). Shale intervals range in thickness from approximately 10 ft (3 m) to less than 1 in. (2.5 cm). Thick shale intervals can be correlated between wells in the WOM and Copper Creek thrust sheets with a high degree of confidence. Dark gray shale is the most common; however, black, dark green, olive green, and brown shale are also present. The darkest shale occurs in intervals where other lithologies are the least abundant (Fig. 4). Adjacent to limestone interbeds, shale color changes to dark green over a thickness of several centimeters. The shale maintains a fairly uniform texture over this interval, suggesting the cause of the color alteration may have been diagenetic. The shale has good fissility, especially in outcrop. Horizontal laminae are well preserved and are defined in thin section by the parallel alignment of clay mineral flakes and, less commonly, intercalated glauconite and quartz-rich layers that are usually less than 1 to 2 mm thick.

Illite, chlorite, and kaolinite are the dominant clay minerals in the Nolichucky Shale (Lee and others, 1987; Weber, 1988). Other minerals include quartz, potassium feldspar, pyrite, glauconite, ferroan calcite, and ferroan dolomite; organic matter is less than 2.0 wt% (Lee and others, 1987). Dolomite, the dominant carbonate mineral in the shale, occurs as euhedral rhombs with iron-depleted cores and iron-rich rims. The most common fossil debris observed in the shale consists of phosphatic, inarticulate brachiopods and trilobite fragments.

Shale: interpretation.--The association of shale with allochthonous deep-water lithologies supports a deep-water origin for the shale in the Nolichucky. Features in the shale that suggest deposition under disaerobic or anaerobic conditions include (1) lack of bioturbation during deposition of the shale, (2) presence of pyrite indicating low Eh conditions (Berner, 1980), and (3) occurrence of glauconite, also indicating Eh conditions that "are not particularly oxidizing" (Odin and Matter, 1981). In the WOM, shale represents the background sedimentation, with deposition of fine-grained material by hemipelagic fallout. This sedimentation was interrupted by allochthonous carbonate debris flows (intraclastic limestone) and carbonate turbidites (oolitic and fossiliferous-peloidal limestone). The rapid deposition of these allochthonous carbonates commonly resulted in soft-sediment deformation. Contorted bedding, in-place pseudo-breccias, and shale injected into overlying carbonate beds are examples of rapid loading features in the Nolichucky. Exceedingly slow deposition and perhaps current winnowing were responsible for the concentration of glauconite in thin layers in the shale.

Calcareous siltstone: description.--Calcareous siltstone is most common in the lower third of the Nolichucky Shale (Fig. 8). Laterally, calcareous siltstone is absent in the Copper Creek thrust sheet and fairly common throughout the lower half of the Nolichucky at Dismal Gap (WOM thrust sheet, Fig. 1). Calcareous siltstone beds are thicker and more abundant at Dismal Gap than in any of the drill core from the ORR. Calcareous siltstone becomes more abundant to the northwest across strike (Rodgers, 1953). In core, beds are usually less than 4 in. (10 cm) thick and have sharp bases and tops (Fig. 5). Calcareous siltstone occurs in coarsening-upward sequences where it usually occurs at the base of the sequence. Carbonate turbidite and intraclastic limestone debris flows typically overlie this lithology. In core, this lithology can be easily mistaken for laminated, fine-grained, peloidal limestone. Internally, calcareous siltstone is finely laminated; phosphatic brachiopod fragments and coarse mica flakes are interlayered with angular quartz silt and feldspar. These interlayers define fining-upward packages, are laterally continuous, and resemble MHCS in outcrop (Fig. 11c and d). The calcareous siltstone beds have erosive bases and commonly have current-rippled tops (Fig. 11c and d). Bed-parallel burrows, filled with calcareous siltstone, are common at the base of many siltstone beds. These burrows do not penetrate more than 1 to 2 mm into the underlying shale.

The carbonate component of the siltstone reflects detrital and diagenetic contributions. Microspar between silt grains may indicate recrystallization of detrital lime mud. Occasionally coarser carbonate resembling peloids was observed; however, most of the carbonate is diagenetic. Euhedral ferroan dolomite, similar to that observed in the shale, is the most abundant carbonate phase. Poikilitic ferroan calcite is less common.

Calcareous siltstone: interpretation.--The internal horizontal laminae in the calcareous siltstone have the morphology characteristic of MHCS (Fig. 11d). Clearly the number of different mechanisms proposed for MHCS suggests that they are multi-genetic. A discussion of the different models for this sedimentary structure is not appropriate for this paper. However, Weber (1988) has interpreted laminated peloidal limestone in the Nolichucky from Copper Creek thrust sheet outcrops, containing MHCS, to represent storm-related deposition. Similar structures in the calcareous siltstone may suggest that these are also storm deposits. However, current ripples at the top of many calcareous siltstone beds (Fig. 11c and d) indicate a dominance of uni-directional flow during the formation of these ripples. The laterally continuous nature of the laminae in the siltstones (Fig. 11c) and absence of shale partings in the laminated siltstone may imply that deposition occurred during a single event. The ripples at the top of many siltstone beds suggest that some of this deposition occurred under uni-directional flow rather than bi-directional flow. This type of deposition may represent a form of antidune stratification in turbidites (Prave and Duke, 1990). Though hummocky cross-stratification is usually considered an indicator of shoreface and shelf environments (Prave and Duke, 1990), a deep-water origin for the calcareous siltstone MHCS units is favored because of its association with deep-water shale, carbonate turbidites, and debris flows. These siltstone units are tentatively classified as incomplete BCE Bouma turbidite deposits.

The lateral trends observed in the calcareous siltstone that indicate a northwest source for the siltstone are the following: (1) siltstone beds become thicker and more abundant to the northwest, (2) absence of siltstone in shallow-water carbonates to the east, and (3) current-ripple orientations on the tops of some siltstone beds that indicate a southeast-directed current (Fig. 11c).

DISCUSSION

Slope-to-Basin Depositional Model

Pronounced lithologic differences recognized in the Nolichucky Shale, between outcrop and core in the WOM and Copper Creek sheets, indicate that water depths were considerably deeper to the northwest. The most obvious difference is the increased amount of shale and siltstone at the expense of carbonate in the WOM sheet. All of the carbonate lithologies described previously are thinner in the WOM sheet than in the Copper Creek sheet to the east. In contrast to several massive oolitic limestone and thrombolitic limestone intervals in the Copper Creek sheet, interpreted to be autochthonous shallow-water deposits (Weber, 1988), limestone deposition in the WOM sheet is the result of down-slope movement of shallower-water carbonate by turbidity currents and debris flows. The intraclastic limestone debris flows are thicker to the east (Copper Creek sheet), and clasts are dominated by shallower-water lime mudstone and fossiliferous limestone; calcareous siltstone clasts are absent in intraclastic limestone in the Copper Creek thrust. Calcareous siltstone and shale become more abundant to the northwest, suggesting a northwest source for the siliciclastic sediment.

Estimated Water Depths

With the exception of MHCS siltstone, which may have formed under the influence of storm currents, the majority of Nolichucky deposition in the WOM sheet probably occurred below normal storm wave base. This is indicated by the absence of oscillation ripples on the tops of carbonate beds and the sharp contact between the carbonate (and siltstone) interbeds and overlying shale (Kreisa, 1981; Weber, 1988). Bottom waters were disaerobic to anaerobic and rarely mixed with overlying oxygenated waters, as indicated by (1) absence of bioturbation expressed by fissile shale and laminated siltstone and carbonate, (2) abundance of pyrite in the shale and some limestone units, and (3) abundance of glauconite in the shale and limestone lithologies.

Approximate minimum water depth in the WOM sheet can be calculated using a minimum palinspastic separation between up-slope lithologies in the Copper Creek and down-slope lithologies in the WOM sheet. Assuming a palinspastic separation of 12.5 miles (20 km) and a depositional slope of 1.0°, water depths in the WOM sheet would have been 1200 ft (365 m) greater than in the Copper Creek thrust sheet. A depositional slope of 1° is very conservative considering the abundance of intraclastic limestone debris-flow deposits and calcarenite turbidite deposits. The presence of glauconite also provides estimates of minimum water depths. Present-day marine sediments contain glauconite at water depths of 160 to 1600 ft (50-500 m); however, glauconite is most common on the upper slope and outer shelf between 650 to 1000 ft (200-300 m) water depth (Odin and Matter, 1981). Clearly water depths of several hundred meters existed in the Nolichucky Basin during deposition.

Contrast With the Nolichucky in Southwestern Virginia

Water depths of several hundred meters or more, which we estimate for Nolichucky lithologies in the vicinity of Oak Ridge, Tennessee, are an order of magnitude greater than published estimates for the Nolichucky in southwestern Virginia (several meters to a few tens of meters; Markello and Read, 1981, 1982). There are some major lithologic differences in the Nolichucky Shale between the two areas. First, a middle limestone member, consisting of shoaling-upward cyclic algal limestone, is present in Virginia (Markello and Read, 1981, 1982) but absent in Tennessee. Shallowing-upward carbonate sequences are also recognized in the middle Nolichucky in the WOM sheet (Fig. 9). However, these consist of allochthonous peloidal and oolitic calcarenite, and intraclastic limestone, indicating deposition from turbidity currents. The presence of this middle limestone member in the Nolichucky in Virginia forced Markello and Read (1981, 1982) to constrain maximum water depths in the intracratonic basin. Second, polymictic conglomerate deposits composed of up-slope lime mudstone and down-slope peloidal limestone and calcareous siltstone occur in the Nolichucky in the WOM sheet, particularly at the base of the Nolichucky. This intraclastic limestone indicates considerable mixing of lithologies during transport. In contrast, the majority of conglomeratic limestone in Virginia contains clasts with similar lithologies to the underlying beds; the similarity of clasts to the underlying lithologies indicates limited down-slope transport on extremely low depositional slopes (Markello and Read, 1981). Finally, cyclic carbonate peritidal facies have not been observed from the Nolichucky in eastern Tennessee (Weber, 1988), in contrast to well-developed cyclic carbonate units in Virginia (Markello and Read, 1981; Read, 1989).

Stratigraphic Model for the Conasauga Group and Nolichucky Shale

Based on our analysis of ORR cores, and outcrop studies by Simmons (1984), Kozar (1986), Weber (1988), and Srinivasan (in preparation), we have developed a preliminary, predictive, depositional model for the Conasauga Group in eastern Tennessee (Fig. 15). The model is based on several key observations concerning the depositional regime and several assumptions about overall depositional patterns. The key observations are as follows:

1. The basin in which the Rogersville and Nolichucky shale formations were deposited varied in water depth from about 100 m to at least 250 m (perhaps substantially more).

2. The carbonate units of the Rutledge and Maryville limestones and the Maynardville Limestone represent progradation of shallow-water platform environments basinward.

3. The lower parts of these carbonate units represent deposition on a very gently sloping ramp at times of the shallowest water depths in the basin.

4. The upper parts of these carbonate units, with the exception of the Maynardville, correspond to times of slope steepening, platform marginal algal buildups, and emplacement of debris-flow, flat-pebble conglomerate, and/or carbonate turbidites in the basin (Srinivasan, in preparation).

5. Thinner, restricted carbonate intervals in the dominantly shale units (e.g., the Nolichucky) represent times of more minor progradations of platform environments (such as oolite shoals) toward the basin, as well as periods of ramp steepening and minor deepening in the basin (see Fig. 15b).

The several assumptions involved in construction of the model are as follows:

1. Times of encroachment of shale onto the upper ramp and platform represent increase in the input of siliciclastic sediment from the northwestern source region.

2. This shale encroachment led to suppression of carbonate production, a retreat of carbonate environments eastward up-slope, and, eventually, establishment of more ramp-like conditions in the area of transition from basin to platform.

3. Times of reduced siliciclastic input led to increased carbonate production in the transition zone, the progradation of shallow-water environments toward the basin, and, eventually, a gradual steepening of the margin (the distally steepened ramp of Read, 1982).

Figure 15. Preliminary model of Conasauga Group stratigraphic relationships. A. Model for the entire Group. Dotted lines are time-lines inferred from stratigraphic relationships; solid lines enclose formation-level units. Datum is the inferred time-line at the top of the Maryville Limestone, which is equivalent to a level about 10 m above the base of the Nolichucky Shale in the ORR area. The part of the succession enlarged in Figure 15b is shown by the dashed rectangle. Note gradual progradation of each carbonate unit toward the basin, and the concomitant conversion of the basin-platform transition from gentle ramp to steeper slope. See text for additional details. B. Stratigraphic model of the Nolichucky Shale and adjacent units in the ORR area, and the inferred transition to the carbonate platform to the east-southeast. Datum is the inferred time-line near the middle of the Nolichucky. Note the smaller-scale platform progradation at the level marked by development of oolite bodies in the transition zones (Copper Creek sheet, I-75 outcrop). Also note how the lithologic succession in the Nolichucky near the ORR (lithologies labeled near left edge of diagram) is in close agreement with the model predictions for the whole group. See text for further discussion.

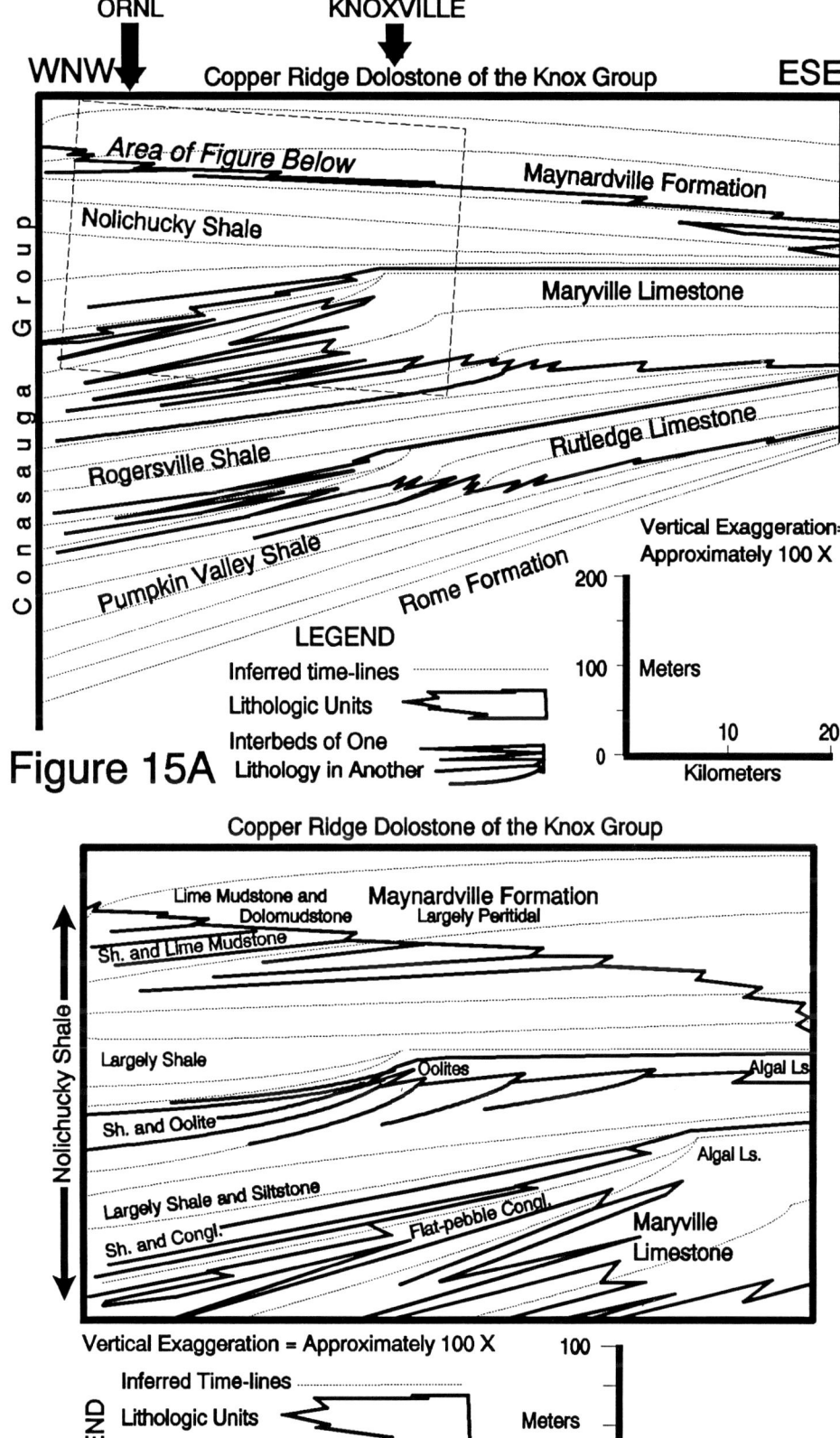

Figure 15A

Figure 15B

Using these observations, assumptions, and unit thickness trends along a west-southwest to east-northeast transect, a cross-section was constructed from the ORR area to northeast Tennessee (Fig. 15). The dotted lines are inferred time-lines (= paleobathymetric profiles). Although they are based on sedimentological evidence, these time-lines are likely more accurate temporally than are biostratigraphically based correlations. In any event, the latter generally agree with the time relationships shown. The time-lines reflect the alternate deepening and shallowing of the basin corresponding to alternate steepening and flattening, respectively, of the transition to the carbonate platform to the east.

Figure 15b shows the Nolichucky-equivalent part of the Conasauga Group enlarged with greater lithologic detail shown (this part of the sequence is shown on Fig. 15a as a dashed rectangle). This figure indicates how the concepts of the model also can be used to explain observations on a smaller scale. Within the shale-dominated parts of the sequence, minor carbonate progradations led at times to establishment of a steeper slope and emplacement of allochthonous carbonate into the basin. Such a situation existed when the middle Nolichucky was being deposited. Oolite shoals developed on the ramp about 7.5 miles (14 km) east of the ORR area (I-75 outcrop). Simultaneously, ooid turbidites were emplaced in the latter area by transportation down the steepened slope. The lithologic composition of the Nolichucky in ORR cores is well explained by logical extension of the larger-scale model.

The controlling influence on the deposition of the entire sequence (Fig. 15a) was changes in siliciclastic input to the basin from its western side, and the effect of increased input as a suppressor of carbonate production in the transition from basin to platform on the east. This resulted in the alternate establishment in the transition zone of true ramp-like geometry and steepened, slope-like geometry. This scenario is the same promulgated by Walker and others (1983) in their "Carbonate Suppression Model" for the development of carbonate to terrigenous-clastic sequences. The control on siliciclastic input is not known, but likely involved sea-level changes. If this is true, suppression of carbonate deposition may have also involved exposure of the platform. We are presently testing this and other hypotheses derived from the model.

CONCLUSIONS

Interbedded limestone, siltstone, and shale of the Nolichucky Shale are the product of slope and basin deposition. Vertical trends indicate an overall shallowing sequence punctuated by numerous smaller-scale, shallowing-upward, and deepening-upward sequences. Dark shale, interbedded with intraclastic limestone and calcareous siltstone, is characteristic of the Nolichucky in the Whiteoak Mountain sheet and represents basinal facies. Maximum water depths in the basin probably ranged from 330 to 1000 ft (100-300 m). These water depths are an order of magnitude greater than those estimated for the Nolichucky basinal lithologies in southwestern Virginia by Markello and Read (1981). Slope facies include shale interbedded with calcarenite, consisting of allochthonous shallow-water carbonate, and intraclastic limestone. Upper-slope facies are predominantly lime mudstone interbedded with shale and calcarenite consisting of shallow-water carbonate grains.

Intraclastic limestone debris-flow deposits and calcarenite turbidite lithologies in the Nolichucky Shale imply that depositional slopes were steep enough to transport shallow-water carbonate at least 12.5 miles (20 km) into the intrashelf basin. Depositional slopes were probably the greatest during deposition of the lower Nolichucky, evidenced by abundant polymictic intraclastic limestone with up-slope-derived lime mudstone clasts and down-slope-derived laminated peloidal limestone and calcareous siltstone. Storms were an important factor for mixing allochthonous carbonate with hemipelagic shale in the intracratonic basin.

ACKNOWLEDGMENTS

This work is part of the senior author's dissertation being completed at the University of Tennessee, Knoxville, while a doctoral research fellow at the Department of Energy, Oak Ridge National Laboratory (ORNL). This fellowship was supervised at ORNL by G. K. Jacobs and funded in part by the ORR Hydrogeology and Geology Studies Project. Additional funding to J. L. Foreman was provided by the Carbonate Research Fund (UT), the UT Discretionary Fund, the Geological Society of America, Appalachian Basin Industrial Associates, and Sigma Xi. K. Srinivasan assisted in the field and provided valuable feedback on many of the ideas presented here.

REFERENCES CITED

COOK, H. E., AND TAYLOR, M. E., 1977, Comparison of continental slope and shelf environments in the Upper Cambrian and lowest Ordovician of Nevada, *in* Cook, H. E., and Enos, P., eds., Deep-Water Carbonate Environments: Society of Economic Paleontologists and Mineralogists Special Publication no. 25, p. 51-81.

COOK, H. E., AND MULLINS, H. T., 1983, Basin margin environment, *in* Scholle, P. A., Bebout, D. G., and Moore, C. H., eds., Carbonate Depositional Environments: American Association of Petroleum Geologists Memoir no. 33, p. 539-617.

FOREMAN, J. L., AND WEBER, L. J., 1989, Diagenesis of an Upper Cambrian oolite, Nolichucky Formation, east Tennessee, *in* Walker, K. R., ed., The Fabric of Cements in Paleozoic Limestones: University of Tennessee Studies in Geology 20, p. 91-104.

HASSON, K. O., AND HAASE, C. S., 1988, Lithofacies and paleogeography of the Conasauga Group (Middle and Late Cambrian) in the Valley and Ridge Province of east Tennessee: Geological Society of America Bulletin, v. 100, p. 234-246.

HOPKINS, J. C., 1977, Production of foreslope breccia by differential submarine cementation and downslope displacement of carbonate sands, *in* Cook, H. E., and Enos, P., eds., Deep-Water Carbonate Environments: Society of Economic Paleontologists and Mineralogists Special Publication no. 25, p. 155-170.

KING, H. L., AND HAASE, C. S., 1987, Sub-surface controlled geologic map for the Y-12 plant and adjacent areas of Bear Creek Valley: ORNL/TM-10112, Martin Marietta Energy Systems, Inc., 21 p.

KOZAR, M. G., 1986, The stratigraphy, petrology, and depositional environments of the Maryville Limestone (Middle Cambrian) in the vicinity of Powell and Oak Ridge, Tennessee: unpublished Master's thesis, University of Tennessee, Knoxville, 242 p.

_____, WEBER, L. J., AND WALKER, K. R., 1986, Transport mechanisms and genesis of limestone-clast conglomerates with examples from the Cambrian of east Tennessee (abstract): American Association of Petroleum Geologists Bulletin, v. 70, p. 609.

KREISA, R. D., 1981, Storm-generated sedimentary structures in the subtidal marine facies, with examples from the Middle and Upper Ordovician of southwest Virginia: Journal of Sedimentary Petrology, v. 51, p. 823-848.

LEE, S. Y., HYDER, L. K., AND ALLEY, P. D., 1987, Mineralogical characterization of selected shale in support of nuclear waste repository studies: ORNL/TM-10567, Martin Marietta Energy Systems, Inc., 84 p.

LEMISZKI, P. J., AND HATCHER, R. D., 1990, Contrasting structural geometries between the Whiteoak Mountain and Copper Creek thrust faults, Bethel Valley Quadrangle, east Tennessee (abstract): Geological Society of America, Abstracts with Programs, v. 22, no. 4, p. 23.

MARKELLO, J. R., 1979, Carbonate ramp to deeper shale shelf transitions of an Upper Cambrian (Dresbachian) shelf embayment, Nolichucky Formation, southwest Virginia: unpublished Master's thesis, Virginia Polytechnic Institute and State University, Blacksburg, 162 p.

MARKELLO, J. R., AND READ, J. F., 1981, Carbonate ramp-to-deeper shale shelf transitions of an Upper Cambrian intrashelf basin, Nolichucky Formation, southwest Virginia Appalachians: Sedimentology, v. 28, p. 573-597.

_____, 1982, Upper Cambrian intrashelf basin, Nolichucky Formation, southwest Virginia Appalachians: American Association of Petroleum Geologists Bulletin, v. 66, p. 860-878.

MCILREATH, I. A., AND JAMES, N. P., 1979, Facies Models 12. Carbonate Slopes, *in* Walker, R. G., ed., Facies Models: Geoscience Canada, Reprint Series 1, p. 133-143.

MIDDLETON, G. V., AND HAMPTON, M. A., 1973, Mechanics of flow and deposition, *in* Middleton, G. V. and Bouma, A. H., eds., Turbidites and Deep Water Sedimentation: Anaheim, California, Society of Economic Paleontologists and Mineralogists Short Course, p. 1-38.

MULLINS, H. T., 1986, Carbonate depositional environments. Part 4: periplatform carbonates: Colorado School of Mines Quarterly, v. 81, no. 2, 63 p.

ODIN, G. S., AND MATTER, A., 1981, De glauconiarum origine: Sedimentology, v. 28, p. 611-641.

PRAVE, A. R., AND DUKE, W. L., 1990, Small-scale hummocky cross-stratification in turbidites: a form of antidune stratification?: Sedimentology, v. 37, p. 531-539.

READ, J. F., 1982, Carbonate platforms of passive (extensional) continental margins: types, characteristics and evolution: Tectonophysics, v. 81, p. 195-212.

_____, 1989, Controls on evolution of Cambrian-Ordovician passive margins, U.S. Appalachians, *in* Crevello, P. D., Wilson, J. L., Sarg, J. F., and Read, J. F., eds., Controls on Carbonate Platform and Basin Development: Society of Economic Paleontologists and Mineralogists Special Publication 44, p. 147-165.

RODGERS, J., 1953, Geologic map of east Tennessee with explanatory text: State of Tennessee, Division of Geology, Bulletin no. 58, Part II, 168 p.

SRINIVASAN, K., Lithofacies analysis of the Middle Cambrian Maryville Limestone, east Tennessee: transition from platform to slope deposition: unpublished Ph.D. dissertation, University of Tennessee, Knoxville (in preparation).

WALKER, K. R., SHANMUGAM, G., AND RUPPEL, S. C., 1983, A model for carbonate to terrigenous clastic sequences: Geological Society of America Bulletin, v. 94, p. 700-712.

WALKER, K. R., FOREMAN, J. L., AND SRINIVASAN, K., 1990, The Cambrian Conasauga Group of eastern Tennessee: a preliminary general stratigraphic model with a more-detailed test for the Nolichucky Formation: Appalachian Basin Industrial Associates, v. 17 (in press).

WALKER, R. G., 1979, Facies models 8. Turbidites and associated coarse clastic deposits, *in* Walker, R. G., ed., Facies Models: Geoscience Canada, Reprint Series 1, p. 91-103.

WEBER, L. J., 1988, Paleoenvironmental analysis and test of stratigraphic cyclicity in the Nolichucky Shale and Maynardville Limestone (Upper Cambrian) in central east Tennessee: unpublished Ph.D. dissertation, University of Tennessee, Knoxville, 389 p.

MIXED CARBONATES AND SILICICLASTICS IN A MISSISSIPPIAN PALEOKARST SETTING, SOUTHWESTERN WYOMING THRUST BELT

J. L. SIEVERDING
Chevron U.S.A., Inc.
Box 1635, Houston, TX 77251
P. M. HARRIS
Chevron Oil Field Research Company
P. O. Box 446, La Habra, CA 90633-0446

ABSTRACT

Cores from the giant Whitney Canyon-Carter Creek gas field in western Wyoming show the nature of a siliciclastic filled karsted carbonate surface in an area with limited outcrop exposure. Regional outcrop studies by others document that a large carbonate platform was subaerially exposed during early Meramecian through Chesterian time across Wyoming, creating a karst surface on the Madison Limestone. From Chesterian to Morrowan time, a west-to-east marine transgression filled solution features with siliciclastic material of the Morgan (Amsden) Formation and equivalents. At Whitney Canyon-Carter Creek Field, the upper Mission Canyon Formation (Madison equivalent) contains this mixed carbonate and siliciclastic facies.

The uppermost 70 to 100 ft (21 to 30 m) of the Mission Canyon at Whitney Canyon-Carter Creek is a correlatable transition zone from preserved evaporite and dolostone beds beneath to the overlying siliciclastics of the Morgan Formation. Cores contain fractured and stylolitized carbonates that underwent solution collapse and brecciation due to dissolution and the subsequent filling of cracks and vugs by siliciclastic sands. Solution-collapse breccias in the cores have a sandstone matrix and dolostone clasts that are rounded, stained to a pinkish hue, locally capped by red clay drapes, and sometimes aligned as if deposited by a current. Breccias interpreted as tectonic in origin have only cement with no matrix and contain angular clasts with matching clast boundaries, i.e., fitted, indicating that fracturing occurred without any extensive dislocation of the clasts.

The mixing of carbonates and siliciclastics within brecciated intervals is significant because these observations suggest that the Mississippian karst plain and associated shoreline extended considerably further west than previously thought, especially when thrust belt structural positions are restored. Evidence presented here suggests that the maximum western shoreline position was at least 20 miles (32.5 km) west of the Utah-Wyoming border.

INTRODUCTION

In southwestern Wyoming, the transition from carbonates of the Mississippian Mission Canyon Formation to siliciclastics of the overlying Mississippian-to-Pennsylvanian Morgan (Amsden) Formation is obscured at the surface because of the structural complexities of the Wyoming thrust belt. Several cores from Whitney Canyon-Carter Creek Field were taken near the Mission

Canyon/Morgan contact and show the nature of this unconventional lithologic mixing. Core slabs and thin sections from two wells in Whitney Canyon-Carter Creek Field were studied. These cores sample only a portion of the interval, but, combined with wireline logs, can be integrated into the regional geologic setting.

GEOLOGIC BACKGROUND

Tectonic Setting

The Wyoming thrust belt is part of the Cordilleran orogenic belt (Armstrong, 1968) that is characterized by large-scale imbricated thrust plates (Fig. 1). A westward-thickening wedge of mostly shallow-water, Precambrian- through Jurassic-age platform rocks were thrust eastward during Early Cretaceous through early Eocene time (Royse and others, 1975). There are four major thrust fault systems; from west to east (and oldest to youngest) they are the Paris-Willard, the Crawford, the Absaroka, and the Hogsback (Darby-Prospect). Palinspastic restorations indicate Absaroka thrust displacement of up to 28 miles (45 km) and total displacement of all thrusts of up to 75 miles (121 km).

The Absaroka thrust sheet Paleozoic section is not exposed in southwestern Wyoming. The closest stratigraphic exposures along depositional strike are in the Salt River Range of north-central Wyoming, approximately 66 miles (106 km) north of Whitney Canyon-Carter Creek Field. The closest exposures along depositional dip are in the Crawford thrust sheet, paleogeographically approximately 42 miles (68 km) west of Whitney Canyon-Carter Creek. Thrust sheet displacements were estimated by palinspastic restoration of a regional cross section by Warner (1982).

Regional Mississippian Paleogeography

During Kinderhookian to Early Meramecian time, carbonate sedimentation occurred on a large cratonic platform that encompassed most of the northern Great Plains, Wyoming, western Colorado, and southeast Utah (Sando, 1976; Rose, 1977; Gutschick and others, 1980; Sandberg and Gutschick, 1980; Sandberg and others, 1982; Gutschick and Sandberg, 1983). The carbonate platform margin trended north-to-south and was located somewhere near the western Wyoming border. Adjacent to the margin was a gentle carbonate slope (1-2°) that graded into a deep, sometimes starved basin (Fig. 2). Although the Antler orogenic highlands trended north-to-south through central Idaho and Nevada, clastic material only sporadically reached the starved basin and never reached the adjacent carbonate slope, possibly because of a submarine rise that restricted transport. Across Wyoming the Madison Limestone was deposited on the carbonate platform. The section thickens to the west where it becomes the Madison Group, consisting of the Mission Canyon and Lodgepole limestones.

Regional uplift started in the Early Meramecian, and the shoreline moved from east to west, exposing subaerially the former depositional carbonate platform. A sabkha, supratidal trend bordered the increasingly exposed carbonate plain (Fig. 3A). The upper part of the Madison

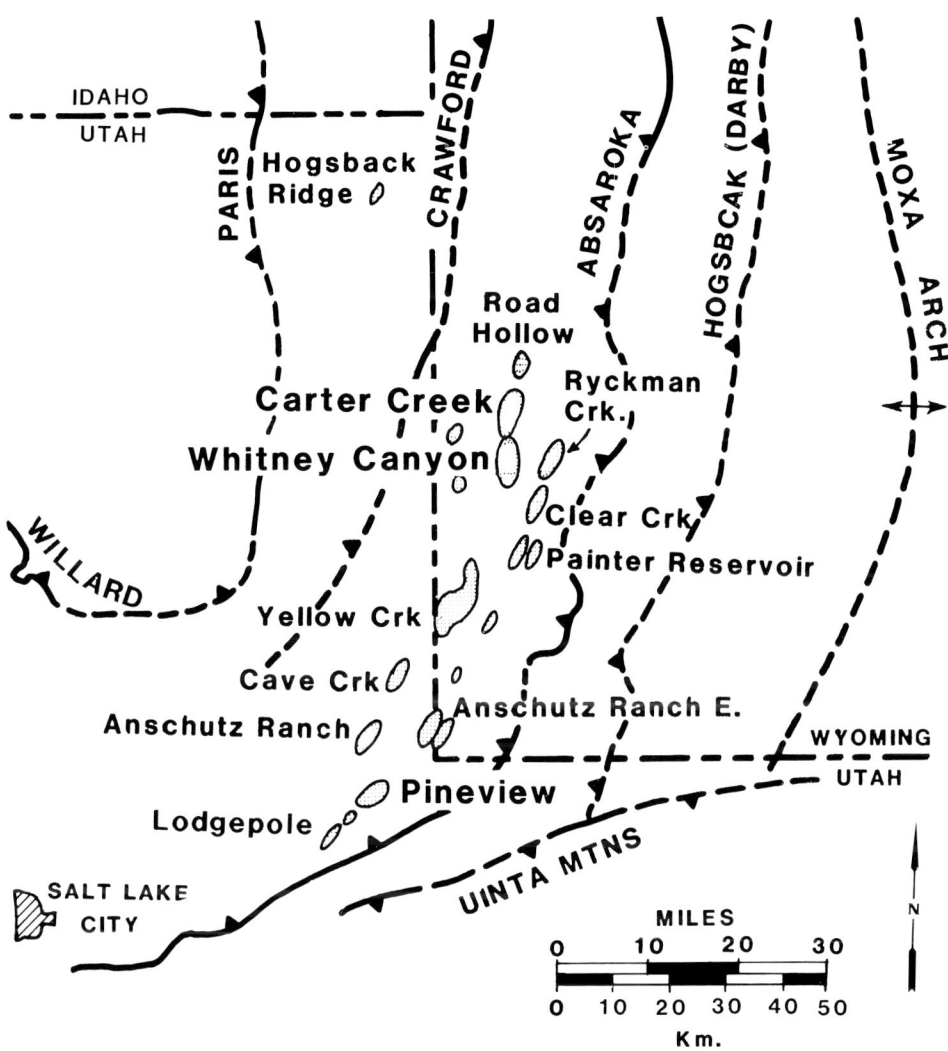

Fig. 1.-- Index map of southwest Wyoming and northern Utah part of the Cordilleran thrust belt showing traces of major thrust faults and location of oil and gas fields.

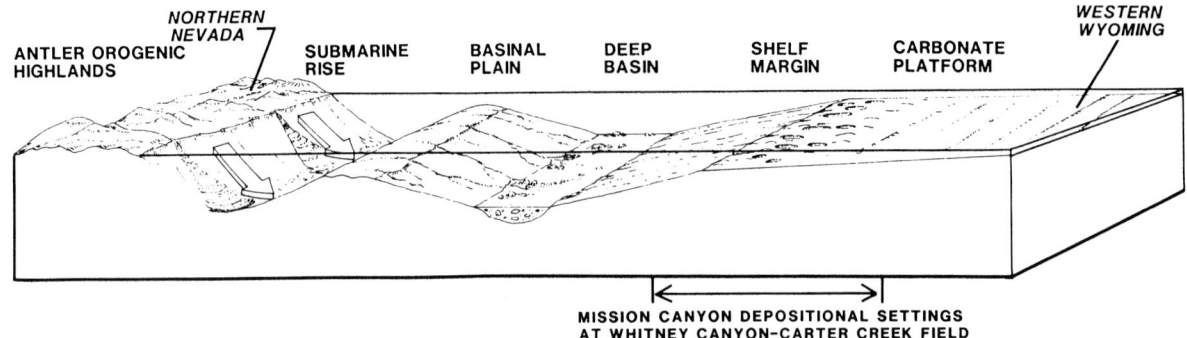

Fig. 2.--Schematic block diagram of interpreted regional geology during Early to Middle Mississippian time along a west-to-east section from northern Nevada to western Wyoming. Mission Canyon depositional settings at Whitney Canyon-Carter Creek field are indicated. Modified from Sandberg and Gutschick, 1980.

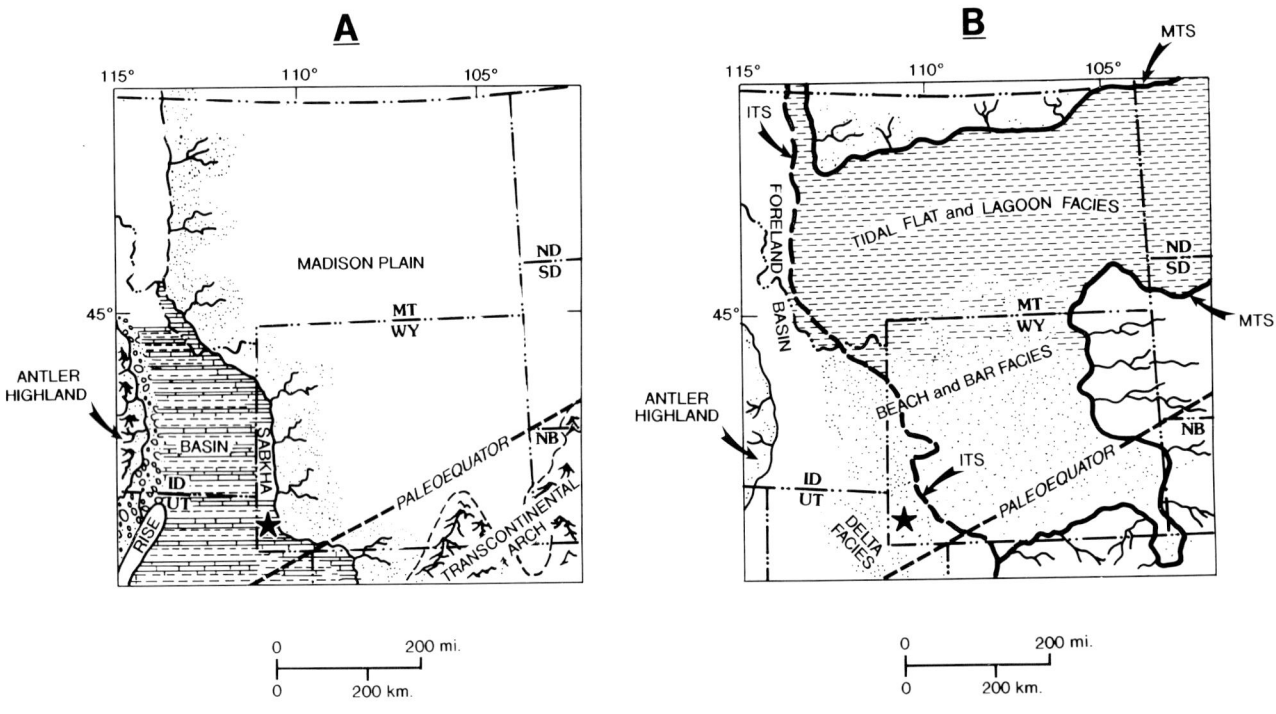

Fig. 3.-- A) Paleogeographic map showing Madison karst plain prior to Morgan/Amsden transgression and B) Map showing distribution of beach and bar facies and tidal flat and lagoon facies of Morgan/Amsden. **ITS** is the initial transgressive strandline and **MTS** is the maximum transgressive strandline. Present day location of Whitney Canyon-Carter Creek field is indicated by a star on both maps. Modified from Sando, 1988.

Limestone (Bull Ridge Member) was deposited as the final regressive facies across Wyoming and contains gypsum deposits, stromatolites, and laterally continuous collapse breccias (Sando, 1968). As subaerial exposure of the carbonate plain continued, karst features developed through evaporite and carbonate solution processes. Features such as enlarged joints, sinkholes, caves, and solution-breccia zones are documented in outcrops from north-central to southeastern Wyoming (Sando, 1988). The infilling by Mississippian and Pennsylvanian terrigenous clastics distinguishes these solution features from any post-Laramide uplift solution features.

During Chesterian through early Morrowan time, a west-to-east marine transgression occurred across Wyoming and deposited siliciclastic material disconformably on Meramecian strata (Fig. 3B). Because the transgression occurred from west to east, the duration of subaerial exposure associated with karsting was greatest in the east and progressively less toward the west. Sando (1988) estimates the maximum duration of karst development at 34 Ma for extreme southeastern Wyoming and interprets nearly constant deposition in westernmost Wyoming. The Darwin Sandstone Member of the Amsden Formation and equivalents were subsequently deposited across the carbonate karst plain. As shown from outcrop studies, karst features were predominantly filled with sand washed into sinkholes and caves by the transgressing shoreline (Gargallo-Quinones, 1985; Sando, 1988; Vice, 1988).

The transition from Meramecian carbonate sedimentation to Chesterian clastic sedimentation implies a significant tectonic event. Detrital material was sourced by the uplift and erosion of the south-central Wyoming and north-central Colorado positive area (ancestral Front Range) and the cratonic transcontinental arch, east of the Front Range (Fig. 3A). The fluvially transported clastics were redeposited as beach, bar, tidal flat, and lagoonal facies by the transgressing sea (Craig, 1972). The preserved thickness of the transgressive clastics is proportional to the topographic relief of the paleokarst surface, which has been estimated to be between 100 and 200 ft (30 and 60 m)(Sando, 1988).

WHITNEY CANYON-CARTER CREEK FIELD

Whitney Canyon-Carter Creek Field is in the Fossil basin area of the Wyoming thrust belt (Fig. 1) and produces mostly sour natural gas from Paleozoic reservoirs. The most significant are dolomitized carbonate reservoirs of the Ordovician Big Horn Dolomite and the Mississippian Mission Canyon and Lodgepole formations (Fig. 4). Hydrocarbons are trapped in large, reverse faulted, anticlinal closures that formed completely within the Absaroka thrust sheet, above the Absaroka fault (Fig. 5). The Absaroka fault cuts the section within the Cambrian; therefore, sandstones, shales, and coals of the Cretaceous (Campanian) Frontier Formation are below the Cambrian Gallatin Limestone. Carter Creek and Whitney Canyon are two separate anticlinal structures, but are considered as one field complex because production testing indicates that they are close to sharing the same gas/water contact in the Mission Canyon Formation main reservoir (Sieverding and Royse, 1990).

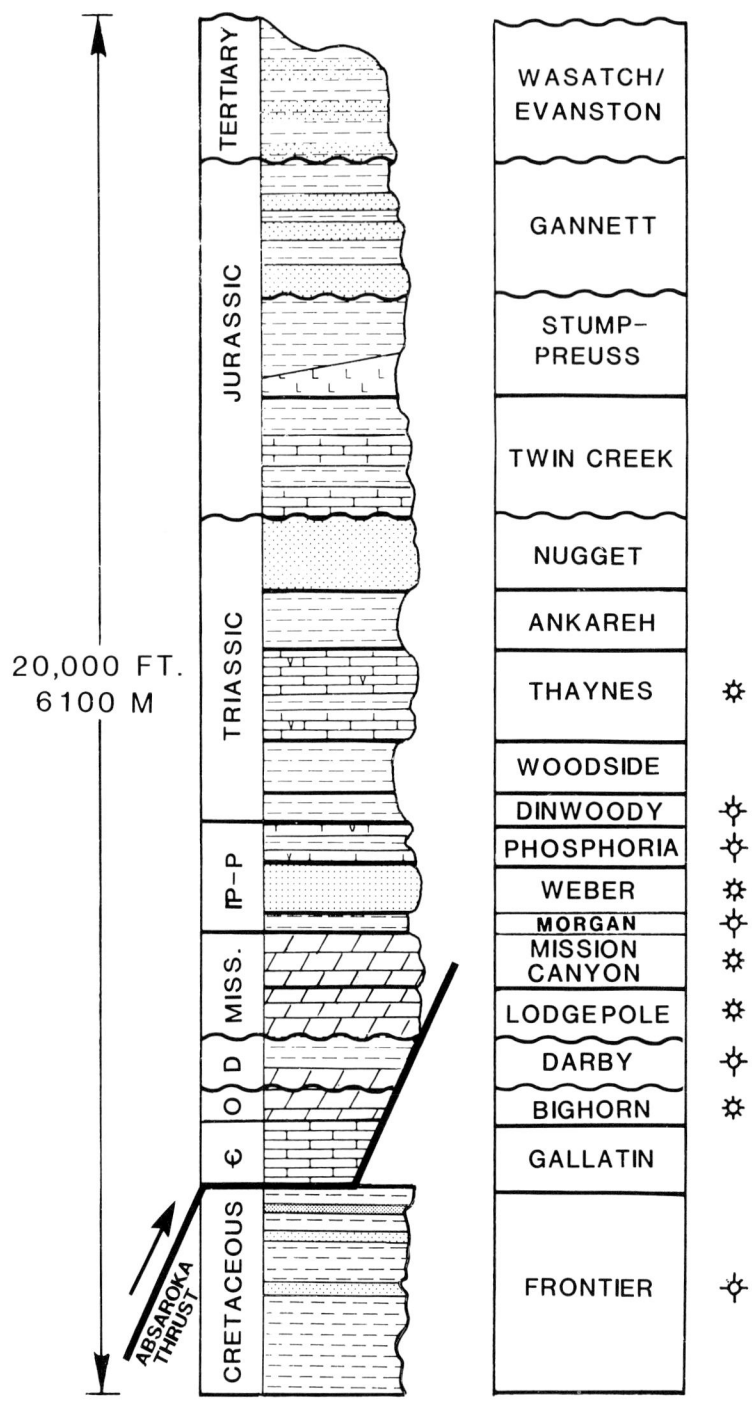

Fig. 4.-- Generalized stratigraphic column for the Whitney Canyon-Carter Creek field area. Productive and potentially productive zones are indicated by gas and gas show symbols. Modified from Hoffman and Kelly, 1981.

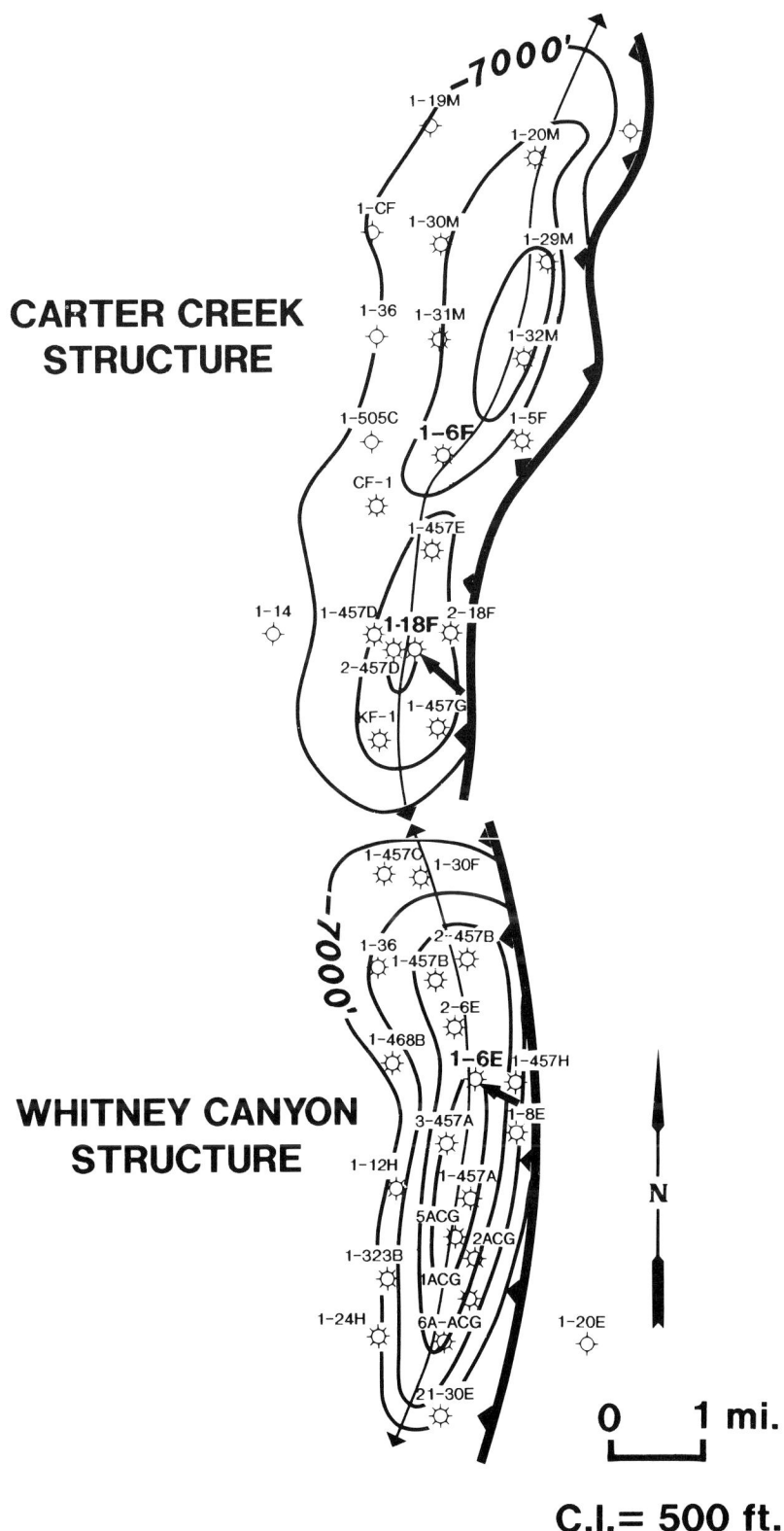

Fig. 5.-- Generalized structure map of Whitney Canyon-Carter Creek field contoured on the top of the Mississippian Mission Canyon Formation. Contours have been smoothed and do not show Chevrons' structural interpretation of the field. Contour interval is 500 ft (150 m). Arrows indicate locations of key wells used in this study.

Mission Canyon Formation

At Whitney Canyon-Carter Creek Field, the Mission Canyon Formation (upper Madison Limestone equivalent) was deposited as an overall shallowing-upward, regressive sequence (Harris and others, 1988). The paleogeographic location was near the Mississippian carbonate platform margin (Fig. 2), and the section indicates shelf margin-to-slope and carbonate shelf deposition. At the base of the Mission Canyon are quiet-water, deeper marine facies. These grade upsection to alternating higher-energy shoals and lower-energy subtidal facies. The upper Mission Canyon is an evaporitic, upper intertidal-to-supratidal mudflat facies (Figs. 6 and 7). Wireline logs indicate that massive anhydrite beds (50+ ft, 15+ m thick) are preserved in all but the uppermost 70 to 100 ft (21 to 30 m) of the upper Mission Canyon (Fig. 8).

The uppermost 70 to 100 ft (21 to 30 m) of the Mission Canyon is a correlatable, yet chaotic unit of mixed carbonate and clastic facies (Figs. 7 and 8). This unit is a transition zone from the preserved massive evaporite and dolostone beds below to the overlying Morgan Formation (Amsden) clastics. Based on both core data and wireline logs, anhydrite is present only as a cement. Wireline log correlations also indicate that the base of the transition zone is a relatively flat horizon, which possibly represents a paleo-groundwater level.

MIXED CARBONATE-SILICICLASTIC LITHOFACIES

Cores of the uppermost Mission Canyon Formation from the 1-6E and 1-18F wells from Whitney Canyon-Carter Creek Field contain fractured and stylolitized carbonates that are intermixed with siliciclastic sands in an unconventional manner. The detailed descriptions of the cores (Figs. 9 and 13) indicate that solution collapse and brecciation of the carbonates created voids that were subsequently filled by siliciclastic sands. There is also evidence that these rocks experienced a considerable tectonic overprint that created tectonic breccias and stylolites. However, several aspects of some brecciated intervals in the cores are definitive criteria for a solution-collapse origin, which corroborates with regional studies of the upper Mission Canyon (Gargallo-Quinones, 1985; Sando, 1988; Vice, 1988).

1-6E Core

Core from the 1-6E well, from 12,477 to 12,503 ft (3802.9 to 3810.9 m), contains prominent intervals of solution-collapse breccia and also sandstone, as well as thinner zones of tectonic breccia and intact dolostone (Fig. 9). The solution-collapse breccia is well developed from 12,488.5 ft (3806.5 m) to the base of the core and also occurs in three thinner intervals higher in the core. This type of breccia is composed of dolostone clasts and minor sandstone clasts that reside within a matrix of sandstone and dolomudstone (Figs. 10-12). The dolostone clasts are commonly a pinkish gray to dark gray, finely crystalline dolomudstone that is often laminated and contains disseminated quartz silt and sand. Locally the clasts are coarsely crystalline dolostone. The matrix around the clasts is dark gray, very fine- to medium-grained, poorly sorted

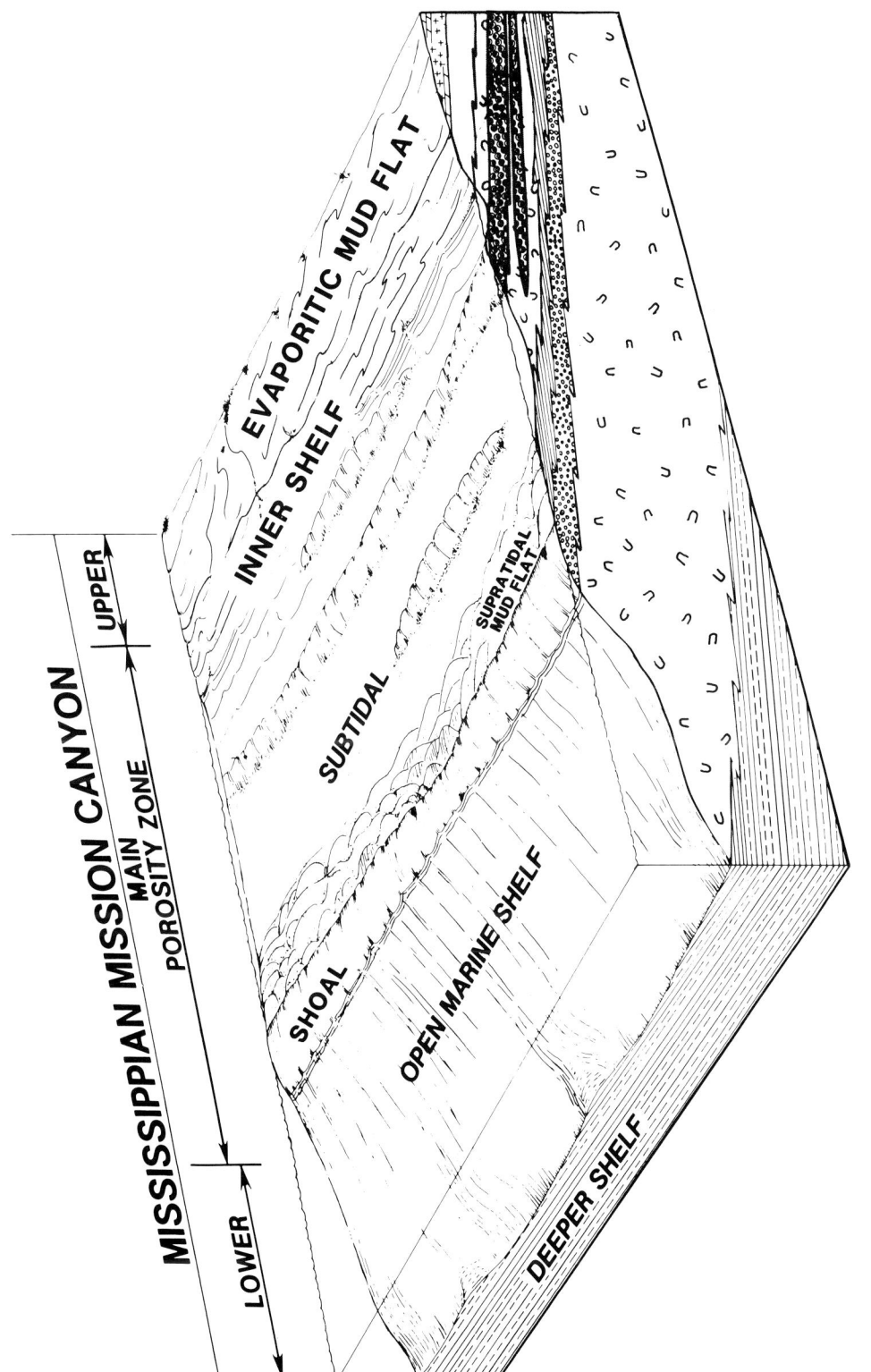

Fig. 6.-- Schematic block diagram of Mission Canyon interpreted depositional facies at Whitney Canyon-Carter Creek field. General depositional environments for the lower Mission Canyon, main porosity zone and Upper Mission Canyon are indicated. Modified from Harris and others, 1988.

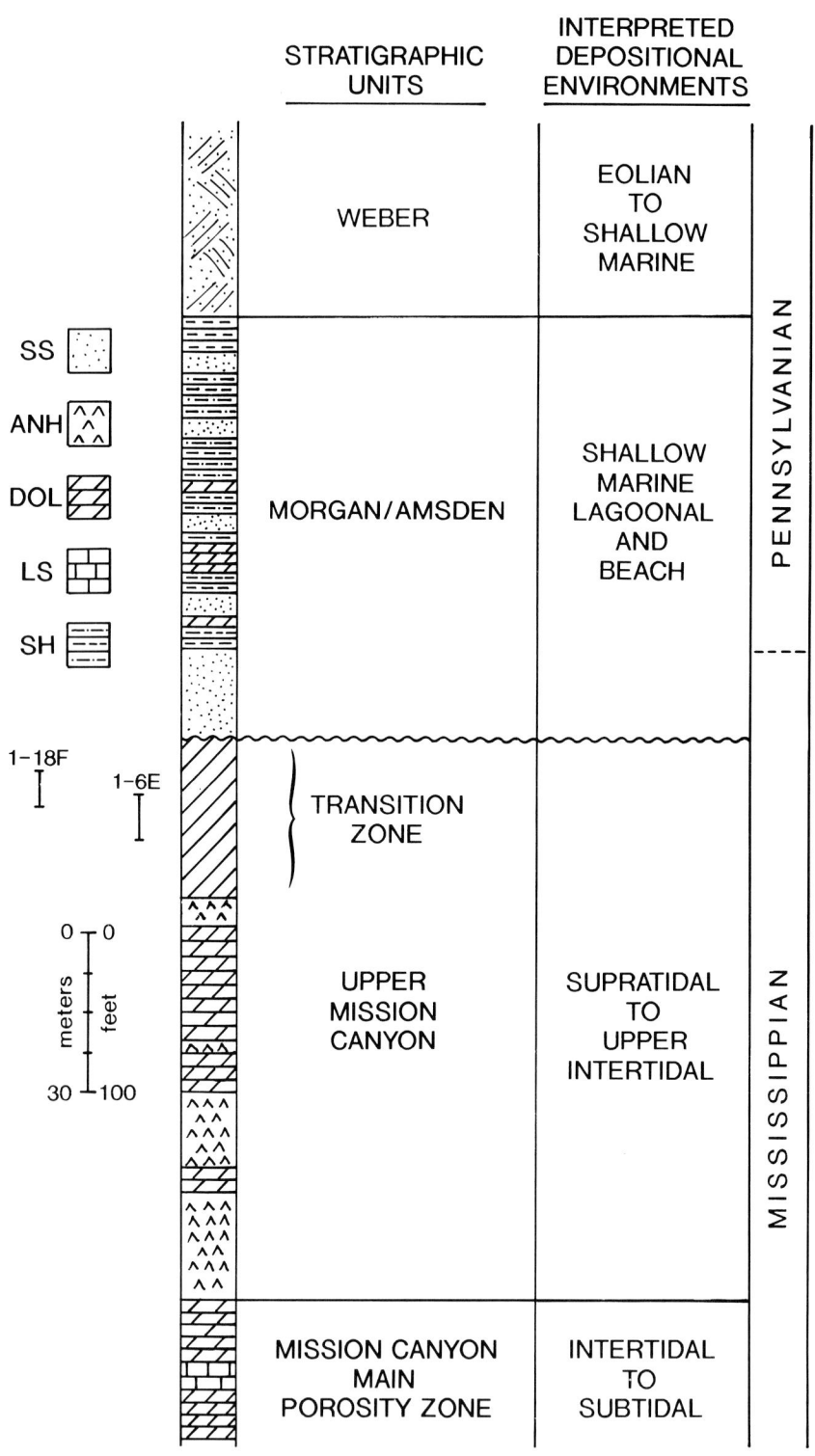

Fig. 7.--Detailed stratigraphic column of the Mississippian to Pennsylvanian transition at Whitney Canyon-Carter Creek field. Approximate stratigraphic location of two cores used in this study are indicated. Core depths could not be accurately adjusted to log depths.

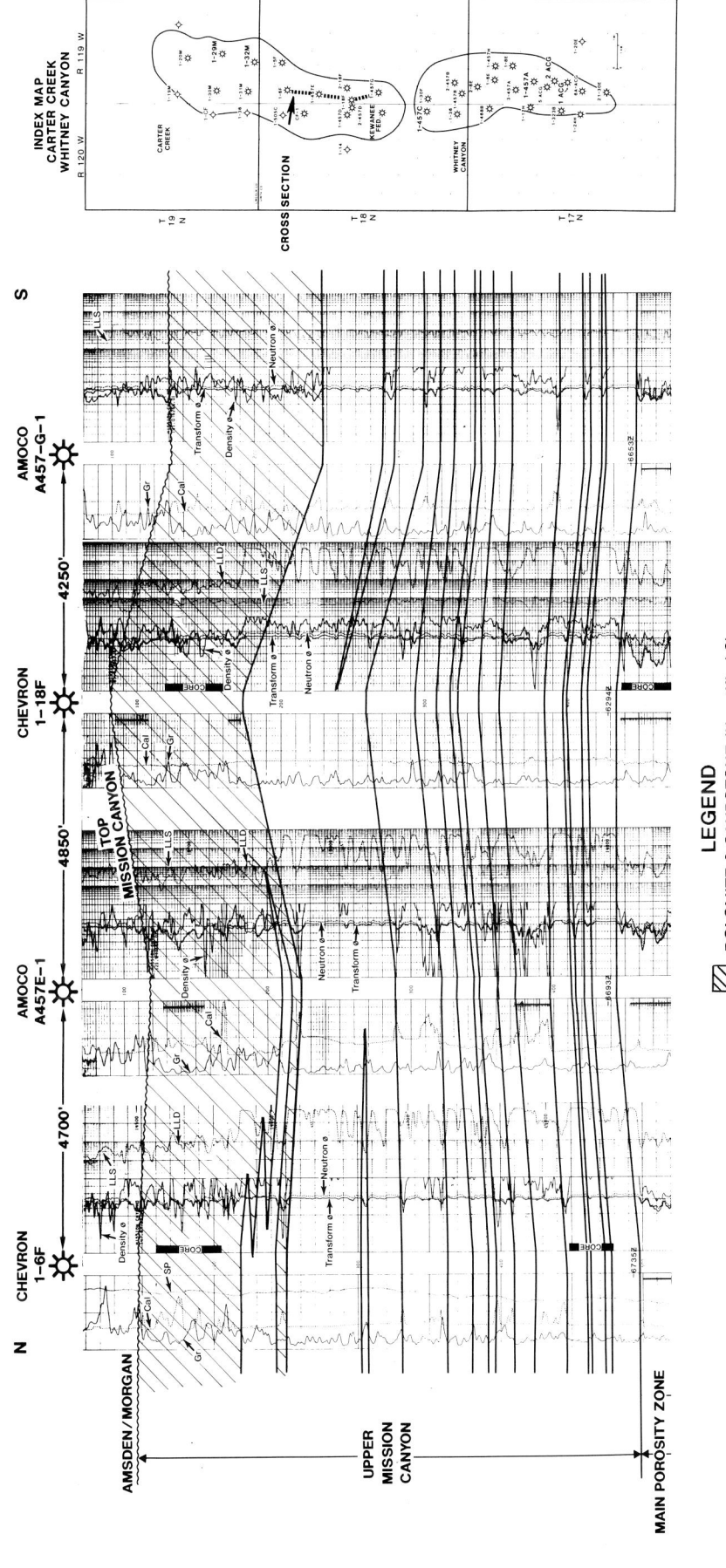

Fig. 8.--True stratigraphic thickness cross-section of the Mississippian Mission Canyon Formation through a portion of Carter Creek field. Well logs show gamma-ray, SP, porosity and resistivity curves and have been computer adjusted for well bore deviation, structural dips, and recognizable faults. Cored intervals are indicated by solid bars on the right side of the logs; perforations are shown on the left. The uppermost Mission Canyon mixed dolomite-sandstone unit is indicated by the hatchured pattern.

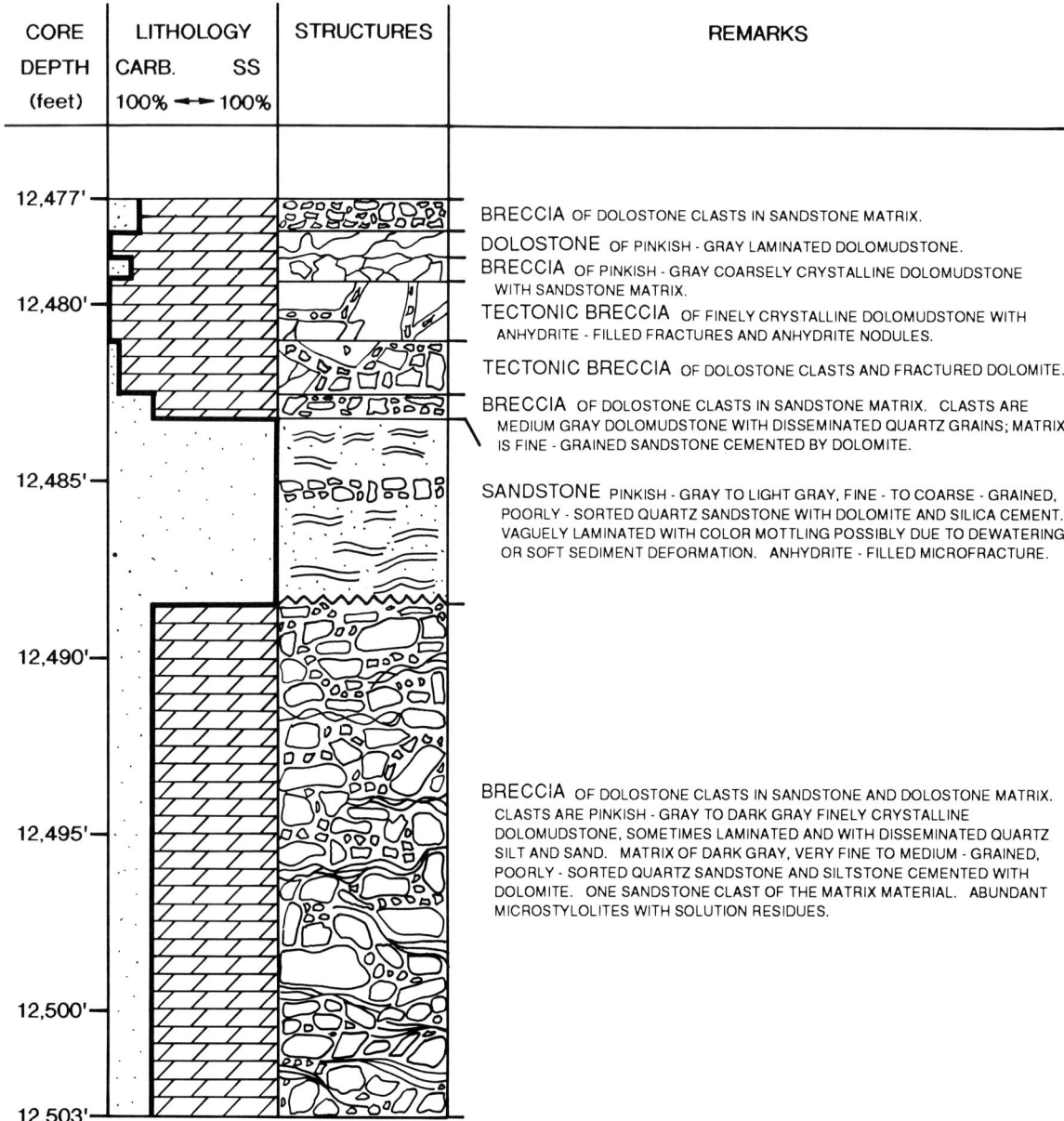

Fig. 9.-- Core description of Chevron 1-6E Federal upper Mission Canyon core (Section 6, T18 N, R119W, Uinta County, Wyoming). Mixing of carbonates and sandstones is shown in the lithology column, and breccia zones are highlighted in the structure column.

Figure 10. Photographs of core from the 1-6E well.

A. Core slab from 12,491 ft (3807.3 m) of solution-collapse breccia containing dolostone and sandstone clasts in a siliciclastic-rich matrix.

B and C. Thin-section photomicrographs of Figure 10A showing clasts of finely crystalline dolomudstone that are vaguely laminated and contain disseminated quartz silt and sand. Matrix is poorly sorted quartz sandstone and siltstone that occurs with a dolomudstone matrix or is cemented by dolomite. Plane polarized light.

Figure 11. Photographs of core from the 1-6E well.

A. Core slab from 12,498 ft (3809.4 m) of solution-collapse breccia that is crosscut by numerous stylolites and pressure-solution seams.

B and C. Thin-section photomicrographs of Figure 11A showing laminated and structureless dolomudstone clasts and a matrix of poorly sorted quartz sandstone and siltstone and dolomudstone. Microfractures crosscut both clasts and matrix. Plane polarized light.

Figure 12. Photographs of core from the 1-6E well.

A. Core slab from 12,499 ft (3809.7 m) of solution-collapse breccia that is crosscut by numerous stylolites.

B and C. Thin-section photomicrographs of Figure 12A showing dolostone clasts in sandstone-siltstone matrix. Clasts are finely crystalline dolomudstone that contains disseminated quartz silt and sand. Locally the clast boundaries are overprinted by stylolites. Plane polarized light.

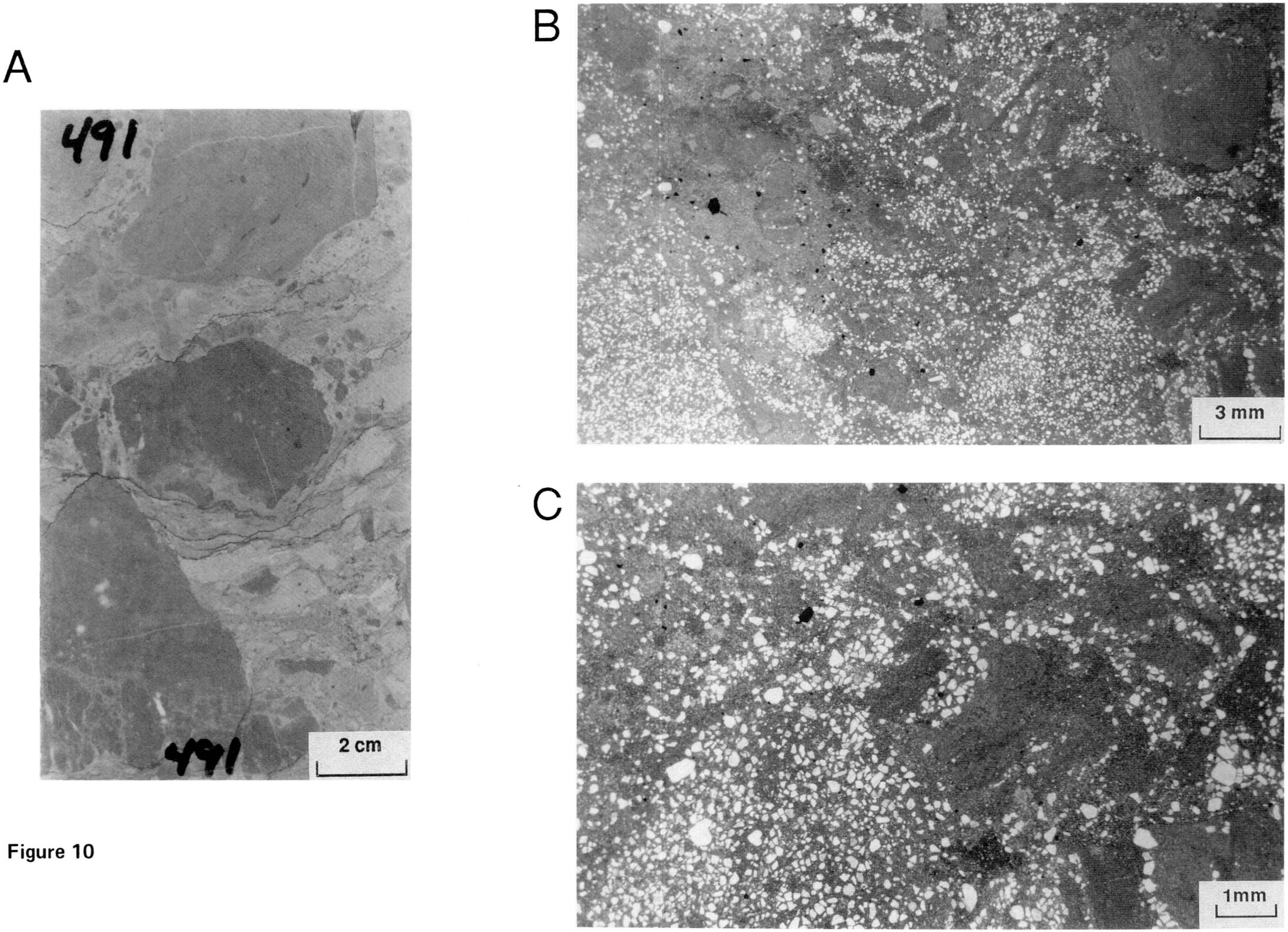

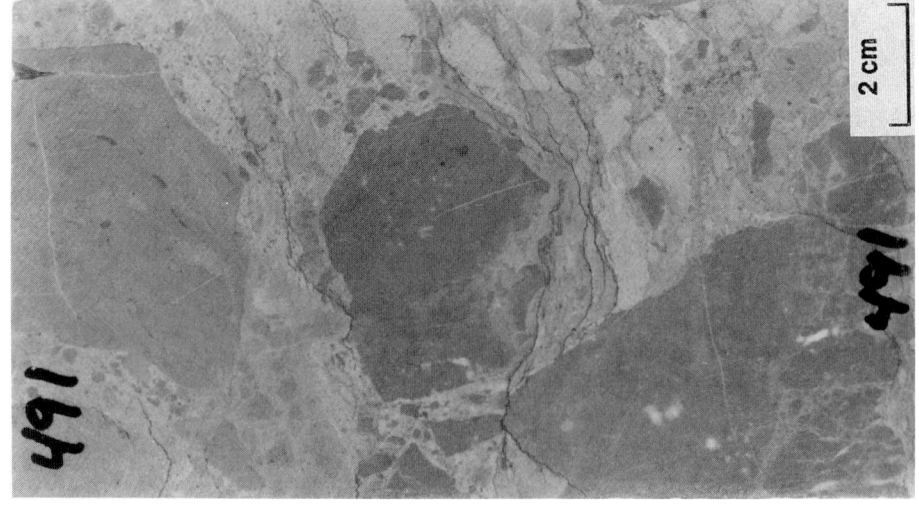

Figure 10

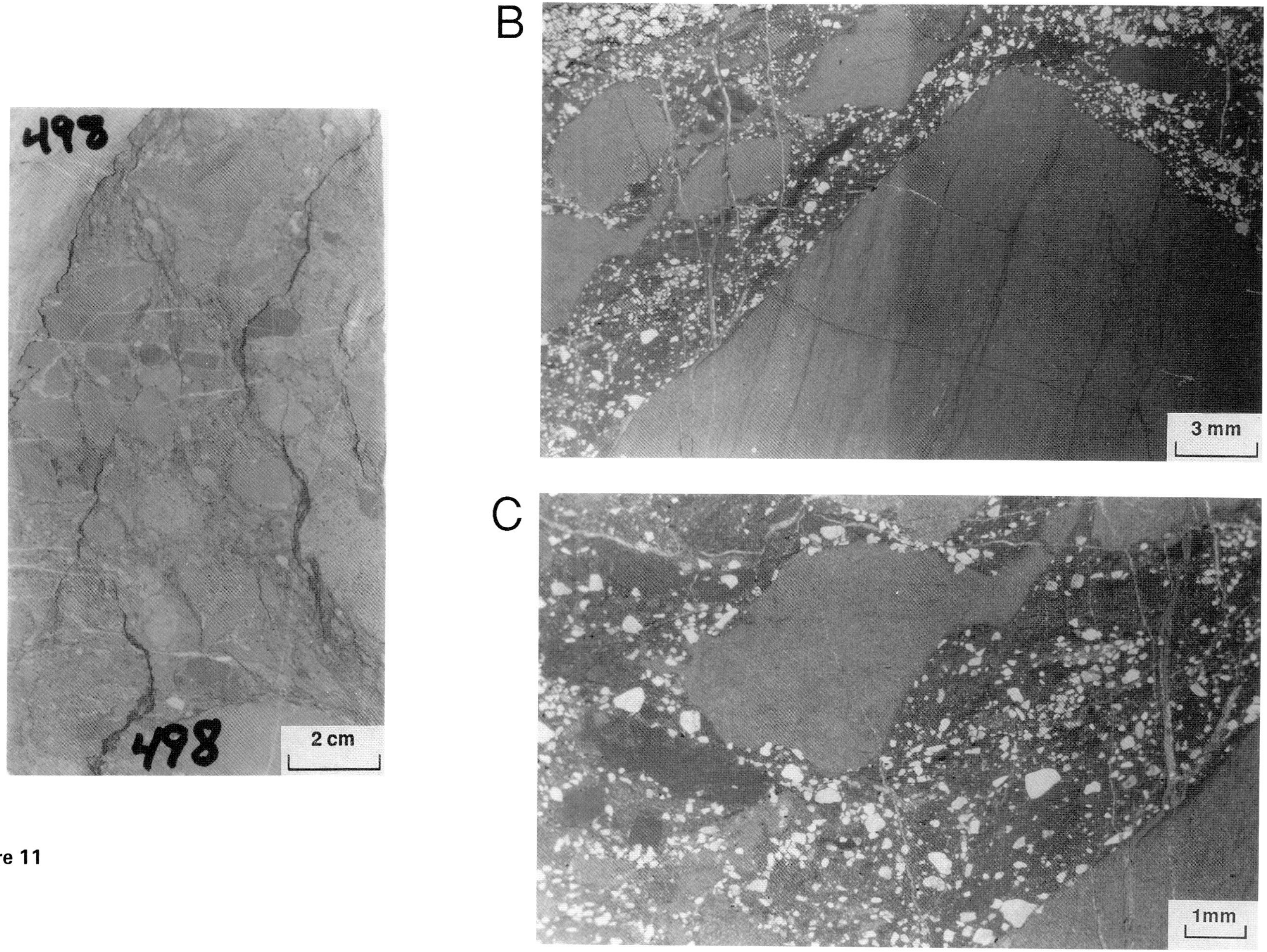

Figure 11

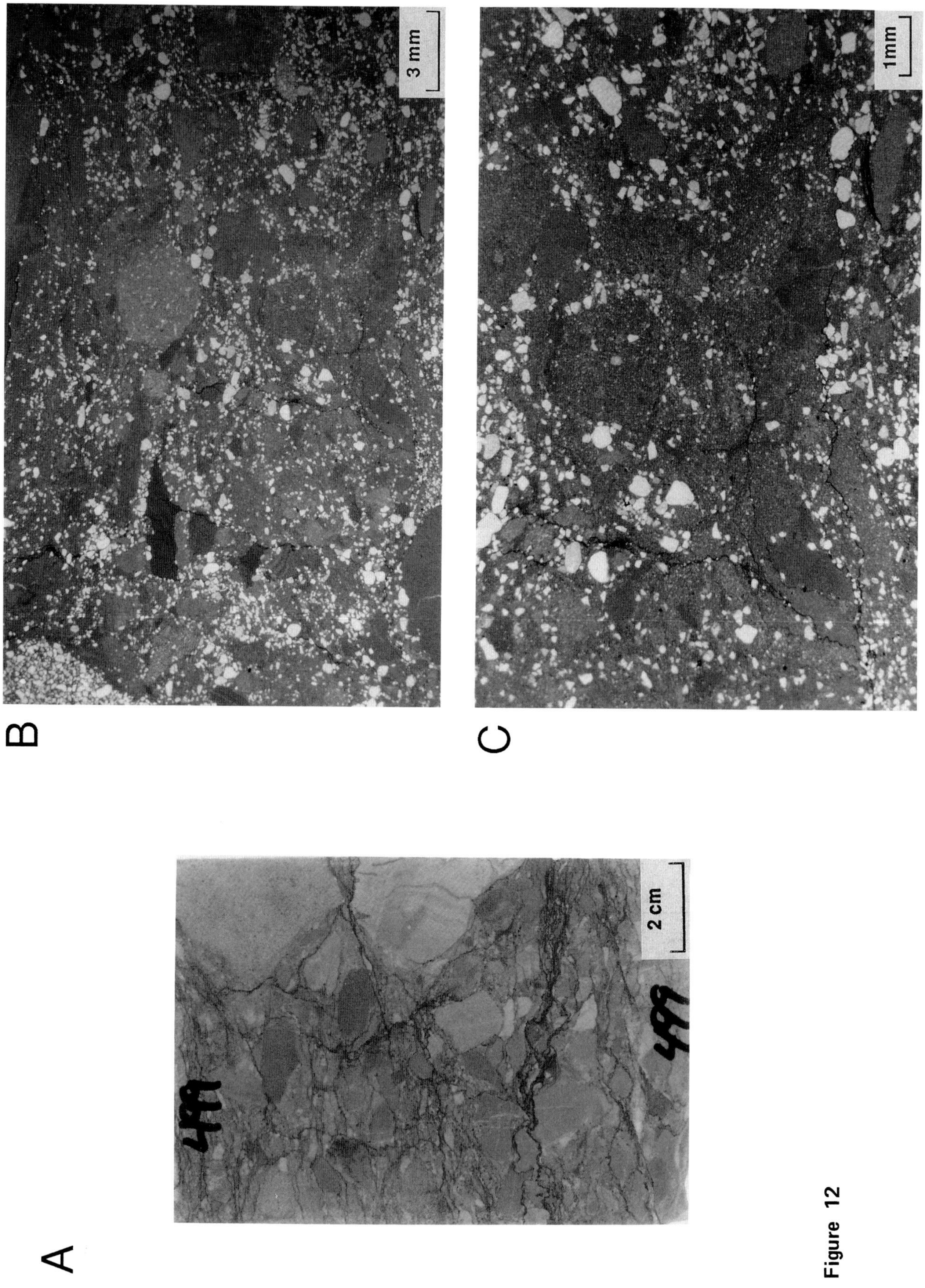

Figure 12

quartz sandstone and siltstone that is locally cemented by dolomite and occurs intermixed with dolomudstone. The breccia intervals commonly contain abundant microstylolites along which variable amounts of insoluble residue have accumulated.

Sandstone from 12,483.5 to 12,488.5 ft (3804.9 to 3806.5 m) is similar in composition and texture to that of the matrix within the breccia intervals. The sandstone contains pinkish gray to light gray, fine- to coarse-grained, poorly sorted quartz grains with both dolomite and silica cement. The sandstone is vaguely laminated and contains a color mottling that may have been produced by dewatering or by soft-sediment deformation.

Two thin intervals, from 12,479.5 to 12,481 ft (3803.7 to 3804.2 m) and from 12,481 to 12,482.5 ft (3804.2 to 3804.7 m), contain breccia that is believed to be tectonic in origin. Clasts of finely crystalline dolomudstone are similar in composition to those of the other brecciated intervals, but here the clasts are angular, have been slightly rotated, and have a more fitted appearance. No matrix surrounds the clasts. The fractures and areas between clasts are filled with anhydrite cement. The general appearance is that of a fractured dolostone, where, in places, the fractures are developed to the point of separating the host rock into separate clasts.

1-18F Core

Cores from the 1-18F well, from 12,822 to 12,854 ft (3908.1 to 3917.9 m), contain the same basic rock types as the 1-6E core, but in a different stratigraphic order (Fig. 13). Four separate intervals of solution-collapse breccia similarly contain dolostone clasts in a sandstone-rich matrix (Figs. 14-16). The clasts and matrix are like those from similar intervals in the other core, with four notable exceptions: (1) the uppermost breccia in the core from 12,822 to 12,827 ft (3908.1 to 3909.7 m) contains chert clasts in addition to dolostone clasts, and, in several cases, it is apparent that the chert is a partial to complete replacement of a carbonate precursor (Figs. 14 and 15); (2) the dolostone clasts vary considerably in color between light gray, greenish gray, and pinkish tan, and in the upper breccia zones of the core also vary in texture from dolomudstone to dolopackstone, with the dolopackstone locally containing fine moldic or microvuggy porosity; (3) clasts in the interval from 12,835 to 12,837.5 ft (3912.1 to 3912.9 m) have thin coatings on their upper surfaces of a reddish brown clay and silt drape; and (4) clasts in the interval from 12,839 to 12,841 ft (3913.3 to 3913.9 m) are aligned as if arranged by current flow during deposition.

Sandstone from 12,850.5 to 12,852.5 ft (3916.8 to 3917.4 m) in the 1-18F core is like that in the other core; whereas, a sandstone just below and extending to the bottom of the cored interval differs slightly in that it is better sorted and contains inclined laminae and thin layers of very fine-grained sand (Fig. 17A). Tectonic breccia from 12,833 to 12,835 ft (3911.5 to 3912.2 m) in the 1-18F core contains anhydrite-cemented fractures as in the other core, but the anhydrite is more coarsely crystalline (Fig. 17B and C). The 1-18F core contains more intervals of intact dolostone that vary from grayish pink silty dolomudstone to dark gray dolomudstone-packstone containing scattered microvuggy and vuggy porosity and anhydrite cement. Chert nodules are developed locally.

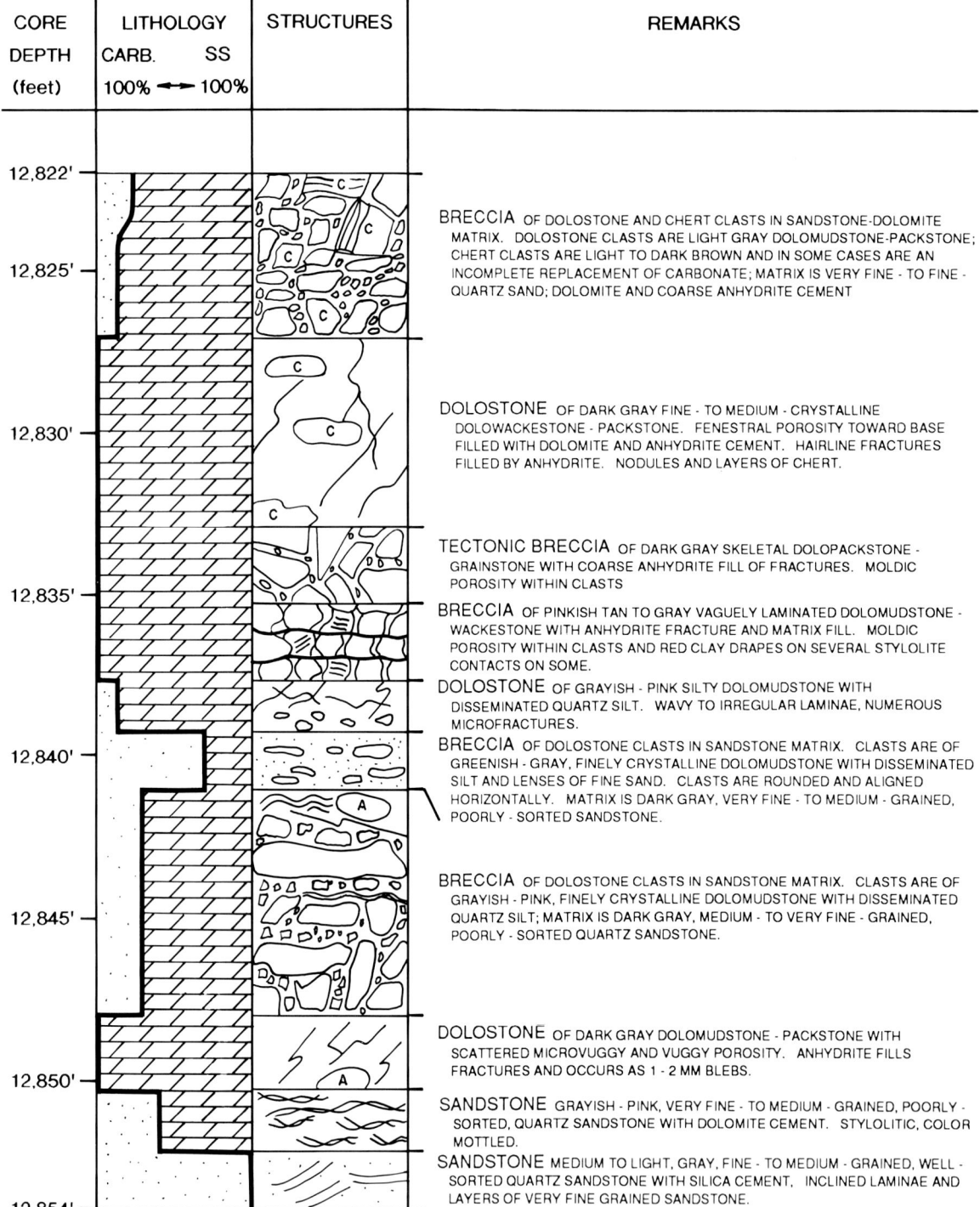

Fig. 13.-- Core description of Chevron 1 - 18F Federal upper Mission Canyon core (Section 6, T18 N, R119W, Unita County, Wyoming). Graphic logs are the same as for Fig. 9.

Figure 14. Photographs of core from the 1-18F well.

A. Core slab from 12,823 ft (3908.4 m) of solution-collapse breccia containing dolostone and chert clasts in a siliciclastic-rich matrix. Portion of large chert clast in upper right.

B and C. Thin-section photomicrographs of Figure 14A showing dolostone (dark color) and chert (light color) clasts in poorly sorted quartz sandstone and siltstone matrix.

Figure 15. Photographs of core from the 1-18F well.

A. Core slab from 12,824 ft (3908.7 m) of solution-collapse breccia containing dolostone and chert (lighter color) clasts in a siliciclastic-rich matrix. Clast boundaries are commonly overprinted by stylolites.

B and C. Thin-section photomicrographs of Figure 15A showing dolostone (dark color) and chert (light color) clasts in sandstone-siltstone matrix. Dolostone clasts vary from dolomudstone to dolopackstone.

Figure 16. Photographs of core from the 1-18F well.

A. Core slab from 12,847 ft (3915.8 m) of solution-collapse breccia containing dolostone clasts in a siliciclastic-rich and dolomudstone matrix. Microfractures and small stylolites crosscut clasts and matrix.

B and C. Thin-section photomicrographs of Figure 16A showing dolomudstone and dolowackestone clasts in a matrix of poorly sorted quartz sandstone and siltstone.

Figure 17. Core slab photographs from the 1-18F well.

A. Core slab from 12,852.8 ft (3917.5 m) of fine- to coarse-grained, moderately sorted quartz sandstone with inclined laminae and thin layers of very fine-grained sandstone.

B and C. Core slabs from 12,834 ft (3911.8 m) of tectonic breccia containing angular clasts of finely crystalline dolomudstone. The clasts have been rotated only slightly, locally have a fitted appearance, and are not surrounded by matrix. The areas between clasts and fractures are filled by coarsely crystalline anhydrite cement.

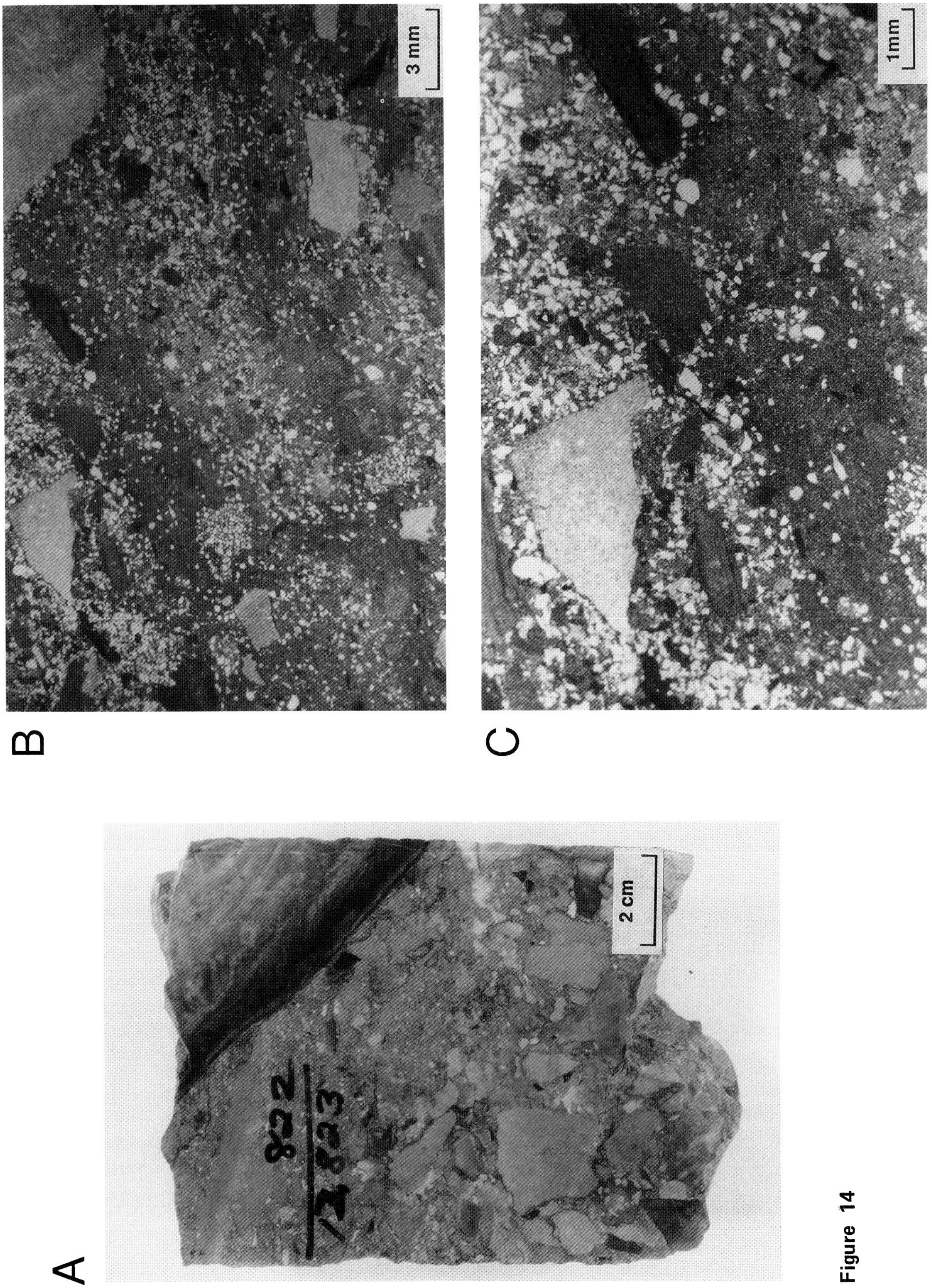

Figure 14

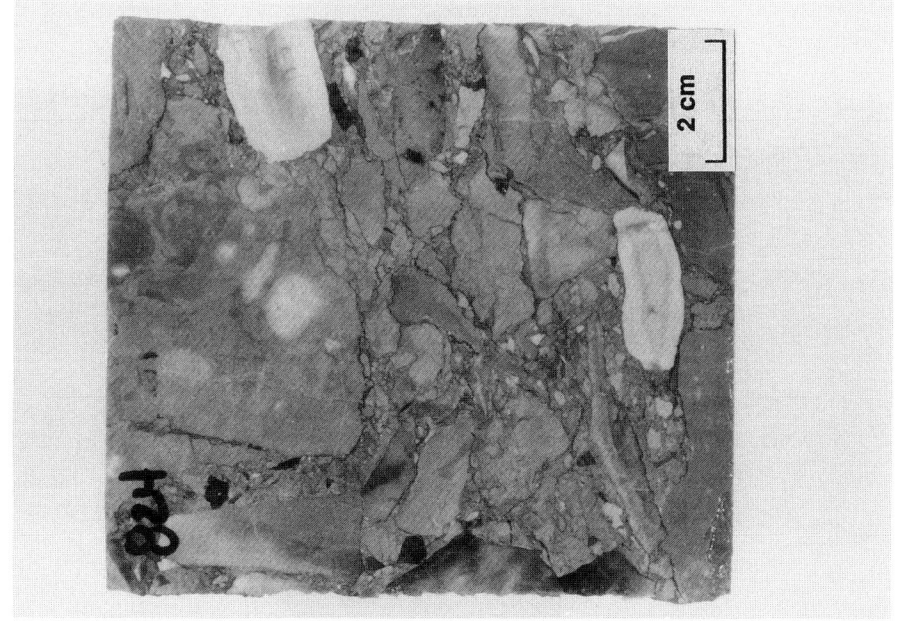

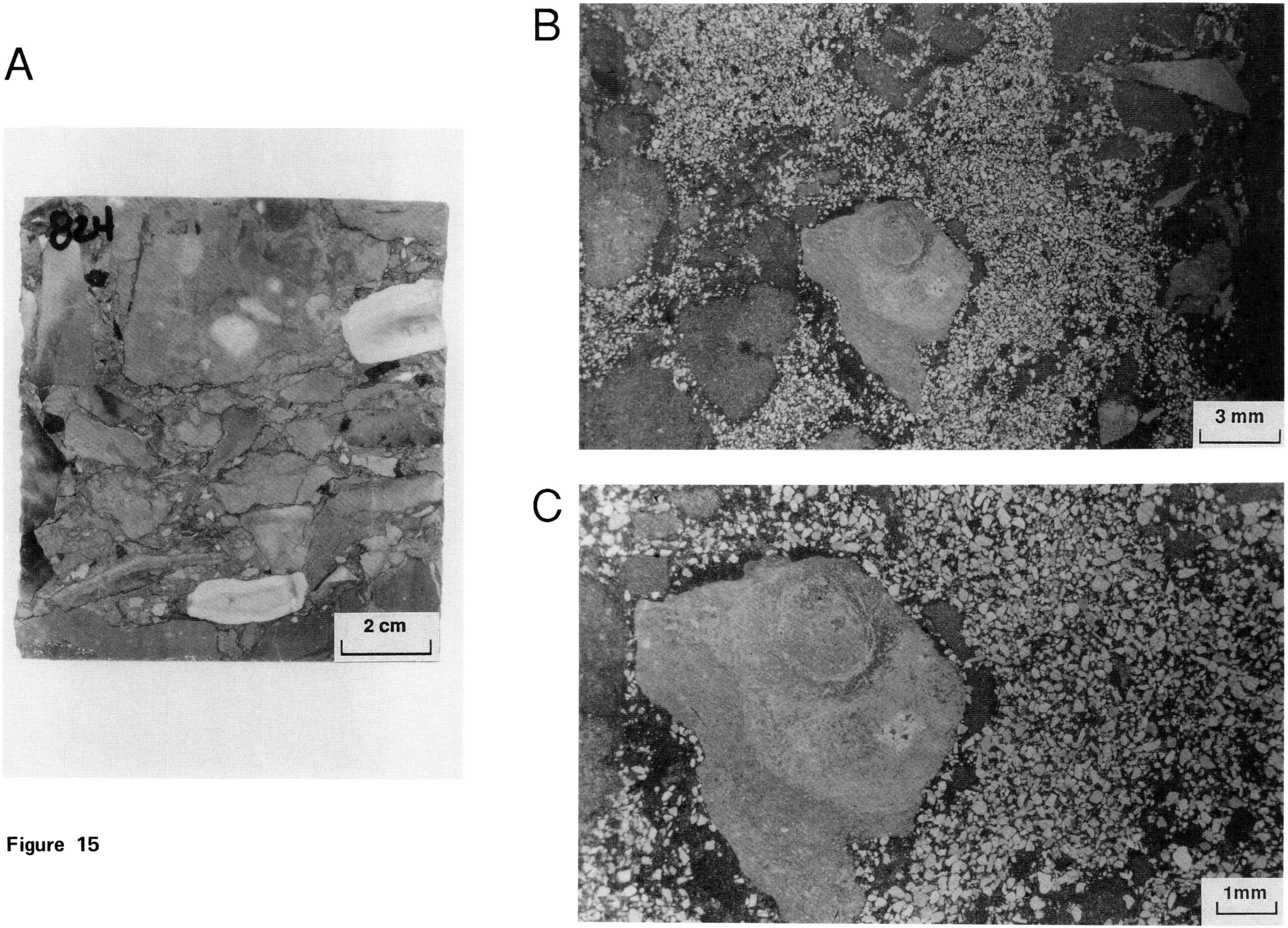

Figure 15

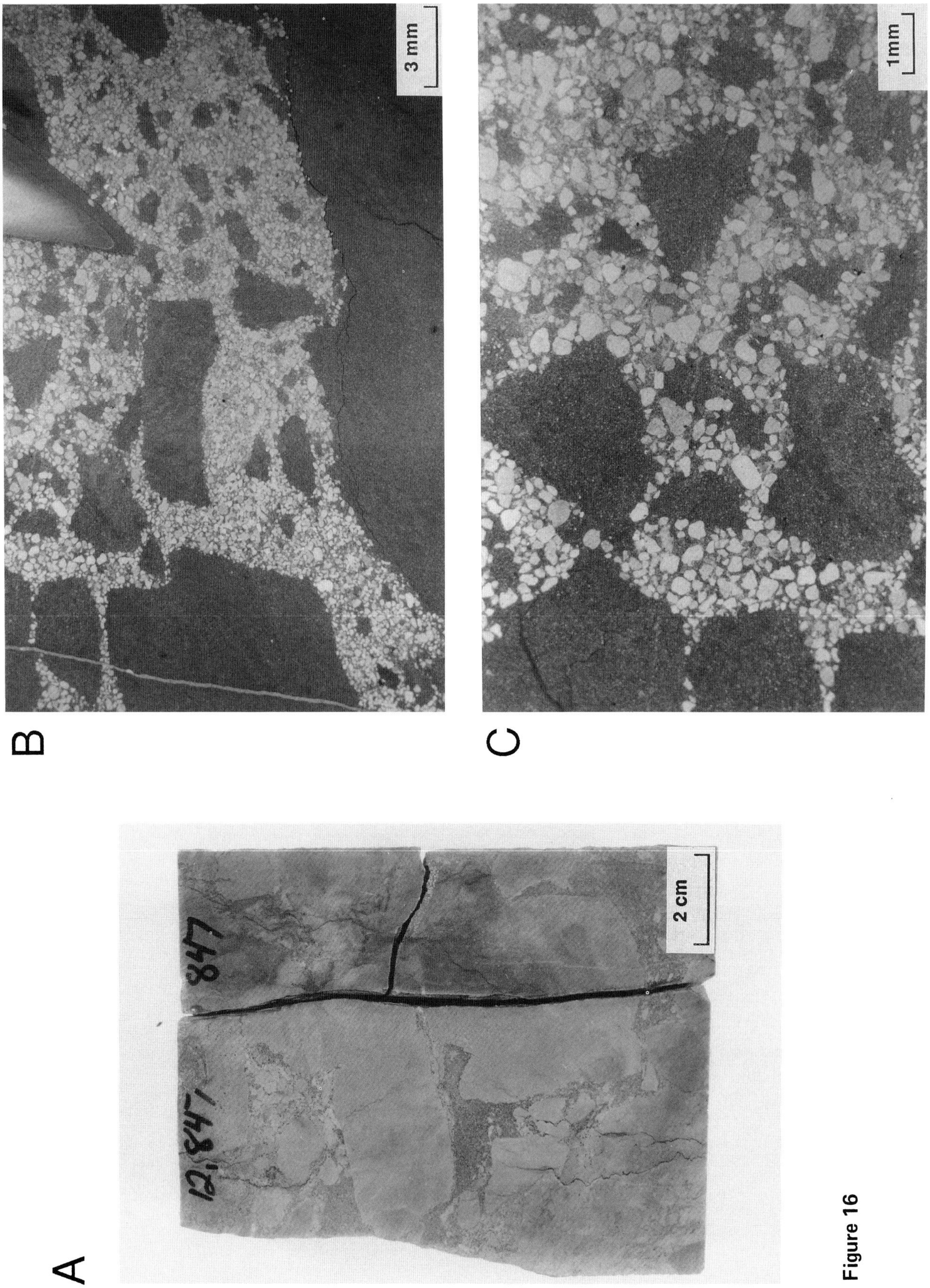

Figure 16

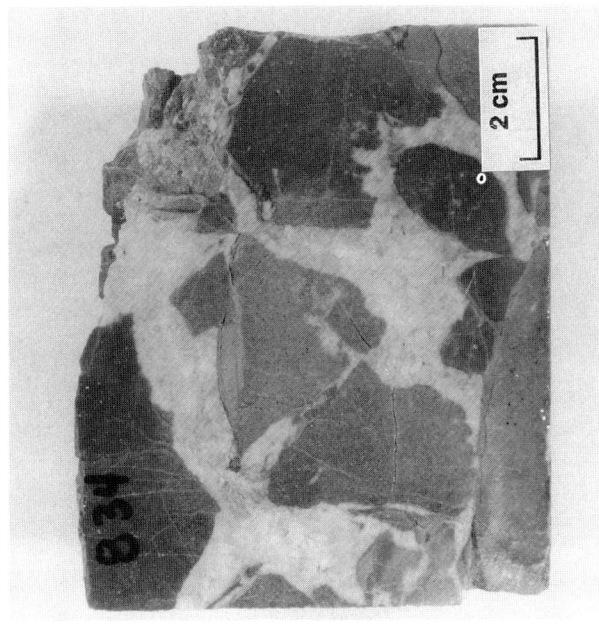
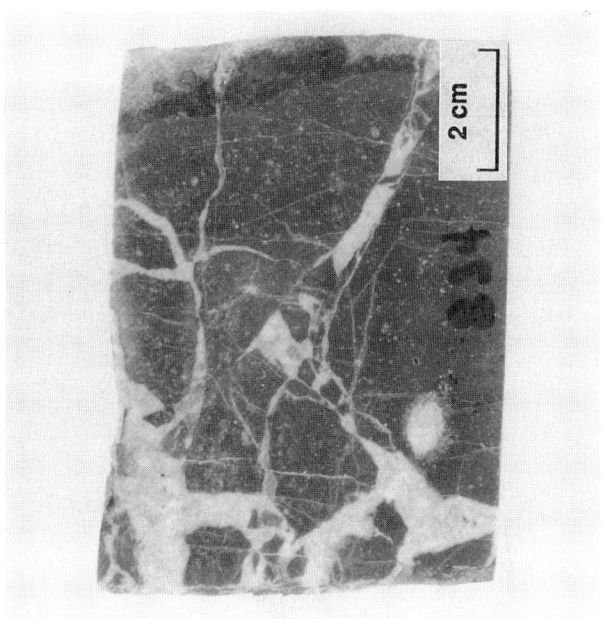
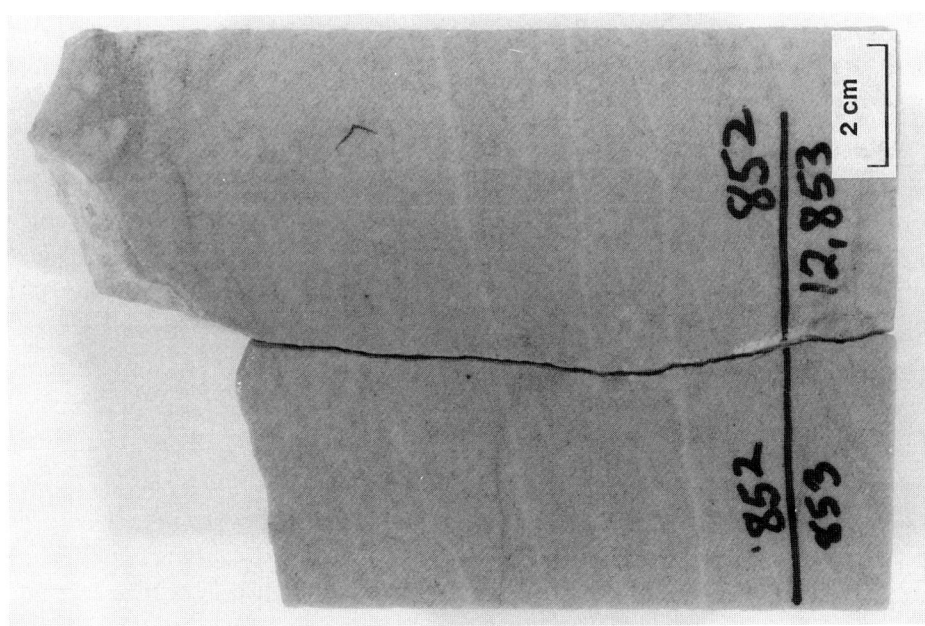

Figure 17

Origin of Breccias

Carbonate breccias can form by solution collapse during karst or evaporite dissolution and also by tectonic processes (Middleton, 1961; Stanton, 1966; Blount and Moore, 1969; Sando, 1974; Wright, 1982; Gargallo-Quinones, 1985). Differentiation of the breccia types is generally based on observations of lateral and vertical variations within the brecciated interval, relationships between breccia and surrounding rocks, the nature of the clasts and matrix within the breccia, and distribution of the brecciated intervals relative to fractures.

The breccias in the two cores that were identified as tectonic contain angular clasts with adjacent matching clast boundaries where fracturing occurred without extensive dislocation of the clasts (Fig. 17C). Matrix is not present between the clasts; instead, the porosity is filled with coarsely crystalline anhydrite that is also present in fractures and small veins (Fig. 17B and C). These tectonic breccias and anhydrite-filled fractures are present throughout the Mississippian section at Whitney Canyon-Carter Creek and have not been associated with a correlatable stratigraphic unit (Harris and others, 1988). They are most likely due to the emplacement of the Absaroka thrust sheet. Associated calcite and dolomite vein and fracture fill cements have been isotopically interpreted as a result of a high-temperature diagenetic event (Budai and others, 1987).

The breccias identified as solution collapse in origin are found in a dominantly sandstone matrix (Figs. 10-12 and 14-16). The clasts are commonly rounded and stained to a pinkish hue and, in one instance, are capped by red clay drapes. Locally the sandstone matrix and intervals of clasts have structures that suggest deposition by flowing waters. These breccias are thought to be related to karstification of the uppermost portions of the Mission Canyon, a process during which the underlying evaporite beds would also have been penetrated and dissolved, leading to the intermixing of karst and evaporite solution-collapse features (Gargallo-Quinones, 1985).

Unconventional Mixing

The mixed carbonate and siliciclastic "transitional" unit of the uppermost Mission Canyon (Figs. 7 and 8) is an example of a laterally correlatable facies that was deposited during two noncontemporaneous stages. Our interpretation presented here indicates that the carbonates were deposited, cemented, and affected by exposure-related solution before siliciclastic filling. Similar examples are found in the Proterozoic Dismal Lakes Group of Northwest Territories, Canada (Kerans and Donaldson, 1988); the Cambrian-Ordovician Arbuckle Formation of eastern Kansas (Merriam and Atkinson, 1956); the Lower Ordovician Ellenburger Group of the Permian Basin (Kerans, 1988); and the Mississippian Leadville Formation of Colorado (DeVoto, 1988). Knowledge of regional unconformities and associated paleontologic dating are critical to distinguish this facies type from a contemporaneous, mixed carbonate-siliciclastic facies, especially with subsurface data.

The filling of solution features by siliciclastics greatly reduced overall potential reservoir quality of the uppermost Mission Canyon. Porosity is limited to massive dolomites that have moldic and intercrystalline porosity types, developed by dolomitization and subsequent solution of small skeletal fragments. Sandstones and breccias are tightly cemented. Wireline log porosity values range from 0% to 4%, but may be as high as 6% to 8% in thin, discontinuous zones. Several wells in the field were perforated and acid-stimulated in these higher-porosity zones, but spinner-flow meter production surveys indicate a relatively insignificant contribution to total flow rates.

REGIONAL SIGNIFICANCE

The presence of solution-collapse breccias and evidence of karstification within the uppermost part of the Mission Canyon Formation in the Absaroka thrust sheet are significant in a regional sense. Previous interpretations of the paleogeographic extent of the Meramecian karst plain place the westernmost limit somewhere to the east of the present-day location of Whitney Canyon-Carter Creek Field (Fig. 3A)(Sando, 1988). These interpretations were based on available outcrop data, from north-central to southeastern Wyoming, and did not include subsurface information from the Wyoming thrust belt.

Evidence presented here indicates that the maximum western shoreline position was approximately in eastern Utah. The shoreline went at least as far west as Whitney Canyon-Carter Creek, but depositionally was even further west when structural deformation is considered. Palinspastic restoration of the thrust belt places the depositional location 28 miles (45.5 km) west of the present location, based on a regional cross section of Warner (1982). This would place the western limit of the karst plain exposure surface at least 20 miles (32.5 km) west of the Utah-Wyoming border. At Hoback Canyon, approximately 120 miles (185 km) north of present-day Whitney Canyon-Carter Creek Field, a preserved evaporite sequence occurs at the surface (Sando, 1977). Sando (1988) interprets this as evidence that Hoback Canyon was just west of the maximum regressive shoreline position and protected from solution during karstification. If this interpretation is correct, then the maximum regressive shoreline may have formed an embayment in northwestern Wyoming, then trended northeast-to-southwest down into north-central Utah.

The Meramecian location of Whitney Canyon-Carter Creek Field would have been very close to the maximum regressive shoreline position and was probably affected by solution processes for a relatively short amount of geologic time. Approximately 3 to 5 Ma of exposure time can be inferred from chronometric charts (Sando, 1988). Because of the duration of exposure, only the upper 70 to 100 ft (21 to 30 m) of section was affected by solution processes. If the exposure time had been longer, groundwater would have been able to remove more anhydrite and more of the upper Mission Canyon would have solution-collapse features. At Elk Basin Field in the northern Big Horn Basin, the subsurface Madison section has solution breccias 300 ft (90 m) below the exposure surface (McCaleb and Wayhan, 1969). Elk Basin Field is 250 miles (406 km) north-northeast of Whitney Canyon-Carter Creek Field, and, according to chronometric charts (Sando, 1988), the karst plain there was exposed for approximately 20 Ma.

ACKNOWLEDGMENTS

We wish to thank Chevron U.S.A., Inc. and Chevron Oil Field Research Company for drafting support and permission to publish this paper. F. J. Conner and J. A. Narbona helped prepare figures, and P. A. Schipperijn offered a helpful review.

REFERENCES

ARMSTRONG, R. C., 1968, Sevier orogenic belt in Nevada and Utah: Geological Society of America Bulletin, v. 79, p. 429-458.

BUDAI, J. M., LOHMANN, K. C., AND WILSON, J. L., 1987, Dolomitization of the Madison Group, Wyoming and Utah overthrust belt: American Association of Petroleum Geologists Bulletin, v. 71, p. 909-924.

BLOUNT, D. H., AND MOORE, C. H., 1969, Depositional and nondepositional carbonate breccias, Chiantla quadrangle, Guatemala: Geological Society of America Bulletin, v. 80, p. 429-442.

CRAIG, L. C., 1972, Mississippian system, *in* Mallory, W. W., ed., Geologic Atlas of the Rocky Mountain Region: Rocky Mountain Association of Geologists, Denver, CO, p. 100-110.

DEVOTO, R. H., 1988, Late Mississippian paleokarst and related mineral deposits, Leadville Formation, central Colorado, *in* James, N. P. and Choquette, P. W., eds., Paleokarst: Springer-Verlag, New York, p. 278-305.

GARGALLO-QUINONES, M., 1985, Study of the ancient karstification and brecciation in the Madison Limestone (Mississippian) in Wyoming: unpublished M.S. thesis, State University of New York at Stonybrook, 208 p.

GUTSCHICK, R. C., AND SANDBERG, C. A., 1983, Mississippian continental margins of the conterminous United States, *in* Stanley, D. J. and Moore, G. T., eds., The Shelfbreak-Critical Interface on Continental Margins: Society of Economic Paleontologists and Mineralogists, Special Publication 33, p. 79-96.

_____, AND SANDO, W. J., 1980, Mississippian shelf margin and carbonate platform from Montana to Nevada, *in* Fouch, T. D. and Magathan, E. R., eds., Paleozoic Paleogeography of the West-Central United States: Society of Economic Paleontologists and Mineralogists - Rocky Mountain Section, Rocky Mountain Paleogeography Symposium 1, p. 111-128.

HARRIS, P. M., FLYNN, P. E., AND SIEVERDING, J. L., 1988, Mission Canyon (Mississippian) reservoir study, Whitney Canyon-Carter Creek Field, southwestern Wyoming, *in* Lomando, A. J. and Harris, P. M., eds., Giant Oil and Gas Fields: Society of Economic Paleontologists and Mineralogists, Core Workshop no. 12, v. 2, p. 695-740.

HOFFMAN, M. E., AND KELLY, J. M., 1981, Whitney Canyon-Carter Creek Field, Uinta and Lincoln Counties, Wyoming, *in* Reid, S. G. and Miller, D. D., eds., Energy Resources of Wyoming: Wyoming Geological Association 32nd Annual Field Conference, p. 99-107.

KERANS, C., 1988, Karst-controlled reservoir heterogeneity in Ellenburger Group carbonates of West Texas: American Association of Petroleum Geologists Bulletin, v. 72, p. 1160-1183.

_____, AND DONALDSON, J. A., 1988, Proterozoic paleokarst profile, Dismal Lakes Group, Northwest Territories, Canada, *in* James, N. P. and Choquette, P. W., eds., Paleokarst: Springer-Verlag, New York, p. 167-182.

McCALEB, J. A., AND WAYHAN, D. A., 1969, Geologic reservoir analysis, Mississippian Madison Formation, Elk Basin Field, Wyoming-Montana: American Association of Petroleum Geologists Bulletin, v. 53, p. 2094-2113.

MERRIAM, D. F., AND ATKINSON, W. R., 1956, Simpson filled sinkholes in eastern Kansas: State Geological Survey of Kansas, Report of Studies Bulletin 119, part 2, p. 61-80.

MIDDLETON, G. V., 1961, Evaporite solution breccias from the Mississippian of Southwest Montana: Journal of Sedimentary Petrology, v. 31, p. 189-195.

ROSE, P. R., 1977, Mississippian carbonate shelf margins, Western United States, *in* Heisey, E. L. and others, eds., Rocky Mountain Thrust Belt Geology and Resources: Joint Wyoming-Montana-Utah Geological Associations Guidebook, p. 155-172.

ROYSE, F., JR., WARNER, M. A., AND REESE, D. L., 1975, Thrust belt of Wyoming, Idaho, and Northern Utah, structural geometry and related stratigraphic problems: Rocky Mountain Association of Geologists Symposium on Deep Drilling Frontiers in Central Rocky Mountains, p. 41-54.

SANDBERG, C. A., AND GUTSCHICK, R. C., 1980, Sedimentation and biostratigraphy of Osagean and Meramecian starved basin and foreslope, Western United States, *in* Fouch, T. D. and Magathan, E. R., eds., Paleozoic Paleogeography of West-Central United States: Society of Economic Paleontologists and Mineralogists - Rocky Mountain Section, West-Central U.S. Paleogeography Symposium 1, p. 129-147.

_____, JOHNSON, J. G., POOLE, F. G., AND SANDO, W. J., 1982, Middle Devonian to Late Mississippian geologic history of the overthrust belt region, Western United States: Rocky Mountain Association of Geologists, Special Publication 1982, Geologic Studies of the Cordilleran Thrust Belt, p. 691-719.

SANDO, W. J., 1968, A new member of the Madison Limestone (Mississippian) in Wyoming: Geological Society of America Bulletin, v. 79, p. 1855-1858.

_____, 1974, Ancient solution phenomena in the Madison Limestone (Mississippian) of North-Central Wyoming: United States Geological Survey, Journal of Research, v. 2, p. 133-141.

SANDO, W. J., 1976, Mississippian history of the northern Rocky Mountains region: United States Geological Survey, Journal of Research, v. 4, p. 317-338.

_____, 1986, Revised Mississippian time scale, western interior region, conterminous United States: United States Geological Survey Bulletin 1605A, 18 p.

_____, 1988, Madison Limestone (Mississippian) paleokarst: a geologic synthesis, *in* James N. P. and Choquette, P. W., eds., Paleokarst: Springer-Verlag, New York, p. 256-277.

SIEVERDING, J. L., AND ROYSE, F., JR., 1990, Whitney Canyon-Carter Creek Field, southwestern Wyoming thrust belt, *in* Beaumont, E. A. and Foster, N. L., eds., Atlas of Oil and Gas Fields: American Association of Petroleum Geologists: Treatise of Petroleum Geology.

STANTON, R. J., 1966, The solution brecciation process: Geological Society of America Bulletin, v. 77, p. 843-848.

VICE, M. A., 1988, Depositional environment and diagenesis in an interval of the Mission Canyon Limestone (Madison Group, Mississippian), south-central Montana and northern Wyoming: unpublished M.S. thesis, Southern Illinois University at Carbondale, 149 p.

WARNER, M. A., 1982, Source and time of generation of hydrocarbons in the Fossil Basin, western Wyoming thrust belt, *in* Powers, R. B., ed., Geologic Studies of the Cordilleran Thrust Belt, Volume II: Rocky Mountain Association of Geologists, Denver, CO, p. 805-815.

WRIGHT, V. P., 1982, The recognition and interpretation of paleokarsts: two examples from the Lower Carboniferous of South Wales: Journal of Sedimentary Petrology, v. 52, p. 83-94.

RESERVOIRS INC.

** GEOLOGY **

Single Well / Field / Regional Studies

** CORE ANALYSIS **

Special / Routine

** ENGINEERING **

Well Completion / Stimulation Studies

FOR MORE INFORMATION CONTACT:

Paul J. Cernock	Randall S. Miller	Philip C. Dupler
President	Vice President	Marketing Representative
	Geology	Staff Geologist

1151-C Brittmore Road
Houston, Texas 77043

PH: (713) 932-7183
FAX: (713) 932-0520
TELEX: 795240